Test Bank

for
McKeague and Turner's

Trigonometry

Fifth Edition

THOMSON

BROOKS/COLE

Australia • Canada • Mexico • Singapore • Spain • United Kingdom • United States

For more information about our products,
contact us at:
Thomson Learning Academic Resource Center
1-800-423-0563

For permission to use material from this text or
product, submit a request online at
http://www.thomsonrights.com.
Any additional questions about permissions can be
submitted by email to **thomsonrights@thomson.com**.

Thomson Brooks/Cole
10 Davis Drive
Belmont, CA 94002-3098
USA

Asia
Thomson Learning
5 Shenton Way #01-01
UIC Building
Singapore 068808

Australia/New Zealand
Thomson Learning
102 Dodds Street
Southbank, Victoria 3006
Australia

Canada
Nelson
1120 Birchmount Road
Toronto, Ontario M1K 5G4
Canada

Europe/Middle East/South Africa
Thomson Learning
High Holborn House
50/51 Bedford Row
London WC1R 4LR
United Kingdom

Latin America
Thomson Learning
Seneca, 53
Colonia Polanco
11560 Mexico D.F.
Mexico

Spain/Portugal
Paraninfo
Calle/Magallanes, 25
28015 Madrid, Spain

Table of Contents

Chapter 5
Test Form A- Free Response
Test Form B- Free Response
Test Form C- Multiple Choice
Test Form D- Multiple Choice
Test Form E- Mixed (Free Response/Multiple Choice)
Test Form F- Mixed (Free Response/Multiple Choice)
Test Form G- Mixed (Free Response/Multiple Choice)
Test Form H- Mixed (Free Response/Multiple Choice)

Chapter 6
Test Form A- Free Response
Test Form B- Free Response
Test Form C- Multiple Choice
Test Form D- Multiple Choice
Test Form E- Mixed (Free Response/Multiple Choice)
Test Form F- Mixed (Free Response/Multiple Choice)
Test Form G- Mixed (Free Response/Multiple Choice)
Test Form H- Mixed (Free Response/Multiple Choice)

Chapter 7
Test Form A- Free Response
Test Form B- Free Response
Test Form C- Multiple Choice
Test Form D- Multiple Choice
Test Form E- Mixed (Free Response/Multiple Choice)
Test Form F- Mixed (Free Response/Multiple Choice)
Test Form G- Mixed (Free Response/Multiple Choice)
Test Form H- Mixed (Free Response/Multiple Choice)

Chapter 8
Test Form A- Free Response
Test Form B- Free Response
Test Form C- Multiple Choice
Test Form D- Multiple Choice
Test Form E- Mixed (Free Response/Multiple Choice)
Test Form F- Mixed (Free Response/Multiple Choice)
Test Form G- Mixed (Free Response/Multiple Choice)
Test Form H- Mixed (Free Response/Multiple Choice)

Final Exams
Final Exam Form A- Free Response
Final Exam Form B- Multiple Choice
Final Exam Form C- Mixed (Free Response/Multiple Choice)

Note: Each test form includes an answer key and list of problem codes for iLrn Testing

1. Indicate if the angle $70°$ is acute or obtuse.

____

Give the complement angle.

_____ °

Give the supplement angle.

_____ °

2. The figure shows a walkway with a handrail. Angle α is the angle between the walkway and the horizontal, while angle β is the angle between the vertical posts of the handrail and the walkway. Assume that the vertical posts are perpendicular to the horizontal.

Find α if $\beta = 60°$.

$\alpha = $ _____ °

3. Suppose ABC is a right triangle with $C = 90°$.

If $a = 9$ and $c = 15$, find b.

4. Find r if $AB = 25$ and $AD = 35$.

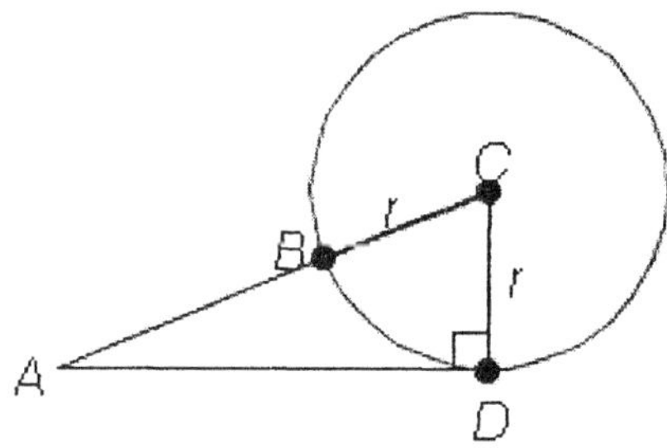

5.

Find the remaining side of a $45° - 45° - 90°$ triangle if the shorter sides are each $\dfrac{3}{4}$.

6. Graph the ordered pair on a rectangular coordinate system:

$(2, -4)$

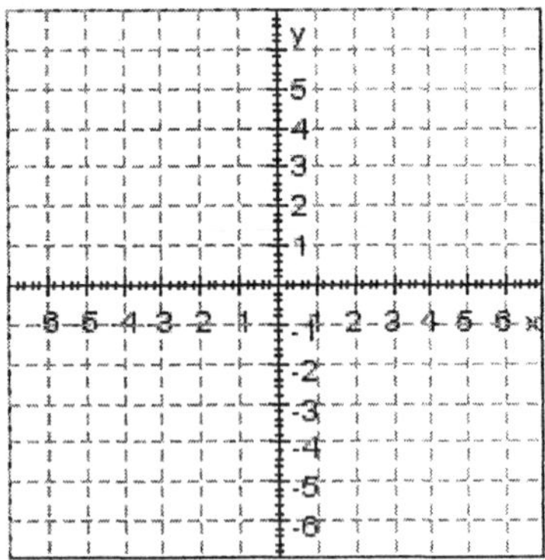

7. Graph the parabola.

$$2x + (x + 1)^2 + 6 = 0$$

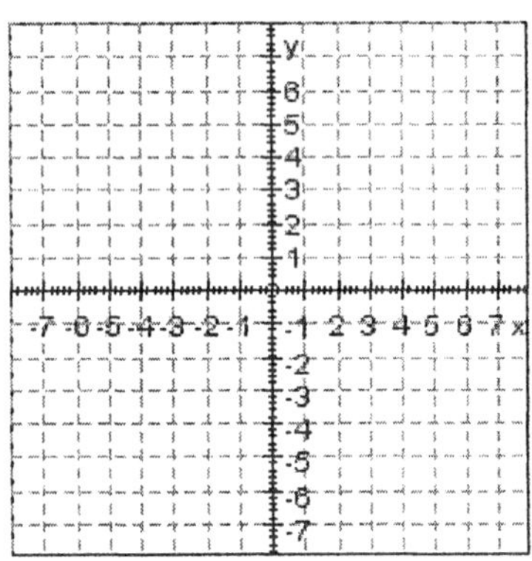

8. Use the given graph of the circle $x^2 + y^2 = 36$ to identify the points at which the line

   $x - y = 6$ will intersect the circle.

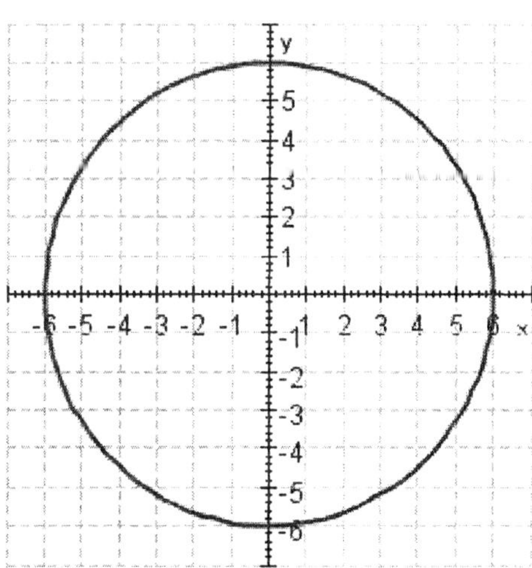

____

9. An airplane is approaching Los Angeles International Airport at an altitude of 4,224 feet. If the horizontal distance from the plane to the runway is 1.5 miles, use the Pythagorean Theorem to find the diagonal distance from the plane to the runway (see the figure below). (5,280 feet equals 1 mile.)

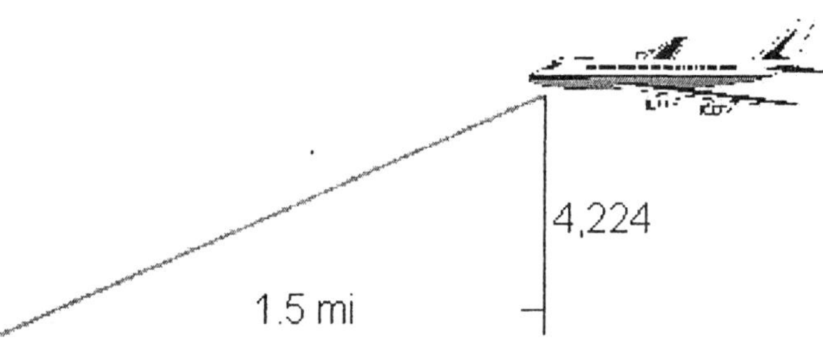

_______ miles

10. Use the diagram to help you name an angle between $0°$ and $360°$ that is coterminal with

the angle $-210°$

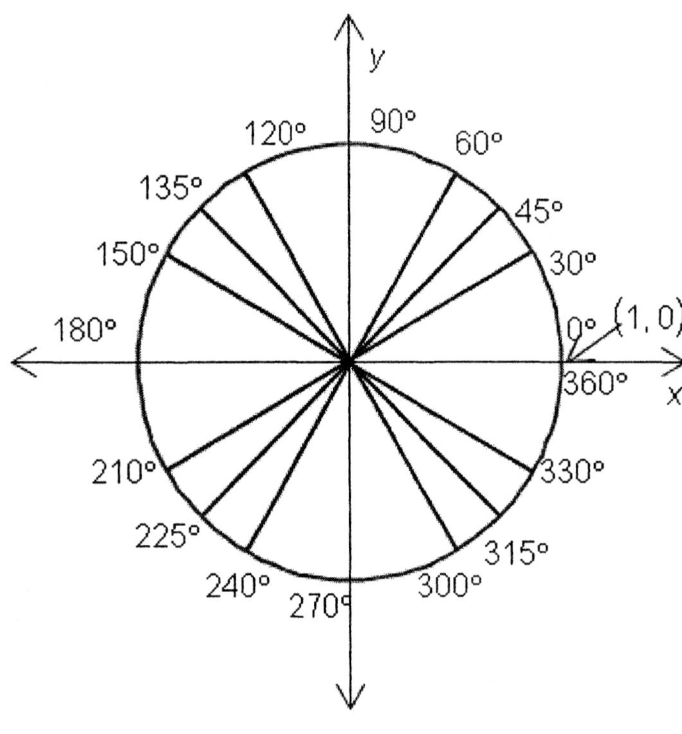

_______   □

11. Find all six trigonometric functions of θ if the given point is on the terminal side of θ.

$(-5, 12)$

Find $\sin \theta$.

Find $\cos \theta$.

Find $\csc \theta$.

Find $\sec \theta$.

Find $\tan \theta$.

Find $\cot \theta$.

12. Find all six trigonometric functions of θ if the given point is on the terminal side of θ.

$$\left(-3,\ \sqrt{7}\right)$$

Find $\sin \theta$.

Find $\cos \theta$.

Find $\csc \theta$.

Find $\sec \theta$.

Find $\tan \theta$.

Find $\cot \theta$.

13. Draw the following angle in standard position.

315°

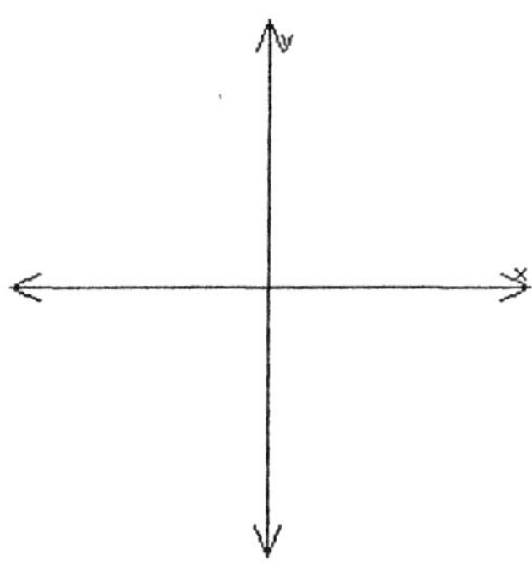

Find a point on the terminal side.   (use a separate sheet to answer if necessary)

Find the sine of the angle.

Find the cosine of the angle.

Find the tangent of the angle.

14.

Find the remaining trigonometric functions of θ if $\sin\theta = \dfrac{12}{13}$ and θ terminates in QI.

Find $\cos\theta$.

Find $\tan\theta$.

Find $\csc\theta$.

Find $\sec\theta$.

Find $\cot\theta$.

15.

Find the remaining trigonometric functions of θ if $\cot\theta = \dfrac{m}{p}$ where m and p are both positive. Assume that θ is between $0°$ and $180°$.

Find $\sin\theta$.

Find $\cos\theta$.

Find $\tan\theta$.

Find $\sec\theta$.

Find $\csc\theta$.

16. Give the reciprocal of the given number.

$$-\dfrac{8}{9}$$

17. Use the reciprocal identities for the following problem.

If $\tan\theta = 5a$, $\left(a \neq 0\right)$ find $\cot\theta$.

18.

For this problem, let $\sin\theta = -\dfrac{60}{61}$, and $\cos\theta = -\dfrac{11}{61}$, and find $\tan\theta$

19. Find $\sec \theta$ if $\tan \theta = \dfrac{4}{3}$ and θ terminates in QIII.

20. If $\cos \theta = \dfrac{3}{\sqrt{13}}$ and $\theta \in$ QIV.

Find $\sin \theta$

Find $\tan \theta$

Find $\cot \theta$

Find $\sec \theta$

Find $\csc \theta$

21. Write the following in terms of $\sin \theta$ only.

$\tan \theta$

22. Write the following in terms of $\sin \theta$ and $\cos \theta$ and then simplify if possible.

$\sec \theta - \tan \theta \sin \theta$

23. Multiply.

$(4 \cos \theta + 7)(5 \cos \theta - 7)$

24. Simplify the expression $\sqrt{x^2 - 4}$ as much as possible after substituting $2 \csc \theta$ for x.

25. Show that the following statement is true by transforming the left side into the right side.

$\cos \theta \cot \theta + \sin \theta = \csc \theta$

(use a separate sheet to answer if necessary)

1. acute

   20

   110

2. 30

3. 12

4. 12

5. $\dfrac{3\sqrt{2}}{4}$

6.

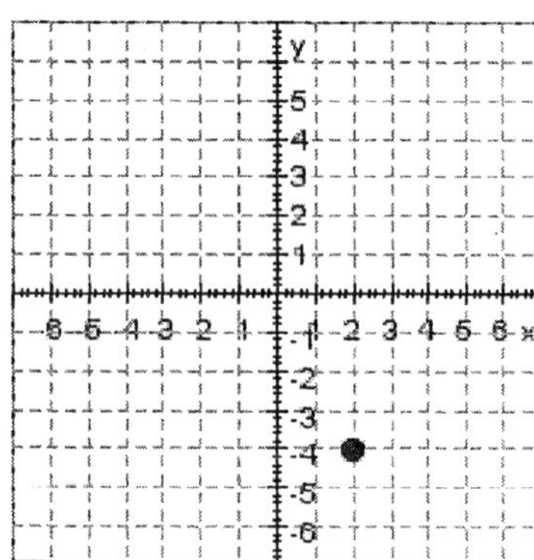

7.

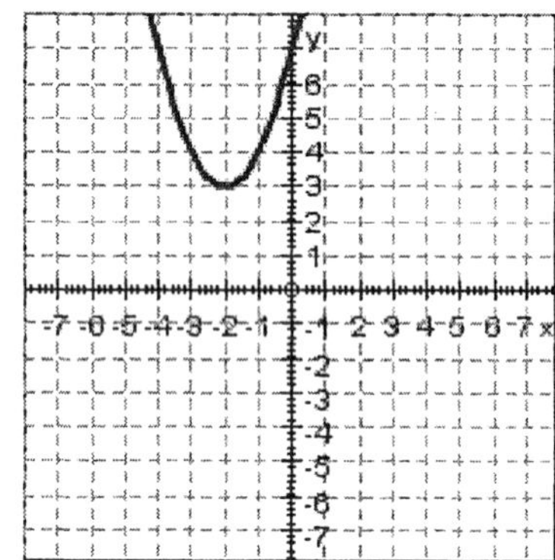

8. $(0, -6), (6, 0)$

9. 1.7

10. 150

11. $\dfrac{12}{13}$,

$-\dfrac{5}{13}$,

$\dfrac{13}{12}$,

$-\dfrac{13}{5}$,

$-\dfrac{12}{5}$,

$-\dfrac{5}{12}$

12. $\dfrac{\sqrt{7}}{4}$,

$-\dfrac{3}{4}$,

$\dfrac{4\sqrt{7}}{7}$,

$-\dfrac{4}{3}$,

$-\dfrac{\sqrt{7}}{3}$,

$-\dfrac{3\sqrt{7}}{7}$

13.

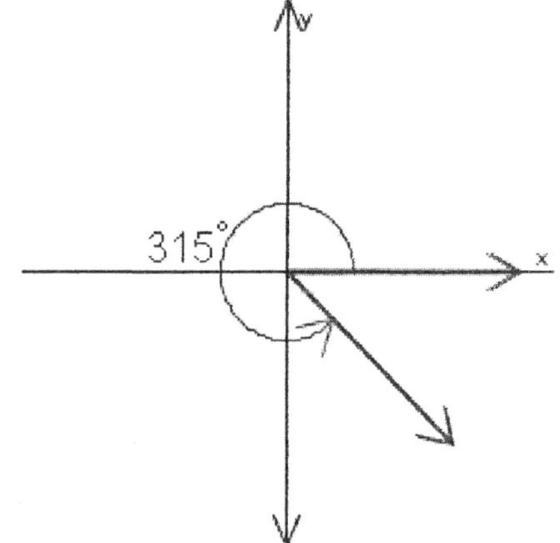

$(1, -1)$ for example

$-\dfrac{\sqrt{2}}{2}$

$\dfrac{\sqrt{2}}{2}$

-1

14. $\dfrac{5}{13}$,

$\dfrac{12}{5}$,

$\dfrac{13}{12}$,

$\dfrac{13}{5}$,

$\dfrac{5}{12}$

15. $\dfrac{p}{\sqrt{p^2+m^2}}$,

$\dfrac{m}{\sqrt{p^2+m^2}}$,

$\dfrac{p}{m}$,

$\dfrac{\sqrt{m^2+p^2}}{m}$,

$$\frac{\sqrt{m^2+p^2}}{p}$$

16. $-\dfrac{9}{8}$

17. $\cot(\theta)=\dfrac{1}{5a}$

18. $\tan(\theta)=\dfrac{60}{11}$

19. $\sec(\theta)=-\dfrac{5}{3}$

20. $\sin(\theta)=-\dfrac{2}{\sqrt{13}}$,

$\tan(\theta)=-\dfrac{2}{3}$,

$\cot(\theta)=-\dfrac{3}{2}$,

$\sec(\theta)=\dfrac{\sqrt{13}}{3}$,

$\csc(\theta)=-\dfrac{\sqrt{13}}{2}$

21. $\pm\dfrac{\sin(\theta)}{\sqrt{1-(\sin(\theta))^2}}$

22. $\cos(\theta)$

23. $20(\cos(\theta))^2+7\cos(\theta)-49$

24. $2|\cot(\theta)|$

25. We begin by writing everything on the left side in terms of $\sin\theta$ and $\cos\theta$.

$$\cos\theta\cot\theta + \sin\theta = \cos\theta\,\frac{\cos\theta}{\sin\theta} + \sin\theta = \frac{\cos^2\theta + \sin^2\theta}{\sin\theta}$$

$$= \frac{1}{\sin\theta} = \csc\theta$$

Since we have succeeded in transforming the left side into the right side, we have shown that the statement $\cos\theta\cot\theta + \sin\theta = \csc\theta$ is an identity.

1. mctr.01.01.02_NoAlgs
2. mctr.01.01.18_NoAlgs
3. mctr.01.01.27_NoAlgs
4. mctr.01.01.40_NoAlgs
5. mctr.01.01.53_NoAlgs
6. mctr.01.02.02_NoAlgs
7. mctr.01.02.19_NoAlgs
8. mctr.01.02.35_NoAlgs
9. mctr.01.02.49_NoAlgs
10. mctr.01.02.68_NoAlgs
11. mctr.01.03.05_NoAlgs
12. mctr.01.03.16_NoAlgs
13. mctr.01.03.30_NoAlgs
14. mctr.01.03.44_NoAlgs
15. mctr.01.03.60_NoAlgs
16. mctr.01.04.03_NoAlgs
17. mctr.01.04.13_NoAlgs
18. mctr.01.04.23_NoAlgs
19. mctr.01.04.39_NoAlgs
20. mctr.01.04.49_NoAlgs
21. mctr.01.05.03_NoAlgs
22. mctr.01.05.25_NoAlgs
23. mctr.01.05.37_NoAlgs
24. mctr.01.05.51_NoAlgs
25. mctr.01.05.72_NoAlgs

1. Suppose ABC is a right triangle with $C = 90°$.

   If $a = 9$ and $c = 15$, find b.

2. Use the given graph of the circle $x^2 + y^2 = 36$ to identify the points at which the line

   $x - y = 6$ will intersect the circle.

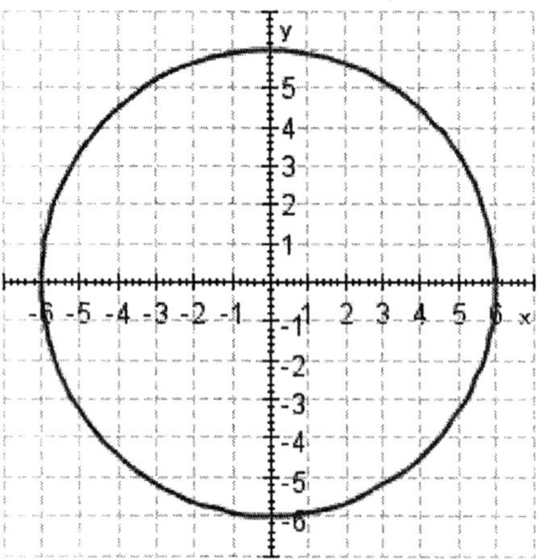

   ____

3. Find $\sec\theta$ if $\tan\theta = \dfrac{4}{3}$ and θ terminates in QIII.

4. Find the remaining trigonometric functions of θ if $\sin\theta = \dfrac{12}{13}$ and θ terminates in QI.

   Find $\cos\theta$.

   Find $\tan\theta$.

   Find $\csc\theta$.

   Find $\sec\theta$.

   Find $\cot\theta$.

5. Show that the following statement is true by transforming the left side into the right side.

$$\cos\theta\cot\theta + \sin\theta = \csc\theta$$

6. Multiply.

$$(4\cos\theta + 7)(5\cos\theta - 7)$$

7. Write the following in terms of $\sin\theta$ only.

$$\tan\theta$$

8. Graph the ordered pair on a rectangular coordinate system.

$$(2, -4)$$

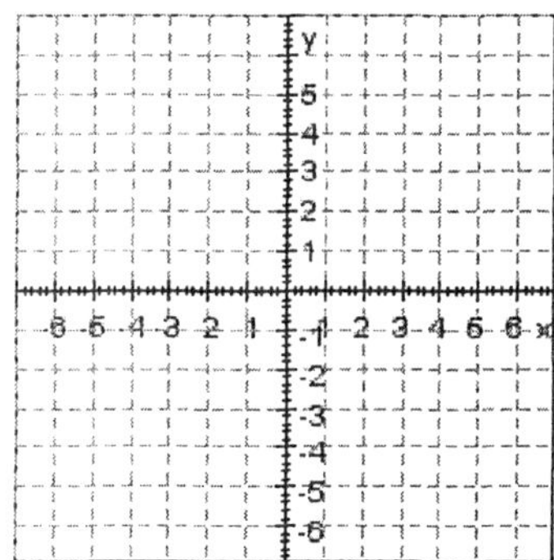

9. The figure shows a walkway with a handrail. Angle α is the angle between the walkway and the horizontal, while angle β is the angle between the vertical posts of the handrail and the walkway. Assume that the vertical posts are perpendicular to the horizontal.

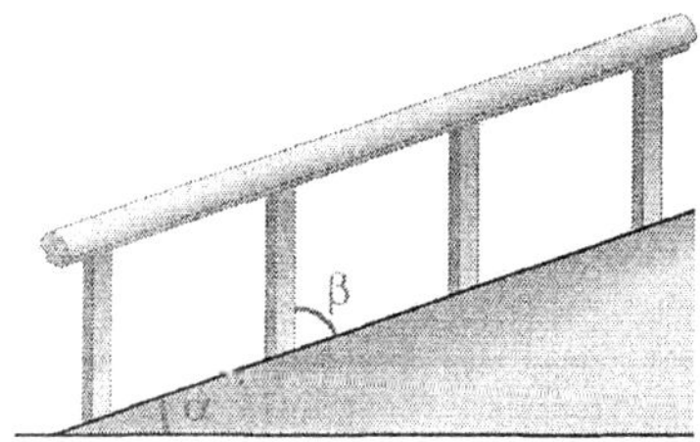

Find α if $\beta = 60°$.

$$\alpha = \underline{\quad\quad}°$$

10.

Find the remaining trigonometric functions of θ if $\cot\theta = \dfrac{m}{p}$ where m and p are both positive. Assume that θ is between $0°$ and $180°$.

Find $\sin\theta$.

Find $\cos\theta$.

Find $\tan\theta$.

Find $\sec\theta$.

Find $\csc\theta$.

11. Indicate if the angle $70°$ is acute or obtuse.

————

Give the complement angle.

———— $°$

Give the supplement angle.

———— $°$

12. Write the following in terms of $\sin\theta$ and $\cos\theta$ and then simplify if possible.

$\sec\theta - \tan\theta\,\sin\theta$

13. Use the diagram to help you name an angle between $0°$ and $360°$ that is coterminal with the angle $-210°$

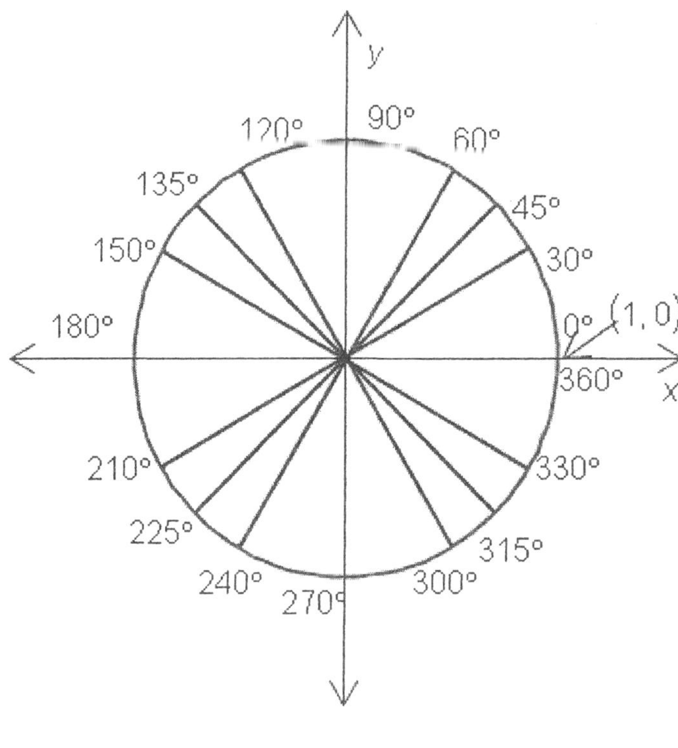

___________ □

14. Find the remaining side of a $45° - 45° - 90°$ triangle if the shorter sides are each $\dfrac{3}{4}$.

15. Give the reciprocal of the given number.

$-\dfrac{8}{9}$

16. Find all six trigonometric functions of θ if the given point is on the terminal side of θ.

$$\left(-3, \sqrt{7}\right)$$

Find $\sin \theta$.

Find $\cos \theta$.

Find $\csc \theta$.

Find $\sec \theta$.

Find $\tan \theta$.

Find $\cot \theta$.

17. Graph the parabola.

$$2x + (x + 1)^2 + 6$$

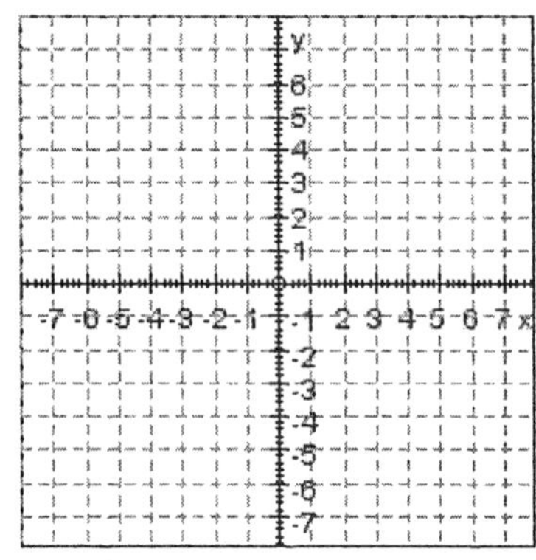

18. Simplify the expression $\sqrt{x^2 - 4}$ as much as possible after substituting $2 \csc \theta$ for x.

19. For this problem, let $\sin \theta = -\dfrac{60}{61}$, and $\cos \theta = -\dfrac{11}{61}$, and find $\tan \theta$

20. Use the reciprocal identities for the following problem.
If $\tan\theta = 5a$, $(a \neq 0)$ find $\cot\theta$.

21. An airplane is approaching Los Angeles International Airport at an altitude of 4,224 feet. If the horizontal distance from the plane to the runway is 1.5 miles, use the Pythagorean Theorem to find the diagonal distance from the plane to the runway (see the figure below). (5,280 feet equals 1 mile.)

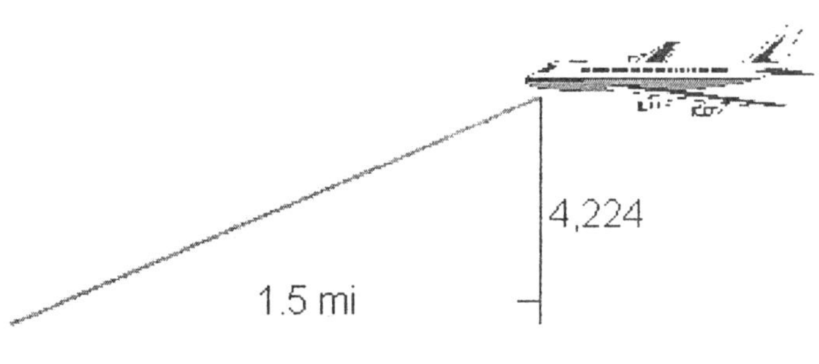

__________ miles

22. Find r if $AB = 25$ and $AD = 35$.

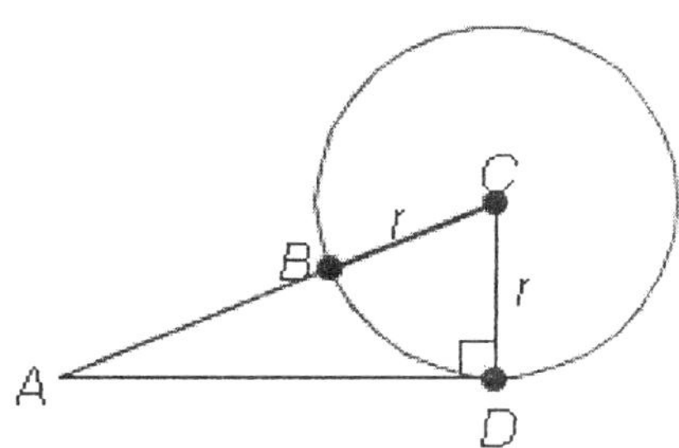

23. Find all six trigonometric functions of θ if the given point is on the terminal side of θ.

$(-5, 12)$

Find $\sin\theta$.

Find $\cos\theta$.

Find $\csc\theta$.

Find $\sec\theta$.

Find $\tan\theta$.

Find $\cot\theta$.

24.

If $\cos\theta = \dfrac{3}{\sqrt{13}}$ and $\theta \in$ QIV.

Find $\sin\theta$

Find $\tan\theta$

Find $\cot\theta$

Find $\sec\theta$

Find $\csc\theta$

25. Draw the following angle in standard position.

$315°$

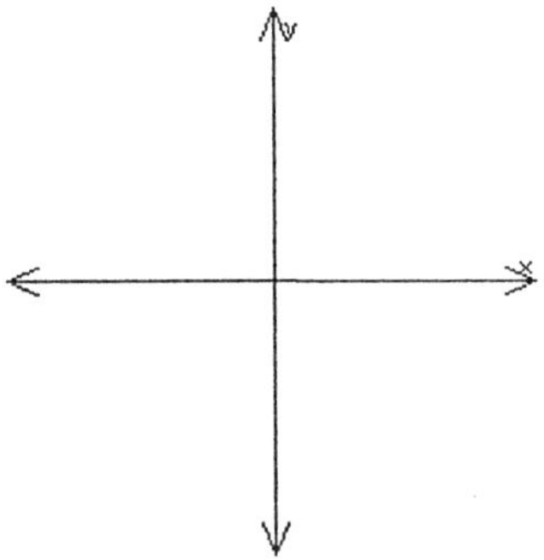

Find a point on the terminal side.

Find the sine of the angle.

Find the cosine of the angle.

Find the tangent of the angle.

1. 12

2. $(0, -6), (6, 0)$

3. $\sec(\theta) = -\dfrac{5}{3}$

4. $\dfrac{5}{13}, \dfrac{12}{5}, \dfrac{13}{12}, \dfrac{13}{5}, \dfrac{5}{12}$

5. We begin by writing everything on the left side in terms of $\sin\theta$ and $\cos\theta$.

$$\cos\theta\cot\theta + \sin\theta = \cos\theta\,\frac{\cos\theta}{\sin\theta} + \sin\theta = \frac{\cos^2\theta + \sin^2\theta}{\sin\theta}$$

$$= \frac{1}{\sin\theta} = \csc\theta$$

Since we have succeeded in transforming the left side into the right side, we have shown that the statement $\cos\theta\cot\theta + \sin\theta = \csc\theta$ is an identity.

6. $20(\cos(\theta))^2 + 7\cos(\theta) - 49$

7. $\pm\,\dfrac{\sin(\theta)}{\sqrt{1 - (\sin(\theta))^2}}$

8. 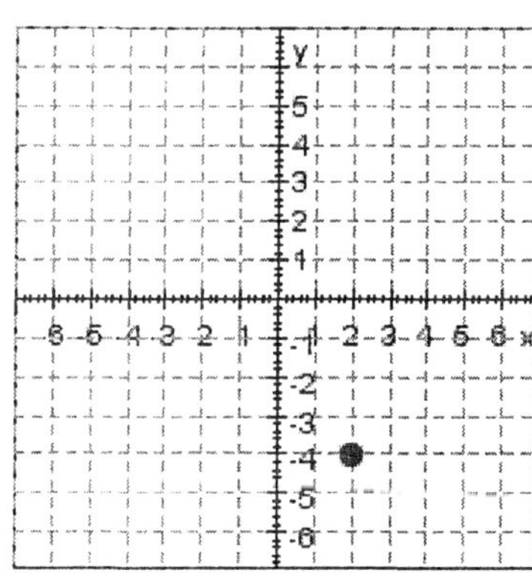

9. 30

10.
$$\frac{p}{\sqrt{p^2+m^2}} \, , \quad \frac{m}{\sqrt{p^2+m^2}} \, , \quad \frac{p}{m} \, , \quad \frac{\sqrt{m^2+p^2}}{m} \, , \quad \frac{\sqrt{m^2+p^2}}{p}$$

11. acute

20

110

12. $\cos(\theta)$

13. 150

14. $\dfrac{3\sqrt{2}}{4}$

15. $-\dfrac{9}{8}$

16. $\dfrac{\sqrt{7}}{4} \, , \ -\dfrac{3}{4} \, , \ \dfrac{4\sqrt{7}}{7} \, , \ -\dfrac{4}{3} \, , \ -\dfrac{\sqrt{7}}{3} \, , \ -\dfrac{3\sqrt{7}}{7}$

17.

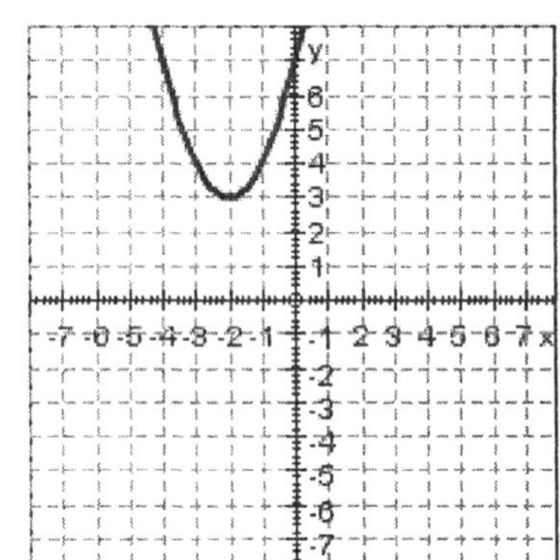

18. $2|\cot(\theta)|$

19. $\tan(\theta)=\dfrac{60}{11}$

20. $\cot(\theta)=\dfrac{1}{5a}$

21. 1.7

22. 12

23. $\dfrac{12}{13}$, $-\dfrac{5}{13}$, $\dfrac{13}{12}$, $-\dfrac{13}{5}$, $-\dfrac{12}{5}$, $-\dfrac{5}{12}$

24.
$$\sin(\theta) = -\dfrac{2}{\sqrt{13}} \, , \; \tan(\theta) = -\dfrac{2}{3} \, , \; \cot(\theta) = -\dfrac{3}{2} \, , \; \sec(\theta) = \dfrac{\sqrt{13}}{3} \, ,$$

$$\csc(\theta) = -\dfrac{\sqrt{13}}{2}$$

25.

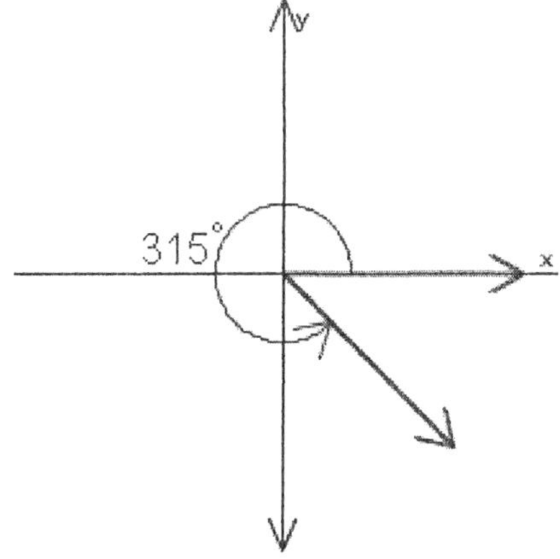

$(1, -1)$ for example

$-\dfrac{\sqrt{2}}{2}$

$\dfrac{\sqrt{2}}{2}$

-1

1. mctr.01.01.27_NoAlgs
2. mctr.01.02.35_NoAlgs
3. mctr.01.04.39_NoAlgs
4. mctr.01.03.44_NoAlgs
5. mctr.01.05.72_NoAlgs
6. mctr.01.05.37_NoAlgs
7. mctr.01.05.03_NoAlgs
8. mctr.01.02.02_NoAlgs
9. mctr.01.01.18_NoAlgs
10. mctr.01.03.60_NoAlgs
11. mctr.01.01.02_NoAlgs
12. mctr.01.05.25_NoAlgs
13. mctr.01.02.68_NoAlgs
14. mctr.01.01.53_NoAlgs
15. mctr.01.04.03_NoAlgs
16. mctr.01.03.16_NoAlgs
17. mctr.01.02.19_NoAlgs
18. mctr.01.05.51_NoAlgs
19. mctr.01.04.23_NoAlgs
20. mctr.01.04.13_NoAlgs
21. mctr.01.02.49_NoAlgs
22. mctr.01.01.40_NoAlgs
23. mctr.01.03.05_NoAlgs
24. mctr.01.04.49_NoAlgs
25. mctr.01.03.30_NoAlgs

1. Find α if $A = 5\alpha$.

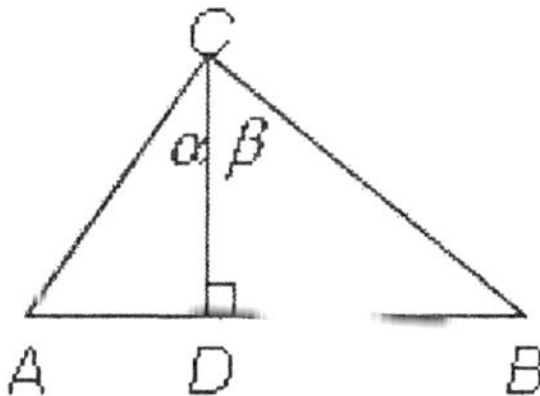

Select the correct answer.

a. $\alpha = 70°$

b. $\alpha = 20°$

c. $\alpha = 15°$

d. $\alpha = 75°$

e. $\alpha = 13°$

2. An isosceles triangle is a triangle in which two sides are equal in length. The angle between the two equal sides is called the vertex angle, while the other two angles are called the base angles. If the vertex angle is $20°$, what is the measure of the base angles?

Select the correct answer.

a. $65°$

b. $70°$

c. $85°$

d. $100°$

e. $80°$

3. Solve for x.

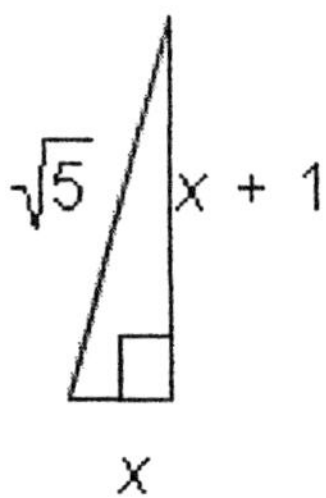

Select the correct answer.

a. $x = 1$

b. $x = 4$

c. $x = 6$

d. $x = 7$

e. $x = 2$

4. Find the remaining sides of a $30° - 60° - 90°$ triangle if the side opposite $60°$ is 18.

Select the correct answer.

a. $\dfrac{16\sqrt{3}}{7}$, $\dfrac{8\sqrt{3}}{7}$

b. $26\sqrt{2}$, $13\sqrt{2}$

c. $\dfrac{36\sqrt{2}}{7}$, $\dfrac{18\sqrt{3}}{7}$

d. $12\sqrt{3}$, $6\sqrt{3}$

e. $\dfrac{16\sqrt{3}}{3}$, $\dfrac{8\sqrt{3}}{3}$

5.   The object shown in the figure is a cube.

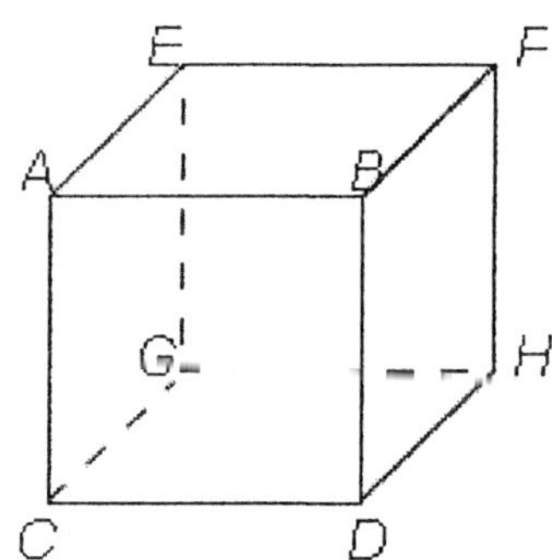

If the length of each edge of the cube shown in the figure is 3 inches, find the length of diagonal *CH*.

Select the correct answer.

a. $\dfrac{3\sqrt{3}}{4}$ inches

b. $3\sqrt{3}$ inches

c. 3 inches

d. $3\sqrt{2}$ inches

e. $\dfrac{3\sqrt{2}}{11}$ inches

6. Graph the ordered pair on a rectangular coordinate system:

$(0, 6)$

Select the correct answer.

a.

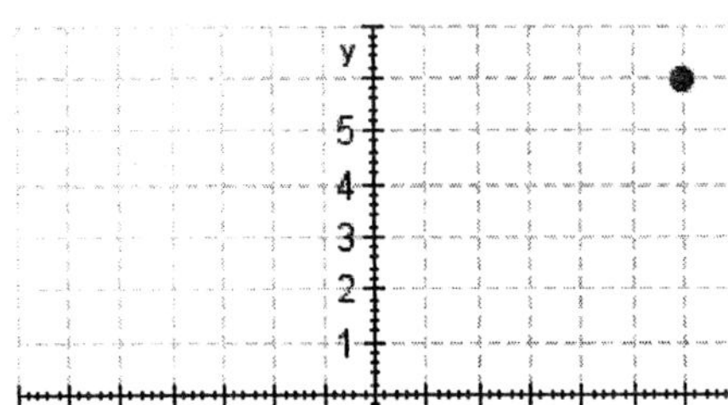

b.

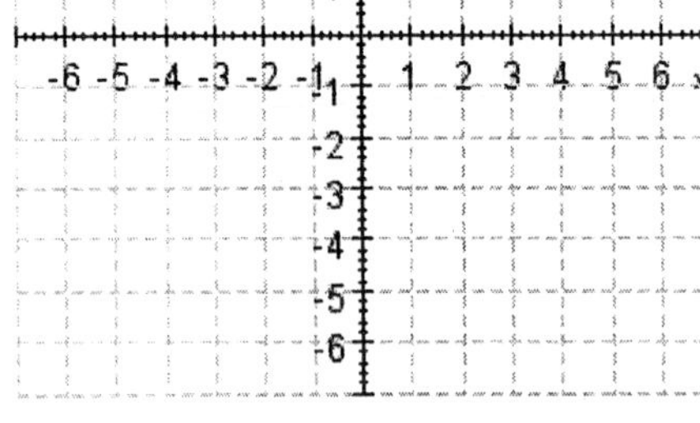

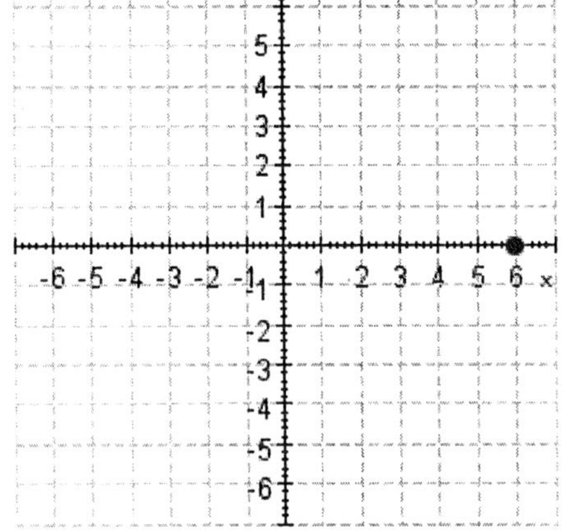

c.

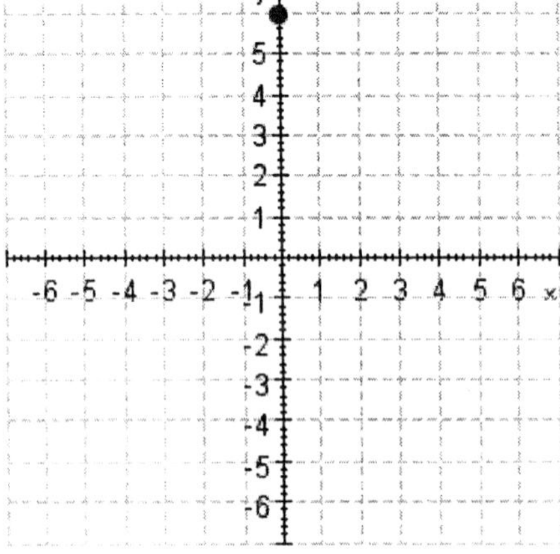

7. Graph the following circle.

$$x^2 + y^2 = 3$$

Select the correct answer.

a.

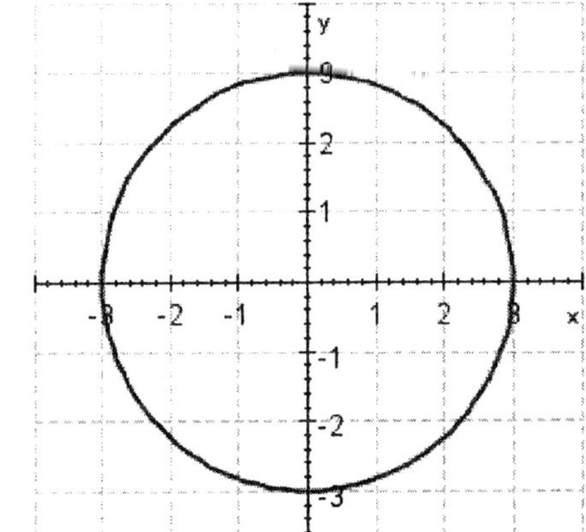

b.

c.

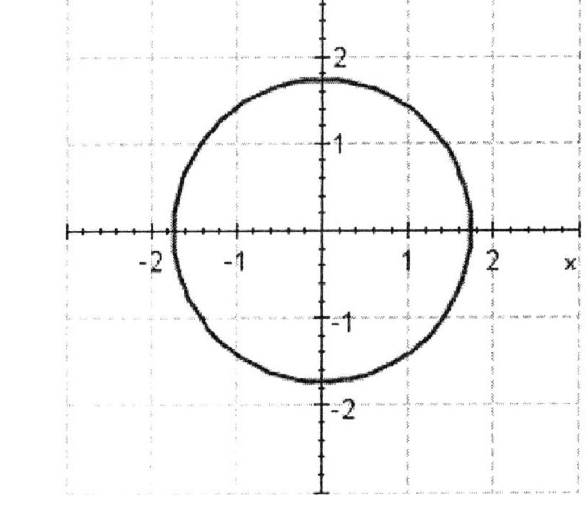

8. Find the distance between the points:

$$(-1, -4), (-11, 7)$$

Select the correct answer.

a. $\sqrt{113}$

b. $\sqrt{185}$

c. $\sqrt{202}$

d. $\sqrt{221}$

e. $\sqrt{181}$

9. Use the diagram to help find the complement of the angle 45°.

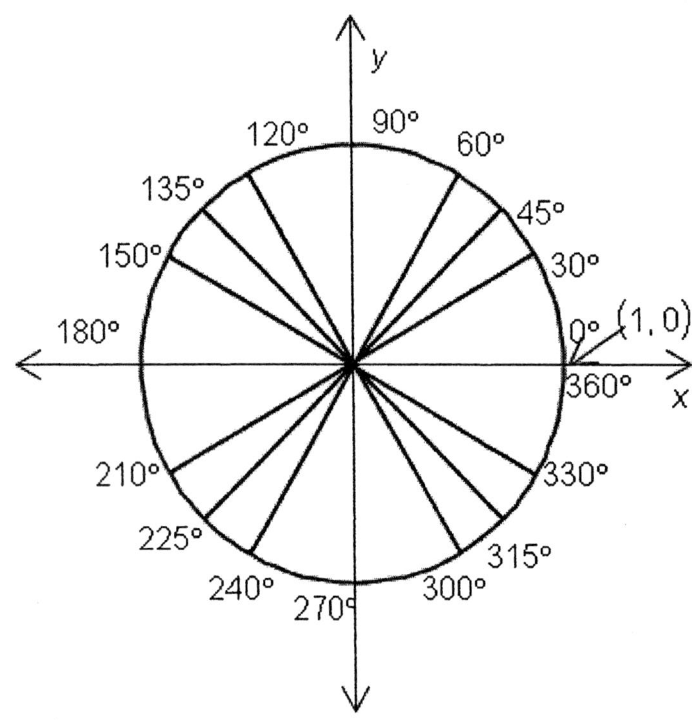

Select the correct answer.

a. 55°

b. 315°

c. 135°

d. 90°

e. 45°

10. Name another angle that is coterminal with $-60°$.

    Select the correct answer(s).

    a. $300°$

    b. $60°$

    c. $420°$

    d. $-420°$

    e. $-300°$

11. Find all six trigonometric functions of θ if the given point is on the terminal side of θ. $(-b, -a)$, assume that b and a are a positive numbers.

    Select the correct answer.

    a. $\sin\theta = -\dfrac{b}{\sqrt{b^2 + a^2}}$; $\cos\theta = -\dfrac{a}{\sqrt{b^2 + a^2}}$; $\csc\theta = -\dfrac{\sqrt{b^2 + a^2}}{a}$

    $\sec\theta = -\dfrac{\sqrt{b^2 + a^2}}{b}$; $\tan\theta = \dfrac{a}{b}$; $\cot\theta = \dfrac{b}{a}$

    b. $\sin\theta = -\dfrac{a}{\sqrt{b^2 + a^2}}$; $\cos\theta = -\dfrac{b}{\sqrt{b^2 + a^2}}$; $\csc\theta = -\dfrac{\sqrt{b^2 + a^2}}{a}$

    $\sec\theta = -\dfrac{\sqrt{b^2 + a^2}}{b}$; $\tan\theta = \dfrac{b}{a}$; $\cot\theta = \dfrac{a}{b}$

    c. $\sin\theta = -\dfrac{a}{\sqrt{b^2 + a^2}}$; $\cos\theta = -\dfrac{b}{\sqrt{b^2 + a^2}}$; $\csc\theta = -\dfrac{\sqrt{b^2 + a^2}}{a}$

    $\sec\theta = -\dfrac{\sqrt{b^2 + a^2}}{b}$; $\tan\theta = \dfrac{a}{b}$; $\cot\theta = \dfrac{b}{a}$

    d. $\sin\theta = -\dfrac{b}{\sqrt{b^2 + a^2}}$; $\cos\theta = -\dfrac{a}{\sqrt{b^2 + a^2}}$; $\csc\theta = -\dfrac{\sqrt{b^2 + a^2}}{a}$

    $\sec\theta = -\dfrac{\sqrt{b^2 + a^2}}{b}$; $\tan\theta = \dfrac{b}{a}$; $\cot\theta = \dfrac{a}{b}$

12. In the diagram below, angle θ is in standard position. Find $\sin \theta$, $\cos \theta$ and $\tan \theta$.

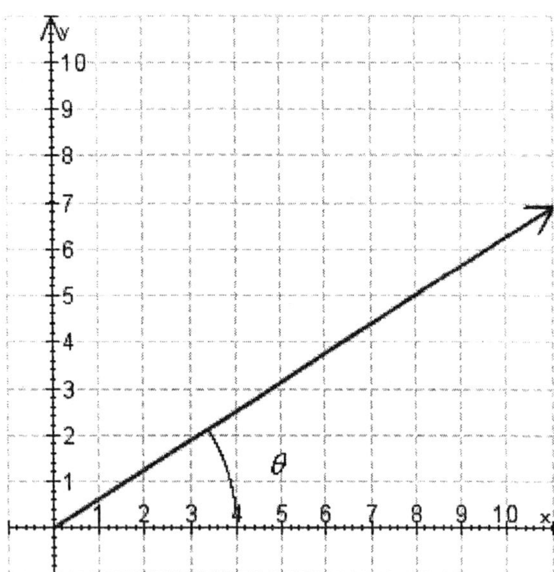

Select the correct answer.

a. $\sin \theta = \dfrac{5\sqrt{89}}{89}$, $\cos \theta = \dfrac{8\sqrt{89}}{89}$, $\tan \theta = \dfrac{5}{8}$

b. $\sin \theta = \dfrac{\sqrt{89}}{5}$, $\cos \theta = \dfrac{\sqrt{89}}{8}$, $\tan \theta = \dfrac{8}{5}$

c. $\sin \theta = \dfrac{8\sqrt{89}}{89}$, $\cos \theta = \dfrac{5\sqrt{89}}{89}$, $\tan \theta = \dfrac{8}{5}$

d. $\sin \theta = \dfrac{8\sqrt{89}}{89}$, $\cos \theta = \dfrac{5\sqrt{89}}{89}$, $\tan \theta = \dfrac{5}{8}$

e. $\sin \theta = \dfrac{\sqrt{89}}{8}$, $\cos \theta = \dfrac{\sqrt{89}}{5}$, $\tan \theta = \dfrac{8}{5}$

13. Indicate the quadrants in which the terminal side of θ must lie in order that $\cos \theta$ is positive and $\tan \theta$ is negative.

Select the correct answer.

a. I
b. IV
c. II
d. III
e. none of the above

14. Find the remaining trigonometric functions of θ if $\csc \theta = \dfrac{25}{7}$ and $\cos \theta < 0$.

Select the correct answer.

a. $\sin \theta = \dfrac{7}{25}$, $\cos \theta = -\dfrac{24}{25}$, $\tan \theta = -\dfrac{7}{24}$,

$\cot \theta = \dfrac{24}{7}$, $\sec \theta = \dfrac{25}{24}$

b. $\sin \theta = \dfrac{24}{25}$, $\cos \theta = \dfrac{7}{25}$, $\tan \theta = \dfrac{7}{24}$,

$\cot \theta = \dfrac{24}{7}$, $\sec \theta = \dfrac{25}{24}$

c. $\sin \theta = \dfrac{24}{25}$, $\cos \theta = -\dfrac{7}{25}$, $\tan \theta = -\dfrac{24}{7}$,

$\cot \theta = -\dfrac{7}{24}$, $\sec \theta = -\dfrac{25}{24}$

d. $\sin \theta = \dfrac{24}{25}$, $\cos \theta = -\dfrac{7}{25}$, $\tan \theta = -\dfrac{24}{7}$,

$\cot \theta = \dfrac{7}{24}$, $\sec \theta = \dfrac{25}{24}$

e. $\sin \theta = \dfrac{7}{25}$, $\cos \theta = -\dfrac{24}{25}$, $\tan \theta = -\dfrac{7}{24}$,

$\cot \theta = -\dfrac{24}{7}$, $\sec \theta = -\dfrac{25}{24}$

15. Find $\sin\theta$ and $\tan\theta$ if the terminal side of θ lies along the line $y = -2x$ in quadrant II.

Select the correct answer.

a. $\quad \sin\theta = \dfrac{2\sqrt{5}}{5},\ \tan\theta = \dfrac{1}{2}$

b. $\quad \sin\theta = -\dfrac{2\sqrt{5}}{5},\ \tan\theta = 2$

c. $\quad \sin\theta = -\dfrac{2\sqrt{5}}{5},\ \tan\theta = \dfrac{1}{2}$

d. $\quad \sin\theta = \dfrac{2\sqrt{5}}{5},\ \tan\theta = -\dfrac{1}{2}$

e. $\quad \sin\theta = \dfrac{2\sqrt{5}}{5},\ \tan\theta = -2$

16. Give the reciprocal of the given number.

$7x$

Select the correct answer.

a. $\quad 1 - 7x$

b. $\quad -7x$

c. $\quad -\dfrac{7}{x}$

d. $\quad 1 + 7x$

e. $\quad \dfrac{1}{7x}$

17. Use a ratio identity to find $\cot \theta$ if

$$\sin \theta = \frac{9}{\sqrt{181}} \quad \text{and}$$

$$\cos \theta = \frac{10}{\sqrt{181}}.$$

Select the correct answer.

a. $\cot \theta = \dfrac{9}{10\sqrt{181}}$

b. $\cot \theta = \dfrac{10}{9\sqrt{181}}$

c. $\cot \theta = \dfrac{10}{9}$

d. $\cot \theta = \dfrac{9}{10}$

e. $\cot \theta = \dfrac{10\sqrt{181}}{9}$

18. If $\sin \theta = -\dfrac{4}{5}$ and θ terminates in QIII, find $\cos \theta$.

a. $\cos \theta = -\dfrac{2}{5}$

b. $\cos \theta = \dfrac{2}{5}$

c. $\cos \theta = \dfrac{3}{5}$

d. $\cos \theta = \dfrac{4}{3}$

e. $\cos \theta = -\dfrac{3}{5}$

19. If $\cos\theta = -\dfrac{1}{9}$ and θ is not in QII, find $\tan\theta$

Select the correct answer.

a. $\tan\theta = 9\sqrt{5}$

b. $\tan\theta = 4\sqrt{2}$

c. $\tan\theta = -4\sqrt{5}$

d. $\tan\theta = 4\sqrt{5}$

e. $\tan\theta = -4\sqrt{2}$

20. Using your calculator and rounding your answers to the nearest hundredth, find $\tan\theta$ if

$\sec\theta = -1.74$ and $\theta \in$ QII.

Select the correct answer.

a. -1.42
b. -1.52
c. -1.82
d. -1.62
e. -1.92

21. Write the following in terms of $\sin\theta$ and $\cos\theta$ and then simplify if possible.

$\sec\theta \tan\theta \csc\theta$

Select the correct answer.

a. $\dfrac{\sin\theta}{\cos^2\theta}$

b. $\dfrac{\cos^2\theta}{\sin\theta}$

c. $\dfrac{1}{\sin\theta\cos\theta}$

d. $\dfrac{\sin^2\theta}{\cos\theta}$

e. $\dfrac{1}{\cos^2\theta}$

22. Add or subtract as indicated. Then simplify your answer if possible.

$$\sin\theta - \frac{1}{\cos\theta}$$

Select the correct answer.

a. $\dfrac{\sin\theta\cos\theta - 1}{\cos\theta}$

b. $\dfrac{\sin\theta}{\cos\theta}$

c. $\sin\theta$

d. $\dfrac{\sin\theta\cos\theta + 1}{\cos\theta}$

e. $\dfrac{\sin\theta\cos\theta + 1}{\sin\theta}$

23. Multiply.

$$(\sin\theta - \cos\theta)^2$$

Select the correct answer.

a. $\sin^2\theta - \cos^2\theta$

b. $1 - 2\sin\theta\cos\theta$

c. 1

d. $\sin^2\theta - \cos\theta$

e. $1 + 2\sin\theta\cos\theta$

24. Find the right part of the identity.

$$\cos \alpha \tan \alpha = \underline{\qquad}$$

Select the correct answer.

   a. $\cos \alpha$

   b. $\sec \alpha$

   c. $\tan \alpha$

   d. $\csc \alpha$

   e. $\sin \alpha$

25. Find the right part of the identity.

$$\sec \theta - \cos \theta = \underline{\qquad}$$

Select the correct answer.

   a. $\dfrac{\sin^2 \theta}{\cos \theta}$

   b. $\cot^2 \theta$

   c. $\dfrac{\cos^2 \theta}{\sin \theta}$

   d. $\tan^2 \theta$

   e. $\csc^2 \theta$

1. c

2. e

3. a

4. d

5. d

6. c

7. c

8. d

9. e

10. a,d

11. c

12. a

13. b

14. e

15. e

16. e

17. c

18. e

19. d

20. a

21. e

22. a

23. b

24. e

25. a

McKeague/Turner - Trigonometry 5e Chapter 1 Form C

1. mctr.01.01.12m_NoAlgs
2. mctr.01.01.24m_NoAlgs
3. mctr.01.01.35m_NoAlgs
4. mctr.01.01.47m_NoAlgs
5. mctr.01.01.61m_NoAlgs
6. mctr.01.02.09m_NoAlgs
7. mctr.01.02.27m_NoAlgs
8. mctr.01.02.43m_NoAlgs
9. mctr.01.02.60m_NoAlgs
10. mctr.01.02.77m_NoAlgs
11. mctr.01.03.09m_NoAlgs
12. mctr.01.03.24m_NoAlgs
13. mctr.01.03.39m_NoAlgs
14. mctr.01.03.52m_NoAlgs
15. mctr.01.03.67m_NoAlgs
16. mctr.01.04.07m_NoAlgs
17. mctr.01.04.18m_NoAlgs
18. mctr.01.04.31m_NoAlgs
19. mctr.01.04.46m_NoAlgs
20. mctr.01.04.57m_NoAlgs
21. mctr.01.05.09m_NoAlgs
22. mctr.01.05.31m_NoAlgs
23. mctr.01.05.43m_NoAlgs
24. mctr.01.05.59m_NoAlgs
25. mctr.01.05.75m_NoAlgs

1. Name another angle that is coterminal with -60°.

   Select the correct answer(s).

   a. 300°

   b. 60°

   c. 420°

   d. -420°

   e. -300°

2. Solve for x.

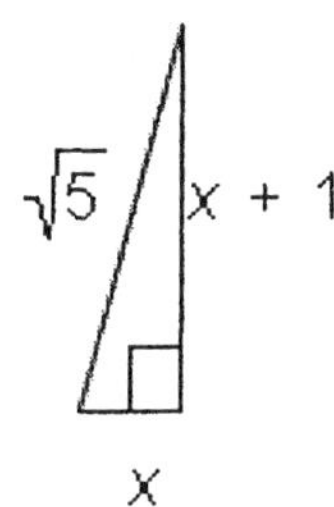

   Select the correct answer.

   a. $x = 1$

   b. $x = 4$

   c. $x = 6$

   d. $x = 7$

   e. $x = 2$

3. If $\cos \theta = -\dfrac{1}{9}$ and θ is not in QII, find $\tan \theta$

   Select the correct answer.

   a. $\tan \theta = 9\sqrt{5}$

   b. $\tan \theta = 4\sqrt{2}$

   c. $\tan \theta = -4\sqrt{5}$

   d. $\tan \theta = 4\sqrt{5}$

   e. $\tan \theta = -4\sqrt{2}$

4. An isosceles triangle is a triangle in which two sides are equal in length. The angle between the two equal sides is called the vertex angle, while the other two angles are called the base angles. If the vertex angle is $20°$, what is the measure of the base angles?

Select the correct answer.

    a. $65°$

    b. $70°$

    c. $85°$

    d. $100°$

    e. $80°$

5. Find all six trigonometric functions of θ if the given point is on the terminal side of θ.

$(-b, -a)$, assume that b and a are a positive numbers

Select the correct answer.

a. $\sin \theta = -\dfrac{b}{\sqrt{b^2 + a^2}}$; $\cos \theta = -\dfrac{a}{\sqrt{b^2 + a^2}}$; $\csc \theta = -\dfrac{\sqrt{b^2 + a^2}}{a}$

$\sec \theta = -\dfrac{\sqrt{b^2 + a^2}}{b}$; $\tan \theta = \dfrac{a}{b}$; $\cot \theta = \dfrac{b}{a}$

b. $\sin \theta = -\dfrac{a}{\sqrt{b^2 + a^2}}$; $\cos \theta = -\dfrac{b}{\sqrt{b^2 + a^2}}$; $\csc \theta = -\dfrac{\sqrt{b^2 + a^2}}{a}$

$\sec \theta = -\dfrac{\sqrt{b^2 + a^2}}{b}$; $\tan \theta = \dfrac{b}{a}$; $\cot \theta = \dfrac{a}{b}$

c. $\sin \theta = -\dfrac{a}{\sqrt{b^2 + a^2}}$; $\cos \theta = -\dfrac{b}{\sqrt{b^2 + a^2}}$; $\csc \theta = -\dfrac{\sqrt{b^2 + a^2}}{a}$

$\sec \theta = -\dfrac{\sqrt{b^2 + a^2}}{b}$; $\tan \theta = \dfrac{a}{b}$; $\cot \theta = \dfrac{b}{a}$

d. $\sin \theta = -\dfrac{b}{\sqrt{b^2 + a^2}}$; $\cos \theta = -\dfrac{a}{\sqrt{b^2 + a^2}}$; $\csc \theta = -\dfrac{\sqrt{b^2 + a^2}}{a}$

$\sec \theta = -\dfrac{\sqrt{b^2 + a^2}}{b}$; $\tan \theta = \dfrac{b}{a}$; $\cot \theta = \dfrac{a}{b}$

e. $\sin \theta = -\dfrac{b}{\sqrt{b^2 + a^2}}$; $\cos \theta = -\dfrac{a}{\sqrt{b^2 + a^2}}$; $\csc \theta = -\dfrac{\sqrt{b^2 + a^2}}{b}$

$\sec \theta = -\dfrac{\sqrt{b^2 + a^2}}{a}$; $\tan \theta = \dfrac{a}{b}$; $\cot \theta = \dfrac{b}{a}$

6. Using your calculator and rounding your answers to the nearest hundredth, find $\tan \theta$ if

$$\sec \theta = -1.74 \text{ and } \theta \in \text{QII.}$$

Select the correct answer.

a. -1.42
b. -1.52
c. -1.82
d. -1.62
e. -1.92

7. Use a ratio identity to find $\cot \theta$ if

$$\sin \theta = \frac{9}{\sqrt{181}} \text{ and}$$

$$\cos \theta = \frac{10}{\sqrt{181}}.$$

Select the correct answer.

a. $\cot \theta = \dfrac{9}{10\sqrt{181}}$

b. $\cot \theta = \dfrac{10}{9\sqrt{181}}$

c. $\cot \theta = \dfrac{10}{9}$

d. $\cot \theta = \dfrac{9}{10}$

e. $\cot \theta = \dfrac{10\sqrt{181}}{9}$

8. Find α if $A = 5\alpha$.

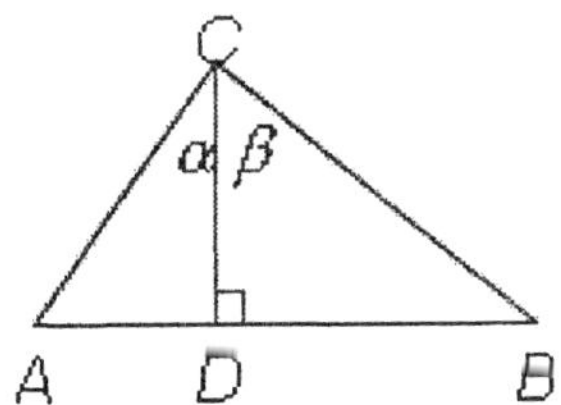

Select the correct answer.

a. $\alpha = 70°$

b. $\alpha = 20°$

c. $\alpha = 15°$

d. $\alpha = 75°$

e. $\alpha = 13°$

9. Give the reciprocal of the given number.

$7x$

Select the correct answer.

a. $1 - 7x$

b. $-7x$

c. $-\dfrac{7}{x}$

d. $1 + 7x$

e. $\dfrac{1}{7x}$

10. Graph the ordered pair on a rectangular coordinate system:

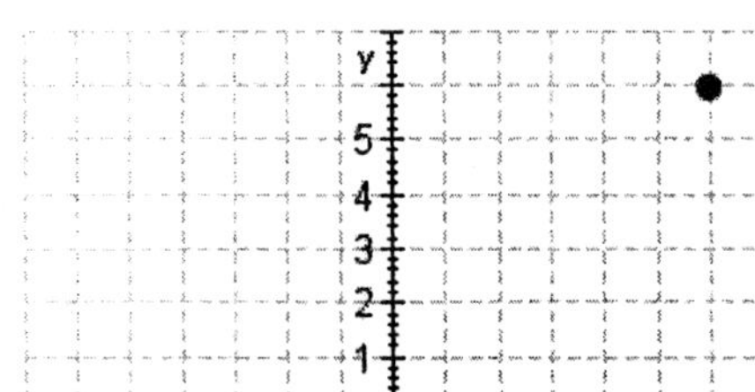

$(0, 6)$

Select the correct answer.

a.

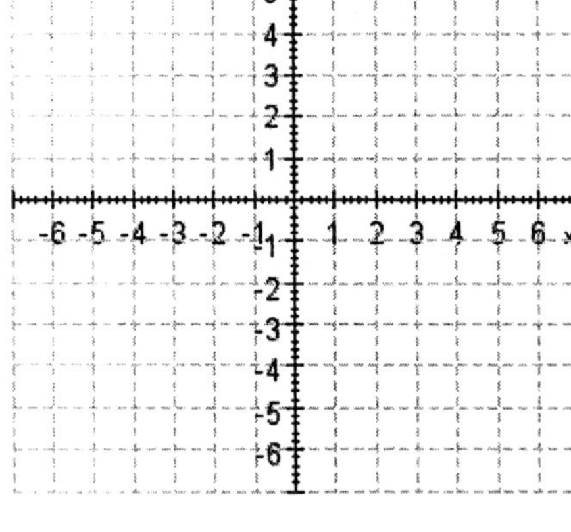

b.

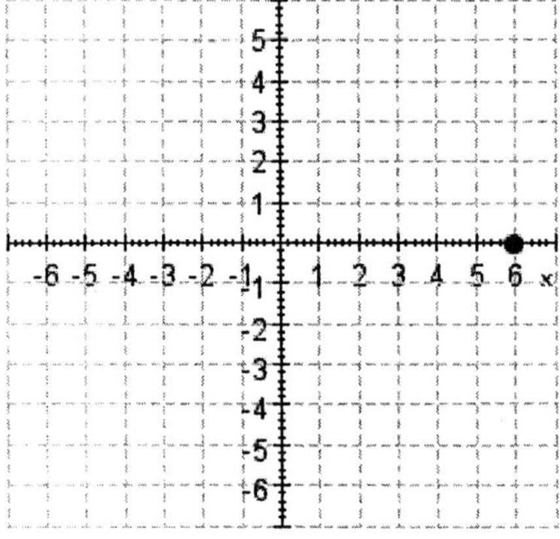

c.

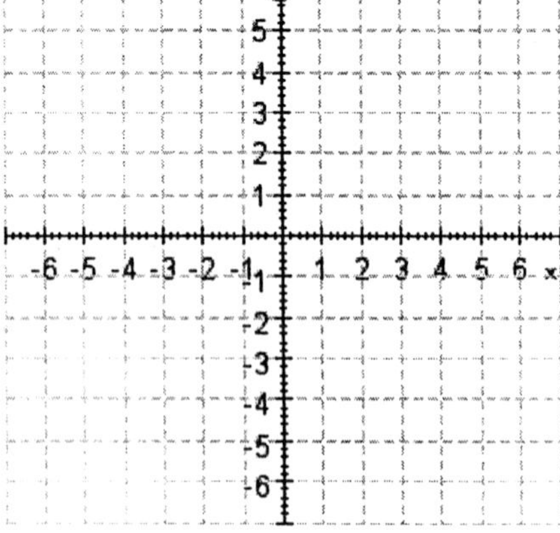

11.

Find the remaining trigonometric functions of θ if $\csc \theta = \dfrac{25}{7}$ and $\cos \theta < 0$.

Select the correct answer.

a. $\sin \theta = \dfrac{7}{25}$, $\cos \theta = -\dfrac{24}{25}$, $\tan \theta = -\dfrac{7}{24}$,

$\cot \theta = \dfrac{24}{7}$, $\sec \theta = \dfrac{25}{24}$

b. $\sin \theta = \dfrac{24}{25}$, $\cos \theta = \dfrac{7}{25}$, $\tan \theta = \dfrac{7}{24}$,

$\cot \theta = \dfrac{24}{7}$, $\sec \theta = \dfrac{25}{24}$

c. $\sin \theta = \dfrac{24}{25}$, $\cos \theta = -\dfrac{7}{25}$, $\tan \theta = -\dfrac{24}{7}$,

$\cot \theta = -\dfrac{7}{24}$, $\sec \theta = -\dfrac{25}{24}$

d. $\sin \theta = \dfrac{24}{25}$, $\cos \theta = -\dfrac{7}{25}$, $\tan \theta = -\dfrac{24}{7}$,

$\cot \theta = \dfrac{7}{24}$, $\sec \theta = \dfrac{25}{24}$

e. $\sin \theta = \dfrac{7}{25}$, $\cos \theta = -\dfrac{24}{25}$, $\tan \theta = -\dfrac{7}{24}$,

$\cot \theta = -\dfrac{24}{7}$, $\sec \theta = -\dfrac{25}{24}$

12. Add or subtract as indicated. Then simplify your answer if possible.

$$\sin\theta - \frac{1}{\cos\theta}$$

Select the correct answer.

a. $\dfrac{\sin\theta\cos\theta - 1}{\cos\theta}$

b. $\dfrac{\sin\theta}{\cos\theta}$

c. $\sin\theta$

d. $\dfrac{\sin\theta\cos\theta + 1}{\cos\theta}$

e. $\dfrac{\sin\theta\cos\theta + 1}{\sin\theta}$

13. Find the distance between the points:

$$(-1, -4), (-11, 7)$$

Select the correct answer.

a. $\sqrt{113}$

b. $\sqrt{185}$

c. $\sqrt{202}$

d. $\sqrt{221}$

e. $\sqrt{181}$

14. Use the diagram to help find the complement of the angle $45°$.

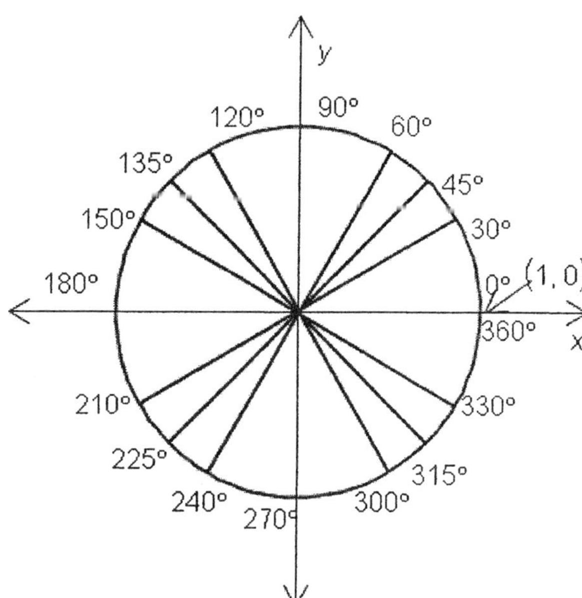

Select the correct answer.

a. $55°$

b. $315°$

c. $135°$

d. $90°$

e. $45°$

15. Find $\sin\theta$ and $\tan\theta$ if the terminal side of θ lies along the line $y = -2x$ in quadrant II.

Select the correct answer.

a. $\sin\theta = \dfrac{2\sqrt{5}}{5}$, $\tan\theta = \dfrac{1}{2}$

b. $\sin\theta = -\dfrac{2\sqrt{5}}{5}$, $\tan\theta = 2$

c. $\sin\theta = -\dfrac{2\sqrt{5}}{5}$, $\tan\theta = \dfrac{1}{2}$

d. $\sin\theta = \dfrac{2\sqrt{5}}{5}$, $\tan\theta = -\dfrac{1}{2}$

e. $\sin\theta = \dfrac{2\sqrt{5}}{5}$, $\tan\theta = -2$

16. Graph the following circle.

$$x^2 + y^2 = 3$$

Select the correct answer.

a.

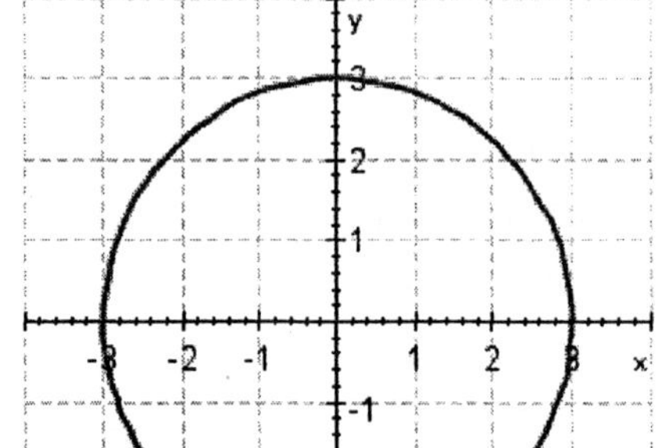

b.

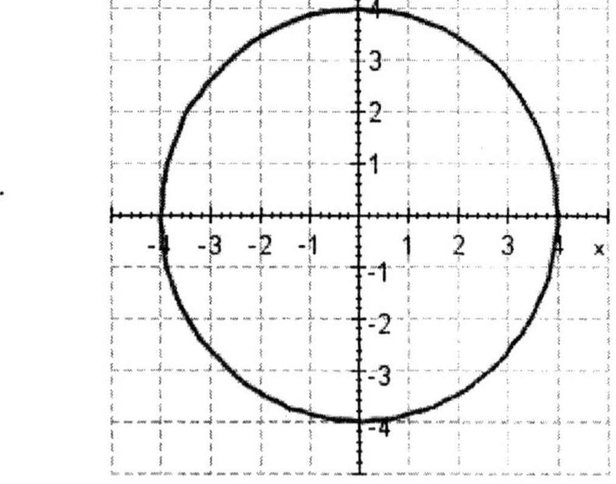

c.

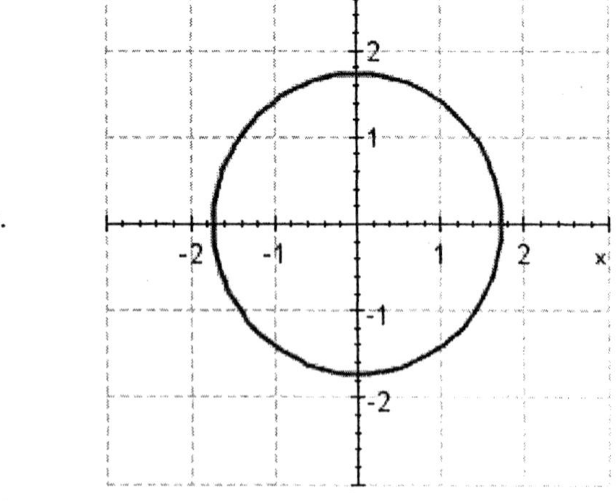

17. In the diagram below, angle θ is in standard position. Find $\sin\theta$, $\cos\theta$ and $\tan\theta$.

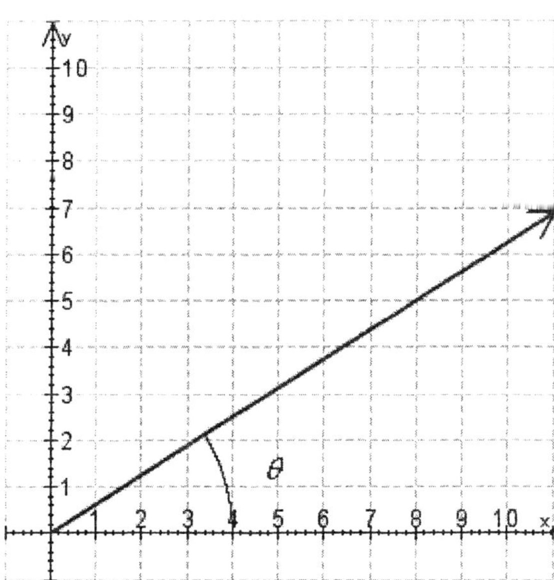

Select the correct answer.

a. $\quad \sin\theta = \dfrac{5\sqrt{89}}{89}$, $\cos\theta = \dfrac{8\sqrt{89}}{89}$, $\tan\theta = \dfrac{5}{8}$

b. $\quad \sin\theta = \dfrac{\sqrt{89}}{5}$, $\cos\theta = \dfrac{\sqrt{89}}{8}$, $\tan\theta = \dfrac{8}{5}$

c. $\quad \sin\theta = \dfrac{8\sqrt{89}}{89}$, $\cos\theta = \dfrac{5\sqrt{89}}{89}$, $\tan\theta = \dfrac{8}{5}$

d. $\quad \sin\theta = \dfrac{8\sqrt{89}}{89}$, $\cos\theta = \dfrac{5\sqrt{89}}{89}$, $\tan\theta = \dfrac{5}{8}$

e. $\quad \sin\theta = \dfrac{\sqrt{89}}{8}$, $\cos\theta = \dfrac{\sqrt{89}}{5}$, $\tan\theta = \dfrac{8}{5}$

18. The object shown in the figure is a cube.

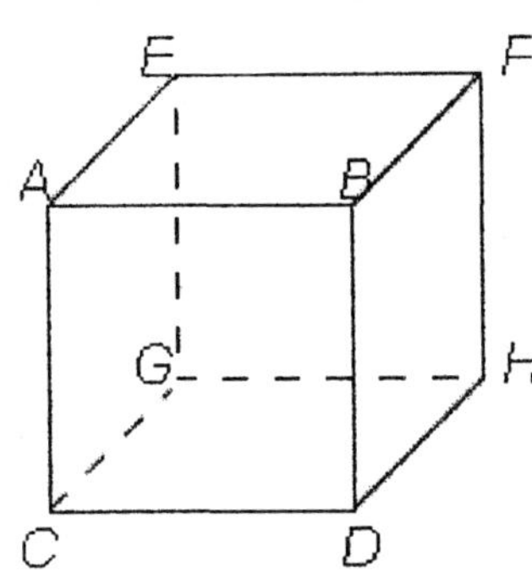

If the length of each edge of the cube shown in the figure is 3 inches, find the length of diagonal *CH*.

Select the correct answer.

a. $\dfrac{3\sqrt{3}}{4}$ inches

b. $3\sqrt{3}$ inches

c. 3 inches

d. $3\sqrt{2}$ inches

e. $\dfrac{3\sqrt{2}}{11}$ inches

19. Find the remaining sides of a $30^\circ - 60^\circ - 90^\circ$ triangle if the side opposite 60° is 18.

Select the correct answer.

a. $\dfrac{16\sqrt{3}}{7} , \dfrac{8\sqrt{3}}{7}$

b. $26\sqrt{2} , 13\sqrt{2}$

c. $\dfrac{36\sqrt{2}}{7} , \dfrac{18\sqrt{3}}{7}$

d. $12\sqrt{3} , 6\sqrt{3}$

e. $\dfrac{16\sqrt{3}}{3} , \dfrac{8\sqrt{3}}{3}$

20. Indicate the quadrants in which the terminal side of θ must lie in order that $\cos\theta$ is positive and $\tan\theta$ is negative.

Select the correct answer.

a. I
b. IV
c. II
d. III
e. none of the above

21. Multiply.

$$\left(\sin\theta - \cos\theta\right)^2$$

Select the correct answer.

a. $\sin^2\theta - \cos^2\theta$

b. $1 - 2\sin\theta\cos\theta$

c. 1

d. $\sin^2\theta - \cos\theta$

e. $1 + 2\sin\theta\cos\theta$

22. Write the following in terms of $\sin\theta$ and $\cos\theta$ and then simplify if possible.

$$\sec\theta\,\tan\theta\,\csc\theta$$

Select the correct answer.

a. $\dfrac{\sin\theta}{\cos^2\theta}$

b. $\dfrac{\cos^2\theta}{\sin\theta}$

c. $\dfrac{1}{\sin\theta\cos\theta}$

d. $\dfrac{\sin^2\theta}{\cos\theta}$

e. $\dfrac{1}{\cos^2\theta}$

23. Find the right part of the identity.

$$\sec \theta - \cos \theta = \underline{\hspace{1cm}}$$

Select the correct answer.

a. $\dfrac{\sin^2 \theta}{\cos \theta}$

b. $\cot^2 \theta$

c. $\dfrac{\cos^2 \theta}{\sin \theta}$

d. $\tan^2 \theta$

24. If $\sin \theta = -\dfrac{4}{5}$ and θ terminates in QIII, find $\cos \theta$.

Select the correct answer.

a. $\cos \theta = -\dfrac{2}{5}$

b. $\cos \theta = \dfrac{2}{5}$

c. $\cos \theta = \dfrac{3}{5}$

d. $\cos \theta = \dfrac{4}{3}$

e. $\cos \theta = -\dfrac{3}{5}$

25. Find the right part of the identity.

$$\cos \alpha \tan \alpha = \underline{\hspace{1cm}}$$

Select the correct answer.

a. $\cos \alpha$

b. $\sec \alpha$

c. $\tan \alpha$

d. $\csc \alpha$

e. $\sin \alpha$

1. a,d
2. a
3. d
4. e
5. c
6. a
7. c
8. c
9. e
10. c
11. e
12. a
13. d
14. e
15. e
16. c
17. a
18. d
19. d
20. b
21. b
22. e
23. a
24. e
25. e

McKeague/Turner - Trigonometry 5e Chapter 1 Form D

1. mctr.01.02.77m_NoAlgs
2. mctr.01.01.35m_NoAlgs
3. mctr.01.04.46m_NoAlgs
4. mctr.01.01.24m_NoAlgs
5. mctr.01.03.09m_NoAlgs
6. mctr.01.04.57m_NoAlgs
7. mctr.01.04.18m_NoAlgs
8. mctr.01.01.12m_NoAlgs
9. mctr.01.04.07m_NoAlgs
10. mctr.01.02.09m_NoAlgs
11. mctr.01.03.52m_NoAlgs
12. mctr.01.05.31m_NoAlgs
13. mctr.01.02.43m_NoAlgs
14. mctr.01.02.60m_NoAlgs
15. mctr.01.03.67m_NoAlgs
16. mctr.01.02.27m_NoAlgs
17. mctr.01.03.24m_NoAlgs
18. mctr.01.01.61m_NoAlgs
19. mctr.01.01.47m_NoAlgs
20. mctr.01.03.39m_NoAlgs
21. mctr.01.05.43m_NoAlgs
22. mctr.01.05.09m_NoAlgs
23. mctr.01.05.75m_NoAlgs
24. mctr.01.04.31m_NoAlgs
25. mctr.01.05.59m_NoAlgs

1. Indicate if the angle 70° is acute or obtuse.

____

Give the complement angle.

____ $^\circ$

Give the supplement angle.

____ $^\circ$

2. Find α if $A = 5\alpha$.

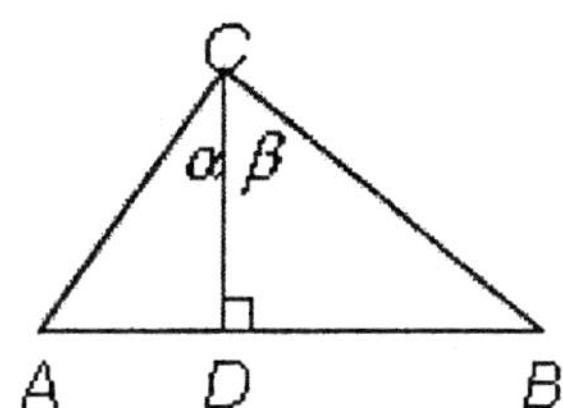

Select the correct answer.

a. $\alpha = 70^\circ$

b. $\alpha = 20^\circ$

c. $\alpha = 15^\circ$

d. $\alpha = 75^\circ$

e. $\alpha = 13^\circ$

3. Suppose ABC is a right triangle with $C = 90^\circ$.

If $a = 9$ and $c = 15$, find b.

4. Solve for x.

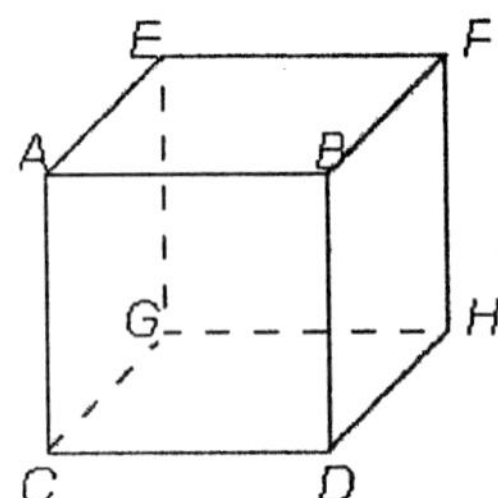

Select the correct answer.

   a. $x = 1$

   b. $x = 4$

   c. $x = 6$

   d. $x = 7$

   e. $x = 2$

5. Find the remaining side of a $45° - 45° - 90°$ triangle if the shorter sides are each $\dfrac{3}{4}$.

6. The object shown in the figure is a cube.

If the length of each edge of the cube shown in the figure is 3 inches, find the length of diagonal CH.

Select the correct answer.

   a. $\dfrac{3\sqrt{3}}{4}$ inches

   b. $3\sqrt{3}$ inches

   c. 3 inches

   d. $3\sqrt{2}$ inches

7. Graph the parabola.

$$2x + (x + 1)^2 + 6$$

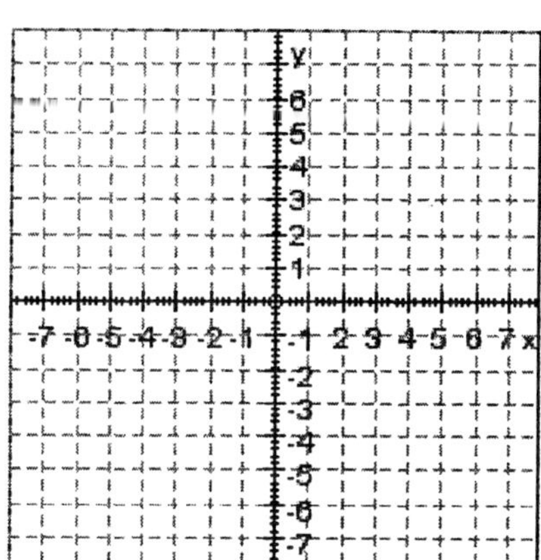

8. Graph the following circle.

$$x^2 + y^2 = 3$$

Select the correct answer.

a.

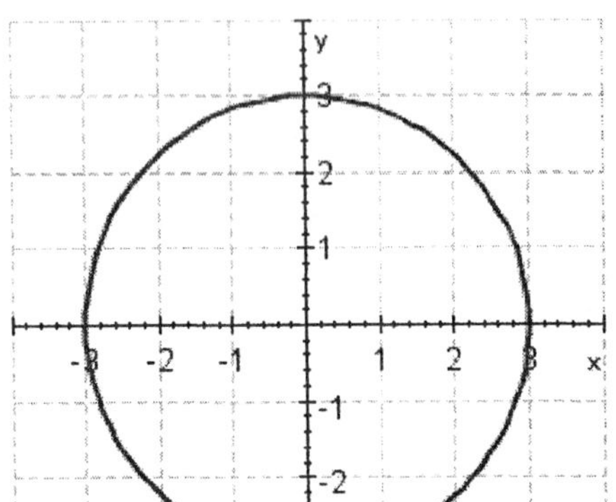

b.

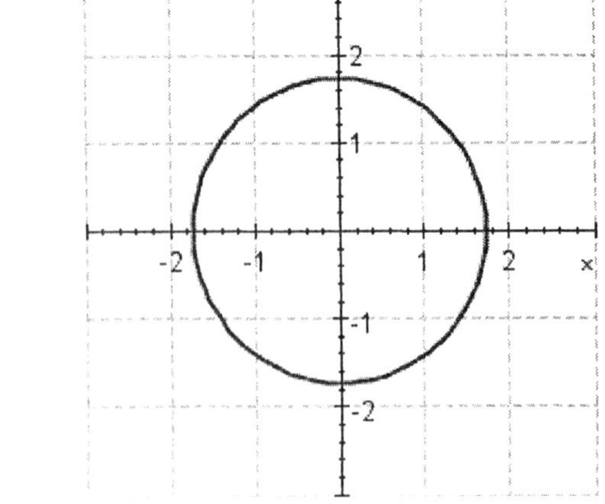

c.

e.

9. An airplane is approaching Los Angeles International Airport at an altitude of 4,224 feet. If the horizontal distance from the plane to the runway is 1.5 miles, use the Pythagorean Theorem to find the diagonal distance from the plane to the runway (see the figure below). (5,280 feet equals 1 mile.)

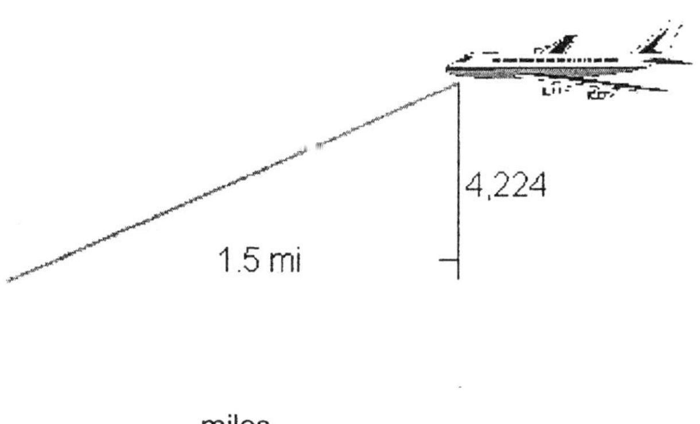

__________ miles

10. Use the diagram to help find the complement of the angle $45°$.

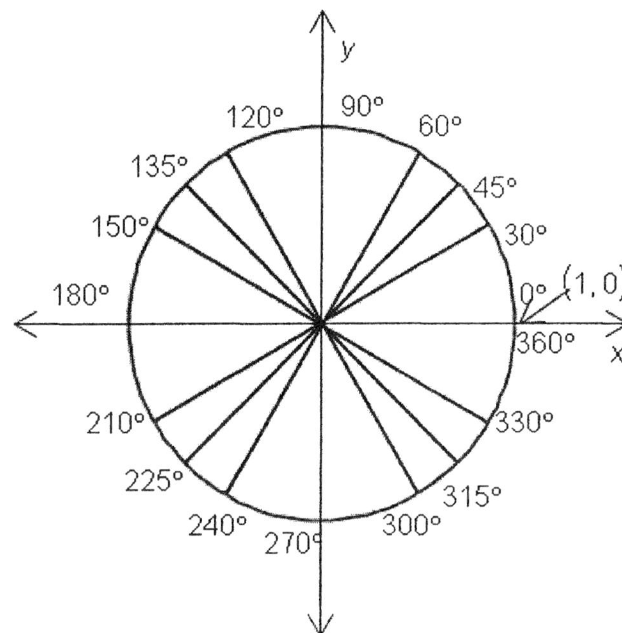

Select the correct answer.

a. $55°$

b. $315°$

c. $135°$

d. $90°$

e. $45°$

11. Find all six trigonometric functions of θ if the given point is on the terminal side of θ.

$(-5, 12)$

Find $\sin \theta$.

Find $\cos \theta$.

Find $\csc \theta$.

Find $\sec \theta$.

Find $\tan \theta$.

Find $\cot \theta$.

12. Find all six trigonometric functions of θ if the given point is on the terminal side of θ.

$(-b, -a)$, assume that b and a are a positive numbers

Select the correct answer.

a. $\quad \sin\theta = -\dfrac{b}{\sqrt{b^2 + a^2}}; \quad \cos\theta = -\dfrac{a}{\sqrt{b^2 + a^2}}; \quad \csc\theta = -\dfrac{\sqrt{b^2 + a^2}}{a}$

$\quad \sec\theta = -\dfrac{\sqrt{b^2 + a^2}}{b}; \quad \tan\theta = \dfrac{a}{b}; \quad \cot\theta = \dfrac{b}{a}$

b. $\quad \sin\theta = -\dfrac{a}{\sqrt{b^2 + a^2}}; \quad \cos\theta = -\dfrac{b}{\sqrt{b^2 + a^2}}; \quad \csc\theta = -\dfrac{\sqrt{b^2 + a^2}}{a}$

$\quad \sec\theta = -\dfrac{\sqrt{b^2 + a^2}}{b}; \quad \tan\theta = \dfrac{b}{a}; \quad \cot\theta = \dfrac{a}{b}$

c. $\quad \sin\theta = -\dfrac{a}{\sqrt{b^2 + a^2}}; \quad \cos\theta = -\dfrac{b}{\sqrt{b^2 + a^2}}; \quad \csc\theta = -\dfrac{\sqrt{b^2 + a^2}}{a}$

$\quad \sec\theta = -\dfrac{\sqrt{b^2 + a^2}}{b}; \quad \tan\theta = \dfrac{a}{b}; \quad \cot\theta = \dfrac{b}{a}$

d. $\quad \sin\theta = -\dfrac{b}{\sqrt{b^2 + a^2}}; \quad \cos\theta = -\dfrac{a}{\sqrt{b^2 + a^2}}; \quad \csc\theta = -\dfrac{\sqrt{b^2 + a^2}}{a}$

$\quad \sec\theta = -\dfrac{\sqrt{b^2 + a^2}}{b}; \quad \tan\theta = \dfrac{b}{a}; \quad \cot\theta = \dfrac{a}{b}$

e. $\quad \sin\theta = -\dfrac{b}{\sqrt{b^2 + a^2}}; \quad \cos\theta = -\dfrac{a}{\sqrt{b^2 + a^2}}; \quad \csc\theta = -\dfrac{\sqrt{b^2 + a^2}}{b}$

$\quad \sec\theta = -\dfrac{\sqrt{b^2 + a^2}}{a}; \quad \tan\theta = \dfrac{a}{b}; \quad \cot\theta = \dfrac{b}{a}$

13. Draw the following angle in standard position.

$315°$

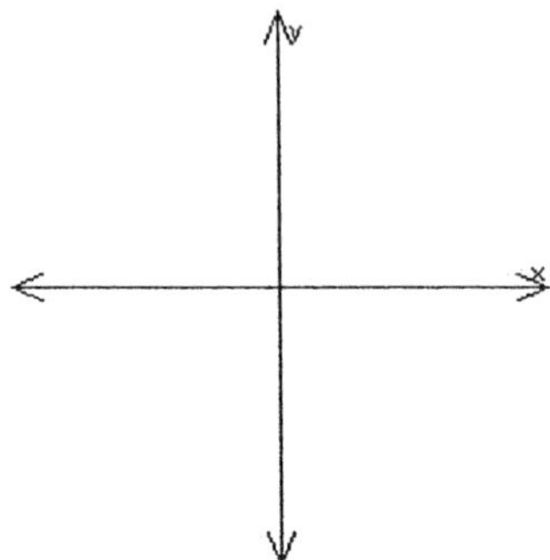

Find a point on the terminal side.

Find the sine of the angle.

Find the cosine of the angle.

Find the tangent of the angle.

14. Indicate the quadrants in which the terminal side of θ must lie in order that $\cos\theta$ is positive and $\tan\theta$ is negative.

Select the correct answer.

a.   I
b.   IV
c.   II
d.   III
e.   none of the above

15. Find the remaining trigonometric functions of θ if $\cot\theta = \dfrac{m}{p}$ where m and p are both positive. Assume that θ is between $0°$ and $180°$.

Find $\sin\theta$.

Find $\cos\theta$.

Find $\tan\theta$.

Find $\sec\theta$.

Find $\csc\theta$.

16. Find $\sin\theta$ and $\tan\theta$ if the terminal side of θ lies along the line $y = -2x$ in quadrant II.

Select the correct answer.

a. $\sin\theta = \dfrac{2\sqrt{5}}{5}$, $\tan\theta = \dfrac{1}{2}$

b. $\sin\theta = -\dfrac{2\sqrt{5}}{5}$, $\tan\theta - 2$

c. $\sin\theta = -\dfrac{2\sqrt{5}}{5}$, $\tan\theta = \dfrac{1}{2}$

d. $\sin\theta = \dfrac{2\sqrt{5}}{5}$, $\tan\theta = -\dfrac{1}{2}$

e. $\sin\theta = \dfrac{2\sqrt{5}}{5}$, $\tan\theta = -2$

17.   Use the reciprocal identities for the following problem.

If $\tan\theta = 5a$, $(a \neq 0)$ find $\cot\theta$.

18.   Use a ratio identity to find $\cot\theta$ if

$$\sin\theta = \dfrac{9}{\sqrt{181}} \quad \text{and} \quad \cos\theta = \dfrac{10}{\sqrt{181}}.$$

Select the correct answer.

a. $\cot\theta = \dfrac{9}{10\sqrt{181}}$

b. $\cot\theta = \dfrac{10}{9\sqrt{181}}$

c. $\cot\theta = \dfrac{10}{9}$

d. $\cot\theta = \dfrac{9}{10}$

e. $\cot\theta = \dfrac{10\sqrt{181}}{9}$

19. Find $\sec \theta$ if $\tan \theta = \dfrac{4}{3}$ and θ terminates in QIII.

20. If $\cos \theta = -\dfrac{1}{9}$ and θ is not in QII, find $\tan \theta$

Select the correct answer.

a. $\tan \theta = 9\sqrt{5}$

b. $\tan \theta = 4\sqrt{2}$

c. $\tan \theta = -4\sqrt{5}$

d. $\tan \theta = 4\sqrt{5}$

e. $\tan \theta = -4\sqrt{2}$

21. Write the following in terms of $\sin \theta$ only.

$\tan \theta$

22. Write the following in terms of $\sin \theta$ and $\cos \theta$ and then simplify if possible.

$\sec \theta \tan \theta \csc \theta$

Select the correct answer.

a. $\dfrac{\sin \theta}{\cos^2 \theta}$

b. $\dfrac{\cos^2 \theta}{\sin \theta}$

c. $\dfrac{1}{\sin \theta \cos \theta}$

d. $\dfrac{\sin^2 \theta}{\cos \theta}$

e. $\dfrac{1}{\cos^2 \theta}$

23. Multiply.

$$(4\cos\theta + 7)(5\cos\theta - 7)$$

24. Multiply.

$$(\sin\theta - \cos\theta)^2$$

Select the correct answer.

a. $\sin^2\theta - \cos^2\theta$

b. $1 - 2\sin\theta\cos\theta$

c. 1

d. $\sin^2\theta - \cos\theta$

e. $1 + 2\sin\theta\cos\theta$

25. Show that the following statement is true by transforming the left side into the right side.

$$\cos\theta\cot\theta + \sin\theta = \csc\theta$$

1. acute

   20

   110

2. c

3. 12

4. a

5. $\dfrac{3\sqrt{2}}{4}$

6. d

7. 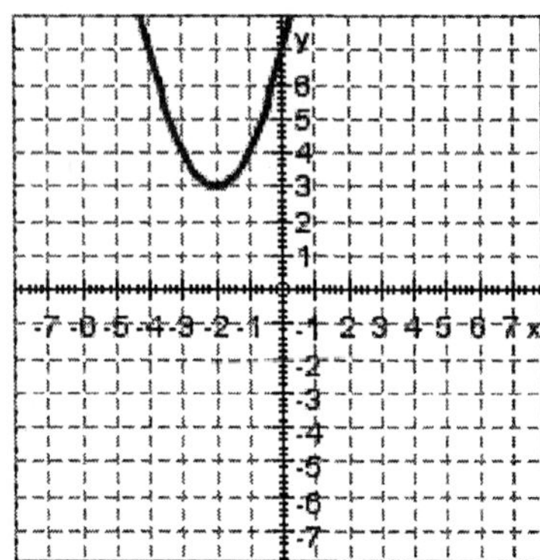

8. c

9. 1.7

10. e

11. $\dfrac{12}{13}, \ -\dfrac{5}{13}, \ \dfrac{13}{12}, \ -\dfrac{13}{5}, \ -\dfrac{12}{5}, \ -\dfrac{5}{12}$

12. c

13.

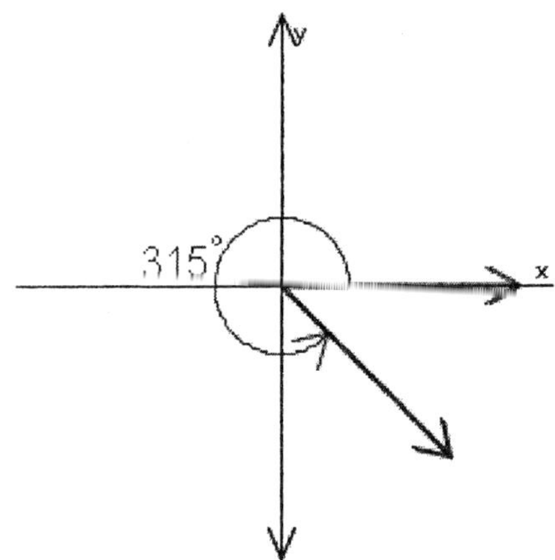

$(1, -1)$ for example

$$-\frac{\sqrt{2}}{2}$$

$$\frac{\sqrt{2}}{2}$$

$$-1$$

14. b

15.
$$\frac{p}{\sqrt{p^2+m^2}},$$

$$\frac{m}{\sqrt{p^2+m^2}},$$

$$\frac{p}{m},$$

$$\frac{\sqrt{m^2+p^2}}{m},$$

$$\frac{\sqrt{m^2+p^2}}{p}$$

16. e

17. $\cot(\theta)=\dfrac{1}{5a}$

18. c

19. $\sec(\theta) = -\dfrac{5}{3}$

20. d

21. $\pm\dfrac{\sin(\theta)}{\sqrt{1-(\sin(\theta))^2}}$

22. e

23. $20(\cos(\theta))^2 + 7\cos(\theta) - 49$

24. b

25. We begin by writing everything on the left side in terms of $\sin\theta$ and $\cos\theta$.

$$\cos\theta\cot\theta + \sin\theta = \cos\theta\,\frac{\cos\theta}{\sin\theta} + \sin\theta = \frac{\cos^2\theta + \sin^2\theta}{\sin\theta}$$

$$= \frac{1}{\sin\theta} = \csc\theta$$

Since we have succeeded in transforming the left side into the right side, we have shown that the statement $\cos\theta\cot\theta + \sin\theta = \csc\theta$ is an identity.

McKeague/Turner - Trigonometry 5e Chapter 1 Form E

1. mctr.01.01.02_NoAlgs
2. mctr.01.01.12m_NoAlgs
3. mctr.01.01.27_NoAlgs
4. mctr.01.01.35m_NoAlgs
5. mctr.01.01.53_NoAlgs
6. mctr.01.01.61m_NoAlgs
7. mctr.01.02.19_NoAlgs
8. mctr.01.02.27m_NoAlgs
9. mctr.01.02.49_NoAlgs
10. mctr.01.02.60m_NoAlgs
11. mctr.01.03.05_NoAlgs
12. mctr.01.03.09m_NoAlgs
13. mctr.01.03.30_NoAlgs
14. mctr.01.03.39m_NoAlgs
15. mctr.01.03.60_NoAlgs
16. mctr.01.03.67m_NoAlgs
17. mctr.01.04.13_NoAlgs
18. mctr.01.04.18m_NoAlgs
19. mctr.01.04.39_NoAlgs
20. mctr.01.04.46m_NoAlgs
21. mctr.01.05.03_NoAlgs
22. mctr.01.05.09m_NoAlgs
23. mctr.01.05.37_NoAlgs
24. mctr.01.05.43m_NoAlgs
25. mctr.01.05.72_NoAlgs

1. Draw the following angle in standard position.

    $315°$

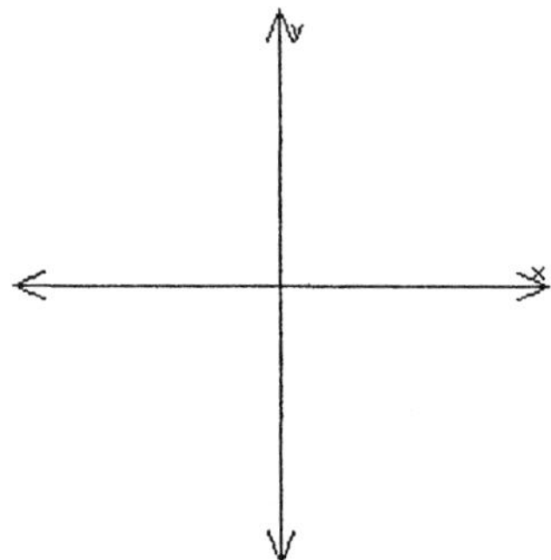

Find a point on the terminal side.

Find the sine of the angle.

Find the cosine of the angle.

Find the tangent of the angle.

2. Use the reciprocal identities for the following problem.

    If $\tan\theta = 5a$, $(a \neq 0)$ find $\cot\theta$.

3. Find $\sin\theta$ and $\tan\theta$ if the terminal side of θ lies along the line $y = -2x$ in quadrant II.

    Select the correct answer.

    a. $\sin\theta = \dfrac{2\sqrt{5}}{5}$, $\tan\theta = \dfrac{1}{2}$

    b. $\sin\theta = -\dfrac{2\sqrt{5}}{5}$, $\tan\theta = 2$

    c. $\sin\theta = -\dfrac{2\sqrt{5}}{5}$, $\tan\theta = \dfrac{1}{2}$

    d. $\sin\theta = \dfrac{2\sqrt{5}}{5}$, $\tan\theta = -\dfrac{1}{2}$

    e. $\sin\theta = \dfrac{2\sqrt{5}}{5}$, $\tan\theta = -2$

4. Find the remaining side of a $45° - 45° - 90°$ triangle if the shorter sides are each $\dfrac{3}{4}$.

5. Indicate if the angle $70°$ is acute or obtuse.

____

Give the complement angle.

_____ $°$

Give the supplement angle.

_____ $°$

6. Use the diagram to help find the complement of the angle $45°$.

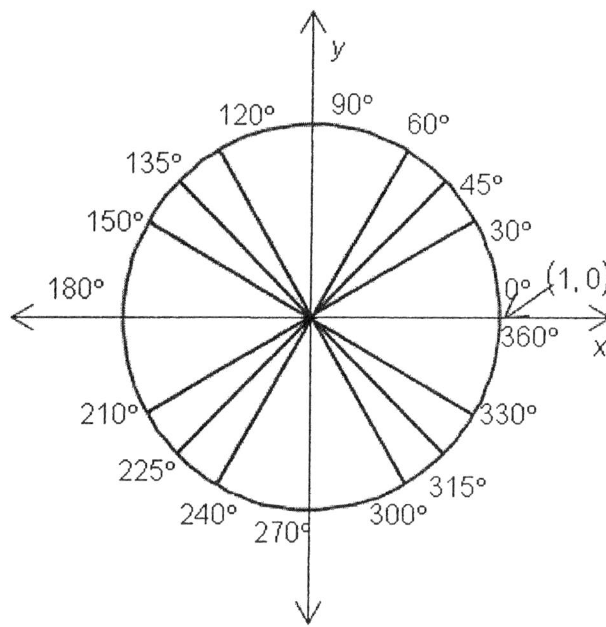

Select the correct answer.

a. $55°$

b. $315°$

c. $135°$

d. $90°$

e. $45°$

7. Indicate the quadrants in which the terminal side of θ must lie in order that $\cos\theta$ is positive and $\tan\theta$ is negative.

   Select the correct answer.

   a.   I
   b.   IV
   c.   II
   d.   III
   e.   none of the above

8. If $\cos\theta = -\dfrac{1}{9}$ and θ is not in QII, find $\tan\theta$

   Select the correct answer.

   a. $\tan\theta = 9\sqrt{5}$

   b. $\tan\theta = 4\sqrt{2}$

   c. $\tan\theta = -4\sqrt{5}$

   d. $\tan\theta = 4\sqrt{5}$

   e. $\tan\theta = -4\sqrt{2}$

9. Multiply.

   $$\left(\sin\theta - \cos\theta\right)^2$$

   Select the correct answer.

   a. $\sin^2\theta - \cos^2\theta$

   b. $1 - 2\sin\theta\cos\theta$

   c. 1

   d. $\sin^2\theta - \cos\theta$

   e. $1 + 2\sin\theta\cos\theta$

10. Find all six trigonometric functions of θ if the given point is on the terminal side of θ.

$(-5, 12)$

Find $\sin \theta$.

Find $\cos \theta$.

Find $\csc \theta$.

Find $\sec \theta$.

Find $\tan \theta$.

Find $\cot \theta$.

11. Suppose ABC is a right triangle with $C = 90^{\circ}$.

If $a = 9$ and $c = 15$, find b.

12. Show that the following statement is true by transforming the left side into the right side.

$\cos \theta \cot \theta + \sin \theta = \csc \theta$

13. Multiply.

$(4 \cos \theta + 7)(5 \cos \theta - 7)$

14. Write the following in terms of $\sin \theta$ only.

$\tan \theta$

15. Find all six trigonometric functions of θ if the given point is on the terminal side of θ.

$(-b,\ -a)$, assume that b and a are a positive numbers

Select the correct answer.

a. $\quad \sin\theta = -\dfrac{b}{\sqrt{b^2+a^2}}; \quad \cos\theta = -\dfrac{a}{\sqrt{b^2+a^2}}; \quad \csc\theta = -\dfrac{\sqrt{b^2+a^2}}{a}$

$\sec\theta = -\dfrac{\sqrt{b^2+a^2}}{b}; \quad \tan\theta = \dfrac{a}{b}; \quad \cot\theta = \dfrac{b}{a}$

b. $\quad \sin\theta = -\dfrac{a}{\sqrt{b^2+a^2}}; \quad \cos\theta = -\dfrac{b}{\sqrt{b^2+a^2}}; \quad \csc\theta = -\dfrac{\sqrt{b^2+a^2}}{a}$

$\sec\theta = -\dfrac{\sqrt{b^2+a^2}}{b}; \quad \tan\theta = \dfrac{b}{a}; \quad \cot\theta = \dfrac{a}{b}$

c. $\quad \sin\theta = -\dfrac{a}{\sqrt{b^2+a^2}}; \quad \cos\theta = -\dfrac{b}{\sqrt{b^2+a^2}}; \quad \csc\theta = -\dfrac{\sqrt{b^2+a^2}}{a}$

$\sec\theta = -\dfrac{\sqrt{b^2+a^2}}{b}; \quad \tan\theta = \dfrac{a}{b}; \quad \cot\theta = \dfrac{b}{a}$

d. $\quad \sin\theta = -\dfrac{b}{\sqrt{b^2+a^2}}; \quad \cos\theta = -\dfrac{a}{\sqrt{b^2+a^2}}; \quad \csc\theta = -\dfrac{\sqrt{b^2+a^2}}{a}$

$\sec\theta = -\dfrac{\sqrt{b^2+a^2}}{b}; \quad \tan\theta = \dfrac{b}{a}; \quad \cot\theta = \dfrac{a}{b}$

e. $\quad \sin\theta = -\dfrac{b}{\sqrt{b^2+a^2}}; \quad \cos\theta = -\dfrac{a}{\sqrt{b^2+a^2}}; \quad \csc\theta = -\dfrac{\sqrt{b^2+a^2}}{b}$

$\sec\theta = -\dfrac{\sqrt{b^2+a^2}}{a}; \quad \tan\theta = \dfrac{a}{b}; \quad \cot\theta = \dfrac{b}{a}$

16. Use a ratio identity to find $\cot\theta$ if

$$\sin\theta = \frac{9}{\sqrt{181}} \quad \text{and}$$

$$\cos\theta = \frac{10}{\sqrt{181}}.$$

Select the correct answer.

a. $\cot\theta = \dfrac{9}{10\sqrt{181}}$

b. $\cot\theta = \dfrac{10}{9\sqrt{181}}$

c. $\cot\theta = \dfrac{10}{9}$

d. $\cot\theta = \dfrac{9}{10}$

e. $\cot\theta = \dfrac{10\sqrt{181}}{9}$

17. An airplane is approaching Los Angeles International Airport at an altitude of 4,224 feet. If the horizontal distance from the plane to the runway is 1.5 miles, use the Pythagorean Theorem to find the diagonal distance from the plane to the runway (see the figure below). (5,280 feet equals 1 mile.)

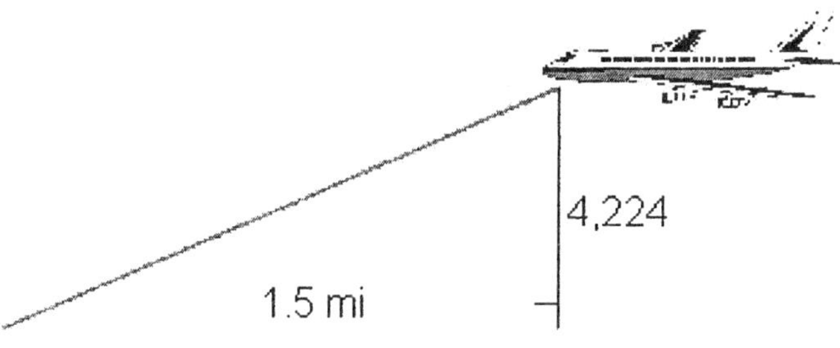

__________ miles

18. Write the following in terms of $\sin\theta$ and $\cos\theta$ and then simplify if possible.

$$\sec\theta\,\tan\theta\,\csc\theta$$

Select the correct answer.

a. $\dfrac{\sin\theta}{\cos^2\theta}$

b. $\dfrac{\cos^2\theta}{\sin\theta}$

c. $\dfrac{1}{\sin\theta\cos\theta}$

d. $\dfrac{\sin^2\theta}{\cos\theta}$

e. $\dfrac{1}{\cos^2\theta}$

19. Solve for x.

$$\sqrt{5}\quad x+1$$
$$x$$

Select the correct answer.

a. $x = 1$

b. $x = 4$

c. $x = 6$

d. $x = 7$

e. $x = 2$

20. Find $\sec\theta$ if $\tan\theta = \dfrac{4}{3}$ and θ terminates in QIII.

21. Graph the following circle.

$$x^2 + y^2 = 3$$

Select the correct answer.

a.
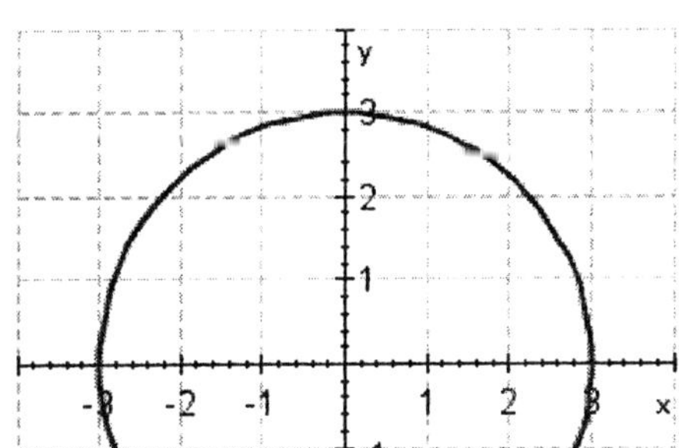

b.
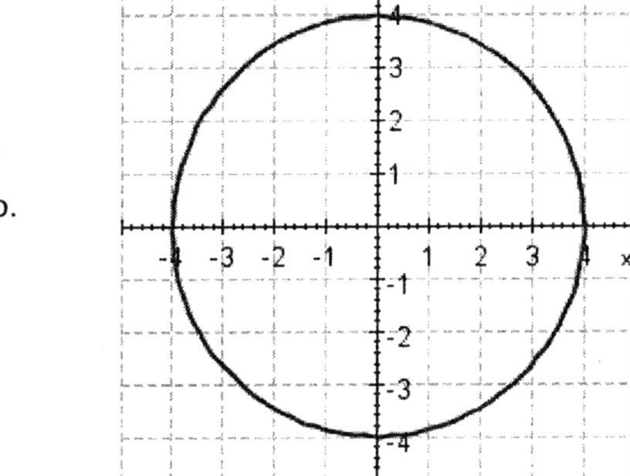

c.
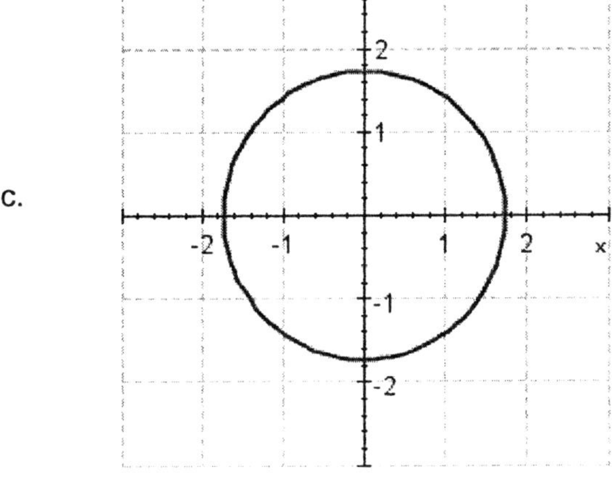

22.

Find the remaining trigonometric functions of θ if $\cot \theta = \dfrac{m}{p}$ where m and p are both positive. Assume that θ is between $0°$ and $180°$.

Find $\sin \theta$.

Find $\cos \theta$.

Find $\tan \theta$.

Find $\sec \theta$.

Find $\csc \theta$.

23. The object shown in the figure is a cube.

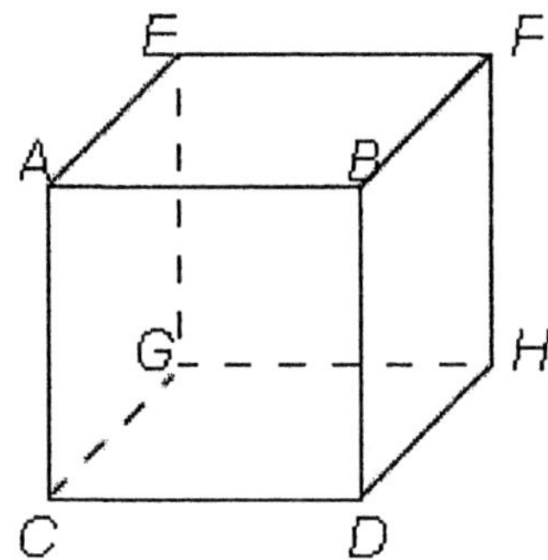

If the length of each edge of the cube shown in the figure is 3 inches, find the length of diagonal CH.

Select the correct answer.

a. $\dfrac{3\sqrt{3}}{4}$ inches

b. $3\sqrt{3}$ inches

c. 3 inches

d. $3\sqrt{2}$ inches

e. $\dfrac{3\sqrt{2}}{11}$ inches

24. Find α if $A = 5\alpha$.

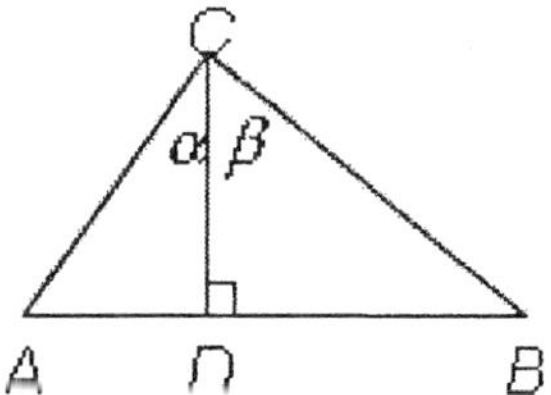

Select the correct answer.

a. $\alpha = 70^\circ$

b. $\alpha = 20^\circ$

c. $\alpha = 15^\circ$

d. $\alpha = 75^\circ$

e. $\alpha = 13^\circ$

25. Graph the parabola.

$$2x + (x + 1)^2 + 6$$

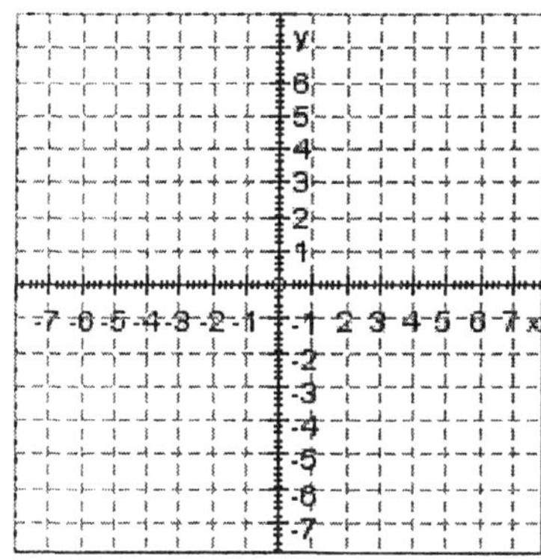

1.

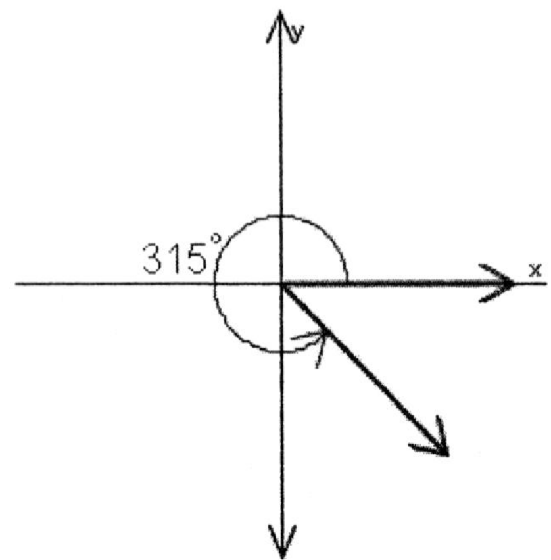

$(1, -1)$ for example

$-\dfrac{\sqrt{2}}{2}$

$\dfrac{\sqrt{2}}{2}$

-1

2. $\cot(\theta) = \dfrac{1}{5a}$

3. e

4. $\dfrac{3\sqrt{2}}{4}$

5. acute
20

110

6. e

7. b

8. d

9. b

10. $\dfrac{12}{13}, \; -\dfrac{5}{13}, \; \dfrac{13}{12}, \; -\dfrac{13}{5}, \; -\dfrac{12}{5}, \; -\dfrac{5}{12}$

11. 12

12. We begin by writing everything on the left side in terms of $\sin\theta$ and $\cos\theta$.

$$\cos\theta\cot\theta + \sin\theta = \cos\theta\,\frac{\cos\theta}{\sin\theta} + \sin\theta = \frac{\cos^2\theta + \sin^2\theta}{\sin\theta}$$

$$= \frac{1}{\sin\theta} = \csc\theta$$

Since we have succeeded in transforming the left side into the right side, we have shown that the statement $\cos\theta\cot\theta + \sin\theta = \csc\theta$ is an identity.

13. $20(\cos(\theta))^2 + 7\cos(\theta) - 49$

14. $\pm\dfrac{\sin(\theta)}{\sqrt{1 - (\sin(\theta))^2}}$

15. c

16. c

17. 1.7

18. e

19. a

20. $\sec(\theta) = -\dfrac{5}{3}$

21. c

22. $\dfrac{p}{\sqrt{p^2+m^2}}\,,\ \dfrac{m}{\sqrt{p^2+m^2}}\,,\ \dfrac{p}{m}\,,\ \dfrac{\sqrt{m^2+p^2}}{m}\,,\ \dfrac{\sqrt{m^2+p^2}}{p}$

23. d

24. c

25.

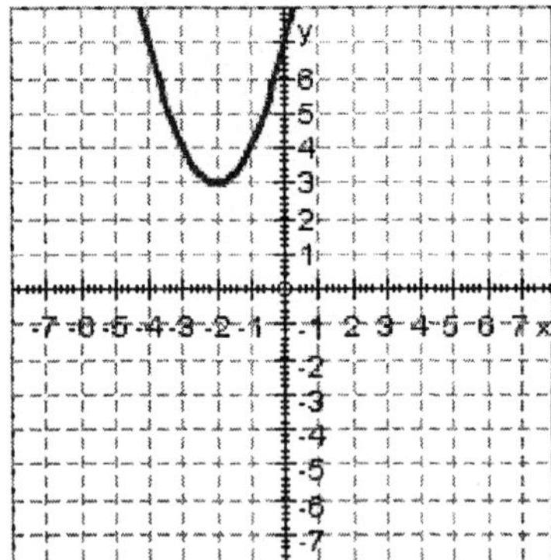

1. mctr.01.03.30_NoAlgs
2. mctr.01.04.13_NoAlgs
3. mctr.01.03.67m_NoAlgs
4. mctr.01.01.53_NoAlgs
5. mctr.01.01.02_NoAlgs
6. mctr.01.02.60m_NoAlgs
7. mctr.01.03.39m_NoAlgs
8. mctr.01.04.46m_NoAlgs
9. mctr.01.05.43m_NoAlgs
10. mctr.01.03.05_NoAlgs
11. mctr.01.01.27_NoAlgs
12. mctr.01.05.72_NoAlgs
13. mctr.01.05.37_NoAlgs
14. mctr.01.05.03_NoAlgs
15. mctr.01.03.09m_NoAlgs
16. mctr.01.04.18m_NoAlgs
17. mctr.01.02.49_NoAlgs
18. mctr.01.05.09m_NoAlgs
19. mctr.01.01.35m_NoAlgs
20. mctr.01.04.39_NoAlgs
21. mctr.01.02.27m_NoAlgs
22. mctr.01.03.60_NoAlgs
23. mctr.01.01.61m_NoAlgs
24. mctr.01.01.12m_NoAlgs
25. mctr.01.02.19_NoAlgs

1. The figure shows a walkway with a handrail. Angle α is the angle between the walkway and the horizontal, while angle β is the angle between the vertical posts of the handrail and the walkway. Assume that the vertical posts are perpendicular to the horizontal.

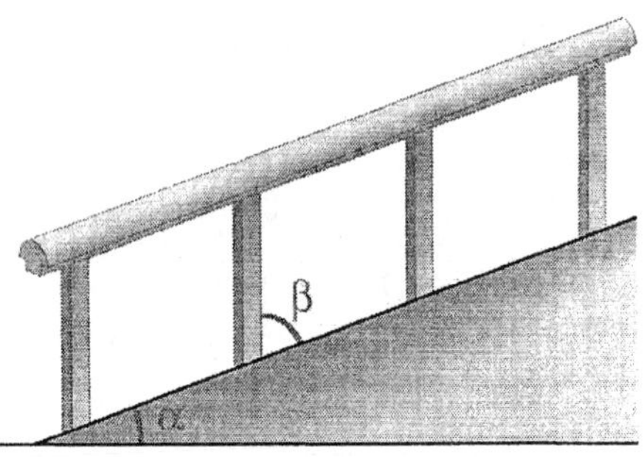

Find α if $\beta = 60°$.

$\alpha = $ _____ °

2. An isosceles triangle is a triangle in which two sides are equal in length. The angle between the two equal sides is called the vertex angle, while the other two angles are called the base angles. If the vertex angle is 20°, what is the measure of the base angles?

Select the correct answer.

    a.   65°

    b.   70°

    c.   85°

    d.   100°

    e.   80°

3. Find r if $AB = 25$ and $AD = 35$.

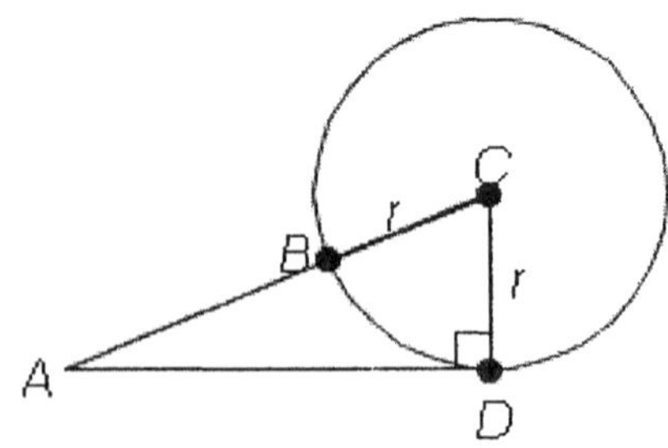

4. Find the remaining sides of a $30^\circ - 60^\circ - 90^\circ$ triangle if the side opposite 60° is 18.

   Select the correct answer.

   a. $\dfrac{16\sqrt{3}}{7}, \dfrac{8\sqrt{3}}{7}$

   b. $26\sqrt{2}, 13\sqrt{2}$

   c. $\dfrac{36\sqrt{2}}{7}, \dfrac{18\sqrt{3}}{7}$

   d. $12\sqrt{3}, 6\sqrt{3}$

   e. $\dfrac{16\sqrt{3}}{3}, \dfrac{8\sqrt{3}}{3}$

5. Graph the ordered pair on a rectangular coordinate system:

$$(4, -5)$$

Select the correct answer.

a.

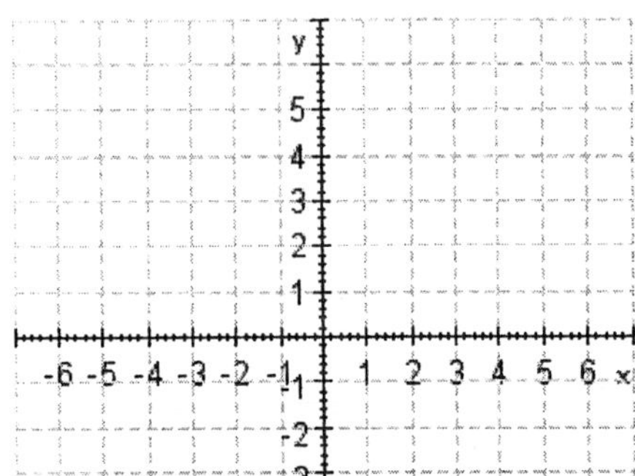

b.

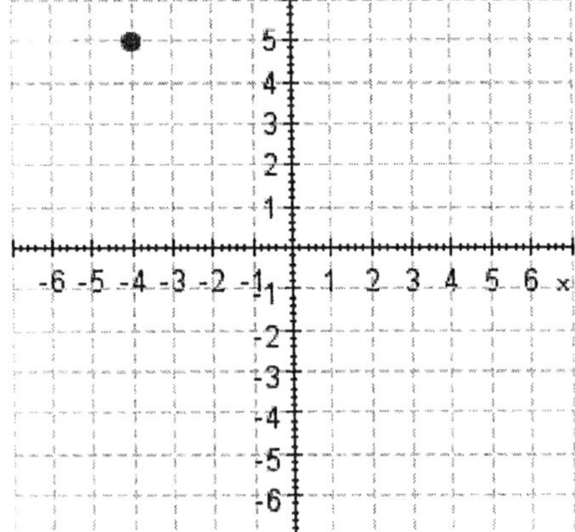

c.

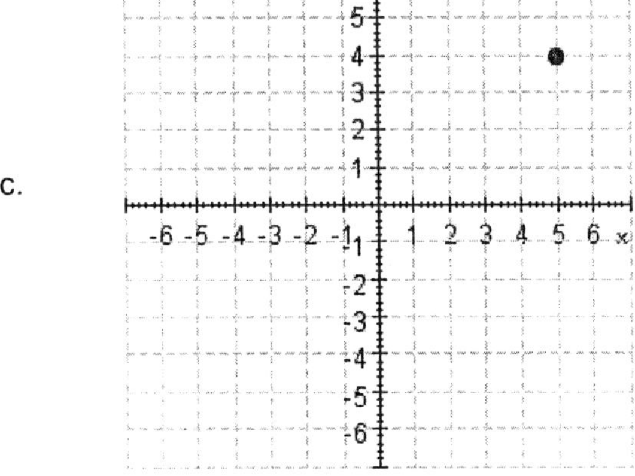

6. Graph the ordered pair on a rectangular coordinate system:

$(0, 6)$

Select the correct answer.

a.

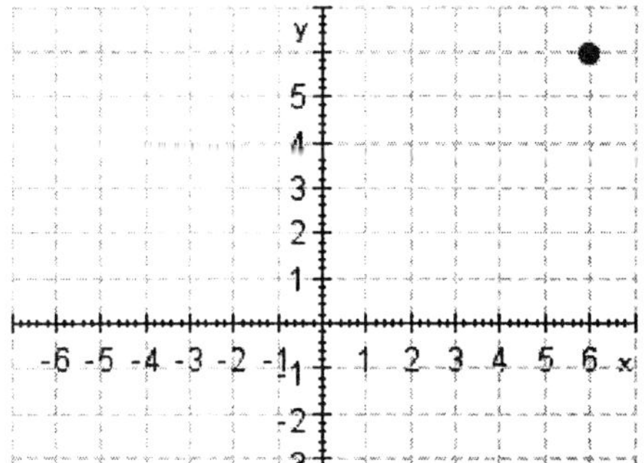

b.

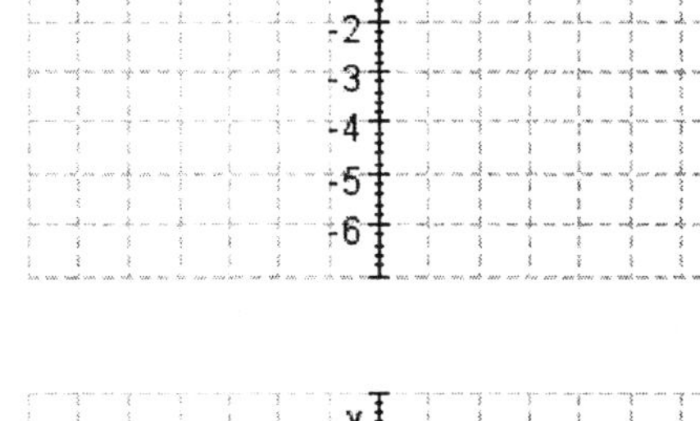

c.

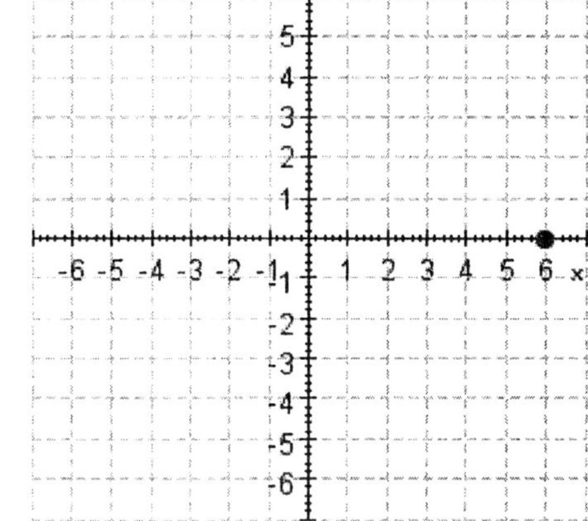

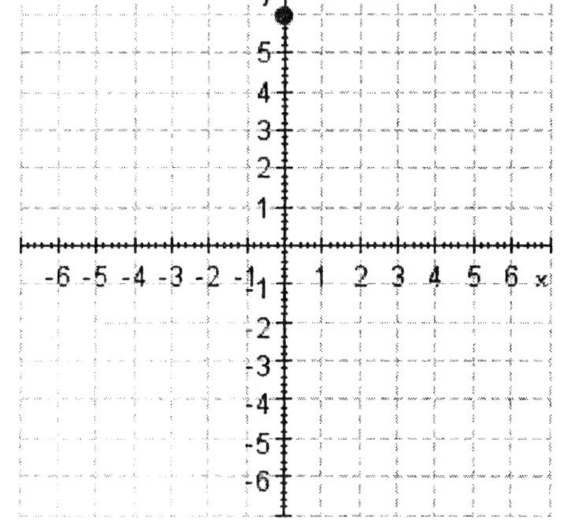

7. Use the given graph of the circle $x^2 + y^2 = 36$ to name the points at which the line

$x - y = 6$ will intersect the circle.

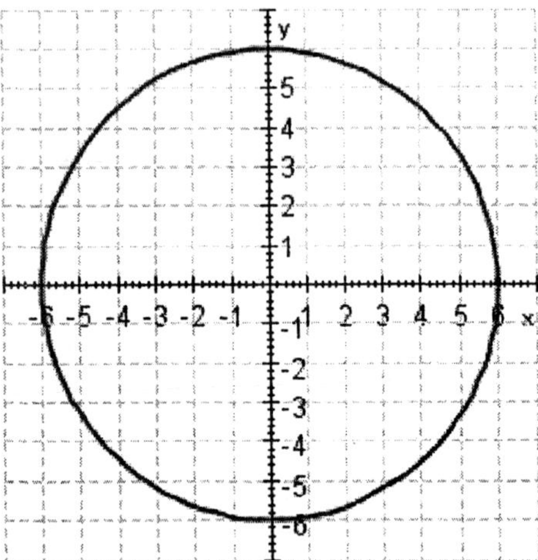

____

8. Find the distance between the points:

$$(-1, -4), (-11, 7)$$

Select the correct answer.

a. $\sqrt{113}$

b. $\sqrt{185}$

c. $\sqrt{202}$

d. $\sqrt{221}$

e. $\sqrt{181}$

9. Use the diagram to help you name an angle between $0°$ and $360°$ that is coterminal with the angle $-210°$

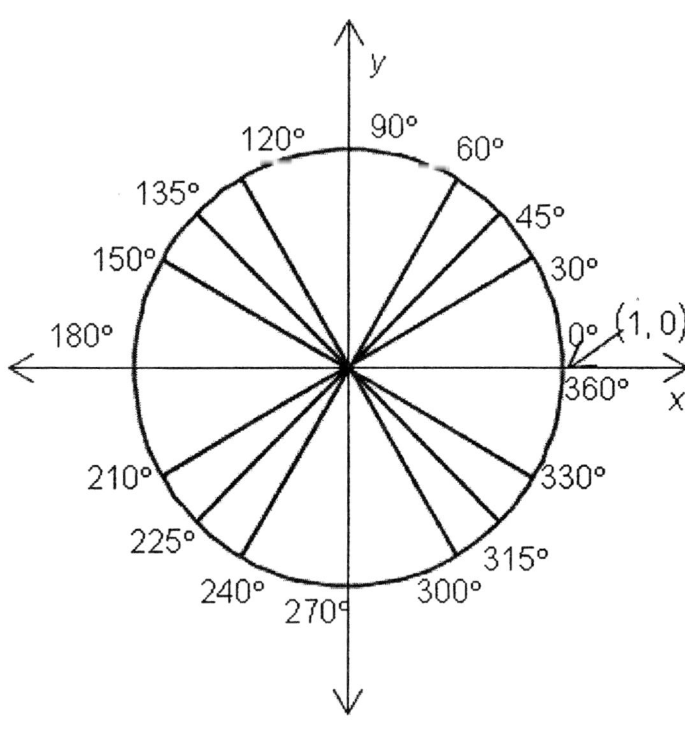

________ °

10. Name another angle that is coterminal with $-60°$.

Select the correct answer(s).

a. $300°$

b. $60°$

c. $420°$

d. $-420°$

e. $-300°$

11. Find all six trigonometric functions of θ if the given point is on the terminal side of θ.

$$\left(-3, \sqrt{7}\right)$$

Find $\sin \theta$.

Find $\cos \theta$.

Find $\csc \theta$.

Find $\sec \theta$.

Find $\tan \theta$.

Find $\cot \theta$.

12. In the diagram below, angle θ is in standard position. Find $\sin\theta$, $\cos\theta$ and $\tan\theta$.

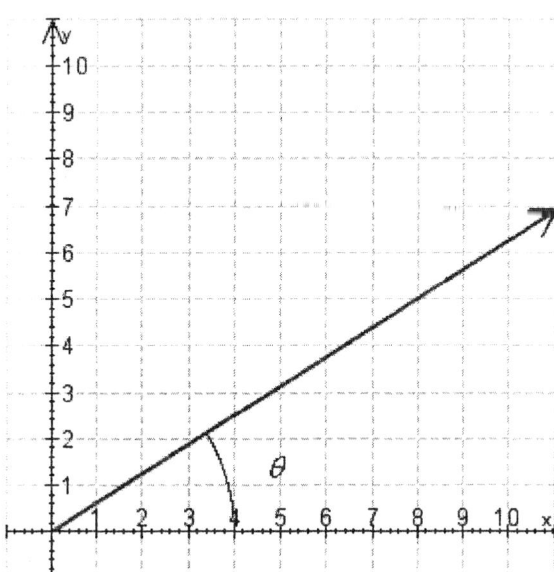

Select the correct answer.

a. $\sin\theta = \dfrac{5\sqrt{89}}{89}$, $\cos\theta = \dfrac{8\sqrt{89}}{89}$, $\tan\theta = \dfrac{5}{8}$

b. $\sin\theta = \dfrac{\sqrt{89}}{5}$, $\cos\theta = \dfrac{\sqrt{89}}{8}$, $\tan\theta = \dfrac{8}{5}$

c. $\sin\theta = \dfrac{8\sqrt{89}}{89}$, $\cos\theta = \dfrac{5\sqrt{89}}{89}$, $\tan\theta = \dfrac{8}{5}$

d. $\sin\theta = \dfrac{8\sqrt{89}}{89}$, $\cos\theta = \dfrac{5\sqrt{89}}{89}$, $\tan\theta = \dfrac{5}{8}$

e. $\sin\theta = \dfrac{\sqrt{89}}{8}$, $\cos\theta = \dfrac{\sqrt{89}}{5}$, $\tan\theta = \dfrac{8}{5}$

13.

Find the remaining trigonometric functions of θ if $\sin \theta = \dfrac{24}{25}$ and θ terminates in QII.

Select the correct answer.

a. $\cos \theta = -\dfrac{7}{25}$, $\tan \theta = -\dfrac{7}{24}$, $\csc \theta = \dfrac{25}{24}$

$\sec \theta = -\dfrac{25}{7}$, $\cot \theta = -\dfrac{24}{7}$

b. $\cos \theta = \dfrac{7}{25}$, $\tan \theta = \dfrac{24}{7}$, $\csc \theta = \dfrac{25}{24}$

$\sec \theta = \dfrac{25}{7}$, $\cot \theta = \dfrac{7}{24}$

c. $\cos \theta = -\dfrac{7}{25}$, $\tan \theta = -\dfrac{24}{7}$, $\csc \theta = \dfrac{25}{24}$

$\sec \theta = -\dfrac{25}{7}$, $\cot \theta = -\dfrac{7}{24}$

d. $\cos \theta = \dfrac{7}{25}$, $\tan \theta = \dfrac{7}{24}$, $\csc \theta = \dfrac{25}{24}$

$\sec \theta = \dfrac{25}{7}$, $\cot \theta = \dfrac{24}{7}$

e. $\cos \theta = -\dfrac{7}{25}$, $\tan \theta = -\dfrac{7}{24}$, $\csc \theta = -\dfrac{25}{7}$

$\sec \theta = \dfrac{25}{24}$, $\cot \theta = -\dfrac{24}{7}$

14. Find the remaining trigonometric functions of θ if $\csc \theta = \dfrac{25}{7}$ and $\cos \theta < 0$.

Select the correct answer.

a. $\sin \theta = \dfrac{7}{25}$, $\cos \theta = -\dfrac{24}{25}$, $\tan \theta = -\dfrac{7}{24}$,

$\cot \theta = \dfrac{24}{7}$, $\sec \theta = \dfrac{26}{24}$

b. $\sin \theta = \dfrac{24}{25}$, $\cos \theta = \dfrac{7}{25}$, $\tan \theta = \dfrac{7}{24}$,

$\cot \theta = \dfrac{24}{7}$, $\sec \theta = \dfrac{25}{24}$

c. $\sin \theta = \dfrac{24}{25}$, $\cos \theta = -\dfrac{7}{25}$, $\tan \theta = -\dfrac{24}{7}$,

$\cot \theta = -\dfrac{7}{24}$, $\sec \theta = -\dfrac{25}{24}$

d. $\sin \theta = \dfrac{24}{25}$, $\cos \theta = -\dfrac{7}{25}$, $\tan \theta = -\dfrac{24}{7}$,

$\cot \theta = \dfrac{7}{24}$, $\sec \theta = \dfrac{25}{24}$

e. $\sin \theta = \dfrac{7}{25}$, $\cos \theta = -\dfrac{24}{25}$, $\tan \theta = -\dfrac{7}{24}$,

$\cot \theta = -\dfrac{24}{7}$, $\sec \theta = -\dfrac{25}{24}$

15. Give the reciprocal of the given number.

$-\dfrac{8}{9}$

16. Give the reciprocal of the given number.

$7x$

Select the correct answer.

a. $1 - 7x$

b. $-7x$

c. $-\dfrac{7}{x}$

d. $1 + 7x$

e. $\dfrac{1}{7x}$

17. For this problem, let $\sin\theta = -\dfrac{60}{61}$, and $\cos\theta = -\dfrac{11}{61}$, and find

$\tan\theta$

18. If $\sin\theta = -\dfrac{4}{5}$ and θ terminates in QIII, find $\cos\theta$.

a. $\cos\theta = -\dfrac{2}{5}$

b. $\cos\theta = \dfrac{2}{5}$

c. $\cos\theta = \dfrac{3}{5}$

d. $\cos\theta = \dfrac{4}{3}$

e. $\cos\theta = -\dfrac{3}{5}$

19. If $\cos \theta = \dfrac{4}{\sqrt{41}}$ and $\theta \in$ QIV, find $\csc \theta$

Select the correct answer.

a. $\csc \theta = \dfrac{\sqrt{41}}{4}$

b. $\csc \theta = -\dfrac{\sqrt{41}}{5}$

c. $\csc \theta = \dfrac{\sqrt{43}}{5}$

d. $\csc \theta = \dfrac{5}{\sqrt{41}}$

e. $\csc \theta = \dfrac{5}{\sqrt{43}}$

20. Using your calculator and rounding your answers to the nearest hundredth, find $\tan \theta$ if

$\sec \theta = -1.74$ and $\theta \in$ QII.

Select the correct answer.

a. $\quad$ - 1.42
b. $\quad$ - 1.52
c. $\quad$ - 1.82
d. $\quad$ - 1.62
e. $\quad$ - 1.92

21. Write the following in terms of $\sin \theta$ and $\cos \theta$ and then simplify if possible.

$\sec \theta - \tan \theta \sin \theta$

22. Add or subtract as indicated. Then simplify your answer if possible.

$$\sin\theta - \frac{1}{\cos\theta}$$

Select the correct answer.

a. $\dfrac{\sin\theta\cos\theta - 1}{\cos\theta}$

b. $\dfrac{\sin\theta}{\cos\theta}$

c. $\sin\theta$

d. $\dfrac{\sin\theta\cos\theta + 1}{\cos\theta}$

e. $\dfrac{\sin\theta\cos\theta + 1}{\sin\theta}$

23. Simplify the expression $\sqrt{x^2 - 4}$ as much as possible after substituting $2\csc\theta$ for x.

24. Find the right part of the identity.

$$\cos\alpha\tan\alpha = \underline{\hspace{1cm}}$$

Select the correct answer.

a. $\cos\alpha$

b. $\sec\alpha$

c. $\tan\alpha$

d. $\csc\alpha$

e. $\sin\alpha$

25. Find the right part of the identity.

$$\sec\theta - \cos\theta = \underline{\hspace{1cm}}$$

Select the correct answer.

a. $\dfrac{\sin^2\theta}{\cos\theta}$

b. $\cot^2\theta$

c. $\dfrac{\cos^2\theta}{\sin\theta}$

d. $\tan^2\theta$

e. $\csc^2\theta$

1. 30

2. e

3. 12

4. d

5. a

6. c

7. $(0, -6), (6, 0)$

8. d

9. 150

10. a,d

11. $\dfrac{\sqrt{7}}{4}$,

$-\dfrac{3}{4}$,

$\dfrac{4\sqrt{7}}{7}$,

$-\dfrac{4}{3}$,

$-\dfrac{\sqrt{7}}{3}$,

$-\dfrac{3\sqrt{7}}{7}$

12. a

13. c

14. e

15. $-\dfrac{9}{8}$

16. e

17. $\tan(\theta) = \dfrac{60}{11}$

18. e

10. b

20. a

21. $\cos(\theta)$

22. a

23. $2|\cot(\theta)|$

24. e

25. a

1. mctr.01.01.18_NoAlgs
2. mctr.01.01.24m_NoAlgs
3. mctr.01.01.40_NoAlgs
4. mctr.01.01.47m_NoAlgs
5. mctr.01.02.02m_NoAlgs
6. mctr.01.02.09m_NoAlgs
7. mctr.01.02.35_NoAlgs
8. mctr.01.02.43m_NoAlgs
9. mctr.01.02.68_NoAlgs
10. mctr.01.02.77m_NoAlgs
11. mctr.01.03.16_NoAlgs
12. mctr.01.03.24m_NoAlgs
13. mctr.01.03.44m_NoAlgs
14. mctr.01.03.52m_NoAlgs
15. mctr.01.04.03_NoAlgs
16. mctr.01.04.07m_NoAlgs
17. mctr.01.04.23_NoAlgs
18. mctr.01.04.31m_NoAlgs
19. mctr.01.04.49m_NoAlgs
20. mctr.01.04.57m_NoAlgs
21. mctr.01.05.25_NoAlgs
22. mctr.01.05.31m_NoAlgs
23. mctr.01.05.51_NoAlgs
24. mctr.01.05.59m_NoAlgs
25. mctr.01.05.75m_NoAlgs

1. Simplify the expression $\sqrt{x^2 - 4}$ as much as possible after substituting $2\csc\theta$ for x.

2. Find the right part of the identity.

$$\cos\alpha\tan\alpha = \underline{\hspace{3em}}$$

Select the correct answer.

    a. $\cos\alpha$

    b. $\sec\alpha$

    c. $\tan\alpha$

    d. $\csc\alpha$

    e. $\sin\alpha$

3. For this problem, let $\sin\theta = -\dfrac{60}{61}$, and $\cos\theta = -\dfrac{11}{61}$, and find

$$\tan\theta$$

4. Find the remaining sides of a $30° - 60° - 90°$ triangle if the side opposite $60°$ is 18.

Select the correct answer.

    a. $\dfrac{16\sqrt{3}}{7}, \dfrac{8\sqrt{3}}{7}$

    b. $26\sqrt{2}, 13\sqrt{2}$

    c. $\dfrac{36\sqrt{2}}{7}, \dfrac{18\sqrt{3}}{7}$

    d. $12\sqrt{3}, 6\sqrt{3}$

    e. $\dfrac{16\sqrt{3}}{3}, \dfrac{8\sqrt{3}}{3}$

5. An isosceles triangle is a triangle in which two sides are equal in length. The angle between the two equal sides is called the vertex angle, while the other two angles are called the base angles. If the vertex angle is 20° , what is the measure of the base angles?

   Select the correct answer.

   a. 65°

   b. 70°

   c. 85°

   d. 100°

   e. 80°

6. Use the given graph of the circle $x^2 + y^2 = 36$ to name the points at which the line

   $x - y = 6$ will intersect the circle.

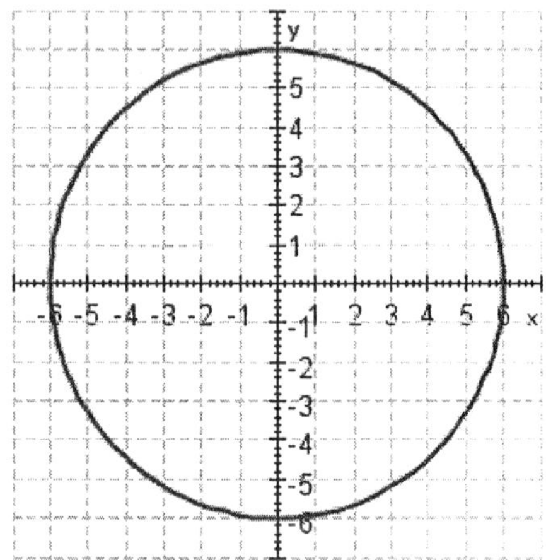

   ____

7. Give the reciprocal of the given number. $7x$

   Select the correct answer.

   a. $1 - 7x$

   b. $-7x$

   c. $-\dfrac{7}{x}$

   d. $\dfrac{1}{7x}$

8. Graph the ordered pair on a rectangular coordinate system:

$$(4, -5)$$

Select the correct answer.

a.
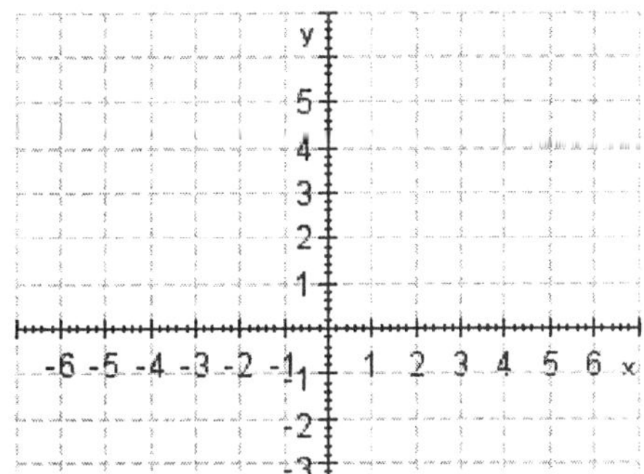

b.
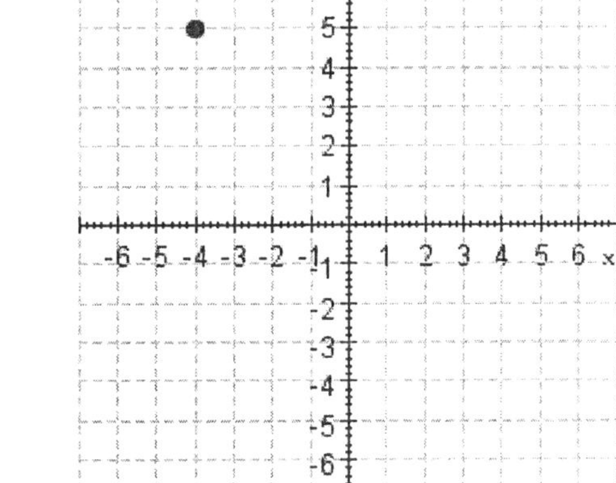

c.

9. Find r if $AB = 25$ and $AD = 35$.

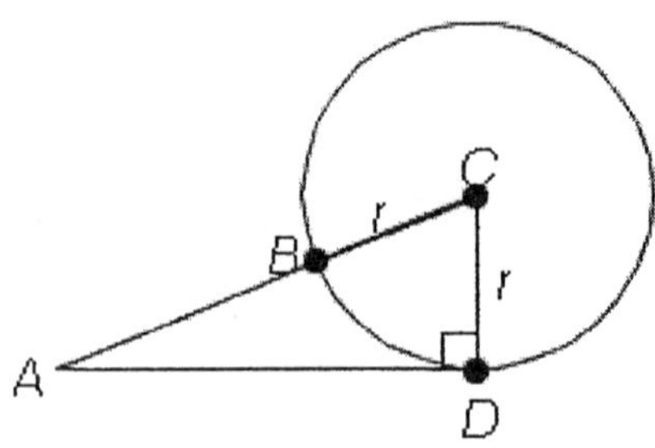

10. Find the distance between the points:

$$\left(-1,-4\right),\ \left(-11,7\right)$$

Select the correct answer.

   a. $\sqrt{113}$

   b. $\sqrt{185}$

   c. $\sqrt{202}$

   d. $\sqrt{221}$

   e. $\sqrt{181}$

11. Name another angle that is coterminal with -60°.

Select the correct answer(s).

   a. 300°

   b. 60°

   c. 420°

   d. -420°

   e. -300°

12. Give the reciprocal of the given number.

$$-\frac{8}{9}$$

13. Graph the ordered pair on a rectangular coordinate system:

$(0, 6)$

Select the correct answer.

a.
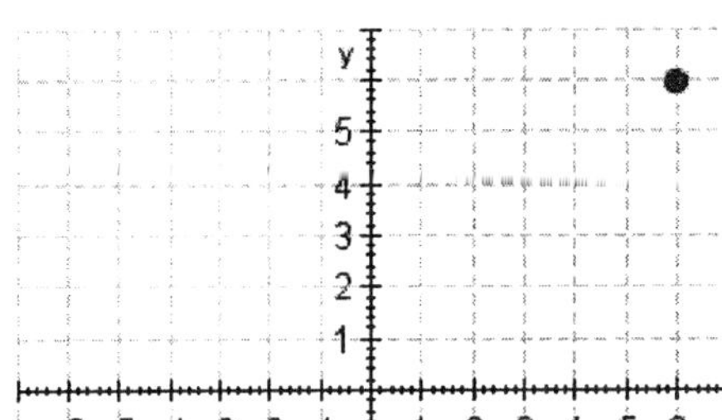

b.
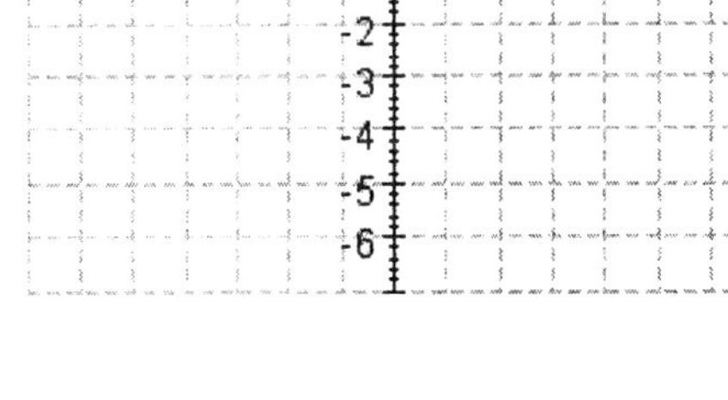

c.
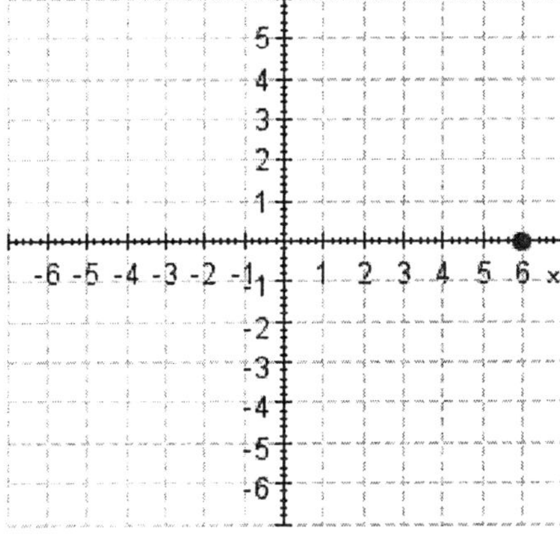

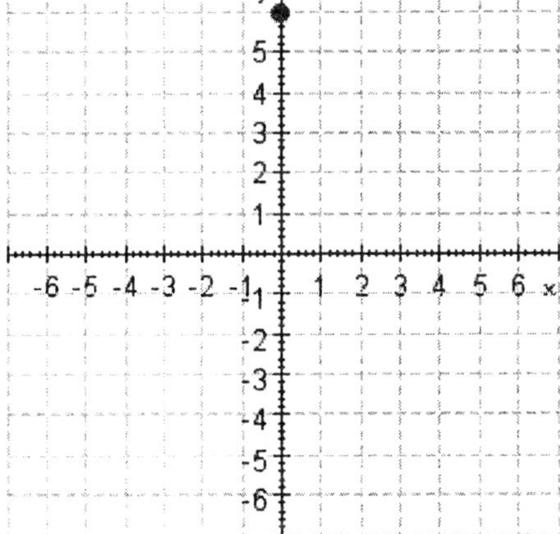

14.

Find the remaining trigonometric functions of θ if $\sin \theta = \dfrac{24}{25}$ and θ terminates in QII.

Select the correct answer.

a. $\cos \theta = -\dfrac{7}{25}$, $\tan \theta = -\dfrac{7}{24}$, $\csc \theta = \dfrac{25}{24}$

$\sec \theta = -\dfrac{25}{7}$, $\cot \theta = -\dfrac{24}{7}$

b. $\cos \theta = \dfrac{7}{25}$, $\tan \theta = \dfrac{24}{7}$, $\csc \theta = \dfrac{25}{24}$

$\sec \theta = \dfrac{25}{7}$, $\cot \theta = \dfrac{7}{24}$

c. $\cos \theta = -\dfrac{7}{25}$, $\tan \theta = -\dfrac{24}{7}$, $\csc \theta = \dfrac{25}{24}$

$\sec \theta = -\dfrac{25}{7}$, $\cot \theta = -\dfrac{7}{24}$

d. $\cos \theta = \dfrac{7}{25}$, $\tan \theta = \dfrac{7}{24}$, $\csc \theta = \dfrac{25}{24}$

$\sec \theta = \dfrac{25}{7}$, $\cot \theta = \dfrac{24}{7}$

e. $\cos \theta = -\dfrac{7}{25}$, $\tan \theta = -\dfrac{7}{24}$, $\csc \theta = -\dfrac{25}{7}$

$\sec \theta = \dfrac{25}{24}$, $\cot \theta = -\dfrac{24}{7}$

15. Use the diagram to help you name an angle between $0°$ and $360°$ that is coterminal with the angle $-210°$

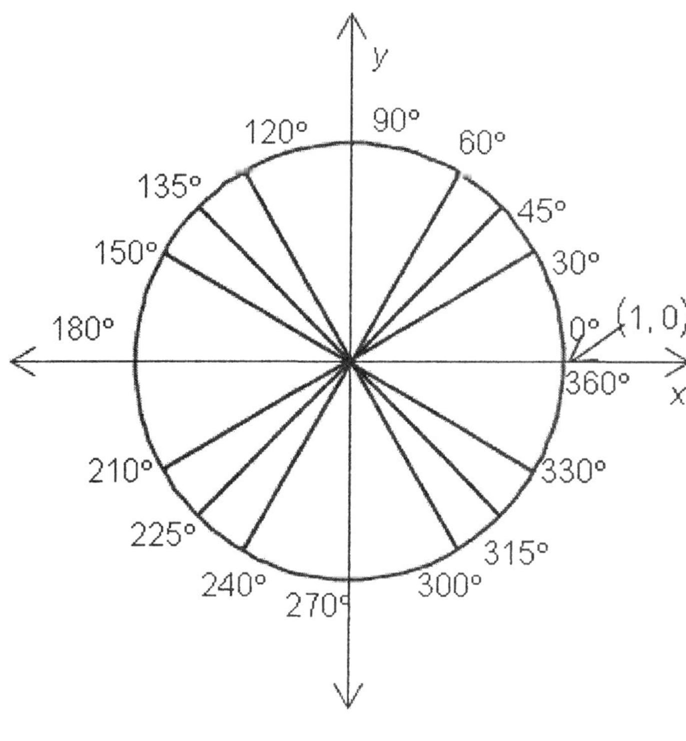

_______ °

16. Using your calculator and rounding your answers to the nearest hundredth, find $\tan \theta$ if

$\sec \theta = -1.74$ and $\theta \in$ QII.

Select the correct answer.

a.　- 1.42
b.　- 1.52
c.　- 1.82
d.　- 1.62
e.　- 1.92

17. Find all six trigonometric functions of θ if the given point is on the terminal side of θ.

$$\left(-3, \ \sqrt{7}\right)$$

Find $\sin \theta$.

Find $\cos \theta$.

Find $\csc \theta$.

Find $\sec \theta$.

Find $\tan \theta$.

Find $\cot \theta$.

18. If $\sin \theta = -\dfrac{4}{5}$ and θ terminates in QIII, find $\cos \theta$.

a. $\cos \theta = -\dfrac{2}{5}$

b. $\cos \theta = \dfrac{2}{5}$

c. $\cos \theta = \dfrac{3}{5}$

d. $\cos \theta = \dfrac{4}{3}$

e. $\cos \theta = -\dfrac{3}{5}$

19.

Find the remaining trigonometric functions of θ if $\csc \theta = \dfrac{25}{7}$ and $\cos \theta < 0$.

Select the correct answer.

a. $\sin \theta = \dfrac{7}{25}$, $\cos \theta = -\dfrac{24}{25}$, $\tan \theta = -\dfrac{7}{24}$,

$\cot \theta = \dfrac{24}{7}$, $\sec \theta = \dfrac{25}{24}$

b. $\sin \theta = \dfrac{24}{25}$, $\cos \theta = \dfrac{7}{25}$, $\tan \theta = \dfrac{7}{24}$,

$\cot \theta = \dfrac{24}{7}$, $\sec \theta = \dfrac{25}{24}$

c. $\sin \theta = \dfrac{24}{25}$, $\cos \theta = -\dfrac{7}{25}$, $\tan \theta = -\dfrac{24}{7}$,

$\cot \theta = -\dfrac{7}{24}$, $\sec \theta = -\dfrac{25}{24}$

d. $\sin \theta = \dfrac{24}{25}$, $\cos \theta = -\dfrac{7}{25}$, $\tan \theta = -\dfrac{24}{7}$,

$\cot \theta = \dfrac{7}{24}$, $\sec \theta = \dfrac{25}{24}$

e. $\sin \theta = \dfrac{7}{25}$, $\cos \theta = -\dfrac{24}{25}$, $\tan \theta = -\dfrac{7}{24}$,

$\cot \theta = -\dfrac{24}{7}$, $\sec \theta = -\dfrac{25}{24}$

20. Find the right part of the identity.

$$\sec \theta - \cos \theta = \underline{\hspace{2cm}}$$

Select the correct answer.

a. $\dfrac{\sin^2 \theta}{\cos \theta}$

b. $\cot^2 \theta$

c. $\dfrac{\cos^2 \theta}{\sin \theta}$

d. $\tan^2 \theta$

e. $\csc^2 \theta$

21. Add or subtract as indicated. Then simplify your answer if possible.

$$\sin \theta - \dfrac{1}{\cos \theta}$$

Select the correct answer.

a. $\dfrac{\sin \theta \cos \theta - 1}{\cos \theta}$

b. $\dfrac{\sin \theta}{\cos \theta}$

c. $\sin \theta$

d. $\dfrac{\sin \theta \cos \theta + 1}{\cos \theta}$

e. $\dfrac{\sin \theta \cos \theta + 1}{\sin \theta}$

22. The figure shows a walkway with a handrail. Angle α is the angle between the walkway and the horizontal, while angle β is the angle between the vertical posts of the handrail and the walkway. Assume that the vertical posts are perpendicular to the horizontal.

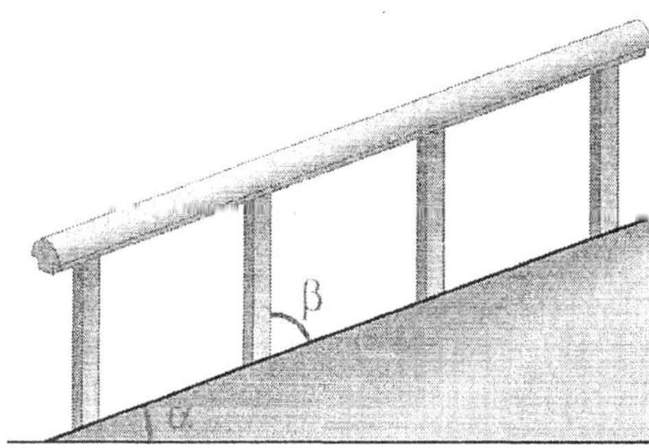

Find α if $\beta = 60°$.

$$\alpha = \underline{\hspace{1cm}} \text{°}$$

23. If $\cos \theta = \dfrac{4}{\sqrt{41}}$ and $\theta \in$ QIV, find $\csc \theta$

Select the correct answer.

a. $\csc \theta = \dfrac{\sqrt{41}}{4}$

b. $\csc \theta = -\dfrac{\sqrt{41}}{5}$

c. $\csc \theta = \dfrac{\sqrt{43}}{5}$

d. $\csc \theta = \dfrac{5}{\sqrt{41}}$

e. $\csc \theta = \dfrac{5}{\sqrt{43}}$

24. In the diagram below, angle θ is in standard position. Find $\sin\theta$, $\cos\theta$ and $\tan\theta$.

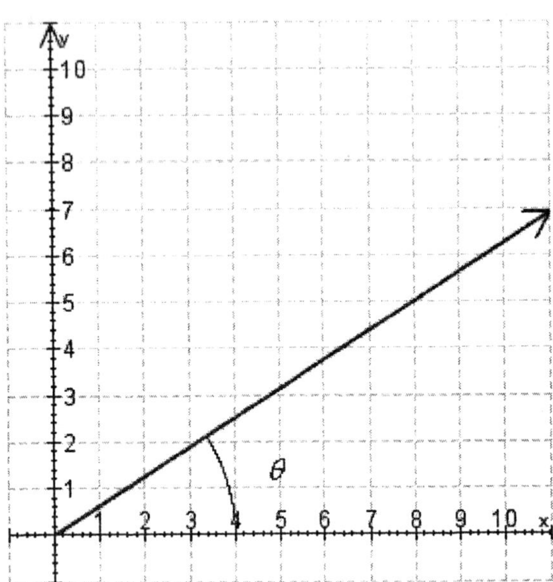

Select the correct answer.

a. $\quad \sin\theta = \dfrac{5\sqrt{89}}{89}, \cos\theta = \dfrac{8\sqrt{89}}{89}, \tan\theta = \dfrac{5}{8}$

b. $\quad \sin\theta = \dfrac{\sqrt{89}}{5}, \cos\theta = \dfrac{\sqrt{89}}{8}, \tan\theta = \dfrac{8}{5}$

c. $\quad \sin\theta = \dfrac{8\sqrt{89}}{89}, \cos\theta = \dfrac{5\sqrt{89}}{89}, \tan\theta = \dfrac{8}{5}$

d. $\quad \sin\theta = \dfrac{8\sqrt{89}}{89}, \cos\theta = \dfrac{5\sqrt{89}}{89}, \tan\theta = \dfrac{5}{8}$

e. $\quad \sin\theta = \dfrac{\sqrt{89}}{8}, \cos\theta = \dfrac{\sqrt{89}}{5}, \tan\theta = \dfrac{8}{5}$

25. Write the following in terms of $\sin\theta$ and $\cos\theta$ and then simplify if possible.

$$\sec\theta - \tan\theta\,\sin\theta$$

1. $2|\cot(\theta)|$

2. e

3. $\tan(\theta) = \dfrac{60}{11}$

4. d

5. e

6. $(0, -6), (6, 0)$

7. d

8. a

9. 12

10. d

11. a,d

12. $-\dfrac{9}{8}$

13. c

14. c

15. 150

16. a

17. $\dfrac{\sqrt{7}}{4}, -\dfrac{3}{4}, \dfrac{4\sqrt{7}}{7}, -\dfrac{4}{3}, -\dfrac{\sqrt{7}}{3}, -\dfrac{3\sqrt{7}}{7}$

18. e

19. e

20. a

21. a

22. 30

23. b

24. a

25. $\cos(\theta)$

McKeague/Turner - Trigonometry 5e Chapter 1 Form H

1. mctr.01.05.51_NoAlgs
2. mctr.01.05.59m_NoAlgs
3. mctr.01.04.23_NoAlgs
4. mctr.01.01.47m_NoAlgs
5. mctr.01.01.24m_NoAlgs
6. mctr.01.02.35_NoAlgs
7. mctr.01.04.07m_NoAlgs
8. mctr.01.02.02m_NoAlgs
9. mctr.01.01.40_NoAlgs
10. mctr.01.02.43m_NoAlgs
11. mctr.01.02.77m_NoAlgs
12. mctr.01.04.03_NoAlgs
13. mctr.01.02.09m_NoAlgs
14. mctr.01.03.44m_NoAlgs
15. mctr.01.02.68_NoAlgs
16. mctr.01.04.57m_NoAlgs
17. mctr.01.03.16_NoAlgs
18. mctr.01.04.31m_NoAlgs
19. mctr.01.03.52m_NoAlgs
20. mctr.01.05.75m_NoAlgs
21. mctr.01.05.31m_NoAlgs
22. mctr.01.01.18_NoAlgs
23. mctr.01.04.49m_NoAlgs
24. mctr.01.03.24m_NoAlgs
25. mctr.01.05.25_NoAlgs

1. Right triangle ABC has $C = 90°$, $b = 9$, and $c = 15$. Use this information to find the six trigonometric functions of A.

   Find $\sin A$

   Find $\cos A$

   Find $\tan A$

   Find $\cot A$

   Find $\sec A$

   Find $\csc A$

2. In the right triangle below

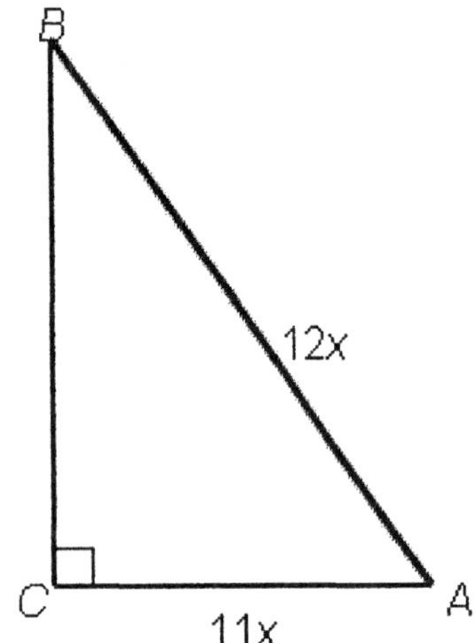

   Find $\sin A$

   Find $\cos A$

   Find $\tan A$

   Find $\sin B$

   Find $\cos B$

   Find $\tan B$

3. Complete the following table using exact values. Do not rationalize the denominator.

x	$\sin x$	$\csc x$
$0°$	0	?
$30°$	$\dfrac{1}{2}$	?
$45°$	$\dfrac{\sqrt{2}}{2}$	?
$60°$	$\dfrac{\sqrt{3}}{2}$	?
$90°$	1	?

Find the value indicated by the question mark. If the value does not exist, enter *undefined*

4. For the expression $6\sin(z - 30°)$, replace z with $60°$, and then simplify as much as possible.

5. Find the exact value of $\cot 60°$.

6. Add.

$$(39°45') + (22°27')$$

_____ ° _____ '

7. Convert the following to degrees and minutes. Round to the nearest minute.

$$99.65°$$

_____ ° _____ '

8. Use a calculator to find the following. Round the answer to four places past the decimal point.

$$\cot 31°$$

____

9. Use a calculator to complete the following table. Round all answers to four digits past the decimal point.

    If an expression is undefined indicate it.

x	$\cos x$	$\sec x$
$0°$	________	________
$30°$	________	________
$45°$	________	________
$60°$	________	________
$90°$	________	________

10. Use a calculator to find a value of θ between $0°$ and $90°$ that satisfies the statement below. Write your answer in degrees and minutes rounded to the nearest minute.

    $\csc \theta = 7.5493$

    $\theta = $ ______ $°$ ______ $'$

11. Refer to right triangle ABC with $C = 90°$.

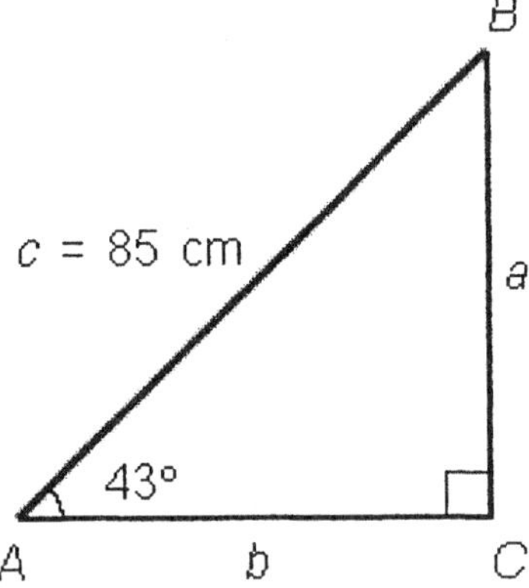

    If $A = 43°$ and $c = 85$ cm, find b.

    Please round the answer to the nearest whole number.

    $b = $ __________ cm

12. Refer to right triangle ABC with $C = 90^\circ$.

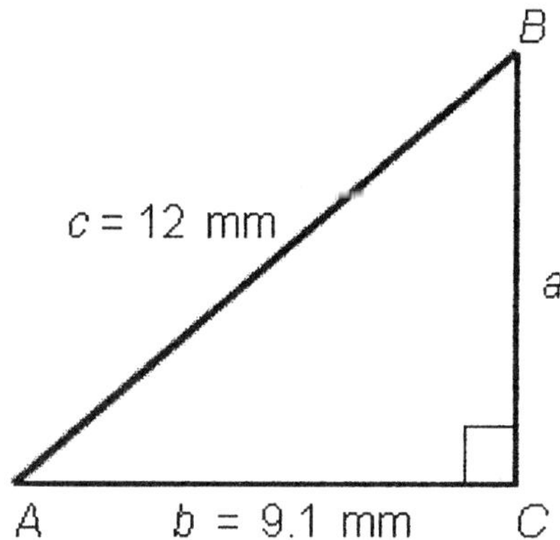

If $b = 9.1$ mm and $c = 12$ mm, find B.

Please round the answer to the nearest whole number.

$B = $ _________ $^\circ$

13. Refer to right triangle ABC with $C = 90^\circ$. Solve for all the missing parts using the given information.

$B = 44.46^\circ$, $a = 5.565$ mi

Please round the answer for the angle to two decimal places, and for the sides, to three decimal places.

$A = $ _________ $^\circ$

$b = $ _________ mi

$c = $ _________ mi

14. The circle in the figure has a radius of *r* and center at *C*.

The distance from *A* to *B* is *x*.

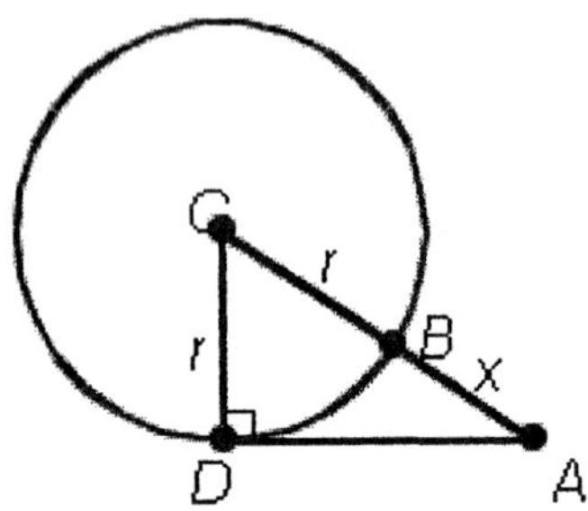

If $A = 32°$ and $r = 13$, find *x*.

Please round the answer to the nearest whole number.

$x =$ _________

15. In the figure, the distance from *A* to *D* is *y*, the distance from *D* to *C* is *x*, and the distance from *C* to *B* is *h*.

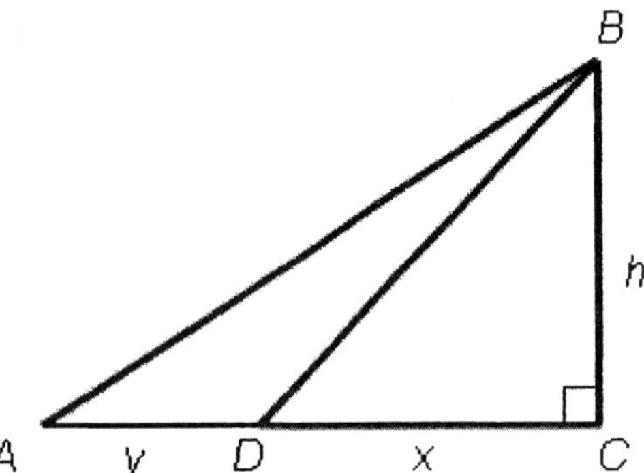

If $A = 32°$, $\angle BDC = 49°$, and $AB = 54$, find *h*, then *x*.

Please round each answer to the nearest whole number.

$h =$ _________

$x =$ _________

16. Solve the problem.

The two equal sides of an isosceles triangle are each 44 centimeters.
The base measures 30 centimeters.

Find the height. Please round your answer to the nearest centimeter.

__________ cm

Find the measure of the two equal angles. Please round your answer to the nearest degree.

__________ °

17. Solve the problem.

A 72.5-foot rope from the top of a circus tent pole is anchored to the ground 42.9 feet from the bottom of the pole. What angle does the rope make with the pole? (Assume the pole is perpendicular to the ground.)

Please round your answer to the nearest tenth of a degree.

__________ °

18. Solve the problem.

A man wandering in the desert walks 2.4 miles in the direction S 32° W. He then turns 90° and walks 3.4 miles in the direction N 58° W. At that time, how far is he from his starting point? Please round your answer to the nearest tenth of a mile.

__________ mi

What is the man's bearing from his starting point? Please round the angle to the nearest degree.
Please enter *S* or *N* (for north or south) in the first field and *E* or *W* (for east or west) in the third field.

___ __________ ° ___

19. From a point on the floor the angle of elevation to the top of a door is $48°$, while the angle of elevation to the ceiling above the door is $57°$.

If the ceiling is 10 feet above the floor, what is the vertical dimension of the door (see the figure)? Please round your answer to the nearest tenth of a foot.

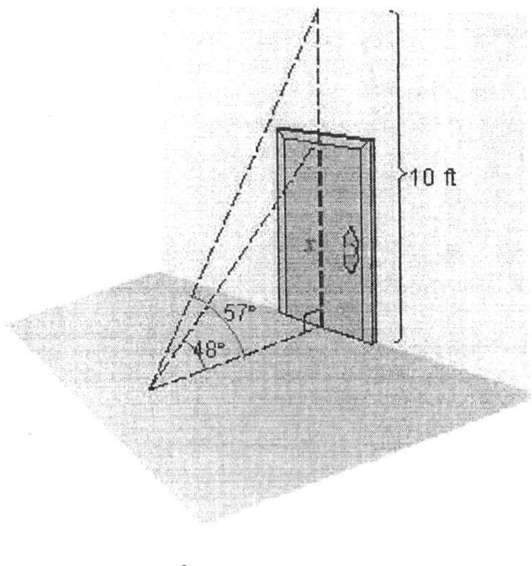

__________ ft

20. Solve the problem.

Two people decide to estimate the height of a flag pole. One person positions himself due north of the pole and the other person stands due east of the pole. If the two people are the same distance from the pole and 24 feet from each other, find the height of the pole if the angle of elevation from the ground at each person's position to the top of the pole is $56°$

(see the figure). Please round your answer to the nearest foot.

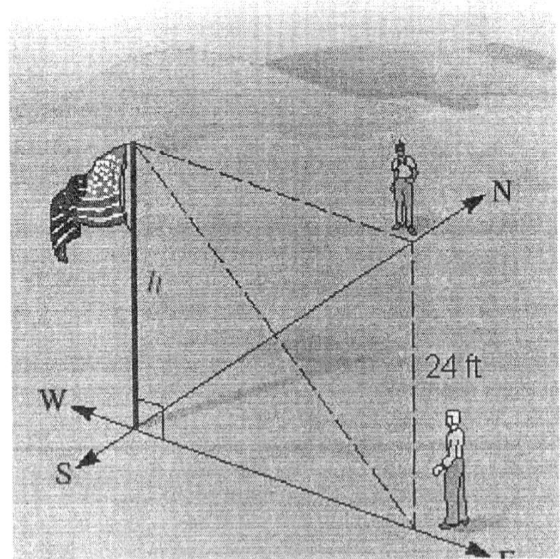

__________ ft

21. Draw vector representing the velocity. 60 mph due south

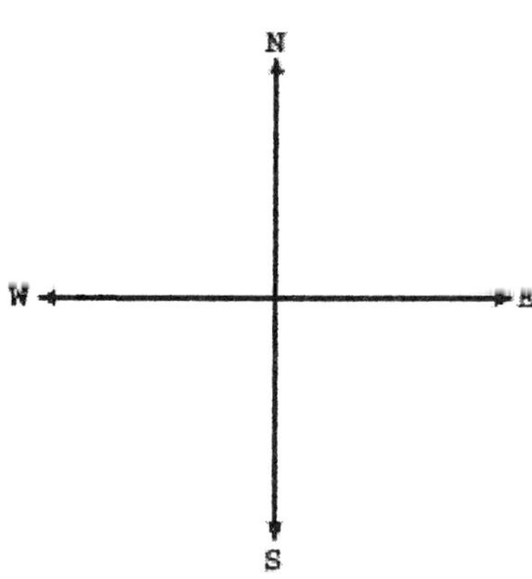

22. Two planes take off at the same time from an airport. The first plane is flying at 260 miles per hour on a bearing of S 45° E. The second plane is flying in the direction S 45° W at

265 miles per hour. If there are no wind currents blowing, how far apart are they after 3 hours? Please round the answer to the nearest whole number.

__________ mi

What is the bearing of the second plane from the first after 3 hours?
Please round the angle to the nearest degree.
Please enter S or N (for north or south) in the first field and E or W (for east or west) in the third field.

___   _______ $^\circ$ ___

23. The problem refers to a vector **V** with magnitude | **V** | that forms an angle with the positive x-axis. Give the magnitude of the horizontal and vertical vector components of **V**, namely V_x and V_y, respectively. Round your answers to the nearest integer.

$|V| = 60,\ \theta = 150^\circ$

$\left|V_x\right| = $ _______

$\left|V_y\right| = $ _______

24. The horizontal and vertical components of the velocity of an arrow shot into the air are 15.0 feet per second and 27.0 feet per second, respectively. Find the velocity of the arrow. Round your answers to the nearest tenth.

__________ ft/sec

25. A 10-pound weight is lying on a sit-up bench at the gym. If the bench is inclined at an angle of $12°$, there are three forces acting on the weight. **N** is called the normal force and it acts in the direction perpendicular to the bench. **F** is the force due to friction that holds the weight on the bench. If the weight does not move, then the sum of these three forces is 0.

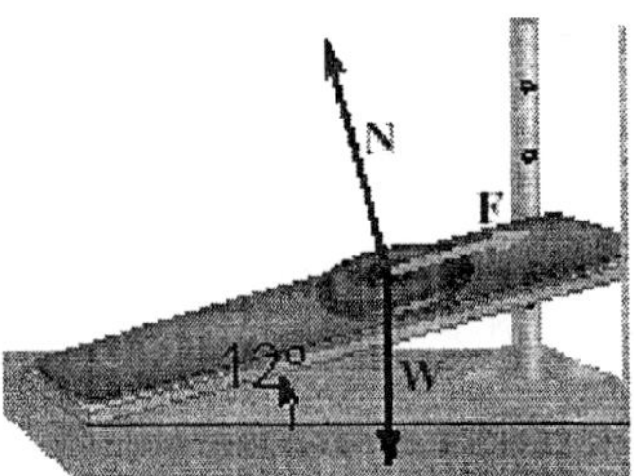

Find the magnitude of **N** and the magnitude of **F**. Please round your answers to the nearest tenth.

| **N** | = __________ lb

| **F** | = __________ lb

1. $\sin(A) = \dfrac{4}{5}$,

   $\cos(A) = \dfrac{3}{5}$,

   $\tan(A) = \dfrac{4}{3}$,

   $\cot(A) = \dfrac{3}{4}$,

   $\sec(A) = \dfrac{5}{3}$,

   $\csc(A) = \dfrac{5}{4}$

2. $\sin(A) = \dfrac{\sqrt{23}}{12}$,

   $\cos(A) = \dfrac{11}{12}$,

   $\tan(A) = \dfrac{\sqrt{23}}{11}$,

   $\sin(B) = \dfrac{11}{12}$,

   $\cos(B) = \dfrac{\sqrt{23}}{12}$,

   $\tan(B) = \dfrac{11}{\sqrt{23}}$

3. undefined

   2

   $\sqrt{2}$

   $\dfrac{2}{\sqrt{3}}$

   1

4. 3

5. $\dfrac{1}{\sqrt{3}}$

6. 62

12

7. 99

39

8. 1.6643

9. 1.0000

1

0.8660

1.1547

0.7071

1.4142

0.5000

2.0000

0.0000

undefined

10. 7

37

11. 62

12. 49

13. 45.54

5.461

7.797

14. 12

15. 29

25

16. 41

70

17. 36.3

18. 4.2

S
87

W

19. 7.2

20. 25

21.

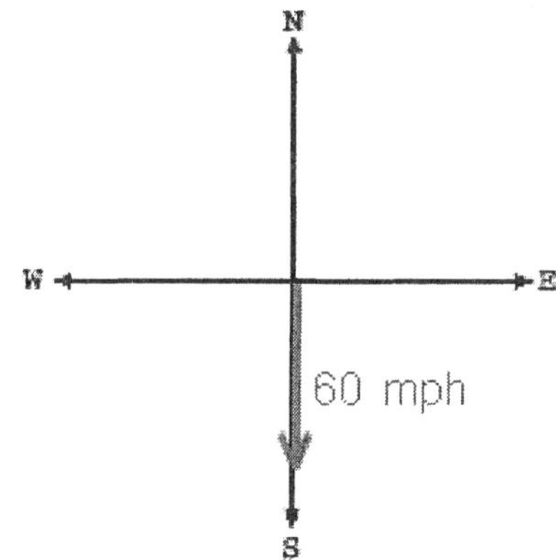

22. 1.114

S
89

W

23. 52

30

24. 30.9

25. 9.8

2.1

McKeague/Turner - Trigonometry 5e Chapter 2 Form A

1. mctr.02.01.01_NoAlgs
2. mctr.02.01.13_NoAlgs
3. mctr.02.01.25_NoAlgs
4. mctr.02.01.39_NoAlgs
5. mctr.02.01.50_NoAlgs
6. mctr.02.02.01_NoAlgs
7. mctr.02.02.19_NoAlgs
8. mctr.02.02.37_NoAlgs
9. mctr.02.02.51_NoAlgs
10. mctr.02.02.74_NoAlgs
11. mctr.02.03.02_NoAlgs
12. mctr.02.03.12_NoAlgs
13. mctr.02.03.26_NoAlgs
14. mctr.02.03.35_NoAlgs
15. mctr.02.03.46_NoAlgs
16. mctr.02.04.01_NoAlgs
17. mctr.02.04.07_NoAlgs
18. mctr.02.04.16_NoAlgs
19. mctr.02.04.21_NoAlgs
20. mctr.02.04.27_NoAlgs
21. mctr.02.05.01_NoAlgs
22. mctr.02.05.10_NoAlgs
23. mctr.02.05.17_NoAlgs
24. mctr.02.05.28_NoAlgs
25. mctr.02.05.35_NoAlgs

1. From a point on the floor the angle of elevation to the top of a door is $48°$, while the angle of elevation to the ceiling above the door is $57°$. If the ceiling is 10 feet above the floor, what is the vertical dimension of the door (see the figure)?

Please round your answer to the nearest tenth of a foot.

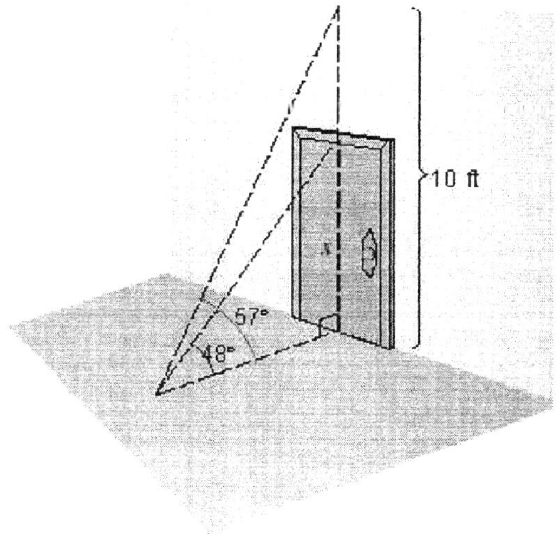

__________ ft

2. Use a calculator to find a value of θ between $0°$ and $90°$ that satisfies the statement below. Write your answer in degrees and minutes rounded to the nearest minute.

$\csc \theta = 7.5493$

$\theta = $ _____ ° _____ '

3. Use a calculator to complete the following table. Round all answers to four digits past the decimal point.

If an expression is undefined indicate it.

x	$\cos x$	$\sec x$
$0°$	__________	__________
$30°$	__________	__________
$45°$	__________	__________
$60°$	__________	__________
$90°$	__________	__________

4. Refer to right triangle *ABC* with $C = 90°$.

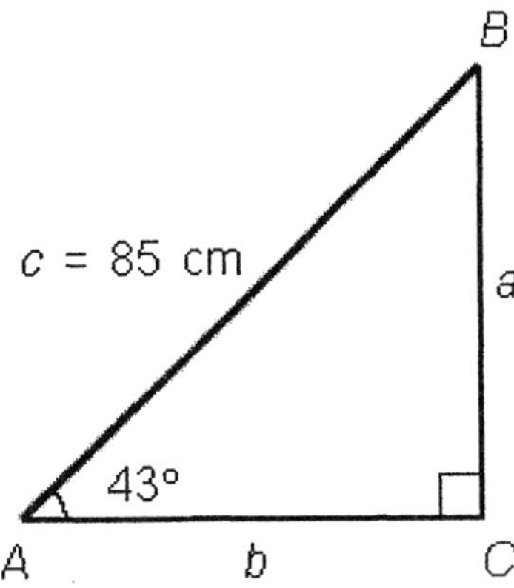

If $A = 43°$ and $c = 85$ cm, find b.

Please round the answer to the nearest whole number.

$b = $ _________ cm

5. Draw vector representing the velocity.

60 mph due south

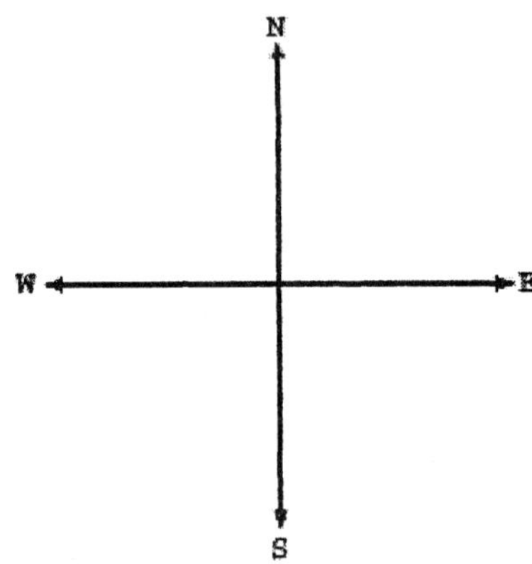

6. Refer to right triangle *ABC* with $C = 90°$.

    Solve for all the missing parts using the given information.

    $B = 44.46°$, $a = 5.565$ mi

    Please round the answer for the angle to two decimal places, and for the sides, to three decimal places.

    $A =$ _________ °

    $b =$ _________ mi

    $c =$ _________ mi

7. Solve the problem.

    A 72.5-foot rope from the top of a circus tent pole is anchored to the ground 42.9 feet from the bottom of the pole. What angle does the rope make with the pole? (Assume the pole is perpendicular to the ground.) Please round your answer to the nearest tenth of a degree.

    _________ °

8. Convert the following to degrees and minutes. Round to the nearest minute.

    $99.65°$

    _____ ° _____ ′

9. In the figure, the distance from *A* to *D* is *y*, the distance from *D* to *C* is *x*, and the distance from *C* to *B* is *h*.

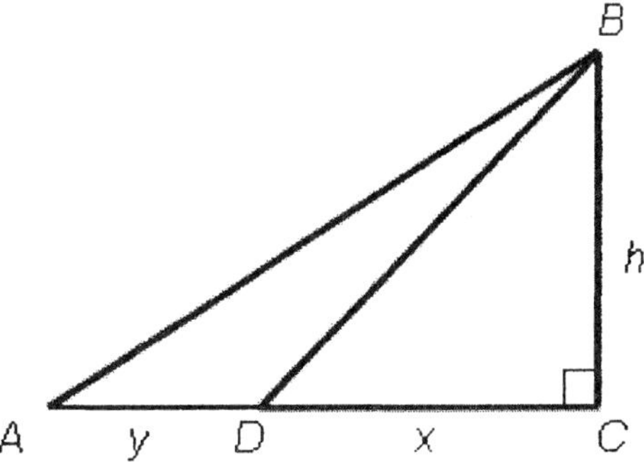

    If $A = 32°$, $\angle BDC = 49°$, and $AB = 54$, find *h*, then *x*.

    Please round each answer to the nearest whole number.

    $h =$ _________

    $x =$ _________

10. In the right triangle below

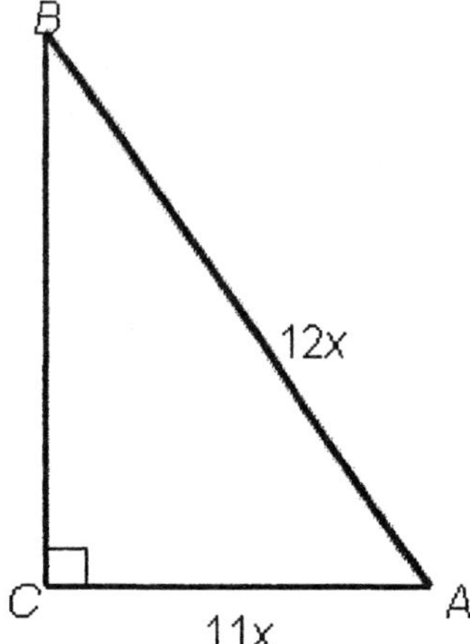

Find sin$A$

Find cos$A$

Find tan$A$

Find sin$B$

Find cos$B$

Find tan$B$

11. Refer to right triangle ABC with $C = 90°$.

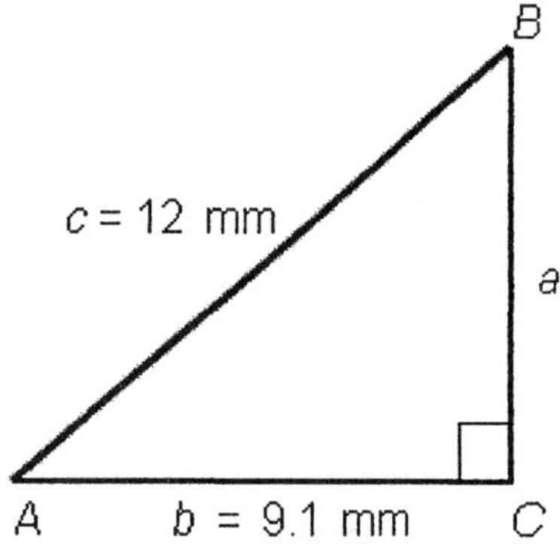

If $b = 9.1$ mm and $c = 12$ mm, find B.

Please round the answer to the nearest whole number.

$B =$ _________ °

12. The circle in the figure has a radius of *r* and center at *C*. The distance from *A* to *B* is *x*.

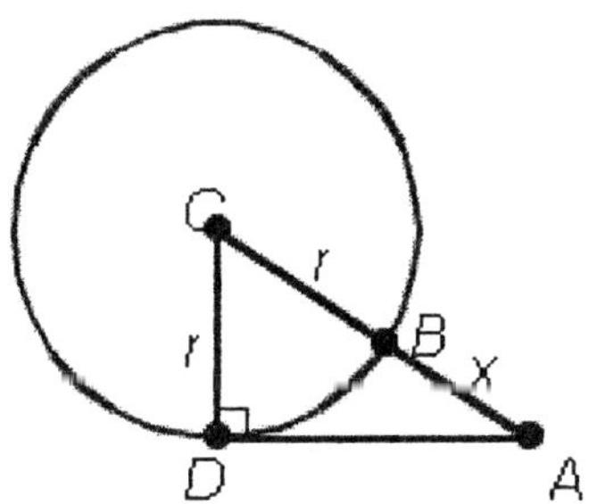

If $A = 32°$ and $r = 13$, find *x*.

Please round the answer to the nearest whole number.

$x =$ _________

13. Solve the problem.

A man wandering in the desert walks 2.4 miles in the direction S 32° W. He then turns 90° and walks 3.4 miles in the direction N 58° W. At that time, how far is he from his starting point? Please round your answer to the nearest tenth of a mile.

_________ mi

What is the man's bearing from his starting point? Please round the angle to the nearest degree.
Please enter *S* or *N* (for north or south) in the first field and *E* or *W* (for east or west) in the third field.

___ _______ ° ___

14. Use a calculator to find the following. Round the answer to four places past the decimal point.

cot 31°

____

15. Solve the problem.

The two equal sides of an isosceles triangle are each 44 centimeters.
The base measures 30 centimeters.

Find the height. Please round your answer to the nearest centimeter.

__________ cm

Find the measure of the two equal angles. Please round your answer to the nearest degree.

__________ °

16. For the expression $6\sin(z - 30°)$, replace z with $60°$, and then simplify as much as possible.

17. Solve the problem.

Two people decide to estimate the height of a flag pole. One person positions himself due north of the pole and the other person stands due east of the pole. If the two people are the same distance from the pole and 24 feet from each other, find the height of the pole if the angle of elevation from the ground at each person's position to the top of the pole is $56°$

(see the figure). Please round your answer to the nearest foot.

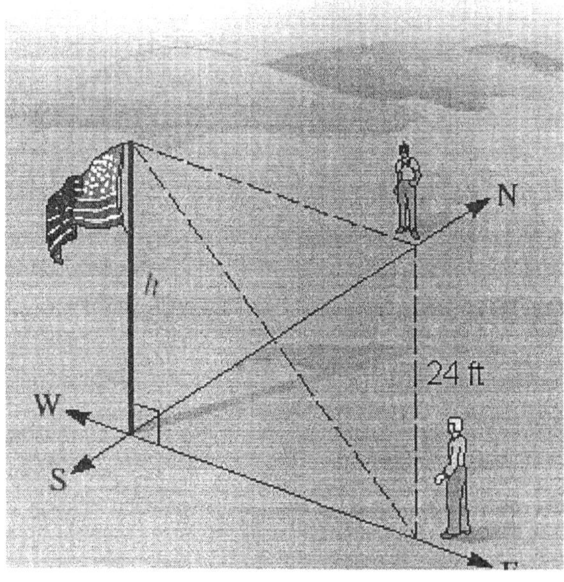

__________ ft

18. Two planes take off at the same time from an airport. The first plane is flying at 260 miles per hour on a bearing of S $45°$ E. The second plane is flying in the direction S $45°$ W at 265 miles per hour. If there are no wind currents blowing, how far apart are they after 3 hours? Please round the answer to the nearest whole number.

    ________ mi

    What is the bearing of the second plane from the first after 3 hours?

    Please round the angle to the nearest degree.

    Please enter S or N (for north or south) in the first field and E or W (for east or west) in the third field.

    __   _____ ° __

19. The problem refers to a vector **V** with magnitude | **V** | that forms an angle with the positive x-axis. Give the magnitude of the horizontal and vertical vector components of **V**, namely $\mathbf{V}_x$ and $\mathbf{V}_y$, respectively. Round your answers to the nearest integer.

    $$|\mathbf{V}| = 60, \ \theta = 150°$$

    $$\left|\mathbf{V}_x\right| = \underline{\hspace{3em}}$$

    $$\left|\mathbf{V}_y\right| = \underline{\hspace{3em}}$$

20. Complete the following table using exact values. Do not rationalize the denominator.

x	$\sin x$	$\csc x$
$0°$	0	?
$30°$	$\dfrac{1}{2}$	?
$45°$	$\dfrac{\sqrt{2}}{2}$	?
$60°$	$\dfrac{\sqrt{3}}{2}$	?
$90°$	1	?

    Find the value indicated by the question mark.

21. Add.

$$(39°45') + (22°27')$$

_____ ° _____ '

22. Find the exact value of $\cot 60°$.

23. Right triangle ABC has $C = 90°$, $b = 9$, and $c = 15$.

Use this information to find the six trigonometric functions of A.

Find $\sin A$

Find $\cos A$

Find $\tan A$

Find $\cot A$

Find $\sec A$

Find $\csc A$

24. The horizontal and vertical components of the velocity of an arrow shot into the air are 15.0 feet per second and 27.0 feet per second, respectively.

Find the velocity of the arrow.

Round your answers to the nearest tenth.

_________ ft/sec

25. A 10-pound weight is lying on a sit-up bench at the gym. If the bench is inclined at an angle of $12°$, there are three forces acting on the weight. **N** is called the normal force and it acts in the direction perpendicular to the bench. **F** is the force due to friction that holds the weight on the bench. If the weight does not move, then the sum of these three forces is 0.

Find the magnitude of **N** and the magnitude of **F**.

Please round your answers to the nearest tenth.

| **N** | = _________ lb

| **F** | = _________ lb

1. 7.2

2. 7

   37

3. 1.0000

   1

   0.8660

   1.1547

   0.7071

   1.4142

   0.5000

   2.0000

   0.0000

   undefined

4. 62

5.

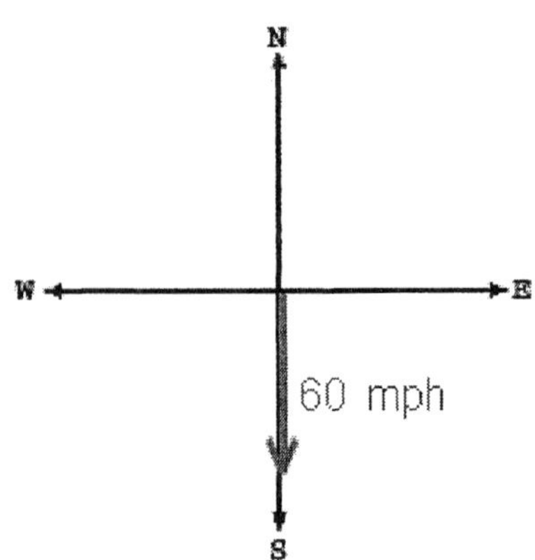

6. 45.54

   5.461

   7.797

7. 36.3

8. 99

   39

McKeague/Turner - Trigonometry 5e Chapter 2 Form B

9. 29

25

10.

$\sin(A) = \dfrac{\sqrt{23}}{12}$,

$\cos(A) = \dfrac{11}{12}$,

$\tan(A) = \dfrac{\sqrt{23}}{11}$,

$\sin(B) = \dfrac{11}{12}$,

$\cos(B) = \dfrac{\sqrt{23}}{12}$,

$\tan(B) = \dfrac{11}{\sqrt{23}}$

11. 4 9

12. 1 2

13. 4 . 2

S
87
W

14. 1 . 664 3

15. 4 1

70

16. 3

17. 25

18. 1 . 11 4

S
89

W

19. 52

30

20. undefined

2

$\sqrt{2}$

$\dfrac{2}{\sqrt{3}}$

1

21. 62

12

22. $\dfrac{1}{\sqrt{3}}$

23. $\sin(A) = \dfrac{4}{5}$,

$\cos(A) = \dfrac{3}{5}$,

$\tan(A) = \dfrac{4}{3}$,

$\cot(A) = \dfrac{3}{4}$,

$\sec(A) = \dfrac{5}{3}$,

$\csc(A) = \dfrac{5}{4}$

24. 30.9

25. 9.8

2.1

McKeague/Turner - Trigonometry 5e Chapter 2 Form B

1. mctr.02.04.21_NoAlgs
2. mctr.02.02.74_NoAlgs
3. mctr.02.02.51_NoAlgs
4. mctr.02.03.02_NoAlgs
5. mctr.02.05.01_NoAlgs
6. mctr.02.03.28_NoAlgs
7. mctr.02.04.07_NoAlgs
8. mctr.02.02.19_NoAlgs
9. mctr.02.03.46_NoAlgs
10. mctr.02.01.13_NoAlgs
11. mctr.02.03.12_NoAlgs
12. mctr.02.03.35_NoAlgs
13. mctr.02.04.16_NoAlgs
14. mctr.02.02.37_NoAlgs
15. mctr.02.04.01_NoAlgs
16. mctr.02.01.39_NoAlgs
17. mctr.02.04.27_NoAlgs
18. mctr.02.05.10_NoAlgs
19. mctr.02.05.17_NoAlgs
20. mctr.02.01.25_NoAlgs
21. mctr.02.02.01_NoAlgs
22. mctr.02.01.50_NoAlgs
23. mctr.02.01.01_NoAlgs
24. mctr.02.05.28_NoAlgs
25. mctr.02.05.35_NoAlgs

1. Find tan *A* in the right triangle below.

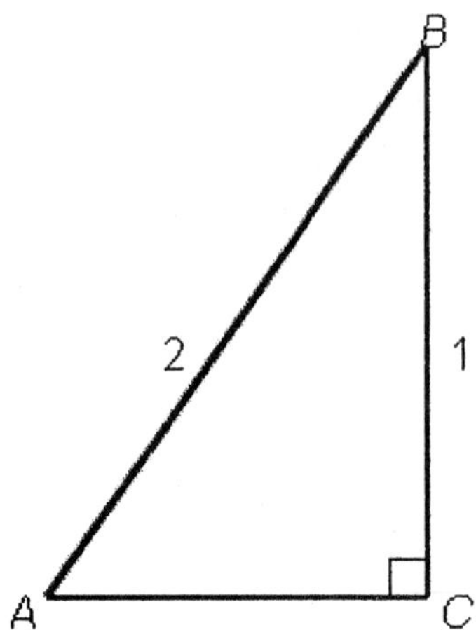

Select the correct answer.

a. $\tan (A) = \dfrac{2}{3}$

b. $\tan (A) = \dfrac{\sqrt{3}}{2}$

c. $\tan (A) = \dfrac{1}{2}$

d. $\tan (A) = \dfrac{1}{3}$

e. $\tan (A) = \dfrac{1}{\sqrt{3}}$

2. Use the Cofunction Theorem to fill in the blank so that the expression becomes a true statement.

$\sin 60^\circ = \cos \underline{\quad}^\circ$

Select the correct answer.

a. 40°

b. 30°

c. 35°

d. 50°

e. 45°

3. Simplify expression by first substituting values from the table of exact values and then simplifying the resulting expression.

$$(\sin 60° + \cos 60°)^2$$

Table of Exact Values

x	$\sin x$	$\cos x$
$0°$	0	1
$30°$	$\dfrac{1}{2}$	$\dfrac{\sqrt{3}}{2}$
$45°$	$\dfrac{1}{\sqrt{2}}$	$\dfrac{1}{\sqrt{2}}$
$60°$	$\dfrac{\sqrt{3}}{2}$	$\dfrac{1}{2}$
$90°$	1	0

Select the correct answer.

a. $\dfrac{2 + \sqrt{3}}{2}$

b. $\dfrac{2 + \sqrt{2}}{2}$

c. $\dfrac{1 - \sqrt{3}}{2}$

d. $\dfrac{2 - \sqrt{2}}{2}$

e. $\dfrac{2 - \sqrt{3}}{2}$

4. Find the exact value for $\csc 45°$.

Select the correct answer.

a. $-5\sqrt{2}$

b. $5\sqrt{2}$

c. $\sqrt{2}$

d. $-\sqrt{2}$

5. There is a right triangle ABC with $C = 90°$, $a = 19.54$, and $b = 5.27$. Find sin B. Round your answers to the nearest hundredth.

    Select the correct answer.

    a.   0.98
    b.   0.97
    c.   0.26
    d.   0.28
    e.   0.25

6. Subtract.

    $(82°38') - (21°68')$

    Select the correct answer.

    a. $30°60'$

    b. $61°31'$

    c. $60°30'$

    d. $60°31'$

    e. $31°60'$

7. Change the following to decimal degrees. If rounding is necessary, round to the nearest hundredth of a degree.

    $43°39'$

    Select the correct answer.

    a. $44.15°$

    b. $42.75°$

    c. $43.65°$

    d. $44.35°$

    e. $42.55°$

8. Use a calculator to find the following.

   Round your answer to four places past the decimal point.

   $\sin 40° 20'$

   Select the correct answer.

   a. 0.7022
   b. 0.5242
   c. 0.6322
   d. 0.6472
   e. 0.8332

9. Find θ if θ is between $0°$ and $90°$. Round your answer to the nearest tenth of a degree.

   $\cot \theta = 0.8732$

   Select the correct answer.

   a. $\theta = 51.7°$

   b. $\theta = 50.0°$

   c. $\theta = 48.8°$

   d. $\theta = 48.9°$

   e. $\theta = 50.4°$

10. Work the following problem on your calculator.

    $\cos^2 36° + \sin^2 36°$

    Select the correct answer.

    a. 1.3
    b. 0.4
    c. 1
    d. 1.2
    e. 2

11. Refer to right triangle ABC with $C = 90°$.

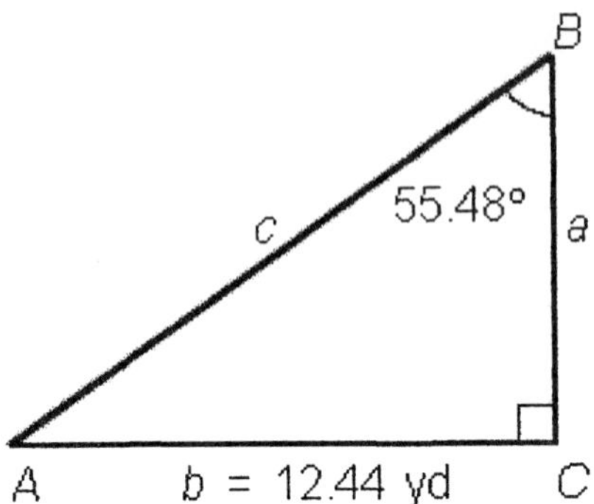

If $B = 55.48°$ and $b = 12.44$ yd, find a.

Please round the answer to the nearest hundredth.

Select the correct answer.

a.   8.54 yd
b.   8.56 yd
c.   8.58 yd
d.   8.60 yd
e.   8.50 yd

12. Refer to right triangle ABC with $C = 90°$.

Solve for all the missing parts using the given information.

$B = 19°,\ c = 4.8$ ft

Please round each answer to the nearest tenth if necessary.

Select the correct answer.

a. $A = 71°,\ a = 4.3$ ft, $b = 1.6$ ft

b. $A = 71°,\ a = 4.5$ ft, $b = 1.8$ ft

c. $A = 71°,\ a = 4.9$ ft, $b = 2.0$ ft

d. $A = 71°,\ a = 4.5$ ft, $b = 1.6$ ft

e. $A = 71°,\ a = 4.3$ ft, $b = 1.8$ ft

13. Refer to right triangle ABC with $C = 90°$.

Solve for all the missing parts using the given information.

$b = 377.6$ inches, $c = 588.5$ inches

Please round the answers for the angles to two decimal places, and for the side, to one decimal place.

Select the correct answer.

a. $A = 50.09°$, $B = 39.91°$, $a = 451.2$ in

b. $A = 50.13°$, $B = 39.87°$, $a = 451.8$ in

c. $A = 50.09°$, $B = 39.91°$, $a = 451.4$ in

d. $A = 50.11°$, $B = 39.89°$, $a = 451.2$ in

e. $A = 50.11°$, $B = 39.89°$, $a = 451.4$ in

14. The figure shows two right triangles drawn at $90°$ to each other.

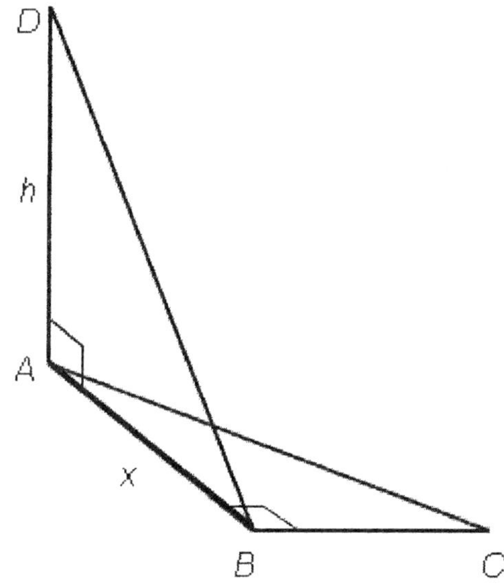

If $\angle ABD = 50°$, $C = 47°$, and $BC = 42$, find x and then find h.

Please round each answer to the nearest whole number.

Select the correct answer.

a. $x = 47$, $h = 56$

b. $x = 45$, $h = 56$

c. $x = 49$, $h = 58$

d. $x = 45$, $h = 54$

e. $x = 47$, $h = 54$

15. The figure shows a simplified model of the first Ferris wheel. The radius of the first Ferris wheel was 125 feet; the distance from the ground to the bottom of the wheel was 14 feet. If θ is the central angle formed as a rider moves from position P_0 to position P_1, find the rider's height above the ground h when θ is 300°.

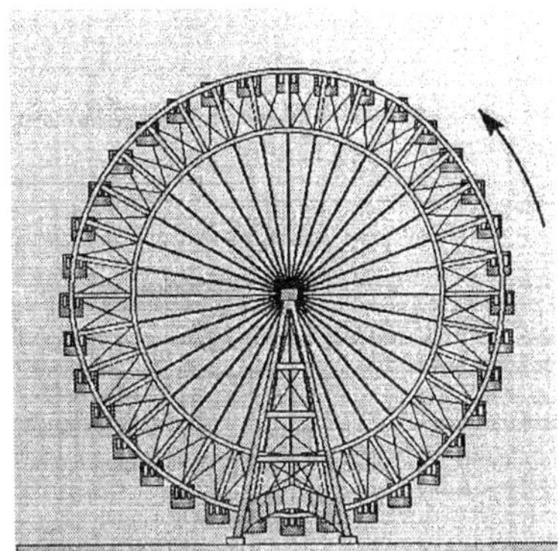

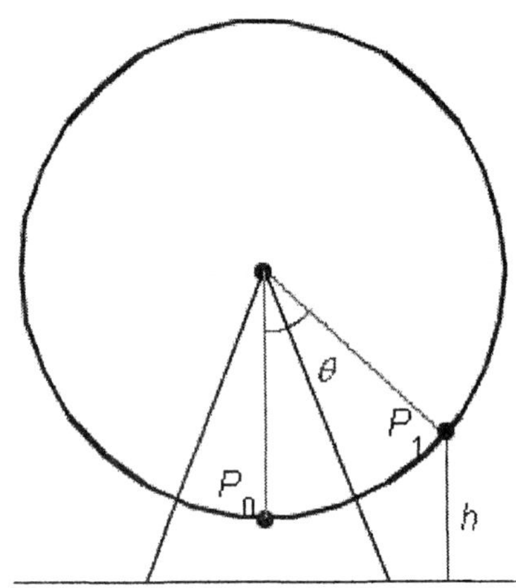

Please round the answer to the nearest whole number.

Select the correct answer.

a.   72 ft
b.   82 ft
c.   79 ft
d.   76 ft
e.   75 ft

16. Solve the problem.

The diagonal of a rectangle is 344 millimeters, while the longer side is 278 millimeters. Find the shorter side of the rectangle and the angles the diagonal makes with the sides.

Round the side length to the nearest millimeter. Round the angles to the nearest degree.

Select the correct answer.

a.   The side is 203 mm; the angles are 54° and 36°.

b.   The side is 203 mm; the angles are 49° and 41°.

c.   The side is 214 mm; the angles are 46° and 44°.

d.   The side is 203 mm; the angles are 46° and 44°.

e.   The side is 214 mm; the angles are 54° and 36°.

17. A person standing 130 centimeters from a mirror notices that the angle of depression from his eyes to the bottom of the mirror is $14°$, while the angle of elevation to the top of the mirror is $12°$.

Find the vertical dimension of the mirror (see the figure).

Please round your answer to the nearest centimeter.

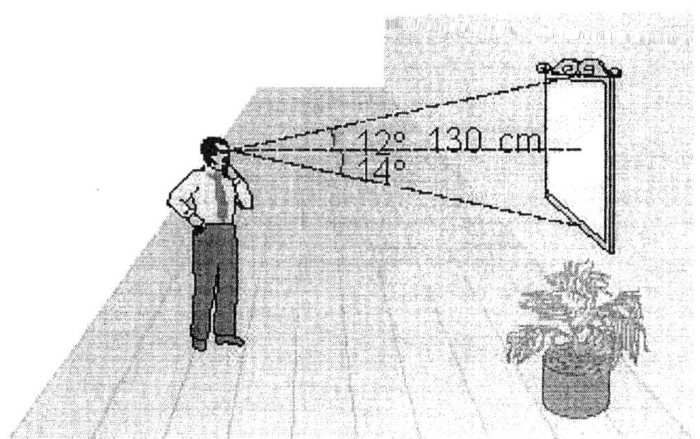

Select the correct answer.

a.   60 cm
b.   62 cm
c.   50 cm
d.   57 cm
e.   61 cm

18. Solve the problem.

A tree on one side of a river is due west of a rock on the other side of the river. From a stake 24 yards north of the rock, the bearing of the tree is $S\,18.4°\,W$.

How far is it from the rock to the tree?

Please round your answer to the nearest tenth of a yard.

Select the correct answer.

a.   6.5 yd
b.   5.6 yd
c.   8.0 yd
d.   13.3 yd
e.   9.0 yd

19. In the figure below, a person standing at point A notices that the angle of elevation to the top of the antenna is $48°\,30'$. A second person standing 37.0 feet farther from the antenna than the person at A finds the angle of elevation to the top of the antenna to be $41°\,15'$.

How far is the person at A from the base of the antenna?

Please round your answer to the nearest foot.

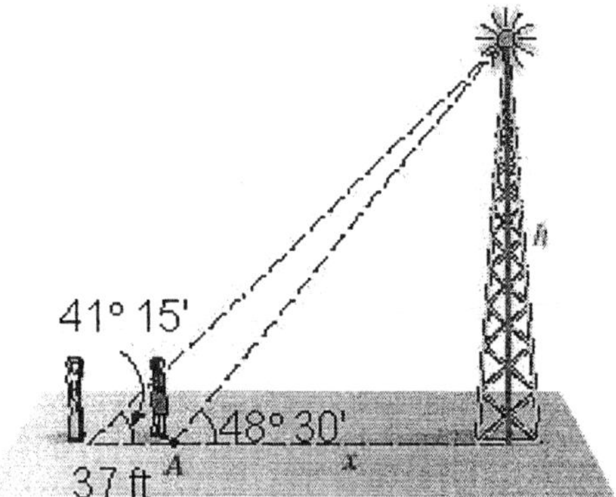

Select the correct answer.

a.   132 ft
b.   129 ft
c.   128 ft
d.   126 ft
e.   125 ft

20. Suppose the figure below is an exaggerated diagram of a plane flying above the earth. The plane is 5.05 miles above the earth and the radius of the earth is 4,000 miles, how far is it from the plane to the horizon?

Please round your answer to the nearest mile.

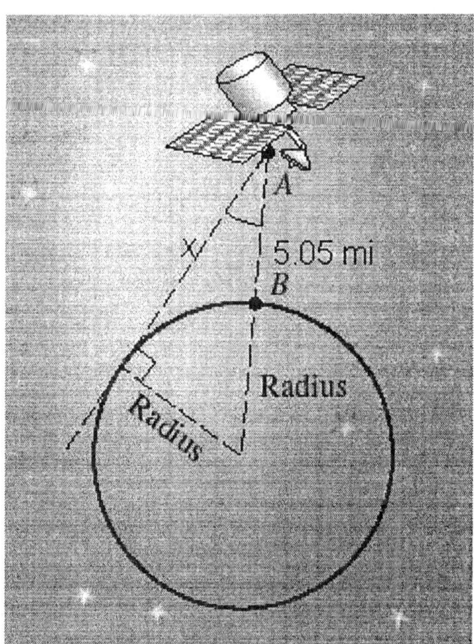

Select the correct answer.

a.   202 mi
b.   201 mi
c.   217 mi
d.   211 mi
e.   197 mi

21. Draw vector representing the velocity.

80 cm/sec N 20° E

Select the correct answer.

a.

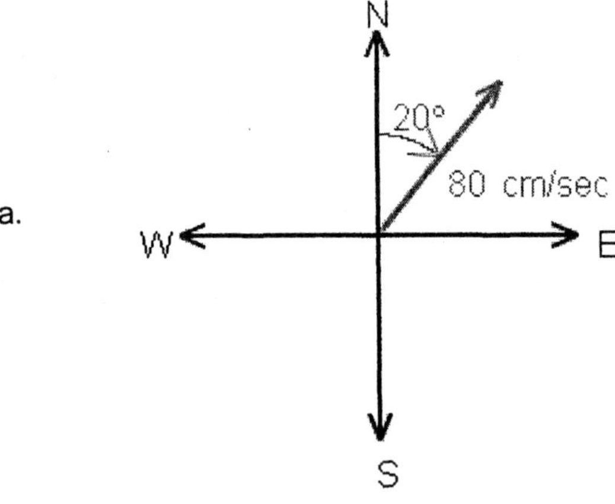

b.

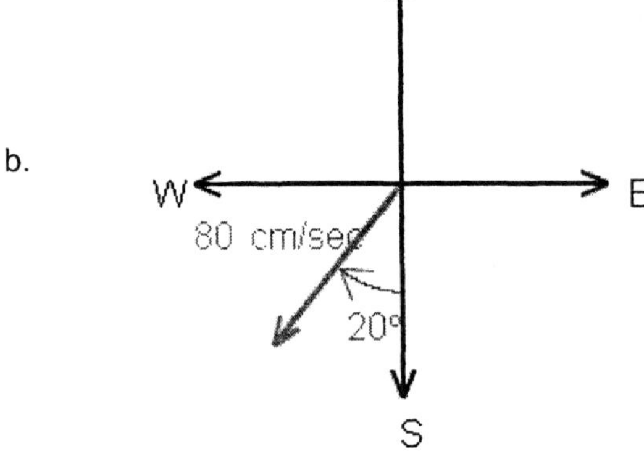

c.

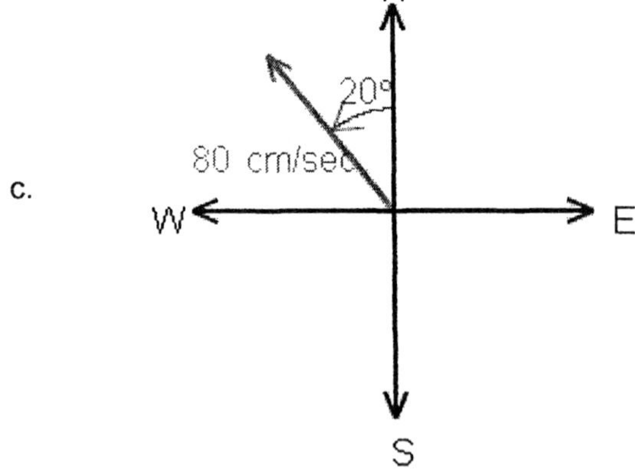

22. The problem refers to a vector $\mathbf{V}$ with magnitude $|\mathbf{V}|$ that forms an angle with the positive x-axis. Give the magnitude of the horizontal and vertical vector components of $\mathbf{V}$, namely $\mathbf{V}_x$ and $\mathbf{V}_y$, respectively.

Please round your answers to the nearest hundredth.

$$|\mathbf{V}| = 13.1, \ \theta = 24.2^\circ$$

Select the correct answer.

a. $\quad |\mathbf{V}_x| = 11.95, \ |\mathbf{V}_y| = 5.37$

b. $\quad |\mathbf{V}_x| = -9.35, \ |\mathbf{V}_y| = 1.17$

c. $\quad |\mathbf{V}_x| = -7.65, \ |\mathbf{V}_y| = -5.37$

d. $\quad |\mathbf{V}_x| = 7.65, \ |\mathbf{V}_y| = -3.67$

e. $\quad |\mathbf{V}_x| = 11.95, \ |\mathbf{V}_y| = -3.67$

23. For the problem, the magnitude of the horizontal and vertical vector components $\mathbf{V}_x$ and $\mathbf{V}_y$, of vector $\mathbf{V}$ are given.

Find the magnitude of $\mathbf{V}$.

Round your answers to the nearest tenth.

$$|\mathbf{V}_x| = 2.2, \ |\mathbf{V}_y| = 5.8$$

Select the correct answer.

a. $\quad |\mathbf{V}| = 8.5$

b. $\quad |\mathbf{V}| = 6.2$

c. $\quad |\mathbf{V}| = 6.1$

d. $\quad |\mathbf{V}| = 8.1$

e. $\quad |\mathbf{V}| = 5.7$

24. A plane travels 145 miles on a bearing of $N \ 11\degree \ E$ and then changes its course to $N \ 48\degree \ E$ and travels another 140 miles. Find the total distance traveled north and the total distance traveled east. Round your answers to the nearest integer.

Select the correct answer.

    a.   plane travels 252 miles to the north and 110 miles to the east

    b.   plane travels 231 miles to the north and 140 miles to the east

    c.   plane travels 236 miles to the north and 110 miles to the east

    d.   plane travels 236 miles to the north and 132 miles to the east

    e.   plane travels 252 miles to the north and 132 miles to the east

25. A package is pushed across a floor for a distance of 60 feet by exerting a force of 35 pounds downward at an angle of $25\degree$ with the horizontal.

How much work is done?

Round your answer to the nearest hundred.

Select the correct answer.

    a.   2,000 ft · lb

    b.   2,200 ft · lb

    c.   2,300 ft · lb

    d.   1,900 ft · lb

    e.   2,100 ft · lb

1. e

2. b

3. a

4. c

5. c

6. c

7. c

8. d

9. d

10. c

11. b

12. d

13. c

14. d

15. d

16. a

17. a

18. c

19. c

20. b

21. a

22. a

23. b

24. d

25. d

1. mctr.02.01.07m_NoAlgs
2. mctr.02.01.17m_NoAlgs
3. mctr.02.01.32m_NoAlgs
4. mctr.02.01.47m_NoAlgs
5. mctr.02.01.55m_NoAlgs
6. mctr.02.02.12m_NoAlgs
7. mctr.02.02.29m_NoAlgs
8. mctr.02.02.44m_NoAlgs
9. mctr.02.02.65m_NoAlgs
10. mctr.02.02.81m_NoAlgs
11. mctr.02.03.07m_NoAlgs
12. mctr.02.03.22m_NoAlgs
13. mctr.02.03.32m_NoAlgs
14. mctr.02.03.40m_NoAlgs
15. mctr.02.03.53m_NoAlgs
16. mctr.02.04.04m_NoAlgs
17. mctr.02.04.11m_NoAlgs
18. mctr.02.04.18m_NoAlgs
19. mctr.02.04.23m_NoAlgs
20. mctr.02.04.30m_NoAlgs
21. mctr.02.05.05m_NoAlgs
22. mctr.02.05.13m_NoAlgs
23. mctr.02.05.22m_NoAlgs
24. mctr.02.05.31m_NoAlgs
25. mctr.02.05.39m_NoAlgs

1. Solve the problem.

   The diagonal of a rectangle is 344 millimeters, while the longer side is 278 millimeters. Find the shorter side of the rectangle and the angles the diagonal makes with the sides.

   Round the side length to the nearest millimeter. Round the angles to the nearest degree.

   Select the correct answer.

   a. The side is 203 mm; the angles are $54°$ and $36°$.

   b. The side is 203 mm; the angles are $49°$ and $41°$.

   c. The side is 214 mm; the angles are $46°$ and $44°$.

   d. The side is 203 mm; the angles are $46°$ and $44°$.

   e. The side is 214 mm; the angles are $54°$ and $36°$.

2. A person standing 130 centimeters from a mirror notices that the angle of depression from his eyes to the bottom of the mirror is $14°$, while the angle of elevation to the top of the mirror is $12°$. Find the vertical dimension of the mirror (see the figure). Please round your answer to the nearest centimeter.

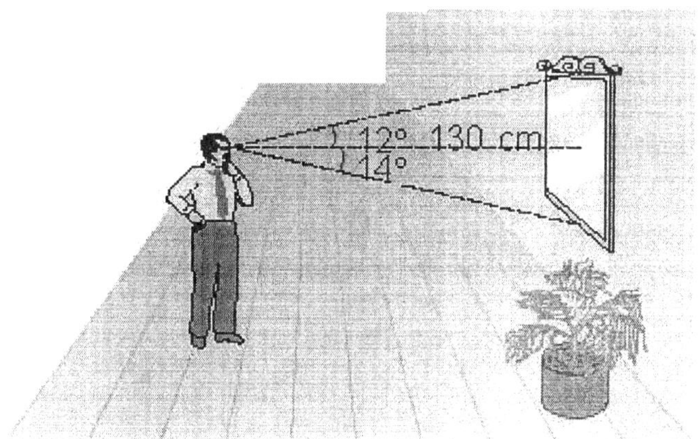

   Select the correct answer.

   a. 60 cm
   b. 62 cm
   c. 50 cm
   d. 57 cm
   e. 61 cm

3. The figure shows two right triangles drawn at 90° to each other.

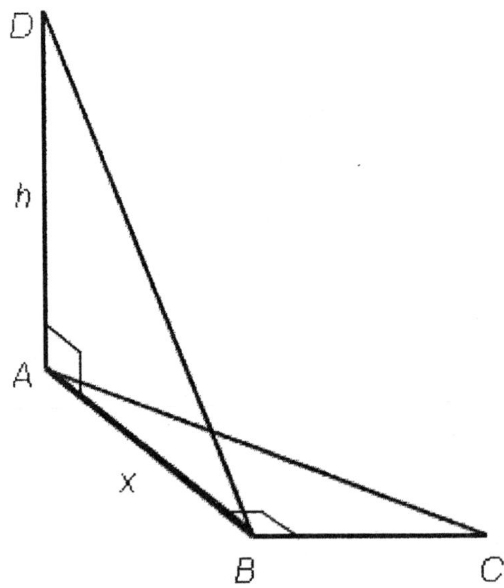

If $\angle ABD = 50°$, $C = 47°$, and $BC = 42$, find x and then find h.

Please round each answer to the nearest whole number.

Select the correct answer.

a. $x = 47,\ h = 56$
b. $x = 45,\ h = 56$
c. $x = 49,\ h = 58$
d. $x = 45,\ h = 54$
e. $x = 47,\ h = 54$

4. Work the following problem on your calculator.

$$\cos^2 36° + \sin^2 36°$$

Select the correct answer.

a.   1.3
b.   0.4
c.   1
d.   1.2
e.   2

5. Refer to right triangle ABC with $C = 90°$.

Solve for all the missing parts using the given information.

$B = 19°$, $c = 4.8$ ft

Please round each answer to the nearest tenth if necessary.

Select the correct answer.

    a. $A = 71°$, $a = 4.3$ ft, $b = 1.6$ ft

    b. $A = 71°$, $a = 4.5$ ft, $b = 1.8$ ft

    c. $A = 71°$, $a = 4.9$ ft, $b = 2.0$ ft

    d. $A = 71°$, $a = 4.5$ ft, $b = 1.6$ ft

    e. $A = 71°$, $a = 4.3$ ft, $b = 1.8$ ft

6. A package is pushed across a floor for a distance of 60 feet by exerting a force of 35 pounds downward at an angle of $25°$ with the horizontal.

How much work is done?

Round your answer to the nearest hundred.

Select the correct answer.

    a. $2{,}000$ ft · lb

    b. $2{,}200$ ft · lb

    c. $2{,}300$ ft · lb

    d. $1{,}900$ ft · lb

    e. $2{,}100$ ft · lb

7. There is a right triangle ABC with $C = 90°$, $a = 19.54$, and $b = 5.27$. Find $\sin B$. Round your answers to the nearest hundredth.

Select the correct answer.

    a.   0.98
    b.   0.97
    c.   0.26
    d.   0.28
    e.   0.25

8. Solve the problem.

    A tree on one side of a river is due west of a rock on the other side of the river. From a stake 24 yards north of the rock, the bearing of the tree is S 18.4° W.

    How far is it from the rock to the tree?

    Please round your answer to the nearest tenth of a yard.

    Select the correct answer.

    a.   6.5 yd
    b.   5.6 yd
    c.   8.0 yd
    d.   13.3 yd
    e.   9.0 yd

9. Refer to right triangle ABC with $C = 90°$.

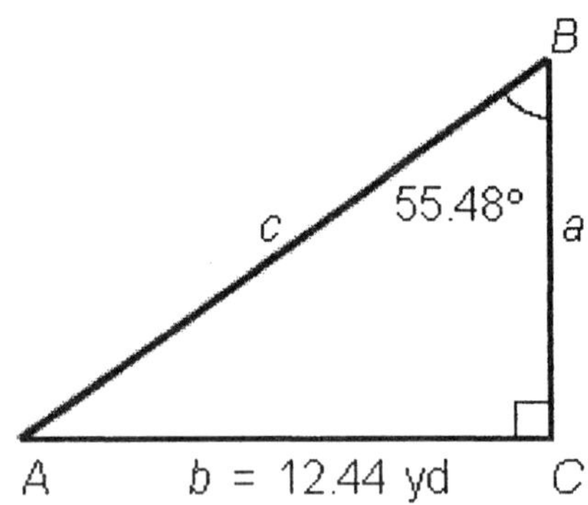

    If $B = 55.48°$ and $b = 12.44$ yd, find a.

    Please round the answer to the nearest hundredth.

    Select the correct answer.

    a.   8.54 yd
    b.   8.56 yd
    c.   8.58 yd
    d.   8.60 yd
    e.   8.50 yd

10. Find tan *A* in the right triangle below.

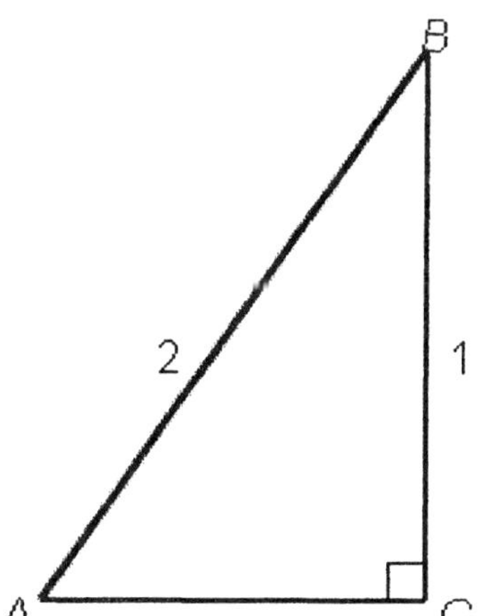

Select the correct answer.

a. $\tan(A) = \dfrac{2}{3}$

b. $\tan(A) = \dfrac{\sqrt{3}}{2}$

c. $\tan(A) = \dfrac{1}{2}$

d. $\tan(A) = \dfrac{1}{3}$

e. $\tan(A) = \dfrac{1}{\sqrt{3}}$

11. Draw vector representing the velocity.

80 cm/sec N 20° E

Select the correct answer.

a.

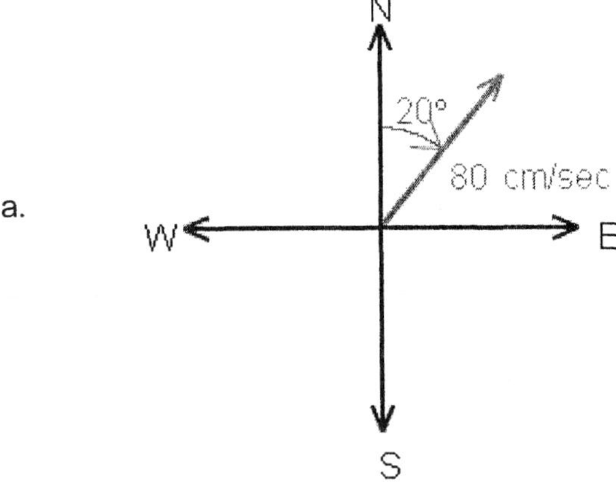

b.

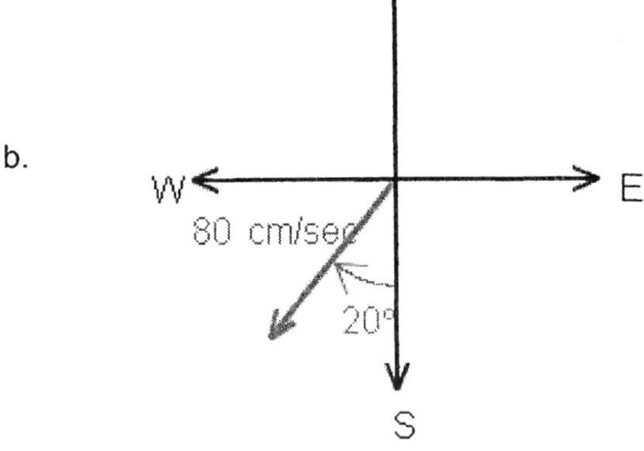

c.

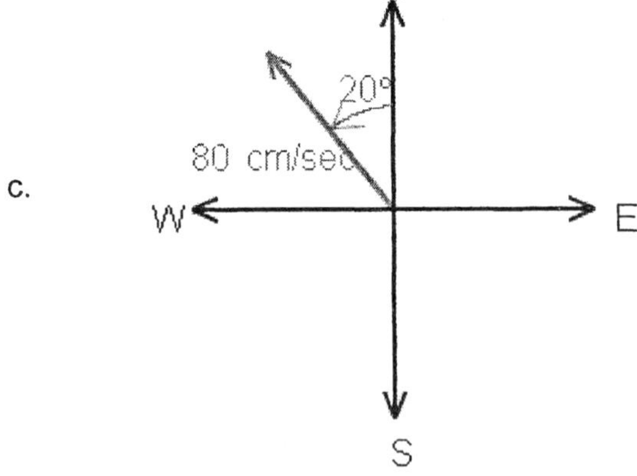

12. Use a calculator to find the following. Round your answer to four places past the decimal point.

$\sin 40^\circ 20'$

Select the correct answer.

    a.   0.7922
    b.   0.5242
    c.   0.6322
    d,   0.6472
    e.   0.8332

13. In the figure below, a person standing at point *A* notices that the angle of elevation to the top of the antenna is $48^\circ 30'$. A second person standing 37.0 feet farther from the antenna than the person at *A* finds the angle of elevation to the top of the antenna to be $41^\circ 15'$.

How far is the person at *A* from the base of the antenna? Please round your answer to the nearest foot.

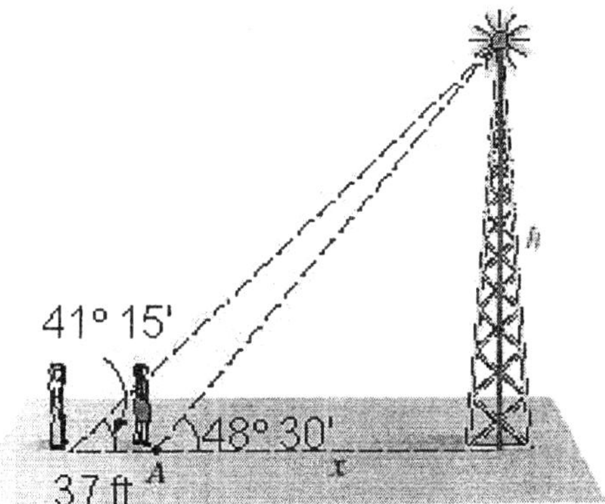

Select the correct answer.

    a.   132 ft
    b.   129 ft
    c.   128 ft
    d.   126 ft
    e.   125 ft

14. Find θ if θ is between $0°$ and $90°$. Round your answer to the nearest tenth of a degree.

$\cot \theta = 0.8732$

Select the correct answer.

    a. $\theta = 51.7°$

    b. $\theta = 50.0°$

    c. $\theta = 48.8°$

    d. $\theta = 48.9°$

    e. $\theta = 50.4°$

15. The figure shows a simplified model of the first Ferris wheel. The radius of the first Ferris wheel was 125 feet; the distance from the ground to the bottom of the wheel was 14 feet. If θ is the central angle formed as a rider moves from position P_0 to position P_1, find the rider's height above the ground h when θ is $300°$.

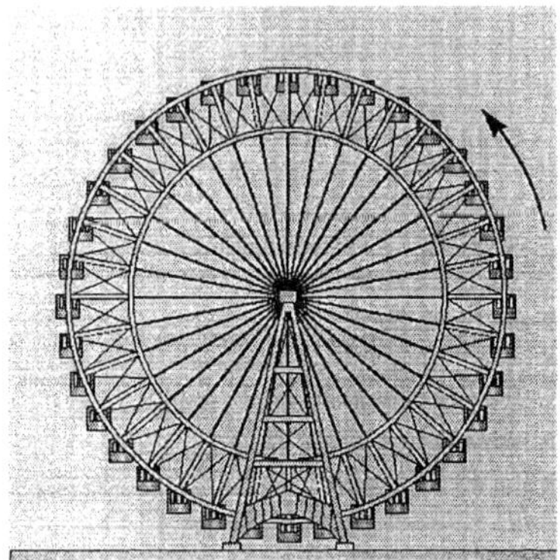
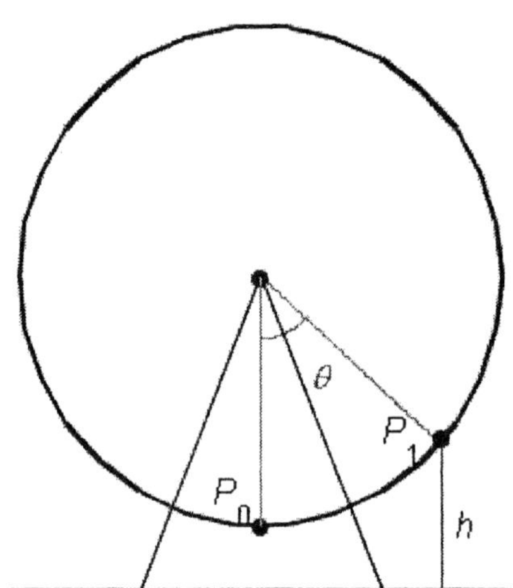

Please round the answer to the nearest whole number.

Select the correct answer.

    a.   72 ft
    b.   82 ft
    c.   79 ft
    d.   76 ft
    e.   75 ft

16. Refer to right triangle ABC with $C = 90°$.

Solve for all the missing parts using the given information.

$b = 377.6$ inches, $c = 588.5$ inches

Please round the answers for the angles to two decimal places, and for the side, to one decimal place.

Select the correct answer.

a. $A = 50.09°$, $B = 39.91°$, $a = 451.2$ in

b. $A = 50.13°$, $B = 39.87°$, $a = 451.8$ in

c. $A = 50.09°$, $B = 39.91°$, $a = 451.4$ in

d. $A = 50.11°$, $B = 39.89°$, $a = 451.2$ in

e. $A = 50.11°$, $B = 39.89°$, $a = 451.4$ in

17. Use the Cofunction Theorem to fill in the blank so that the expression becomes a true statement.

$\sin 60° = \cos \underline{\quad}°$

Select the correct answer.

a. $40°$

b. $30°$

c. $35°$

d. $50°$

e. $45°$

18. The problem refers to a vector $\mathbf{V}$ with magnitude $|\mathbf{V}|$ that forms an angle with the positive x-axis. Give the magnitude of the horizontal and vertical vector components of $\mathbf{V}$, namely $\mathbf{V}_x$ and $\mathbf{V}_y$, respectively. Please round your answers to the nearest hundredth.

$|\mathbf{V}| = 13.1$, $\theta = 24.2°$

Select the correct answer.

a. $\left|\mathbf{V}_x\right| = 11.95$, $\left|\mathbf{V}_y\right| = 5.37$

b. $\left|\mathbf{V}_x\right| = -9.35$, $\left|\mathbf{V}_y\right| = 1.17$

c. $\left|\mathbf{V}_x\right| = -7.65$, $\left|\mathbf{V}_y\right| = -5.37$

d. $\left|\mathbf{V}_x\right| = 7.65$, $\left|\mathbf{V}_y\right| = -3.67$

e. $\left|\mathbf{V}_x\right| = 11.95$, $\left|\mathbf{V}_y\right| = -3.67$

19. Change the following to decimal degrees. If rounding is necessary, round to the nearest hundredth of a degree.

$43°39'$

Select the correct answer.

a. $44.15°$

b. $42.75°$

c. $43.65°$

d. $44.35°$

e. $42.55°$

20. Find the exact value for $\csc 45°$.

Select the correct answer.

a. $-5\sqrt{2}$

b. $5\sqrt{2}$

c. $\sqrt{2}$

d. $-\sqrt{2}$

21. Simplify expression by first substituting values from the table of exact values and then simplifying the resulting expression.

$$(\sin 60° + \cos 60°)^2$$

Table of Exact Values

x	$\sin x$	$\cos x$
0°	0	1
30°	$\dfrac{1}{2}$	$\dfrac{\sqrt{3}}{2}$
45°	$\dfrac{1}{\sqrt{2}}$	$\dfrac{1}{\sqrt{2}}$
60°	$\dfrac{\sqrt{3}}{2}$	$\dfrac{1}{2}$
90°	1	0

Select the correct answer.

a. $\dfrac{2 + \sqrt{3}}{2}$

b. $\dfrac{2 + \sqrt{2}}{2}$

c. $\dfrac{1 - \sqrt{3}}{2}$

d. $\dfrac{2 - \sqrt{2}}{2}$

e. $\dfrac{2 - \sqrt{3}}{2}$

22. A plane travels 145 miles on a bearing of N 11 ° E and then changes its course to N 48° E and travels another 140 miles. Find the total distance traveled north and the total distance traveled east. Round your answers to the nearest integer.

Select the correct answer.

a.   plane travels 252 miles to the north and 110 miles to the east

b.   plane travels 231 miles to the north and 140 miles to the east

c.   plane travels 236 miles to the north and 110 miles to the east

d.   plane travels 236 miles to the north and 132 miles to the east

e.   plane travels 252 miles to the north and 132 miles to the east

23. Suppose the figure below is an exaggerated diagram of a plane flying above the earth. The plane is 5.05 miles above the earth and the radius of the earth is 4,000 miles, how far is it from the plane to the horizon?

Please round your answer to the nearest mile.

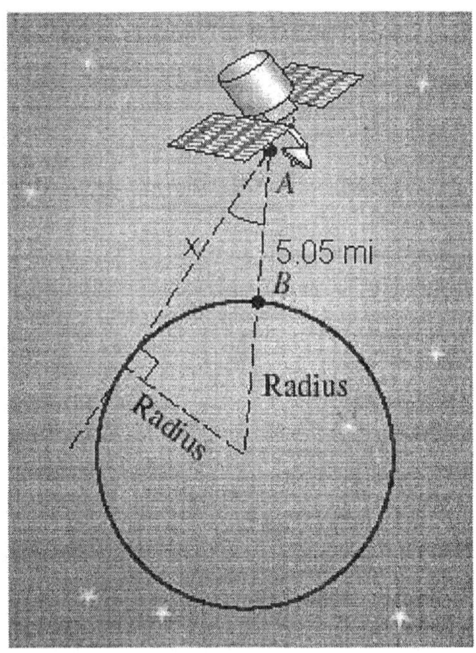

Select the correct answer.

a.   202 mi
b.   201 mi
c.   217 mi
d.   211 mi
e.   197 mi

24. For the problem, the magnitude of the horizontal and vertical vector components V_x and V_y, of vector V are given.

Find the magnitude of V.

Round your answers to the nearest tenth.

$$\left|V_x\right| = 2.2, \quad \left|V_y\right| = 5.8$$

Select the correct answer.

a. $\left|V\right| = 8.5$

b. $\left|V\right| = 6.2$

c. $\left|V\right| = 6.1$

d. $\left|V\right| = 8.1$

e. $\left|V\right| = 5.7$

25. Subtract.

$$(82°38') - (21°68')$$

Select the correct answer.

a. $30°60'$

b. $61°31'$

c. $60°30'$

d. $60°31'$

e. $31°60'$

1. a
2. a
3. d
4. c
5. d
6. d
7. c
8. c
9. b
10. e
11. a
12. d
13. c
14. d
15. d
16. c
17. b
18. a
19. c
20. c
21. a
22. d
23. b
24. b
25. c

McKeague/Turner - Trigonometry 5e Chapter 2 Form D

1. mctr.02.04.04m_NoAlgs
2. mctr.02.04.11m_NoAlgs
3. mctr.02.03.40m_NoAlgs
4. mctr.02.02.81m_NoAlgs
5. mctr.02.03.22m_NoAlgs
6. mctr.02.05.39m_NoAlgs
7. mctr.02.01.55m_NoAlgs
8. mctr.02.04.18m_NoAlgs
9. mctr.02.03.07m_NoAlgs
10. mctr.02.01.07m_NoAlgs
11. mctr.02.05.05m_NoAlgs
12. mctr.02.02.44m_NoAlgs
13. mctr.02.04.23m_NoAlgs
14. mctr.02.02.65m_NoAlgs
15. mctr.02.03.53m_NoAlgs
16. mctr.02.03.32m_NoAlgs
17. mctr.02.01.17m_NoAlgs
18. mctr.02.05.13m_NoAlgs
19. mctr.02.02.29m_NoAlgs
20. mctr.02.01.47m_NoAlgs
21. mctr.02.01.32m_NoAlgs
22. mctr.02.05.31m_NoAlgs
23. mctr.02.04.30m_NoAlgs
24. mctr.02.05.22m_NoAlgs
25. mctr.02.02.12m_NoAlgs

1. In the right triangle below

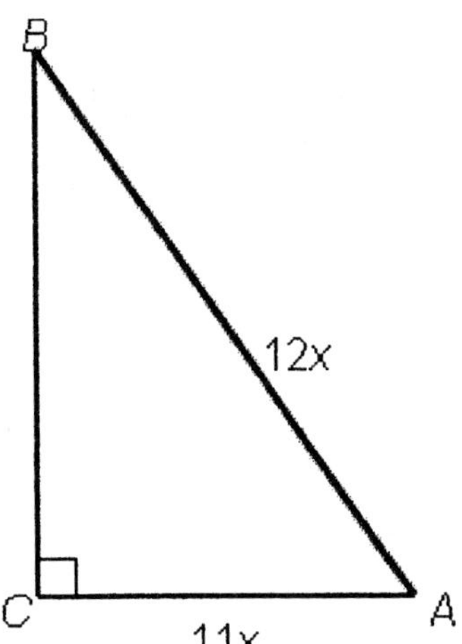

Find $\sin A$

Find $\cos A$

Find $\tan A$

Find $\sin B$

Find $\cos B$

Find $\tan B$

2. Use the Cofunction Theorem to fill in the blank so that the expression becomes a true statement.

$$\sin 60° = \cos \underline{\quad}°$$

Select the correct answer.

a. $40°$

b. $30°$

c. $35°$

d. $50°$

e. $45°$

3. For the expression $6\sin(z - 30°)$, replace z with $60°$, and then simplify.

4. Find the exact value for $\csc 45°$.

   Select the correct answer.

   a. $-5\sqrt{2}$

   b. $5\sqrt{2}$

   c. $\sqrt{2}$

   d. 0

   e. $-\sqrt{2}$

5. Add.

   $$(39°45') + (22°27')$$

   _____ ° _____ '

6. Subtract.

   $$(82°38') - (21°68')$$

   Select the correct answer.

   a. $30°60'$

   b. $61°31'$

   c. $60°30'$

   d. $60°31'$

   e. $31°60'$

7. Use a calculator to find the following. Round the answer to four places past the decimal point.

   $\cot 31°$

   ____

8. Use a calculator to find the following. Round your answer to four places past the decimal point.

$$\sin 40^\circ 20'$$

Select the correct answer.

a. 0.7922
b. 0.5242
c. 0.6322
d. 0.6472
e. 0.8332

9. Use a calculator to find a value of θ between 0° and 90° that satisfies the statement below. Write your answer in degrees and minutes rounded to the nearest minute.

$$\csc \theta = 7.5493$$

$$\theta = \underline{\quad\quad}^\circ \underline{\quad\quad}'$$

10. Work the following problem on your calculator. Do not write down or round off any intermediate answers.

$$\cos^2 36^\circ + \sin^2 36^\circ$$

Select the correct answer.

a. 1.3
b. 0.4
c. 1
d. 1.2
e. 2

11. Refer to right triangle ABC with $C = 90^\circ$.

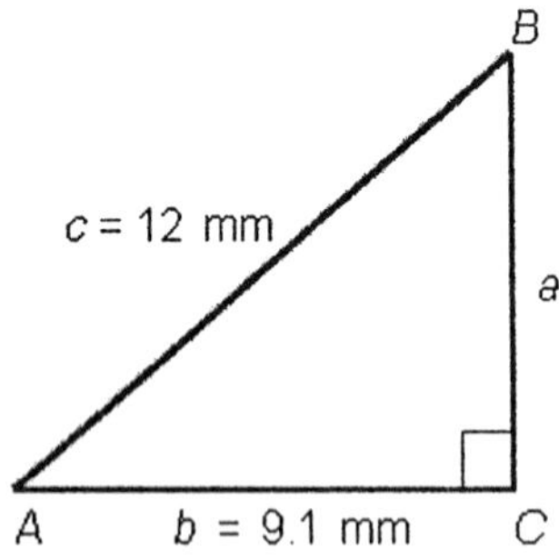

If $b = 9.1$ mm and $c = 12$ mm, find B. Please round the answer to the nearest whole number.

$$B = \underline{\quad\quad\quad}^\circ$$

12. Refer to right triangle *ABC* with $C = 90°$.

Solve for all the missing parts using the given information.

$B = 19°$, $c = 4.8$ ft

Please round each answer to the nearest tenth if necessary.

Select the correct answer.

a. $A = 71°$, $a = 4.3$ ft, $b = 1.6$ ft

b. $A = 71°$, $a = 4.5$ ft, $b = 1.8$ ft

c. $A = 71°$, $a = 4.9$ ft, $b = 2.0$ ft

d. $A = 71°$, $a = 4.5$ ft, $b = 1.6$ ft

e. $A = 71°$, $a = 4.3$ ft, $b = 1.8$ ft

13. The circle in the figure has a radius of *r* and center at *C*. The distance from *A* to *B* is *x*.

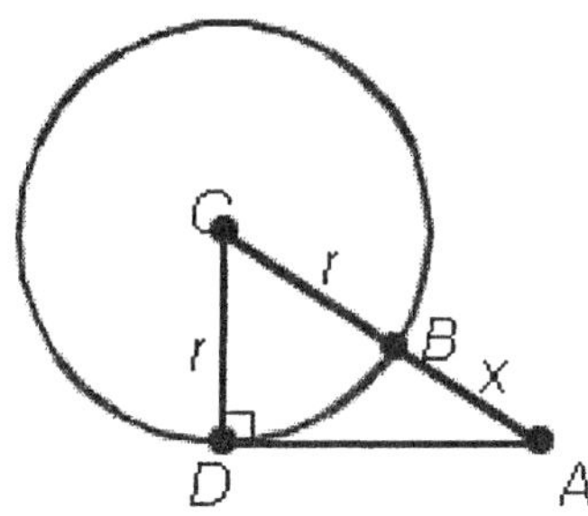

If $A = 32°$ and $r = 13$, find *x*.

Please round the answer to the nearest whole number.

$x = $ _________

14. The figure shows two right triangles drawn at 90° to each other.

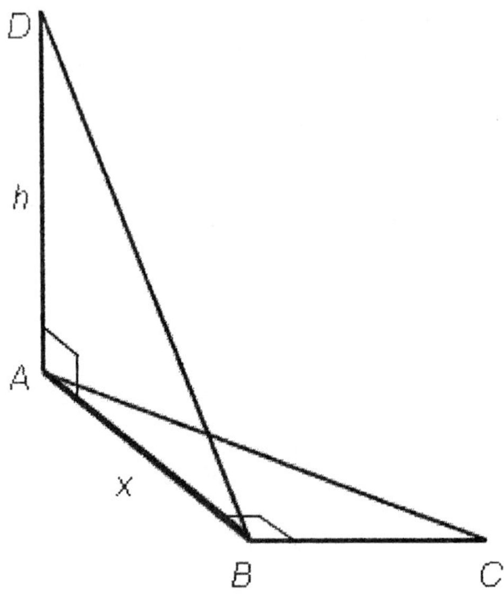

If $\angle ABD = 50°$, $C = 47°$, and $BC = 42$, find x and then find h.

Please round each answer to the nearest whole number.

Select the correct answer.

a. $x = 47, h = 56$
b. $x = 45, h = 56$
c. $x = 49, h = 58$
d. $x = 45, h = 54$
e. $x = 47, h = 54$

15. Solve the problem.

The two equal sides of an isosceles triangle are each 44 centimeters.
The base measures 30 centimeters.

Find the height. Please round your answer to the nearest centimeter.

__________ cm

Find the measure of the two equal angles. Please round your answer to the nearest degree.

__________ °

16. Solve the problem.

The diagonal of a rectangle is 344 millimeters, while the longer side is 278 millimeters. Find the shorter side of the rectangle and the angles the diagonal makes with the sides.

Round the side length to the nearest millimeter. Round the angles to the nearest degree.

Select the correct answer.

a.   The side is 203 mm; the angles are 54° and 36°.

b.   The side is 203 mm; the angles are 49° and 41°.

c.   The side is 214 mm; the angles are 46° and 44°.

d.   The side is 203 mm; the angles are 46° and 44°.

e.   The side is 214 mm; the angles are 54° and 36°.

17. Solve the problem.

A man wandering in the desert walks 2.4 miles in the direction S 32° W. He then turns 90° and walks 3.4 miles in the direction N 58° W. At that time, how far is he from his starting point? Please round your answer to the nearest tenth of a mile.

__________ mi

What is the man's bearing from his starting point? Please round the angle to the nearest degree.
Please enter *S* or *N* (for north or south) in the first field and *E* or *W* (for east or west) in the third field.

___   ______ $^\circ$ ___

18. Solve the problem.

A tree on one side of a river is due west of a rock on the other side of the river. From a stake 24 yards north of the rock, the bearing of the tree is S 18.4° W. How far is it from the rock to the tree? Please round your answer to the nearest tenth of a yard.

Select the correct answer.

a.   6.5 yd
b.   5.6 yd
c.   8.0 yd
d.   13.3 yd
e.   9.0 yd

19. Solve the problem.

Two people decide to estimate the height of a flag pole. One person positions himself due north of the pole and the other person stands due east of the pole. If the two people are the same distance from the pole and 24 feet from each other, find the height of the pole if the angle of elevation from the ground at each person's position to the top of the pole is $56°$

(see the figure). Please round your answer to the nearest foot.

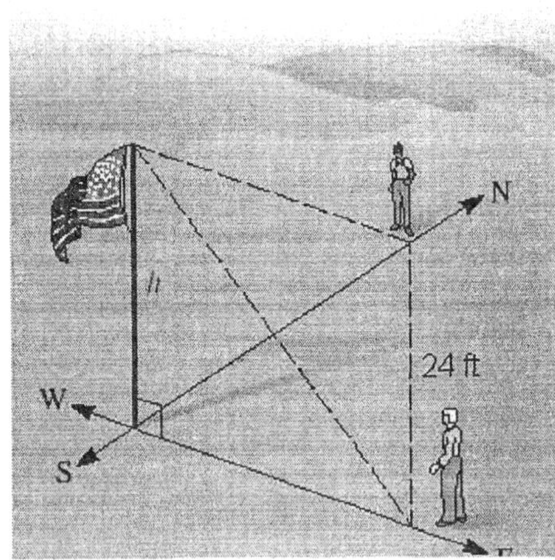

__________ ft

20. Suppose the figure below is an exaggerated diagram of a plane flying above the earth. The plane is 5.05 miles above the earth and the radius of the earth is 4,000 miles, how far is it from the plane to the horizon? Please round your answer to the nearest mile.

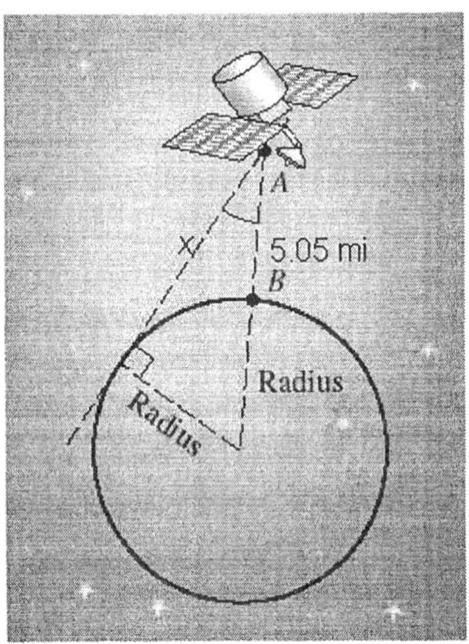

Select the correct answer.

a.   202 mi
b.   201 mi
c.   217 mi
d.   211 mi
e.   197 mi

21. Two planes take off at the same time from an airport. The first plane is flying at 260 miles per hour on a bearing of S $45°$ E. The second plane is flying in the direction S $45°$ W at 265 miles per hour. If there are no wind currents blowing, how far apart are they after 3 hours? Please round the answer to the nearest whole number.

_________ mi

What is the bearing of the second plane from the first after 3 hours?

Please round the angle to the nearest degree.

Please enter S or N (for north or south) in the first field and E or W (for east or west) in the third field.

___ ___ ° ___

22. The problem refers to a vector **V** with magnitude | **V** | that forms an angle with the positive x-axis. Give the magnitude of the horizontal and vertical vector components of **V**, namely $\mathbf{V}_x$

and $\mathbf{V}_y$, respectively.

Please round your answers to the nearest hundredth.

$$|\mathbf{V}| = 13.1, \ \theta = 24.2°$$

Select the correct answer.

a. $\left|\mathbf{V}_x\right| = 11.95, \ \left|\mathbf{V}_y\right| = 5.37$

b. $\left|\mathbf{V}_x\right| = -9.35, \ \left|\mathbf{V}_y\right| = 1.17$

c. $\left|\mathbf{V}_x\right| = -7.65, \ \left|\mathbf{V}_y\right| = -5.37$

d. $\left|\mathbf{V}_x\right| = 7.65, \ \left|\mathbf{V}_y\right| = -3.67$

e. $\left|\mathbf{V}_x\right| = 11.95, \ \left|\mathbf{V}_y\right| = -3.67$

23. The horizontal and vertical components of the velocity of an arrow shot into the air are 15.0 feet per second and 27.0 feet per second, respectively.

Find the velocity of the arrow.

Round your answers to the nearest tenth.

_________ ft/sec

24. A plane travels 145 miles on a bearing of N 11 $^{\circ}$ E and then changes its course to N 48° E and travels another 140 miles. Find the total distance traveled north and the total distance traveled east. Round your answers to the nearest integer.

Select the correct answer.

a.   plane travels 252 miles to the north and 110 miles to the east

b.   plane travels 231 miles to the north and 140 miles to the east

c.   plane travels 236 miles to the north and 110 miles to the east

d.   plane travels 236 miles to the north and 132 miles to the east

e.   plane travels 252 miles to the north and 132 miles to the east

25. A package is pushed across a floor for a distance of 60 feet by exerting a force of 35 pounds downward at an angle of 25° with the horizontal.

How much work is done?

Round your answer to the nearest hundred.

Select the correct answer.

a.   2,000 ft · lb

b.   2,200 ft · lb

c.   2,300 ft · lb

d.   1,900 ft · lb

e.   2,100 ft · lb

McKeague/Turner - Trigonometry 5e Chapter 2 Form E

1.
$$\sin(A)= \frac{\sqrt{23}}{12},$$

$$\cos(A)= \frac{11}{12},$$

$$\tan(A)= \frac{\sqrt{23}}{11},$$

$$\sin(B)= \frac{11}{12},$$

$$\cos(B)= \frac{\sqrt{23}}{12},$$

$$\tan(B)= \frac{11}{\sqrt{23}}$$

2. b

3. 3

4. c

5. 62

   12

6. c

7. 1.6643

8. d

9. 7

   37

10. c

11. 49

12. d

13. 12

14. d

15. 4 1

70

16. a

17. 4 . 2

S
87

W

18. c

19. 25

20. b

21. 1 . 11 4

S
89

W

22. a

23. 30 . 9

24. d

25. d

McKeague/Turner - Trigonometry 5e Chapter 2 Form E

1. mctr.02.01.13_NoAlgs
2. mctr.02.01.17m_NoAlgs
3. mctr.02.01.39_NoAlgs
4. mctr.02.01.47m_NoAlgs
5. mctr.02.02.01_NoAlgs
6. mctr.02.02.12m_NoAlgs
7. mctr.02.02.37_NoAlgs
8. mctr.02.02.44m_NoAlgs
9. mctr.02.02.74_NoAlgs
10. mctr.02.02.81m_NoAlgs
11. mctr.02.03.12_NoAlgs
12. mctr.02.03.22m_NoAlgs
13. mctr.02.03.35_NoAlgs
14. mctr.02.03.40m_NoAlgs
15. mctr.02.04.01_NoAlgs
16. mctr.02.04.04m_NoAlgs
17. mctr.02.04.16_NoAlgs
18. mctr.02.04.18m_NoAlgs
19. mctr.02.04.27_NoAlgs
20. mctr.02.04.30m_NoAlgs
21. mctr.02.05.10_NoAlgs
22. mctr.02.05.13m_NoAlgs
23. mctr.02.05.28_NoAlgs
24. mctr.02.05.31m_NoAlgs
25. mctr.02.05.39m_NoAlgs

1.   The figure shows two right triangles drawn at 90° to each other.

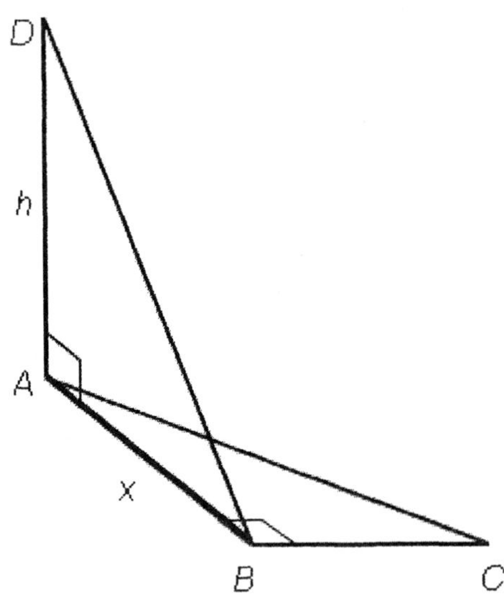

If $\angle ABD = 50°$, $C = 47°$, and $BC = 42$, find x and then find h.

Please round each answer to the nearest whole number.

Select the correct answer.

  a. $x = 47,\ h = 56$
  b. $x = 45,\ h = 56$
  c. $x = 49,\ h = 58$
  d. $x = 45,\ h = 54$
  e. $x = 47,\ h = 54$

2.   Solve the problem.

The diagonal of a rectangle is 344 millimeters, while the longer side is 278 millimeters. Find the shorter side of the rectangle and the angles the diagonal makes with the sides.

Round the side length to the nearest millimeter. Round the angles to the nearest degree.

Select the correct answer.

  a.   The side is 203 mm; the angles are 54° and 36°.

  b.   The side is 203 mm; the angles are 49° and 41°.

  c.   The side is 214 mm; the angles are 46° and 44°.

  d.   The side is 203 mm; the angles are 46° and 44°.

  e.   The side is 214 mm; the angles are 54° and 36°.

3. The problem refers to a vector **V** with magnitude | **V** | that forms an angle with the positive *x*-axis. Give the magnitude of the horizontal and vertical vector components of **V**, namely V_x and V_y, respectively. Please round your answers to the nearest hundredth.

$$|V| = 13.1, \ \theta = 24.2°$$

Select the correct answer.

a. $|V_x| = 11.95, \ |V_y| = 5.37$

b. $|V_x| = -9.35, \ |V_y| = 1.17$

c. $|V_x| = -7.65, \ |V_y| = -5.37$

d. $|V_x| = 7.65, \ |V_y| = -3.67$

e. $|V_x| = 11.95, \ |V_y| = -3.67$

4. Subtract.

$$(82°38') - (21°68')$$

Select the correct answer.

a. $30°60'$

b. $61°31'$

c. $60°30'$

d. $60°31'$

e. $31°60'$

5. Use a calculator to find a value of θ between $0°$ and $90°$ that satisfies the statement below. Write your answer in degrees and minutes rounded to the nearest minute.

$$\csc \theta = 7.5493$$

$$\theta = \underline{\quad}° \ \underline{\quad}'$$

6. Use a calculator to find the following. Round the answer to four places past the decimal point.

$$\cot 31°$$

____

7. For the expression $6 \sin (z - 30°)$, replace z with $60°$, and then simplify as much as possible.

8. Solve the problem.

   Two people decide to estimate the height of a flag pole. One person positions himself due north of the pole and the other person stands due east of the pole. If the two people are the same distance from the pole and 24 feet from each other, find the height of the pole if the angle of elevation from the ground at each person's position to the top of the pole is $56°$

   (see the figure). Please round your answer to the nearest foot.

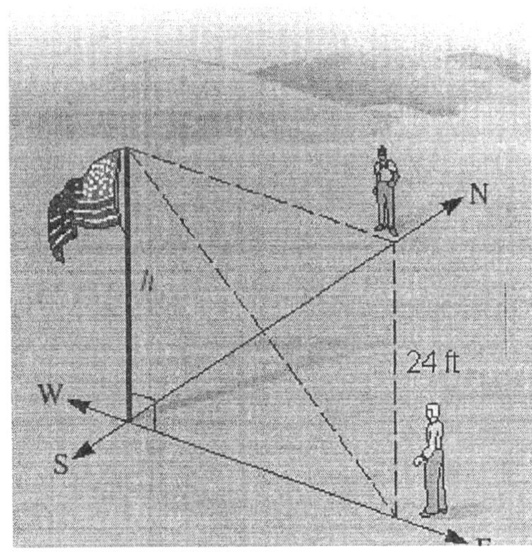

   _________ ft

9. Two planes take off at the same time from an airport. The first plane is flying at 260 miles per hour on a bearing of S $45°$ E. The second plane is flying in the direction S $45°$ W at 265 miles per hour. If there are no wind currents blowing, how far apart are they after 3 hours? Please round the answer to the nearest whole number.

   _________ mi

   What is the bearing of the second plane from the first after 3 hours?
   Please round the angle to the nearest degree.

   Please enter S or N (for north or south) in the first field and E or W (for east or west) in the third field.

   ___ _____ ° ___

10. Use a calculator to find the following. Round your answer to four places past the decimal point.

$$\sin 40^\circ 20'$$

Select the correct answer.

a.   0.7922
b.   0.5242
c.   0.6322
d.   0.6472
e.   0.8332

11. Solve the problem.

A tree on one side of a river is due west of a rock on the other side of the river. From a stake 24 yards north of the rock, the bearing of the tree is $S\ 18.4^\circ\ W$. How far is it from the rock to the tree? Please round your answer to the nearest tenth of a yard.

Select the correct answer.

a.   6.5 yd
b.   5.6 yd
c.   8.0 yd
d.   13.3 yd
e.   9.0 yd

12. The circle in the figure has a radius of r and center at C. The distance from A to B is x.

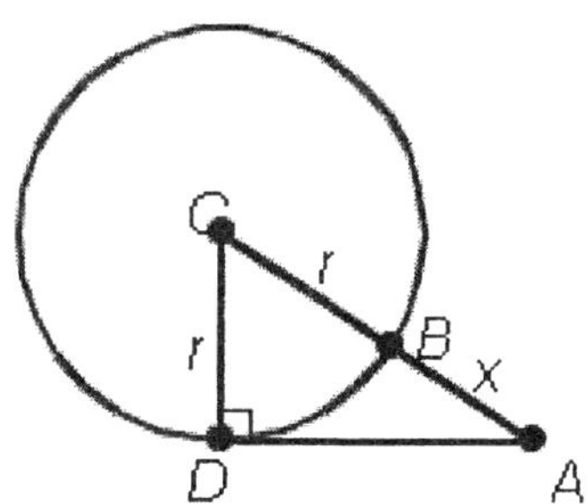

If $A = 32^\circ$ and $r = 13$, find x.

Please round the answer to the nearest whole number.

$x =$ _________

13. Solve the problem.

A man wandering in the desert walks 2.4 miles in the direction S 32° W. He then turns 90° and walks 3.4 miles in the direction N 58° W. At that time, how far is he from his starting point? Please round your answer to the nearest tenth of a mile.

__________ mi

What is the man's bearing from his starting point? Please round the angle to the nearest degree.

Please enter S or N (for north or south) in the first field and E or W (for east or west) in the third field.

__ _____ ° __

14. Solve the problem.

The two equal sides of an isosceles triangle are each 44 centimeters.
The base measures 30 centimeters.

Find the height. Please round your answer to the nearest centimeter.

__________ cm

Find the measure of the two equal angles. Please round your answer to the nearest degree.

__________ °

15. In the right triangle below

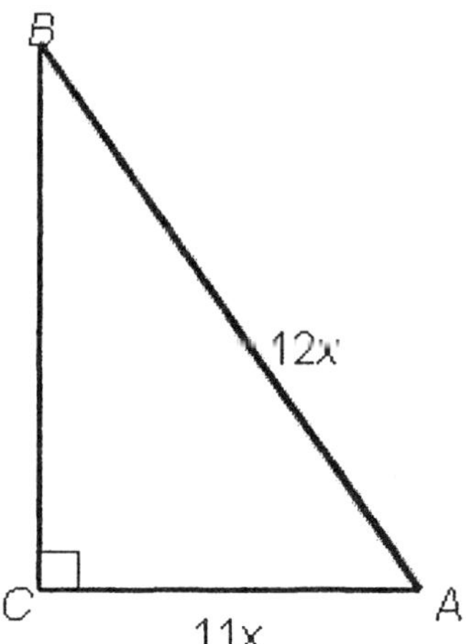

Find sin$A$

Find cos$A$

Find tan$A$

Find sin$B$

Find cos$B$

Find tan$B$

16. The horizontal and vertical components of the velocity of an arrow shot into the air are 15.0 feet per second and 27.0 feet per second, respectively. Find the velocity of the arrow. Round your answers to the nearest tenth.

__________ ft/sec

17. A package is pushed across a floor for a distance of 60 feet by exerting a force of 35 pounds downward at an angle of $25°$ with the horizontal. How much work is done?

Round your answer to the nearest hundred.

Select the correct answer.

    a.   2,000 ft · lb

    b.   2,200 ft · lb

    c.   2,300 ft · lb

    d.   1,900 ft · lb

    e.   2,100 ft · lb

18. Refer to right triangle ABC with $C = 90^\circ$.

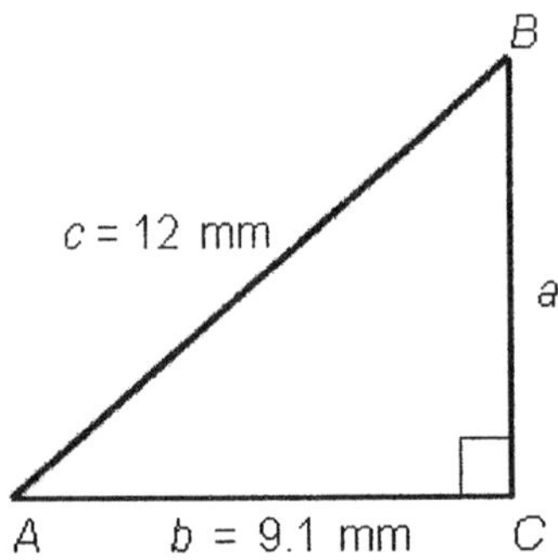

If $b = 9.1$ mm and $c = 12$ mm, find B.

Please round the answer to the nearest whole number.

$B = $ _________ $^\circ$

19. Suppose the figure below is an exaggerated diagram of a plane flying above the earth. The plane is 5.05 miles above the earth and the radius of the earth is 4,000 miles, how far is it from the plane to the horizon? Please round your answer to the nearest mile.

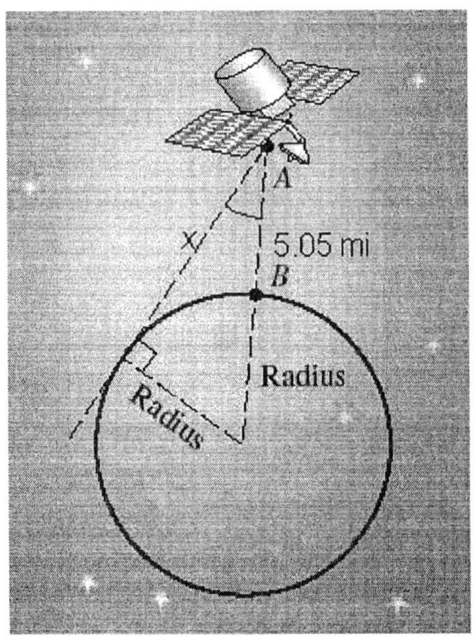

Select the correct answer.

a.   202 mi
b.   201 mi
c.   217 mi
d.   211 mi
e.   197 mi

20. Refer to right triangle ABC with $C = 90°$. Solve for all the missing parts using the given information.

$B = 19°$, $c = 4.8$ ft

Please round each answer to the nearest tenth if necessary.

Select the correct answer.

a. $A = 71°$, $a = 4.3$ ft, $b = 1.6$ ft

b. $A = 71°$, $a = 4.5$ ft, $b = 1.8$ ft

c. $A = 71°$, $a = 4.9$ ft, $b = 2.0$ ft

d. $A = 71°$, $a = 4.5$ ft, $b = 1.6$ ft

e. $A = 71°$, $a = 4.3$ ft, $b = 1.8$ ft

21. Find the exact value for $\csc 45°$.

Select the correct answer.

a. $-5\sqrt{2}$

b. $5\sqrt{2}$

c. $\sqrt{2}$

d. 0

e. $-\sqrt{2}$

22. Add.

$(39°45') + (22°27')$

_____ ° _____ '

23. Use the Cofunction Theorem to fill in the blank so that the expression becomes a true statement.

$$\sin 60^\circ = \cos \underline{\quad}^\circ$$

Select the correct answer.

a. 40°

b. 30°

c. 35°

d. 50°

e. 45°

24. Work the following problem on your calculator.

$$\cos^2 36^\circ + \sin^2 36^\circ$$

Select the correct answer.

a. 1.3
b. 0.4
c. 1
d. 1.2
e. 2

25. A plane travels 145 miles on a bearing of N 11° E and then changes its course to N 48° E and travels another 140 miles. Find the total distance traveled north and the total distance traveled east. Round your answers to the nearest integer.

Select the correct answer.

a. plane travels 252 miles to the north and 110 miles to the east

b. plane travels 231 miles to the north and 140 miles to the east

c. plane travels 236 miles to the north and 110 miles to the east

d. plane travels 236 miles to the north and 132 miles to the east

e. plane travels 252 miles to the north and 132 miles to the east

1. d

2. a

3. a

4. c

5. 7

   37

6. 1.6643

7. 3

8. 25

9. 1,114

   S
   89
   W

10. d

11. c

12. 12

13. 4.2

    S
    87

    W

14. 41

    70

15.

$$\sin(A) = \frac{\sqrt{23}}{12},$$

$$\cos(A) = \frac{11}{12},$$

$$\tan(A) = \frac{\sqrt{23}}{11},$$

$$\sin(B) = \frac{11}{12},$$

$$\cos(B) = \frac{\sqrt{23}}{12},$$

$$\tan(B) = \frac{11}{\sqrt{23}}$$

16. 30.9

17. d

18. 49

19. b

20. d

21. c

22. 62

12

23. b

24. c

25. d

1. mctr.02.03.40m_NoAlgs
2. mctr.02.04.04m_NoAlgs
3. mctr.02.05.13m_NoAlgs
4. mctr.02.02.12m_NoAlgs
5. mctr.02.02.74_NoAlgs
6. mctr.02.02.37_NoAlgo
7. mctr.02.01.39_NoAlgs
8. mctr.02.04.27_NoAlgs
9. mctr.02.05.10_NoAlgs
10. mctr.02.02.44m_NoAlgs
11. mctr.02.04.18m_NoAlgs
12. mctr.02.03.35_NoAlgs
13. mctr.02.04.16_NoAlgs
14. mctr.02.04.01_NoAlgs
15. mctr.02.01.13_NoAlgs
16. mctr.02.05.28_NoAlgs
17. mctr.02.05.39m_NoAlgs
18. mctr.02.03.12_NoAlgs
19. mctr.02.04.30m_NoAlgs
20. mctr.02.03.22m_NoAlgs
21. mctr.02.01.47m_NoAlgs
22. mctr.02.02.01_NoAlgs
23. mctr.02.01.17m_NoAlgs
24. mctr.02.02.81m_NoAlgs
25. mctr.02.05.31m_NoAlgs

1. Right triangle ABC has $C = 90°$, $b = 9$, and $c = 15$.

   Use this information to find the six trigonometric functions of A.

   Find $\sin A$

   Find $\cos A$

   Find $\tan A$

   Find $\cot A$

   Find $\sec A$

   Find $\csc A$

2. Find tan A in the right triangle below.

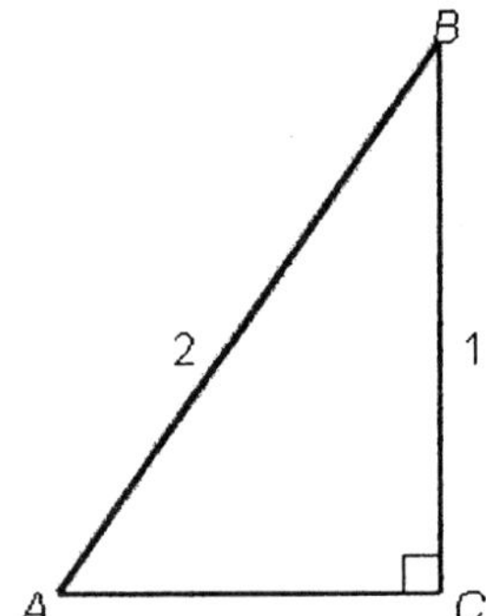

   Select the correct answer.

   a. $\tan(A) = \dfrac{2}{3}$

   b. $\tan(A) = \dfrac{\sqrt{3}}{2}$

   c. $\tan(A) = \dfrac{1}{2}$

   d. $\tan(A) = \dfrac{1}{3}$

   e. $\tan(A) = \dfrac{1}{\sqrt{3}}$

3. Complete the following table using exact values. Do not rationalize the denominator.

x	$\sin x$	$\csc x$
$0°$	0	?
$30°$	$\dfrac{1}{2}$	?
$45°$	$\dfrac{\sqrt{2}}{2}$	?
$60°$	$\dfrac{\sqrt{3}}{2}$	?
$90°$	1	?

Find the value indicated by the question mark.

4. Simplify expression by first substituting values from the table of exact values and then simplifying the resulting expression.

$$(\sin 60° + \cos 60°)^2$$

Table of Exact Values

x	$\sin x$	$\cos x$
$0°$	0	1
$30°$	$\dfrac{1}{2}$	$\dfrac{\sqrt{3}}{2}$
$45°$	$\dfrac{1}{\sqrt{2}}$	$\dfrac{1}{\sqrt{2}}$
$60°$	$\dfrac{\sqrt{3}}{2}$	$\dfrac{1}{2}$
$90°$	1	0

Select the correct answer.

a. $\dfrac{2 + \sqrt{3}}{2}$

b. $\dfrac{2 + \sqrt{2}}{2}$

c. $\dfrac{1 - \sqrt{3}}{2}$

d. $\dfrac{2 - \sqrt{2}}{2}$

e. $\dfrac{2 - \sqrt{3}}{2}$

5. Find the exact value of $\cot 60°$.

6. There is a right triangle ABC with $C = 90°$, $a = 19.54$, and $b = 5.27$. Find sin B.

   Round your answers to the nearest hundredth.

   Select the correct answer.

   a. 0.98
   b. 0.97
   c. 0.26
   d. 0.28
   e. 0.25

7. Convert the following to degrees and minutes. Round to the nearest minute.

   $99.45°$

   Select the correct answer.

   a. $99°24'$

   b. $99°45'$

   c. $99°26'$

   d. $99°32'$

   e. $99°27'$

8. Change the following to decimal degrees.

   $43°39'$

   Select the correct answer.

   a. $44.15°$

   b. $42.75°$

   c. $43.65°$

   d. $44.35°$

   e. $42.55°$

9. Use a calculator to complete the following table. Round all answers to four digits past the decimal point.

x	$\cos x$	$\sec x$
$0°$	_______	_______
$30°$	_______	_______
$45°$	_______	_______
$60°$	_______	_______
$90°$	_______	_______

10. Find θ if θ is between $0°$ and $90°$. Round your answer to the nearest tenth of a degree.

$\cot \theta = 0.8732$

Select the correct answer.

a. $\theta = 51.7°$

b. $\theta = 50.0°$

c. $\theta = 48.8°$

d. $\theta = 48.9°$

e. $\theta = 50.4°$

11. Refer to right triangle ABC with $C = 90°$.

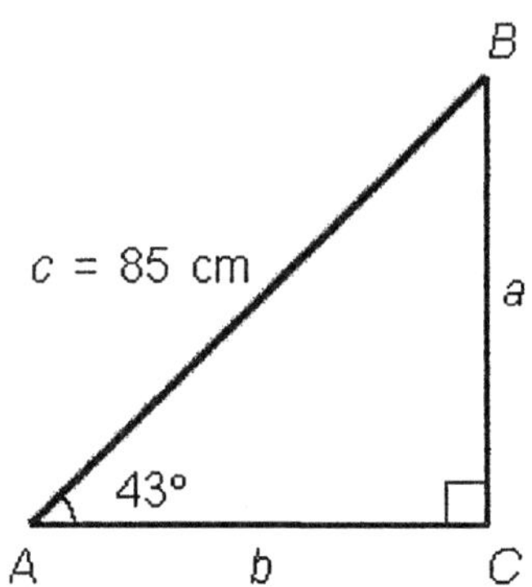

If $A = 43°$ and $c = 85$ cm, find b.

Please round the answer to the nearest whole number.

$b =$ _______ cm

12. Refer to right triangle ABC with $C = 90^\circ$.

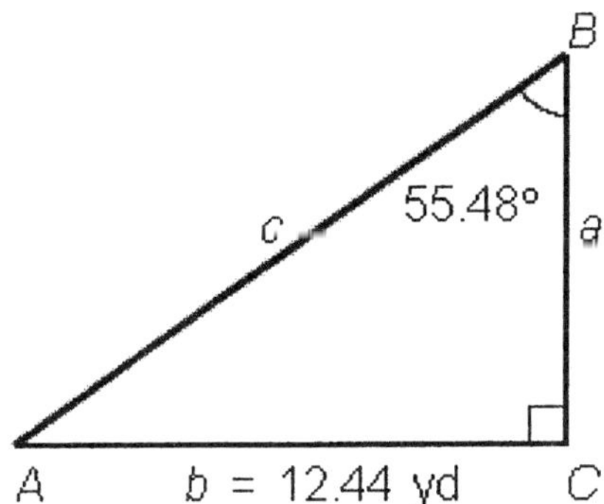

If $B = 55.48^\circ$ and $b = 12.44$ yd, find a.

Please round the answer to the nearest hundredth.

Select the correct answer.

a.   8.54 yd
b.   8.56 yd
c.   8.58 yd
d.   8.60 yd
e.   8.50 yd

13. Refer to right triangle ABC with $C = 90^\circ$.

Solve for all the missing parts using the given information.

$$B = 44.45^\circ, \quad a = 5.585 \text{ mi}$$

Please round the answer for the angle to two decimal places, and for the sides, to three decimal places.

Select the correct answer.

a. $A = 45.55^\circ, b = 5.479$ mi, $c = 7.824$ mi

b. $A = 45.55^\circ, b = 5.477$ mi, $c = 7.824$ mi

c. $A = 45.55^\circ, b = 5.479$ mi, $c = 7.826$ mi

d. $A = 45.55^\circ, b = 5.483$ mi, $c = 7.828$ mi

e. $A = 45.55^\circ, b = 5.477$ mi, $c = 7.826$ mi

14. Refer to right triangle ABC with $C = 90°$.

Solve for all the missing parts using the given information.

$b = 377.6$ inches, $c = 588.5$ inches

Please round the answers for the angles to two decimal places, and for the side, to one decimal place.

Select the correct answer.

a. $A = 50.09°$, $B = 39.91°$, $a = 451.2$ in

b. $A = 50.13°$, $B = 39.87°$, $a = 451.8$ in

c. $A = 50.09°$, $B = 39.91°$, $a = 451.4$ in

d. $A = 50.11°$, $B = 39.89°$, $a = 451.2$ in

e. $A = 50.11°$, $B = 39.89°$, $a = 451.4$ in

15. In the figure, the distance from A to D is y, the distance from D to C is x, and the distance from C to B is h.

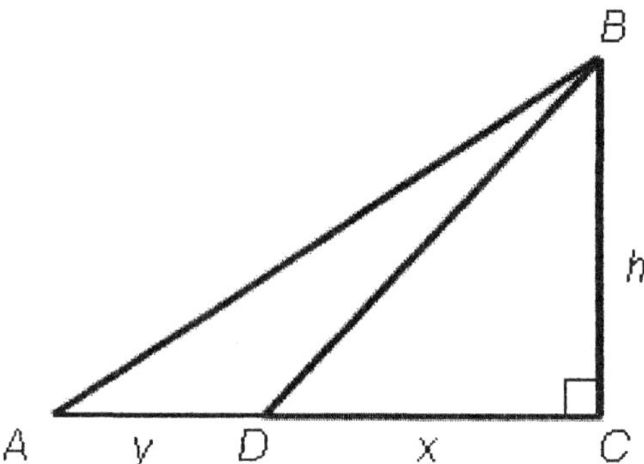

If $A = 32°$, $\angle BDC = 49°$, and $AB = 54$, find h, then x.

Please round each answer to the nearest whole number.

$h =$ ________

$x =$ ________

16. The figure shows a simplified model of the first Ferris wheel. The radius of the first Ferris wheel was 125 feet; the distance from the ground to the bottom of the wheel was 14 feet. If θ is the central angle formed as a rider moves from position P_0 to position P_1, find the rider's height above the ground h when θ is $300°$.

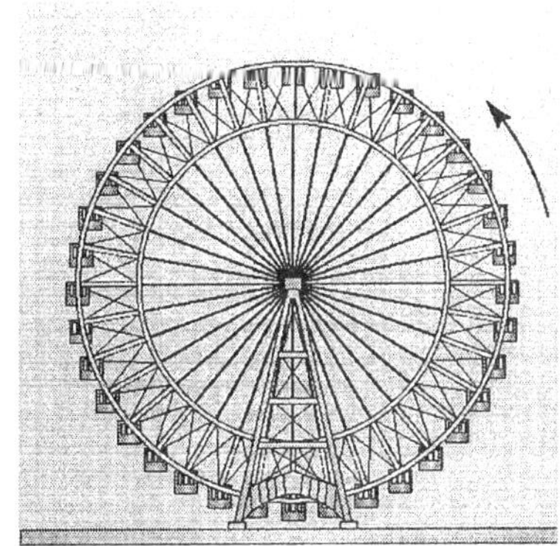
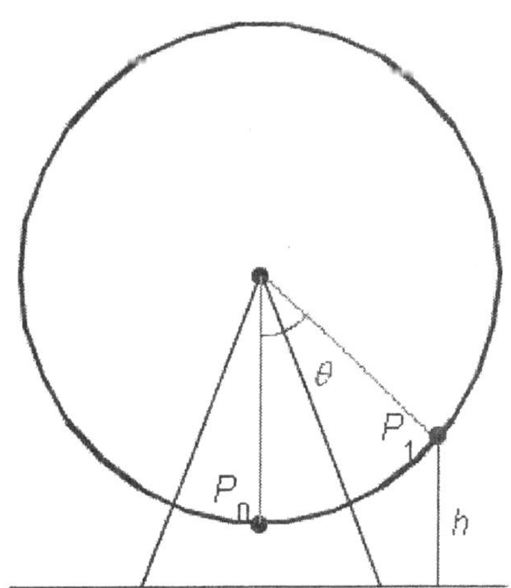

Please round the answer to the nearest whole number.

Select the correct answer.

a.   72 ft
b.   82 ft
c.   79 ft
d.   76 ft
e.   75 ft

17. Solve the problem.

A 72.5-foot rope from the top of a circus tent pole is anchored to the ground 42.9 feet from the bottom of the pole. What angle does the rope make with the pole? (Assume the pole is perpendicular to the ground.)

Please round your answer to the nearest tenth of a degree.

_________ °

18. A person standing 130 centimeters from a mirror notices that the angle of depression from his eyes to the bottom of the mirror is $14°$, while the angle of elevation to the top of the mirror is $12°$. Find the vertical dimension of the mirror (see the figure). Please round your answer to the nearest centimeter.

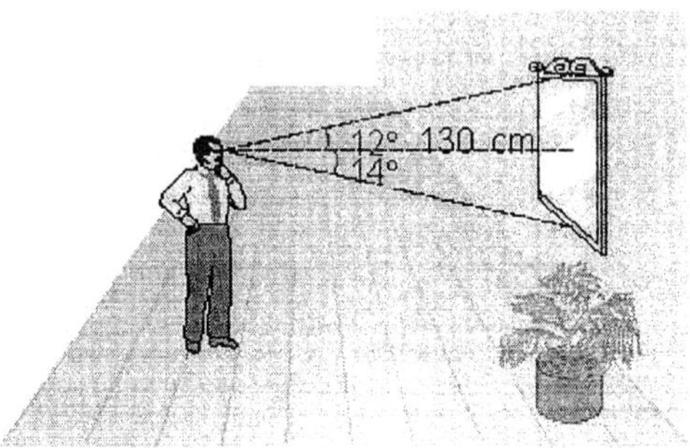

Select the correct answer.

a. 60 cm
b. 62 cm
c. 50 cm
d. 57 cm
e. 61 cm

19. From a point on the floor the angle of elevation to the top of a door is $46°$, while the angle of elevation to the ceiling above the door is $62°$. If the ceiling is 12 feet above the floor, what is the vertical dimension of the door (see the figure). Please round your answer to the nearest tenth of a foot.

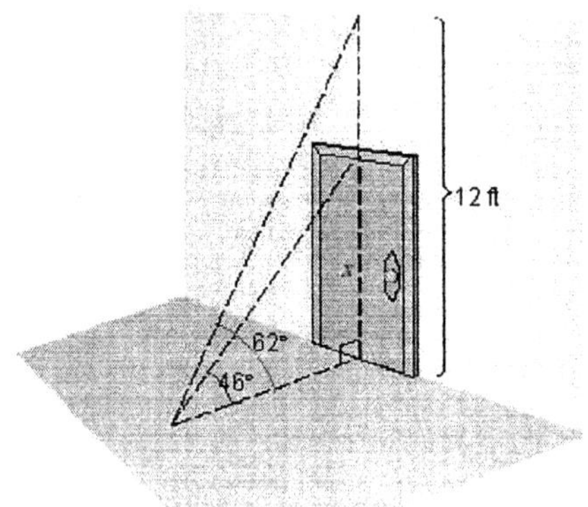

Select the correct answer.

a. 5.1 ft
b. 6.6 ft
c. 7.9 ft
d. 5.6 ft

20. In the figure below, a person standing at point *A* notices that the angle of elevation to the top of the antenna is $48°\,30'$. A second person standing 37.0 feet farther from the antenna than the person at *A* finds the angle of elevation to the top of the antenna to be $41°\,15'$. How far is the person at *A* from the base of the antenna? Please round your answer to the nearest foot.

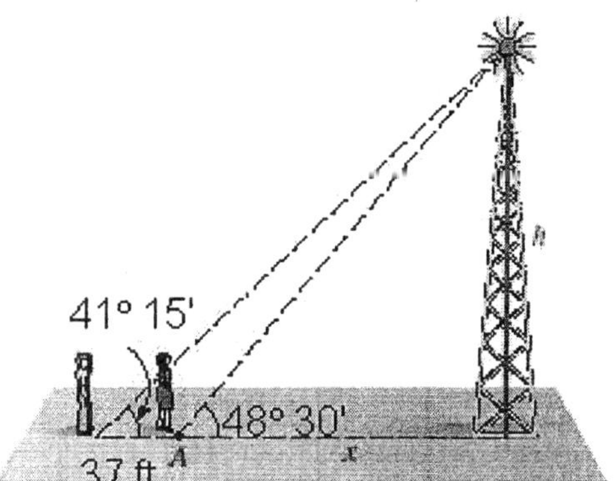

Select the correct answer.

a.   132 ft
b.   129 ft
c.   128 ft
d.   126 ft
e.   125 ft

21. Draw vector representing the velocity.

60 mph due south

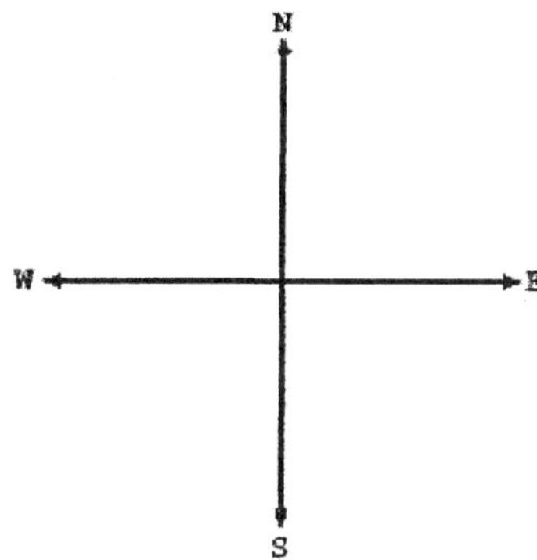

22. Draw vector representing the velocity.

80 cm/sec N 20° E

Select the correct answer.

a.

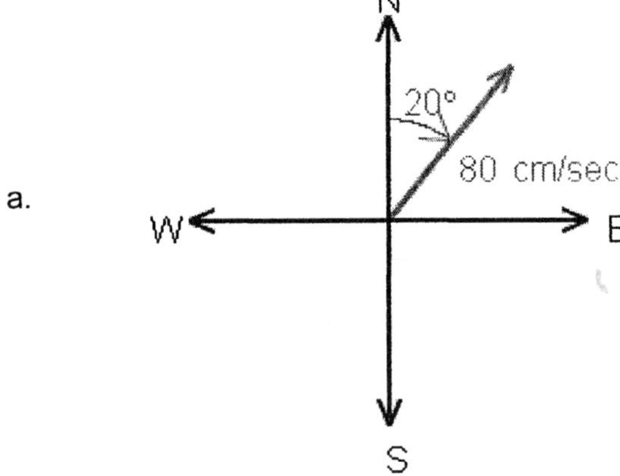

b.

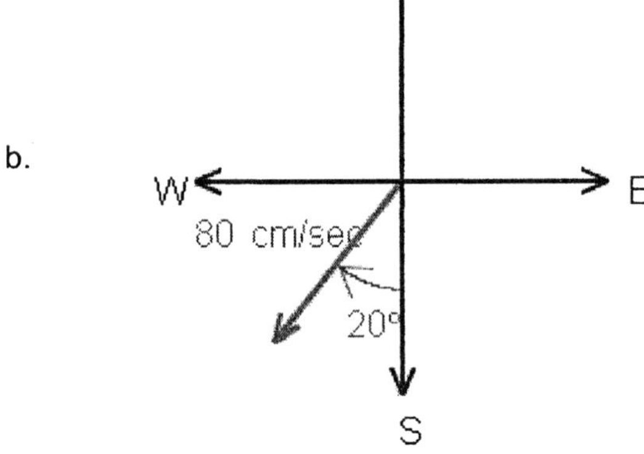

c.

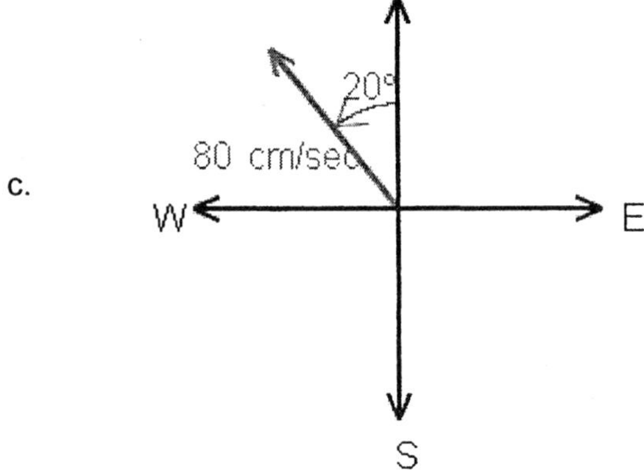

23. The problem refers to a vector **V** with magnitude | **V** | that forms an angle with the positive x-axis. Give the magnitude of the horizontal and vertical vector components of **V**, namely $\mathbf{V}_x$

    and $\mathbf{V}_y$, respectively.

    Round your answers to the nearest integer.

    $$|\mathbf{V}| = 66, \; \theta = 120°$$

    Selcct the oorroct ancwor.

    a. $\left|\mathbf{V}_x\right| = -36, \; \left|\mathbf{V}_y\right| = -56$

    b. $\left|\mathbf{V}_x\right| = -29, \; \left|\mathbf{V}_y\right| = 57$

    c. $\left|\mathbf{V}_x\right| = 33, \; \left|\mathbf{V}_y\right| = 59$

    d. $\left|\mathbf{V}_x\right| = 33, \; \left|\mathbf{V}_y\right| = 57$

    e. $\left|\mathbf{V}_x\right| = 29, \; \left|\mathbf{V}_y\right| = 59$

24. For the problem, the magnitude of the horizontal and vertical vector components $\mathbf{V}_x$ and $\mathbf{V}_y$, of vector **V** are given. Find the magnitude of **V**.

    Round your answers to the nearest tenth.

    $$\left|\mathbf{V}_x\right| = 2.2, \; \left|\mathbf{V}_y\right| = 5.8$$

    Select the correct answer.

    a. $|\mathbf{V}| = 8.5$

    b. $|\mathbf{V}| = 6.2$

    c. $|\mathbf{V}| = 6.1$

    d. $|\mathbf{V}| = 8.1$

    e. $|\mathbf{V}| = 5.7$

25. A 10-pound weight is lying on a sit-up bench at the gym. If the bench is inclined at an angle of $12°$, there are three forces acting on the weight. **N** is called the normal force and it acts in the direction perpendicular to the bench. **F** is the force due to friction that holds the weight on the bench. If the weight does not move, then the sum of these three forces is 0.

Find the magnitude of **N** and the magnitude of **F**. Please round your answers to the nearest tenth.

| **N** | = ________ lb

| **F** | = ________ lb

1. $\sin(A) = \dfrac{4}{5}$,

   $\cos(A) = \dfrac{3}{5}$,

   $\tan(A) = \dfrac{4}{3}$,

   $\cot(A) = \dfrac{3}{4}$,

   $\sec(A) = \dfrac{5}{3}$,

   $\csc(A) = \dfrac{5}{4}$

2. e

3. undefined

   2

   $\sqrt{2}$

   $\dfrac{2}{\sqrt{3}}$

   1

4. a

5. $\dfrac{1}{\sqrt{3}}$

6. c

7. e

8. c

9. 1.0000

   1

   0.8660

   1.1547

   0.7071

   1.4142

   0.5000

  2.0000

  0.0000

  undefined

10. d

11. 62

12. b

13. a

14. c

15. 29

  25

16. d

17. 36.3

18. a

19. b

20. c

21.

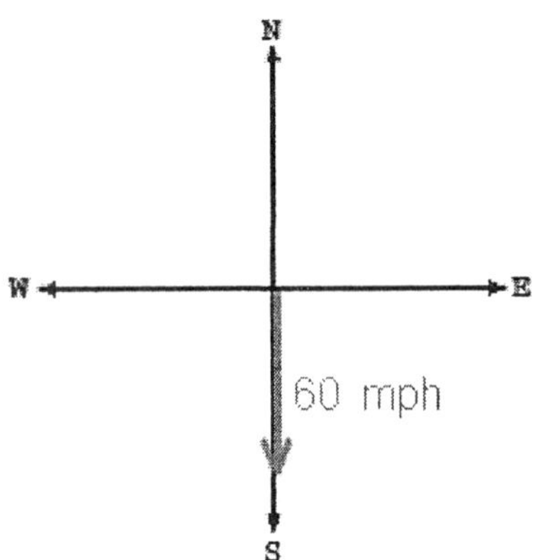

22. a

23. d

24. b

25. 9.8

  2.1

1. mctr.02.01.01_NoAlgs
2. mctr.02.01.07m_NoAlgs
3. mctr.02.01.25_NoAlgs
4. mctr.02.01.32m_NoAlgs
5. mctr.02.01.50_NoAlgs
6. mctr.02.01.55m_NoAlgs
7. mctr.02.02.19m_NoAlgs
8. mctr.02.02.29m_NoAlgs
9. mctr.02.02.51_NoAlgs
10. mctr.02.02.65m_NoAlgs
11. mctr.02.03.02_NoAlgs
12. mctr.02.03.07m_NoAlgs
13. mctr.02.03.26m_NoAlgs
14. mctr.02.03.32m_NoAlgs
15. mctr.02.03.46_NoAlgs
16. mctr.02.03.53m_NoAlgs
17. mctr.02.04.07_NoAlgs
18. mctr.02.04.11m_NoAlgs
19. mctr.02.04.21m_NoAlgs
20. mctr.02.04.23m_NoAlgs
21. mctr.02.05.01_NoAlgs
22. mctr.02.05.05m_NoAlgs
23. mctr.02.05.17m_NoAlgs
24. mctr.02.05.22m_NoAlgs
25. mctr.02.05.35_NoAlgs

1. Find tan A in the right triangle below.

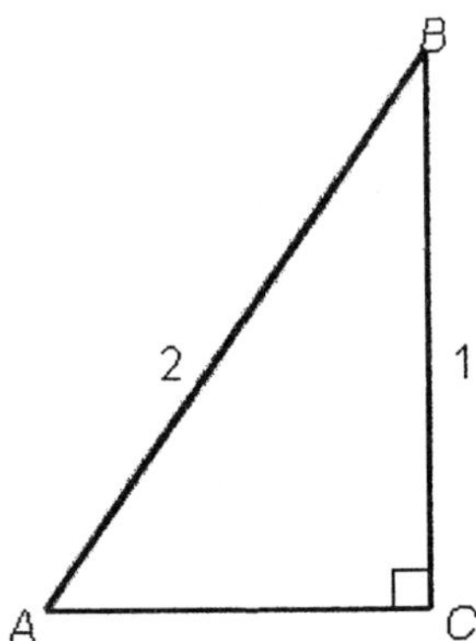

Select the correct answer.

a. $\tan(A) = \dfrac{2}{3}$

b. $\tan(A) = \dfrac{\sqrt{3}}{2}$

c. $\tan(A) = \dfrac{1}{2}$

d. $\tan(A) = \dfrac{1}{\sqrt{3}}$

2. Refer to right triangle ABC with $C = 90^\circ$.

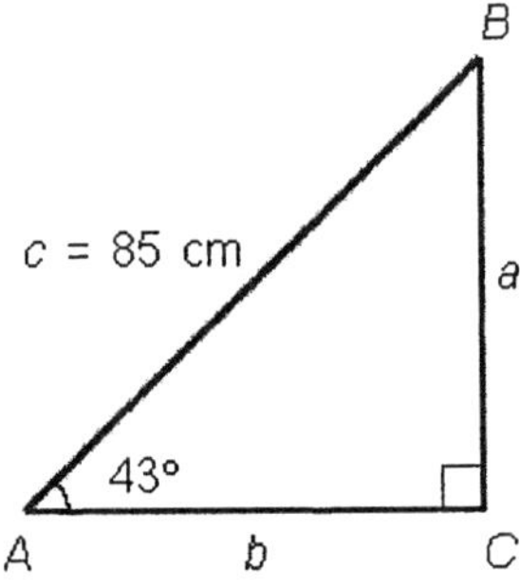

If $A = 43^\circ$ and $c = 85$ cm, find b.

Please round the answer to the nearest whole number.

$b =$ _________ cm

3. Right triangle ABC has $C = 90°$, $b = 9$, and $c = 15$. Use this information to find the six trigonometric functions of A.

Find $\sin A$

Find $\cos A$

Find $\tan A$

Find $\cot A$

Find $\sec A$

Find $\csc A$

4. Refer to right triangle ABC with $C = 90°$. Solve for all the missing parts using the given information.

$b = 377.6$ inches, $c = 588.5$ inches

Please round the answers for the angles to two decimal places, and for the side, to one decimal place.

Select the correct answer.

a. $A = 50.09°$, $B = 39.91°$, $a = 451.2$ in
b. $A = 50.13°$, $B = 39.87°$, $a = 451.8$ in
c. $A = 50.09°$, $B = 39.91°$, $a = 451.4$ in
d. $A = 50.11°$, $B = 39.89°$, $a = 451.2$ in
e. $A = 50.11°$, $B = 39.89°$, $a = 451.4$ in

5. Complete the following table using exact values. Do not rationalize the denominator.

x	$\sin x$	$\csc x$
$0°$	0	?
$30°$	$\dfrac{1}{2}$	?
$45°$	$\dfrac{\sqrt{2}}{2}$	?
$60°$	$\dfrac{\sqrt{3}}{2}$	?
$90°$	1	?

Find the value indicated by the question mark.

6. For the problem, the magnitude of the horizontal and vertical vector components V_x and V_y, of vector **V** are given. Find the magnitude of **V**.

Round your answers to the nearest tenth.

$$\left|V_x\right| = 2.2, \quad \left|V_y\right| = 5.8$$

Select the correct answer.

a. $\left|V\right| = 8.5$

b. $\left|V\right| = 6.2$

c. $\left|V\right| = 6.1$

d. $\left|V\right| = 8.1$

e. $\left|V\right| = 5.7$

7. There is a right triangle ABC with $C = 90°$, $a = 19.54$, and $b = 5.27$. Find $\sin B$.

Select the correct answer.

a.   0.98
b.   0.97
c.   0.26
d.   0.28
e.   0.25

8. Simplify expression by first substituting values from the table of exact values and then simplifying the resulting expression.

$$\left(\sin 60° + \cos 60°\right)^2$$

Table of Exact Values

x	$\sin x$	$\cos x$
$0°$	0	1
$30°$	$\dfrac{1}{2}$	$\dfrac{\sqrt{3}}{2}$
$45°$	$\dfrac{1}{\sqrt{2}}$	$\dfrac{1}{\sqrt{2}}$
$60°$	$\dfrac{\sqrt{3}}{2}$	$\dfrac{1}{2}$
$90°$	1	0

Select the correct answer.

a. $\dfrac{2 + \sqrt{3}}{2}$

b. $\dfrac{2 + \sqrt{2}}{2}$

c. $\dfrac{1 - \sqrt{3}}{2}$

d. $\dfrac{2 - \sqrt{2}}{2}$

e. $\dfrac{2 - \sqrt{3}}{2}$

9. The figure shows a simplified model of the first Ferris wheel. The radius of the first Ferris wheel was 125 feet; the distance from the ground to the bottom of the wheel was 14 feet. If θ is the central angle formed as a rider moves from position P_0 to position P_1, find the rider's height above the ground h when θ is $300°$.

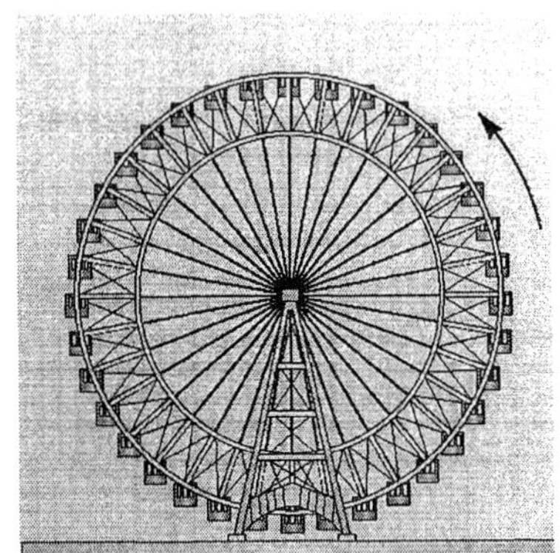
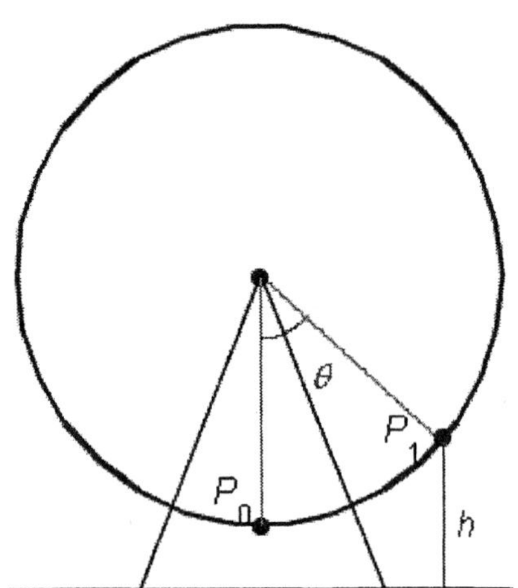

Please round the answer to the nearest whole number.

Select the correct answer.

a. 72 ft
b. 82 ft
c. 79 ft
d. 76 ft
e. 75 ft

10. A person standing 130 centimeters from a mirror notices that the angle of depression from his eyes to the bottom of the mirror is $14°$, while the angle of elevation to the top of the mirror is $12°$. Find the vertical dimension of the mirror (see the figure). Please round your answer to the nearest centimeter.

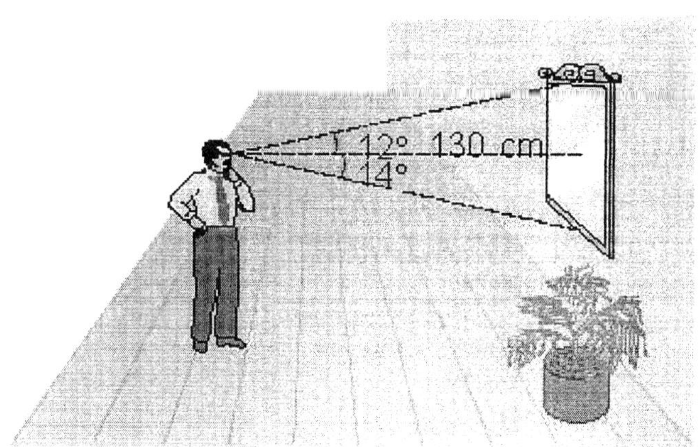

Select the correct answer.

a.   60 cm
b.   62 cm
c.   50 cm
d.   57 cm
e.   61 cm

11. In the figure below, a person standing at point *A* notices that the angle of elevation to the top of the antenna is $48°\,30'$. A second person standing 37.0 feet farther from the antenna than the person at *A* finds the angle of elevation to the top of the antenna to be $41°\,15'$. How far is the person at *A* from the base of the antenna? Please round your answer to the nearest foot.

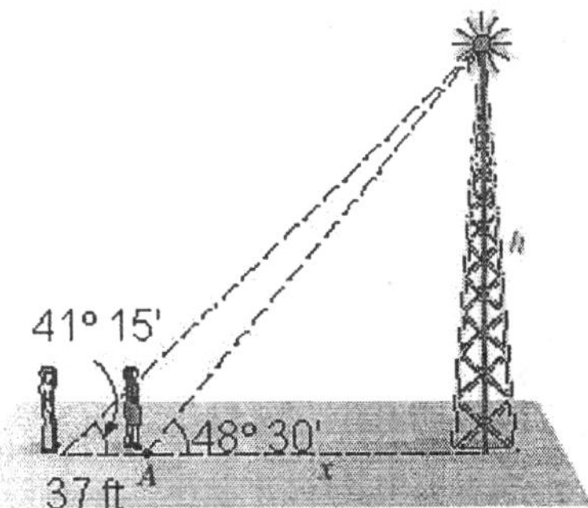

Select the correct answer.

   a.   132 ft
   b.   129 ft
   c.   128 ft
   d.   126 ft
   e.   125 ft

12. Refer to right triangle *ABC* with $C = 90°$.

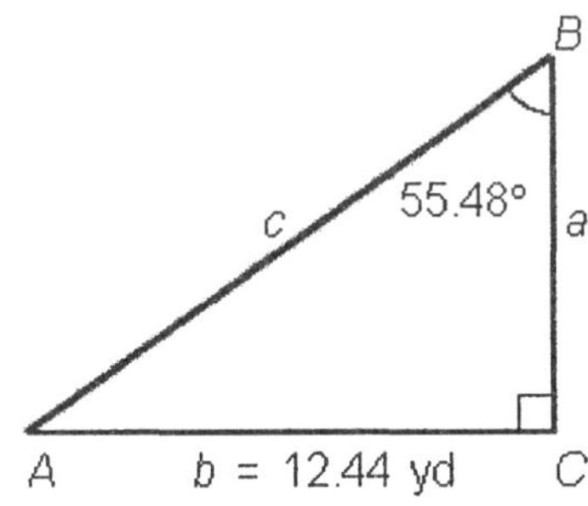

If $B = 55.48°$ and $b = 12.44$ yd, find *a*.

Please round the answer to the nearest hundredth.

Select the correct answer.

   a.   8.54 yd
   b.   8.56 yd
   c.   8.58 yd
   d.   8.60 yd

13. Refer to right triangle *ABC* with $C = 90^\circ$. Solve for all the missing parts using the given information.

$B = 44.45^\circ$, $a = 5.585$ mi

Please round the answer for the angle to two decimal places, and for the sides, to three decimal places.

Select the correct answer.

a. $A = 45.55^\circ$, $b = 5.479$ mi, $c = 7.824$ mi

b. $A = 45.55^\circ$, $b = 5.477$ mi, $c = 7.824$ mi

c. $A = 45.55^\circ$, $b = 5.479$ mi, $c = 7.826$ mi

d. $A = 45.55^\circ$, $b = 5.483$ mi, $c = 7.828$ mi

e. $A = 45.55^\circ$, $b = 5.477$ mi, $c = 7.826$ mi

14. Change the following to decimal degrees. If rounding is necessary, round to the nearest hundredth of a degree.

$43^\circ 39'$

Select the correct answer.

a. 44.15°

b. 42.75°

c. 43.65°

d. 44.35°

e. 42.55°

15. A 10-pound weight is lying on a sit-up bench at the gym. If the bench is inclined at an angle of $12°$, there are three forces acting on the weight. **N** is called the normal force and it acts in

the direction perpendicular to the bench. **F** is the force due to friction that holds the weight on the bench. If the weight does not move, then the sum of these three forces is 0.

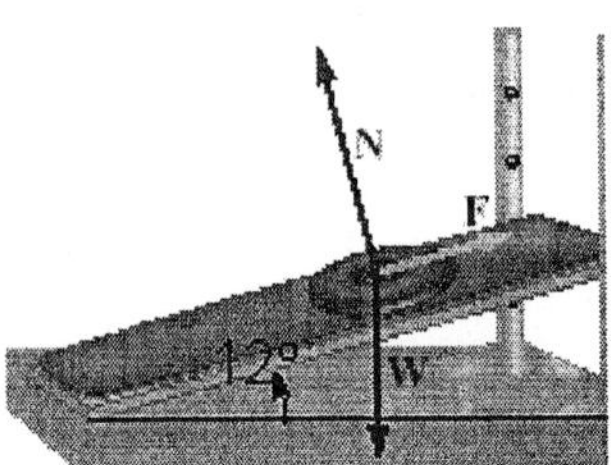

Find the magnitude of **N** and the magnitude of **F**. Please round answers to the nearest tenth.

| **N** | = ________   lb

| **F** | = ________   lb

16. From a point on the floor the angle of elevation to the top of a door is $46°$, while the angle

of elevation to the ceiling above the door is $62°$. If the ceiling is 12 feet above the floor,

what is the vertical dimension of the door (see the figure). Please round your answer to the nearest tenth of a foot.

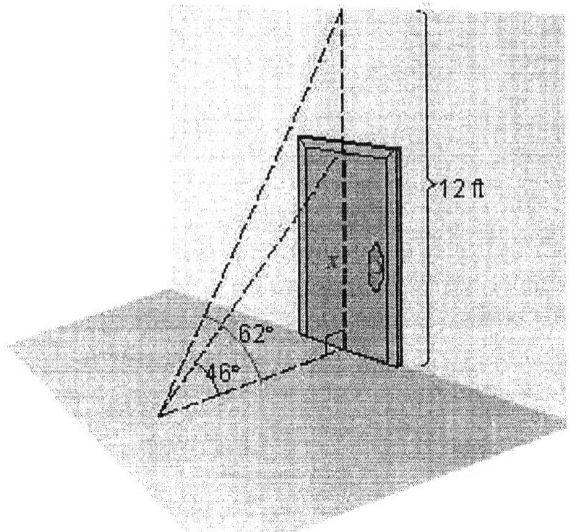

Select the correct answer.

a.   5.1 ft
b.   6.6 ft
c.   7.9 ft
d.   5.6 ft
e.   4.6 ft

17. Convert the following to degrees and minutes. Round to the nearest minute.

$99.45°$

Select the correct answer.

    a. $99°24'$

    b. $99°45'$

    c. $99°26'$

    d. $99°32'$

    e. $99°27'$

18. Solve the problem.

A 72.5-foot rope from the top of a circus tent pole is anchored to the ground 42.9 feet from the bottom of the pole. What angle does the rope make with the pole? (Assume the pole is perpendicular to the ground.) Please round your answer to the nearest tenth of a degree.

__________ $°$

19. Find the exact value of $\cot 60°$.

20. The problem refers to a vector **V** with magnitude | **V** | that forms an angle with the positive x-axis. Give the magnitude of the horizontal and vertical vector components of **V**, namely V_x and V_y, respectively. Round your answers to the nearest integer.

$$|V| = 66, \; \theta = 120°$$

Select the correct answer.

    a. $|V_x| = -36, \; |V_y| = -56$

    b. $|V_x| = -29, \; |V_y| = 57$

    c. $|V_x| = 33, \; |V_y| = 59$

    d. $|V_x| = 33, \; |V_y| = 57$

    e. $|V_x| = 29, \; |V_y| = 59$

21. Use a calculator to complete the following table. Round all answers to four digits past the decimal point.

If an expression is undefined indicate it.

x	$\cos x$	$\sec x$
$0°$	_______	_______
$30°$	_______	_______
$45°$	_______	_______
$60°$	_______	_______
$90°$	_______	_______

22. In the figure, the distance from A to D is y, the distance from D to C is x, and the distance from C to B is h.

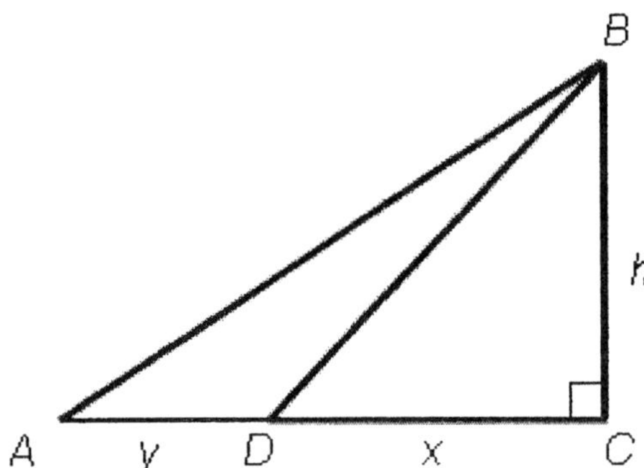

If $A = 32°$, $\angle BDC = 49°$, and $AB = 54$, find h, then x.

Please round each answer to the nearest whole number.

$h =$ _______

$x =$ _______

23. Find θ if θ is between 0° and 90°. Round your answer to the nearest tenth of a degree.

$$\cot \theta = 0.8732$$

Select the correct answer.

a. $\theta = 51.7^\circ$

b. $\theta = 50.0^\circ$

c. $\theta = 48.8^\circ$

d. $\theta = 48.9^\circ$

e. $\theta = 50.4^\circ$

24. Draw vector representing the velocity.

60 mph due south

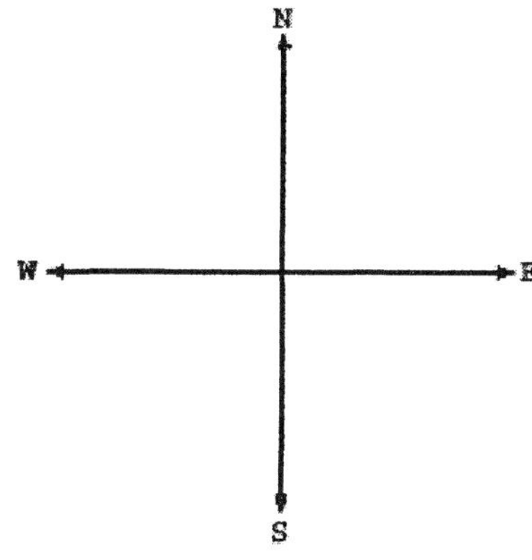

25. Draw vector representing the velocity.

80 cm/sec N 20° E

Select the correct answer.

a.

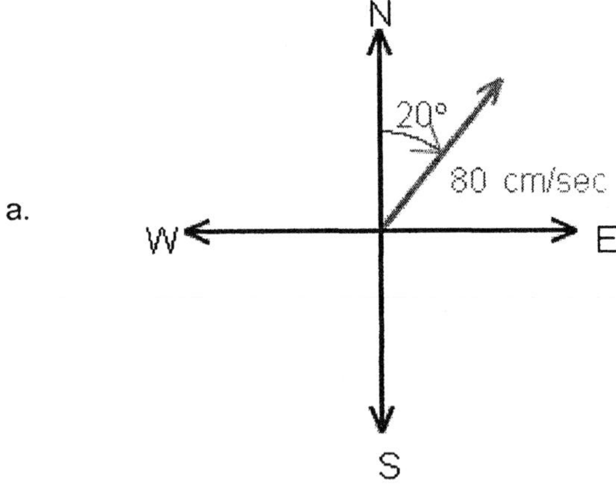

b.

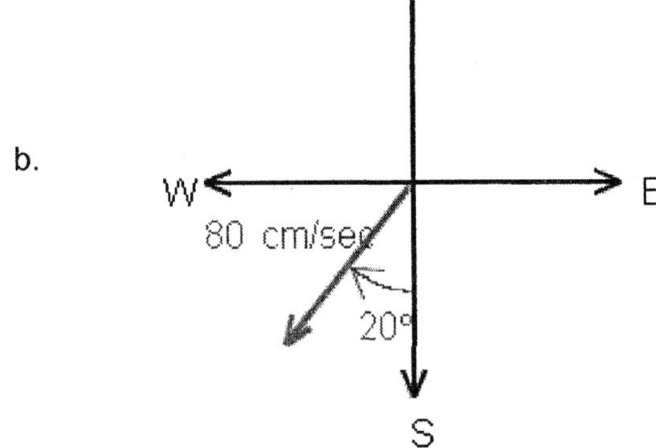

c.

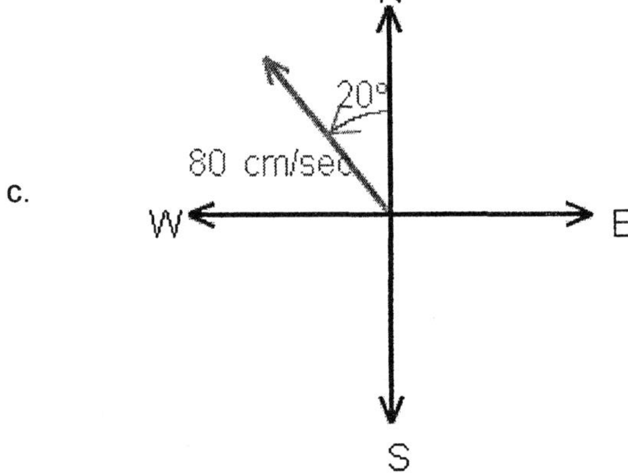

1. d

2. 62

3. $\sin(A) = \dfrac{4}{5}$,

   $\cos(A) = \dfrac{3}{5}$,

   $\tan(A) = \dfrac{4}{3}$,

   $\cot(A) = \dfrac{3}{4}$,

   $\sec(A) = \dfrac{5}{3}$,

   $\csc(A) = \dfrac{5}{4}$

4. c

5. undefined

   2

   $\sqrt{2}$

   $\dfrac{2}{\sqrt{3}}$

   1

6. b

7. c

8. a

9. d

10. a

11. c

12. b

13. a

14. c

15. 9.8

    2.1

16. b

17. e

18. 36.3

19. $\dfrac{1}{\sqrt{3}}$

20. d

21. 1.0000

1

0.8660

1.1547

0.7071

1.4142

0.5000

2.0000

0.0000

undefined

22. 29

25

23. d

24.

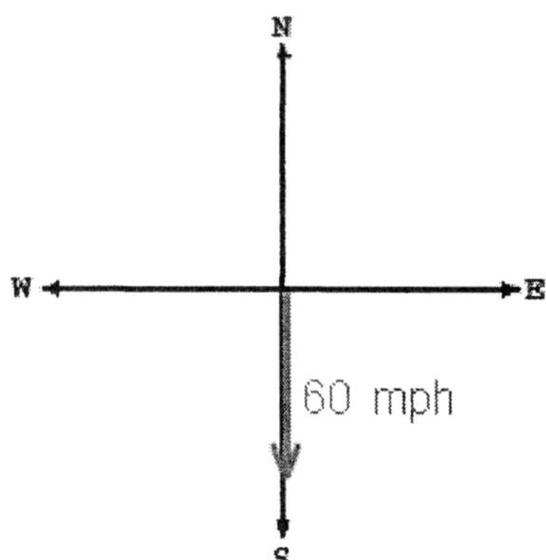

25. a

1. mctr.02.01.07m_NoAlgs
2. mctr.02.03.02_NoAlgs
3. mctr.02.01.01_NoAlgs
4. mctr.02.03.32m_NoAlgs
5. mctr.02.01.25_NoAlgs
6. mctr.02.05.22m_NoAlgs
7. mctr.02.01.55m_NoAlgs
8. mctr.02.01.32m_NoAlgs
9. mctr.02.03.53m_NoAlgs
10. mctr.02.04.11m_NoAlgs
11. mctr.02.04.23m_NoAlgs
12. mctr.02.03.07m_NoAlgs
13. mctr.02.03.26m_NoAlgs
14. mctr.02.02.29m_NoAlgs
15. mctr.02.05.35_NoAlgs
16. mctr.02.04.21m_NoAlgs
17. mctr.02.02.19m_NoAlgs
18. mctr.02.04.07_NoAlgs
19. mctr.02.01.50_NoAlgs
20. mctr.02.05.17m_NoAlgs
21. mctr.02.02.51_NoAlgs
22. mctr.02.03.46_NoAlgs
23. mctr.02.02.65m_NoAlgs
24. mctr.02.05.01_NoAlgs
25. mctr.02.05.05m_NoAlgs

1. Draw the following angle in standard position and then name the reference angle.

    240°

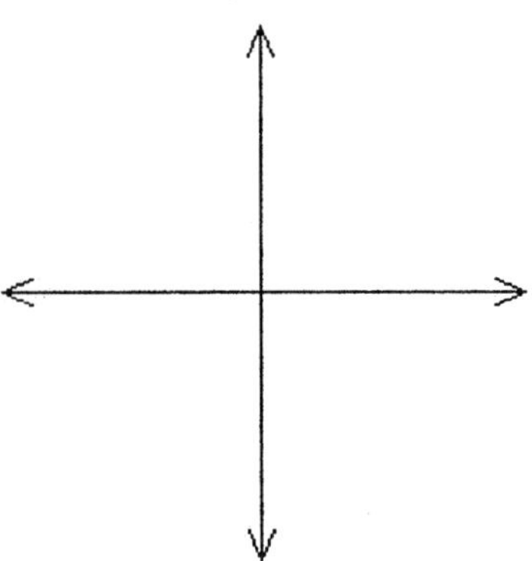

2. Find the exact value of $\cos 210^\circ$.

3. Use a calculator to find $\sec 100.9^\circ$.

    Please round the answer to the nearest ten-thousandth.

    ____

4. Use a calculator to find θ to the nearest tenth of a degree, if $0^\circ < \theta < 360^\circ$ and
    $\sin \theta = -0.3010$ with θ in QIII

    $\theta =$ ________ $^\circ$

5. Use a calculator to find θ to the nearest tenth of a degree, if $0^\circ < \theta < 360^\circ$ and

    $\cot \theta = -0.7357$ with θ in QII

    $\theta =$ ________ $^\circ$

6. Find the radian measure of angle θ, if θ is a central angle in a circle of radius r, and θ cuts off an arc of length s.

$r = 3$ inches, $s = 6$ inches

$\theta = $ _________ radians

7. Draw the angle in standard position.

$\theta = 200°$

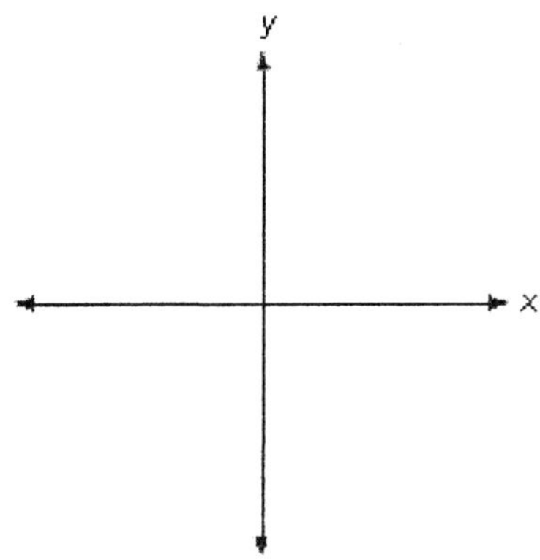

Convert to radian measure using exact values.

Name the reference angle in both degrees and radians.

8.

Convert to degree measure. $\theta = \dfrac{8\pi}{9}$

$\theta = $ _________ °

Draw the angle in standard position.

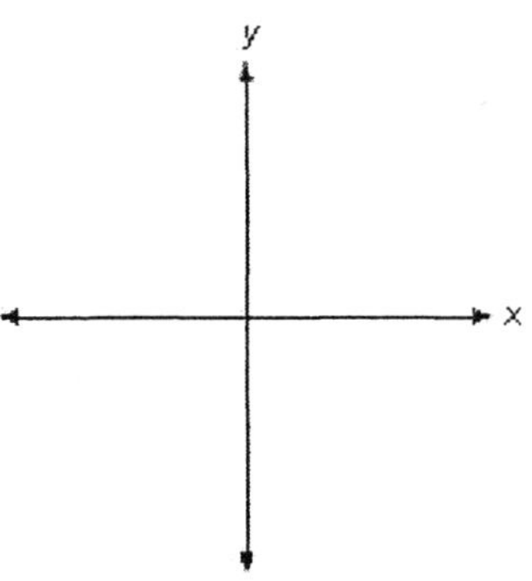

Label the reference angle in both degrees and radians.

9. Use a calculator to convert 1.35 to degree measure to tenth of a degree.

_________ °

10.

Evaluate the expression when x is $\dfrac{\pi}{12}$. Use exact values.

4 cos 6x

11. Use the unit circle to find the six trigonometric functions of 330 °.

Find sin 330 °.

Find cos 330 °.

Find tan 330 °.

Find cot 330 °.

Find sec 330 °.

Find csc 330 °.

12.

Use the unit circle and the fact that sine is an odd function to find $\sin\left(\dfrac{3\pi}{4}\right)$.

13.

If angle θ is in standard position and intersects the unit circle at $\left(\dfrac{1}{\sqrt{17}}, -\dfrac{4}{\sqrt{17}}\right)$ find

$\sin\theta$, $\cos\theta$ and $\tan\theta$.

14. If we start at the point (1, 0) and travel once around the unit circle, we travel a distance of 2π units and arrive back where we started at the point (1, 0). If we continue around the unit circle a second time, we will repeat all the values of x and y that occurred during our first trip around. Use this discussion to evaluate the expression.

$$\cos\left(\dfrac{7\pi}{3}\right)$$

15. Prove the identity.

$$\cos\left(-\theta\right)\csc\left(-\theta\right)\tan\left(-\theta\right) = 1$$

16. For the problem below, θ is a central angle in a circle of radius r. Find the length of arc s cut off by θ.

$\theta = 4$, $r = 3$ inches.

If the answer needs rounding, round it to three significant digits.

$s =$ _________ inches

17. The pendulum on a grandfather clock swings from side to side once every second. If the length of the pendulum is 5 feet and the angle through which it swings is $16°$, how far does the tip of the pendulum travel in 1 second?

If the answer needs rounding, round it to three significant digits.

_________ feet

18. The figure is a model of George Ferris's Ferris wheel. The diameter of the wheel is 248 feet; and θ is the central angle formed as a rider travels from his or her initial position P_0 to position P_1. Find the distance traveled by the rider if $\theta = 211°$. Round your answer to the nearest tenth.

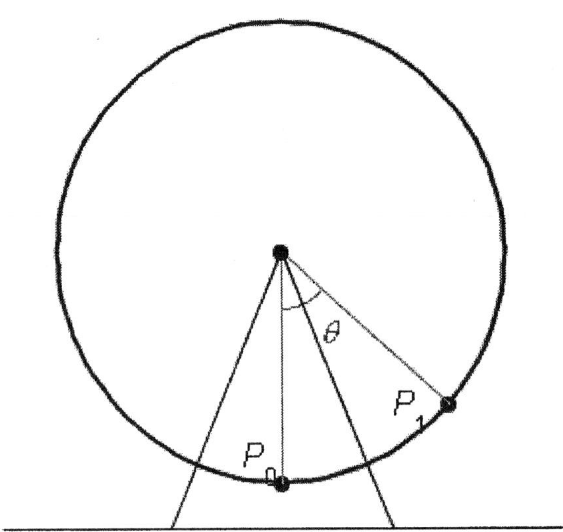

$s =$ _________ ft

19. θ is a central angle that cuts off an arc of length s. Find the radius of the circle if

$$\theta = 6°, \quad s = \frac{\pi}{2} \text{ m}.$$

If the answer needs rounding, round it to three significant digits.

$r =$ _________ m

20. Find the area of the sector formed by central angle $\theta = 15°$ in a circle of radius $r = 5$ m.

If the answer needs rounding, round it to three significant digits.

$A =$ _________ m^2

21. Find the linear velocity of a point moving with uniform circular motion, if the point covers a distance s in an amount of time t, where

$s = 20$ cm and $t = 4$ sec

If the answer needs rounding, round it to three significant digits.

__________ cm/sec

22. Point P sweeps out central angle θ as it rotates on a circle of radius r.

Find the angular velocity of point P.

$$\theta = \frac{2\pi}{5}\,,\ t = 5 \text{ sec}$$

If the answer needs rounding, round it to three significant digits.

__________ rad/sec

23. Point P moves with angular velocity ω on a circle of radius r.

Find the distance s traveled by the point in time t.

$\omega = 5 \text{ rad/sec},\ r = 3 \text{ ft},\ t = 5 \text{ min}$

If the answer needs rounding, round it to three significant digits.

__________ ft

24. A point is rotating with uniform circular motion on a circle of radius r.

Find v if $r = 6$ ft and the point rotates at 7 rpm.

If the answer needs rounding, round it to three significant digits.

__________ ft/min

25. The figure below is a model of the Ferris wheel. The diameter of the wheel is 165 feet, and one complete revolution takes 16 minutes.

Find the linear velocity of a person riding on the wheel.

Give your answer in miles per hour and round to the nearest hundredth.

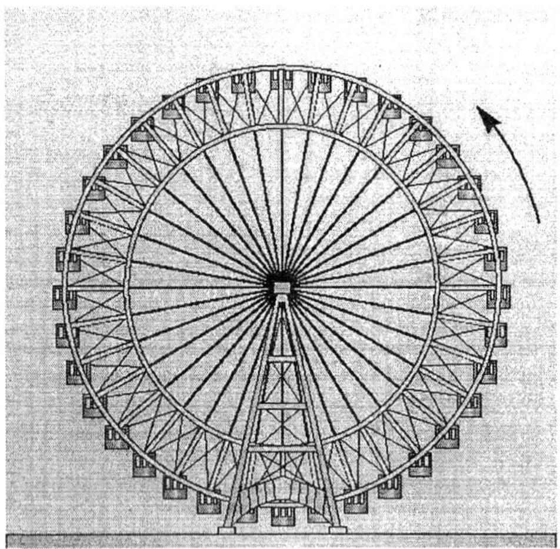

_________ mph

1.

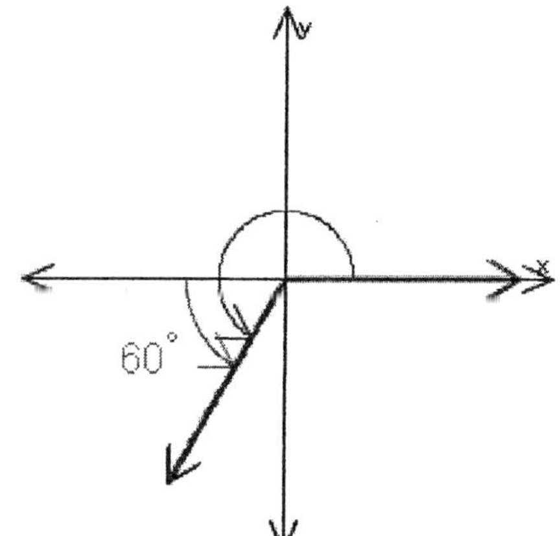

2. $-\dfrac{\sqrt{3}}{2}$

3. -5.2883

4. 197.5

5. 126.3

6. 2

7.

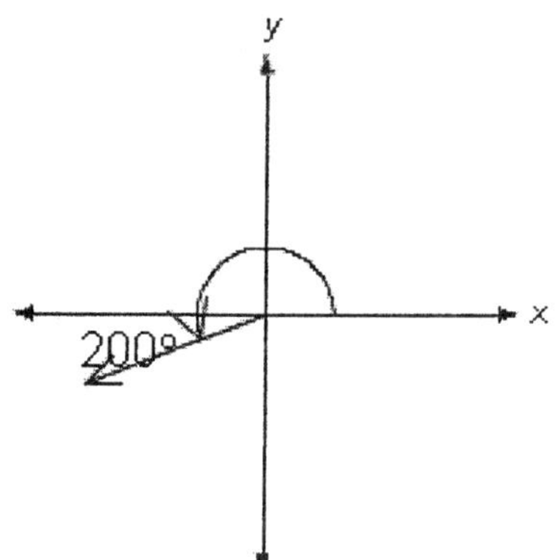

$\dfrac{10\pi}{9}$

$\dfrac{\pi}{9}, 20°$

8. 160

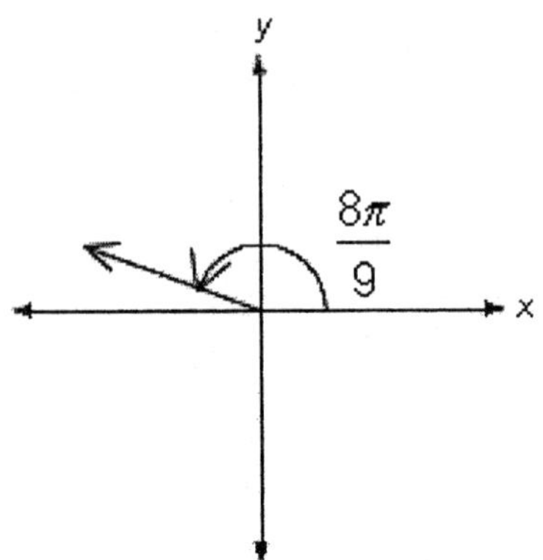

$\dfrac{\pi}{9}$, 20°

9. 77.3

10. 0

11. $-\dfrac{1}{2}$,

$\dfrac{\sqrt{3}}{2}$,

$-\dfrac{\sqrt{3}}{3}$,

$-\sqrt{3}$,

$\dfrac{2\sqrt{3}}{3}$,

-2

12. $\dfrac{\sqrt{2}}{2}$

13. $-\dfrac{4}{\sqrt{17}}$,

$\dfrac{1}{\sqrt{17}}$,

-4

14. $\dfrac{1}{2}$

15. $\cos(-\theta)\,\csc(-\theta)\,\tan(-\theta) = \cos(-\theta)\,\dfrac{1}{\sin(-\theta)}\,\dfrac{\sin(-\theta)}{\cos(-\theta)} = 1$

16. 12

17. 1.40

18. 456.6

19. 15

20. 3.27

21. 5

22. 0.251

23. 4,500

24. 264

25. 0.368

1. mctr.03.01.02_NoAlgs
2. mctr.03.01.16_NoAlgs
3. mctr.03.01.31_NoAlgs
4. mctr.03.01.49_NoAlgs
5. mctr.03.01.63_NoAlgs
6. mctr.03.02.06_NoAlgs
7. mctr.03.02.16_NoAlgs
8. mctr.03.02.33_NoAlgs
9. mctr.03.02.47_NoAlgs
10. mctr.03.02.67_NoAlgs
11. mctr.03.03.01_NoAlgs
12. mctr.03.03.16_NoAlgs
13. mctr.03.03.39_NoAlgs
14. mctr.03.03.50_NoAlgs
15. mctr.03.03.57_NoAlgs
16. mctr.03.04.01_NoAlgs
17. mctr.03.04.15_NoAlgs
18. mctr.03.04.25_NoAlgs
19. mctr.03.04.35_NoAlgs
20. mctr.03.04.45_NoAlgs
21. mctr.03.05.04_NoAlgs
22. mctr.03.05.13_NoAlgs
23. mctr.03.05.28_NoAlgs
24. mctr.03.05.39_NoAlgs
25. mctr.03.05.51_NoAlgs

1. Point P sweeps out central angle θ as it rotates on a circle of radius r.

   Find the angular velocity of point P.

   $$\theta = \frac{2\pi}{5}, \ t = 5 \text{ sec}$$

   If the answer needs rounding, round it to three significant digits.

   __________ rad/sec

2. Use a calculator to find sec 100.9°.

   Please round the answer to the nearest ten-thousandth.

   ____

3. Find the radian measure of angle θ, if θ is a central angle in a circle of radius r, and θ cuts off an arc of length s.

   $r = 3$ inches, $s = 6$ inches

   $\theta = $ __________ radians

4. If angle θ is in standard position and intersects the unit circle at $\left(\dfrac{1}{\sqrt{17}}, -\dfrac{4}{\sqrt{17}} \right)$ find $\sin \theta$, $\cos \theta$ and $\tan \theta$.

   Find $\sin \theta$.

   Find $\cos \theta$.

   Find $\tan \theta$.

5. Evaluate the expression when x is $\dfrac{\pi}{12}$. Use exact values.

   $4 \cos 6x$

6. Point P moves with angular velocity w on a circle of radius r.

    Find the distance s traveled by the point in time t.

    $w = 5$ rad/sec, $r = 3$ ft, $t = 5$ min

    If the answer needs rounding, round it to three significant digits.

    _________ ft

7. The figure is a model of George Ferris's Ferris wheel. The diameter of the wheel is 248 feet; and θ is the central angle formed as a rider travels from his or her initial position P_0 to position P_1. Find the distance traveled by the rider if $\theta = 211°$.

    Round your answer to the nearest tenth.

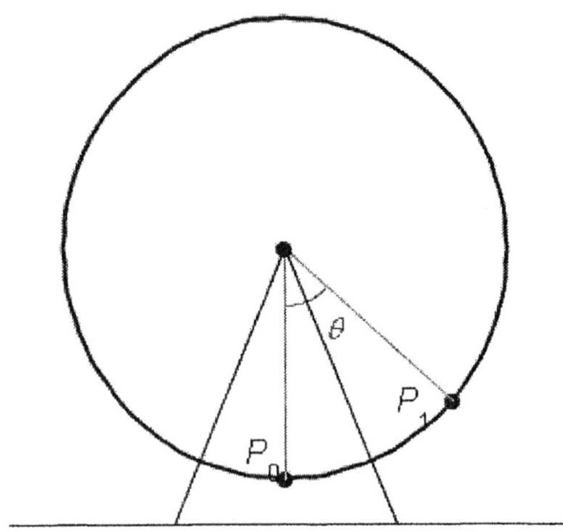

    $s =$ _________ ft

8. Find the exact value of $\cos 210°$.

9. For the problem below, θ is a central angle in a circle of radius r.

    Find the length of arc s cut off by θ .

    $\theta = 4$, $r = 3$ inches.

    If the answer needs rounding, round it to three significant digits.

    $s =$ _________ inches

10. Draw the angle in standard position.

$$\theta = 200°$$

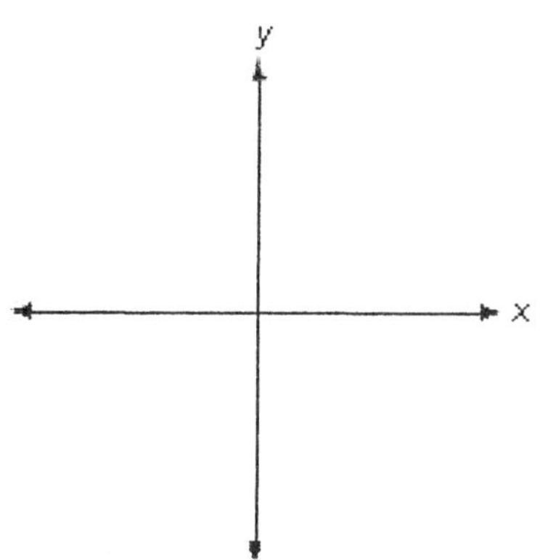

Convert to radian measure using exact values.

Name the reference angle in both degrees and radians.

11. Use a calculator to find θ to the nearest tenth of a degree, if $0° < \theta < 360°$ and

$\sin \theta = -0.3010$ with θ in QIII

$\theta =$ _______ °

12. A point is rotating with uniform circular motion on a circle of radius r. Find v if $r = 6$ ft and the point rotates at 7 rpm.

If the answer needs rounding, round it to three significant digits.

_______ ft/min

13. Draw the following angle in standard position and then name the reference angle.

240°

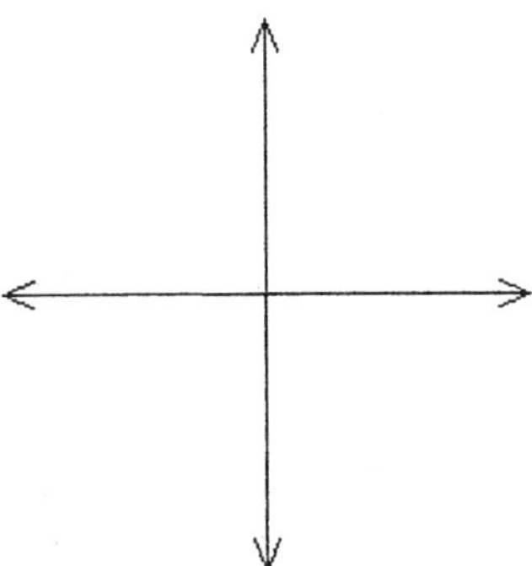

14. The figure below is a model of the Ferris wheel. The diameter of the wheel is 165 feet, and one complete revolution takes 16 minutes. Find the linear velocity of a person riding on the wheel.
Give your answer in miles per hour and round to the nearest hundredth.

__________ mph

15. θ is a central angle that cuts off an arc of length s. Find the radius of the circle if

$$\theta = 6^\circ, \; s = \frac{\pi}{2} \text{ m}.$$

If the answer needs rounding, round it to three significant digits.

$r =$ __________ m

16. Use the unit circle and the fact that sine is an odd function to find $\sin\left(\dfrac{3\pi}{4}\right)$.

17. If we start at the point (1, 0) and travel once around the unit circle, we travel a distance of 2π units and arrive back where we started at the point (1, 0). If we continue around the unit circle a second time, we will repeat all the values of x and y that occurred during our first trip around. Use this discussion to evaluate the expression.

$$\cos\left(\dfrac{7\pi}{3}\right)$$

18. Prove the identity.

$$\cos(-\theta)\csc(-\theta)\tan(-\theta) = 1$$

19. Use a calculator to find θ to the nearest tenth of a degree, if $0° < \theta < 360°$ and $\cot\theta = -0.7357$ with θ in QII

$$\theta = \underline{\qquad}°$$

20. Use the unit circle to find the six trigonometric functions of $330°$.

Find $\sin 330°$.

Find $\cos 330°$.

Find $\tan 330°$.

Find $\cot 330°$.

Find $\sec 330°$.

Find $\csc 330°$.

21. Convert to degree measure.

$$\theta = \frac{8\pi}{9}$$

$\theta =$ _________ °

Draw the angle in standard position.

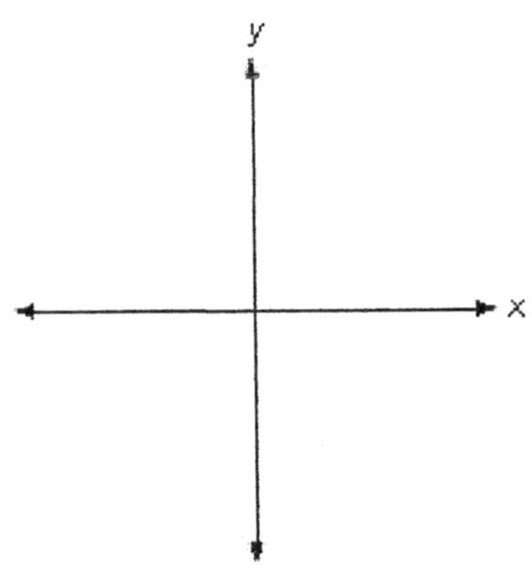

Label the reference angle in both degrees and radians.

22. Find the area of the sector formed by central angle $\theta = 15°$ in a circle of radius $r = 5$ m. If the answer needs rounding, round it to three significant digits.

$A =$ _________ m^2

23. Find the linear velocity of a point moving with uniform circular motion, if the point covers a distance s in an amount of time t, where

$s = 20$ cm and $t = 4$ sec

If the answer needs rounding, round it to three significant digits.

_________ cm/sec

24. Use a calculator to convert 1.35 to degree measure to tenth of a degree.

__________ °

25. The pendulum on a grandfather clock swings from side to side once every second. If the length of the pendulum is 5 feet and the angle through which it swings is 16°, how far does the tip of the pendulum travel in 1 second?

If the answer needs rounding, round it to three significant digits.

__________ feet

1. 0.251

2. -5.2883

3. 2

4. $-\dfrac{4}{\sqrt{17}}$,

$\dfrac{1}{\sqrt{17}}$,

-4

5. 0

6. 4,500

7. 456.6

8. $-\dfrac{\sqrt{3}}{2}$

9. 12

10.

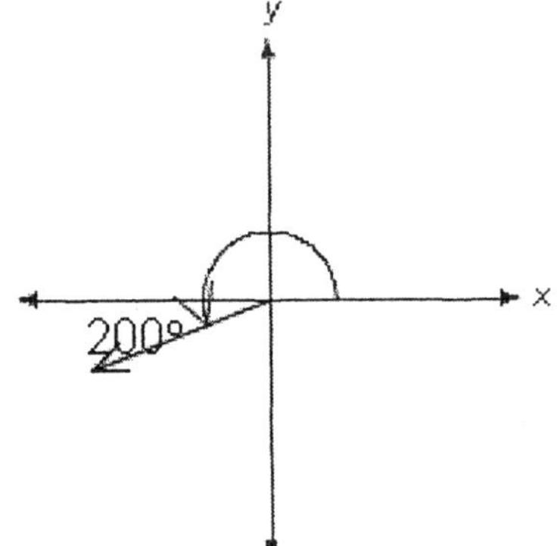

$\dfrac{10\pi}{9}$

$\dfrac{\pi}{9}$, $20°$

11. 197.5

12. 264

13.

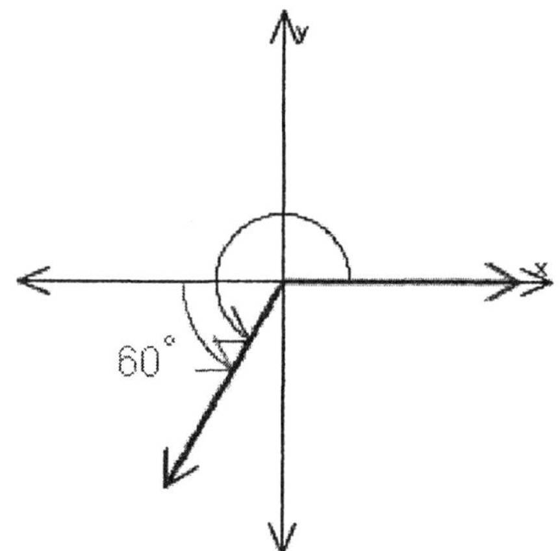

14. 0.368

15. 15

16. $\dfrac{\sqrt{2}}{2}$

17. $\dfrac{1}{2}$

18.
$$\cos(-\theta)\,\csc(-\theta)\,\tan(-\theta) = \cos(-\theta)\,\dfrac{1}{\sin(-\theta)}\,\dfrac{\sin(-\theta)}{\cos(-\theta)} = 1$$

19. 126.3

20. $-\dfrac{1}{2}$,

$\dfrac{\sqrt{3}}{2}$,

$-\dfrac{\sqrt{3}}{3}$,

$-\sqrt{3}$,

$\dfrac{2\sqrt{3}}{3}$,

-2

21. 160

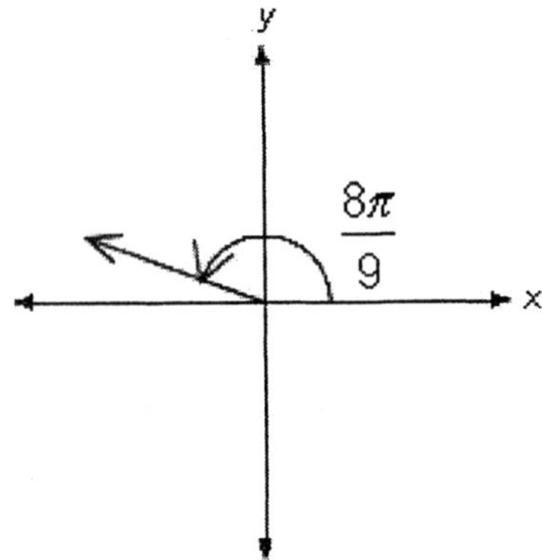

$\dfrac{\pi}{9}$, 20°

22. 3. 27

23. 5

24. 77. 3

25. 1. 4 0

McKeague/Turner - Trigonometry 5e Chapter 3 Form B

1. mctr.03.05.13_NoAlgs
2. mctr.03.01.31_NoAlgs
3. mctr.03.02.06_NoAlgs
4. mctr.03.03.39_NoAlgs
5. mctr.03.02.67_NoAlgs
6. mctr.03.05.28_NoAlgs
7. mctr.03.04.25_NoAlgs
8. mctr.03.01.16_NoAlgs
9. mctr.03.04.01_NoAlgs
10. mctr.03.02.16_NoAlgs
11. mctr.03.01.49_NoAlgs
12. mctr.03.05.39_NoAlgs
13. mctr.03.01.02_NoAlgs
14. mctr.03.05.51_NoAlgs
15. mctr.03.04.35_NoAlgs
16. mctr.03.03.16_NoAlgs
17. mctr.03.03.50_NoAlgs
18. mctr.03.03.57_NoAlgs
19. mctr.03.01.63_NoAlgs
20. mctr.03.03.01_NoAlgs
21. mctr.03.02.33_NoAlgs
22. mctr.03.04.45_NoAlgs
23. mctr.03.05.04_NoAlgs
24. mctr.03.02.47_NoAlgs
25. mctr.03.04.15_NoAlgs

1. Draw the following angle in standard position and then name the reference angle.

    92.4°

    Select the correct answer.

a.

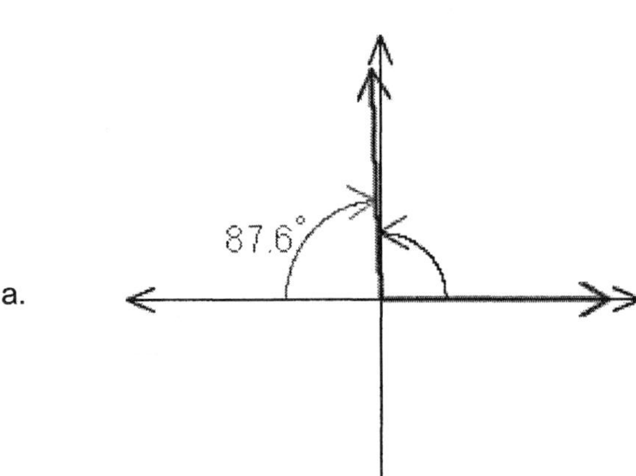

b.

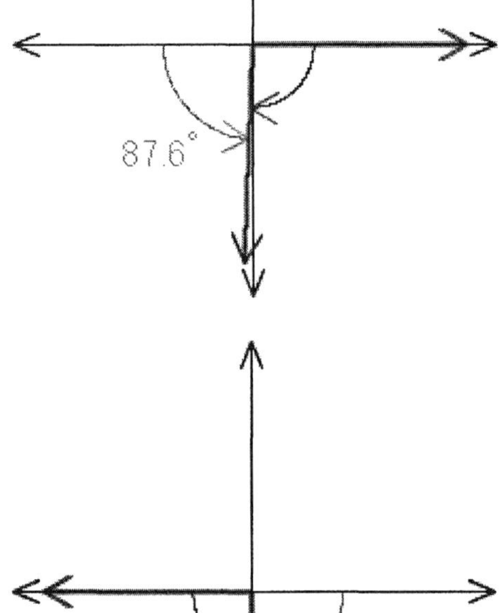

c.

2. Find the exact value of $\cot 690°$.

   Select the correct answer.

   a. $\dfrac{\sqrt{3}}{3}$

   b. $\sqrt{3}$

   c. $-\sqrt{3}$

   d. $\dfrac{\sqrt{3}}{2}$

   e. $-\dfrac{\sqrt{3}}{2}$

3. Use a calculator to find $\sin\left(-235\right)°$.

   Please round the answer to the nearest ten-thousandth.

   Select the correct answer.

   a. 0.8193
   b. 0.7972
   c. 0.7792
   d. 0.8192
   e. 0.7992

4. Use a calculator to find θ to the nearest tenth of a degree, if $0° < \theta < 360°$ and

   $\sin\theta = 0.9636$ with θ in QII

   Select the correct answer.

   a. $106.5°$

   b. $108.5°$

   c. $103.2°$

   d. $109.0°$

   e. $105.5°$

5. Find θ, if $0° < \theta < 360°$ and

$$\tan \theta = -\frac{\sqrt{3}}{3} \quad \text{and } \theta \text{ in QII}$$

Select the correct answer.

    a. $210°$

    b. $150°$

    c. $30°$

    d. $135°$

    e. $315°$

6. Two cities are approximately 350 miles apart on the surface of the earth. Assuming that the radius of the earth is 4,000 miles, find the radian measure of the central angle with its vertex at the center of the earth that has one city on one side and another one on the other side.

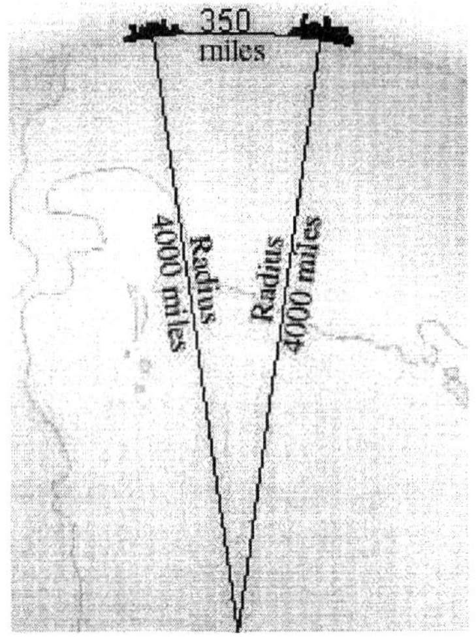

Select the correct answer.

    a.   0.0875 radians
    b.   0.1075 radians
    c.   0.0925 radians
    d.   0.0725 radians
    e.   0.0775 radians

7. Use a calculator to convert $110°10'$ to radians. Round your answer to the nearest hundredth.

    Select the correct answer.

    a.   1.92
    b.   1.98
    c.   2.00
    d.   1.96
    e.   1.94

8. Label the reference angle in both degrees and radians.

    $$\theta = \frac{9\pi}{4}$$

    Select the correct answer.

    a. $30° = \dfrac{\pi}{6}$

    b. $45° = \dfrac{\pi}{2}$

    c. $90° = \dfrac{\pi}{4}$

    d. $90° = \dfrac{\pi}{2}$

    e. $45° = \dfrac{\pi}{4}$

9. Give the exact value of $\sec \dfrac{3\pi}{4}$.

    Select the correct answer.

    a. $-\dfrac{2}{\sqrt{3}}$

    b. -2

    c. $\dfrac{2}{\sqrt{3}}$

    d. $\sqrt{2}$

    e. $-\sqrt{2}$

10. For the following expression, find the value of y that corresponds to each value of x.

$$y = \sin 2x \quad \text{for } x = 0,\ \frac{\pi}{4},\ \frac{\pi}{2},\ \frac{3\pi}{4},\ \pi$$

Select the correct answer.

a. $(0,0),\ \left(\dfrac{\pi}{4}, \dfrac{1}{\sqrt{2}}\right),\ \left(\dfrac{\pi}{2}, 1\right),\ \left(\dfrac{3\pi}{4}, \dfrac{1}{\sqrt{2}}\right),\ (\pi, 0)$

b. $(0,0),\ \left(\dfrac{\pi}{4}, 1\right),\ \left(\dfrac{\pi}{2}, 0\right),\ \left(\dfrac{3\pi}{4}, 1\right),\ (\pi, 0)$

c. $(0,0),\ \left(\dfrac{\pi}{4}, 1\right),\ \left(\dfrac{\pi}{2}, 0\right),\ \left(\dfrac{3\pi}{4}, -1\right),\ (\pi, 0)$

d. $(0,0),\ \left(\dfrac{\pi}{4}, \dfrac{9}{\sqrt{2}}\right),\ \left(\dfrac{\pi}{2}, 9\right),\ \left(\dfrac{3\pi}{4}, \dfrac{9}{\sqrt{2}}\right),\ (\pi, 0)$

11. Use the unit circle to find the six trigonometric functions of $\dfrac{7\pi}{4}$.

Select the correct answer.

a. $\sin \dfrac{7\pi}{4} = -\dfrac{\sqrt{2}}{2},\ \cos \dfrac{7\pi}{4} = \dfrac{\sqrt{2}}{2},\ \tan \dfrac{7\pi}{4} = -1$

   $\cot \dfrac{7\pi}{4} = -1,\ \sec \dfrac{7\pi}{4} = \sqrt{2},\ \csc \dfrac{7\pi}{4} = -\sqrt{2}$

b. $\sin \dfrac{7\pi}{4} = \dfrac{\sqrt{2}}{2},\ \cos \dfrac{7\pi}{4} = -\dfrac{\sqrt{2}}{2},\ \tan \dfrac{7\pi}{4} = 1$

   $\cot \dfrac{7\pi}{4} = 1,\ \sec \dfrac{7\pi}{4} = -\sqrt{2},\ \csc \dfrac{7\pi}{4} = \sqrt{2}$

c. $\sin \dfrac{7\pi}{4} = -1,\ \cos \dfrac{7\pi}{4} = 1,\ \tan \dfrac{7\pi}{4} = -\dfrac{\sqrt{2}}{2}$

   $\cot \dfrac{7\pi}{4} = -\sqrt{2},\ \sec \dfrac{7\pi}{4} = 1,\ \csc \dfrac{7\pi}{4} = -1$

d. $\sin \dfrac{7\pi}{4} = -1,\ \cos \dfrac{7\pi}{4} = \dfrac{\sqrt{2}}{2},\ \tan \dfrac{7\pi}{4} = -\sqrt{2}$

   $\cot \dfrac{7\pi}{4} = -\dfrac{\sqrt{2}}{2},\ \sec \dfrac{7\pi}{4} = \sqrt{2},\ \csc \dfrac{7\pi}{4} = -1$

12.

Use the unit circle to find all values of θ between 0 and 2π for which $\tan\theta = \dfrac{\sqrt{3}}{3}$.

Select the correct answer.

a. $\dfrac{5\pi}{6}, \dfrac{11\pi}{3}$

b. $\dfrac{2\pi}{3}, \dfrac{5\pi}{3}$

c. $\dfrac{2\pi}{3}, \dfrac{\pi}{3}$

d. $\dfrac{\pi}{6}, \dfrac{5\pi}{6}$

e. $\dfrac{\pi}{6}, \dfrac{7\pi}{6}$

13. If we start at the point (1, 0) and travel once around the unit circle, we travel a distance of 2π units and arrive back where we started at the point (1, 0).

If we continue around the unit circle a second time, we will repeat all the values of x and y that occurred during our first trip around. Use this discussion to evaluate the expression.

$$\sin\left(2\pi - \frac{\pi}{2}\right)$$

Select the correct answer.

a.   1

b. $\dfrac{1}{2}$

c.   0

d. $\dfrac{\sqrt{2}}{2}$

e. -1

14. The pendulum on a grandfather clock swings from side to side once every second. If the length of the pendulum is 5 feet and the angle through which it swings is $16°$, how far does the tip of the pendulum travel in 1 second?

Select the correct answer.

    a.   1.42
    b.   1.48
    c.   2.30
    d.   1.40
    e.   2.94

15. Find the algebraic expression that is equal to $\sin \alpha \sec \alpha \cot \alpha$.

Select the correct answer.

    a. 1

    b. $\cos \alpha$

    c. $\sec \alpha$

    d. $\sin \alpha$

    e. $\csc \alpha$

16. For the problem below, θ is a central angle in a circle of radius r.

Find the length of arc s cut off by θ.

$\theta = 330°$, $r = 5$ inches.

Select the correct answer.

    a. $s = 33$ inches
    b. $s = 29.8$ inches
    c. $s = 28.8$ inches
    d. $s = 31$ inches
    e. $s = 26.8$ inches

17. The light truck with manual transmission has a circular brake drum with a diameter of 320 millimeters. Each brake pad, which presses against the drum, is 309 millimeters long. What central angle is subtended by one of the brake pads?

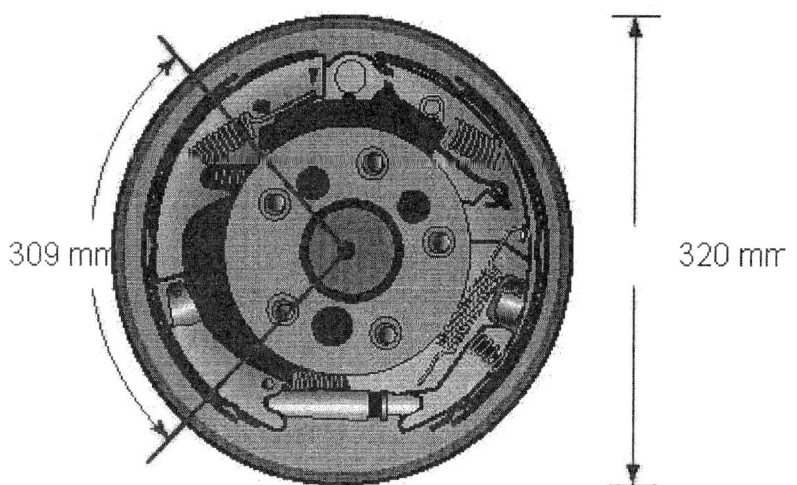

Select the correct answer.

a. 1.63 radians, 108°

b. 1.93 radians, 111°

c. 2.13 radians, 113°

d. 1.83 radians, 110°

e. 1.53 radians, 107°

18. θ is a central angle that cuts off an arc of length s. Find the radius of the circle if

$\theta = 10$, $s = 5$ ft.

Select the correct answer.

a. 0.5 ft
b. 0.47 ft
c. 0.48 ft
d. 0.46 ft
e. 0.49 ft

19. Find the area of the sector formed by central angle $\theta = 2.4$

in a circle of radius $r = 6$ inches.

Select the correct answer.

    a.   43.4 inches^2

    b.   42.9 inches^2

    c.   43.2 inches^2

    d.   43.6 inches^2

    e.   43.1 inches^2

20. A sector of area $\dfrac{2\pi}{3}$ in^2 is formed by a central angle of 45°.

What is the radius of the circle?

Select the correct answer.

    a.   2.30 inches
    b.   2.29 inches
    c.   2.27 inches
    d.   2.31 inches
    e.   2.34 inches

21. Find the distance s covered by a point moving with linear velocity v for a time t if

$v = 20$ ft/sec and $t = 4$ sec

Select the correct answer.

    a.   60 ft
    b.   55.8 ft
    c.   90 ft
    d.   66.5 ft
    e.   80 ft

22. Point P sweeps out central angle θ as it rotates on a circle of radius r.

Find the angular velocity of point P.

$\theta = 26\pi,\ t = 0.8\ hr$

Select the correct answer.

   a.   197 rad/hr
   b.   187 rad/hr
   c.   150 rad/hr
   d.   102 rad/hr
   e.   167 rad/hr

23. Find the angular velocity associated with the given rpm.

5.1 rpm

Select the correct answer.

   a.   28.9 rad/min
   b.   41.5 rad/min
   c.   19 rad/min
   d.   26 rad/min
   e.   32.0 rad/min

24. The San Francisco cable cars travel by clamping onto a steel cable that circulates in a channel beneath the streets. This cable is driven by a large 13-foot-diameter pulley, called a *sheave* (see the figure). The sheave turns at a rate of 18 revolutions per minute.
Find the speed of the cable car, in miles per hour (rounded to three significant digits), by determining the linear velocity of the cable. (1 mi = 5,280 ft).

Select the correct answer.

   a.   7.94 mph
   b.   8.91 mph
   c.   8.35 mph
   d.   8.33 mph
   e.   9.19 mph

25. Lance Armstrong, four-time winner of the Tour de France, rides a Trek 5,900 bicycle equipped with Dura-Ace components. (see the figure). When Lance pedals, he turns a gear, called a chainring. The angular velocity of the chainring will determine the linear speed at which the chain travels. The chain connects the chainring to a smaller gear, called a sprocket, which is attached to the rear wheel (see the scheme). The angular velocity of the sprocket depends upon the linear speed of the chain. The sprocket and rear wheel rotate at the same rate, and the diameter of the rear wheel is 700 millimeters. The speed at which Lance travels is determined by the angular velocity of his rear wheel.

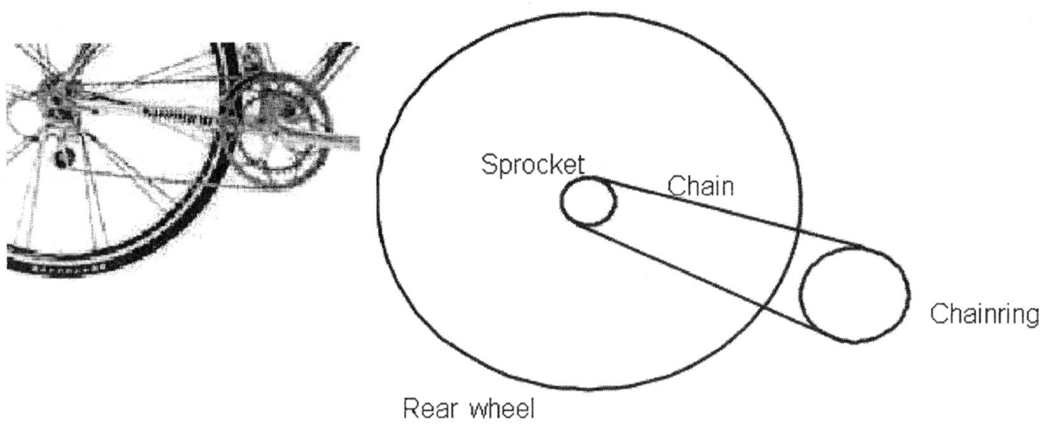

If Lance was using his 150-millimeter-diameter chainring and pedaling at a rate of 80 revolutions per minute, what diameter sprocket would he need in order to maintain a speed of 40 kilometers per hour?

Select the correct answer.

a.   49.5 mm
b.   36.8 mm
c.   33.2 mm
d.   54.7 mm
e.   39.6 mm

1. a

2. c

3. d

4. e

5. b

6. a

7. a

8. e

9. e

10. c

11. a

12. e

13. e

14. d

15. a

16. c

17. b

18. a

19. c

20. d

21. e

22. d

23. e

24. c

25. e

1. mctr.03.01.06m_NoAlgs
2. mctr.03.01.28m_NoAlgs
3. mctr.03.01.42m_NoAlgs
4. mctr.03.01.57m_NoAlgs
5. mctr.03.01.73m_NoAlgs
6. mctr.03.02.09m_NoAlgs
7. mctr.03.02.23m_NoAlgs
8. mctr.03.02.38m_NoAlgs
9. mctr.03.02.58m_NoAlgs
10. mctr.03.02.73m_NoAlgs
11. mctr.03.03.07m_NoAlgs
12. mctr.03.03.21m_NoAlgs
13. mctr.03.03.43m_NoAlgs
14. mctr.03.03.50m_NoAlgs
15. mctr.03.03.57m_NoAlgs
16. mctr.03.04.10m_NoAlgs
17. mctr.03.04.19m_NoAlgs
18. mctr.03.04.29m_NoAlgs
19. mctr.03.04.41m_NoAlgs
20. mctr.03.04.51m_NoAlgs
21. mctr.03.05.08m_NoAlgs
22. mctr.03.05.19m_NoAlgs
23. mctr.03.05.34m_NoAlgs
24. mctr.03.05.45m_NoAlgs
25. mctr.03.05.57m_NoAlgs

1. For the problem below, θ is a central angle in a circle of radius r.

   Find the length of arc s cut off by θ.

   $\theta = 330°$, $r = 5$ inches.

   Select the correct answer.

   a. $s = 33$ inches
   b. $s = 29.8$ inches
   c. $s = 28.8$ inches
   d. $s = 31$ inches
   e. $s = 26.8$ inches

2. Give the exact value of $\sec \dfrac{3\pi}{4}$.

   Select the correct answer.

   a. $-\dfrac{2}{\sqrt{3}}$

   b. -2

   c. $\dfrac{2}{\sqrt{3}}$

   d. $\sqrt{2}$

   e. $-\sqrt{2}$

3. A sector of area $\dfrac{2\pi}{3}$ in^2 is formed by a central angle of $45°$.

   What is the radius of the circle?

   Select the correct answer.

   a.   2.30 inches
   b.   2.29 inches
   c.   2.27 inches
   d.   2.31 inches
   e.   2.34 inches

4. Find the algebraic expression that is equal to $\sin \alpha \sec \alpha \cot \alpha$.

Select the correct answer.

a. 1

b. $\cos \alpha$

c. $\sec \alpha$

d. $\sin \alpha$

e. $\csc \alpha$

5. Use the unit circle to find all values of θ between 0 and 2π for which $\tan \theta = \dfrac{\sqrt{3}}{3}$.

Select the correct answer.

a. $\dfrac{5\pi}{6}, \dfrac{11\pi}{3}$

b. $\dfrac{2\pi}{3}, \dfrac{5\pi}{3}$

c. $\dfrac{2\pi}{3}, \dfrac{\pi}{3}$

d. $\dfrac{\pi}{6}, \dfrac{5\pi}{6}$

e. $\dfrac{\pi}{6}, \dfrac{7\pi}{6}$

6. Find the exact value of $\cot 690^\circ$.

Select the correct answer.

a. $\dfrac{\sqrt{3}}{3}$

b. $\sqrt{3}$

c. $-\sqrt{3}$

d. $\dfrac{\sqrt{3}}{2}$

e. $-\dfrac{\sqrt{3}}{2}$

7. Label the reference angle in both degrees and radians.

$$\theta = \frac{9\pi}{4}$$

Select the correct answer.

a. $30° = \dfrac{\pi}{6}$

b. $45° = \dfrac{\pi}{2}$

c. $90° = \dfrac{\pi}{4}$

d. $90° = \dfrac{\pi}{2}$

e. $45° = \dfrac{\pi}{4}$

8. For the following expression, find the value of y that corresponds to each value of x.

$$y = \sin 2x \quad \text{for } x = 0, \ \frac{\pi}{4}, \ \frac{\pi}{2}, \ \frac{3\pi}{4}, \ \pi$$

Select the correct answer.

a. $(0,0), \ \left(\dfrac{\pi}{4}, 9\right), \ \left(\dfrac{\pi}{2}, 0\right), \ \left(\dfrac{3\pi}{4}, -9\right), \ (\pi, 0)$

b. $(0,0), \ \left(\dfrac{\pi}{4}, 1\right), \ \left(\dfrac{\pi}{2}, 0\right), \ \left(\dfrac{3\pi}{4}, 1\right), \ (\pi, 0)$

c. $(0,0), \ \left(\dfrac{\pi}{4}, 1\right), \ \left(\dfrac{\pi}{2}, 0\right), \ \left(\dfrac{3\pi}{4}, -1\right), \ (\pi, 0)$

d. $(0,0), \ \left(\dfrac{\pi}{4}, \dfrac{9}{\sqrt{2}}\right), \ \left(\dfrac{\pi}{2}, 9\right), \ \left(\dfrac{3\pi}{4}, \dfrac{9}{\sqrt{2}}\right), \ (\pi, 0)$

e. $(0,0), \ \left(\dfrac{\pi}{4}, \dfrac{1}{\sqrt{2}}\right), \ \left(\dfrac{\pi}{2}, 1\right), \ \left(\dfrac{3\pi}{4}, \dfrac{1}{\sqrt{2}}\right), \ (\pi, 0)$

9. Point P sweeps out central angle θ as it rotates on a circle of radius r.

   Find the angular velocity of point P.

   $\theta = 26\pi, \; t = 0.8 \, hr$

   Select the correct answer.

   a. 197 rad/hr
   b. 187 rad/hr
   c. 156 rad/hr
   d. 102 rad/hr
   e. 167 rad/hr

10. The pendulum on a grandfather clock swings from side to side once every second. If the length of the pendulum is 5 feet and the angle through which it swings is $16°$, how far does

    the tip of the pendulum travel in 1 second?
    If the answer needs rounding, round it to three significant digits.

    Select the correct answer.

    a. 1.42
    b. 1.40
    c. 2.00
    d. 1.46
    e. 2.94

11. Use a calculator to convert $110°10'$ to radians. Round your answer to the nearest

    hundredth.

    Select the correct answer.

    a. 1.92
    b. 1.98
    c. 2.00
    d. 1.96
    e. 1.94

12. Draw the following angle in standard position and then name the reference angle.

92.4 °

Select the correct answer.

a. 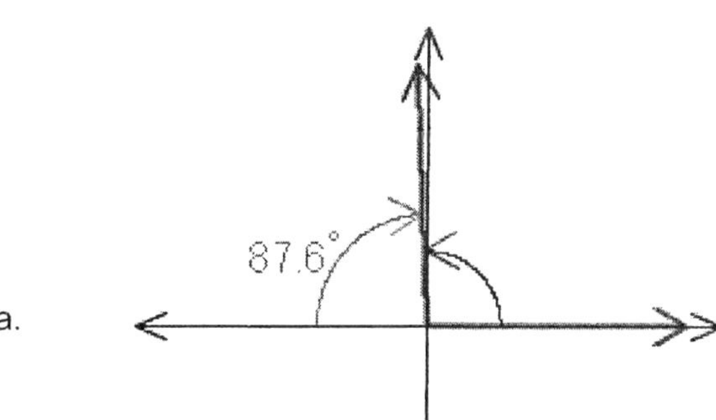

b. 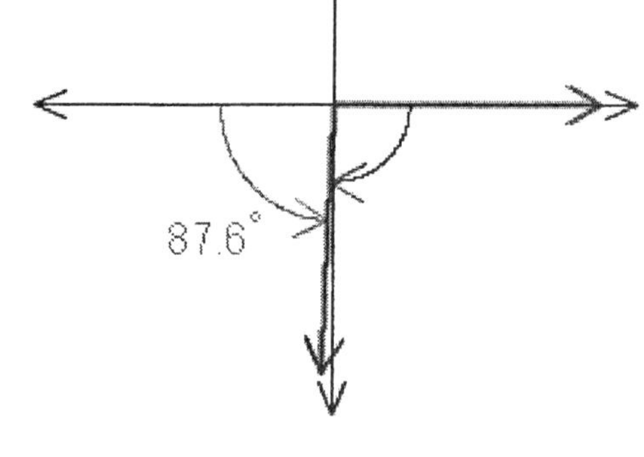

c. 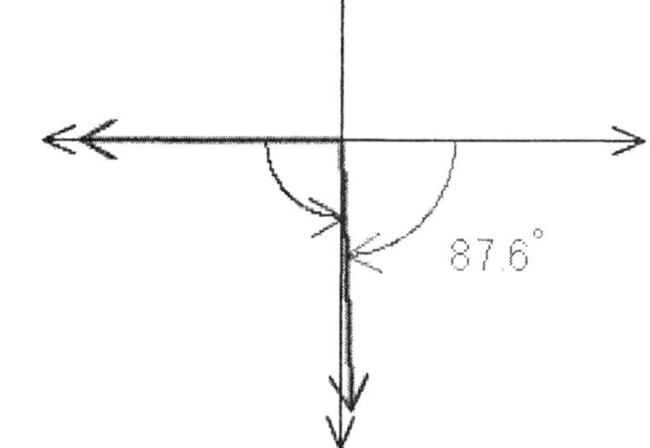

13. If we start at the point (1, 0) and travel once around the unit circle, we travel a distance of 2π units and arrive back where we started at the point (1, 0). If we continue around the unit circle a second time, we will repeat all the values of x and y that occurred during our first trip around. Use this discussion to evaluate the expression.

$$\cos\left(\frac{17\pi}{6}\right)$$

Select the correct answer.

a. 1

b. $\dfrac{\sqrt{3}}{2}$

c. $-\dfrac{1}{2}$

d. $\dfrac{\sqrt{2}}{2}$

e. $-\dfrac{\sqrt{3}}{2}$

14. Two cities are approximately 350 miles apart on the surface of the earth. Assuming that the radius of the earth is 4,000 miles, find the radian measure of the central angle with its vertex at the center of the earth that has one city on one side and another one on the other side.

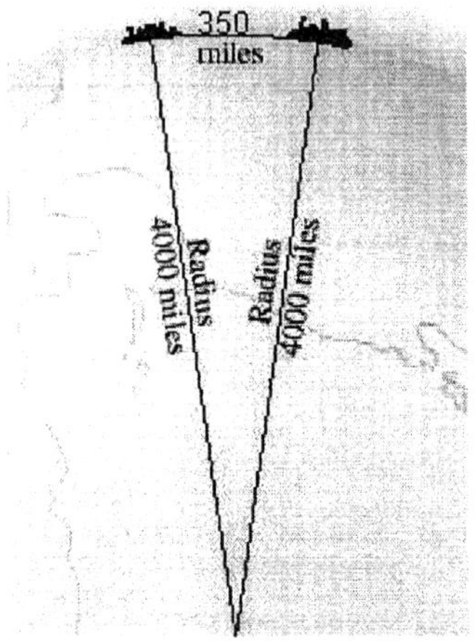

Select the correct answer.

a. 0.0875 radians
b. 0.1075 radians
c. 0.0925 radians
d. 0.0725 radians
e. 0.0775 radians

15. Find the angular velocity associated with the given rpm.

5.1 rpm

Select the correct answer.

a. 28.9 rad/min
b. 41.5 rad/min
c. 19 rad/min
d. 26 rad/min
e. 32.0 rad/min

16. Use the unit circle to find the six trigonometric functions of $\dfrac{7\pi}{4}$.

Select the correct answer.

a. $\sin \dfrac{7\pi}{4} = -\dfrac{\sqrt{2}}{2}$, $\cos \dfrac{7\pi}{4} = \dfrac{\sqrt{2}}{2}$, $\tan \dfrac{7\pi}{4} = -1$

$\cot \dfrac{7\pi}{4} = -1$, $\sec \dfrac{7\pi}{4} = \sqrt{2}$, $\csc \dfrac{7\pi}{4} = -\sqrt{2}$

b. $\sin \dfrac{7\pi}{4} = \dfrac{\sqrt{2}}{2}$, $\cos \dfrac{7\pi}{4} = -\dfrac{\sqrt{2}}{2}$, $\tan \dfrac{7\pi}{4} = 1$

$\cot \dfrac{7\pi}{4} = 1$, $\sec \dfrac{7\pi}{4} = -\sqrt{2}$, $\csc \dfrac{7\pi}{4} = \sqrt{2}$

c. $\sin \dfrac{7\pi}{4} = -1$, $\cos \dfrac{7\pi}{4} = 1$, $\tan \dfrac{7\pi}{4} = -\dfrac{\sqrt{2}}{2}$

$\cot \dfrac{7\pi}{4} = -\sqrt{2}$, $\sec \dfrac{7\pi}{4} = 1$, $\csc \dfrac{7\pi}{4} = -1$

d. $\sin \dfrac{7\pi}{4} = -1$, $\cos \dfrac{7\pi}{4} = \dfrac{\sqrt{2}}{2}$, $\tan \dfrac{7\pi}{4} = -\sqrt{2}$

$\cot \dfrac{7\pi}{4} = -\dfrac{\sqrt{2}}{2}$, $\sec \dfrac{7\pi}{4} = \sqrt{2}$, $\csc \dfrac{7\pi}{4} = -1$

e. $\sin \dfrac{7\pi}{4} = -\dfrac{\sqrt{2}}{2}$, $\cos \dfrac{7\pi}{4} = 1$, $\tan \dfrac{7\pi}{4} = -\dfrac{\sqrt{2}}{2}$

$\cot \dfrac{7\pi}{4} = -\sqrt{2}$, $\sec \dfrac{7\pi}{4} = 1$, $\csc \dfrac{7\pi}{4} = -\sqrt{2}$

17. θ is a central angle that cuts off an arc of length s. Find the radius of the circle if

$\theta = 10$, $s = 5$ ft.

Select the correct answer.

    a.   0.5 ft
    b.   0.47 ft
    c.   0.48 ft
    d.   0.46 ft
    e.   0.49 ft

18. The San Francisco cable cars travel by clamping onto a steel cable that circulates in a channel beneath the streets. This cable is driven by a large 13-foot-diameter pulley, called a *sheave* (see the figure). The sheave turns at a rate of 18 revolutions per minute. Find the speed of the cable car, in miles per hour (rounded to three significant digits), by determining the linear velocity of the cable. (1 mi = 5,280 ft).

Select the correct answer.

    a.   7.94 mph
    b.   8.91 mph
    c.   8.35 mph
    d.   8.33 mph
    e.   9.19 mph

19. Use a calculator to find θ to the nearest tenth of a degree, if $0° < \theta < 360°$ and

$\sin \theta = 0.9636$ with θ in QII

Select the correct answer.

    a.   106.5°

    b.   108.5°

    c.   103.2°

    d.   109.0°

    e.   105.5°

20. Find θ, if $0° < \theta < 360°$ and

$$\tan \theta = -\frac{\sqrt{3}}{3} \quad \text{and } \theta \text{ in QII}$$

Select the correct answer.

    a. $210°$

    h $150°$

    c. $30°$

    d. $135°$

    e. $315°$

21. Find the area of the sector formed by central angle $\theta = 2.4$ in a circle of radius $r = 6$ inches.

Select the correct answer.

    a. 43.4 inches^2

    b. 42.9 inches^2

    c. 43.2 inches^2

    d. 43.6 inches^2

    e. 43.1 inches^2

22. Lance Armstrong, four-time winner of the Tour de France, rides a Trek 5,900 bicycle equipped with Dura-Ace components. (see the figure). When Lance pedals, he turns a gear, called a chainring. The angular velocity of the chainring will determine the linear speed at which the chain travels. The chain connects the chainring to a smaller gear, called a sprocket, which is attached to the rear wheel (see the scheme). The angular velocity of the sprocket depends upon the linear speed of the chain. The sprocket and rear wheel rotate at the same rate, and the diameter of the rear wheel is 700 millimeters. The speed at which Lance travels is determined by the angular velocity of his rear wheel.

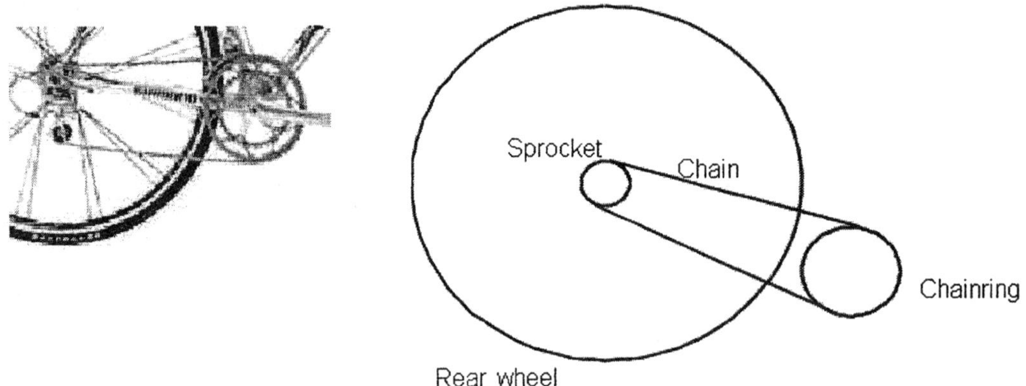

If Lance was using his 150-millimeter-diameter chainring and pedaling at a rate of 80 revolutions per minute, what diameter sprocket would he need in order to maintain a speed of 40 kilometers per hour?

Select the correct answer.

a.   49.5 mm
b.   36.8 mm
c.   33.2 mm
d.   54.7 mm
e.   39.6 mm

23. Use a calculator to find $\sin\left(-235\right)^{\circ}$.

Please round the answer to the nearest ten-thousandth.

Select the correct answer.

a.   0.8193
b.   0.7972
c.   0.7792
d.   0.8192
e.   0.7992

24. Find the distance s covered by a point moving with linear velocity v for a time t if

$v = 20$ ft/sec and $t = 4$ sec

Select the correct answer.

a. 60 ft
b. 55.8 ft
c. 90 ft
d. 66.5 ft
e. 80 ft

25. The light truck with manual transmission has a circular brake drum with a diameter of 320 millimeters. Each brake pad, which presses against the drum, is 309 millimeters long. What central angle is subtended by one of the brake pads? Write your answer in both radians and in degrees.

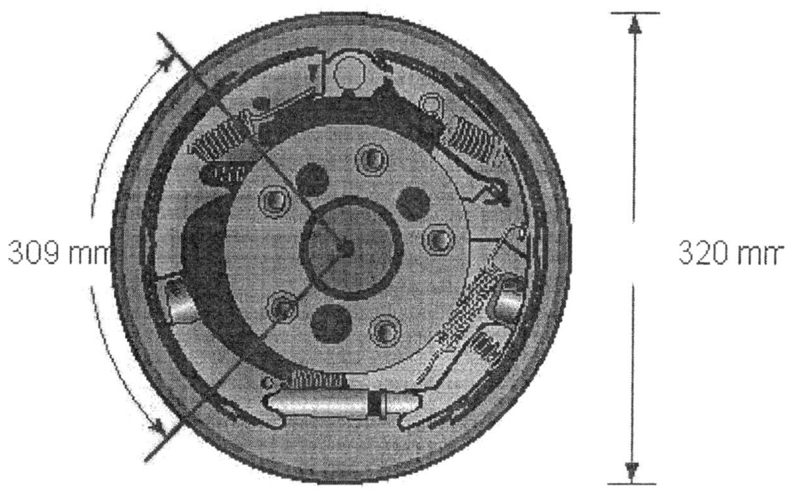

Select the correct answer.

a. 1.63 radians, 108°

b. 1.93 radians, 111°

c. 2.13 radians, 113°

d. 1.83 radians, 110°

e. 1.53 radians, 107°

McKeague/Turner - Trigonometry 5e Chapter 3 Form D

1. c

2. e

3. d

4. a

5. e

6. c

7. e

8. c

9. d

10. b

11. a

12. a

13. e

14. a

15. e

16. a

17. a

18. c

19. e

20. b

21. c

22. e

23. d

24. e

25. b

1. mctr.03.04.10m_NoAlgs
2. mctr.03.02.58m_NoAlgs
3. mctr.03.04.51m_NoAlgs
4. mctr.03.03.57m_NoAlgs
5. mctr.03.03.21m_NoAlgs
6. mctr.03.01.28m_NoAlgs
7. mctr.03.02.38m_NoAlgs
8. mctr.03.02.73m_NoAlgs
9. mctr.03.05.19m_NoAlgs
10. mctr.03.03.43m_NoAlgs
11. mctr.03.02.23m_NoAlgs
12. mctr.03.01.06m_NoAlgs
13. mctr.03.03.50m_NoAlgs
14. mctr.03.02.09m_NoAlgs
15. mctr.03.05.34m_NoAlgs
16. mctr.03.03.07m_NoAlgs
17. mctr.03.04.29m_NoAlgs
18. mctr.03.05.45m_NoAlgs
19. mctr.03.01.57m_NoAlgs
20. mctr.03.01.73m_NoAlgs
21. mctr.03.04.41m_NoAlgs
22. mctr.03.05.57m_NoAlgs
23. mctr.03.01.42m_NoAlgs
24. mctr.03.05.08m_NoAlgs
25. mctr.03.04.19m_NoAlgs

1. Draw the following angle in standard position and then name the reference angle.

   92.4 °

   Select the correct answer.

a. 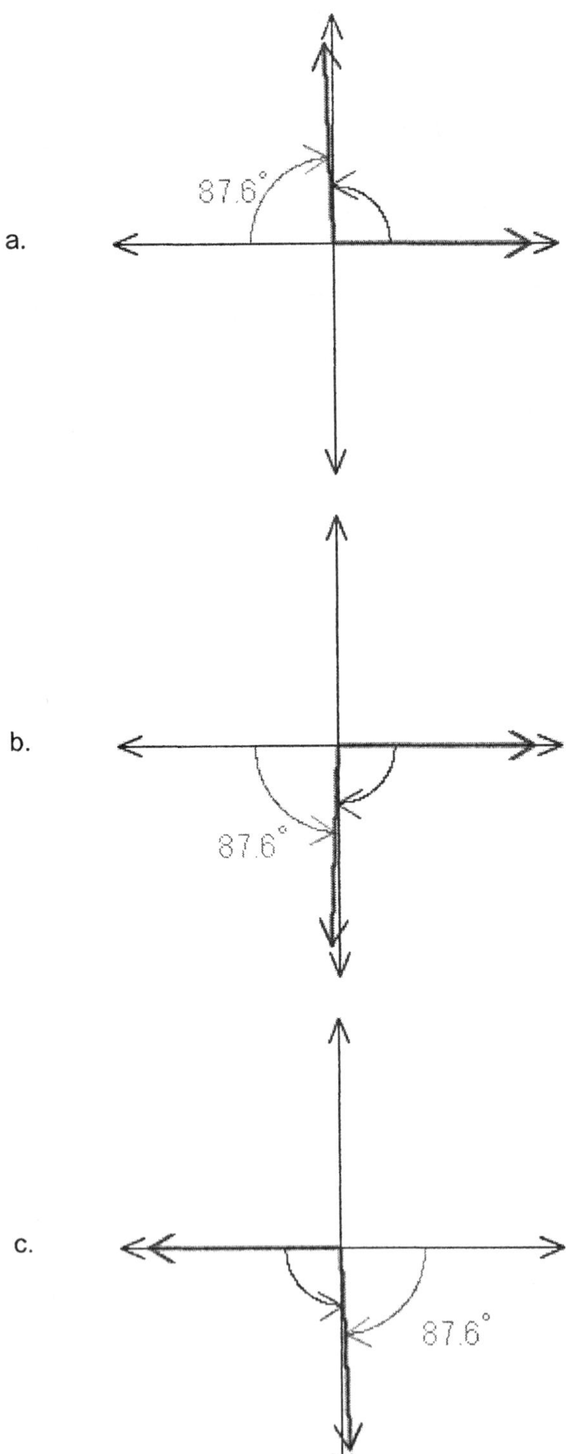

b.

c.

2. Draw the following angle in standard position and then name the reference angle.

   240°

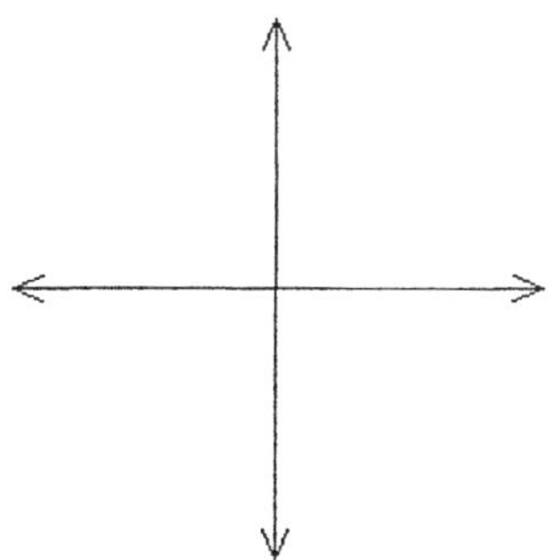

3. Use a calculator to find sec 100.9° .

   Please round the answer to the nearest ten-thousandth.

   ____

4. Use a calculator to find $\sin\left(-235\right)^\circ$.

   Please round the answer to the nearest ten-thousandth.

   Select the correct answer.

   a. 0.8193
   b. 0.7972
   c. 0.7792
   d. 0.8192
   e. 0.7992

5. Use a calculator to find θ to the nearest tenth of a degree, if $0^\circ < \theta < 360^\circ$ and

   $\cot \theta = -0.7357$ with θ in QII

   $\theta = $ ________ $^\circ$

6. Find θ, if $0° < \theta < 360°$ and

$$\tan\theta = -\frac{\sqrt{3}}{3} \quad \text{and } \theta \text{ in QII}$$

Select the correct answer.

    a. $210°$

    b. $150°$

    c. $30°$

    d. $135°$

    e. $315°$

7. Draw the angle in standard position.

$$\theta = 200°$$

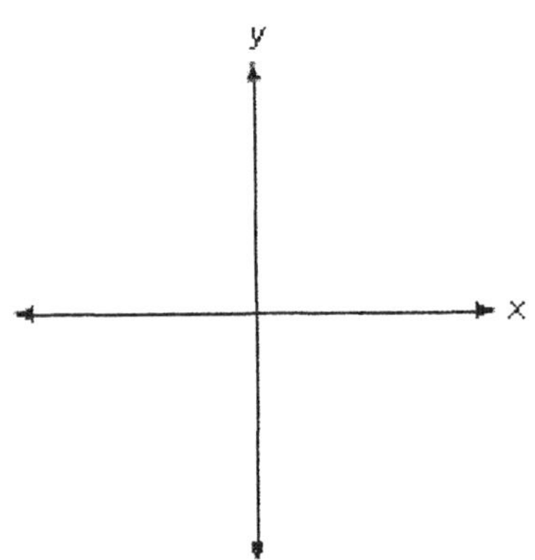

Convert to radian measure using exact values.

Name the reference angle in both degrees and radians.

8. Use a calculator to convert $110°10'$ to radians.

    Round your answer to the nearest hundredth.

    Select the correct answer.

    a.   1.92
    b.   1.98
    c.   2.00
    d.   1.96
    e.   1.94

9. Use a calculator to convert 1.5 to degree measure to the nearest tenth of a degree.

    __________ °

10.
    Give the exact value of $\sec \dfrac{3\pi}{4}$.

    Select the correct answer.

    a. $-\dfrac{2}{\sqrt{3}}$

    b. -2

    c. $\dfrac{2}{\sqrt{3}}$

    d. $\sqrt{2}$

    e. $-\sqrt{2}$

11. Use the unit circle to find the six trigonometric functions of $330°$.

    Find $\sin 330°$.

    Find $\cos 330°$.

    Find $\tan 330°$.

    Find $\cot 330°$.

    Find $\sec 330°$.

    Find $\csc 330°$.

12. Use the unit circle to find the six trigonometric functions of $\dfrac{7\pi}{4}$.

Select the correct answer.

a. $\sin \dfrac{7\pi}{4} = -\dfrac{\sqrt{2}}{2}$, $\cos \dfrac{7\pi}{4} = \dfrac{\sqrt{2}}{2}$, $\tan \dfrac{7\pi}{4} = -1$

   $\cot \dfrac{7\pi}{4} = -1$, $\sec \dfrac{7\pi}{4} = \sqrt{2}$, $\csc \dfrac{7\pi}{4} = -\sqrt{2}$

b. $\sin \dfrac{7\pi}{4} = \dfrac{\sqrt{2}}{2}$, $\cos \dfrac{7\pi}{4} = -\dfrac{\sqrt{2}}{2}$, $\tan \dfrac{7\pi}{4} = 1$

   $\cot \dfrac{7\pi}{4} = 1$, $\sec \dfrac{7\pi}{4} = -\sqrt{2}$, $\csc \dfrac{7\pi}{4} = \sqrt{2}$

c. $\sin \dfrac{7\pi}{4} = -1$, $\cos \dfrac{7\pi}{4} = 1$, $\tan \dfrac{7\pi}{4} = -\dfrac{\sqrt{2}}{2}$

   $\cot \dfrac{7\pi}{4} = -\sqrt{2}$, $\sec \dfrac{7\pi}{4} = 1$, $\csc \dfrac{7\pi}{4} = -1$

d. $\sin \dfrac{7\pi}{4} = -1$, $\cos \dfrac{7\pi}{4} = \dfrac{\sqrt{2}}{2}$, $\tan \dfrac{7\pi}{4} = -\sqrt{2}$

   $\cot \dfrac{7\pi}{4} = -\dfrac{\sqrt{2}}{2}$, $\sec \dfrac{7\pi}{4} = \sqrt{2}$, $\csc \dfrac{7\pi}{4} = -1$

e. $\sin \dfrac{7\pi}{4} = -\dfrac{\sqrt{2}}{2}$, $\cos \dfrac{7\pi}{4} = 1$, $\tan \dfrac{7\pi}{4} = -\dfrac{\sqrt{2}}{2}$

   $\cot \dfrac{7\pi}{4} = -\sqrt{2}$, $\sec \dfrac{7\pi}{4} = 1$, $\csc \dfrac{7\pi}{4} = -\sqrt{2}$

13. If angle θ is in standard position and intersects the unit circle at $\left(\dfrac{1}{\sqrt{17}}, -\dfrac{4}{\sqrt{17}} \right)$

find $\sin\theta$, $\cos\theta$ and $\tan\theta$.

Find $\sin\theta$.

Find $\cos\theta$.

Find $\tan\theta$.

14. If we start at the point (1, 0) and travel once around the unit circle, we travel a distance of 2π units and arrive back where we started at the point (1, 0). If we continue around the unit circle a second time, we will repeat all the values of x and y that occurred during our first trip around. Use this discussion to evaluate the expression.

$$\sin\left(2\pi - \dfrac{\pi}{2} \right)$$

Select the correct answer.

a. 1

b. $\dfrac{1}{2}$

c. 0

d. $\dfrac{\sqrt{2}}{2}$

e. -1

15. If we start at the point (1, 0) and travel once around the unit circle, we travel a distance of 2π units and arrive back where we started at the point (1, 0). If we continue around the unit circle a second time, we will repeat all the values of x and y that occurred during our first trip around. Use this discussion to evaluate the expression.

$$\cos\left(\dfrac{7\pi}{3} \right)$$

16. Find the algebraic expression that is equal to $\sin \alpha \sec \alpha \cot \alpha$.

Select the correct answer.

a.   1

b. $\cos \alpha$

c. $\sec \alpha$

d. $\sin \alpha$

e. $\csc \alpha$

17. The pendulum on a grandfather clock swings from side to side once every second. If the length of the pendulum is 5 feet and the angle through which it swings is $16°$, how far

does the tip of the pendulum travel in 1 second?

If the answer needs rounding, round it to three significant digits.

__________ feet

18. The light truck with manual transmission has a circular brake drum with a diameter of 320 millimeters. Each brake pad, which presses against the drum, is 309 millimeters long. What central angle is subtended by one of the brake pads? Write your answer in both radians and in degrees.

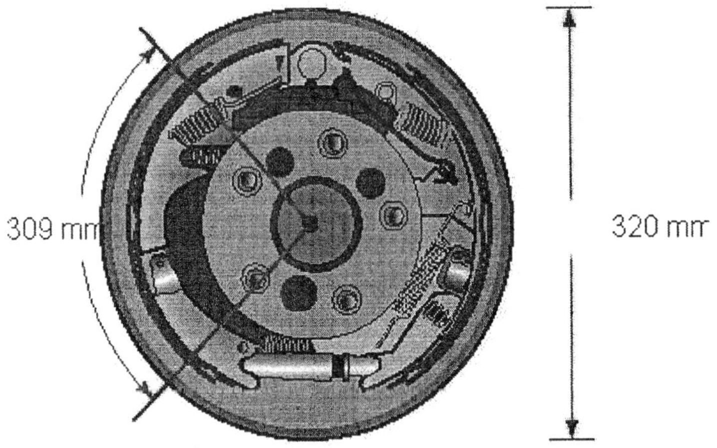

Select the correct answer.

a.   1.63 radians, 108°

b.   1.93 radians, 111°

c.   2.13 radians, 113°

d.   1.83 radians, 110°

e.   1.53 radians, 107°

19. θ is a central angle that cuts off an arc of length s. Find the radius of the circle if

$$\theta = 6°, \ s = \frac{\pi}{2} \ \text{m} .$$

If the answer needs rounding, round it to three significant digits.

$r =$ _________ m

20. Find the area of the sector formed by central angle $\theta = 2.4$ in a circle of radius $r = 6$ inches.

Select the correct answer.

    a.   43.4 inches^2

    b.   42.9 inches^2

    c.   43.2 inches^2

    d.   43.6 inches^2

    e.   43.1 inches^2

21. Find the linear velocity of a point moving with uniform circular motion, if the point covers a distance s in an amount of time t, where

$s = 20$ cm and $t = 4$ sec

If the answer needs rounding, round it to three significant digits.

_________ cm/sec

22. Find the distance s covered by a point moving with linear velocity v for a time t if

$v = 20$ ft/sec and $t = 4$ sec

Select the correct answer.

    a.   60 ft
    b.   55.8 ft
    c.   90 ft
    d.   66.5 ft
    e.   80 ft

23. Point *P* moves with angular velocity w on a circle of radius *r*. Find the distance *s* traveled by the point in time *t*.

$$w = 5 \text{ rad/sec}, r = 3 \text{ ft}, t = 5 \text{ min}$$

If the answer needs rounding, round it to three significant digits.

__________ ft

24. Find the angular velocity associated with the given rpm.

5.1 rpm

Select the correct answer.

a. 28.9 rad/min
b. 41.5 rad/min
c. 19 rad/min
d. 26 rad/min
e. 32.0 rad/min

25. The figure below is a model of the Ferris wheel. The diameter of the wheel is 165 feet, and one complete revolution takes 16 minutes.

Find the linear velocity of a person riding on the wheel.

Give your answer in miles per hour and round to the nearest hundredth.

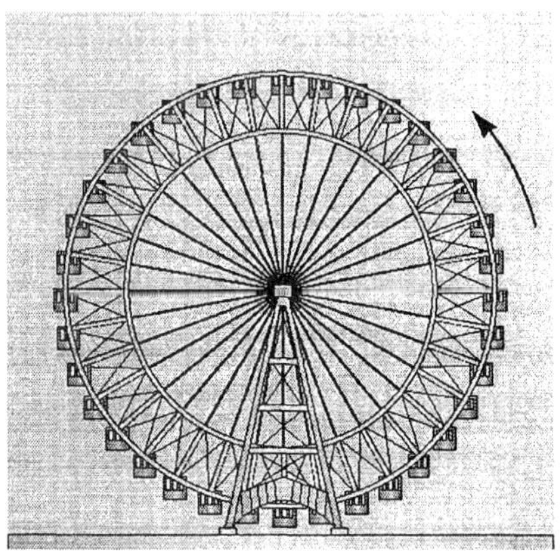

__________ mph

1. a

2.

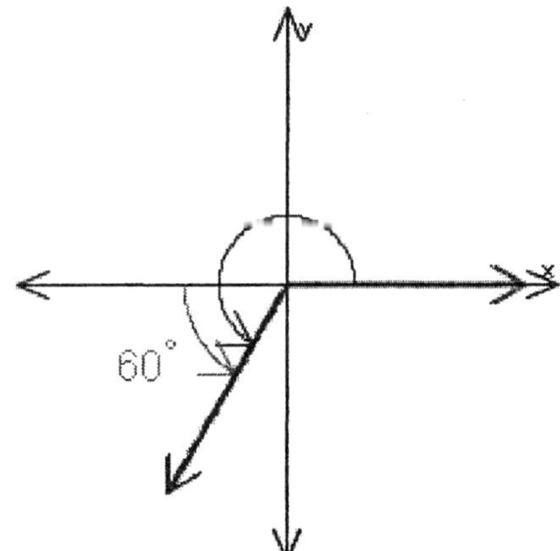

3. −5. 2883

4. d

5. 126. 3

6. b

7.

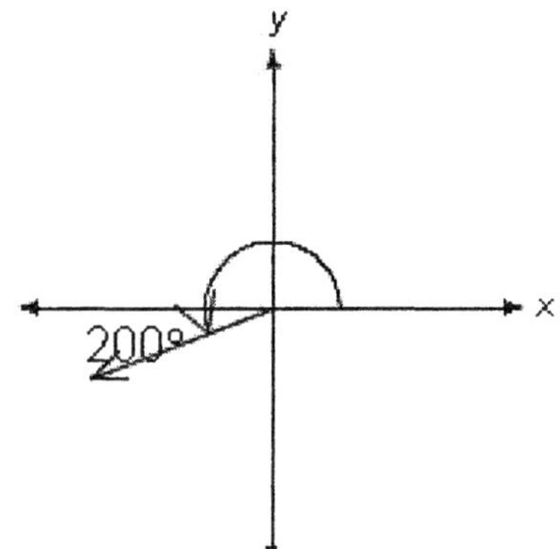

$$\frac{10\pi}{9}$$

$$\frac{\pi}{9}, 20°$$

8. a

9. 85. 9

10. e

11. $-\dfrac{1}{2}$,

$\dfrac{\sqrt{3}}{2}$,

$-\dfrac{\sqrt{3}}{3}$,

$-\sqrt{3}$,

$\dfrac{2\sqrt{3}}{3}$,

-2

12. a

13. $-\dfrac{4}{\sqrt{17}}$,

$\dfrac{1}{\sqrt{17}}$,

-4

14. e

15. $\dfrac{1}{2}$

16. a

17. 1.40

18. b

19. 15

20. c

21. 5

22. e

23. 4.500

24. e

25. 0.368

1. mctr.03.01.02_NoAlgs
2. mctr.03.01.06m_NoAlgs
3. mctr.03.01.31_NoAlgs
4. mctr.03.01.42m_NoAlgs
5. mctr.03.01.63_NoAlgs
6. mctr.03.01.73m_NoAlgs
7. mctr.03.02.16_NoAlgs
8. mctr.03.02.23m_NoAlgs
9. mctr.03.02.44_NoAlgs
10. mctr.03.02.58m_NoAlgs
11. mctr.03.03.01_NoAlgs
12. mctr.03.03.07m_NoAlgs
13. mctr.03.03.39_NoAlgs
14. mctr.03.03.43m_NoAlgs
15. mctr.03.03.50_NoAlgs
16. mctr.03.03.57m_NoAlgs
17. mctr.03.04.15_NoAlgs
18. mctr.03.04.19m_NoAlgs
19. mctr.03.04.35_NoAlgs
20. mctr.03.04.41m_NoAlgs
21. mctr.03.05.04_NoAlgs
22. mctr.03.05.08m_NoAlgs
23. mctr.03.05.28_NoAlgs
24. mctr.03.05.34m_NoAlgs
25. mctr.03.05.51_NoAlgs

1. Find the area of the sector formed by central angle $\theta = 2.4$

    in a circle of radius $r = 6$ inches.

    Select the correct answer.

    a.   43.4 inches^2

    b.   42.9 inches^2

    c.   43.2 inches^2

    d.   43.6 inches^2

    e.   43.1 inches^2

2. Give the exact value of $\sec \dfrac{3\pi}{4}$.

    Select the correct answer.

    a. $-\dfrac{2}{\sqrt{3}}$

    b. -2

    c. $\dfrac{2}{\sqrt{3}}$

    d. $\sqrt{2}$

    e. $-\sqrt{2}$

3. Find the linear velocity of a point moving with uniform circular motion, if the point covers a distance s in an amount of time t, where

    $s = 20$ cm and $t = 4$ sec

    If the answer needs rounding, round it to three significant digits.

    _________   cm/sec

4. Use the unit circle to find the six trigonometric functions of $\dfrac{7\pi}{4}$.

Select the correct answer.

a. $\sin \dfrac{7\pi}{4} = -\dfrac{\sqrt{2}}{2}$, $\cos \dfrac{7\pi}{4} = \dfrac{\sqrt{2}}{2}$, $\tan \dfrac{7\pi}{4} = -1$

   $\cot \dfrac{7\pi}{4} = -1$, $\sec \dfrac{7\pi}{4} = \sqrt{2}$, $\csc \dfrac{7\pi}{4} = -\sqrt{2}$

b. $\sin \dfrac{7\pi}{4} = \dfrac{\sqrt{2}}{2}$, $\cos \dfrac{7\pi}{4} = -\dfrac{\sqrt{2}}{2}$, $\tan \dfrac{7\pi}{4} = 1$

   $\cot \dfrac{7\pi}{4} = 1$, $\sec \dfrac{7\pi}{4} = -\sqrt{2}$, $\csc \dfrac{7\pi}{4} = \sqrt{2}$

c. $\sin \dfrac{7\pi}{4} = -1$, $\cos \dfrac{7\pi}{4} = 1$, $\tan \dfrac{7\pi}{4} = -\dfrac{\sqrt{2}}{2}$

   $\cot \dfrac{7\pi}{4} = -\sqrt{2}$, $\sec \dfrac{7\pi}{4} = 1$, $\csc \dfrac{7\pi}{4} = -1$

d. $\sin \dfrac{7\pi}{4} = -1$, $\cos \dfrac{7\pi}{4} = \dfrac{\sqrt{2}}{2}$, $\tan \dfrac{7\pi}{4} = -\sqrt{2}$

   $\cot \dfrac{7\pi}{4} = -\dfrac{\sqrt{2}}{2}$, $\sec \dfrac{7\pi}{4} = \sqrt{2}$, $\csc \dfrac{7\pi}{4} = -1$

e. $\sin \dfrac{7\pi}{4} = -\dfrac{\sqrt{2}}{2}$, $\cos \dfrac{7\pi}{4} = 1$, $\tan \dfrac{7\pi}{4} = -\dfrac{\sqrt{2}}{2}$

   $\cot \dfrac{7\pi}{4} = -\sqrt{2}$, $\sec \dfrac{7\pi}{4} = 1$, $\csc \dfrac{7\pi}{4} = -\sqrt{2}$

5. The pendulum on a grandfather clock swings from side to side once every second. If the length of the pendulum is 5 feet and the angle through which it swings is $16°$, how far does the tip of the pendulum travel in 1 second?

   If the answer needs rounding, round it to three significant digits.

   _________ feet

6. Find the angular velocity associated with the given rpm.

    5.1 rpm

    Select the correct answer.

    a.   28.9 rad/min
    b.   41.5 rad/min
    c.   19 rad/min
    d.   26 rad/min
    e.   32.0 rad/min

7. Use the unit circle to find the six trigonometric functions of $330°$.

    Find $\sin 330°$.

    Find $\cos 330°$.

    Find $\tan 330°$.

    Find $\cot 330°$.

    Find $\sec 330°$.

    Find $\csc 330°$.

8. Draw the following angle in standard position and then name the reference angle.

    $240°$

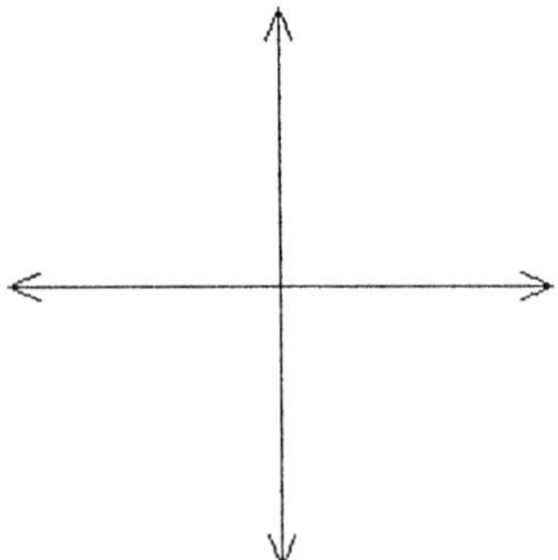

9. θ is a central angle that cuts off an arc of length s. Find the radius of the circle if

$$\theta = 6^\circ, \ s = \frac{\pi}{2} \ m \, .$$

If the answer needs rounding, round it to three significant digits.

$r = $ _________ m

10. Draw the following angle in standard position and then name the reference angle.

92.4°

Select the correct answer.

a.
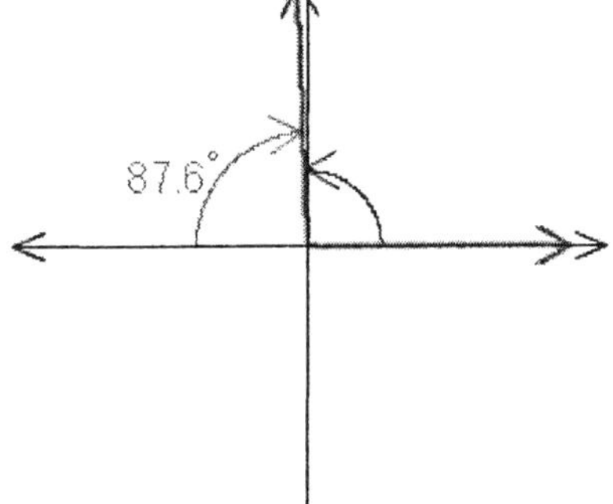

b.
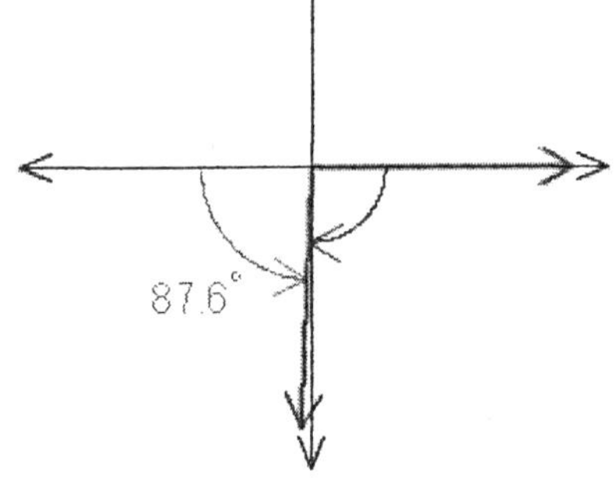

c.
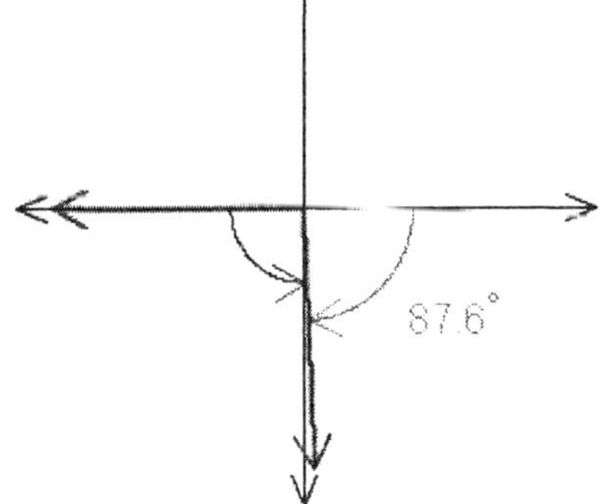

11. Find the algebraic expression that is equal to $\sin \alpha \sec \alpha \cot \alpha$.

Select the correct answer.

    a.   1

    b. $\cos \alpha$

    c. $\sec \alpha$

    d. $\sin \alpha$

    e. $\csc \alpha$

12. Find the distance s covered by a point moving with linear velocity v for a time t if

$v = 20$ ft/sec and $t = 4$ sec

Select the correct answer.

    a.   60 ft
    b.   55.8 ft
    c.   90 ft
    d.   66.5 ft
    e.   80 ft

13. The light truck with manual transmission has a circular brake drum with a diameter of 320 millimeters. Each brake pad, which presses against the drum, is 309 millimeters long. What central angle is subtended by one of the brake pads? Write your answer in both radians and in degrees.

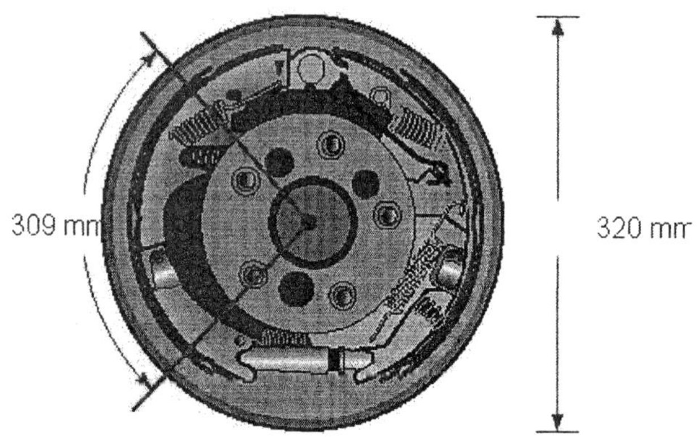

Select the correct answer.

    a.   1.63 radians, 108°

    b.   1.93 radians, 111°

    c.   2.13 radians, 113°

    d.   1.83 radians, 110°

    e.   1.53 radians, 107°

14. Use a calculator to find θ to the nearest tenth of a degree, if $0^\circ < \theta < 360^\circ$ and

$\cot \theta = -0.7357$ with θ in QII

$$\theta = \underline{\hspace{2cm}}^\circ$$

15. If we start at the point (1, 0) and travel once around the unit circle, we travel a distance of 2π units and arrive back where we started at the point (1, 0), If we continue around the unit circle a second time, we will repeat all the values of x and y that occurred during our first trip around. Use this discussion to evaluate the expression.

$$\cos\left(\frac{7\pi}{3}\right)$$

16. Use a calculator to find $\sin\left(-235\right)^\circ$.

Please round the answer to the nearest ten-thousandth.

Select the correct answer.

a.   0.8193
b.   0.7972
c.   0.7792
d.   0.8192
e.   0.7992

17. If we start at the point (1, 0) and travel once around the unit circle, we travel a distance of 2π units and arrive back where we started at the point (1, 0). If we continue around the unit circle a second time, we will repeat all the values of x and y that occurred during our first trip around. Use this discussion to evaluate the expression.

$$\sin\left(2\pi - \frac{\pi}{2}\right)$$

Select the correct answer.

a.   1

b. $\dfrac{1}{2}$

c.   0

d. $\dfrac{\sqrt{2}}{2}$

e. -1

18. Draw the angle in standard position.

$\theta = 200°$

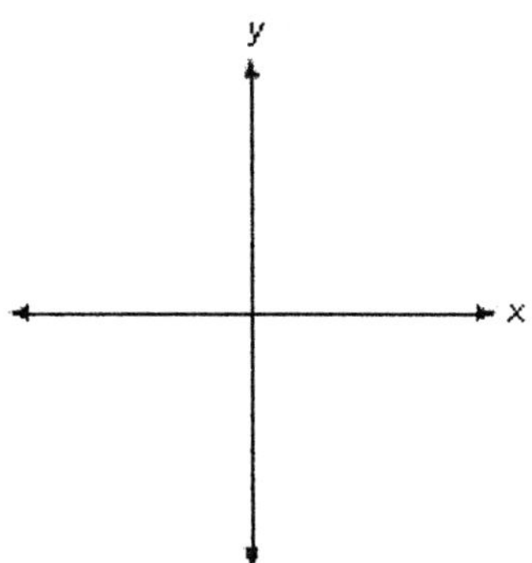

Convert to radian measure using exact values.

Name the reference angle in both degrees and radians.

19. Point P moves with angular velocity w on a circle of radius r.

Find the distance s traveled by the point in time t.

$w = 5 \ \text{rad/sec}, \ r = 3 \ \text{ft}, \ t = 5 \ \text{min}$

If the answer needs rounding, round it to three significant digits.

__________ ft

20.

If angle θ is in standard position and intersects the unit circle at $\left(\dfrac{1}{\sqrt{17}}, \ -\dfrac{4}{\sqrt{17}} \right)$ find

$\sin\theta, \ \cos\theta$ and $\tan\theta$.

Find $\sin\theta$.

Find $\cos\theta$.

Find $\tan\theta$.

21. Use a calculator to convert $110°10'$ to radians. Round your answer to the nearest hundredth.

Select the correct answer.

a. 1.92
b. 1.98
c. 2.00
d. 1.96
e. 1.94

22. Use a calculator to find sec $100.9°$.

Please round the answer to the nearest ten-thousandth.

____

23. Use a calculator to convert 1.5 to degree measure to the nearest tenth of a degree.

___________ $°$

24.

Find θ , if $0° < \theta < 360°$ and $\tan \theta = -\dfrac{\sqrt{3}}{3}$ and θ in QII

Select the correct answer.

a. $210°$

b. $150°$

c. $30°$

d. $135°$

e. $315°$

25. The figure below is a model of the Ferris wheel. The diameter of the wheel is 165 feet, and one complete revolution takes 16 minutes. Find the linear velocity of a person riding on the wheel. Give your answer in miles per hour and round to the nearest hundredth.

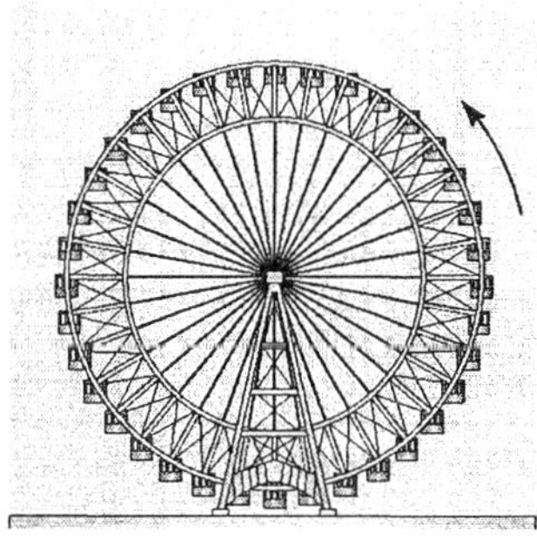

___________ mph

1. c

2. e

3. 5

4. a

5. 1 . 4 0

6. e

7. $-\dfrac{1}{2}$,

$\dfrac{\sqrt{3}}{2}$,

$-\dfrac{\sqrt{3}}{3}$,

$-\sqrt{3}$,

$\dfrac{2\sqrt{3}}{3}$,

-2

8.

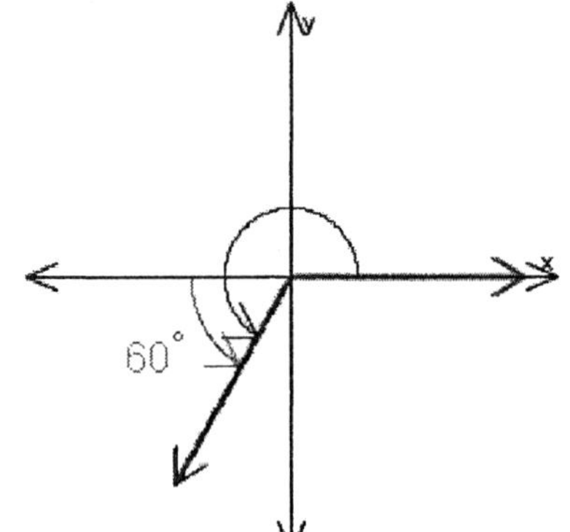

9. 15

10. a

11. a

12. e

13. b

14. 126.3

15. $\dfrac{1}{2}$

16. d

17. e

18.

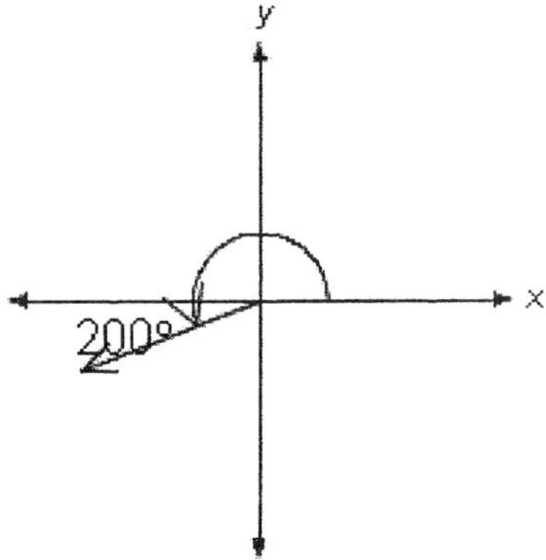

$\dfrac{10\pi}{9}$

$\dfrac{\pi}{9}, 20°$

19. $4,500$

20. $-\dfrac{4}{\sqrt{17}}, \ \dfrac{1}{\sqrt{17}}, \ -4$

21. a

22. -5.2883

23. 85.9

24. b

25. 0.368

1. mctr.03.04.41m_NoAlgs
2. mctr.03.02.58m_NoAlgs
3. mctr.03.05.04_NoAlgs
4. mctr.03.03.07m_NoAlgs
5. mctr.03.04.15_NoAlgs
6. mctr.03.05.34m_NoAlgs
7. mctr.03.03.01_NoAlgs
8. mctr.03.01.02_NoAlgs
9. mctr.03.04.35_NoAlgs
10. mctr.03.01.06m_NoAlgs
11. mctr.03.03.57m_NoAlgs
12. mctr.03.05.08m_NoAlgs
13. mctr.03.04.19m_NoAlgs
14. mctr.03.01.63_NoAlgs
15. mctr.03.03.50_NoAlgs
16. mctr.03.01.42m_NoAlgs
17. mctr.03.03.43m_NoAlgs
18. mctr.03.02.16_NoAlgs
19. mctr.03.05.28_NoAlgs
20. mctr.03.03.39_NoAlgs
21. mctr.03.02.23m_NoAlgs
22. mctr.03.01.31_NoAlgs
23. mctr.03.02.44_NoAlgs
24. mctr.03.01.73m_NoAlgs
25. mctr.03.05.51_NoAlgs

1. Find the exact value of $\cos 210^\circ$.

2. Find the exact value of $\cot 690^\circ$.

    Select the correct answer.

    a. $\dfrac{\sqrt{3}}{3}$

    b. $\overset{00}{\sqrt{3}}$

    c. $-\sqrt{3}$

    d. $\dfrac{\sqrt{3}}{2}$

    e. $-\dfrac{\sqrt{3}}{2}$

3. Use a calculator to find θ to the nearest tenth of a degree, if $0^\circ < \theta < 360^\circ$ and $\sin \theta = -0.3070$ with θ in QIII

    Select the correct answer.

    a. 198.9°

    b. 197.9°

    c. 200.9°

    d. 194.4°

    e. 200.2°

4. Use a calculator to find θ to the nearest tenth of a degree, if $0^\circ < \theta < 360^\circ$ and $\sin \theta = 0.9636$ with θ in QII

    Select the correct answer.

    a. 106.5°

    b. 108.5°

    c. 103.2°

    d. 109.0°

    e. 105.5°

5. Find the radian measure of angle θ, if θ is a central angle in a circle of radius r, and θ cuts off an arc of length s.

$r = 3$ inches, $s = 6$ inches

$\theta = $ _________ radians

6. Two cities are approximately 350 miles apart on the surface of the earth. Assuming that the radius of the earth is 4,000 miles, find the radian measure of the central angle with its vertex at the center of the earth that has one city on one side and another one on the other side.

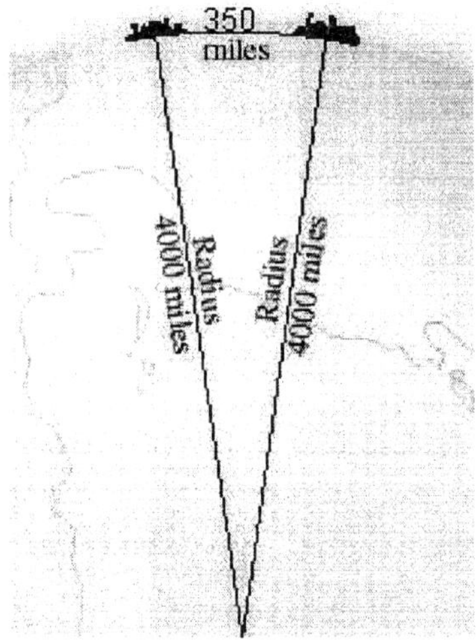

Select the correct answer.

a. 0.0875 radians
b. 0.1075 radians
c. 0.0925 radians
d. 0.0725 radians
e. 0.0775 radians

7. Convert to degree measure. $\theta = \dfrac{8\pi}{9}$

$\theta = $ _________ °

Draw the angle in standard position.

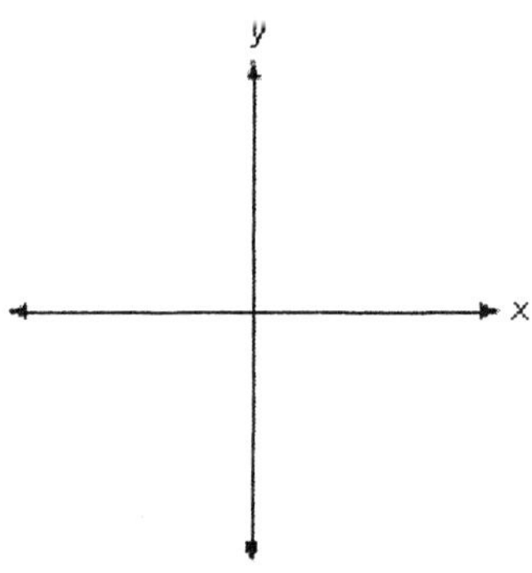

Label the reference angle in both degrees and radians.

8. Label the reference angle in both degrees and radians.

$\theta = \dfrac{9\pi}{4}$

Select the correct answer.

a. $30° = \dfrac{\pi}{6}$

b. $45° = \dfrac{\pi}{2}$

c. $90° = \dfrac{\pi}{4}$

d. $90° = \dfrac{\pi}{2}$

e. $45° = \dfrac{\pi}{4}$

9. Evaluate the expression when x is $\dfrac{\pi}{10}$.

$6 \cos 5x$

Select the correct answer.

    a.   - 6
    b.   - 2
    c.   2
    d.   0
    e.   6

10. For the following expression, find the value of y that corresponds to each value of x.

$$y = \sin 2x \quad \text{for } x = 0,\ \frac{\pi}{4},\ \frac{\pi}{2},\ \frac{3\pi}{4},\ \pi$$

Select the correct answer.

a. $(0,0),\ \left(\dfrac{\pi}{4},9\right),\ \left(\dfrac{\pi}{2},0\right),\ \left(\dfrac{3\pi}{4},-9\right),\ (\pi,0)$

b. $(0,0),\ \left(\dfrac{\pi}{4},1\right),\ \left(\dfrac{\pi}{2},0\right),\ \left(\dfrac{3\pi}{4},1\right),\ (\pi,0)$

c. $(0,0),\ \left(\dfrac{\pi}{4},1\right),\ \left(\dfrac{\pi}{2},0\right),\ \left(\dfrac{3\pi}{4},-1\right),\ (\pi,0)$

d. $(0,0),\ \left(\dfrac{\pi}{4},\dfrac{9}{\sqrt{2}}\right),\ \left(\dfrac{\pi}{2},9\right),\ \left(\dfrac{3\pi}{4},\dfrac{9}{\sqrt{2}}\right),\ (\pi,0)$

e. $(0,0),\ \left(\dfrac{\pi}{4},\dfrac{1}{\sqrt{2}}\right),\ \left(\dfrac{\pi}{2},1\right),\ \left(\dfrac{3\pi}{4},\dfrac{1}{\sqrt{2}}\right),\ (\pi,0)$

11. Use the unit circle and the fact that sine is an odd function to find $\sin\left(\dfrac{3\pi}{4}\right)$.

0

12. Use the unit circle to find all values of θ between 0 and 2π for which $\tan \theta = \dfrac{\sqrt{3}}{3}$.

Select the correct answer.

a. $\dfrac{5\pi}{6}, \dfrac{11\pi}{3}$

b. $\dfrac{2\pi}{3}, \dfrac{5\pi}{3}$

c. $\dfrac{2\pi}{3}, \dfrac{\pi}{3}$

d. $\dfrac{\pi}{6}, \dfrac{5\pi}{6}$

e. $\dfrac{\pi}{6}, \dfrac{7\pi}{6}$

13. If we start at the point (1, 0) and travel once around the unit circle, we travel a distance of 2π units and arrive back where we started at the point (1, 0). If we continue

    around the unit circle a second time, we will repeat all the values of x and y that occurred during our first trip around. Use this discussion to evaluate the expression.

$$\cos\left(\frac{7\pi}{3}\right)$$

14. Prove the identity.

$$\cos(-\theta)\,\csc(-\theta)\,\tan(-\theta) = 1$$

15. For the problem below, θ is a central angle in a circle of radius r. Find the length of arc s cut off by θ.

$\theta = 4$, $r = 3$ inches.

If the answer needs rounding, round it to three significant digits.

$s =$ _________ inches

16. For the problem below, θ is a central angle in a circle of radius r. Find the length of arc s cut off by θ.

$\theta = 330°$, $r = 5$ inches.

Select the correct answer.

a. $s = 33$ inches
b. $s = 29.8$ inches
c. $s = 28.8$ inches
d. $s = 31$ inches
e. $s = 26.8$ inches

17. The figure is a model of George Ferris's Ferris wheel. The diameter of the wheel is 250 feet; and θ is the central angle formed as a rider travels from his or her initial position P_0 to position P_1.

Find the distance traveled by the rider if $\theta = 212°$.

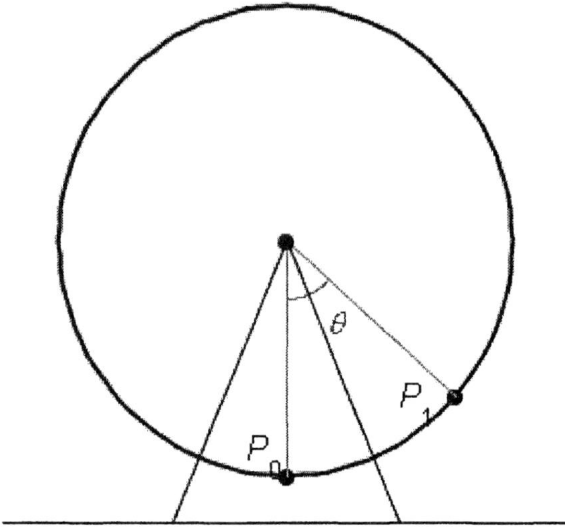

Select the correct answer.

a. 463.5 ft
b. 459.5 ft
c. 462.5 ft
d. 466.5 ft
e. 464.5 ft

18. θ is a central angle that cuts off an arc of length s. Find the radius of the circle if

$\theta = 10$, $s = 5$ ft.

Select the correct answer.

a.   0.5 ft
b.   0.47 ft
c.   0.48 ft
d.   0.46 ft
e.   0.49 ft

19. Find the area of the sector formed by central angle $\theta = 15°$

in a circle of radius $r = 5$ m.

If the answer needs rounding, round it to three significant digits.

$A = \underline{\hspace{2cm}}$ m^2

20.

A sector of area $\dfrac{2\pi}{3}$ in^2 is formed by a central angle of $45°$.

What is the radius of the circle?

Select the correct answer.

a.   2.30 inches
b.   2.29 inches
c.   2.27 inches
d.   2.31 inches
e.   2.34 inches

21. Point P sweeps out central angle θ as it rotates on a circle of radius r.

Find the angular velocity of point P.

$$\theta = \dfrac{2\pi}{5}, \; t = 5 \text{ sec}$$

If the answer needs rounding, round it to three significant digits.

$\underline{\hspace{2cm}}$ rad/sec

22. Point P sweeps out central angle θ as it rotates on a circle of radius r.

Find the angular velocity of point P.

$\theta = 26\pi, \ t = 0.8 \ hr$

Select the correct answer.

   a.   197 rad/hr
   b.   187 rad/hr
   c.   156 rad/hr
   d.   102 rad/hr
   e.   167 rad/hr

23. A point is rotating with uniform circular motion on a circle of radius r. Find v if $r = 6$ ft and the point rotates at 8 rpm.

Select the correct answer.

   a.   232 ft/min
   b.   286 ft/min
   c.   267 ft/min
   d.   214 ft/min
   e.   302 ft/min

24. The San Francisco cable cars travel by clamping onto a steel cable that circulates in a channel beneath the streets. This cable is driven by a large 13-foot-diameter pulley, called a *sheave* (see the figure). The sheave turns at a rate of 18 revolutions per minute.

Find the speed of the cable car, in miles per hour, by determining the linear velocity of the cable. (1 mi = 5,280 ft)

If the answer needs rounding, round it to three significant digits.

__________   mph

25. Lance Armstrong, four-time winner of the Tour de France, rides a Trek 5,900 bicycle equipped with Dura-Ace components. (see the figure). When Lance pedals, he turns a gear, called a chainring. The angular velocity of the chainring will determine the linear speed at which the chain travels. The chain connects the chainring to a smaller gear, called a sprocket, which is attached to the rear wheel (see the scheme).

The angular velocity of the sprocket depends upon the linear speed of the chain. The sprocket and rear wheel rotate at the same rate, and the diameter of the rear wheel is 700 millimeters. The speed at which Lance travels is determined by the angular velocity of his rear wheel.

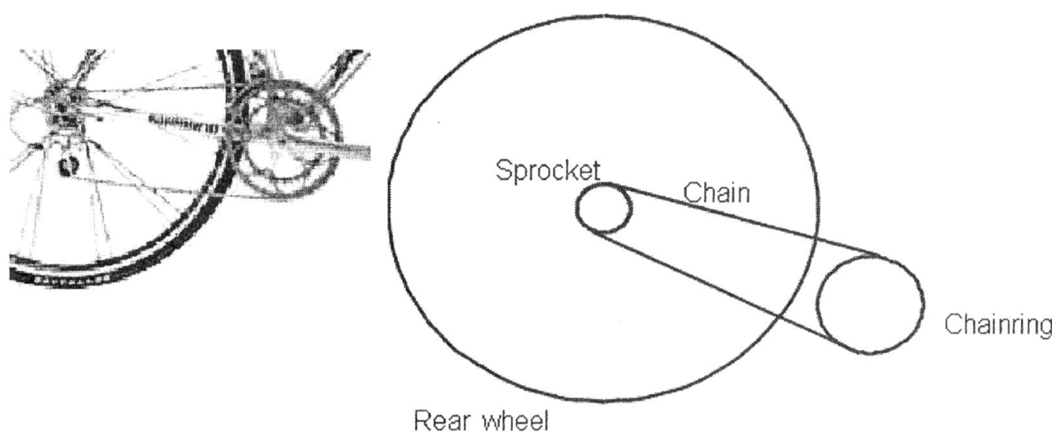

If Lance was using his 160-millimeter-diameter chainring and pedaling at a rate of 85 revolutions per minute, what diameter sprocket would he need in order to maintain a speed of 45 kilometers per hour?

If the answer needs rounding round it to three significant digits.

__________ mm

1. $-\dfrac{\sqrt{3}}{2}$

2. c

3. b

4. e

5. 2

6. a

7. 160

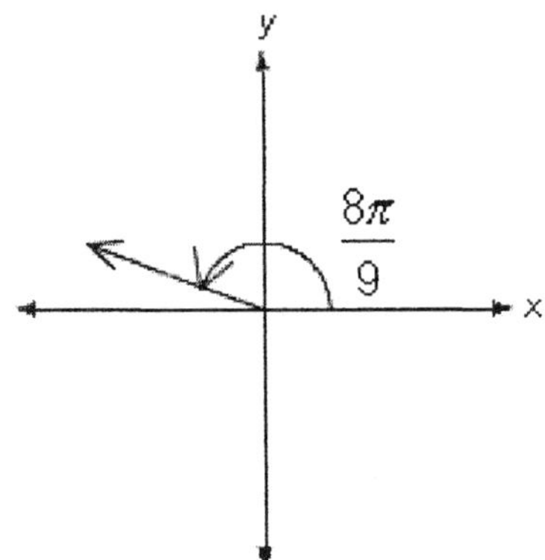

$\dfrac{\pi}{9}$, 20°

8. e

9. d

10. c

11. $\dfrac{\sqrt{2}}{2}$

12. e

13. $\dfrac{1}{2}$

14.

$$\cos(-\theta)\csc(-\theta)\tan(-\theta) = \cos(-\theta)\ \frac{1}{\sin(-\theta)}\ \frac{\sin(-\theta)}{\cos(-\theta)} = 1$$

15. 12

16. c

17. c

18. a

19. 3. 27

20. d

21. 0. 251

22. d

23. e

24. 8. 35

25. 39. 9

1. mctr.03.01.16_NoAlgs
2. mctr.03.01.28m_NoAlgs
3. mctr.03.01.49m_NoAlgs
4. mctr.03.01.57m_NoAlgs
5. mctr.03.02.06_NoAlgs
6. mctr.03.02.09m_NoAlgs
7. mctr.03.02.33_NoAlgs
8. mctr.03.02.38m_NoAlgs
9. mctr.03.02.67m_NoAlgs
10. mctr.03.02.73m_NoAlgs
11. mctr.03.03.16_NoAlgs
12. mctr.03.03.21m_NoAlgs
13. mctr.03.03.50_NoAlgs
14. mctr.03.03.57_NoAlgs
15. mctr.03.04.01_NoAlgs
16. mctr.03.04.10m_NoAlgs
17. mctr.03.04.25m_NoAlgs
18. mctr.03.04.29m_NoAlgs
19. mctr.03.04.45_NoAlgs
20. mctr.03.04.51m_NoAlgs
21. mctr.03.05.13_NoAlgs
22. mctr.03.05.19m_NoAlgs
23. mctr.03.05.39m_NoAlgs
24. mctr.03.05.45_NoAlgs
25. mctr.03.05.57_NoAlgs

1. Evaluate the expression when x is $\dfrac{\pi}{10}$.

$6 \cos 5x$

Select the correct answer.

a. -6
b. -2
c. 2
d. 0
e. 6

2. Use the unit circle to find all values of θ between 0 and 2π for which $\tan \theta = \dfrac{\sqrt{3}}{3}$.

Select the correct answer.

a. $\dfrac{5\pi}{6}, \dfrac{11\pi}{3}$

b. $\dfrac{2\pi}{3}, \dfrac{5\pi}{3}$

c. $\dfrac{2\pi}{3}, \dfrac{\pi}{3}$

d. $\dfrac{\pi}{6}, \dfrac{5\pi}{6}$

e. $\dfrac{\pi}{6}, \dfrac{7\pi}{6}$

3. If we start at the point (1, 0) and travel once around the unit circle, we travel a distance of 2π units and arrive back where we started at the point (1, 0). If we continue

around the unit circle a second time, we will repeat all the values of x and y that occurred during our first trip around. Use this discussion to evaluate the expression.

$\cos\left(\dfrac{7\pi}{3}\right)$

4. Point P sweeps out central angle θ as it rotates on a circle of radius r.

Find the angular velocity of point P.

$\theta = \dfrac{2\pi}{5}, t = 5 \text{ sec}$

If the answer needs rounding, round it to three significant digits.

__________ rad/sec

5. Use a calculator to find θ to the nearest tenth of a degree, if $0° < \theta < 360°$ and

    $\sin \theta = 0.9636$ with θ in QII

    Select the correct answer.

    a. $106.5°$

    b. $108.5°$

    c. $103.2°$

    d. $109.0°$

    e. $105.5°$

6. Find the radian measure of angle θ, if θ is a central angle in a circle of radius r, and θ cuts off an arc of length s.

    $r = 3$ inches, $s = 6$ inches

    $\theta = $ _________ radians

7. Find the area of the sector formed by central angle $\theta = 15°$
    in a circle of radius $r = 5$ m.

    If the answer needs rounding, round it to three significant digits.

    $A = $ _________ m^2

8. For the problem below, θ is a central angle in a circle of radius r.
    Find the length of arc s cut off by θ.

    $\theta = 330°,\ r = 5$ inches.

    Select the correct answer.

    a. $s = 33$ inches
    b. $s = 29.8$ inches
    c. $s = 28.8$ inches
    d. $s = 31$ inches
    e. $s = 26.8$ inches

9. Lance Armstrong, four-time winner of the Tour de France, rides a Trek 5,900 bicycle equipped with Dura-Ace components. (see the figure). When Lance pedals, he turns a gear, called a chainring. The angular velocity of the chainring will determine the linear speed at which the chain travels. The chain connects the chainring to a smaller gear, called a sprocket, which is attached to the rear wheel (see the scheme). The angular velocity of the sprocket depends upon the linear speed of the chain. The sprocket and rear wheel rotate at the same rate, and the diameter of the rear wheel is 700 millimeters. The speed at which Lance travels is determined by the angular velocity of his rear wheel.

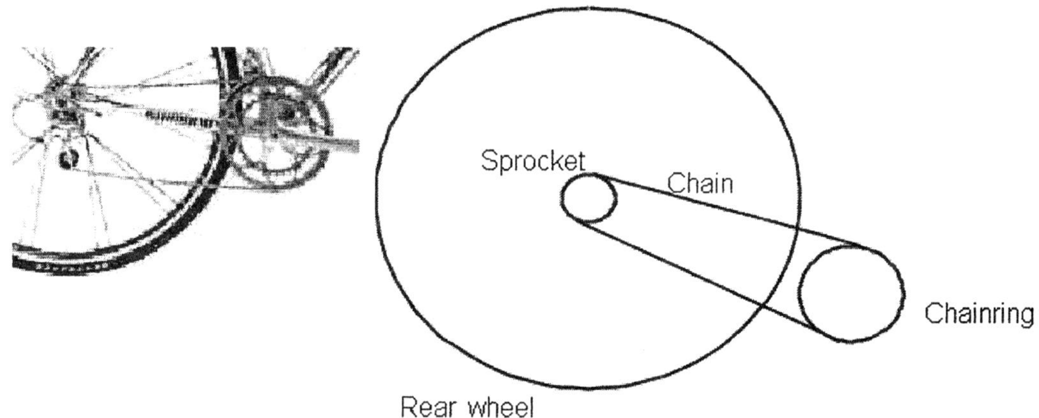

If Lance was using his 160-millimeter-diameter chainring and pedaling at a rate of 85 revolutions per minute, what diameter sprocket would he need in order to maintain a speed of 45 kilometers per hour?

If the answer needs rounding round it to three significant digits.

__________ mm

10. Label the reference angle in both degrees and radians.

$$\theta = \frac{9\pi}{4}$$

Select the correct answer.

a. $\quad 30° = \dfrac{\pi}{6}$

b. $\quad 45° = \dfrac{\pi}{2}$

c. $\quad 90° = \dfrac{\pi}{4}$

d. $\quad 90° = \dfrac{\pi}{2}$

e. $\quad 45° = \dfrac{\pi}{4}$

11. Find the exact value of cot 690° .

Select the correct answer.

a. $\dfrac{\sqrt{3}}{3}$

b. $\sqrt{3}$

c. $-\sqrt{3}$

d. $\dfrac{\sqrt{3}}{2}$

e. $-\dfrac{\sqrt{3}}{2}$

12. Use a calculator to find θ to the nearest tenth of a degree, if $0^\circ < \theta < 360^\circ$ and

$\sin \theta = -0.3070$ with θ in QIII

Select the correct answer.

a. 198.9°

b. 197.9°

c. 200.9°

d. 194.4°

e. 200.2°

13. A sector of area $\dfrac{2\pi}{3}$ in^2 is formed by a central angle of 45°.

What is the radius of the circle?

Select the correct answer.

a. 2.30 inches
b. 2.29 inches
c. 2.27 inches
d. 2.31 inches
e. 2.34 inches

14. A point is rotating with uniform circular motion on a circle of radius r. Find v if $r = 6$ ft and the point rotates at 8 rpm.

Select the correct answer.

a.   232 ft/min
b.   286 ft/min
c.   267 ft/min
d.   214 ft/min
e.   302 ft/min

15. Find the exact value of cos 210° .

16. The San Francisco cable cars travel by clamping onto a steel cable that circulates in a channel beneath the streets. This cable is driven by a large 13-foot-diameter pulley, called a *sheave* (see the figure). The sheave turns at a rate of 18 revolutions per minute.

Find the speed of the cable car, in miles per hour, by determining the linear velocity of the cable. (1 mi = 5,280 ft)

If the answer needs rounding, round it to three significant digits.

__________ mph

17.
Use the unit circle and the fact that sine is an odd function to find $\sin\left(\dfrac{3\pi}{4}\right)$.

18. For the following expression, find the value of *y* that corresponds to each value of *x*.

$$y = \sin 2x \quad \text{for } x = 0,\ \frac{\pi}{4},\ \frac{\pi}{2},\ \frac{3\pi}{4},\ \pi$$

Select the correct answer.

a. $(0,0),\ \left(\dfrac{\pi}{4},9\right),\ \left(\dfrac{\pi}{2},0\right),\ \left(\dfrac{3\pi}{4},-9\right),\ (\pi,0)$

b. $(0,0),\ \left(\dfrac{\pi}{4},1\right),\ \left(\dfrac{\pi}{2},0\right),\ \left(\dfrac{3\pi}{4},1\right),\ (\pi,0)$

c. $(0,0),\ \left(\dfrac{\pi}{4},1\right),\ \left(\dfrac{\pi}{2},0\right),\ \left(\dfrac{3\pi}{4},-1\right),\ (\pi,0)$

d. $(0,0),\ \left(\dfrac{\pi}{4},\dfrac{9}{\sqrt{2}}\right),\ \left(\dfrac{\pi}{2},9\right),\ \left(\dfrac{3\pi}{4},\dfrac{9}{\sqrt{2}}\right),\ (\pi,0)$

e. $(0,0),\ \left(\dfrac{\pi}{4},\dfrac{1}{\sqrt{2}}\right),\ \left(\dfrac{\pi}{2},1\right),\ \left(\dfrac{3\pi}{4},\dfrac{1}{\sqrt{2}}\right),\ (\pi,0)$

19. Convert to degree measure.

$$\theta = \frac{8\pi}{9}$$

$\theta = \underline{\hspace{2cm}}°$

Draw the angle in standard position.

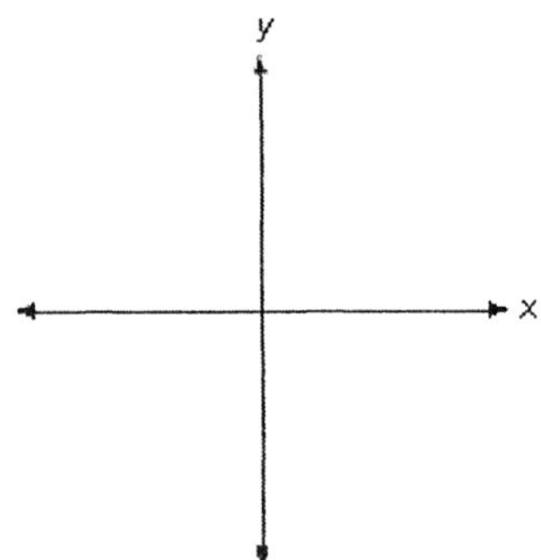

Label the reference angle in both degrees and radians.

20. Two cities are approximately 350 miles apart on the surface of the earth. Assuming that the radius of the earth is 4,000 miles, find the radian measure of the central angle with its vertex at the center of the earth that has one city on one side and another one on the other side.

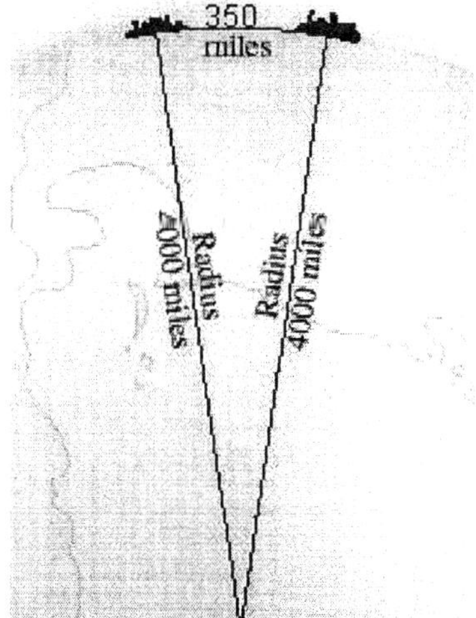

Select the correct answer.

a.   0.0875 radians
b.   0.1075 radians
c.   0.0925 radians
d.   0.0725 radians
e.   0.0775 radians

21. θ is a central angle that cuts off an arc of length s. Find the radius of the circle if

$\theta = 10$, $s = 5$ ft.

Select the correct answer.

a.   0.5 ft
b.   0.47 ft
c.   0.48 ft
d.   0.46 ft
e.   0.49 ft

22. Prove the identity.

$$\cos(-\theta)\csc(-\theta)\tan(-\theta) = 1$$

23. The figure is a model of George Ferris's Ferris wheel. The diameter of the wheel is 250 feet; and θ is the central angle formed as a rider travels from his or her initial position P_0 to position P_1. Find the distance traveled by the rider if $\theta = 212°$.

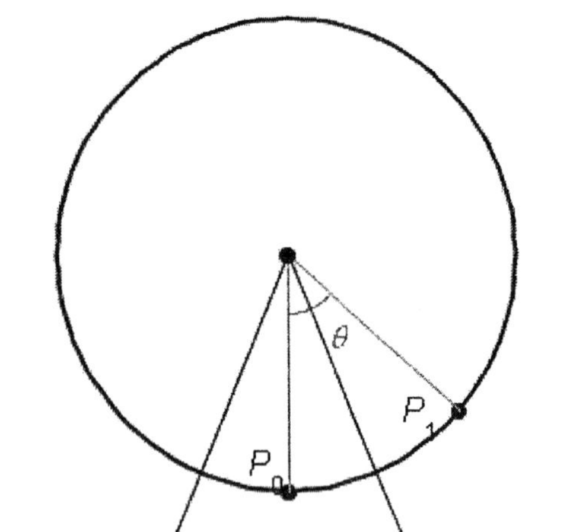

Select the correct answer.

    a.   463.5 ft
    b.   459.5 ft
    c.   462.5 ft
    d.   466.5 ft
    e.   464.5 ft

24. For the problem below, θ is a central angle in a circle of radius r.

Find the length of arc s cut off by θ.

$\theta = 4$, $r = 3$ inches.

If the answer needs rounding, round it to three significant digits.

$s =$ ________ inches

25. Point P sweeps out central angle θ as it rotates on a circle of radius r.

Find the angular velocity of point P.

$\theta = 26\pi$, $t = 0.8$ hr

Select the correct answer.

    a.   197 rad/hr
    b.   187 rad/hr
    c.   156 rad/hr
    d.   102 rad/hr
    e.   167 rad/hr

1. d

2. e

3. $\dfrac{1}{2}$

4. 0.251

5. e

6. 2

7. 3.27

8. c

9. 39.9

10. e

11. c

12. b

13. d

14. e

15. $-\dfrac{\sqrt{3}}{2}$

16. 8.35

17. $\dfrac{\sqrt{2}}{2}$

18. c

19. 160

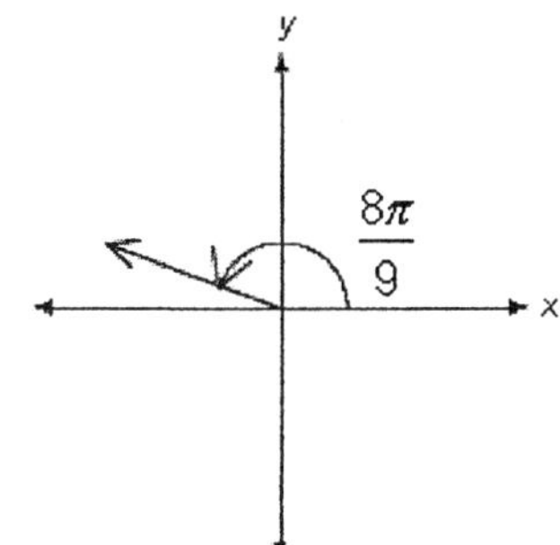

$\dfrac{\pi}{9}$. 20°

20. a

21. a

22.

$$\cos(-\theta)\,\csc(-\theta)\,\tan(-\theta) = \cos(-\theta)\,\dfrac{1}{\sin(-\theta)}\,\dfrac{\sin(-\theta)}{\cos(-\theta)} = 1$$

23. c

24. 1 2

25. d

McKeague/Turner - Trigonometry 5e Chapter 3 Form H

1. mctr.03.02.67m_NoAlgs
2. mctr.03.03.21m_NoAlgs
3. mctr.03.03.50_NoAlgs
4. mctr.03.05.13_NoAlgs
5. mctr.03.01.57m_NoAlgs
6. mctr.03.02.06_NoAlgs
7. mctr.03.04.45_NoAlgs
8. mctr.03.04.10m_NoAlgs
9. mctr.03.05.57_NoAlgs
10. mctr.03.02.38m_NoAlgs
11. mctr.03.01.28m_NoAlgs
12. mctr.03.01.49m_NoAlgs
13. mctr.03.04.51m_NoAlgs
14. mctr.03.05.39m_NoAlgs
15. mctr.03.01.16_NoAlgs
16. mctr.03.05.45_NoAlgs
17. mctr.03.03.16_NoAlgs
18. mctr.03.02.73m_NoAlgs
19. mctr.03.02.33_NoAlgs
20. mctr.03.02.09m_NoAlgs
21. mctr.03.04.29m_NoAlgs
22. mctr.03.03.57_NoAlgs
23. mctr.03.04.25m_NoAlgs
24. mctr.03.04.01_NoAlgs
25. mctr.03.05.19m_NoAlgs

1. Make a table of values for $y = \sin x$ using multiples of $\dfrac{\pi}{4}$ for x.

x	$\sin x$
0	
$\dfrac{\pi}{4}$	
$\dfrac{\pi}{2}$	
$\dfrac{3\pi}{4}$	
π	
$\dfrac{5\pi}{4}$	
$\dfrac{3\pi}{2}$	
$\dfrac{7\pi}{4}$	
2π	

Use the entries in the table to sketch the graph of the function for x between 0 and 2π

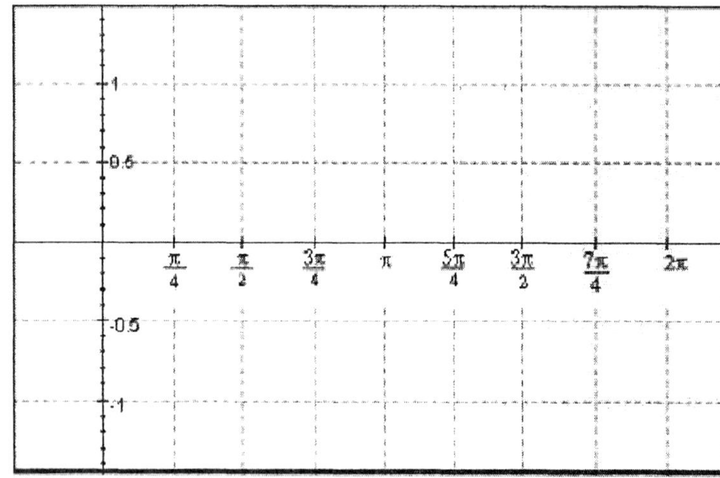

2. Match each equation in the left column with all the corresponding values of x in the right column which make the equation true.

$\sec x = 1$

$\dfrac{\pi}{2} + k\pi$

$\cot x = 0$

$2k\pi$

3. Give the amplitude and period of the graph.

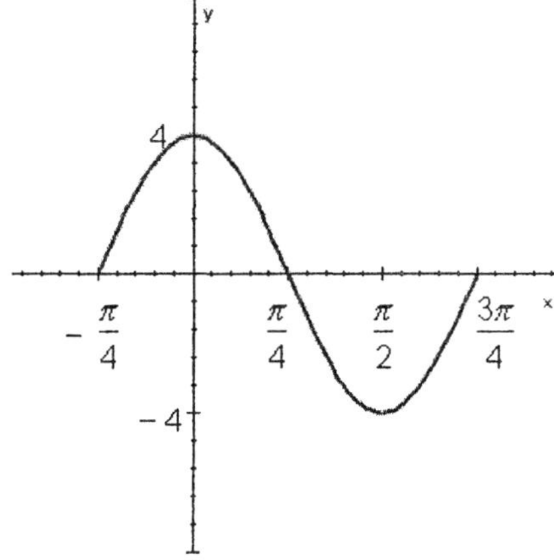

amplitude.

period.

4. Graph one complete cycle of $y = \dfrac{3}{10} \cos x$. Label the axes accurately.

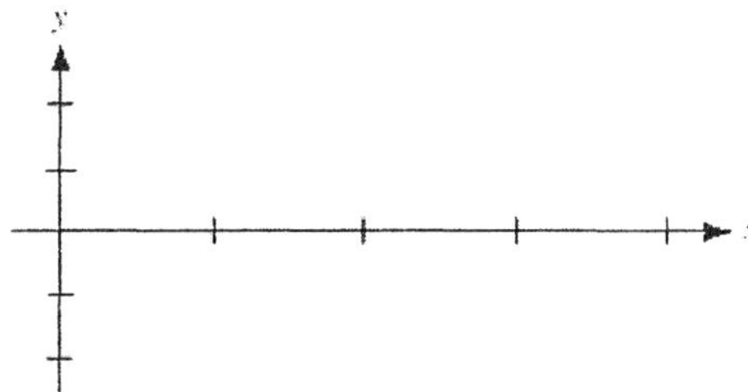

Identify the amplitude for the graph (if defined).

5. Graph one complete cycle of the graph. Label the axes accurately.

$$y = \cos 5x$$

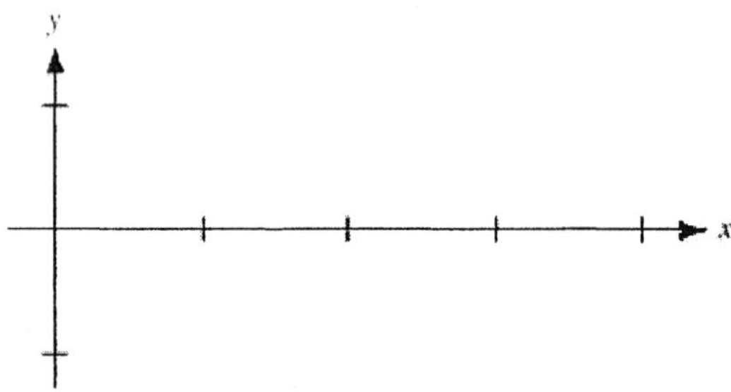

Identify the period.

6. Graph one complete cycle for the function. Label the axes so that the amplitude (if defined) and period are easy to read.

$$y = 3\sin 2x$$

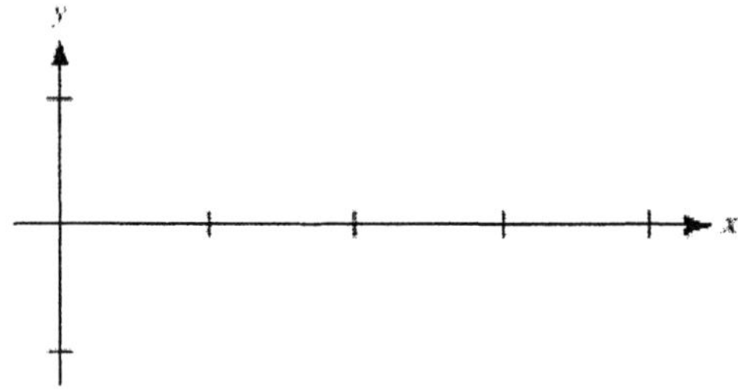

7. Use the graph of $y = \dfrac{3}{2} \cos 3x$ for reference and graph one complete cycle of the

equation $y = 1 + \dfrac{3}{2} \cos 3x$.

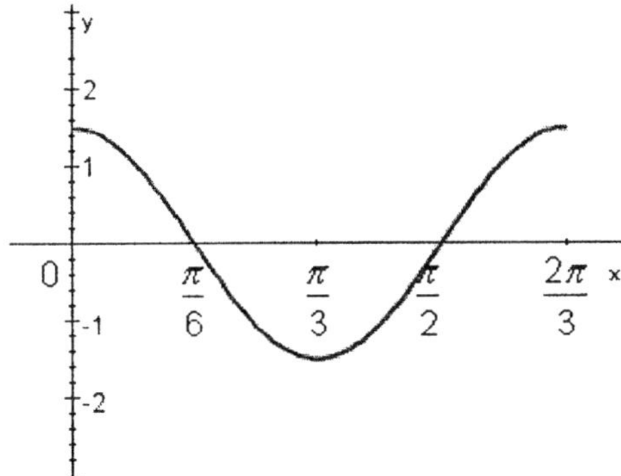

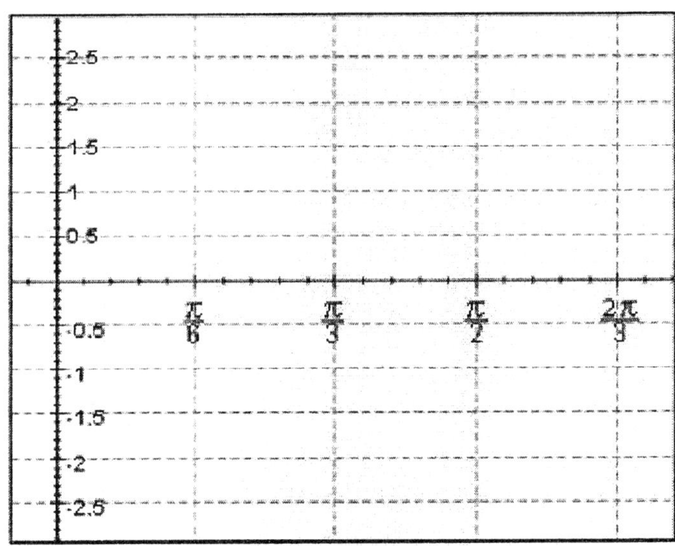

8. The current in an alternating circuit varies in intensity with time. If I represents the intensity of the current and t represents time, then the relationship between I and t is given by

$$I = 10 \cos 140\pi t$$

where I is measured in amperes and t is measured in seconds.

Find the maximum value of I.

Find the time it takes for I to go through one complete cycle.

9. Identify the phase shift for the equation.

$$y = \sin\left(x + \frac{\pi}{2}\right)$$

Sketch one complete cycle of the graph. Graph $y = \sin x$ on the same coordinate system.

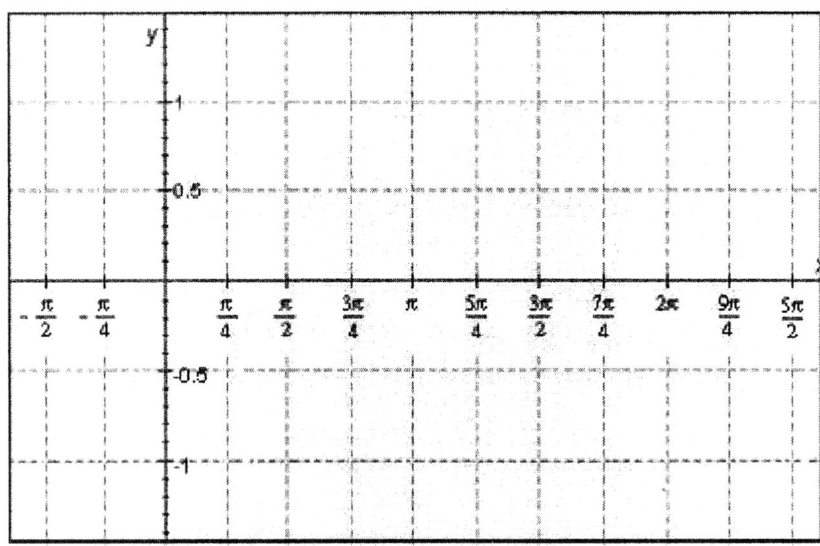

10. Identify the amplitude for the equation.

$$y = 2\sin\left(\pi x + \frac{\pi}{3}\right)$$

____

Identify the period for the equation.

Identify the phase shift for the equation.

Label the axes accordingly and sketch one complete cycle of the curve.

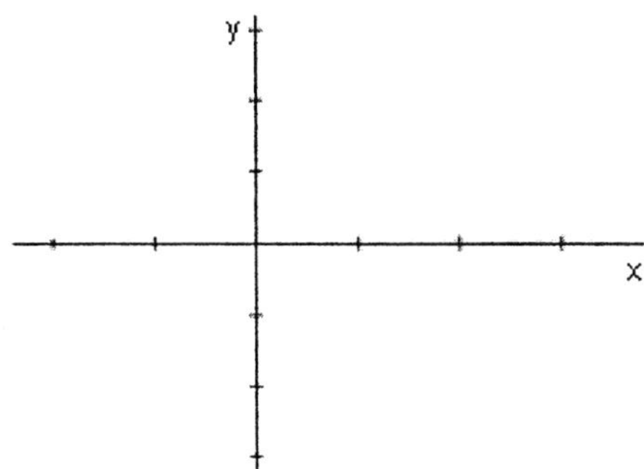

11. Use the graph of the equation $y = -\sin\left(2x + \dfrac{\pi}{2}\right)$ shown below to graph one

complete cycle of the equation $y = 2 - \sin\left(2x + \dfrac{\pi}{2}\right)$.

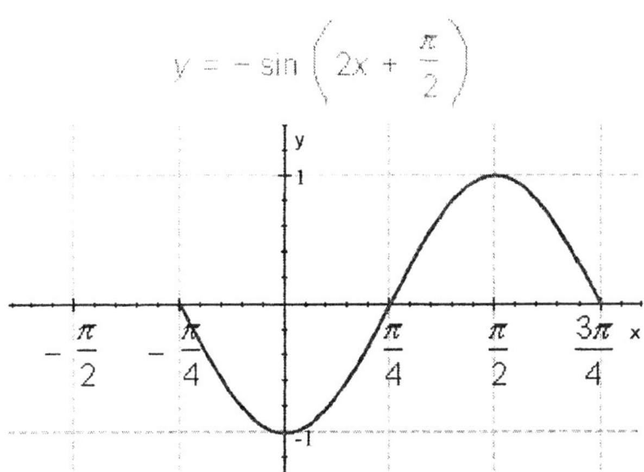

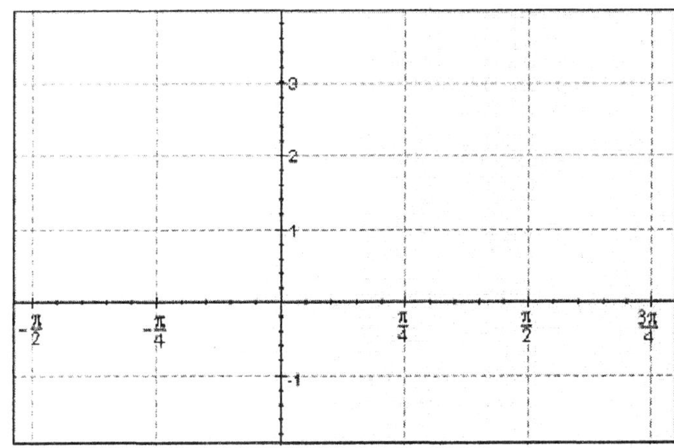

12. Graph the equation over the given interval. Be sure to label the axes so that the amplitude, period, and phase shift are easy to read.

$$y = -\frac{2}{3} \cos\left(3x + \frac{\pi}{2}\right), \quad -\pi \le x \le \pi$$

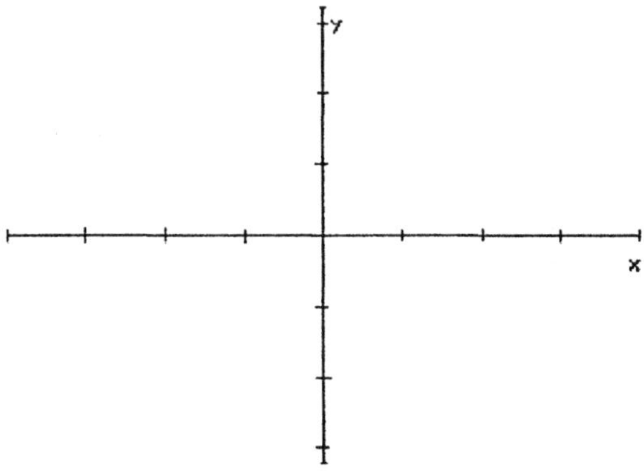

13. Find the equation of the line. Write your answer in slope-intercept form, $y = mx + b$.

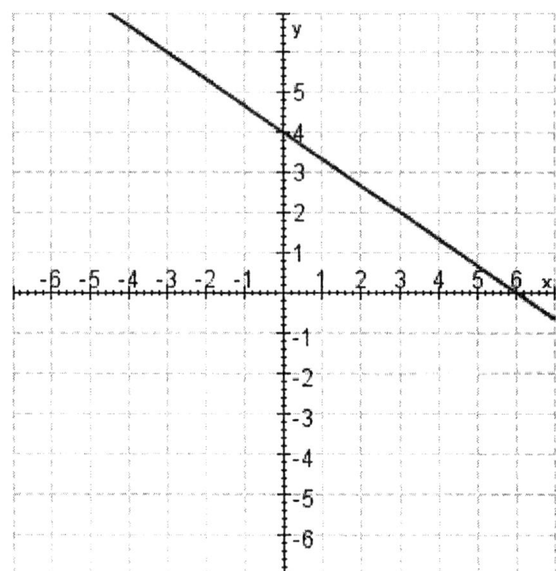

14. The graph below is one complete cycle of the graph of an equation containing a trigonometric function. Find an equation to match the graph.

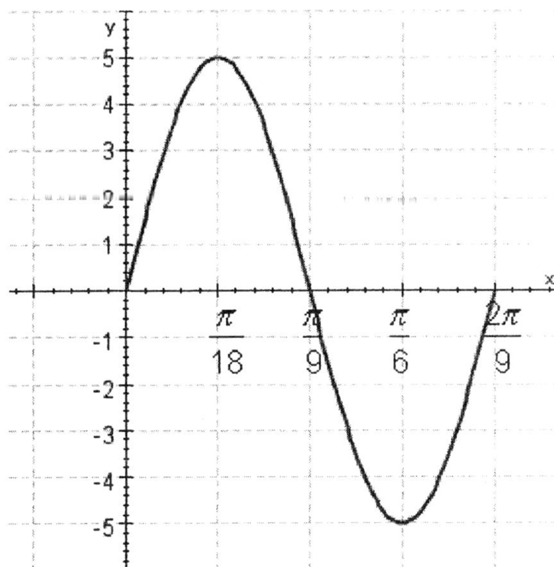

15. The graph below is one complete cycle of the graph of an equation containing a trigonometric function. Find an equation to match the graph.

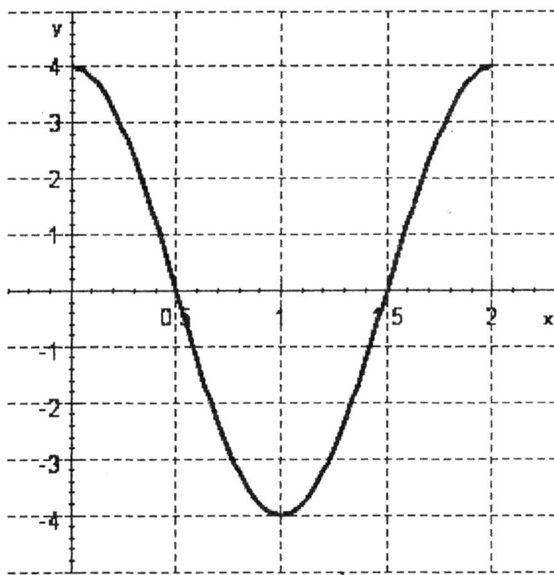

16. The graph below is one complete cycle of the graph of an equation containing a trigonometric function. Find an equation to match the graph.

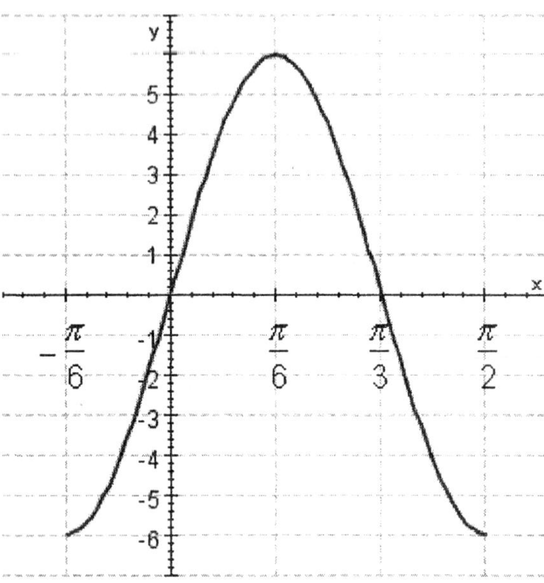

17. Use addition of y-coordinates to sketch the graph of the function between

$x = 0$ and $x = 4\pi$.

$y = 4 + \cos x$

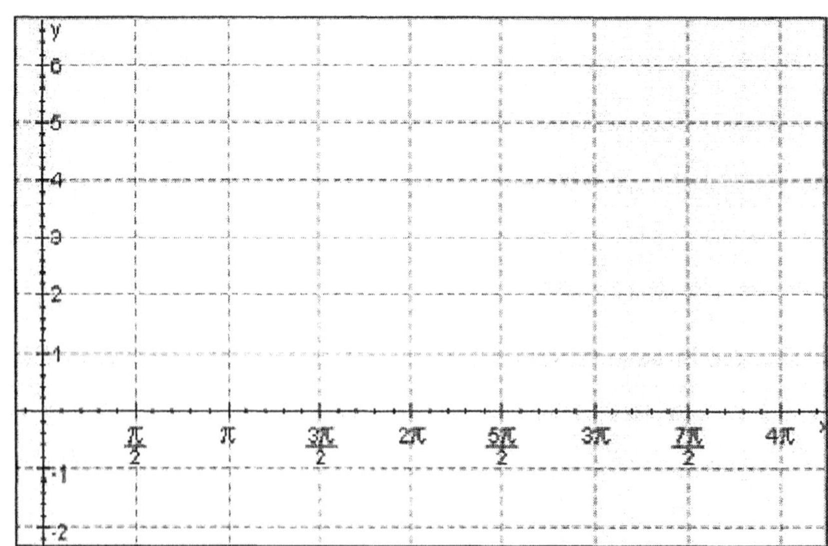

18. Use addition of *y*-coordinates to sketch the graph of the function between

$x = 0$ and $x = 4\pi$.

$y = 4 + 2\sin x$

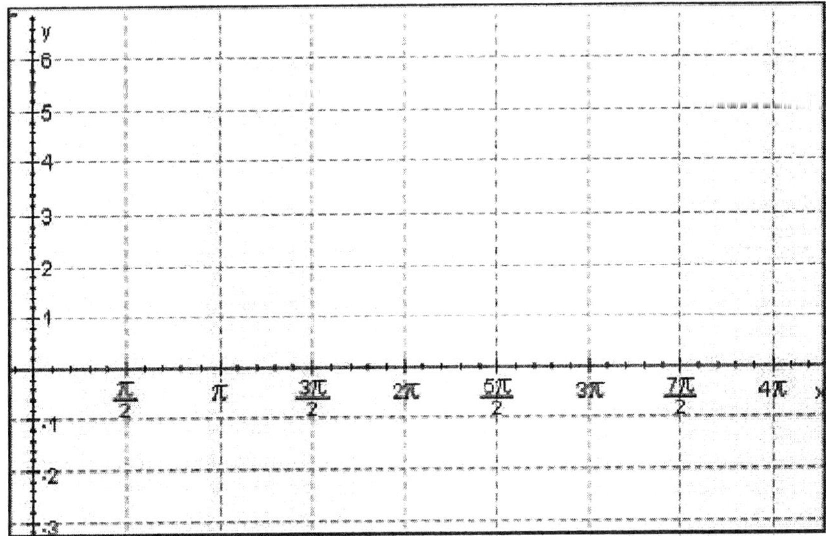

19. Use addition of *y*-coordinates to sketch the graph of the function between

$x = 0$ and $x = 4\pi$.

$y = \dfrac{1}{6}x + \cos x$

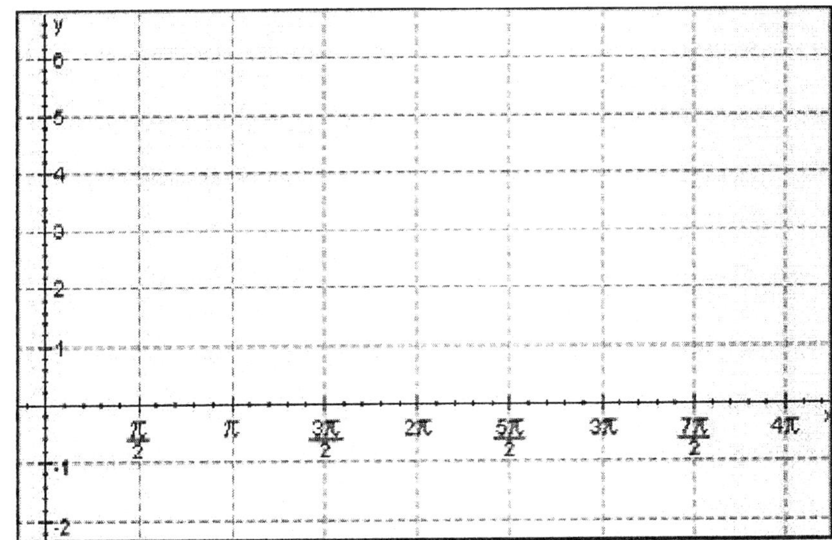

20. Sketch the graph from

$x = 0$ to $x = 4\pi$.

$$y = \cos x + \sin \frac{x}{2}$$

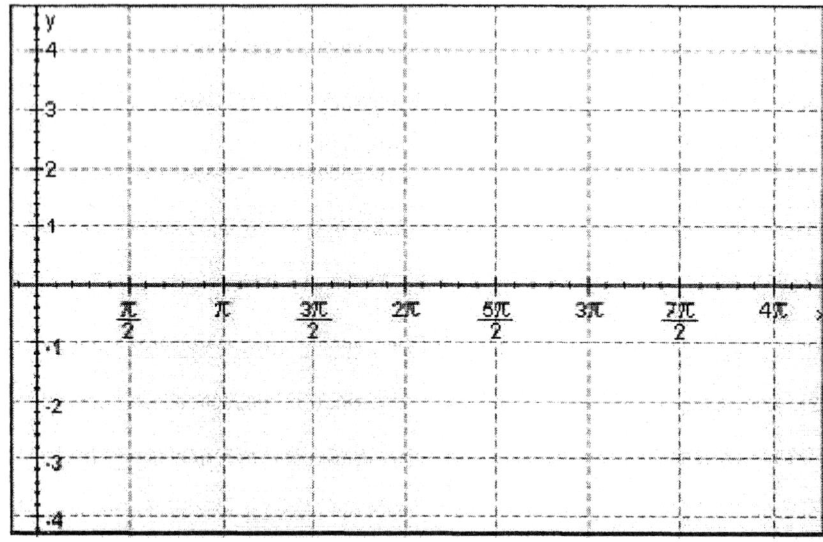

21. Graph

$y = \sin x$ between $-\dfrac{3\pi}{2}$ and $\dfrac{3\pi}{2}$, and then reflect the graph about the line

$y = x$ to obtain the graph of $x = \sin y$.

22. Evaluate the expression without using a calculator, and write your answer in radians.

$$\sin^{-1}(1)$$

23. Simplify

$3\left|\sin\theta\right|$ if $\theta = \cos^{-1}\dfrac{x}{3}$ for some real number x .

24. Evaluate without using a calculator.

$$\tan\left(\cos^{-1}\frac{3}{5}\right)$$

25. Simplify

$$\sin^{-1}(\sin x) \quad \text{If} \quad \frac{\pi}{2} \leq x \leq \pi.$$

1.

$$0, \ \frac{\sqrt{2}}{2}, \ 1, \ \frac{\sqrt{2}}{2}, \ 0, \ -\frac{\sqrt{2}}{2}, \ -1, \ -\frac{\sqrt{2}}{2}, \ 0$$

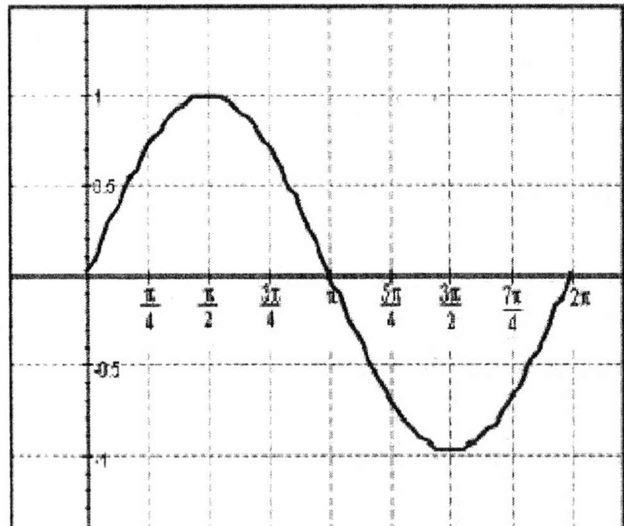

2.

$$\cot x = 0 \rightarrow \frac{\pi}{2} + k\pi,$$

$$\sec x = 1 \rightarrow 2k\pi$$

3. $4,$

π

4.

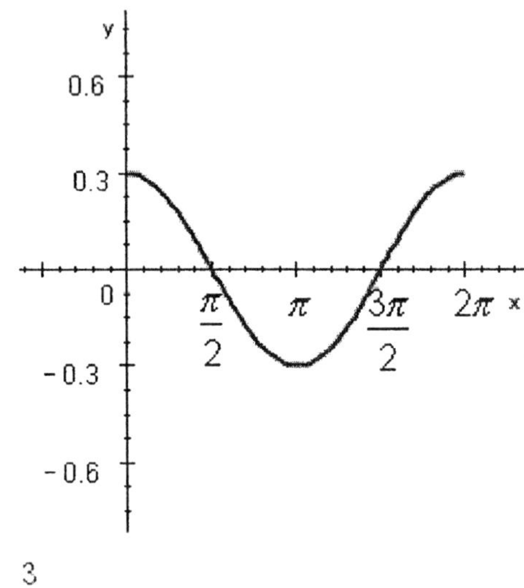

$$\frac{3}{10}$$

5.

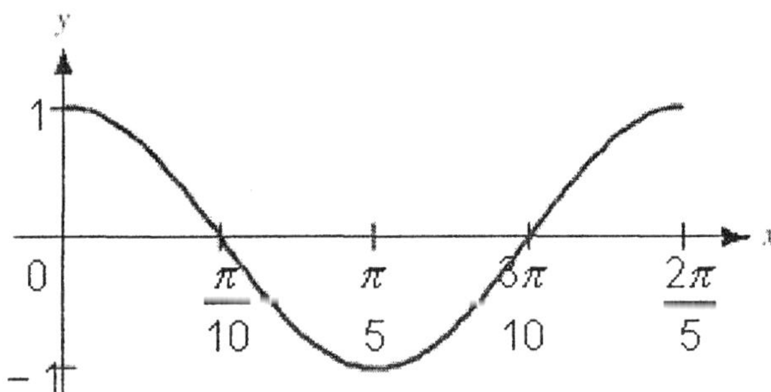

$$\frac{2\pi}{5}$$

6.

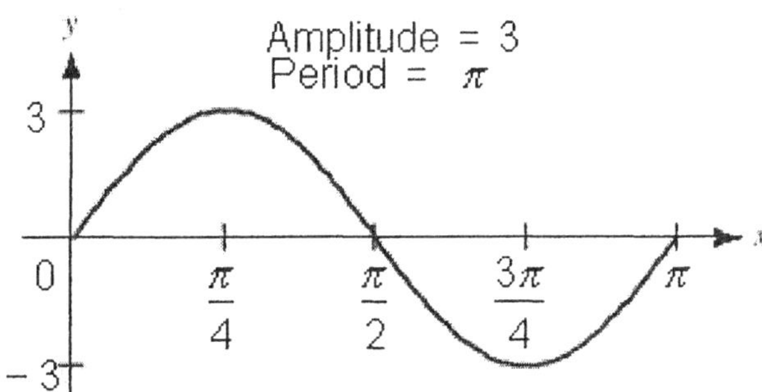

7.

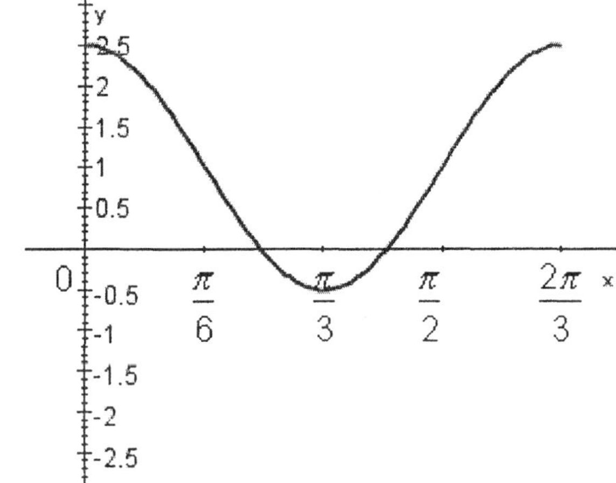

8. 10,

$\dfrac{1}{70}$

9. $-\dfrac{\pi}{2}$

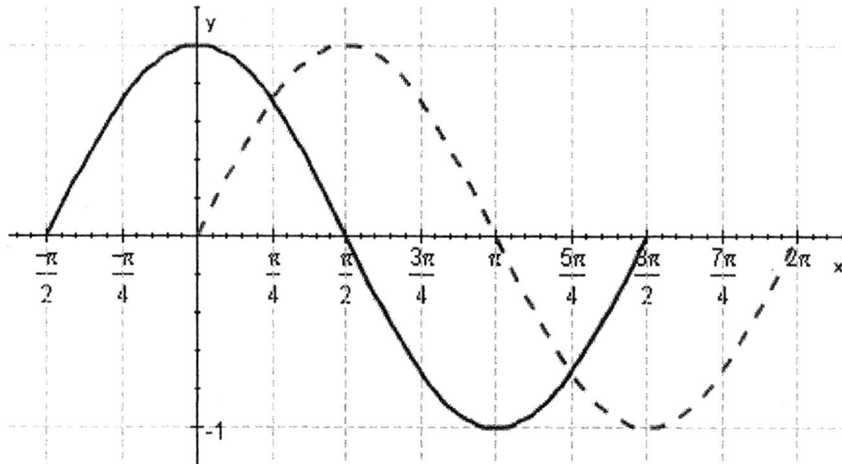

10. 2

2

$-\dfrac{1}{3}$

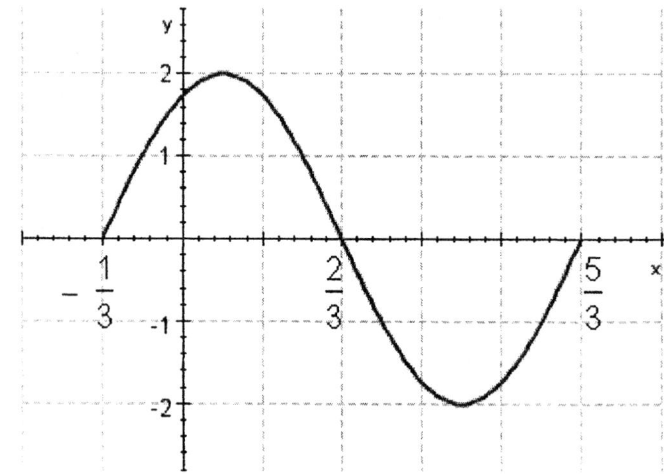

11.

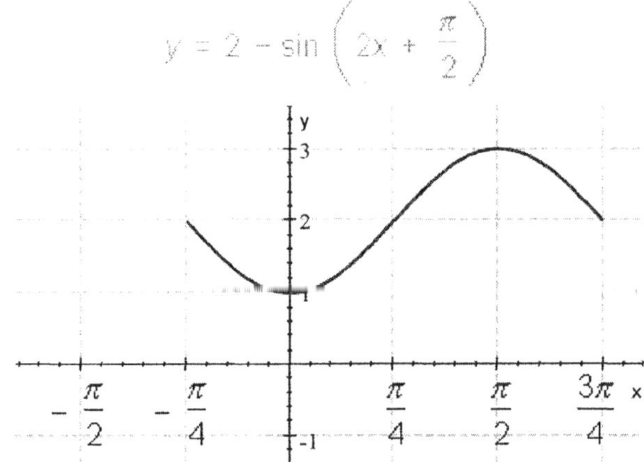

12.

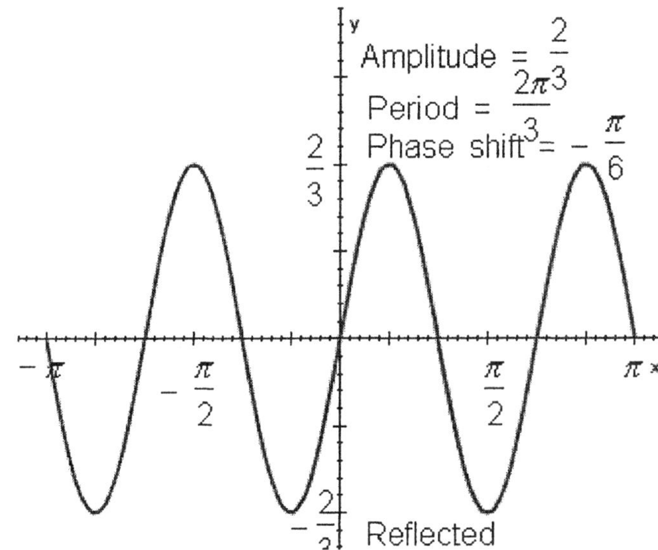

13.
$$y = -\frac{2}{3} \cdot x + 4$$

14. $y = 5\sin(9x)$

15. $y = 4\cos(\pi \cdot x)$

16.
$$y = -6\cos\left(3x + \frac{\pi}{2}\right)$$

McKeague/Turner - Trigonometry 5e Chapter 4 Form A

17.

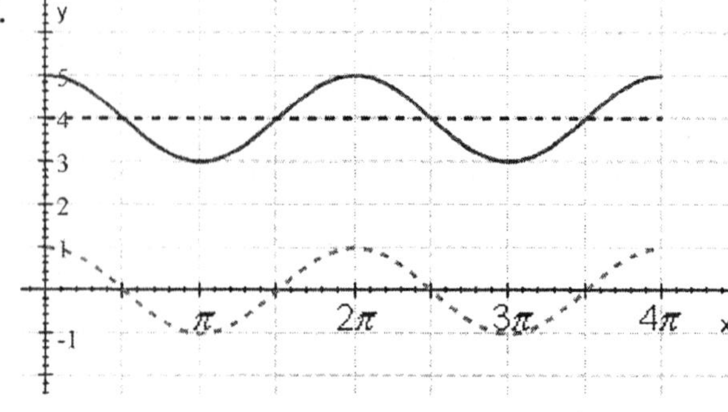

18.

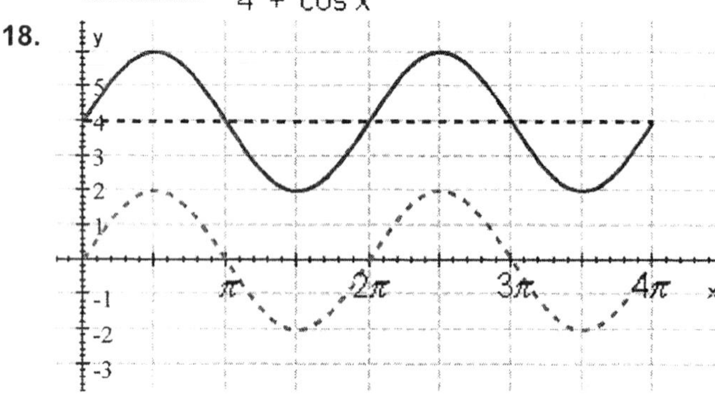

19.

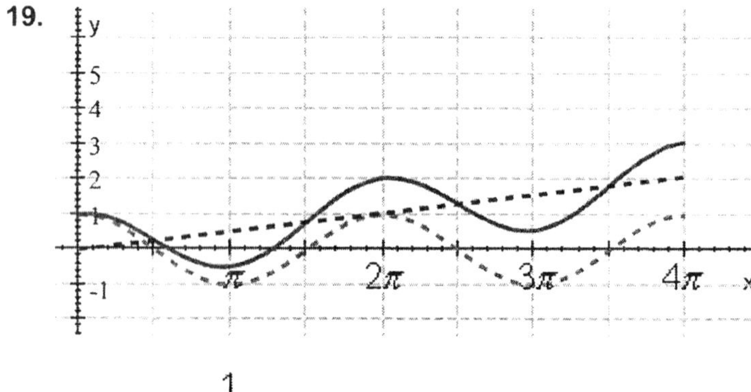

20.

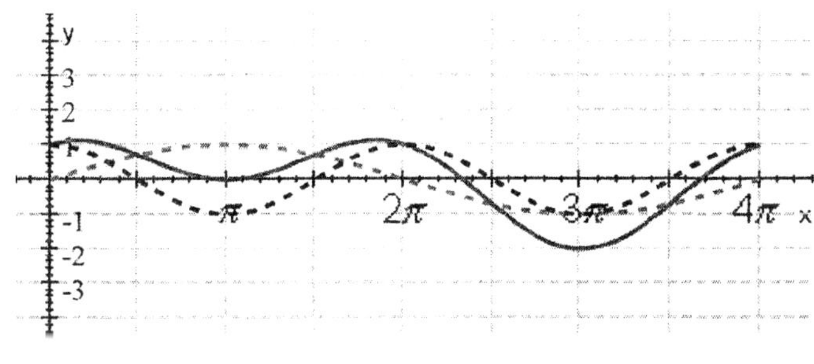

21.

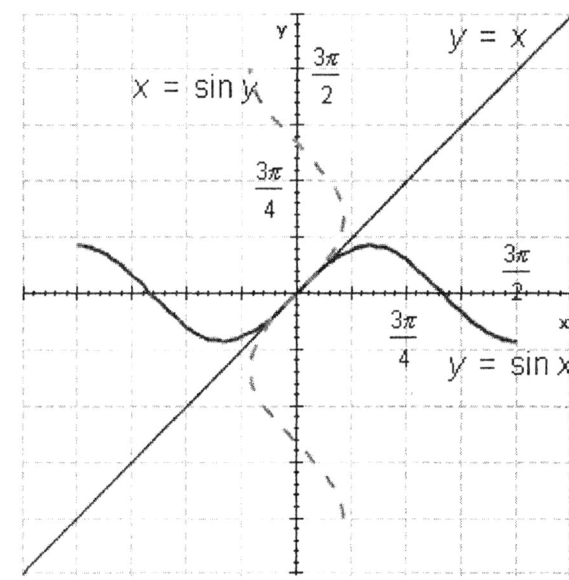

22. $\dfrac{\pi}{2}$

23. $\sqrt{9 - x^2}$

24. $\dfrac{4}{3}$

25. $\pi - x$

McKeague/Turner - Trigonometry 5e Chapter 4 Form A

1. mctr.04.01.01_NoAlgs
2. mctr.04.01.18_NoAlgs
3. mctr.04.01.29_NoAlgs
4. mctr.04.01.53_NoAlgs
5. mctr.04.02.01_NoAlgs
6. mctr.04.02.17_NoAlgs
7. mctr.04.02.36_NoAlgs
8. mctr.04.02.53_NoAlgs
9. mctr.04.03.01_NoAlgs
10. mctr.04.03.13_NoAlgs
11. mctr.04.03.27_NoAlgs
12. mctr.04.03.39_NoAlgs
13. mctr.04.04.02_NoAlgs
14. mctr.04.04.11_NoAlgs
15. mctr.04.04.18_NoAlgs
16. mctr.04.04.25_NoAlgs
17. mctr.04.05.02_NoAlgs
18. mctr.04.05.05_NoAlgs
19. mctr.04.05.10_NoAlgs
20. mctr.04.05.17_NoAlgs
21. mctr.04.06.01_NoAlgs
22. mctr.04.06.22_NoAlgs
23. mctr.04.06.42_NoAlgs
24. mctr.04.06.66_NoAlgs
25. mctr.04.06.73_NoAlgs

1. The graph below is one complete cycle of the graph of an equation containing a trigonometric function.

   Find an equation to match the graph.

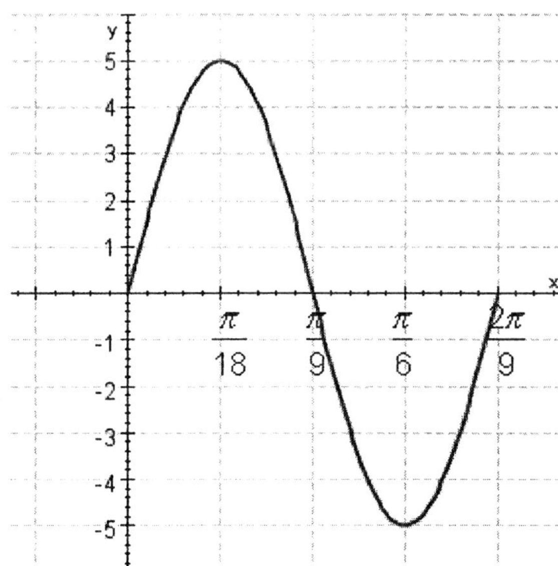

2. Use the graph of $y = \dfrac{3}{2}\cos 3x$ for reference and graph one complete cycle of the equation $y = 1 + \dfrac{3}{2}\cos 3x$.

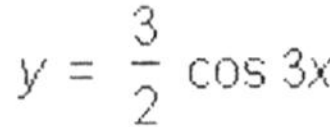

$$y = \dfrac{3}{2}\cos 3x$$

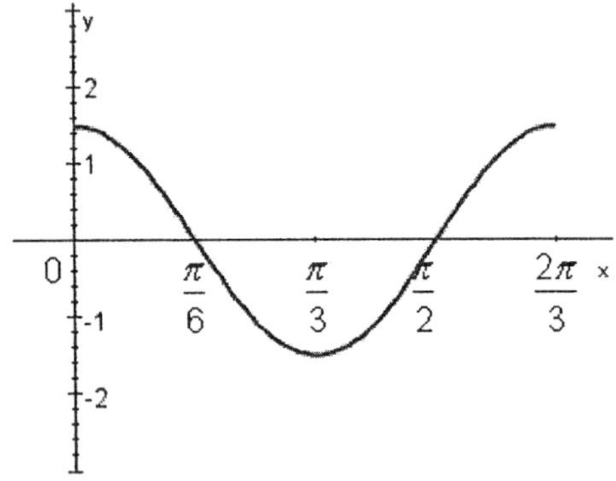

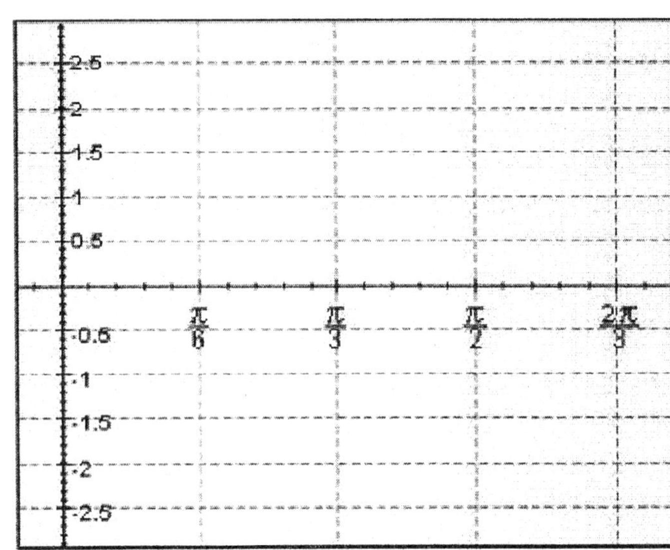

3. Identify the amplitude for the equation.

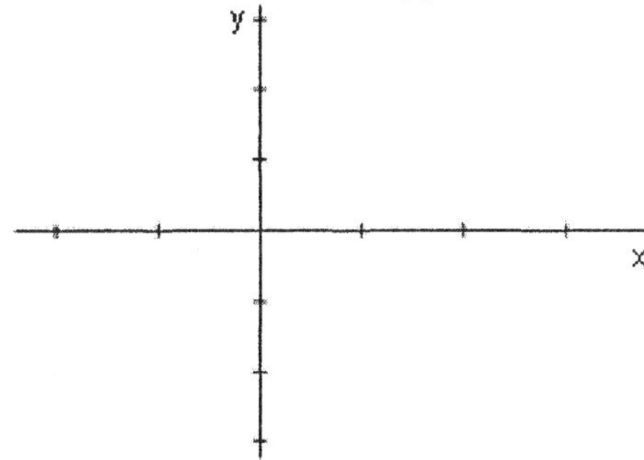

$$y = 2\sin\left(\pi x + \frac{\pi}{3}\right)$$

———————

Identify the period for the equation.

Identify the phase shift for the equation.

Label the axes accordingly and sketch one complete cycle of the curve.

4. The current in an alternating circuit varies in intensity with time. If I represents the intensity of the current and t represents time, then the relationship between I and t is given by

$$I = 10\cos 140\pi t$$

where I is measured in amperes and t is measured in seconds.

Find the maximum value of I.

Find the time it takes for I to go through one complete cycle.

5. Match each equation in the left column with all the corresponding values of x in the right column which make the equation true.

$\sec x = 1$ $\qquad\qquad\qquad\qquad\qquad\qquad \dfrac{\pi}{2} + k\pi$

$\cot x = 0$ $\qquad\qquad\qquad\qquad\qquad\qquad 2k\pi$

6. Evaluate without using a calculator.

$$\tan\left(\cos^{-1}\frac{3}{5}\right)$$

7. Use addition of *y*-coordinates to sketch the graph of the function between

    $x = 0$ and $x = 4\pi$.

    $y = 4 + 2\sin x$

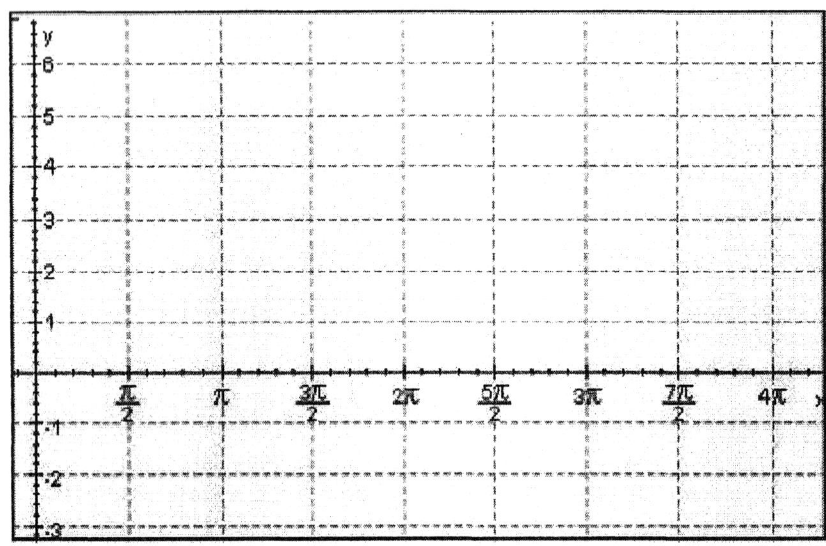

8. Find the equation of the line. Write your answer in slope-intercept form, $y = mx + b$.

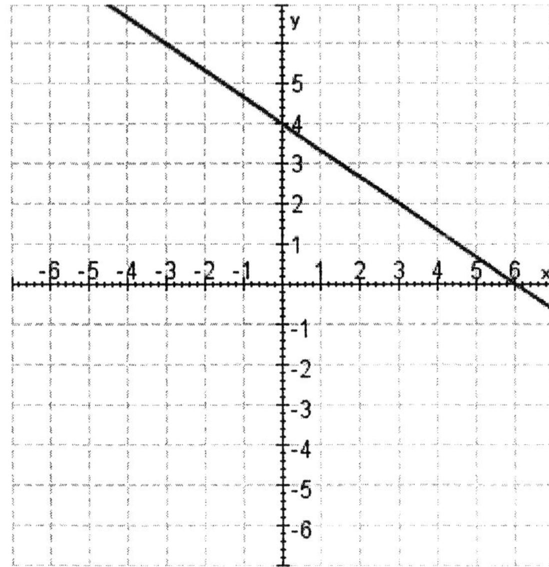

9. Graph one complete cycle for the function. Label the axes so that the amplitude (if defined) and period are easy to read.

$$y = 3\sin 2x$$

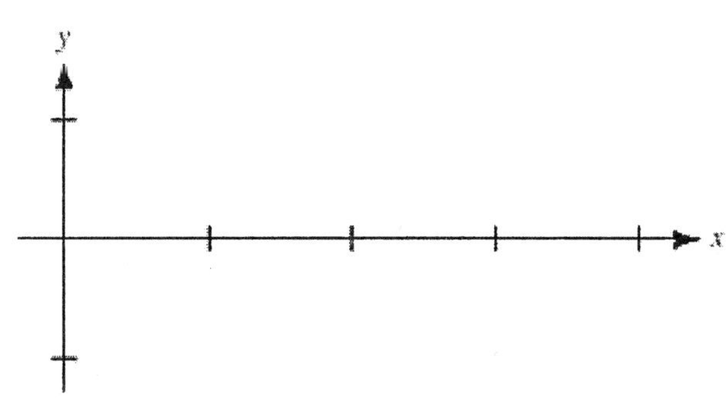

10. Use addition of y-coordinates to sketch the graph of the function between

$x = 0$ and $x = 4\pi$.

$$y = \frac{1}{6}x + \cos x$$

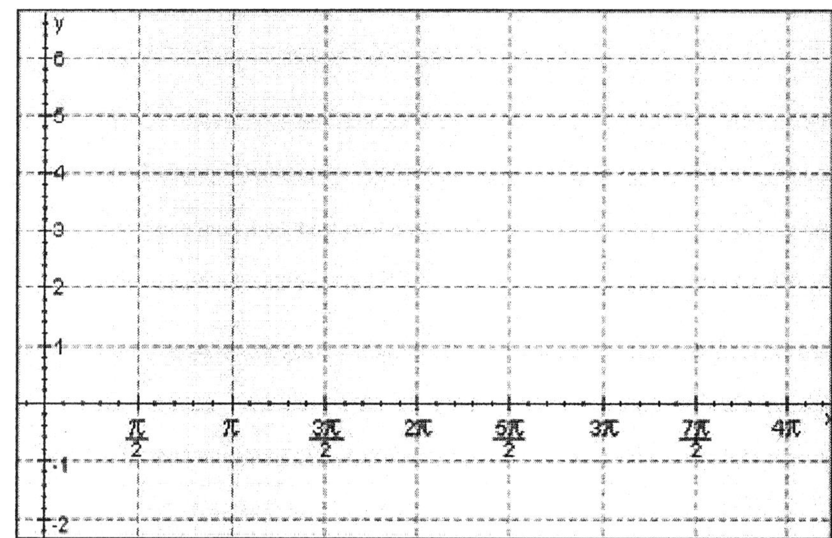

11. Graph

$y = \sin x$ between $-\dfrac{3\pi}{2}$ and $\dfrac{3\pi}{2}$, and then reflect the graph about the line

$y = x$ to obtain the graph of $x = \sin y$.

12. Graph the equation over the given interval. Be sure to label the axes so that the amplitude, period, and phase shift are easy to read.

$$y = -\frac{2}{3}\cos\left(3x + \frac{\pi}{2}\right), \ -\pi \le x \le \pi$$

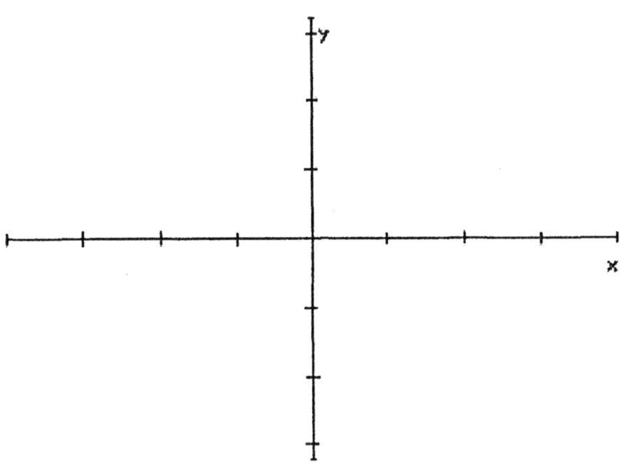

13. Evaluate the expression without using a calculator, and write your answer in radians.

$\sin^{-1}(1)$

14. The graph below is one complete cycle of the graph of an equation containing a trigonometric function. Find an equation to match the graph.

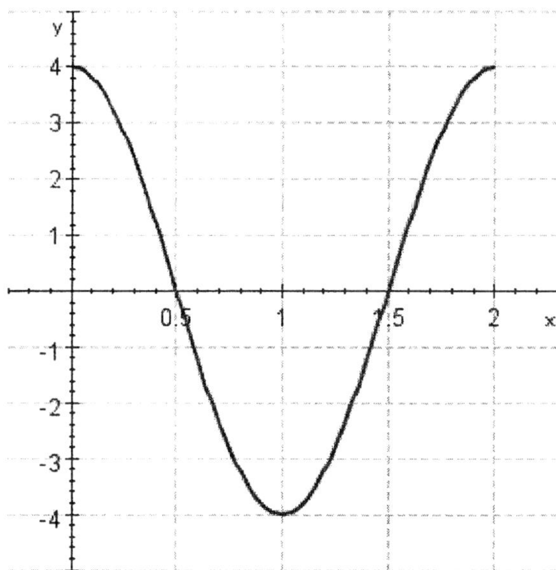

15. The graph below is one complete cycle of the graph of an equation containing a trigonometric function. Find an equation to match the graph.

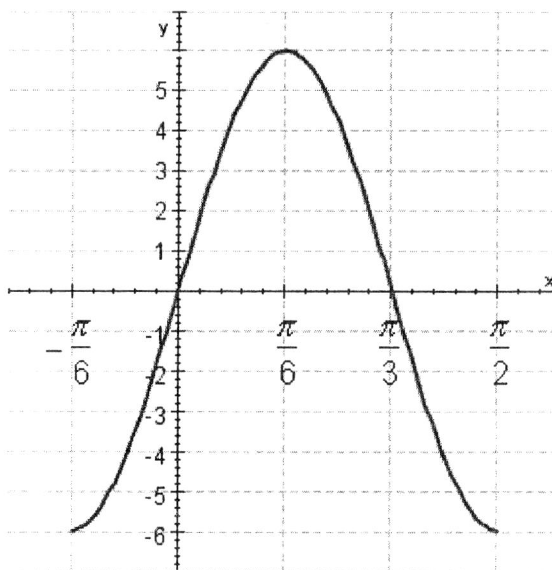

16. Use the graph of the equation $y = -\sin\left(2x + \dfrac{\pi}{2}\right)$ shown below to graph one complete cycle of the equation $y = 2 - \sin\left(2x + \dfrac{\pi}{2}\right)$.

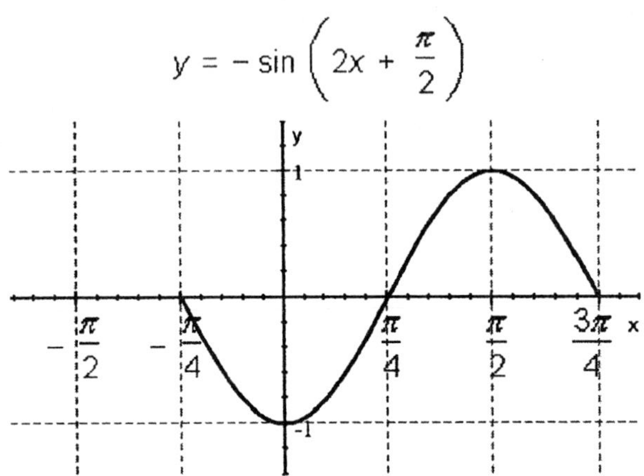

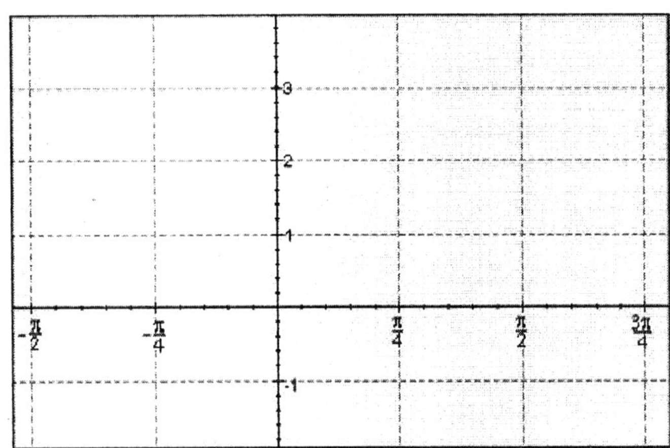

17.
Graph one complete cycle of $y = \dfrac{3}{10} \cos x$. Label the axes accurately.

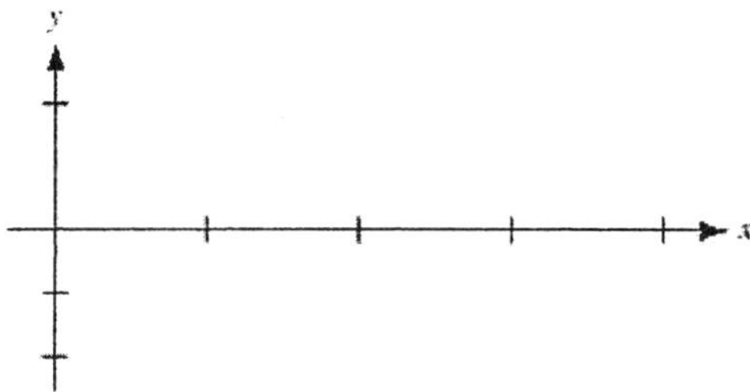

Identify the amplitude for the graph (if defined).

18.

Make a table of values for $y = \sin x$ using multiples of $\dfrac{\pi}{4}$ for x.

x	$\sin x$
0	
$\dfrac{\pi}{4}$	
$\dfrac{\pi}{2}$	
$\dfrac{3\pi}{4}$	
π	
$\dfrac{5\pi}{4}$	
$\dfrac{3\pi}{2}$	
$\dfrac{7\pi}{4}$	
2π	

Use the entries in the table to sketch the graph of the function for x between 0 and 2π

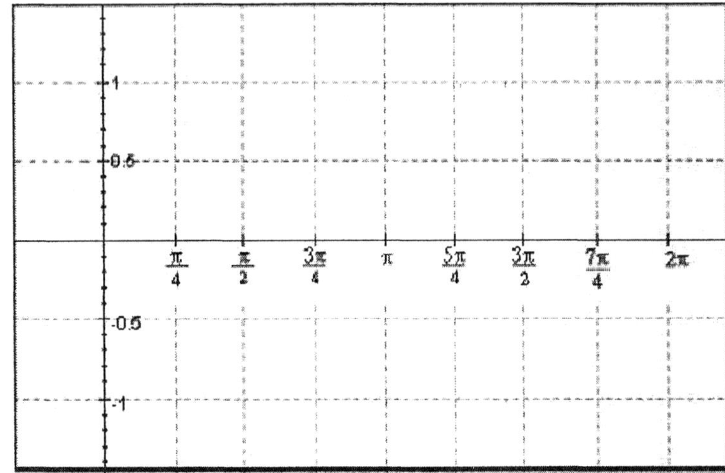

19. Identify the phase shift for the equation.

$$y = \sin\left(x + \frac{\pi}{2}\right)$$

Sketch one complete cycle of the graph. Graph $y = \sin x$ on the same coordinate system.

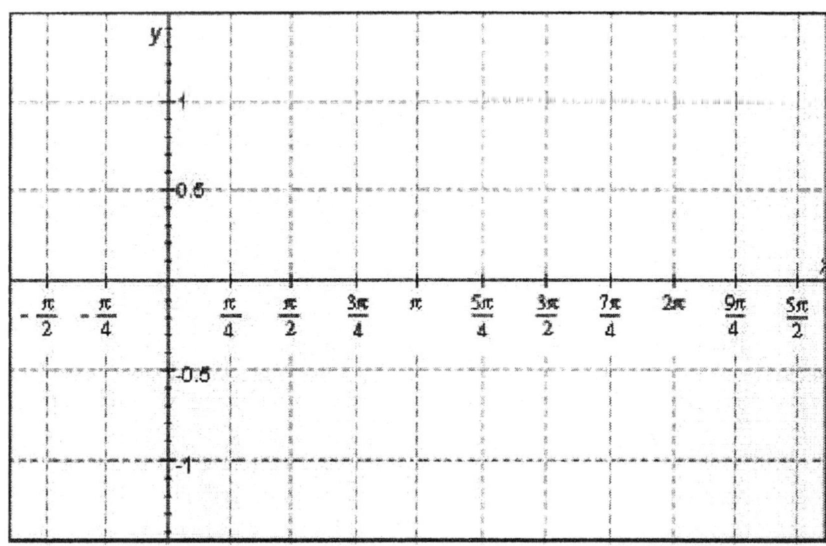

20. Graph one complete cycle of the graph. Label the axes accurately.

$$y = \cos 5x$$

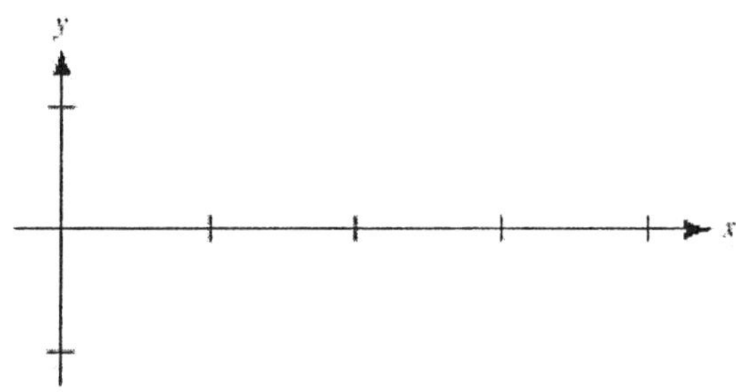

Identify the period.

21. Give the amplitude and period of the graph.

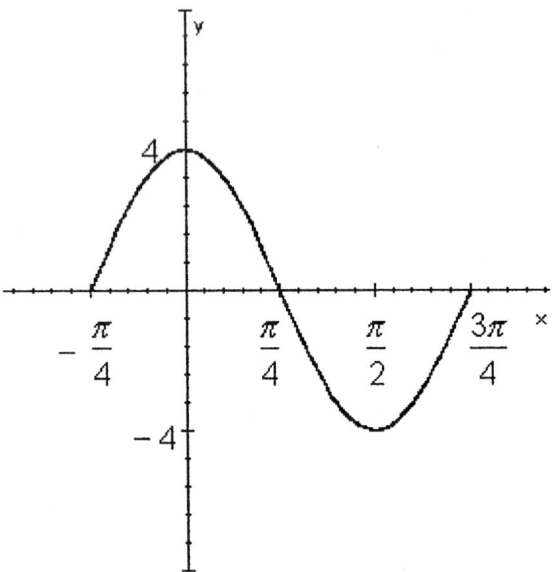

amplitude.

period.

22. Sketch the graph from

$x = 0$ to $x = 4\pi$.

$$y = \cos x + \sin \frac{x}{2}$$

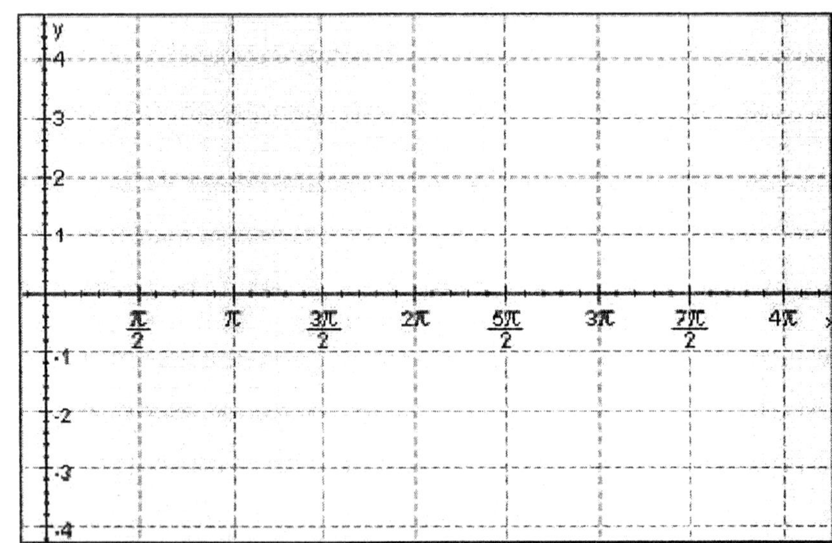

23. Use addition of y-coordinates to sketch the graph of the function between

$x = 0$ and $x = 4\pi$.

$y = 4 + \cos x$

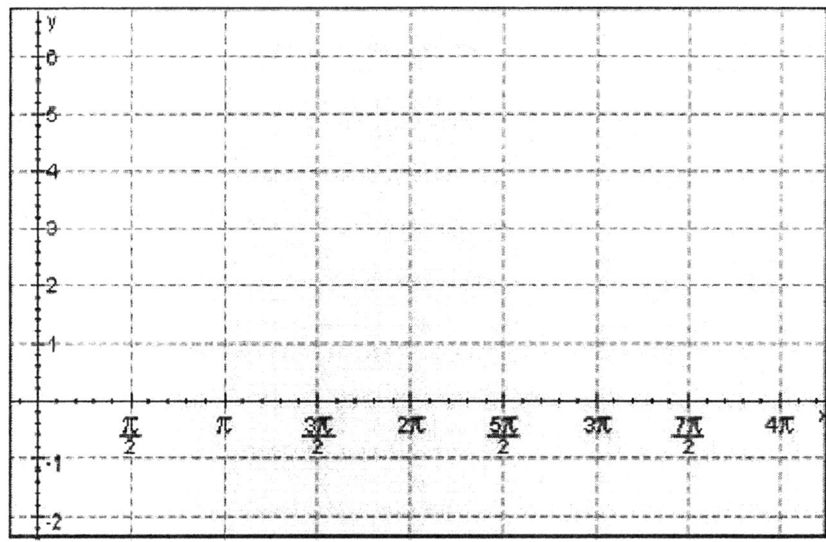

24. Simplify

$\sin^{-1}(\sin x)$ if $\dfrac{\pi}{2} \leq x \leq \pi$.

25. Simplify

$3\left|\sin\theta\right|$ if $\theta = \cos^{-1}\dfrac{x}{3}$ for some real number x.

1. $y = 5\sin(9x)$

2.

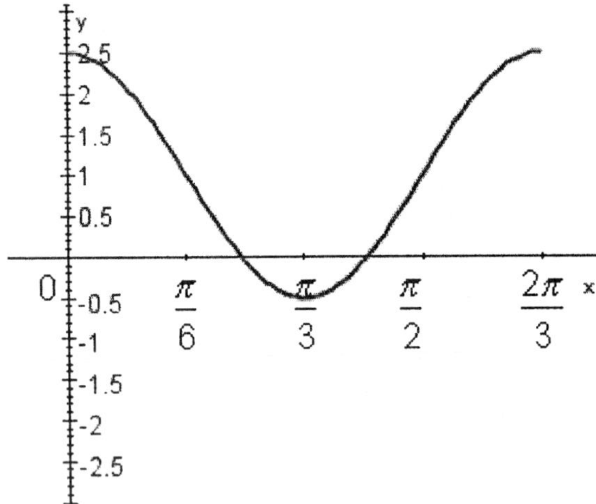

3. 2

2

$-\dfrac{1}{3}$

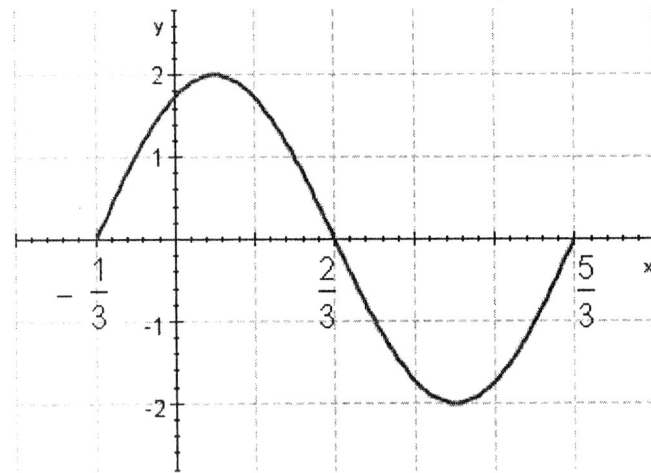

4. 10,

$\dfrac{1}{70}$

5. $\cot x = 0 \rightarrow \dfrac{\pi}{2} + k\pi,$

   $\sec x = 1 \rightarrow 2k\pi$

6. $\dfrac{4}{3}$

7. 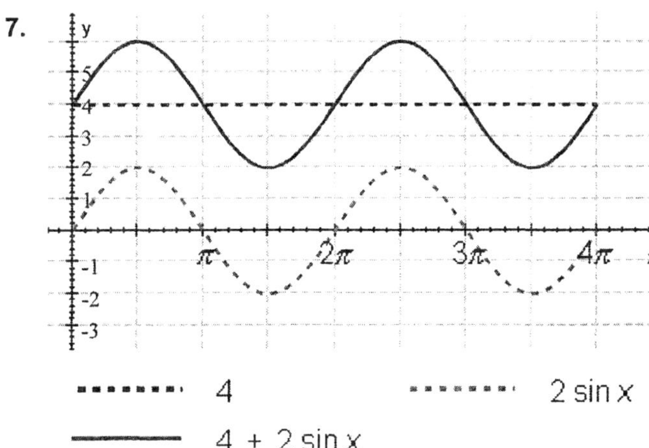

········· 4		········· $2 \sin x$
——— $4 + 2 \sin x$		

8. $y = -\dfrac{2}{3} \cdot x + 4$

9.

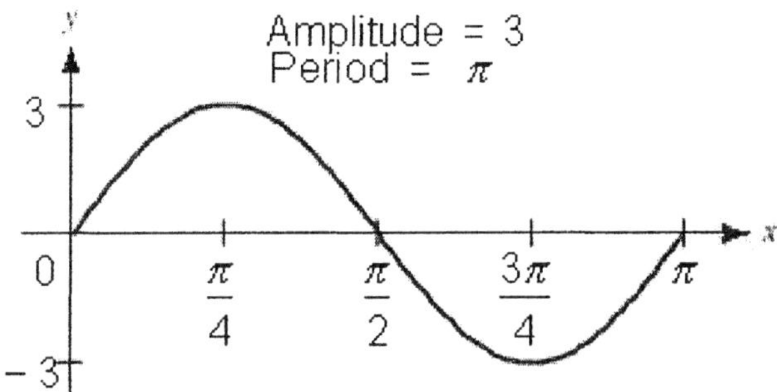

10.

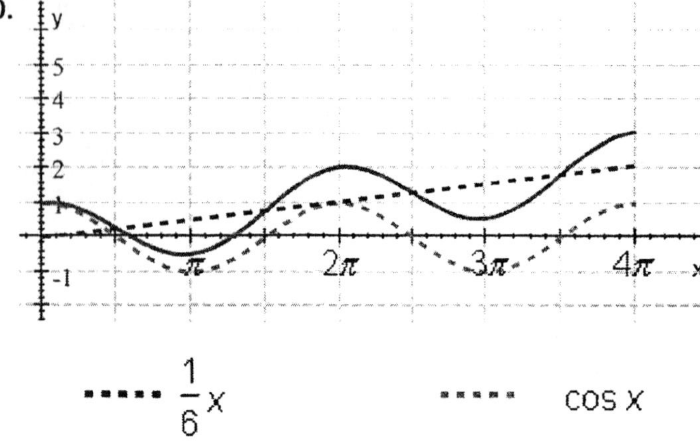

$$\cdots\cdots \quad \frac{1}{6}x \qquad \cdots\cdots \quad \cos x$$

$$\overline{} \quad \frac{1}{6}x + \cos x$$

11.

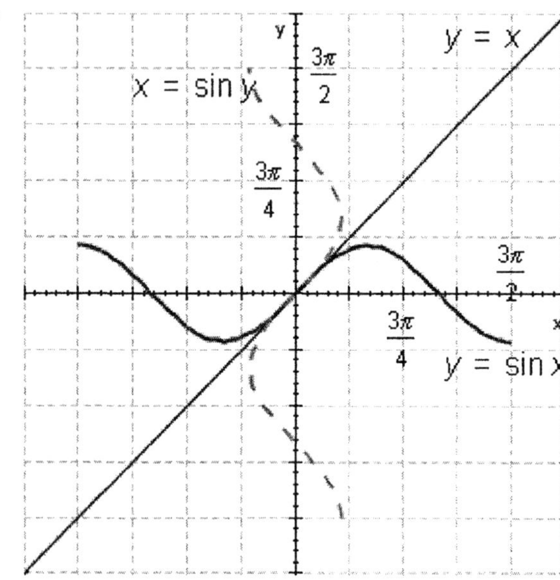

12.

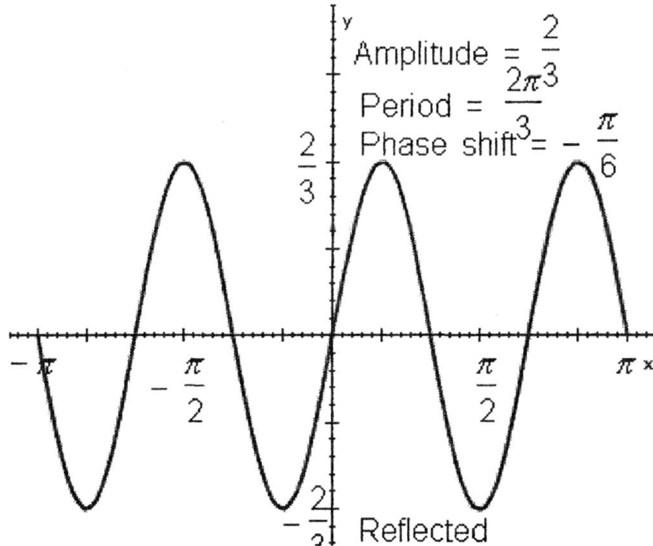

13. $\dfrac{\pi}{2}$

14. $y = 4\cos(\pi \cdot x)$

15. $y = -6\cos\left(3x + \dfrac{\pi}{2}\right)$

16.

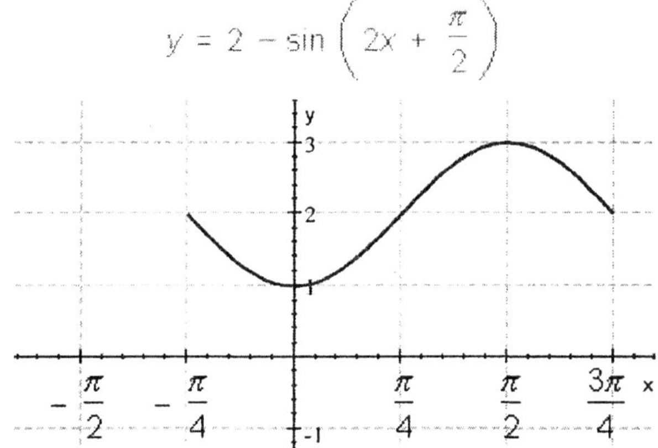

17.

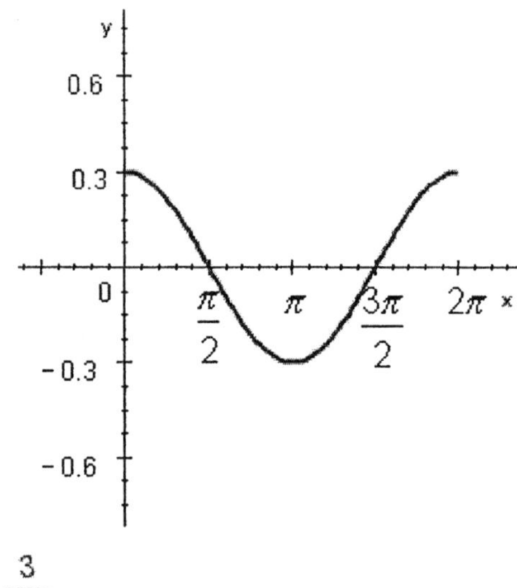

$$\dfrac{3}{10}$$

18. 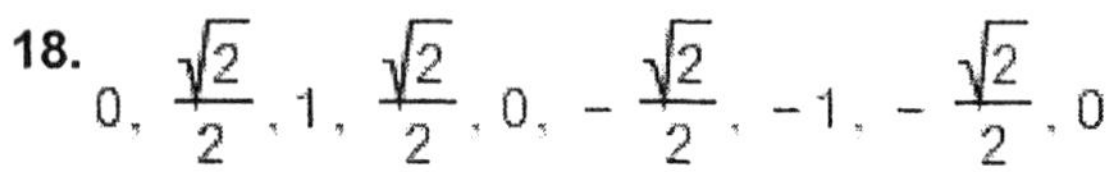
$$0,\ \dfrac{\sqrt{2}}{2},\ 1,\ \dfrac{\sqrt{2}}{2},\ 0,\ -\dfrac{\sqrt{2}}{2},\ -1,\ -\dfrac{\sqrt{2}}{2},\ 0$$

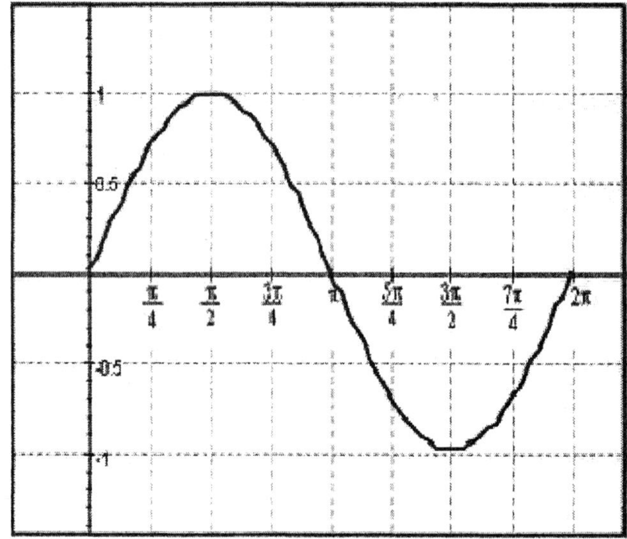

19. $-\dfrac{\pi}{2}$

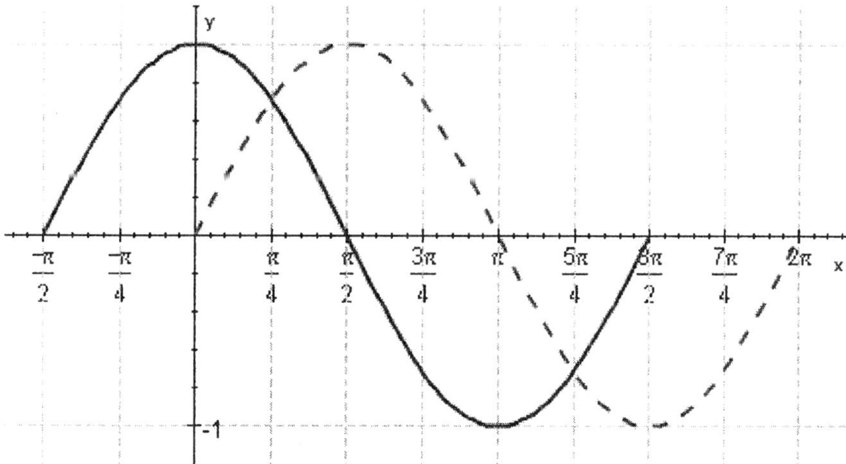

20.

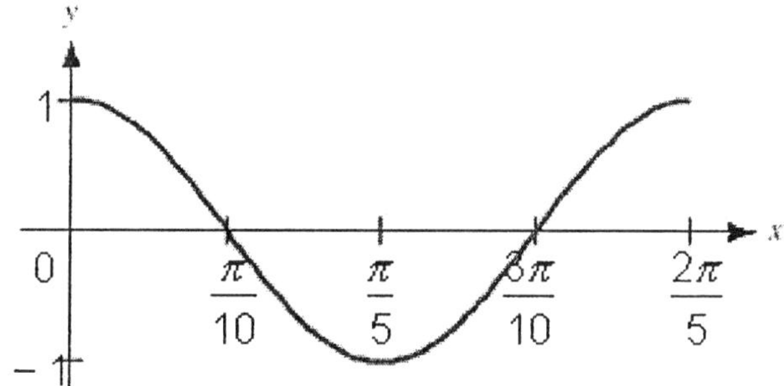

$\dfrac{2\pi}{5}$

21. 4 ,

π

McKeague/Turner - Trigonometry 5e Chapter 4 Form B

22.

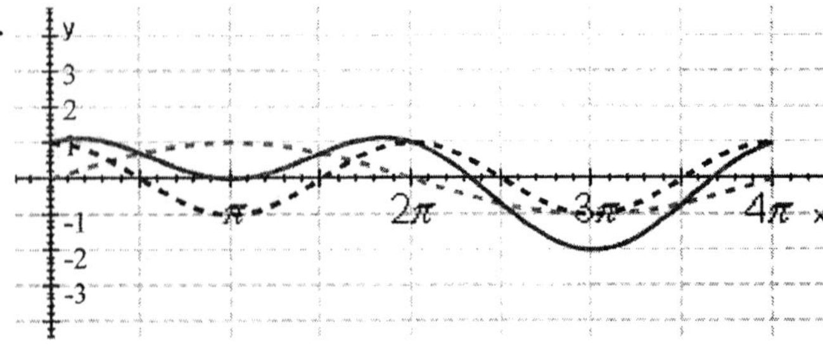

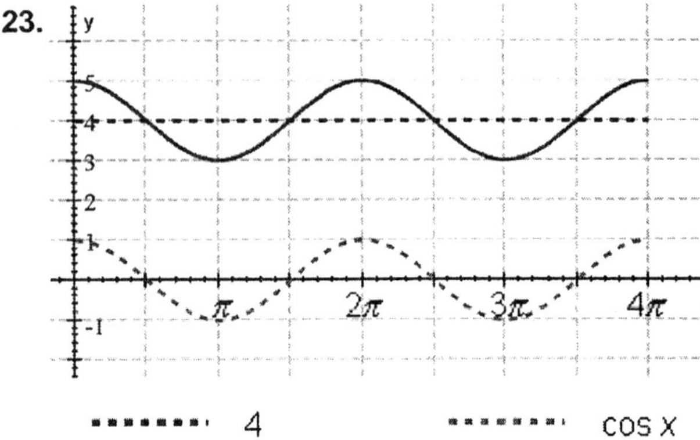

$$\cdots\cdots\cdots \quad \cos x \qquad \cdots\cdots\cdots \quad \sin \frac{x}{2}$$

$$\text{———} \quad \cos x + \sin \frac{x}{2}$$

23.

$$\cdots\cdots\cdots \quad 4 \qquad \cdots\cdots\cdots \quad \cos x$$

$$\text{———} \quad 4 + \cos x$$

24. $\pi - x$

25. $\sqrt{9 - x^2}$

1. mctr.04.04.11_NoAlgs
2. mctr.04.02.36_NoAlgs
3. mctr.04.03.13_NoAlgs
4. mctr.04.02.53_NoAlgs
5. mctr.04.01.18_NoAlgs
6. mctr.04.06.66_NoAlgs
7. mctr.04.05.05_NoAlgs
8. mctr.04.04.02_NoAlgs
9. mctr.04.02.17_NoAlgs
10. mctr.04.05.10_NoAlgs
11. mctr.04.06.01_NoAlgs
12. mctr.04.03.39_NoAlgs
13. mctr.04.06.22_NoAlgs
14. mctr.04.04.18_NoAlgs
15. mctr.04.04.25_NoAlgs
16. mctr.04.03.27_NoAlgs
17. mctr.04.01.53_NoAlgs
18. mctr.04.01.01_NoAlgs
19. mctr.04.03.01_NoAlgs
20. mctr.04.02.01_NoAlgs
21. mctr.04.01.29_NoAlgs
22. mctr.04.05.17_NoAlgs
23. mctr.04.05.02_NoAlgs
24. mctr.04.06.73_NoAlgs
25. mctr.04.06.42_NoAlgs

1. Sketch the graph of $y = \cos x$ for x between 0 and 2π.

Select the correct answer.

a.

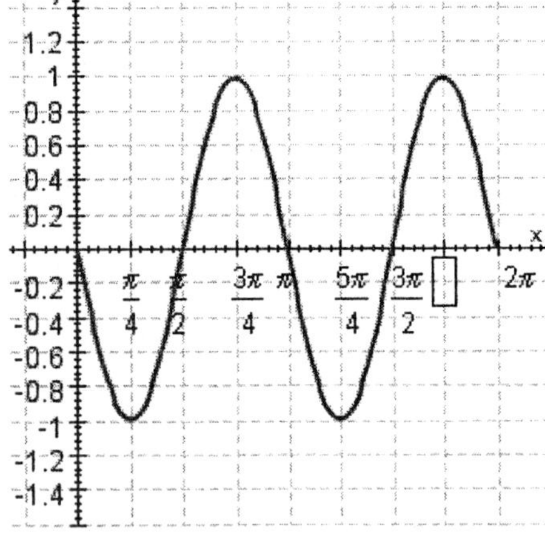

b.

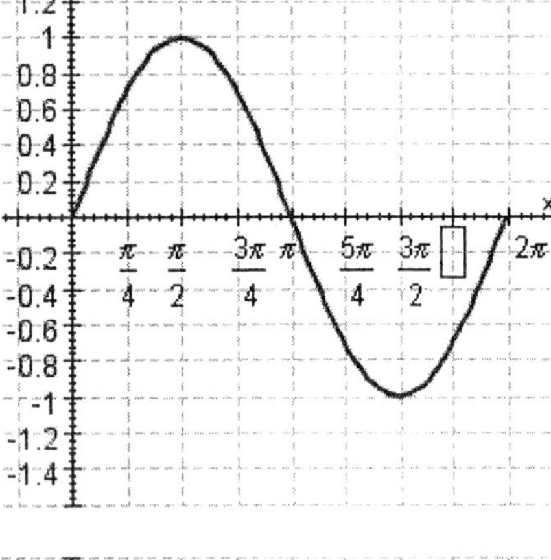

c.

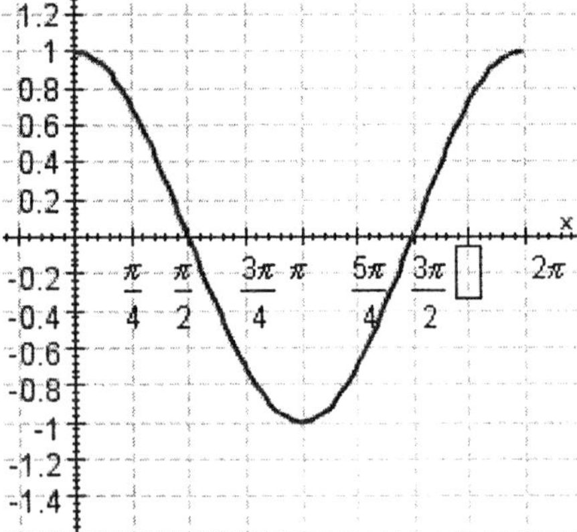

2. Sketch the graph of $y = \cos x$ between $x = -4\pi$ and $x = 4\pi$ by extending the graph for x between 0 and 2π.

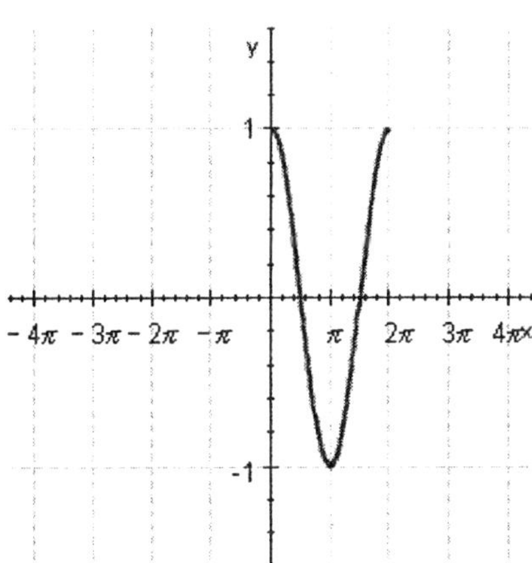

Select the correct answer.

a.
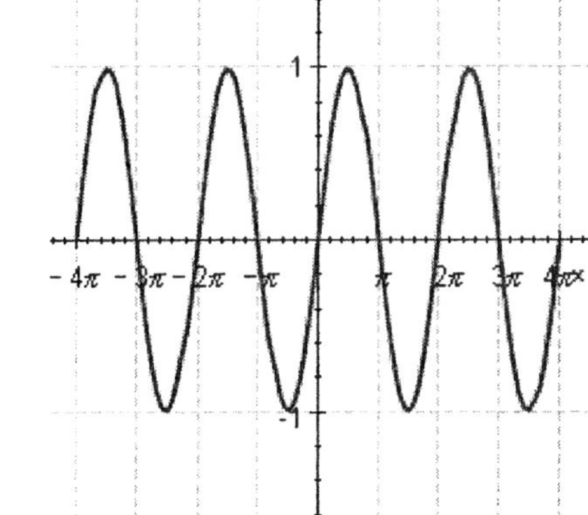

b.
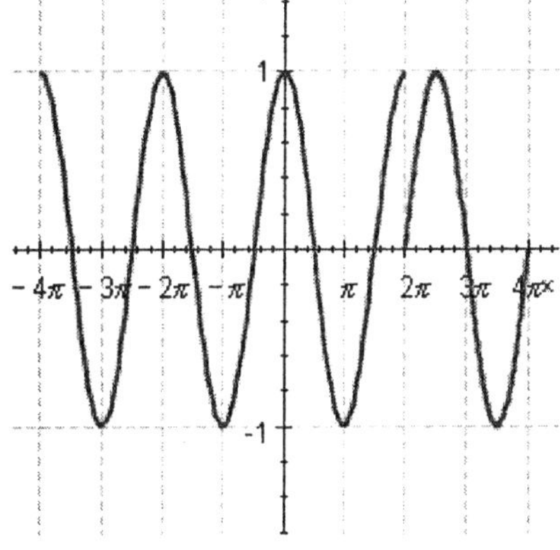

c.
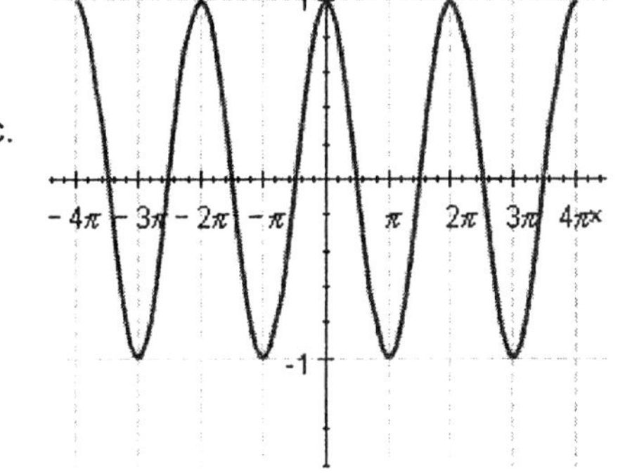

d.
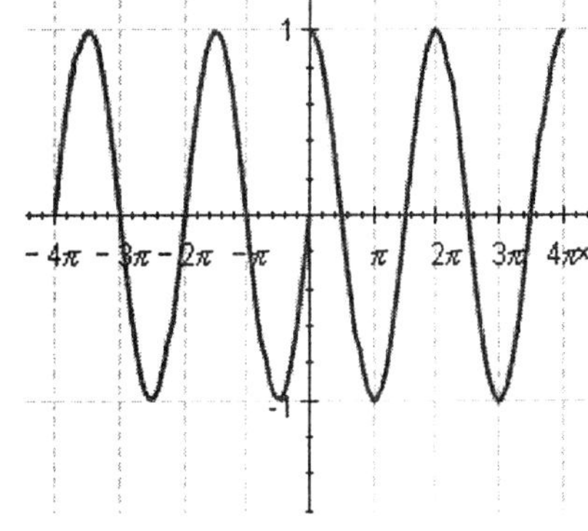

3. Find all values of *x* for which the following are true.

$$\sec x = 1$$

Select the correct answer.

a. $\dfrac{\pi}{3} + 2k\pi$

b. $2k\pi$

c. $\dfrac{\pi}{4} + 2k\pi$

d. $-\dfrac{\pi}{3} + 2k\pi$

e. $\dfrac{\pi}{2} + k\pi$

4. Give the amplitude and period of the graph.

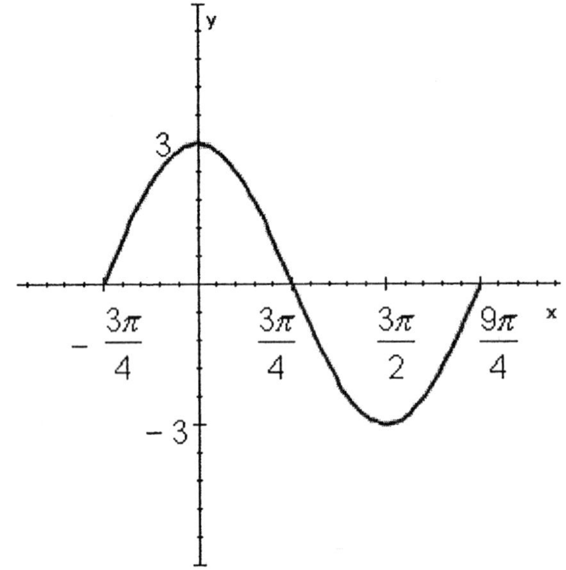

Select the correct answer.

a. Amplitude = **3**, period = π.

b. Amplitude = **6**, period = π.

c. Amplitude = **6**, period = $\dfrac{5\pi}{2}$.

d. Amplitude = **3**, period = 3π.

e. Amplitude = **6**, period = 3π.

5. Graph one complete cycle of the graph. Label the axes accurately and identify the period.

$$y = \sin \frac{\pi}{4} x$$

Select the correct answer.

a.

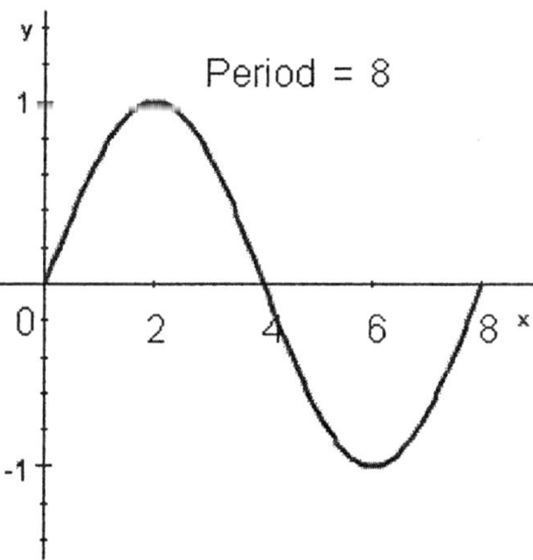

b.

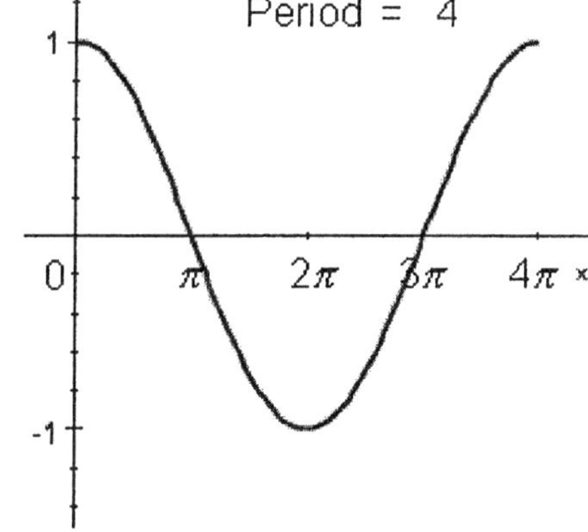

c.

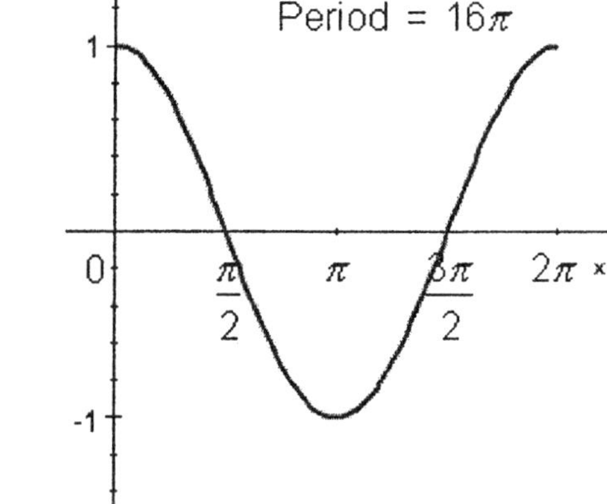

6. Graph one complete cycle for the function.

$$y = 4 \csc \frac{1}{6} x$$

Select the correct answer.

a.

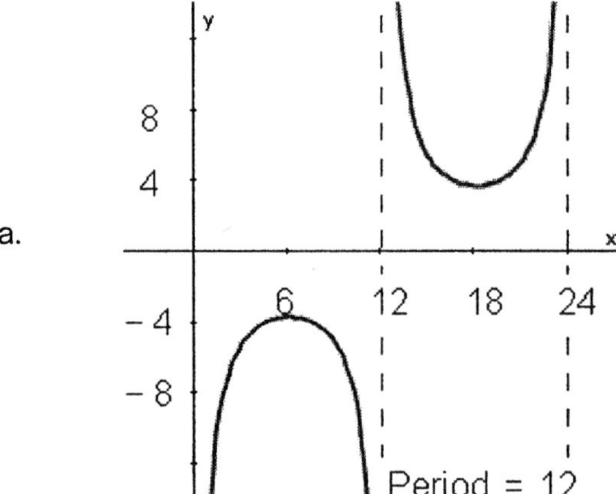

b.

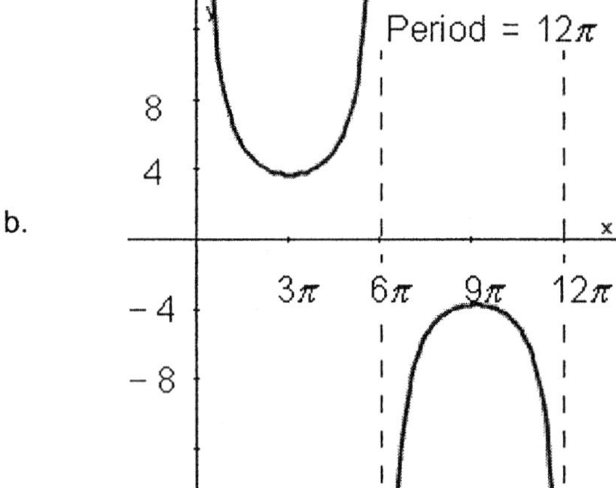

c.

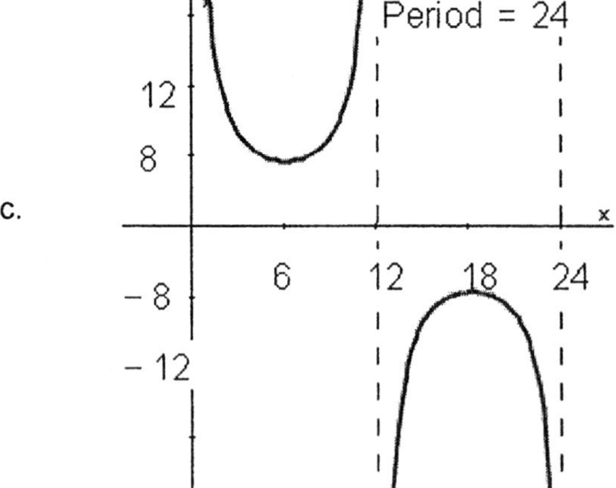

7. Graph the function over the given interval.

$$y = 3\cos \pi x, \quad -2 \le x \le 4$$

Select the correct answer.

a.

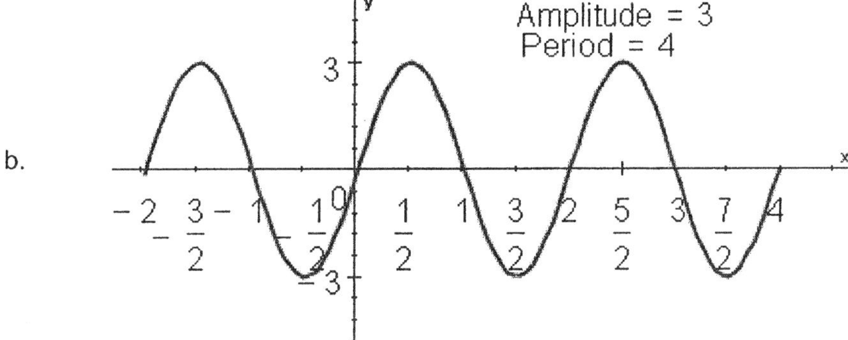

b.

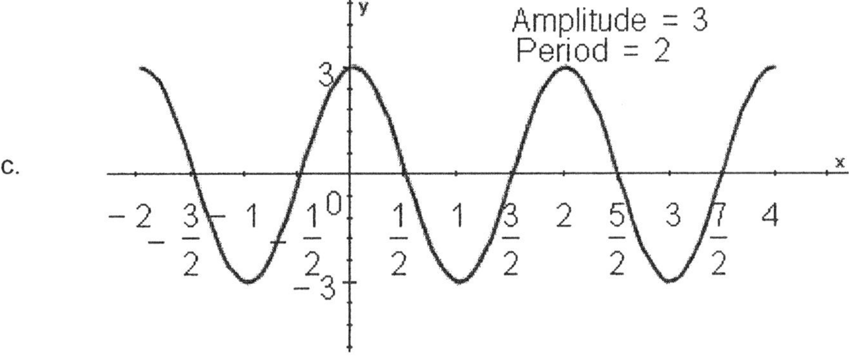

c.

8. The current in an alternating circuit varies in intensity with time. If I represents the intensity of the current and t represents time, then the relationship between I and t is given by

$$I = 16 \sin 160\pi t$$

where I is measured in amperes and t is measured in seconds. Find the maximum value of I and the time it takes for I to go through one complete cycle.

Select the correct answer.

a.   The maximum value of I is 16 amperes; one complete cycle takes $\dfrac{1}{160}$ seconds.

b.   The maximum value of I is 16 amperes; one complete cycle takes $\dfrac{1}{80}$ seconds.

c.   The maximum value of I is 32 amperes; one complete cycle takes $\dfrac{1}{80}$ seconds.

d.   The maximum value of I is 32 amperes; one complete cycle takes $\dfrac{1}{92}$ seconds.

e.   The maximum value of I is 64 amperes; one complete cycle takes $\dfrac{1}{160}$ seconds.

9. Sketch one complete cycle of the graph. Graph $y = \cos x$ on the same coordinate system.

$$y = \cos\left(x + \frac{\pi}{4}\right)$$

Select the correct answer.

a.

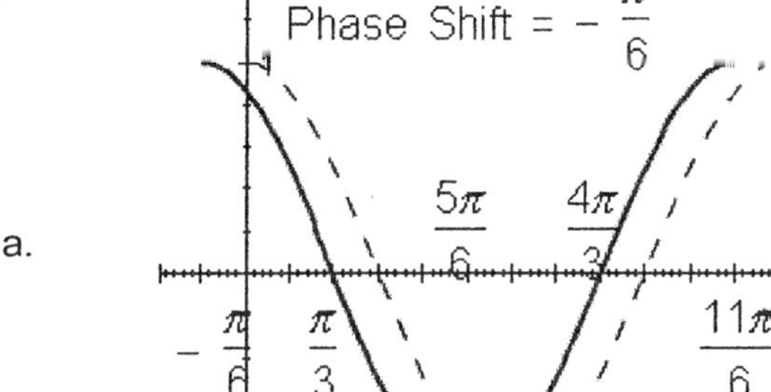

b.

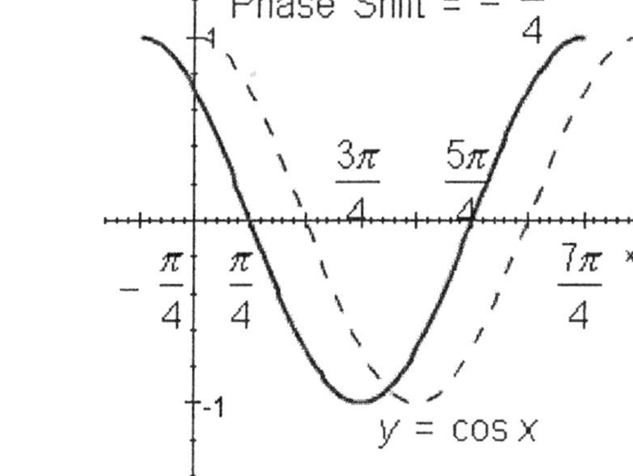

c.

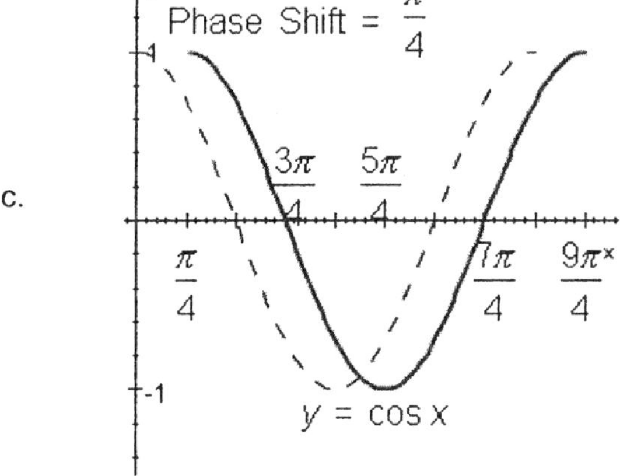

10.

Use the graph of the equation $y = \cos\left(2x - \dfrac{\pi}{2}\right)$ shown below to graph one complete cycle of the equation $y = 2 + \cos\left(2x - \dfrac{\pi}{2}\right)$.

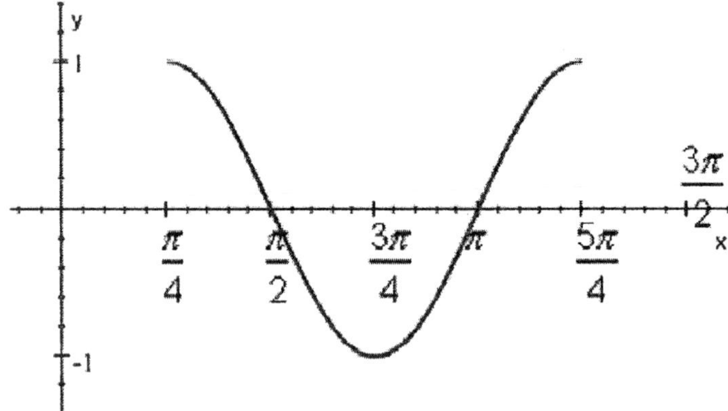

Select the correct answer.

a.

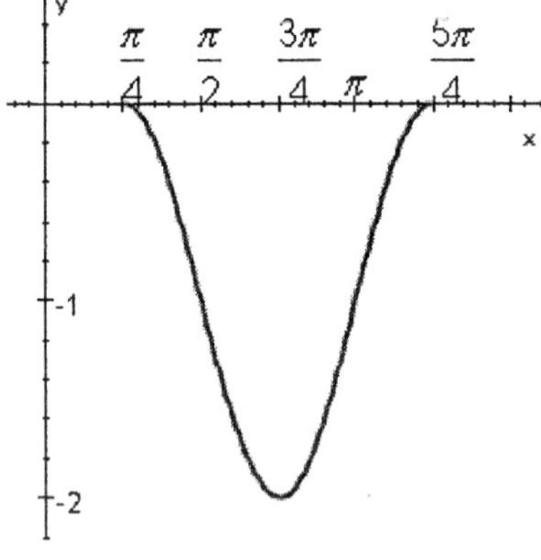

b.

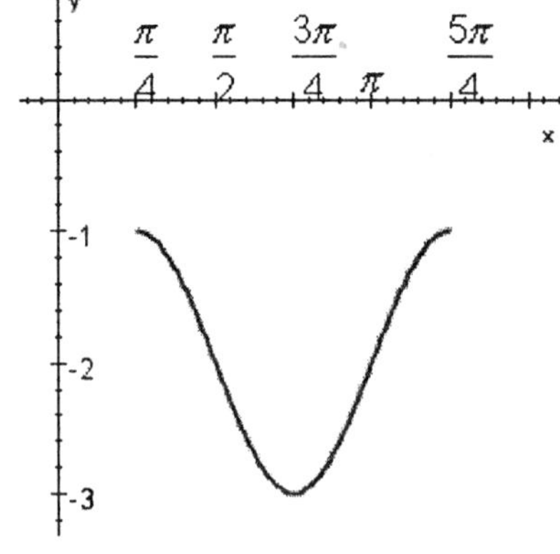

c.

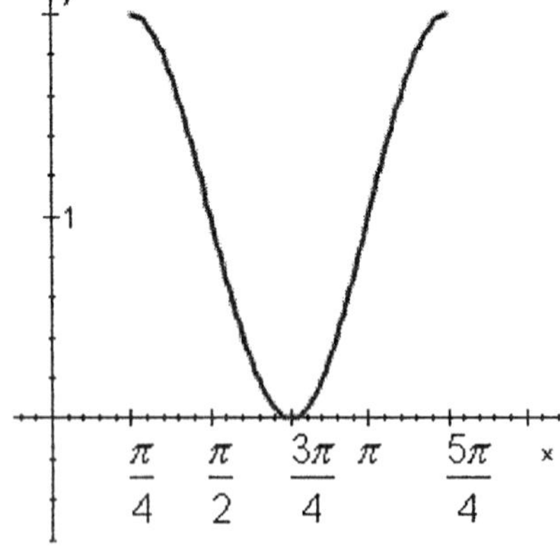

d.

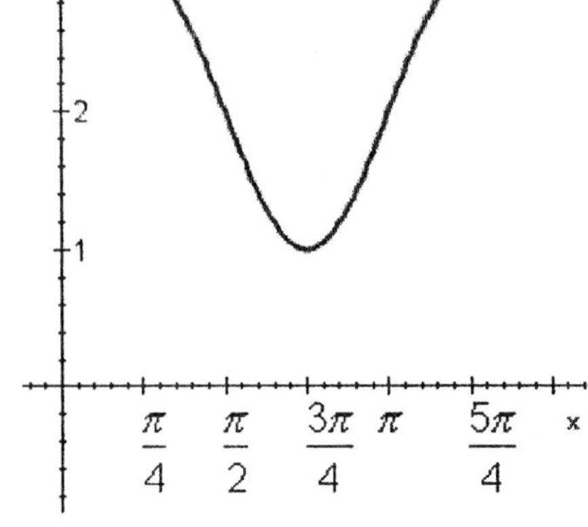

11. Graph the equation over the given interval.

$$y = 3\cos(2x - \pi),\ -\frac{\pi}{4} \le x \le \frac{3\pi}{2}$$

Select the correct answer.

a.

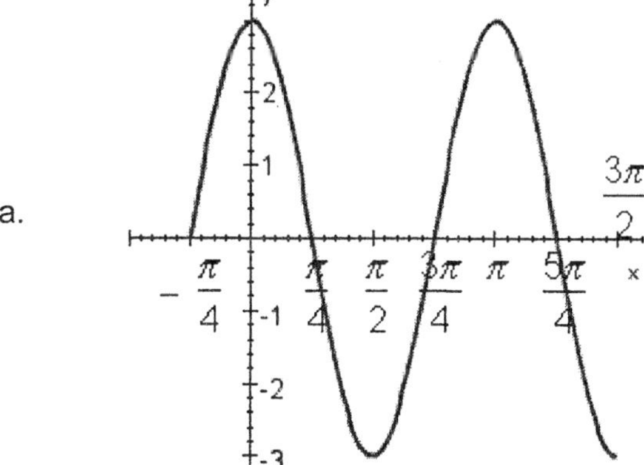

b.

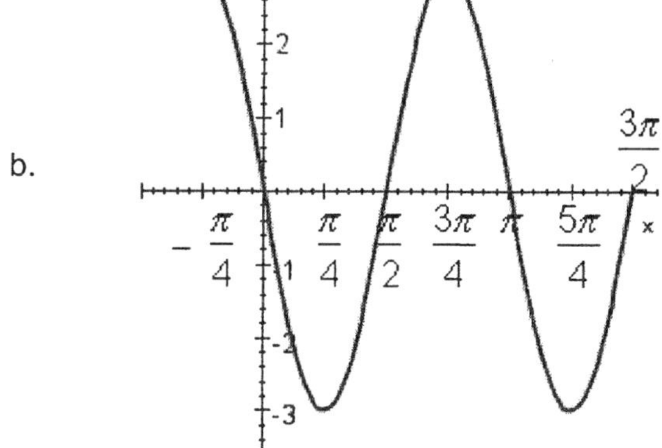

c.

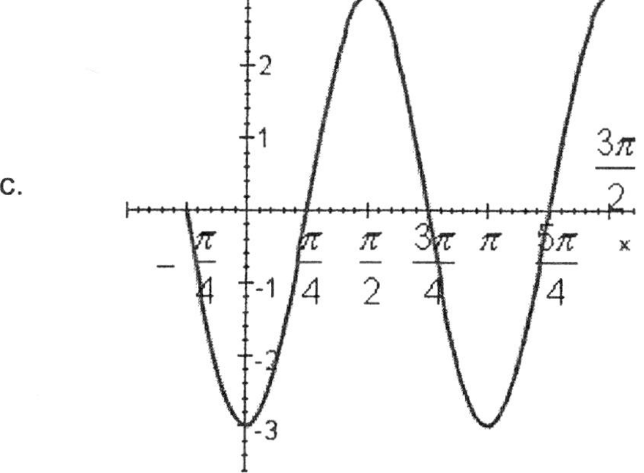

12. Sketch the graph of the equation below.

$$y = \tan\left(x - \frac{\pi}{4}\right)$$

Select the correct answer.

a.

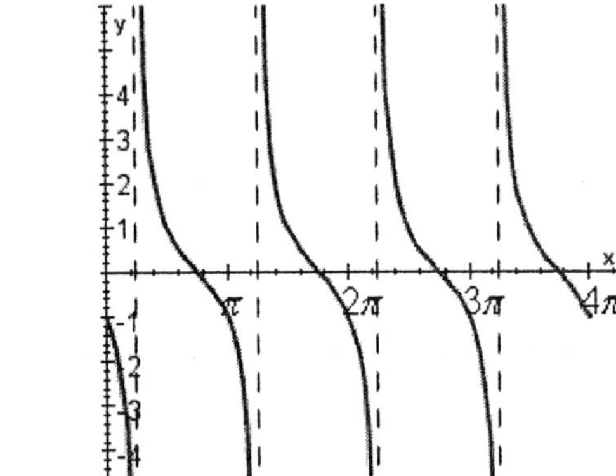

b.

c.

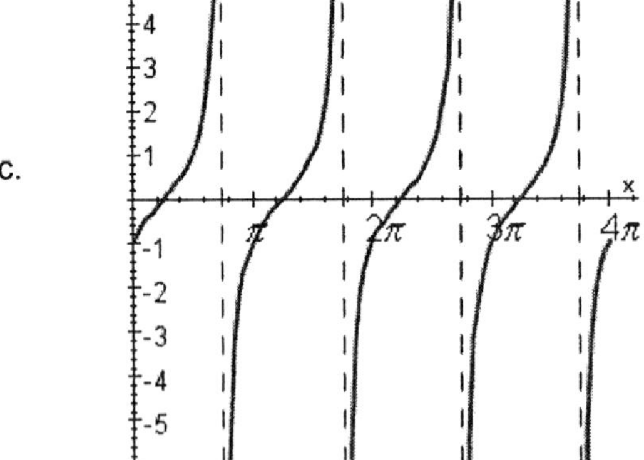

13. The graph below is one complete cycle of the graph of an equation containing a trigonometric function. Find an equation to match the graph.

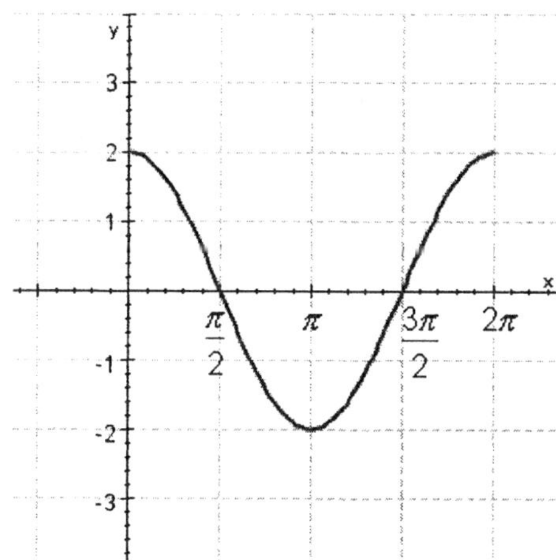

Select the correct answer.

a. $y = 2\cos x$

b. $y = 4\cos x$

c. $y = 6\cos x$

d. $y = 3\cos x$

14. The graph below is one complete cycle of the graph of an equation containing a trigonometric function. Find an equation to match the graph.

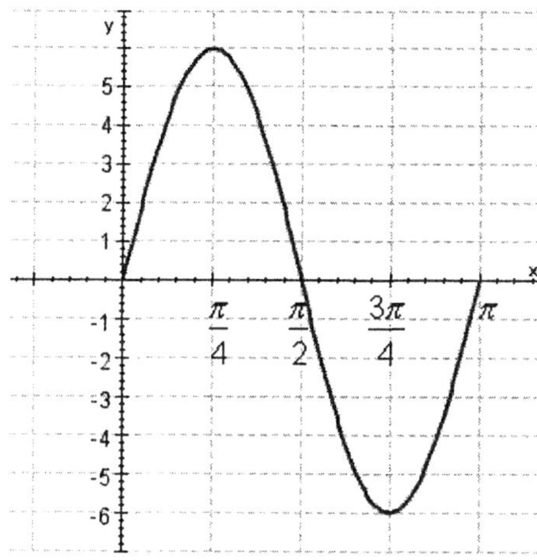

Select the correct answer.

a. $y = 6\cos 2x$

b. $y = 6\sin 2x$

c. $y = 6\sin x$

d. $y = 4\sin x$

15. The graph below is one complete cycle of the graph of an equation containing a trigonometric function. Find an equation to match the graph.

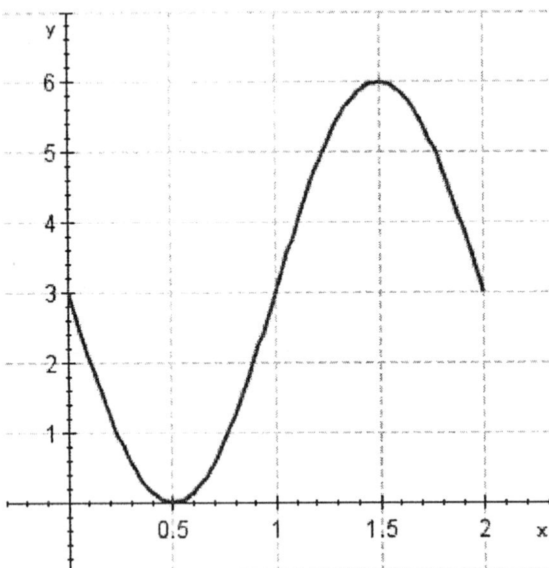

Select the correct answer.

a. $y = 3 - 3\sin \pi x$

b. $y = -1 + \sin \pi x$

c. $y = 1 + 2\sin \pi x$

d. $y = -3 - 4\sin \pi x$

e. $y = 4\sin \pi x$

16. The graph below is one complete cycle of the graph of an equation containing a trigonometric function. Find an equation to match the graph.

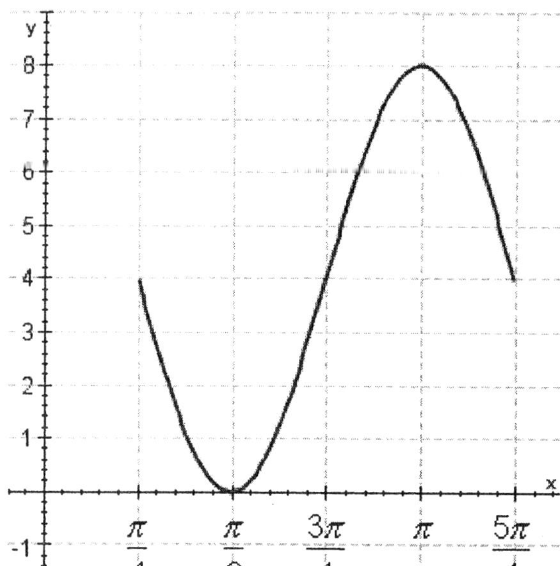

Select the correct answer.

a. $\quad y = 4 - \sin\left(2x - \dfrac{\pi}{2}\right)$

b. $\quad y = 5 - 2\sin\left(2x - \dfrac{\pi}{2}\right)$

c. $\quad y = 4 - 4\sin\left(2x - \dfrac{\pi}{2}\right)$

d. $\quad y = 4 - 3\sin\left(2x - \dfrac{\pi}{2}\right)$

e. $\quad y = 4 - 6\sin\left(2x - \dfrac{\pi}{2}\right)$

17. Use addition of *y*-coordinates to sketch the graph of the function between

$x = 0$ and $x = 4\pi$.

$y = 4 - 2\sin x$

Select the correct answer.

a.

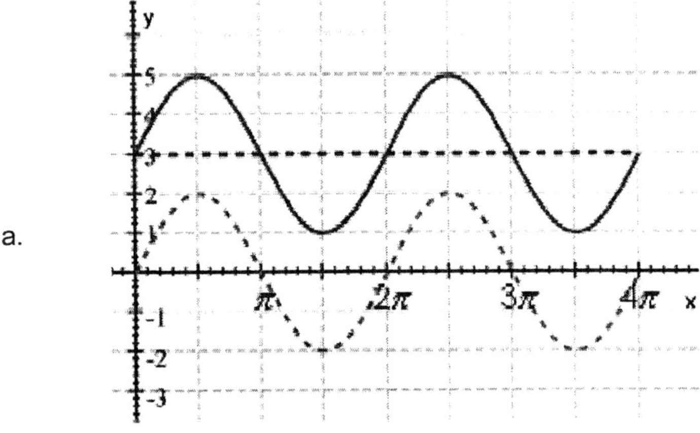

$$\cdots\cdots\ 4\ \cdots\cdots\ -2\sin x \qquad 4 - 2\sin x$$

b.

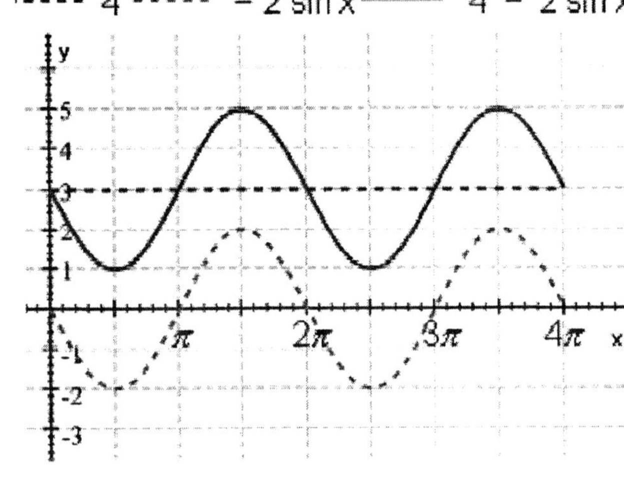

$$\cdots\cdots\ 4\ \cdots\cdots\ -2\sin x \qquad 4 - 2\sin x$$

c.

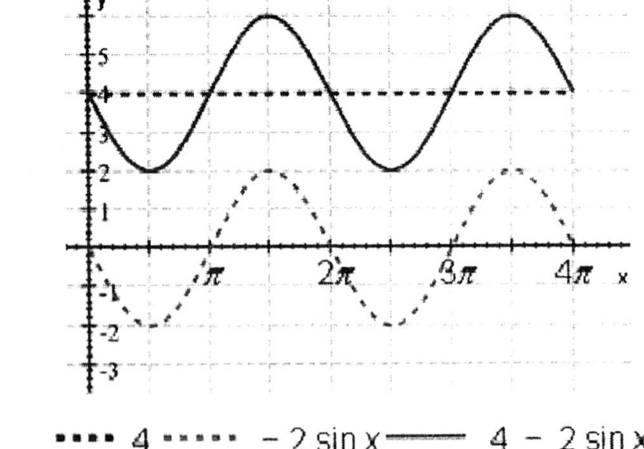

$$\cdots\cdots\ 4\ \cdots\cdots\ -2\sin x \qquad 4 - 2\sin x$$

18. Use addition of *y*-coordinates to sketch the graph of the function between

$x = 0$ and $x = 4\pi$.

$$y = \frac{1}{6}x + \cos x$$

Select the correct answer.

a.

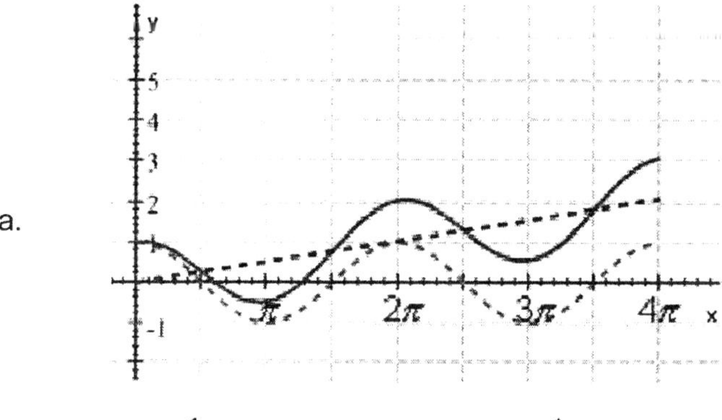

$$\cdots\cdots \ \frac{1}{6}x \qquad \cdots\cdots \ \cos x \qquad \underline{\quad\quad} \ \frac{1}{6}x + \cos x$$

b.

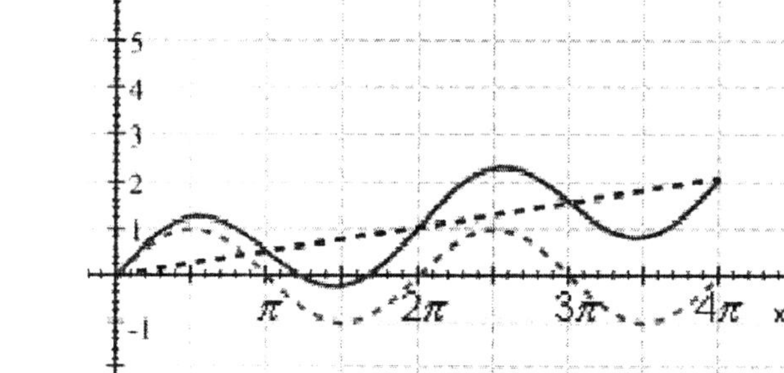

$$\cdots\cdots \ \frac{1}{6}x \qquad \cdots\cdots \ \cos x \qquad \underline{\quad\quad} \ \frac{1}{6}x + \cos x$$

c.

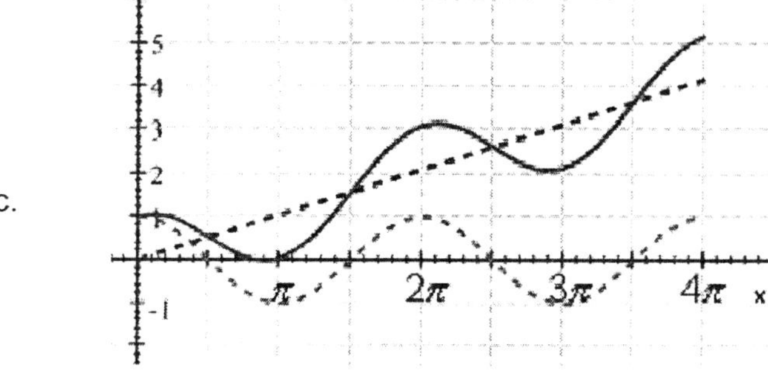

$$\cdots\cdots \ \frac{1}{6}x \qquad \cdots\cdots \ \cos x \qquad \underline{\quad\quad} \ \frac{1}{6}x + \cos x$$

19. Sketch the graph from

$x = 0$ to $x = 4\pi$.

$y = 2\sin x + \sin 2x$

Select the correct answer.

a.

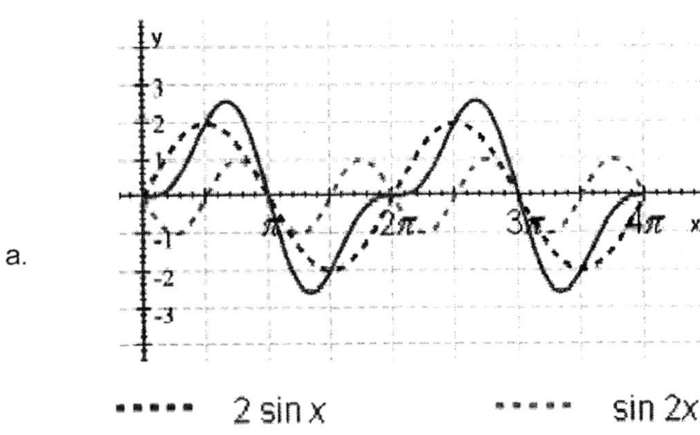

······   2 sin x   ·····   sin 2x

————   2 sin x + sin 2x

b.

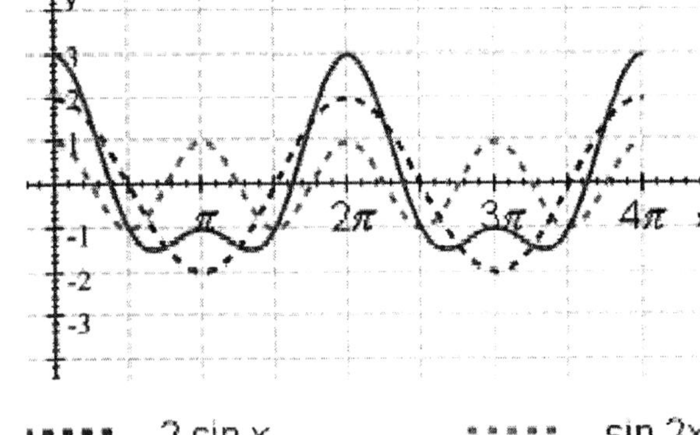

······   2 sin x   ·····   sin 2x

————   2 sin x + sin 2x

c.

······   2 sin x   ·····   sin 2x

————   2 sin x + sin 2x

20. Sketch the graph from

$x = 0$ to $x = 4\pi$.

$$y = \sin x - \frac{1}{2}\cos 2x$$

Select the correct answer.

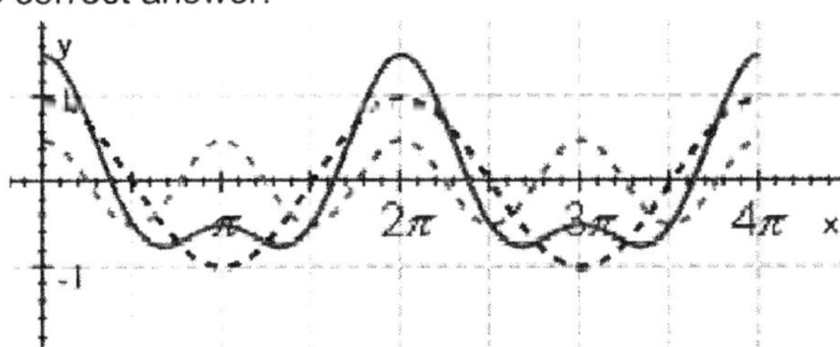

a.

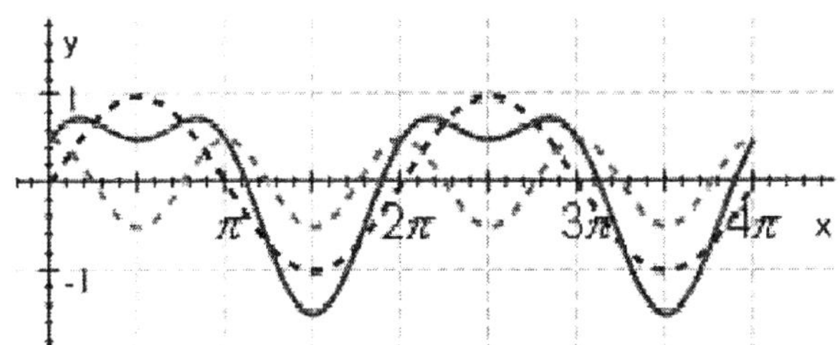

b.

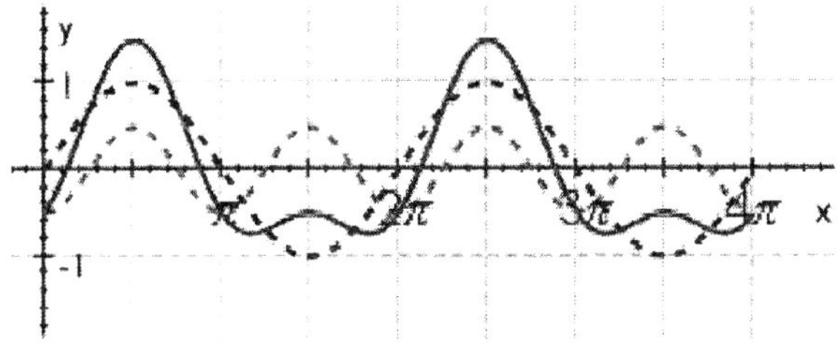

c.

21. Evaluate the expression without using a calculator, and write your answer in radians.

$$\sin^{-1}\left(\frac{1}{2}\right)$$

Select the correct answer.

a. $-\dfrac{\pi}{2}$

b. $-\dfrac{\pi}{6}$

c. $-\dfrac{\pi}{3}$

d. 0

e. $\dfrac{\pi}{6}$

22. Use a calculator to evaluate the expression to the nearest tenth of a degree.

$$\cos^{-1}(-0.7210)$$

Select the correct answer.

a. 136.1°

b. 130.6°

c. 131.7°

d. 138.4°

e. 137.3°

23. Evaluate without using a calculator.

$$\sin^{-1}\left(\sin 330°\right)$$

Select the correct answer.

a. $-45°$

b. $-180°$

c. $-90°$

d. $-30°$

e. $-150°$

24. Evaluate without using a calculator.

$$\csc\left(\tan^{-1}\frac{3}{4}\right)$$

Select the correct answer.

a. $\dfrac{8}{9}$

b. $\dfrac{5}{7}$

c. $\dfrac{1}{2}$

d. $\dfrac{5}{3}$

25. Simplify $\sin^{-1}\left(\sin x\right)$

If $0 \le x \le \dfrac{\pi}{2}$.

Select the correct answer.

a. $\pi - x$

b. $-x$

c. $x - \pi$

d. $-x - \pi$

e. x

McKeague/Turner - Trigonometry 5e Chapter 4 Form C

1. c

2. c

3. b

4. d

5. a

6. b

7. c

8. b

9. b

10. d

11. c

12. c

13. a

14. b

15. a

16. c

17. c

18. a

19. c

20. c

21. e

22. a

23. d

24. d

25. e

1. mctr.04.01.01m_NoAlgs
2. mctr.04.01.08m_NoAlgs
3. mctr.04.01.18m_NoAlgs
4. mctr.04.01.29m_NoAlgs
5. mctr.04.02.06m_NoAlgs
6. mctr.04.02.28m_NoAlgs
7. mctr.04.02.42m_NoAlgs
8. mctr.04.02.53m_NoAlgs
9. mctr.04.03.09m_NoAlgs
10. mctr.04.03.23m_NoAlgs
11. mctr.04.03.33m_NoAlgs
12. mctr.04.03.47m_NoAlgs
13. mctr.04.04.07m_NoAlgs
14. mctr.04.04.16m_NoAlgs
15. mctr.04.04.21m_NoAlgs
16. mctr.04.04.30m_NoAlgs
17. mctr.04.05.05m_NoAlgs
18. mctr.04.05.10m_NoAlgs
19. mctr.04.05.14m_NoAlgs
20. mctr.04.05.22m_NoAlgs
21. mctr.04.06.13m_NoAlgs
22. mctr.04.06.38m_NoAlgs
23. mctr.04.06.52m_NoAlgs
24. mctr.04.06.66m_NoAlgs
25. mctr.04.06.73m_NoAlgs

1. Graph one complete cycle for the function.

$$y = 4 \csc \frac{1}{6} x$$

Select the correct answer.

a.

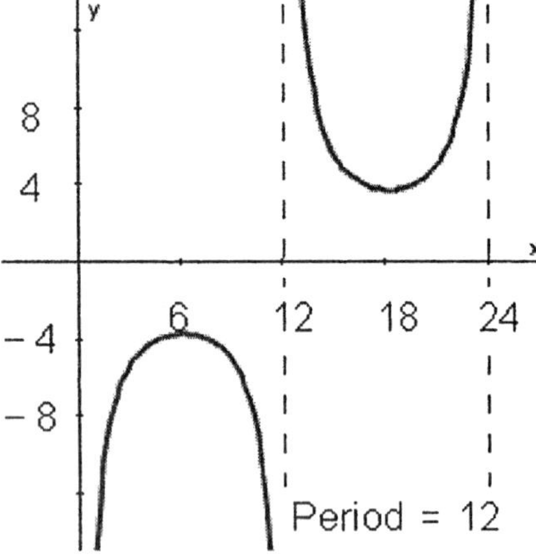

b.

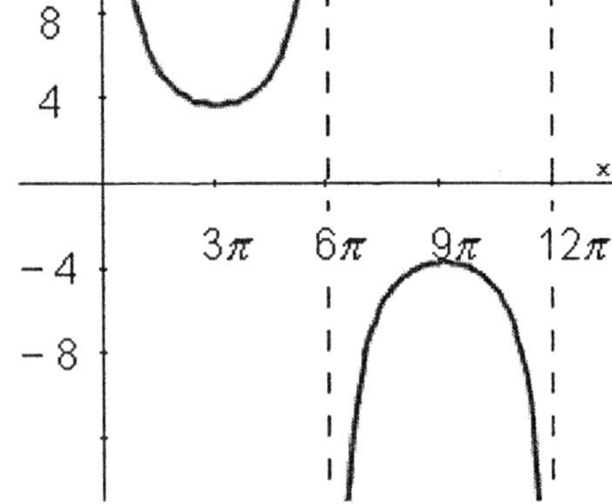

c.

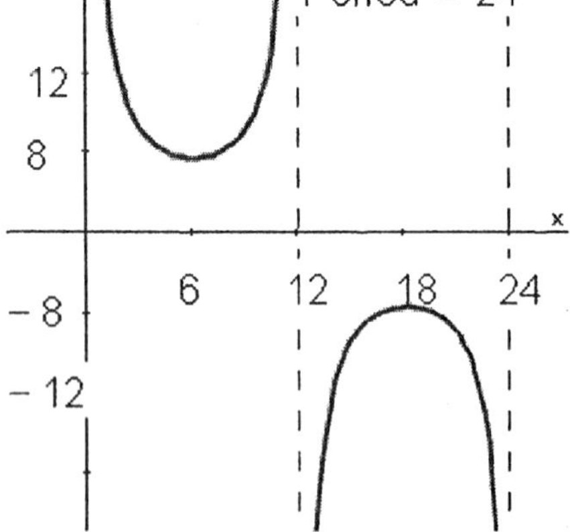

2. Sketch the graph from

$x = 0$ to $x = 4\pi$.

$y = \sin x - \dfrac{1}{2}\cos 2x$

Select the correct answer.

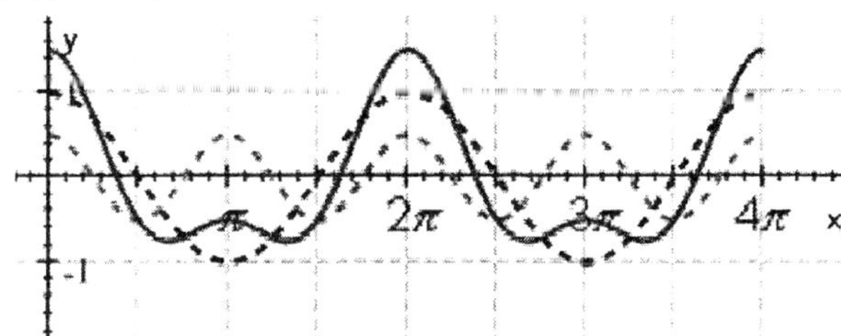

a.

    `....` $\sin x$ `......` $-\dfrac{1}{2}\cos 2x$

    `——` $\sin x - \dfrac{1}{2}\cos 2x$

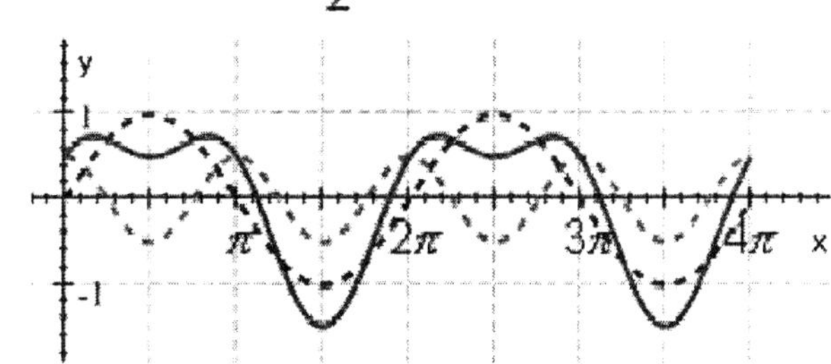

b.

    `....` $\sin x$ `.....` $-\dfrac{1}{2}\cos 2x$

    `——` $\sin x - \dfrac{1}{2}\cos 2x$

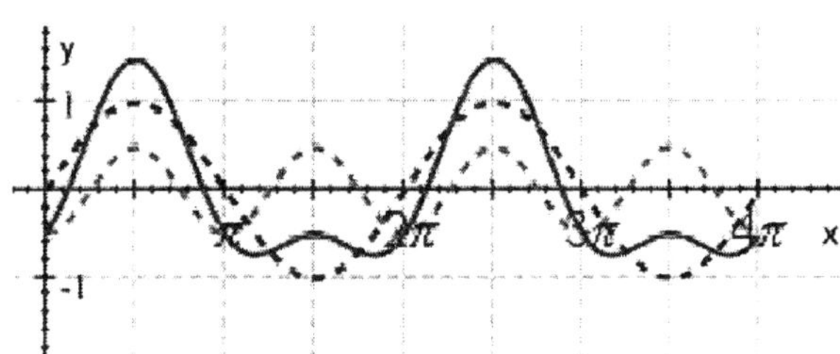

c.

    `- - -` $\sin x$ `......` $-\dfrac{1}{2}\cos 2x$

    `——` $\sin x - \dfrac{1}{2}\cos 2x$

3. Graph the function over the given interval.

$$y = 3\cos \pi x, \quad -2 \le x \le 4$$

Select the correct answer.

a.

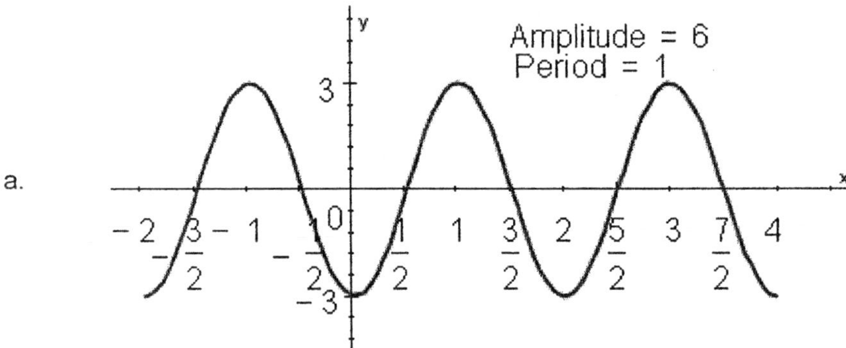

b.

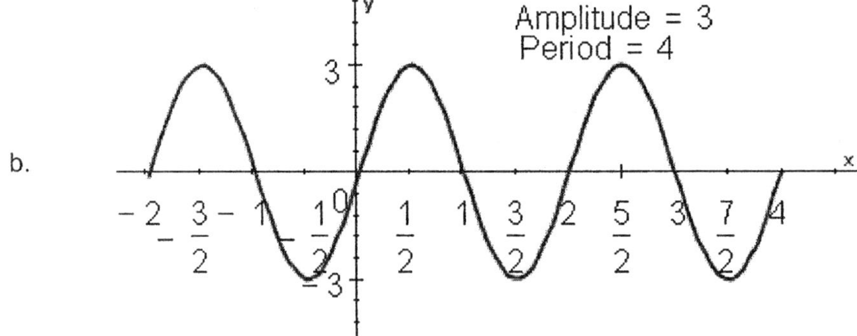

c.

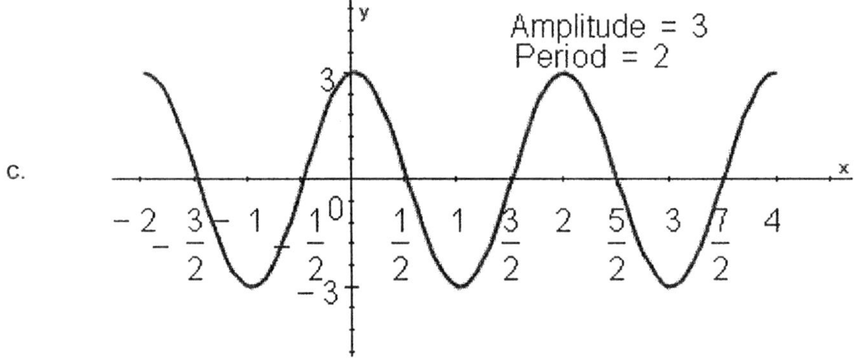

4. Graph the equation over the given interval.

$$y = 3\cos(2x - \pi), \quad -\frac{\pi}{4} \leq x \leq \frac{3\pi}{2}$$

Select the correct answer.

a.

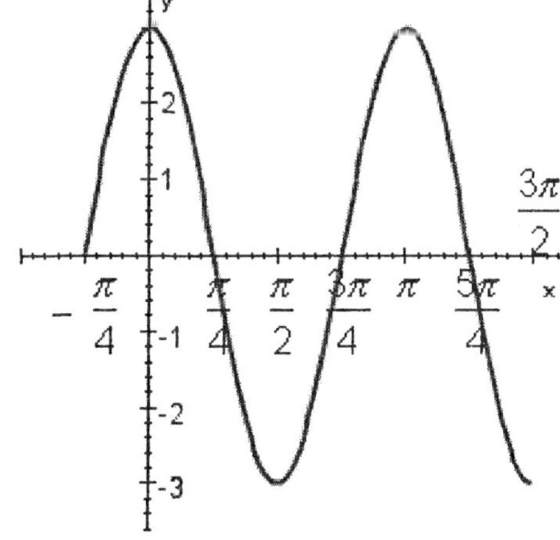

b.

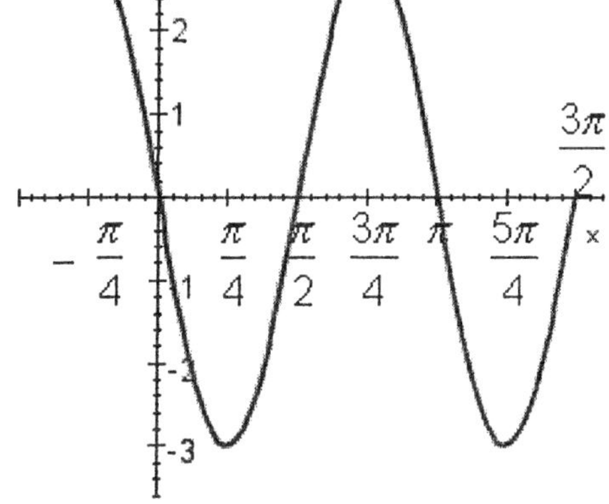

c.

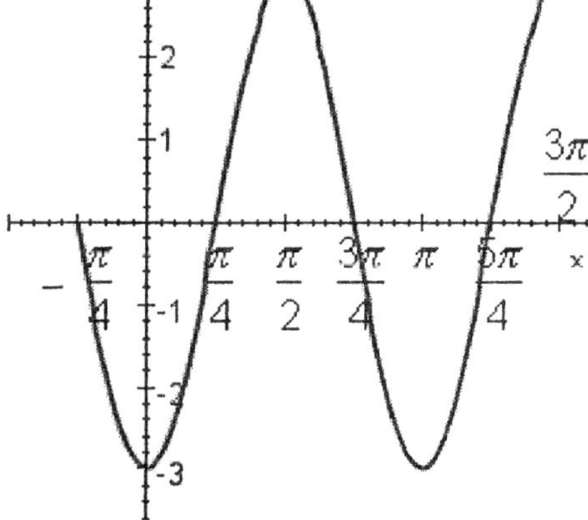

5. The graph below is one complete cycle of the graph of an equation containing a trigonometric function. Find an equation to match the graph.

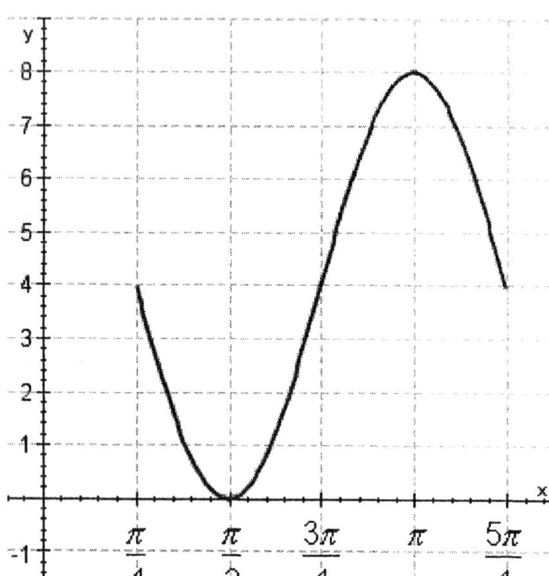

Select the correct answer.

a. $y = 4 - \sin\left(2x - \dfrac{\pi}{2}\right)$

b. $y = 5 - 2\sin\left(2x - \dfrac{\pi}{2}\right)$

c. $y = 4 - 4\sin\left(2x - \dfrac{\pi}{2}\right)$

d. $y = 4 - 3\sin\left(2x - \dfrac{\pi}{2}\right)$

e. $y = 4 - 6\sin\left(2x - \dfrac{\pi}{2}\right)$

6. Use the graph of the equation $y = \cos\left(2x - \dfrac{\pi}{2}\right)$ shown below to graph one complete cycle of the equation $y = 2 + \cos\left(2x - \dfrac{\pi}{2}\right)$.

$$y = \cos\left(2x - \frac{\pi}{2}\right)$$

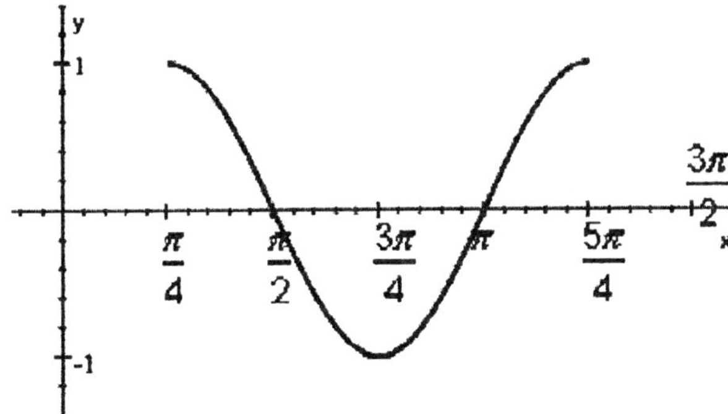

Select the correct answer.

a.
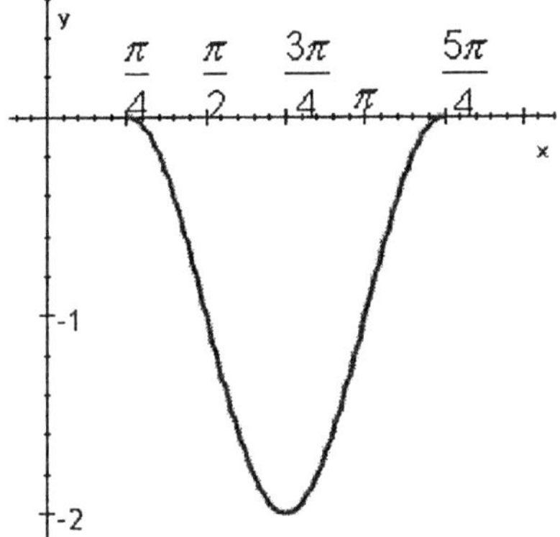

b.
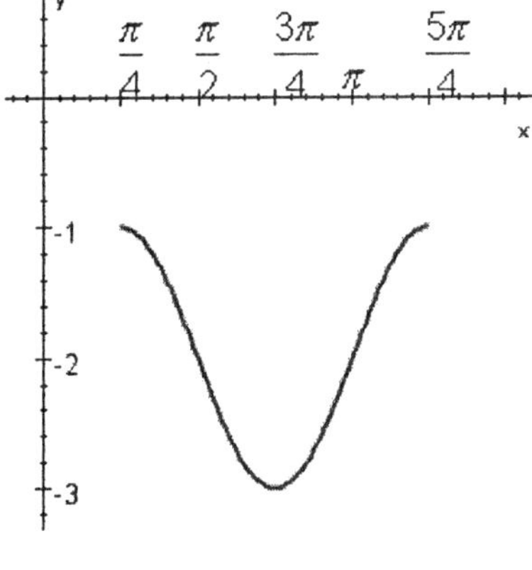

c.
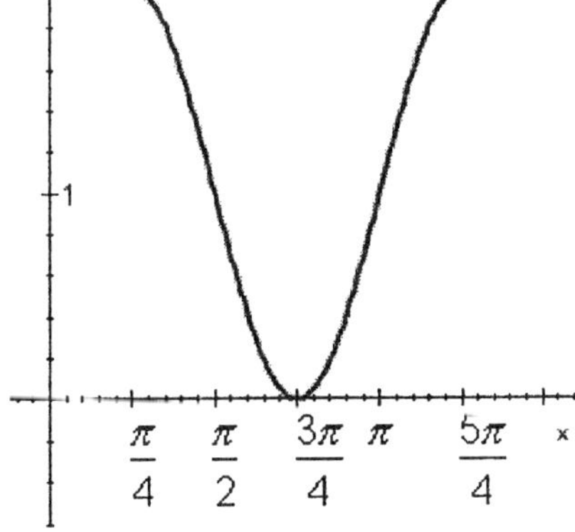

d.
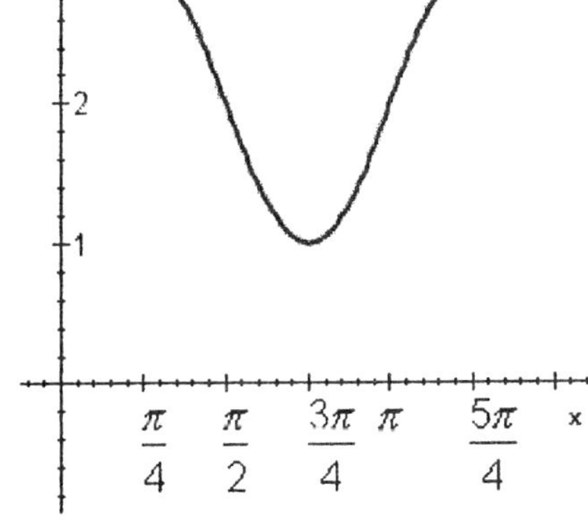

7. Sketch the graph of $y = \cos x$ for x between 0 and 2π.

   Select the correct answer.

a.

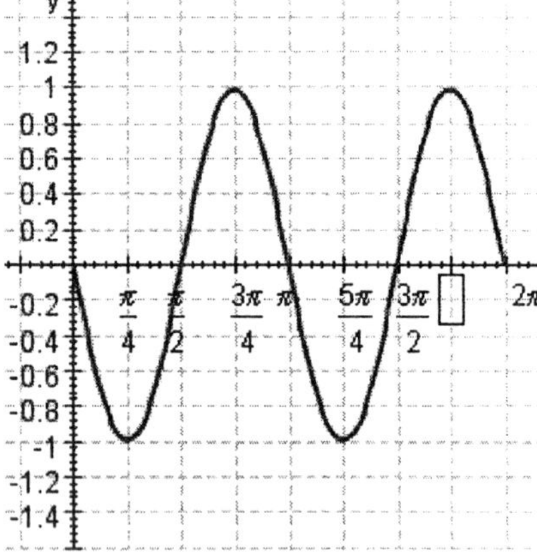

b.

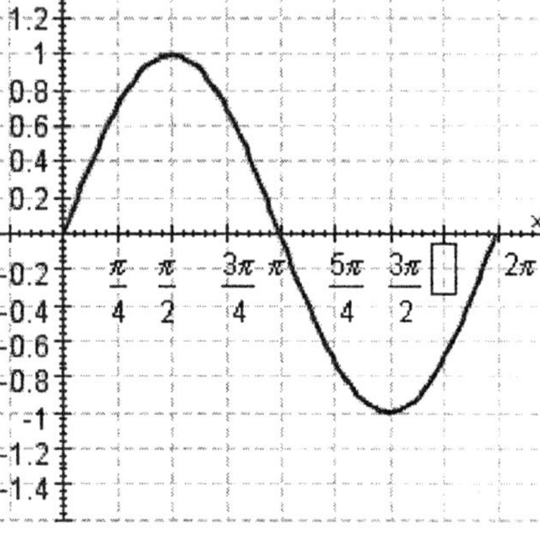

c.

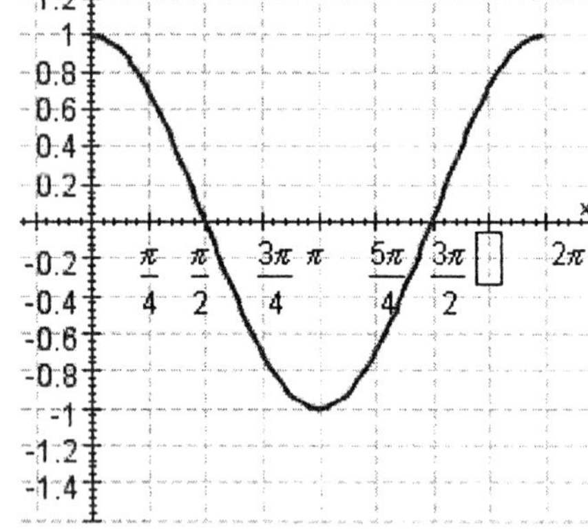

8. Evaluate without using a calculator.

$$\csc\left(\tan^{-1}\frac{3}{4}\right)$$

Select the correct answer.

a. $\dfrac{8}{9}$

b. $\dfrac{5}{7}$

c. $\dfrac{1}{2}$

d. $\dfrac{5}{3}$

9. Give the amplitude and period of the graph.

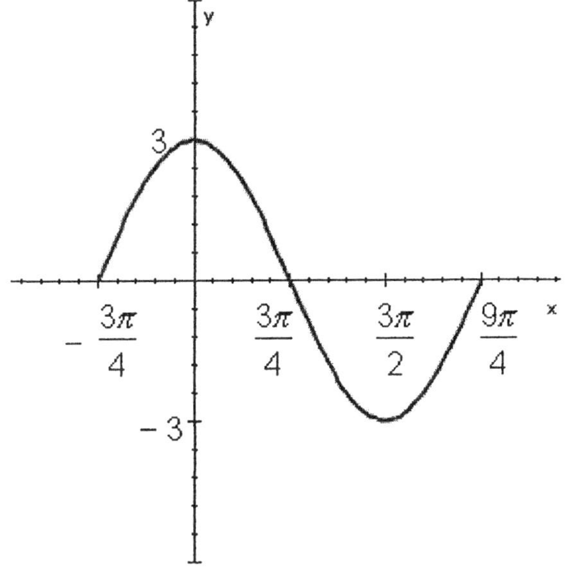

Select the correct answer.

a. Amplitude = 3, period = π.

b. Amplitude = 6, period = π.

c. Amplitude = 6, period = $\dfrac{5\pi}{2}$.

d. Amplitude = 3, period = 3π.

10. Sketch the graph from

$x = 0$ to $x = 4\pi$.

$y = 2\sin x + \sin 2x$

Select the correct answer.

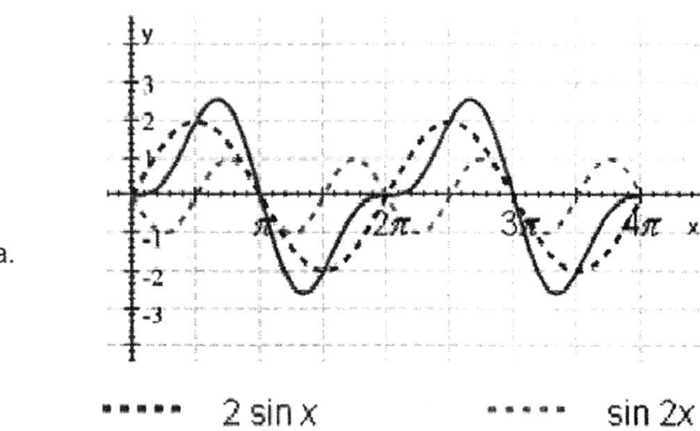

a.

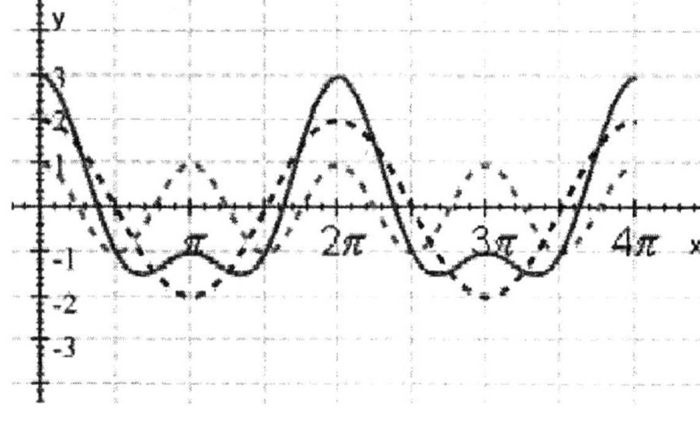

b.

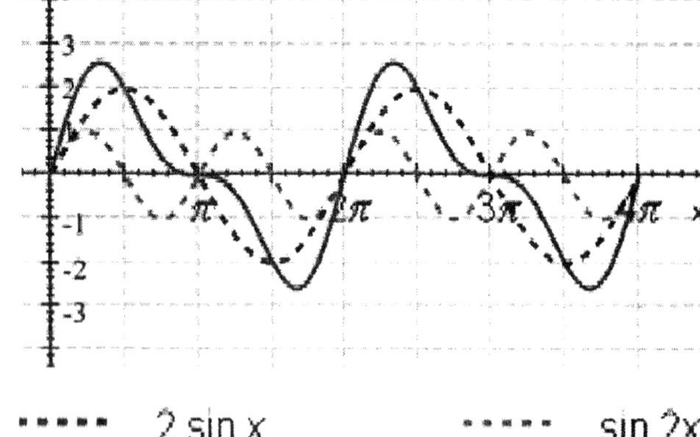

c.

11. For the given equation, first identify the phase shift and then sketch one complete cycle of the graph. Graph $y = \cos x$ on the same coordinate system.

$$y = \cos\left(x + \frac{\pi}{4}\right)$$

Select the correct answer.

a.
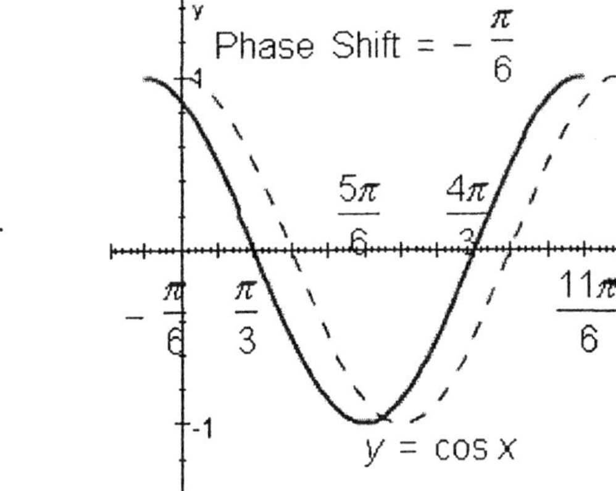

b.
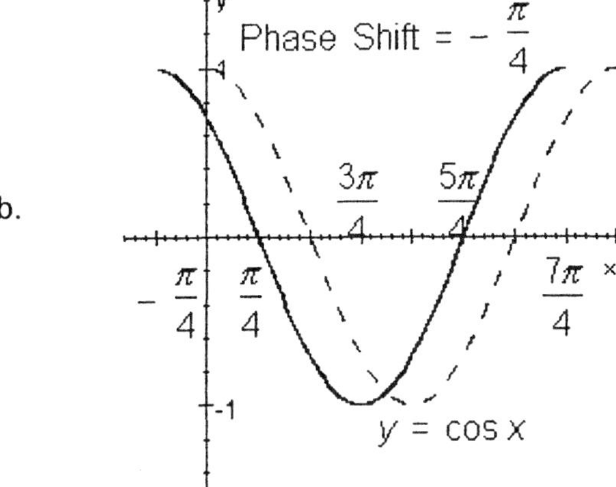

c.
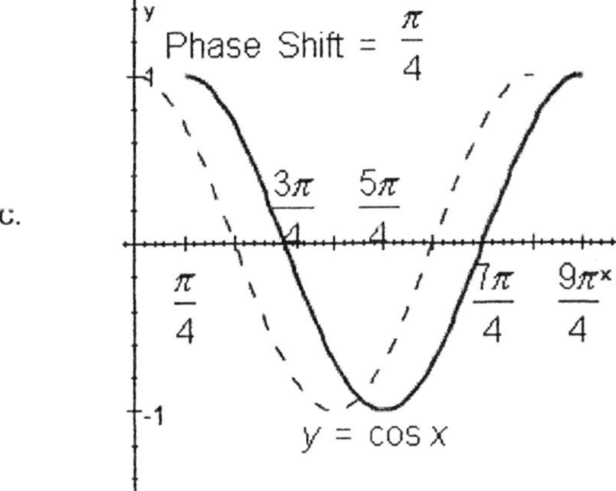

12. The graph below is one complete cycle of the graph of an equation containing a trigonometric function. Find an equation to match the graph.

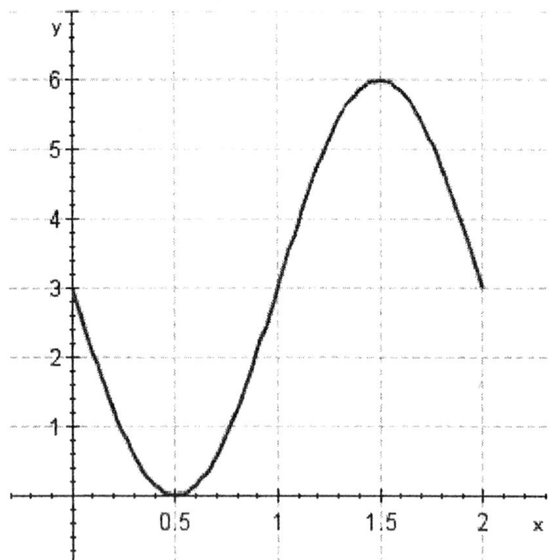

Select the correct answer.

a. $y = 3 - 3\sin \pi x$

b. $y = -1 + \sin \pi x$

c. $y = 1 + 2\sin \pi x$

d. $y = -3 - 4\sin \pi x$

e. $y = 4\sin \pi x$

13. Find all values of x for which the following are true.

$\sec x = 1$

Select the correct answer.

a. $\dfrac{\pi}{3} + 2k\pi$

b. $2k\pi$

c. $\dfrac{\pi}{4} + 2k\pi$

d. $-\dfrac{\pi}{3} + 2k\pi$

e. $\dfrac{\pi}{2} + k\pi$

14. The current in an alternating circuit varies in intensity with time. If I represents the intensity of the current and t represents time, then the relationship between I and t is given by

$$I = 16 \sin 160\pi t$$

Where I is measured in amperes and t is measured in seconds.

Find the maximum value of I and the time it takes for I to go through one complete cycle.

Select the correct answer.

a.   The maximum value of I is 16 amperes; one complete cycle takes $\dfrac{1}{160}$ seconds.

b.   The maximum value of I is 16 amperes; one complete cycle takes $\dfrac{1}{80}$ seconds.

c.   The maximum value of I is 32 amperes; one complete cycle takes $\dfrac{1}{80}$ seconds.

d.   The maximum value of I is 32 amperes; one complete cycle takes $\dfrac{1}{92}$ seconds.

e.   The maximum value of I is 64 amperes; one complete cycle takes $\dfrac{1}{160}$ seconds.

15. Sketch the graph of the equation below.

$$y = \tan\left(x - \frac{\pi}{4}\right)$$

Select the correct answer.

a.

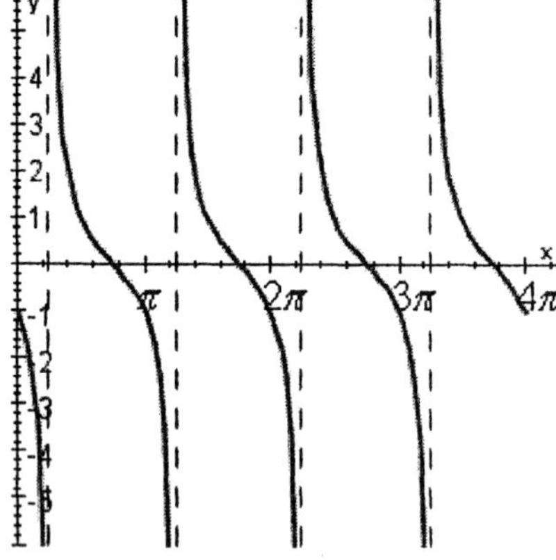

b.

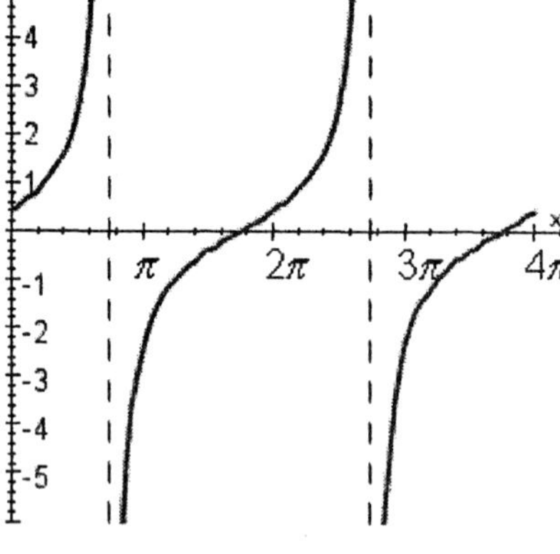

c.

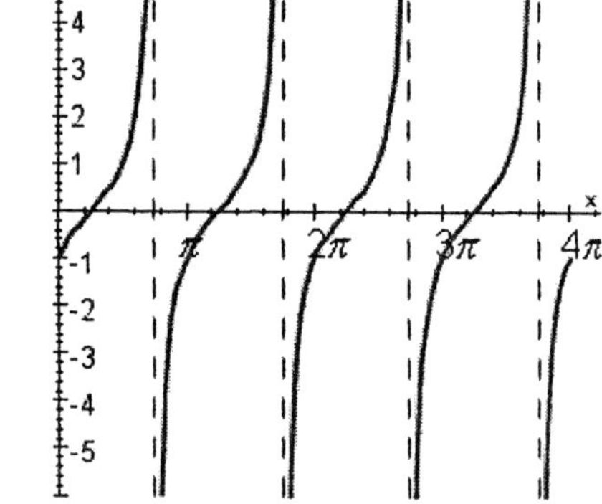

16. Use addition of *y*-coordinates to sketch the graph of the function between

$$x = 0 \text{ and } x = 4\pi.$$

$$y = \frac{1}{6}x + \cos x$$

Select the correct answer.

a.

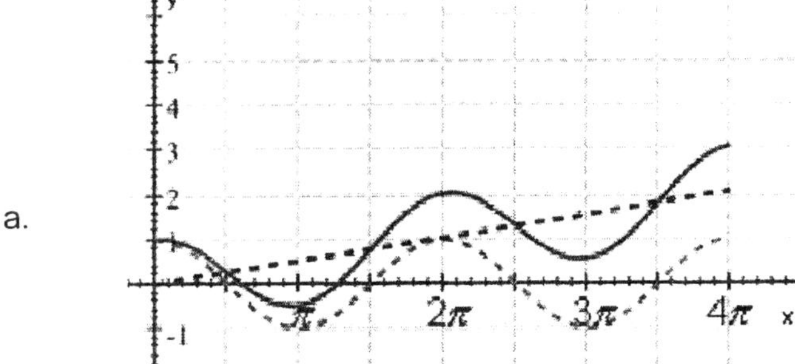

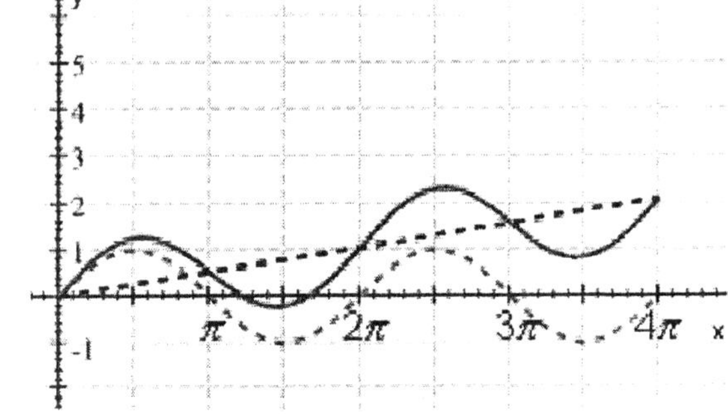

b.

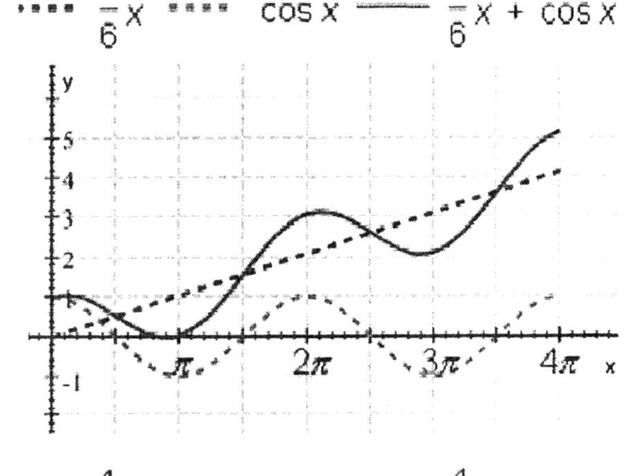

c.

17. Evaluate the expression without using a calculator, and write your answer in radians.

$$\sin^{-1}\left(\frac{1}{2}\right)$$

Select the correct answer.

a. $-\dfrac{\pi}{2}$

b. $-\dfrac{\pi}{6}$

c. $-\dfrac{\pi}{3}$

d.   0

e. $\dfrac{\pi}{6}$

18. Evaluate without using a calculator.

$$\sin^{-1}\left(\sin 330°\right)$$

Select the correct answer.

a. $-45°$

b. $-180°$

c. $-90°$

d. $-30°$

e. $-150°$

19. The graph below is one complete cycle of the graph of an equation containing a trigonometric function. Find an equation to match the graph.

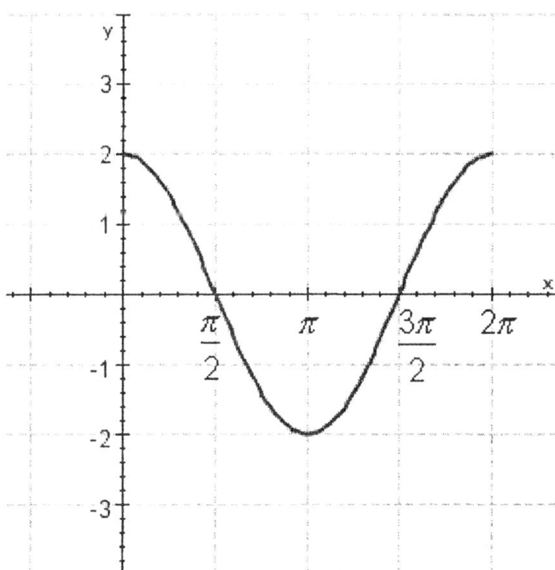

Select the correct answer.

a. $y = 2\cos x$

b. $y = 4\cos x$

c. $y = 6\cos x$

d. $y = 3\cos x$

e. $y = 5\cos x$

20. Graph one complete cycle of the graph.

$$y = \sin \frac{\pi}{4} x$$

Select the correct answer.

a.

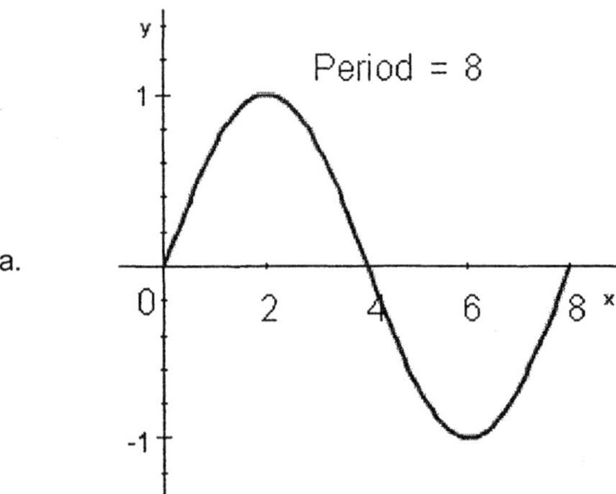

b.

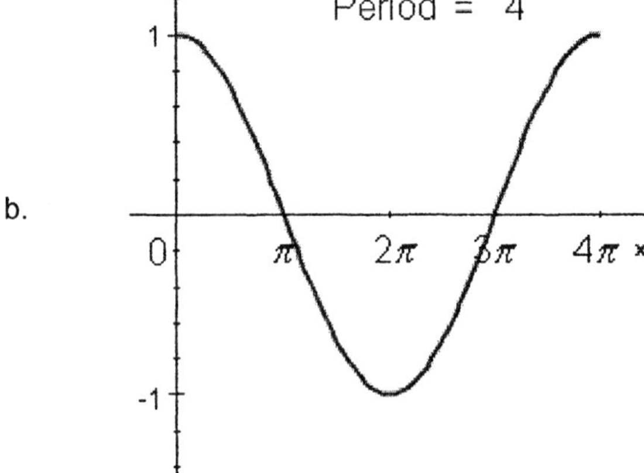

c.

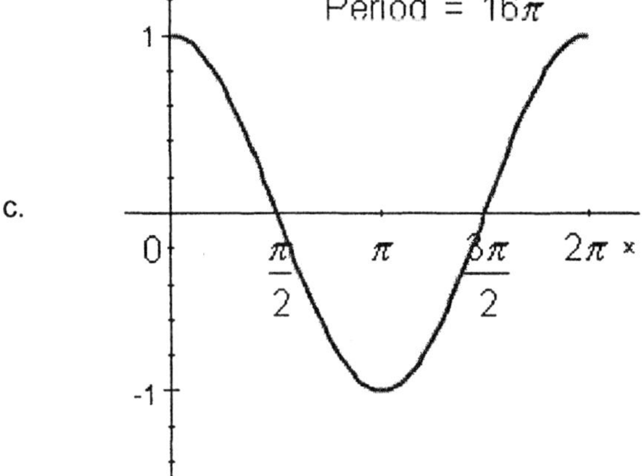

21. Sketch the graph of $y = \cos x$ between $x = -4\pi$ and $x = 4\pi$.

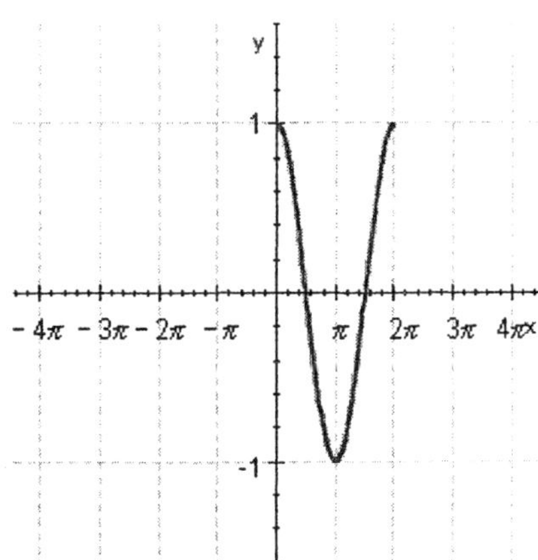

Select the correct answer.

a.

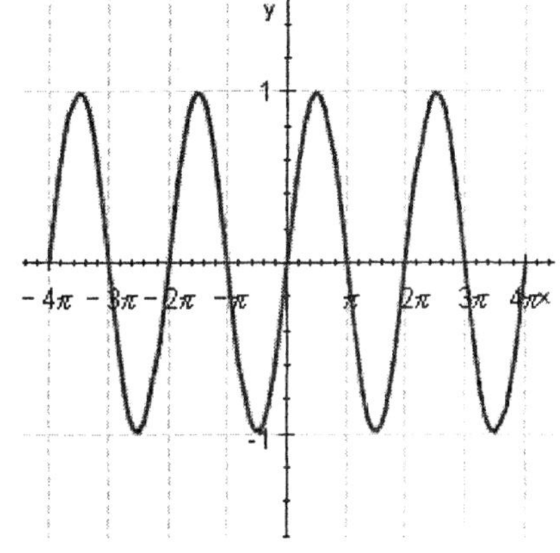

b.

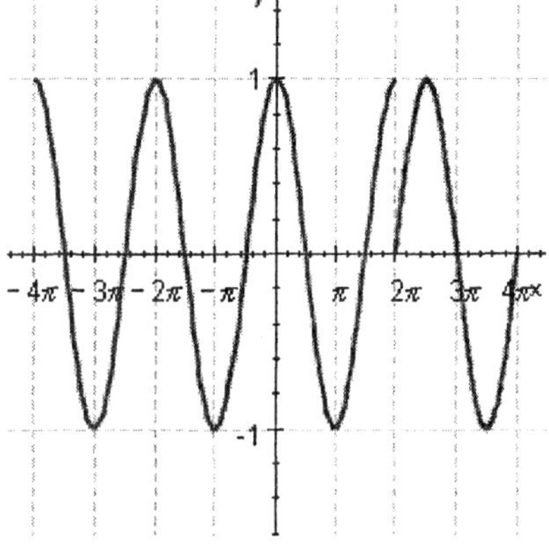

c.

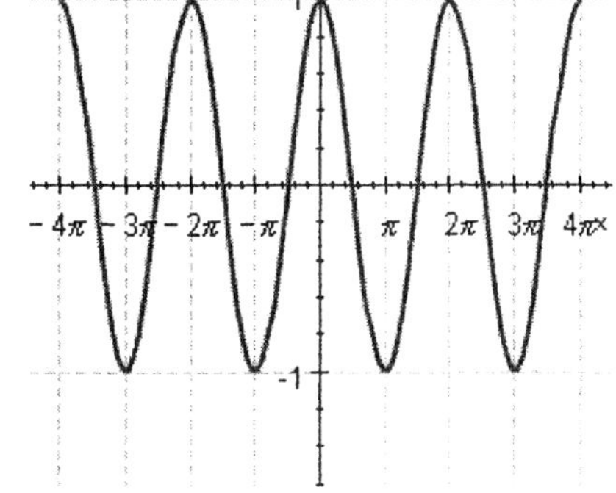

d.

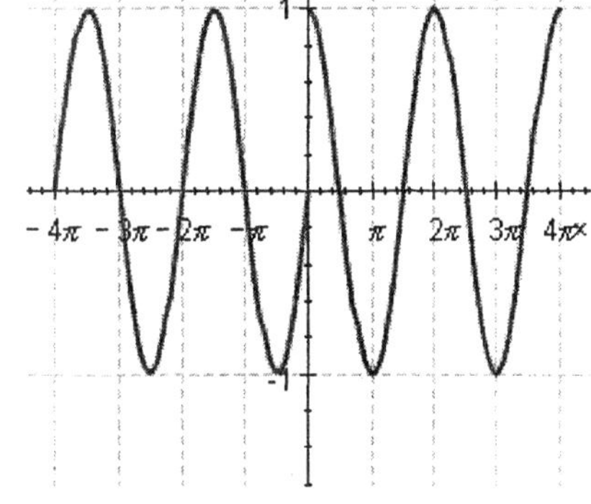

22. Use addition of *y*-coordinates to sketch the graph of the function between

$x = 0$ and $x = 4\pi$.

$y = 4 - 2\sin x$

Select the correct answer.

a.

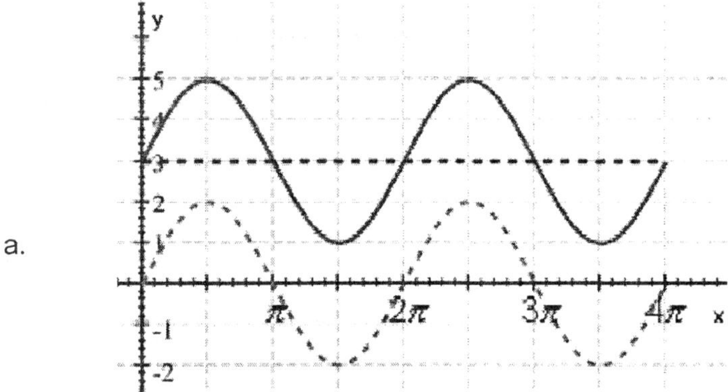

b.

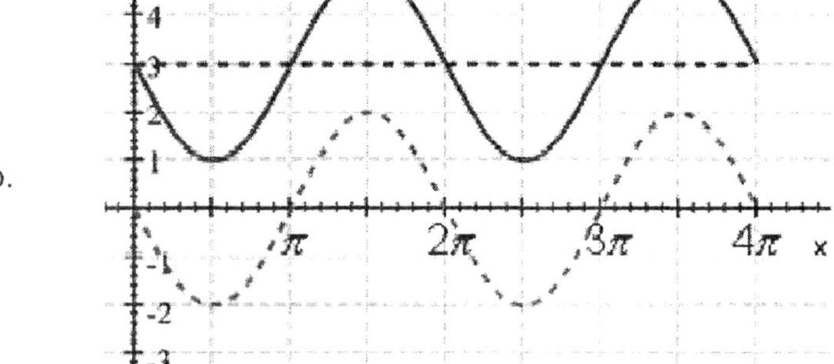

c.

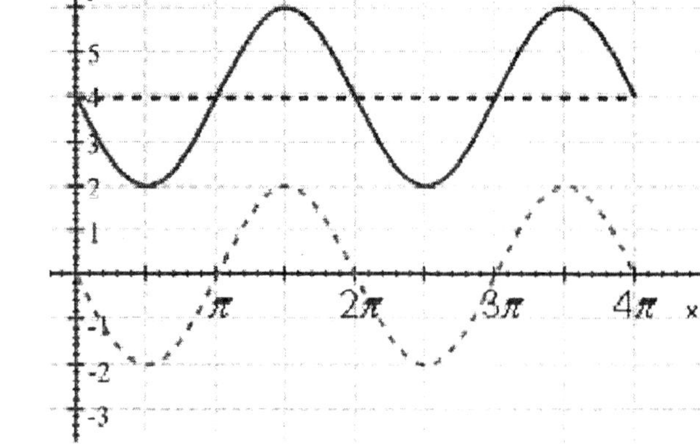

23. The graph below is one complete cycle of the graph of an equation containing a trigonometric function. Find an equation to match the graph.

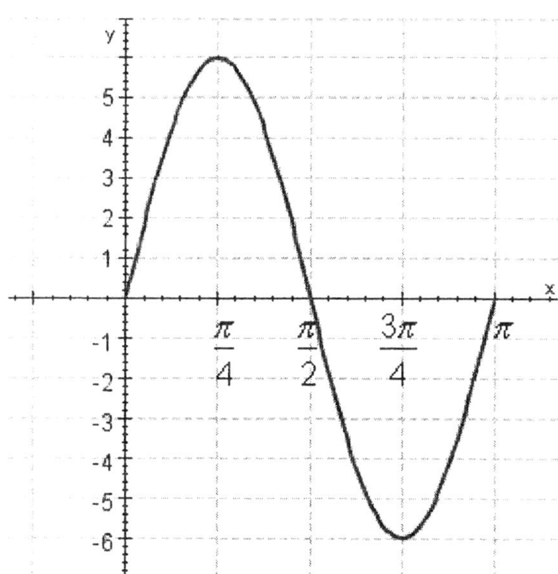

Select the correct answer.

a. $y = 6\cos 2x$

b. $y = 6\sin 2x$

c. $y = 6\sin x$

d. $y = 4\sin x$

e. $y = 4\sin 2x$

24. Use a calculator to evaluate the expression to the nearest tenth of a degree.

$$\cos^{-1}(-0.7210)$$

Select the correct answer.

a. 136.1°

b. 130.6°

c. 131.7°

d. 138.4°

e. 137.3°

25. Simplify

$$\sin^{-1}(\sin x) \quad \text{if} \quad 0 \le x \le \frac{\pi}{2}.$$

Select the correct answer.

a. $\pi - x$

b. $-x$

c. $x - \pi$

d. $-x - \pi$

e. x

1. b

2. c

3. c

4. c

5. c

6. d

7. c

8. d

9. d

10. c

11. b

12. a

13. b

14. b

15. c

16. a

17. e

18. d

19. a

20. a

21. c

22. c

23. b

24. a

25. e

1. mctr.04.02.28m_NoAlgs
2. mctr.04.05.22m_NoAlgs
3. mctr.04.02.42m_NoAlgs
4. mctr.04.03.33m_NoAlgs
5. mctr.04.04.30m_NoAlgs
6. mctr.04.03.23m_NoAlgs
7. mctr.04.01.01m_NoAlgs
8. mctr.04.06.66m_NoAlgs
9. mctr.04.01.29m_NoAlgs
10. mctr.04.05.14m_NoAlgs
11. mctr.04.03.09m_NoAlgs
12. mctr.04.04.21m_NoAlgs
13. mctr.04.01.18m_NoAlgs
14. mctr.04.02.53m_NoAlgs
15. mctr.04.03.47m_NoAlgs
16. mctr.04.05.10m_NoAlgs
17. mctr.04.06.13m_NoAlgs
18. mctr.04.06.52m_NoAlgs
19. mctr.04.04.07m_NoAlgs
20. mctr.04.02.06m_NoAlgs
21. mctr.04.01.08m_NoAlgs
22. mctr.04.05.05m_NoAlgs
23. mctr.04.04.16m_NoAlgs
24. mctr.04.06.38m_NoAlgs
25. mctr.04.06.73m_NoAlgs

1.

Make a table of values for $y = \sin x$ using multiples of $\dfrac{\pi}{4}$ for x.

x	$\sin x$
0	
$\dfrac{\pi}{4}$	
$\dfrac{\pi}{2}$	
$\dfrac{3\pi}{4}$	
π	
$\dfrac{5\pi}{4}$	
$\dfrac{3\pi}{2}$	
$\dfrac{7\pi}{4}$	
2π	

Use the entries in the table to sketch the graph of the function for x between 0 and 2π

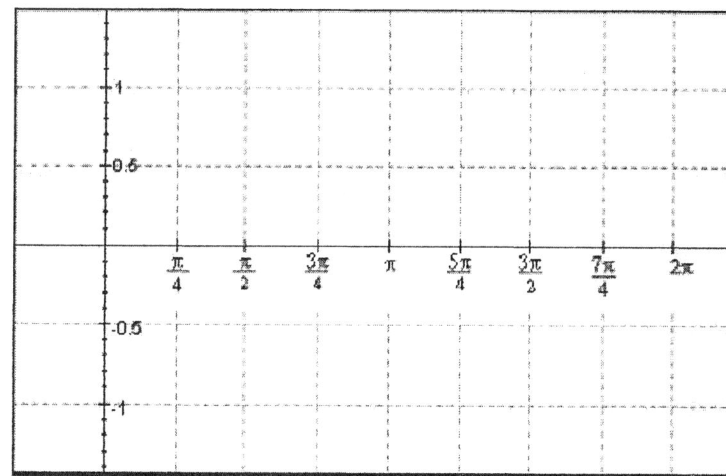

2. Find all values of x for which the following are true.

   $\sec x = 1$

   Select the correct answer.

   a. $\dfrac{\pi}{3} + 2k\pi$

   b. $2k\pi$

   c. $\dfrac{\pi}{4} + 2k\pi$

   d. $-\dfrac{\pi}{3} + 2k\pi$

   e. $\dfrac{\pi}{2} + k\pi$

3. Give the amplitude and period of the graph.

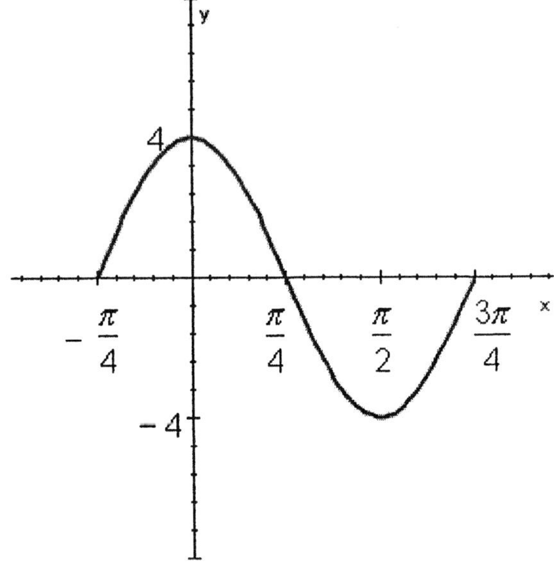

   amplitude.

   period.

4.

Graph one complete cycle of $y = \dfrac{3}{10}\cos x$. Label the axes accurately.

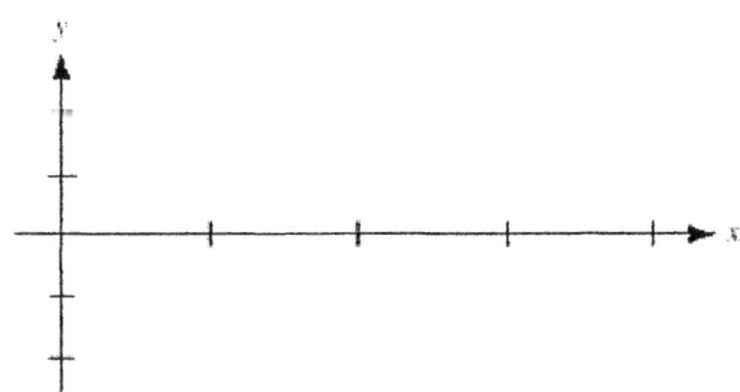

Identify the amplitude for the graph (if defined).

5. Graph one complete cycle for the function. Label the axes so that the amplitude (if defined) and period are easy to read.

$y = 3\sin 2x$

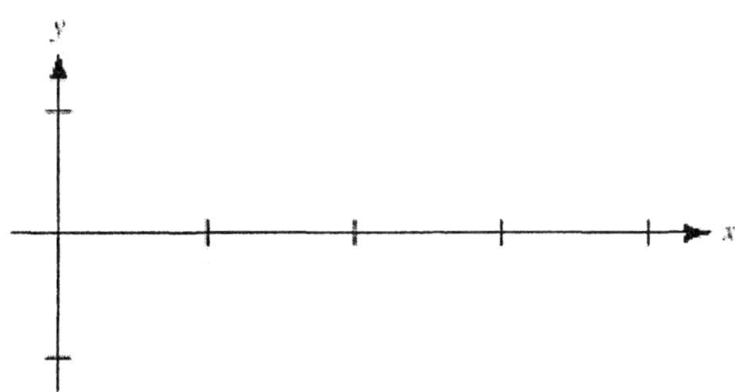

6. Graph one complete cycle for the function.

$$y = 4 \csc \frac{1}{6} x$$

Select the correct answer.

a.

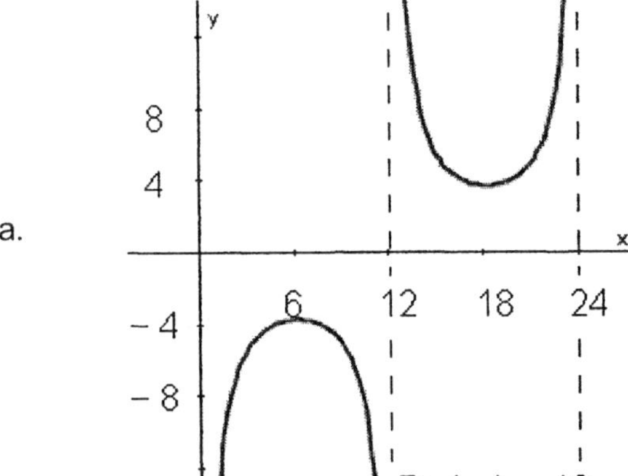

b.

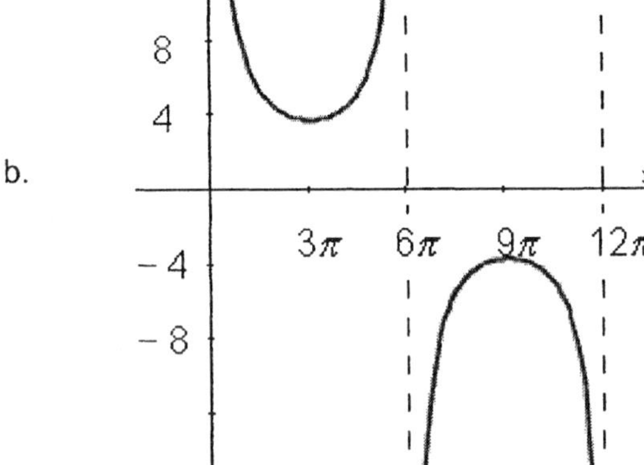

c.

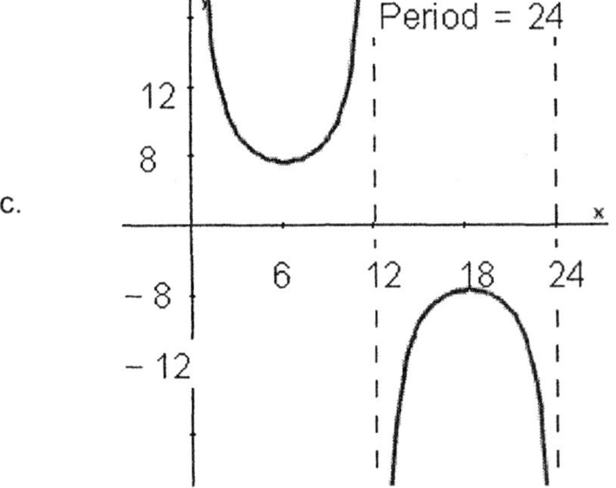

7. Use the graph of $y = \dfrac{3}{2} \cos 3x$ for reference, and graph one complete cycle of the equation $y = 1 + \dfrac{3}{2} \cos 3x$.

$$y = \frac{3}{2} \cos 3x$$

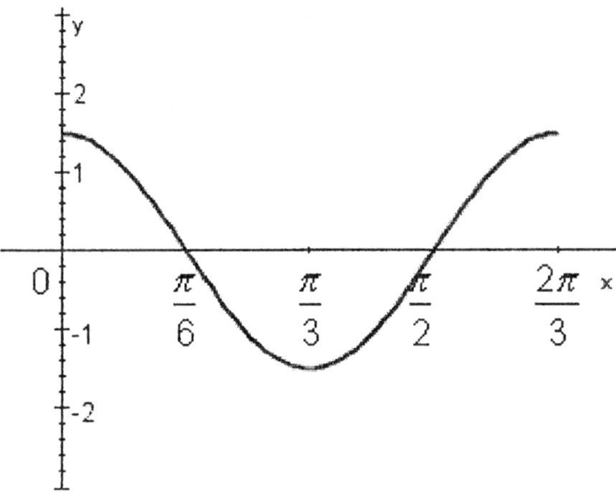

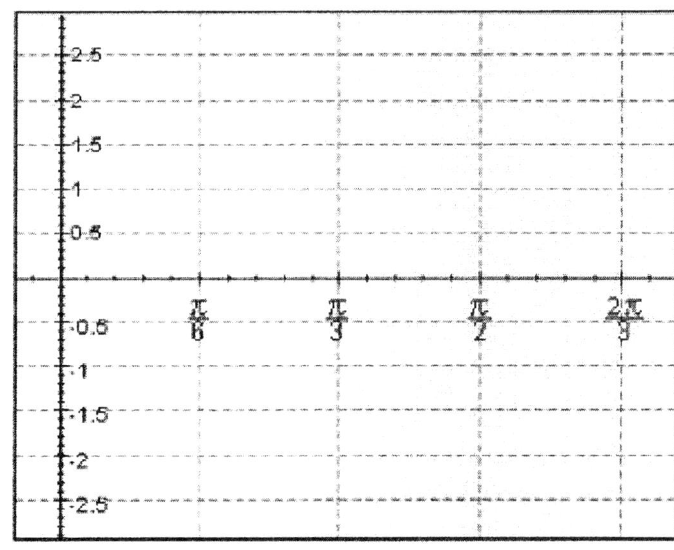

8. Graph the function over the given interval.

$$y = 3\cos \pi x, \quad -2 \le x \le 4$$

Select the correct answer.

a.

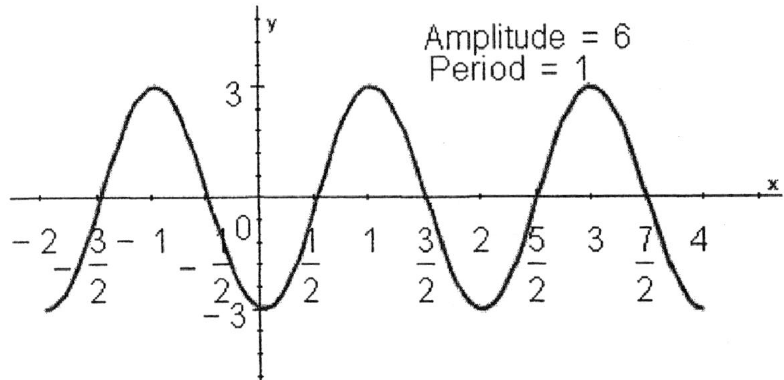

b.

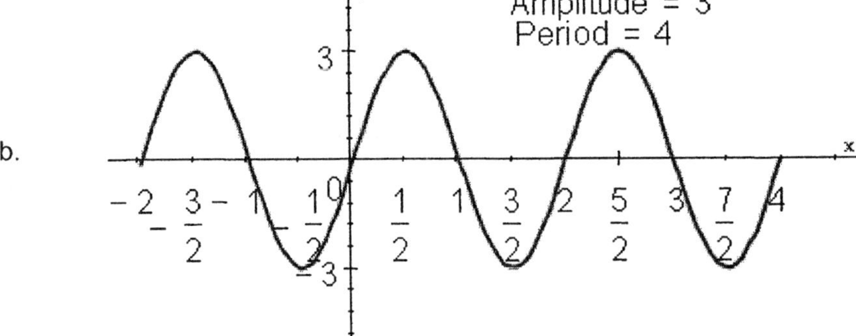

c.

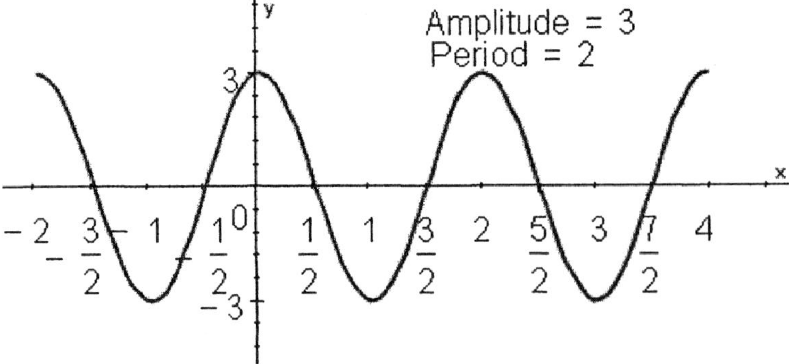

9. The current in an alternating circuit varies in intensity with time. If I represents the intensity of the current and t represents time, then the relationship between I and t is given by

$$I = 16 \sin 160\pi t$$

Where I is measured in amperes and t is measured in seconds. Find the maximum value of I and the time it takes for I to go through one complete cycle.

Select the correct answer.

a. The maximum value of I is 16 amperes; one complete cycle takes $\dfrac{1}{160}$ seconds.

b. The maximum value of I is 16 amperes; one complete cycle takes $\dfrac{1}{80}$ seconds.

c. The maximum value of I is 32 amperes; one complete cycle takes $\dfrac{1}{80}$ seconds.

d. The maximum value of I is 32 amperes; one complete cycle takes $\dfrac{1}{92}$ seconds.

e. The maximum value of I is 64 amperes; one complete cycle takes $\dfrac{1}{160}$ seconds.

10. Identify the amplitude for the equation.

$$y = 2 \sin \left(\pi x + \frac{\pi}{3} \right)$$

____

Identify the period for the equation.

Identify the phase shift for the equation.

Label the axes accordingly and sketch one complete cycle of the curve.

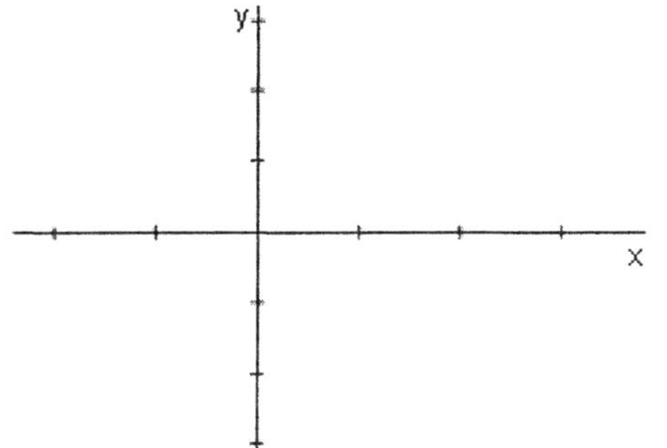

11. Use the graph of the equation $y = \cos\left(2x - \dfrac{\pi}{2}\right)$ shown below to graph one complete cycle of the equation $y = 2 + \cos\left(2x - \dfrac{\pi}{2}\right)$.

$$y = \cos\left(2x - \frac{\pi}{2}\right)$$

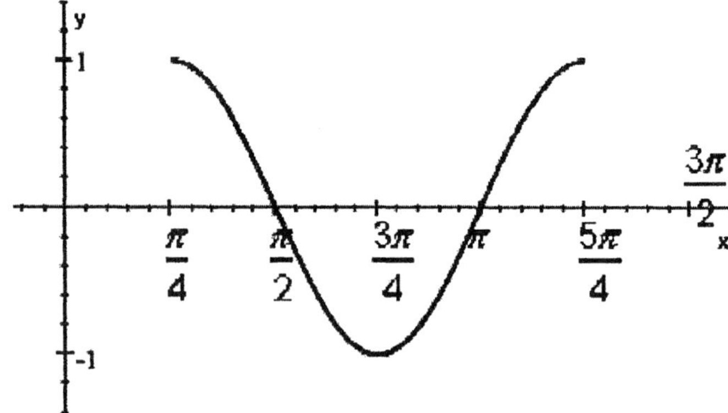

Select the correct answer.

a.
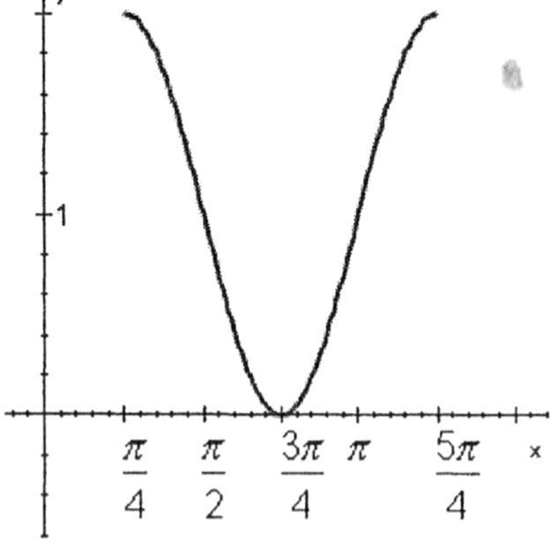

b.

c.

d.
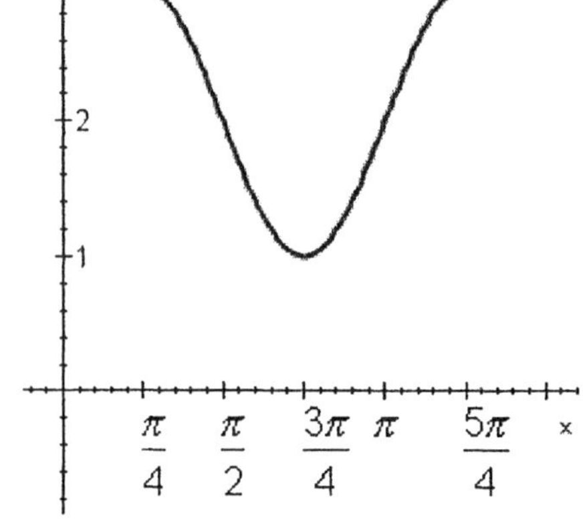

12. Graph the equation over the given interval. Be sure to label the axes so that the amplitude, period, and phase shift are easy to read.

$$y = -\frac{2}{3} \cos\left(3x + \frac{\pi}{2}\right), \quad -\pi \leq x \leq \pi$$

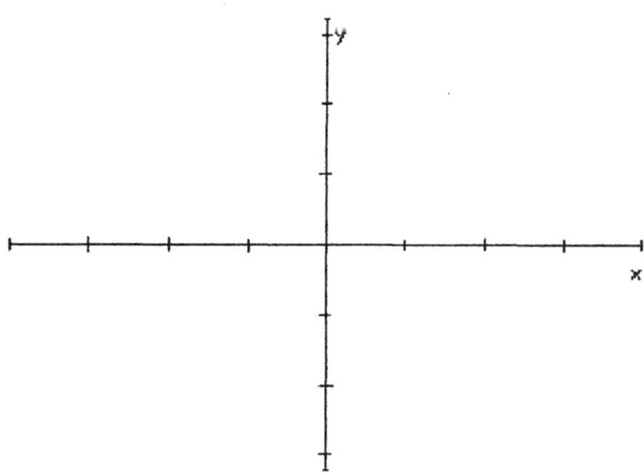

13. Sketch the graph of the equation below.

$$y = \tan\left(x - \frac{\pi}{4}\right)$$

Select the correct answer.

a.

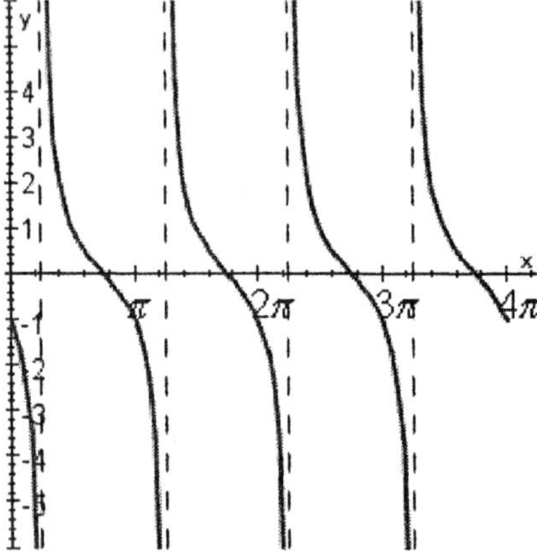

b.

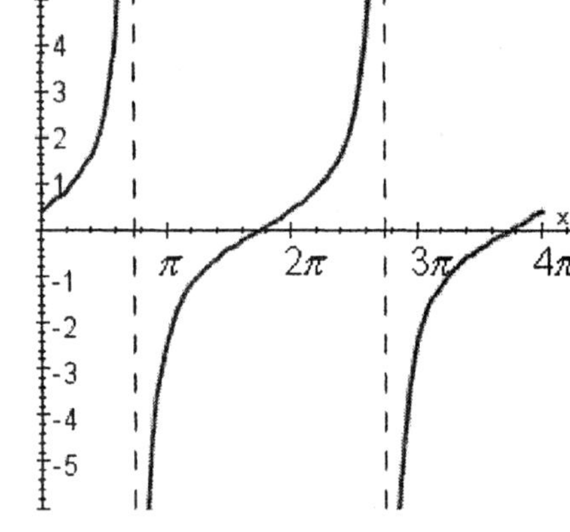

c.

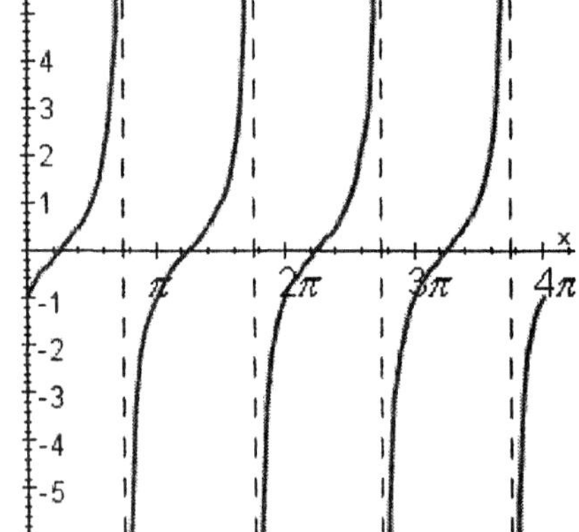

14. The graph below is one complete cycle of the graph of an equation containing a trigonometric function. Find an equation to match the graph.

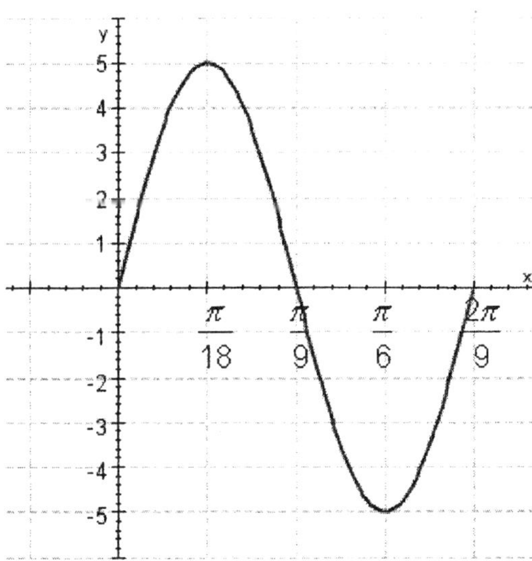

15. The graph below is one complete cycle of the graph of an equation containing a trigonometric function. Find an equation to match the graph.

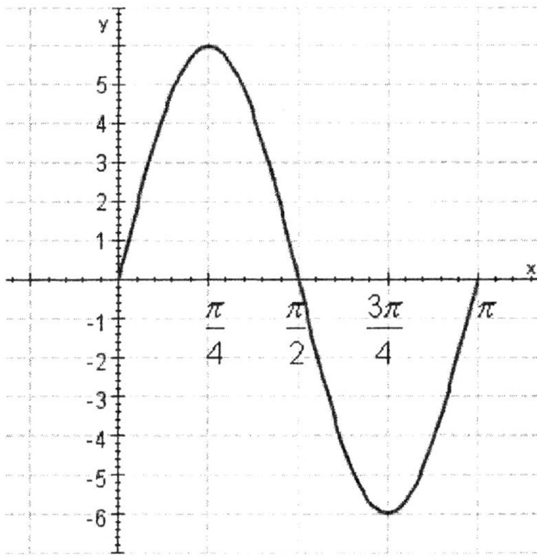

Select the correct answer.

a. $y = 6\cos 2x$

b. $y = 6\sin 2x$

c. $y = 6\sin x$

d. $y = 4\sin x$

e. $y = 4\sin 2x$

16. The graph below is one complete cycle of the graph of an equation containing a trigonometric function. Find an equation to match the graph.

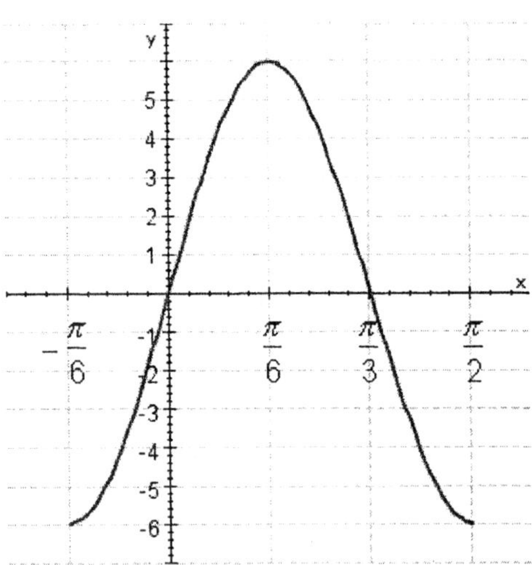

17. The graph below is one complete cycle of the graph of an equation containing a trigonometric function. Find an equation to match the graph.

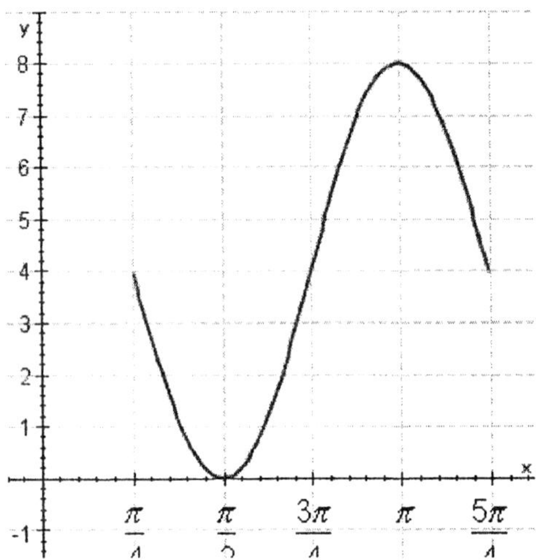

Select the correct answer.

a. $\quad y = 4 - \sin\left(2x - \dfrac{\pi}{2}\right)$

b. $\quad y = 5 - 2\sin\left(2x - \dfrac{\pi}{2}\right)$

c. $\quad y = 4 - 4\sin\left(2x - \dfrac{\pi}{2}\right)$

d. $\quad y = 4 - 3\sin\left(2x - \dfrac{\pi}{2}\right)$

18. Use addition of *y*-coordinates to sketch the graph of the function between

$x = 0$ and $x = 4\pi$.

$y = 4 + \cos x$

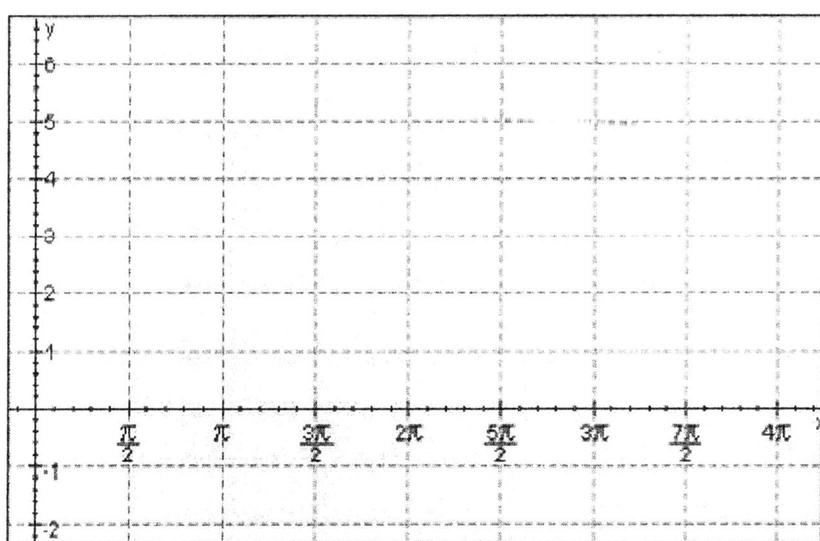

19. Use addition of *y*-coordinates to sketch the graph of the function between

$x = 0$ and $x = 4\pi$.

$y = 4 - 2\sin x$

Select the correct answer.

a.

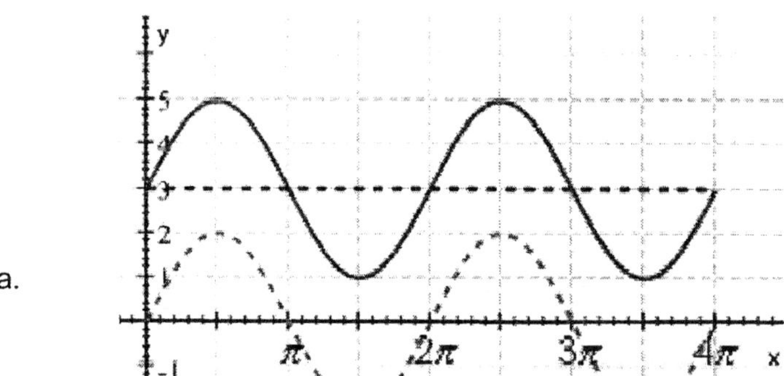

b.

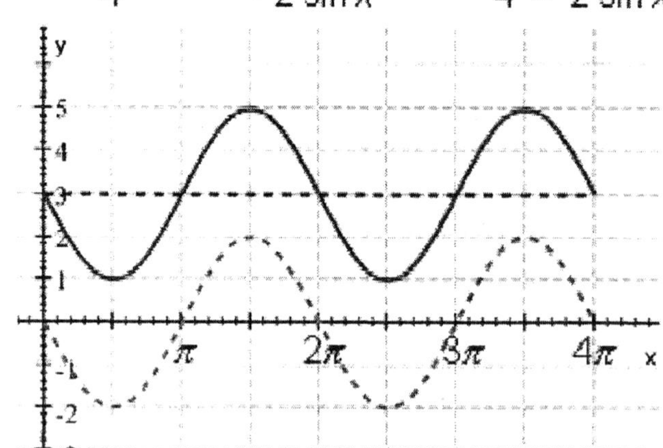

c.

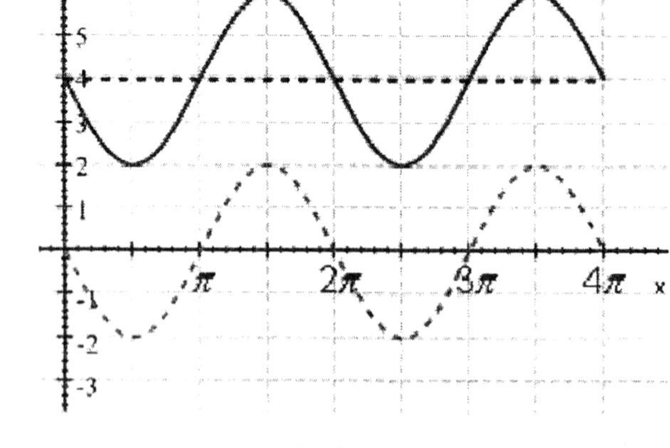

20. Use addition of *y*-coordinates to sketch the graph of the function between

$x = 0$ and $x = 4\pi$.

$$y = \frac{1}{6}x + \cos x$$

Select the correct answer.

a.

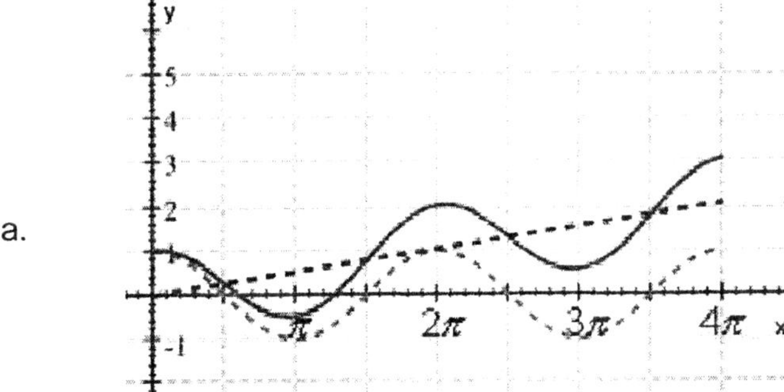

$$\cdots\cdots \frac{1}{6}X \quad \cdots\cdots \cos X \quad \text{——} \quad \frac{1}{6}X + \cos X$$

b.

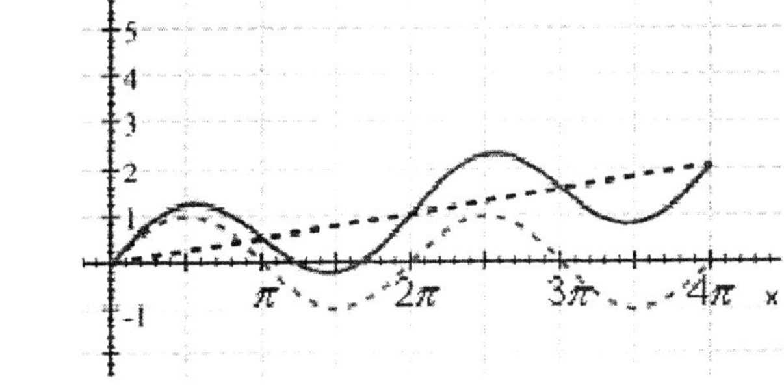

$$\cdots\cdots \frac{1}{6}X \quad \cdots\cdots \cos X \quad \text{——} \quad \frac{1}{6}X + \cos X$$

c.

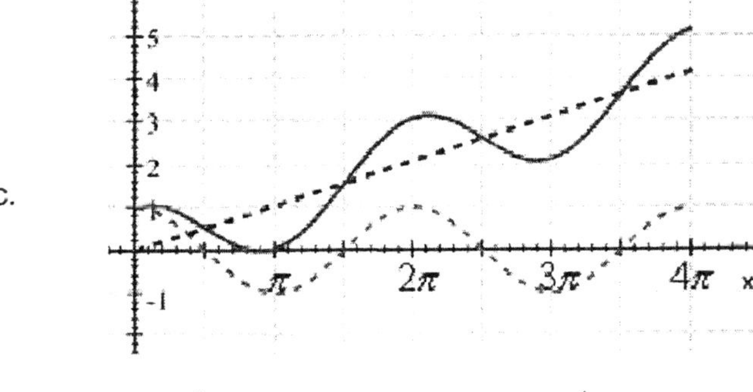

$$\cdots\cdots \frac{1}{6}X \quad \cdots\cdots \cos X \quad \text{——} \quad \frac{1}{6}X + \cos X$$

21. Sketch the graph from

$x = 0$ to $x = 4\pi$.

$y = 2\sin x + \sin 2x$

Select the correct answer.

a.

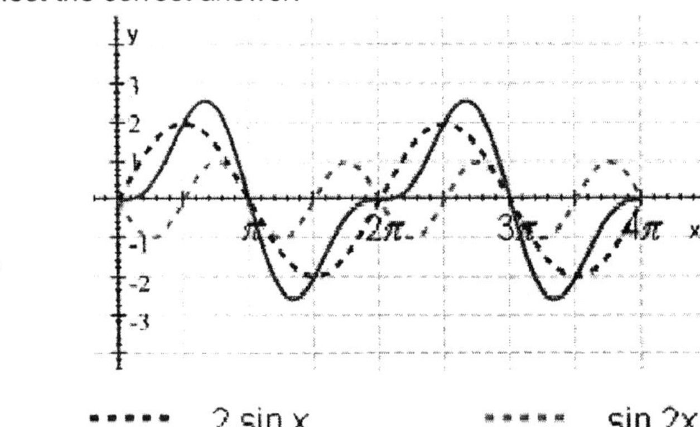

······ 2 sin x   ······ sin 2x

—— 2 sin x + sin 2x

b.

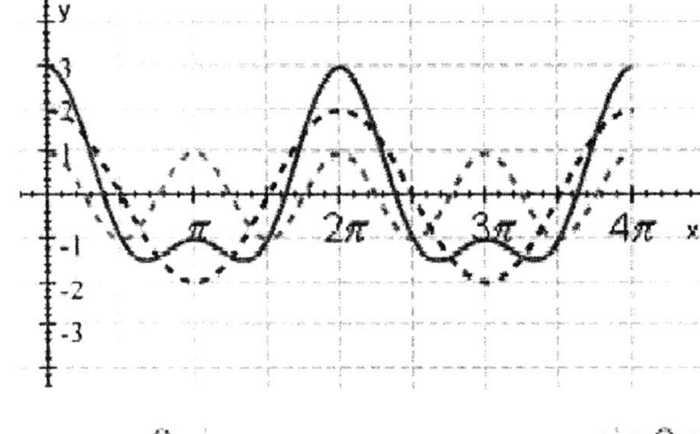

······ 2 sin x   ······ sin 2x

—— 2 sin x + sin 2x

c.

······ 2 sin x   ······ sin 2x

—— 2 sin x + sin 2x

22. Graph $y = \sin x$ between $-\dfrac{3\pi}{2}$ and $\dfrac{3\pi}{2}$,

and then reflect the graph about the line $y = x$ to obtain the graph of $x = \sin y$.

23. Evaluate the expression without using a calculator, and write your answer in radians.

$$\sin^{-1}\left(\frac{1}{2}\right)$$

Select the correct answer.

a. $\quad -\dfrac{\pi}{2}$

b. $\quad -\dfrac{\pi}{6}$

c. $\quad -\dfrac{\pi}{3}$

d. $\quad 0$

e. $\quad \dfrac{\pi}{6}$

24. Simplify

$$3\left|\sin\theta\right| \text{ if } \theta = \cos^{-1}\frac{x}{3} \text{ for some real number } x.$$

25. Evaluate without using a calculator.

$$\sin^{-1}\left(\sin 330°\right)$$

Select the correct answer.

a. $\quad -45°$

b. $\quad -180°$

c. $\quad -90°$

d. $\quad -30°$

e. $\quad -150°$

1.
$$0, \ \frac{\sqrt{2}}{2}, \ 1, \ \frac{\sqrt{2}}{2}, \ 0, \ -\frac{\sqrt{2}}{2}, \ -1, \ -\frac{\sqrt{2}}{2}, \ 0$$

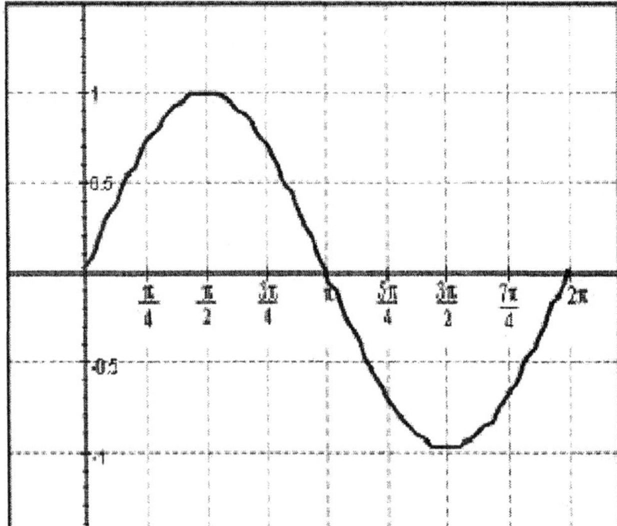

2. b

3. $4,$

π

4.

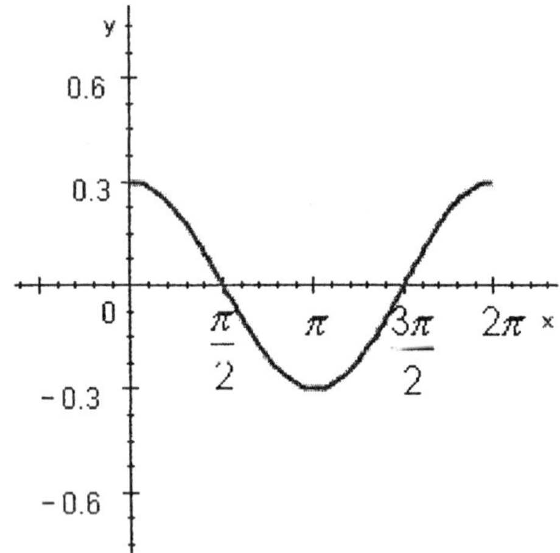

$$\frac{3}{10}$$

5.

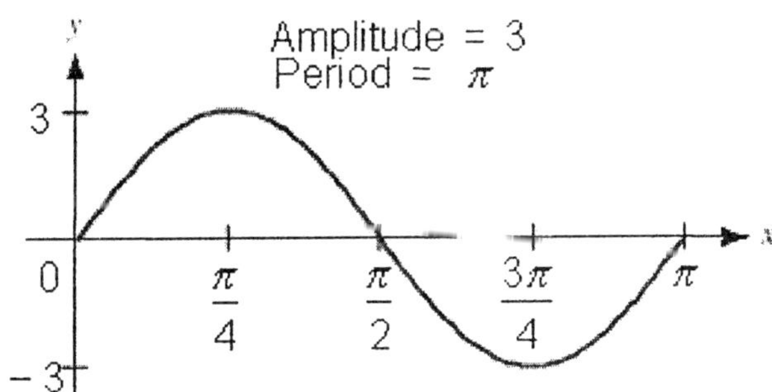

6. b

7.

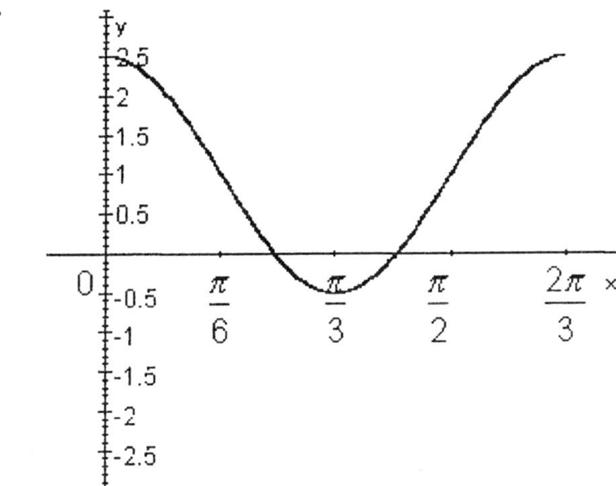

8. c

9. b

10. 2

2

$-\dfrac{1}{3}$

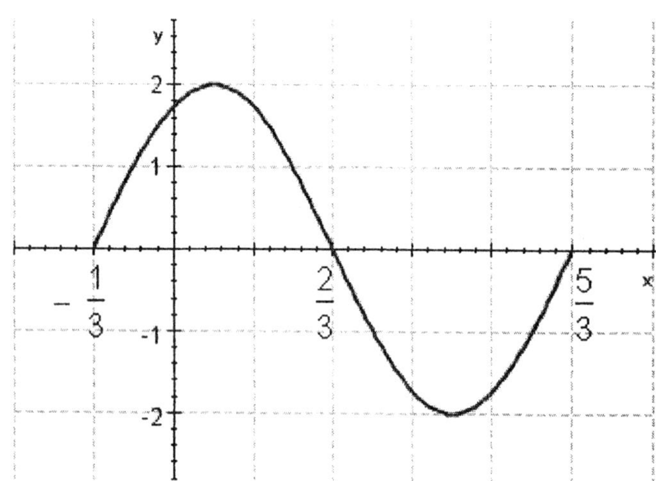

11. d

12.

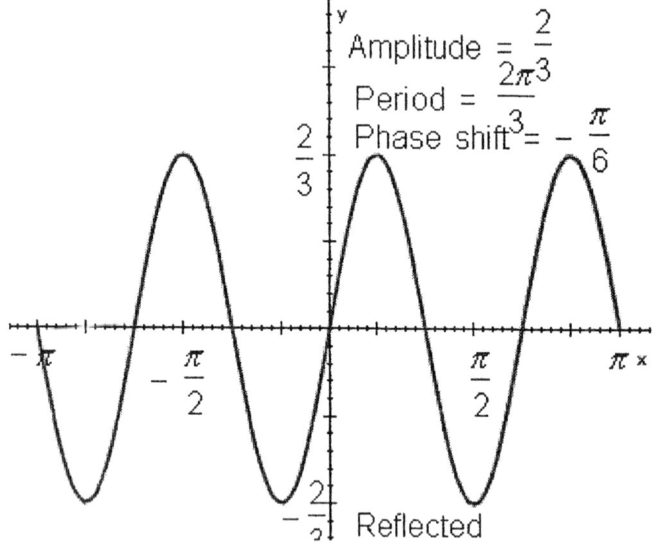

13. c

14. $y = 5\sin(9x)$

15. b

16.
$$y = -6\cos\left(3x + \frac{\pi}{2}\right)$$

17. c

18.

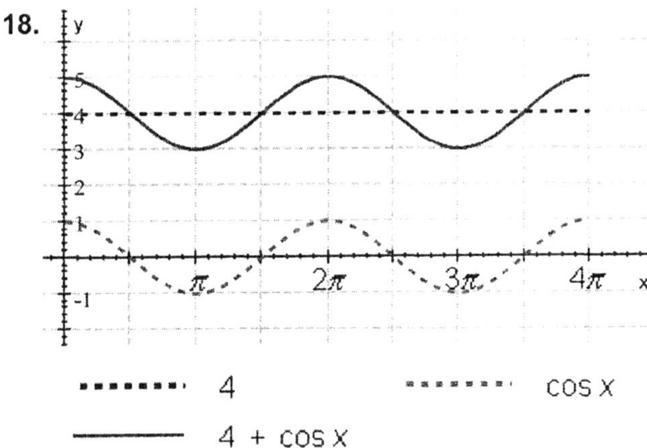

19. c

20. a

21. c

22.

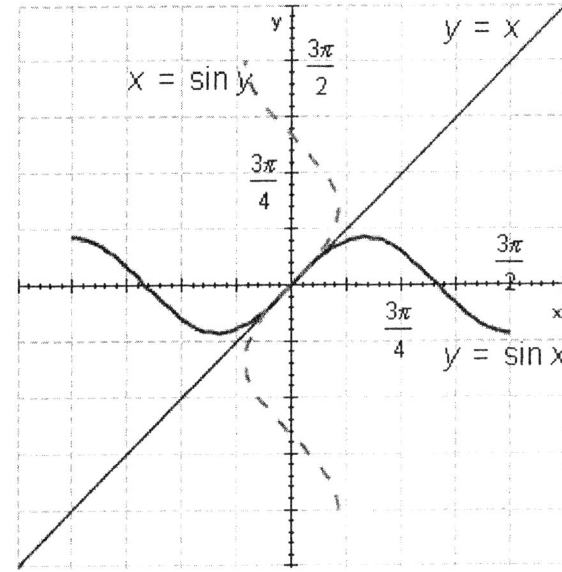

23. e

24. $\sqrt{9 - x^2}$

25. d

McKeague/Turner - Trigonometry 5e Chapter 4 Form E

1. mctr.04.01.01_NoAlgs
2. mctr.04.01.18m_NoAlgs
3. mctr.04.01.29_NoAlgs
4. mctr.04.01.53_NoAlgs
5. mctr.04.02.17_NoAlgs
6. mctr.04.02.28m_NoAlgs
7. mctr.04.02.36_NoAlgs
8. mctr.04.02.42m_NoAlgs
9. mctr.04.02.53m_NoAlgs
10. mctr.04.03.13_NoAlgs
11. mctr.04.03.23m_NoAlgs
12. mctr.04.03.39_NoAlgs
13. mctr.04.03.47m_NoAlgs
14. mctr.04.04.11_NoAlgs
15. mctr.04.04.16m_NoAlgs
16. mctr.04.04.25_NoAlgs
17. mctr.04.04.30m_NoAlgs
18. mctr.04.05.02_NoAlgs
19. mctr.04.05.05m_NoAlgs
20. mctr.04.05.10m_NoAlgs
21. mctr.04.05.14m_NoAlgs
22. mctr.04.06.01_NoAlgs
23. mctr.04.06.13m_NoAlgs
24. mctr.04.06.42_NoAlgs
25. mctr.04.06.52m_NoAlgs

1. Evaluate the expression without using a calculator, and write your answer in radians.

$$\sin^{-1}\left(\frac{1}{2}\right)$$

Select the correct answer.

a. $-\dfrac{\pi}{2}$

b. $-\dfrac{\pi}{6}$

c. $-\dfrac{\pi}{3}$

d. 0

e. $\dfrac{\pi}{6}$

2. Graph the equation over the given interval. Be sure to label the axes so that the amplitude, period, and phase shift are easy to read.

$$y = -\frac{2}{3}\cos\left(3x + \frac{\pi}{2}\right), \quad -\pi \le x \le \pi$$

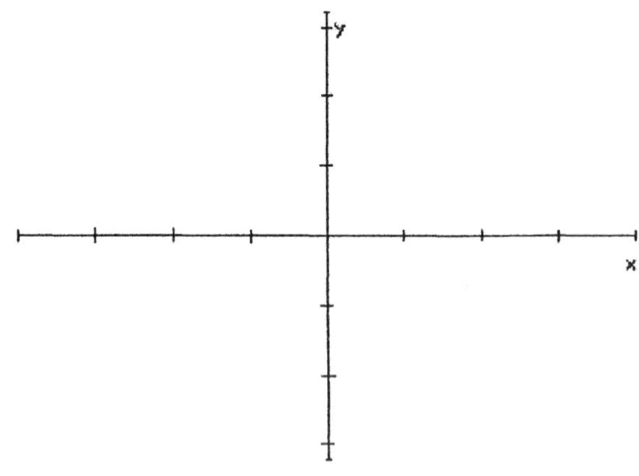

3. Find all values of x for which the following are true:

$\sec x = 1$

Select the correct answer.

a. $\dfrac{\pi}{3} + 2k\pi$

b. $2k\pi$

c. $\dfrac{\pi}{4} + 2k\pi$

d. $-\dfrac{\pi}{3} + 2k\pi$

e. $\dfrac{\pi}{2} + k\pi$

4. Graph one complete cycle for the function.

$$y = 4 \csc \frac{1}{6}x$$

Select the correct answer.

a.

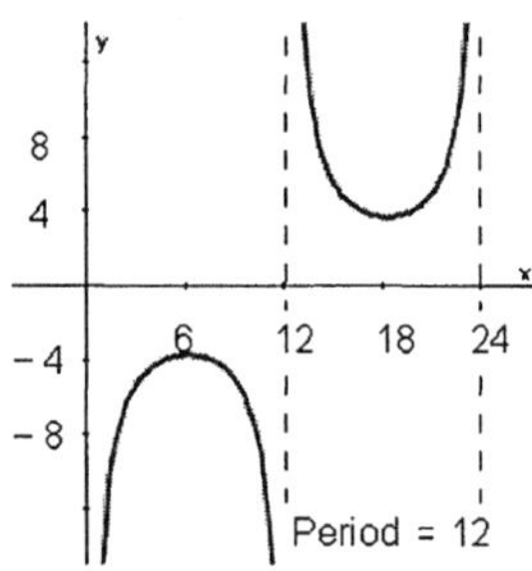

c.

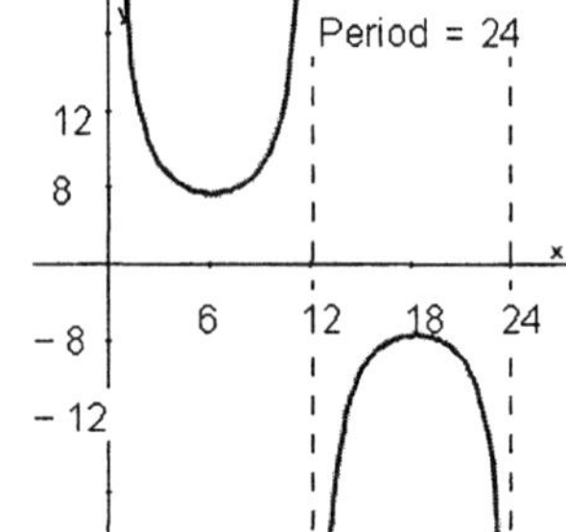

b.

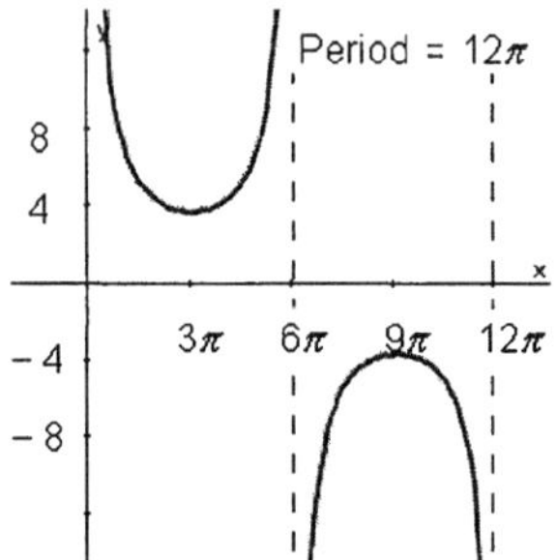

d.

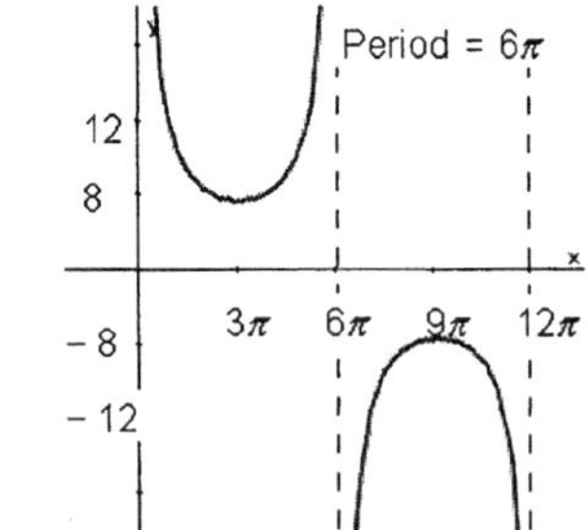

5. Use the graph of the equation $y = \cos\left(2x - \dfrac{\pi}{2}\right)$ shown below to graph one complete cycle of the equation $y = 2 + \cos\left(2x - \dfrac{\pi}{2}\right)$.

$$y = \cos\left(2x - \dfrac{\pi}{2}\right)$$

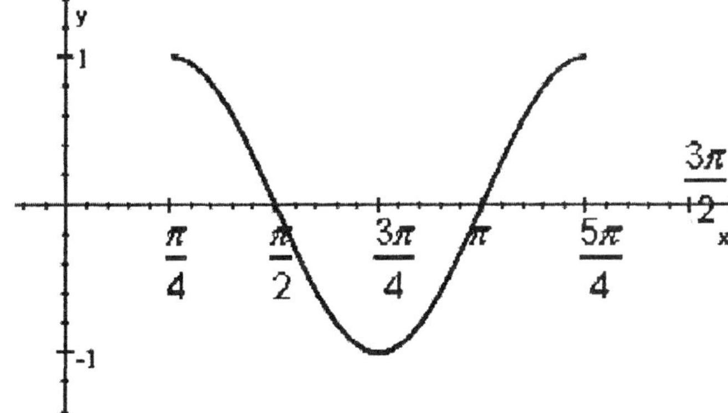

Select the correct answer.

a.

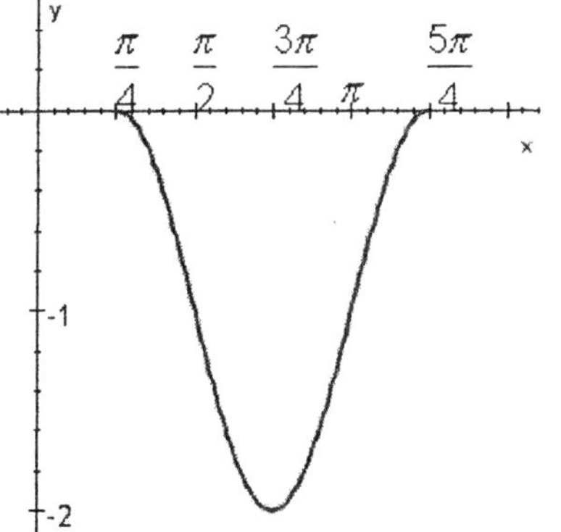

b.

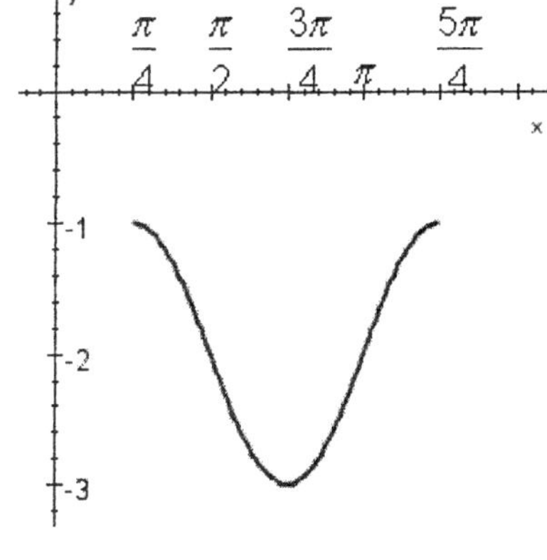

c.

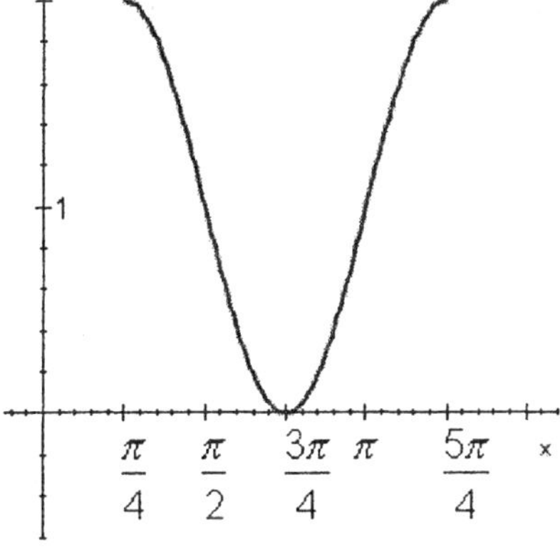

d.

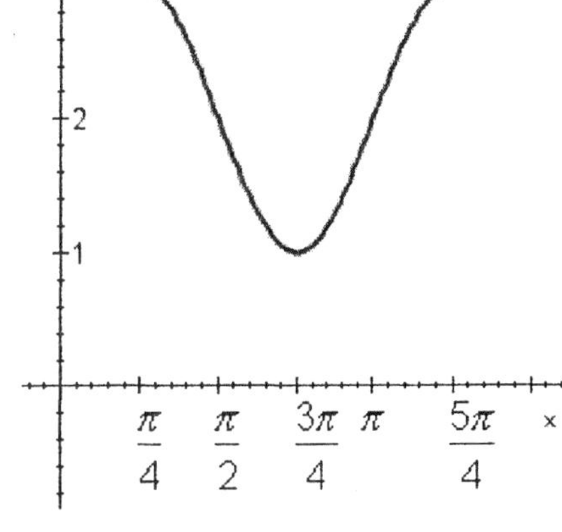

6. Sketch the graph from

$x = 0$ to $x = 4\pi$.

$y = 2\sin x + \sin 2x$

Select the correct answer.

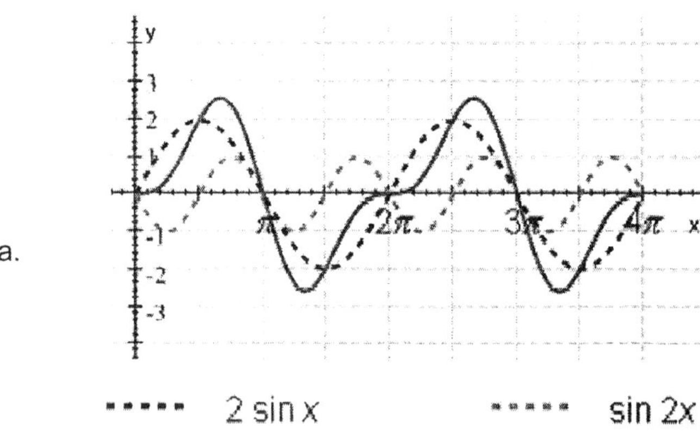

a.

······ $2\sin x$ ······ $\sin 2x$

——— $2\sin x + \sin 2x$

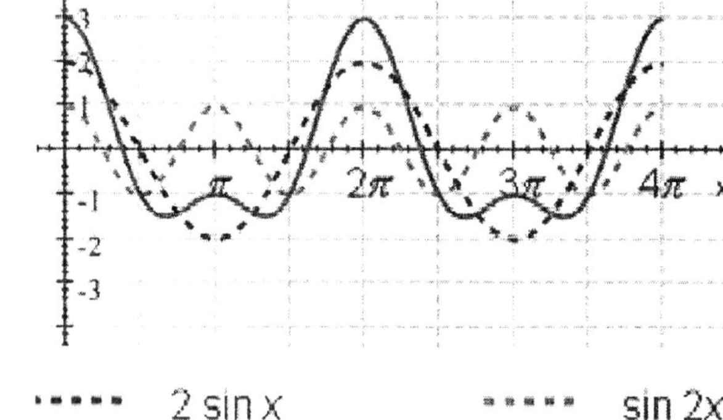

b.

······ $2\sin x$ ······ $\sin 2x$

——— $2\sin x + \sin 2x$

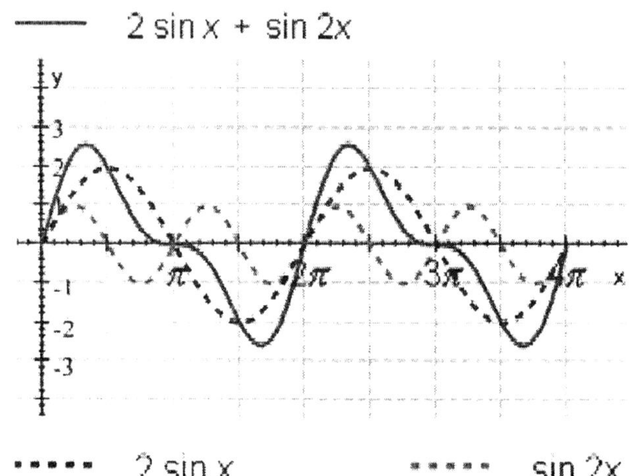

c.

······ $2\sin x$ ······ $\sin 2x$

——— $2\sin x + \sin 2x$

7. The current in an alternating circuit varies in intensity with time. If I represents the intensity of the current and t represents time, then the relationship between I and t is given by

$$I = 16 \sin 160\pi t$$

Where I is measured in amperes and t is measured in seconds. Find the maximum value of I and the time it takes for I to go through one complete cycle.

Select the correct answer.

a. The maximum value of I is 16 amperes; one complete cycle takes $\dfrac{1}{160}$ seconds.

b. The maximum value of I is 16 amperes; one complete cycle takes $\dfrac{1}{80}$ seconds.

c. The maximum value of I is 32 amperes; one complete cycle takes $\dfrac{1}{80}$ seconds.

d. The maximum value of I is 32 amperes; one complete cycle takes $\dfrac{1}{92}$ seconds.

e. The maximum value of I is 64 amperes; one complete cycle takes $\dfrac{1}{160}$ seconds.

8. Give the amplitude and period of the graph.

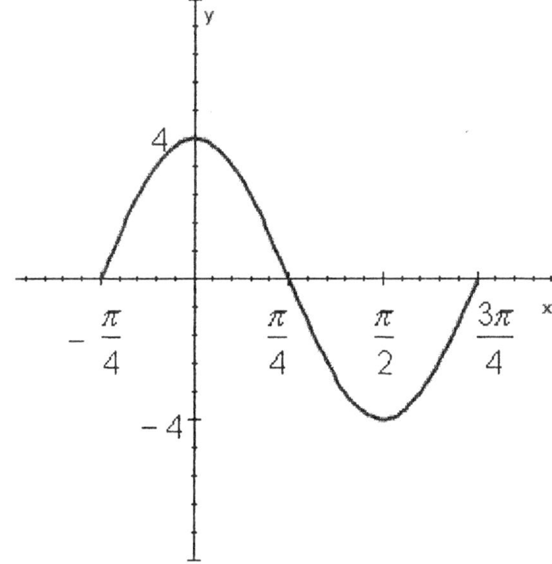

amplitude.

period.

9. The graph below is one complete cycle of the graph of an equation containing a trigonometric function. Find an equation to match the graph.

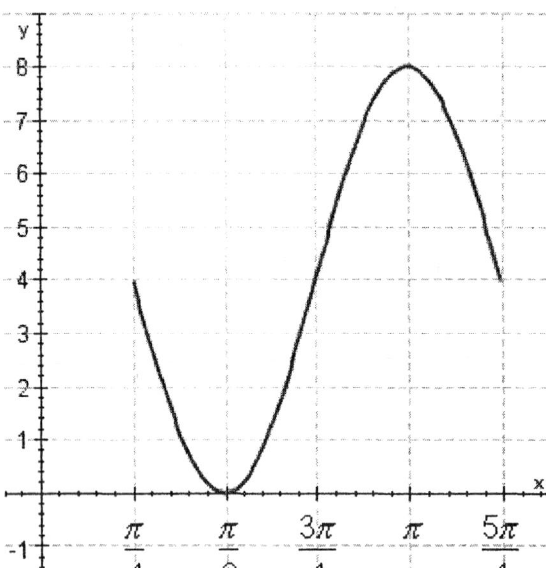

Select the correct answer.

a. $y = 4 - \sin\left(2x - \dfrac{\pi}{2}\right)$

b. $y = 5 - 2\sin\left(2x - \dfrac{\pi}{2}\right)$

c. $y = 4 - 4\sin\left(2x - \dfrac{\pi}{2}\right)$

d. $y = 4 - 3\sin\left(2x - \dfrac{\pi}{2}\right)$

e. $y = 4 - 6\sin\left(2x - \dfrac{\pi}{2}\right)$

10. Use addition of y-coordinates to sketch the graph of the function between

$x = 0$ and $x = 4\pi$.

$$y = \frac{1}{6}x + \cos x$$

Select the correct answer.

a.

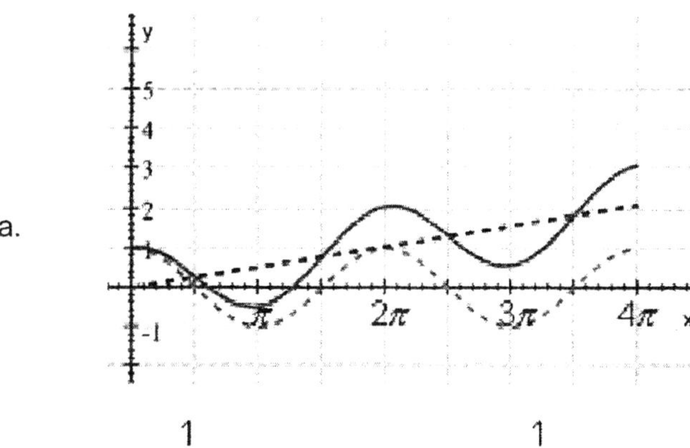

$$\cdots\cdots \frac{1}{6}x \qquad \cdots\cdots \cos x \qquad \text{———} \frac{1}{6}x + \cos x$$

b.

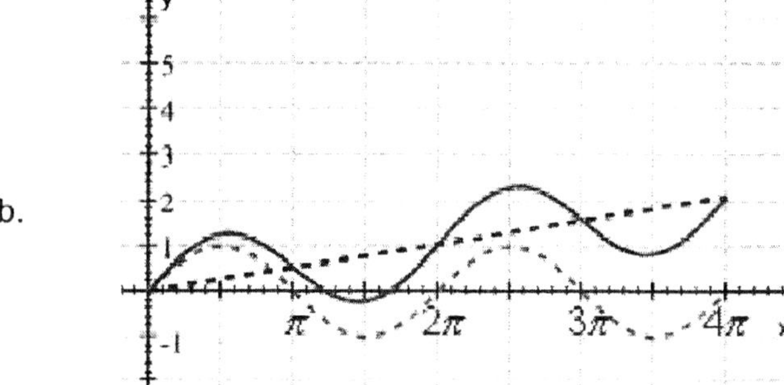

$$\cdots\cdots \frac{1}{6}x \qquad \cdots\cdots \cos x \qquad \text{———} \frac{1}{6}x + \cos x$$

c.

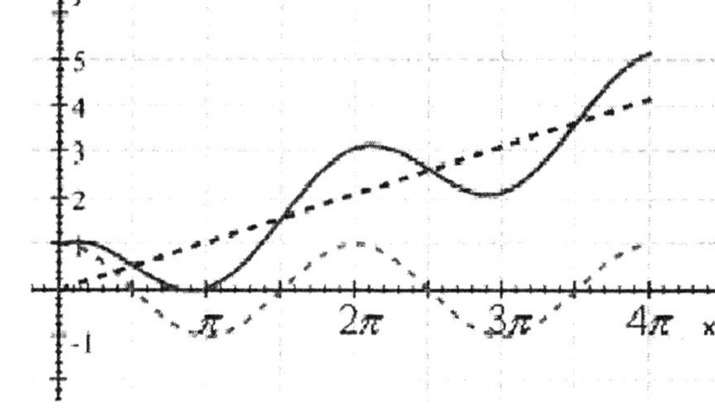

$$\cdots\cdots \frac{1}{6}x \qquad \cdots\cdots \cos x \qquad \text{———} \frac{1}{6}x + \cos x$$

11. Use the graph of $y = \dfrac{3}{2}\cos 3x$ for reference and graph one complete cycle of the equation

$$y = 1 + \frac{3}{2}\cos 3x.$$

$$y = \frac{3}{2}\cos 3x$$

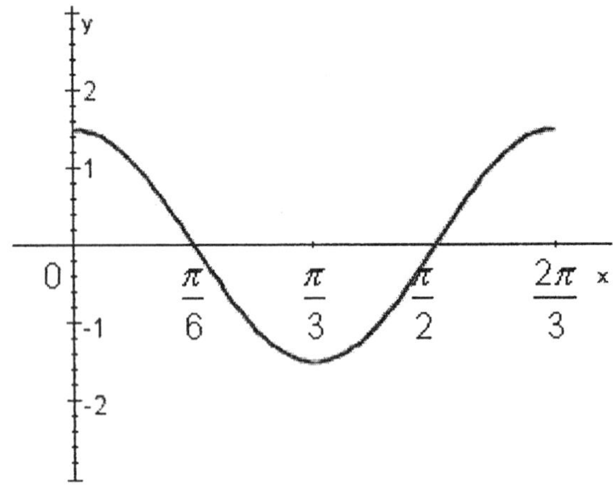

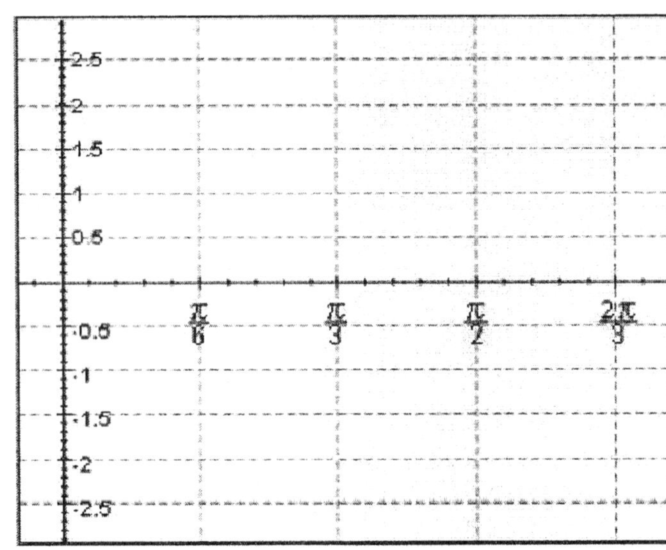

12.

Make a table of values for $y = \sin x$ using multiples of $\dfrac{\pi}{4}$ for x.

x	$\sin x$
0	
$\dfrac{\pi}{4}$	
$\dfrac{\pi}{2}$	
$\dfrac{3\pi}{4}$	
π	
$\dfrac{5\pi}{4}$	
$\dfrac{3\pi}{2}$	
$\dfrac{7\pi}{4}$	
2π	

Use the entries in the table to sketch the graph of the function for x between 0 and 2π

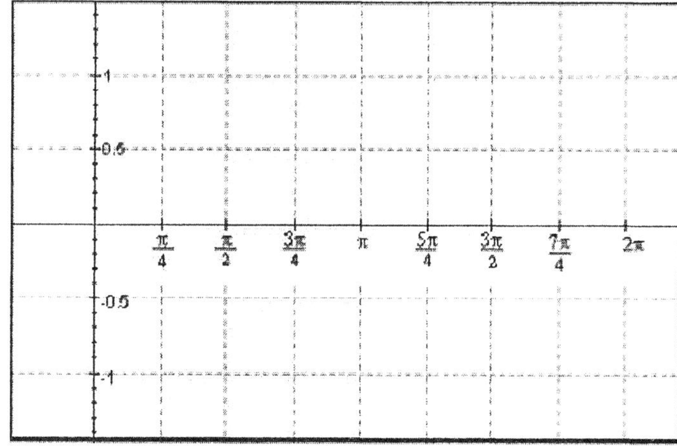

13.

Graph $y = \sin x$ between $-\dfrac{3\pi}{2}$ and $\dfrac{3\pi}{2}$, and then reflect the graph about the line

$y = x$ to obtain the graph of $x = \sin y$.

14. Sketch the graph of the equation below.

$$y = \tan\left(x - \frac{\pi}{4}\right)$$

Select the correct answer.

a.

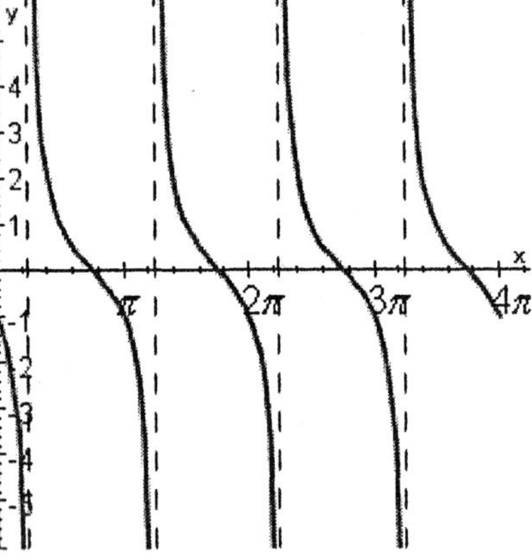

b.

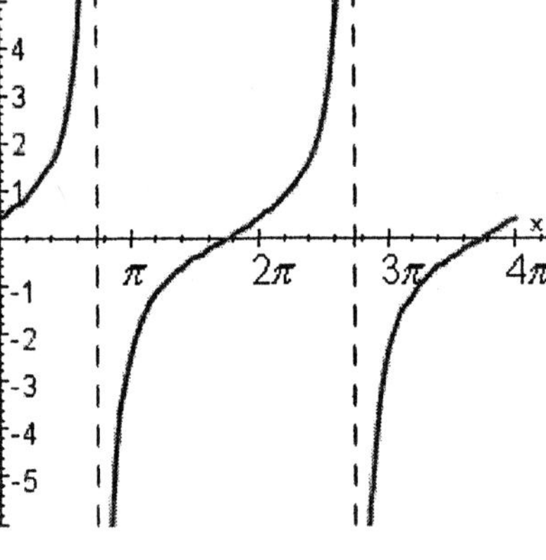

c.

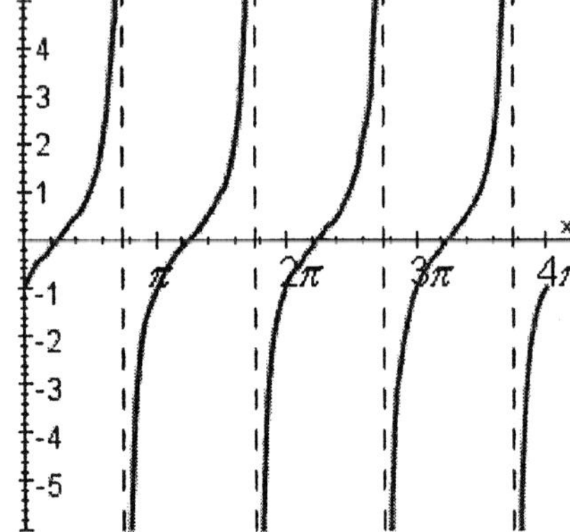

15. Simplify

$3\left|\sin\theta\right|$ if $\theta = \cos^{-1}\dfrac{x}{3}$ for some real number x.

16.

Graph one complete cycle of $y = \dfrac{3}{10}\cos x$. Label the axes accurately.

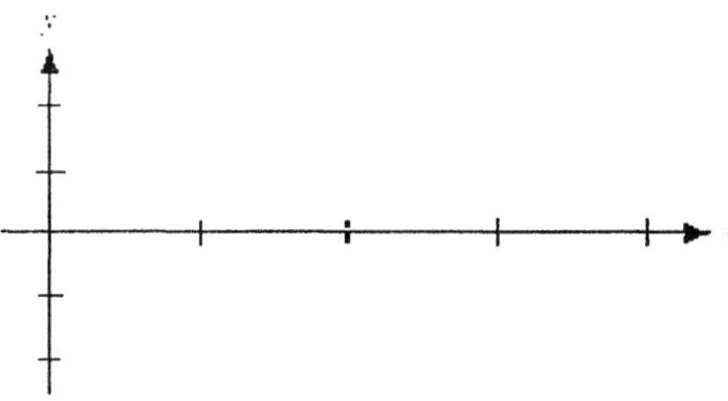

Identify the amplitude for the graph.

17. Use addition of *y*-coordinates to sketch the graph of the function between

$x = 0$ and $x = 4\pi$.

$y = 4 - 2\sin x$

Select the correct answer.

a.

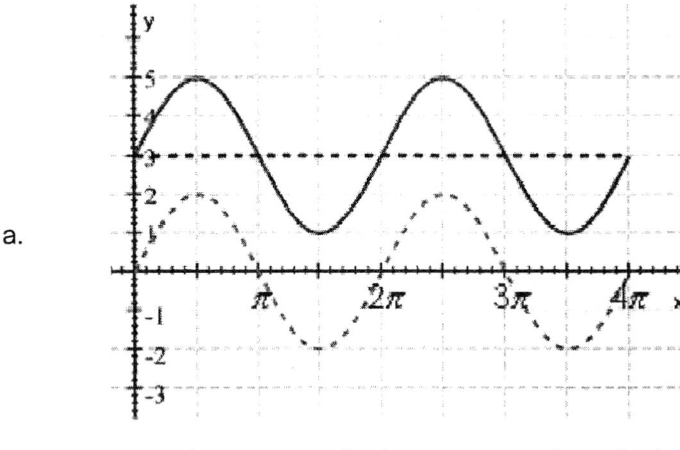

$$\cdots\cdots 4 \cdots\cdots \quad -2\sin x \quad\text{————}\quad 4 - 2\sin x$$

b.

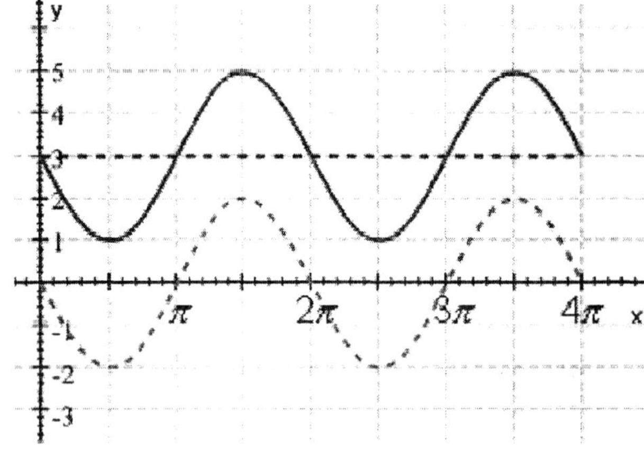

$$\cdots\cdots 4 \cdots\cdots \quad -2\sin x \quad\text{————}\quad 4 - 2\sin x$$

c.

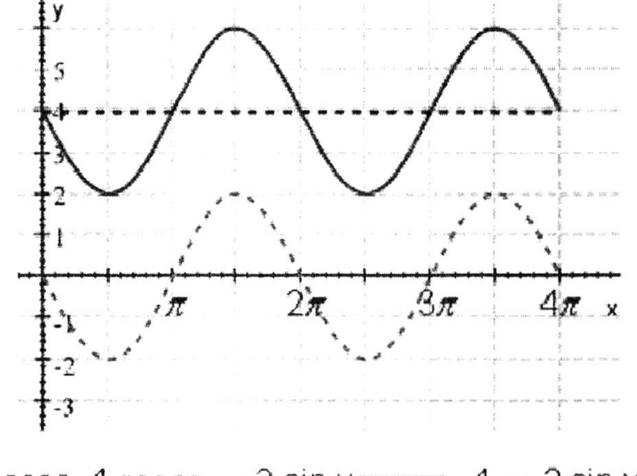

$$\cdots\cdots 4 \cdots\cdots \quad -2\sin x \quad\text{————}\quad 4 - 2\sin x$$

18. Evaluate without using a calculator.

$$\sin^{-1}\left(\sin 330^\circ\right)$$

Select the correct answer.

a. -45°

b. -180°

c. -90°

d. -30°

e. -150°

19. Graph the function over the given interval.

$$y = 3\cos\pi x, \quad -2 \le x \le 4$$

Select the correct answer.

a.

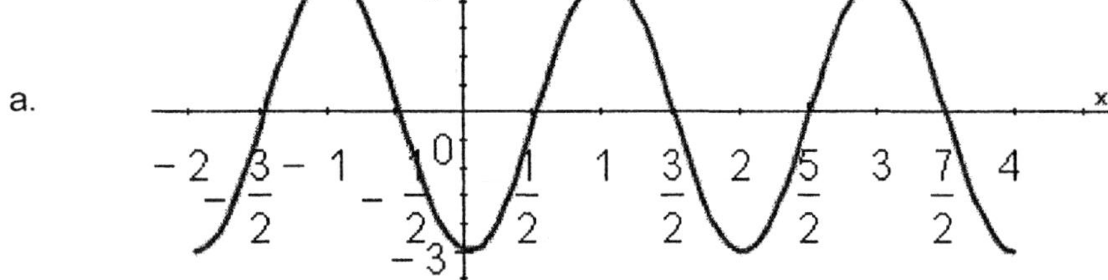

b.

c.

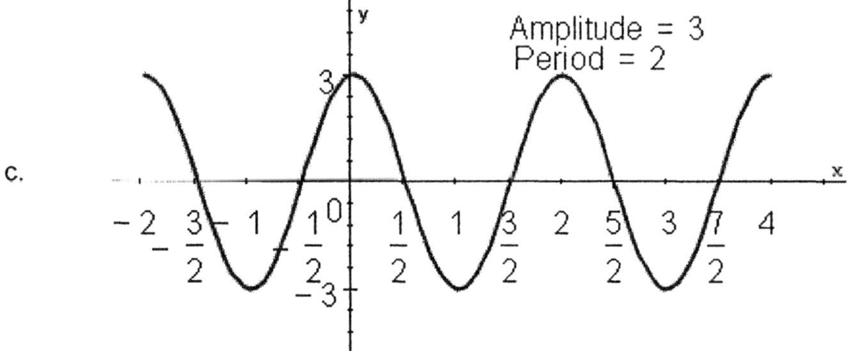

20. Graph one complete cycle for the function. Label the axes so that the amplitude (if defined) and period are easy to read.

$$y = 3\sin 2x$$

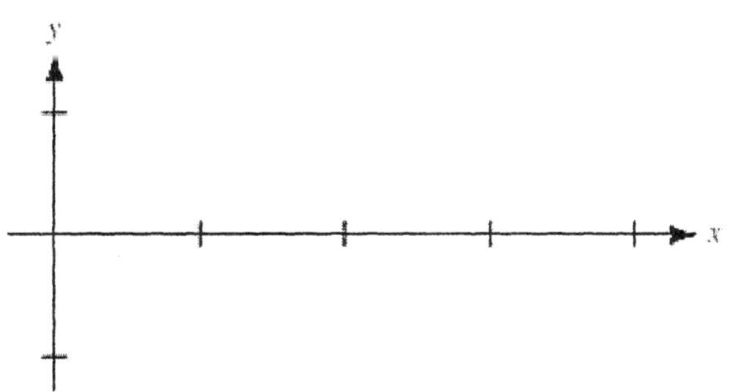

21. The graph below is one complete cycle of the graph of an equation containing a trigonometric function. Find an equation to match the graph.

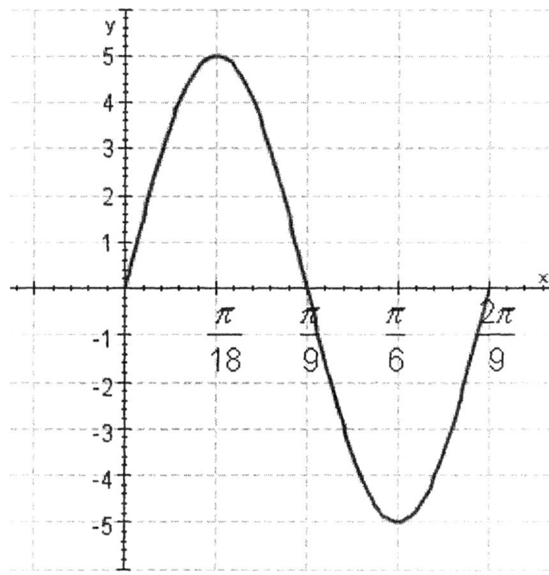

22. Use addition of *y*-coordinates to sketch the graph of the function between

$x = 0$ and $x = 4\pi$.

$y = 4 + \cos x$

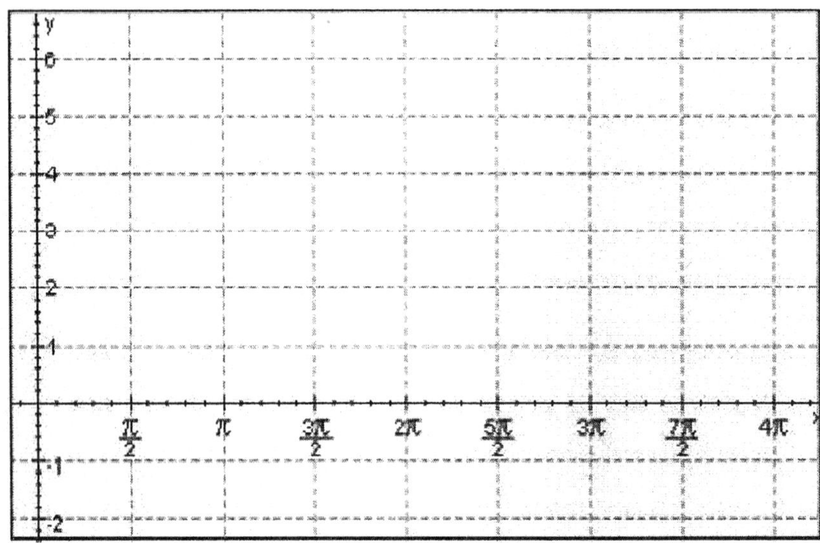

23. Identify the amplitude for the equation.

$$y = 2\sin\left(\pi x + \frac{\pi}{3}\right)$$

____

Identify the period for the equation.

Identify the phase shift for the equation.

Label the axes accordingly and sketch one complete cycle of the curve.

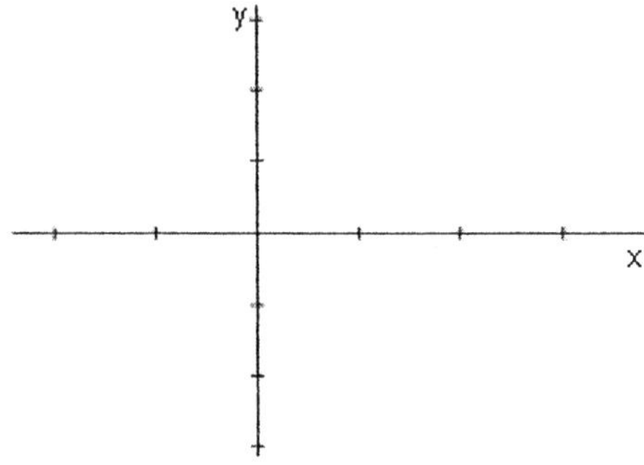

24. The graph below is one complete cycle of the graph of an equation containing a trigonometric function. Find an equation to match the graph.

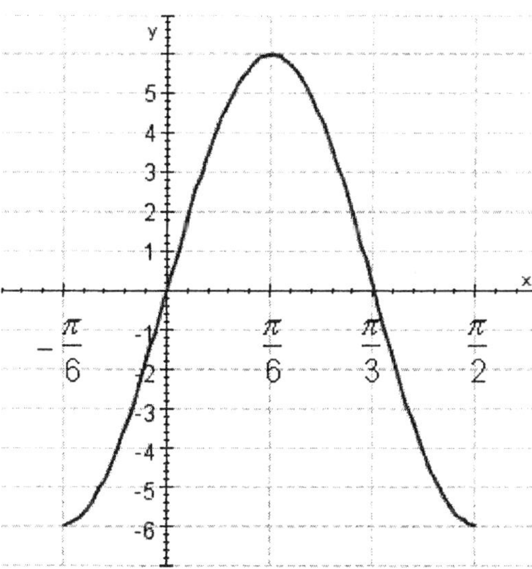

25. The graph below is one complete cycle of the graph of an equation containing a trigonometric function. Find an equation to match the graph.

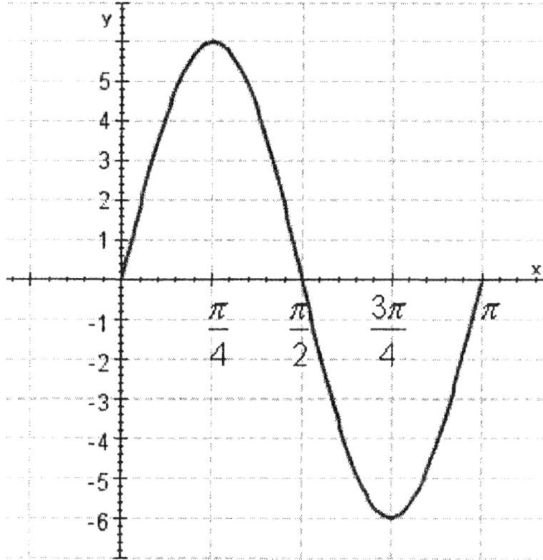

Select the correct answer.

a. $y = 6\cos 2x$

b. $y = 6\sin 2x$

c. $y = 6\sin x$

d. $y = 4\sin x$

1. e

2. 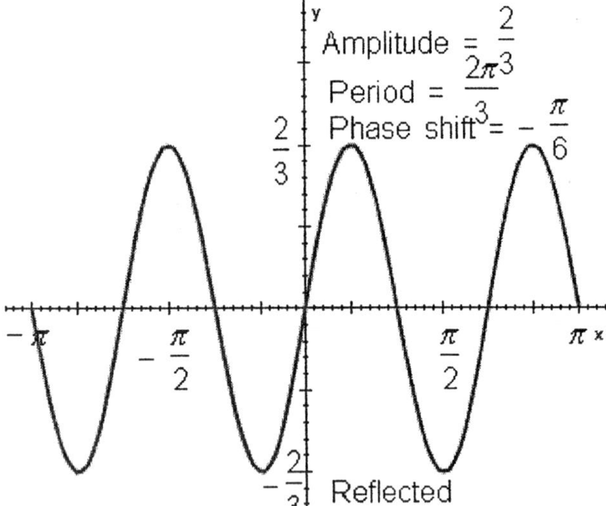

3. b

4. b

5. d

6. c

7. b

8. 4,

    π

9. c

10. a

11. 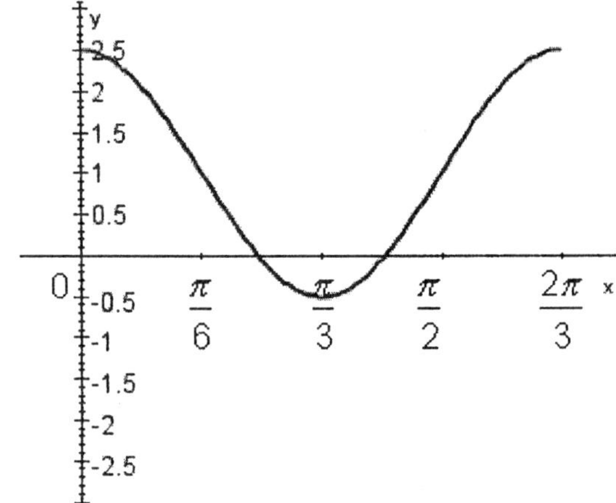

12. $0,\ \dfrac{\sqrt{2}}{2},\ 1,\ \dfrac{\sqrt{2}}{2},\ 0,\ -\dfrac{\sqrt{2}}{2},\ -1,\ -\dfrac{\sqrt{2}}{2},\ 0$

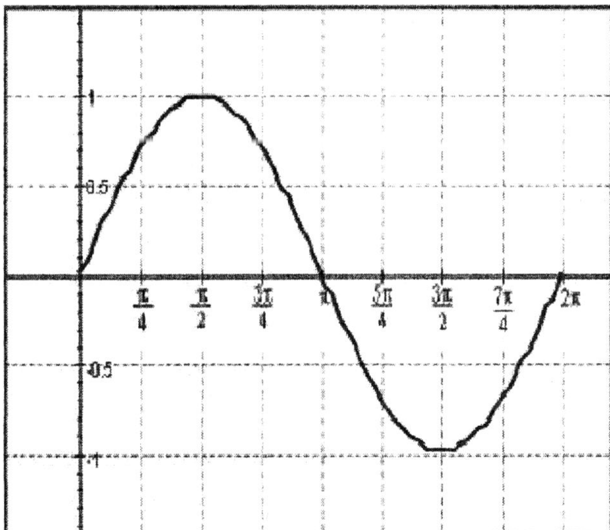

13.

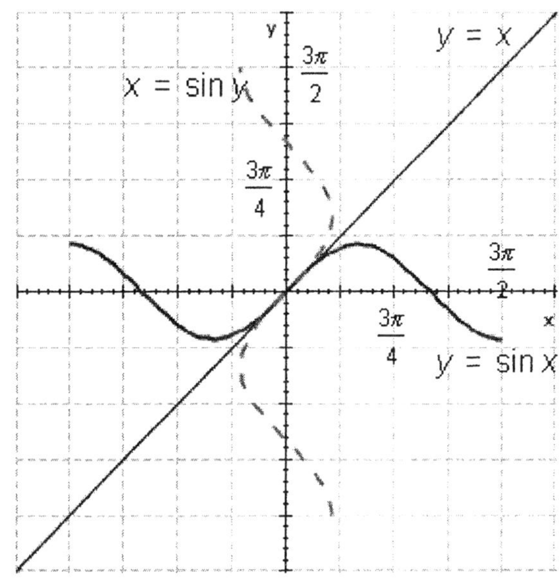

14. c

15. $\sqrt{9 - x^2}$

16.

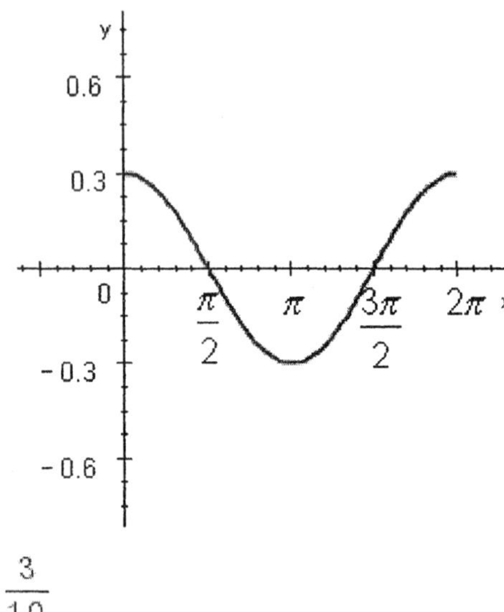

$$\frac{3}{10}$$

17. c

18. d

19. c

20.

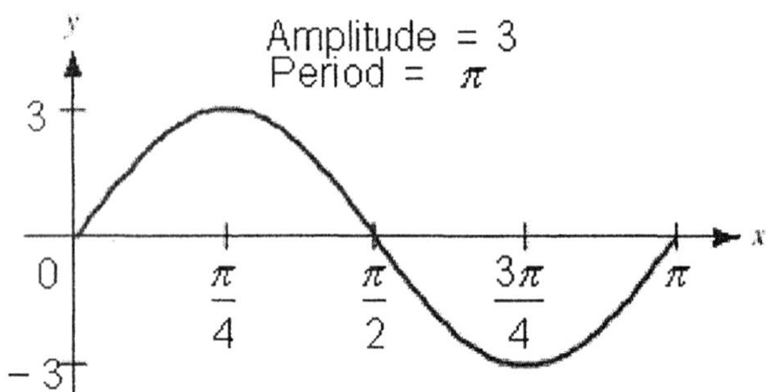

21. $y = 5\sin(9x)$

22.

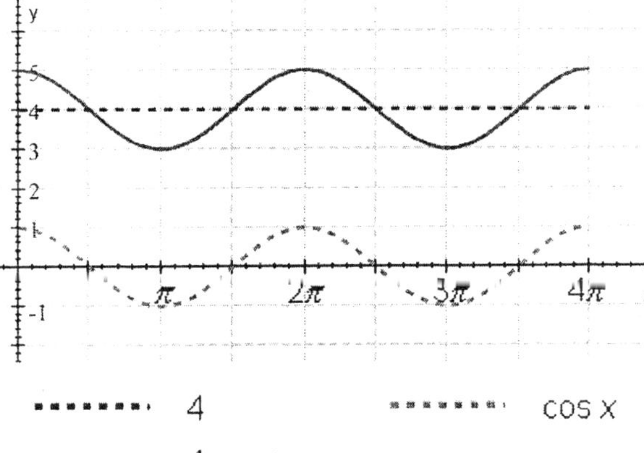

23. 2

2

$-\dfrac{1}{3}$

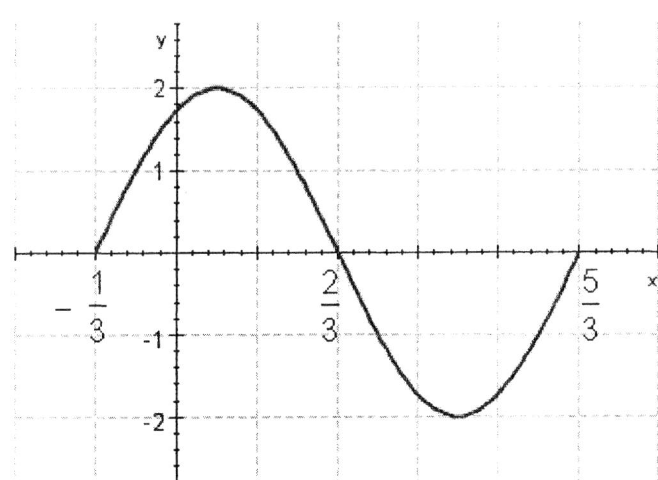

24. $y = -6\cos\left(3x + \dfrac{\pi}{2}\right)$

25. b

1. mctr.04.06.13m_NoAlgs
2. mctr.04.03.39_NoAlgs
3. mctr.04.01.18m_NoAlgs
4. mctr.04.02.28m_NoAlgs
5. mctr.04.03.23m_NoAlgs
6. mctr.04.05.14m_NoAlgs
7. mctr.04.02.53m_NoAlgs
8. mctr.04.01.29_NoAlgs
9. mctr.04.04.30m_NoAlgs
10. mctr.04.05.10m_NoAlgs
11. mctr.04.02.36_NoAlgs
12. mctr.04.01.01_NoAlgs
13. mctr.04.06.01_NoAlgs
14. mctr.04.03.47m_NoAlgs
15. mctr.04.06.42_NoAlgs
16. mctr.04.01.53_NoAlgs
17. mctr.04.05.05m_NoAlgs
18. mctr.04.06.52m_NoAlgs
19. mctr.04.02.42m_NoAlgs
20. mctr.04.02.17_NoAlgs
21. mctr.04.04.11_NoAlgs
22. mctr.04.05.02_NoAlgs
23. mctr.04.03.13_NoAlgs
24. mctr.04.04.25_NoAlgs
25. mctr.04.04.16m_NoAlgs

1. Sketch the graph of $y = \cos x$ between $x = -4\pi$ and $x = 4\pi$.

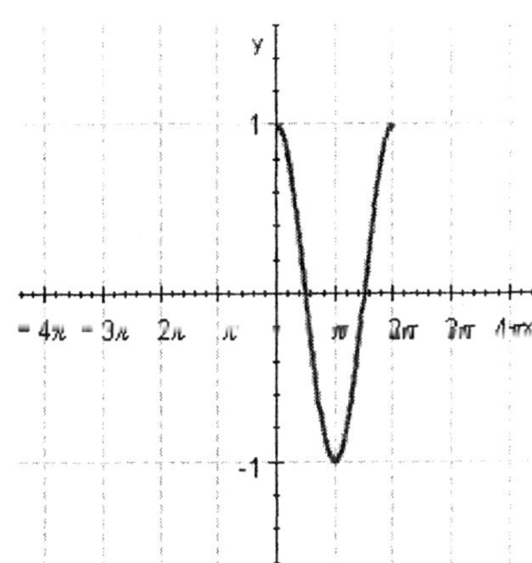

Select the correct answer.

a. 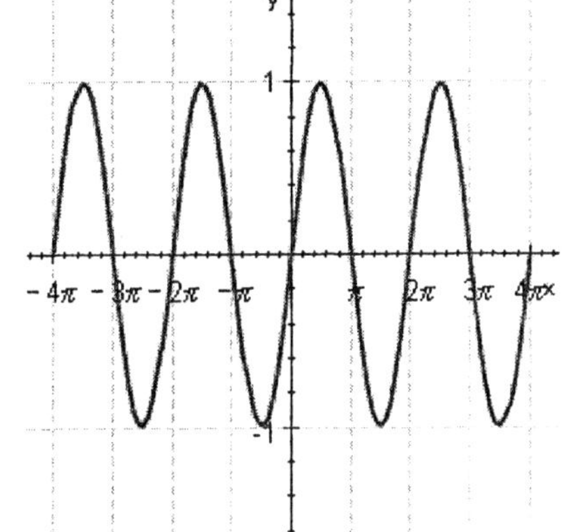

b. 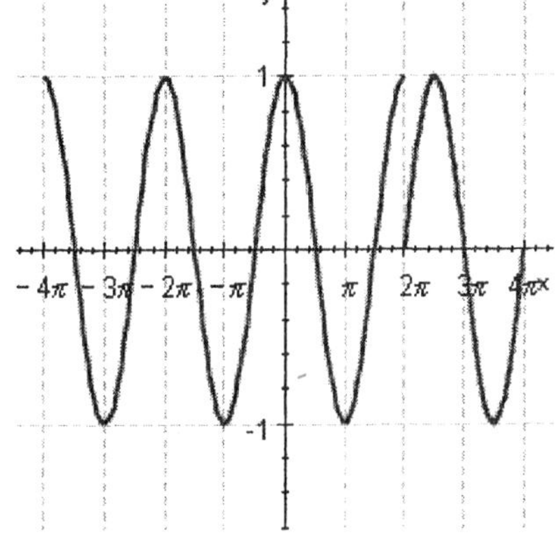

c. 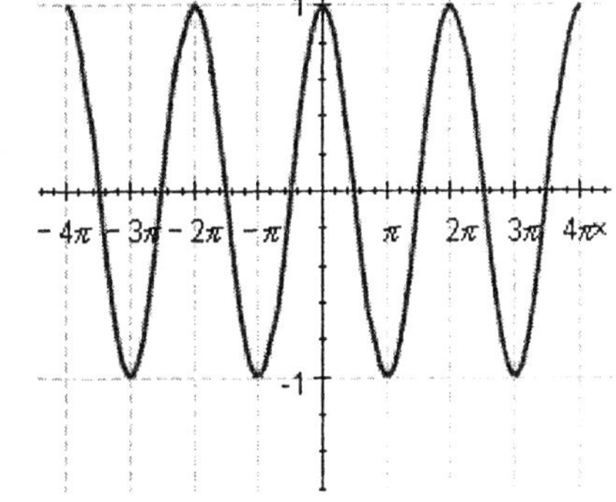

d. 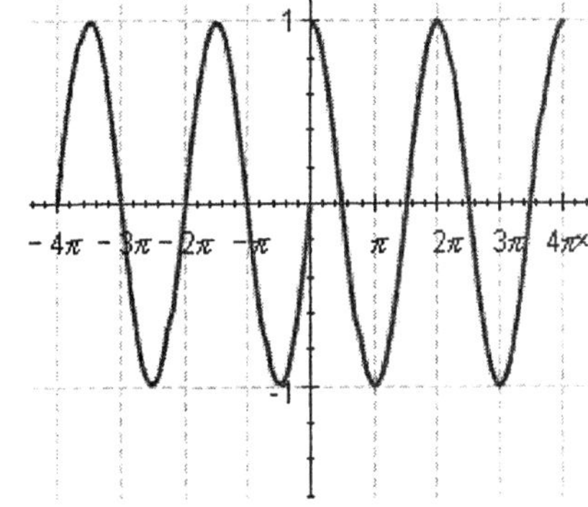

2. Use your graphing calculator to graph the family of functions for $-2\pi \leq x \leq 2\pi$ together on a single coordinate system.

$y = A \cos x$ for $A = 1, 4, 6$

Select the correct answer.

a.

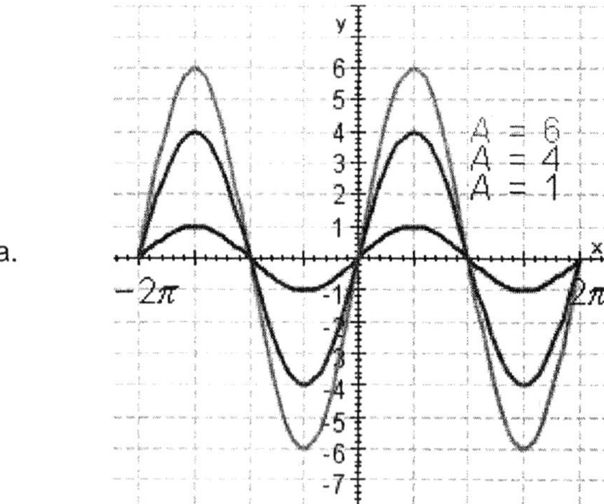

b.

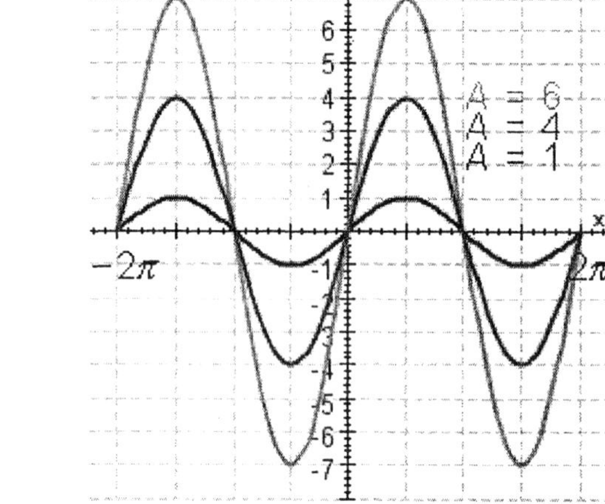

c.

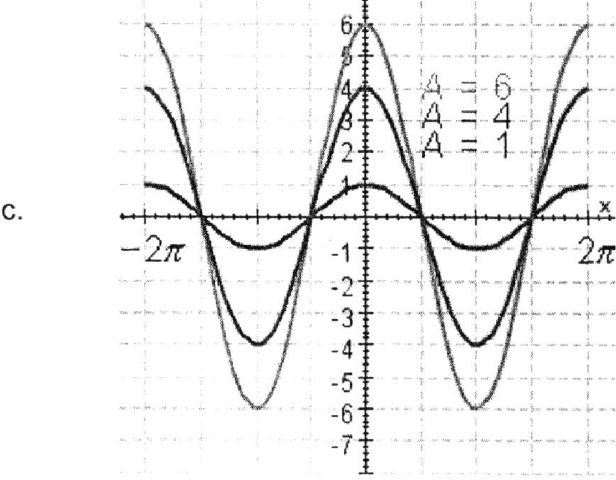

3. Use your graphing calculator to graph the pair of functions for $0 \leq x \leq 4\pi$.

$y = \cos Bx$ for $B = 1, 3$

Select the correct answer.

a.

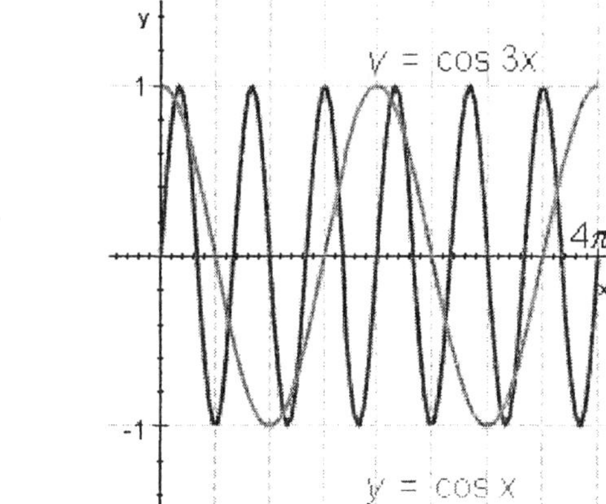

b.

c.

4. Use your graphing calculator to graph the pair of functions for $-2\pi \leq x \leq 2\pi$ together on a single coordinate system.

$$y = 3\cos x, \quad -3\cos x$$

Select the correct answer.

a.

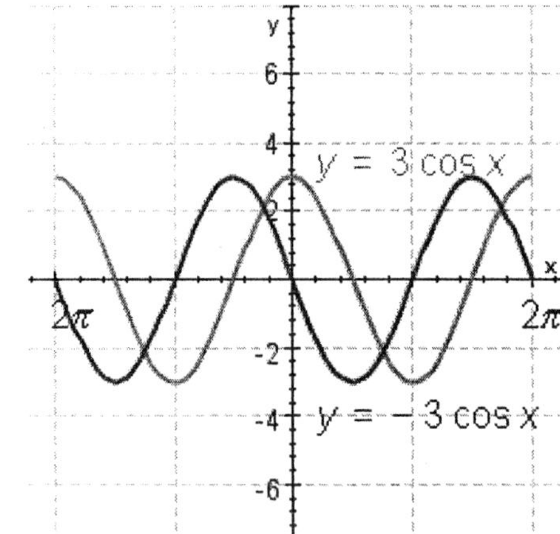

b.

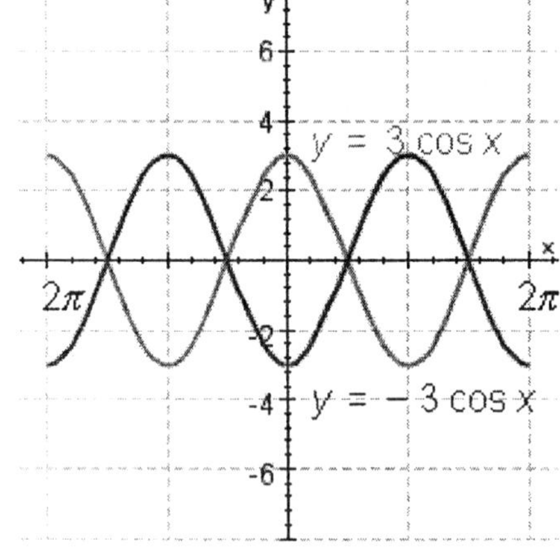

c.

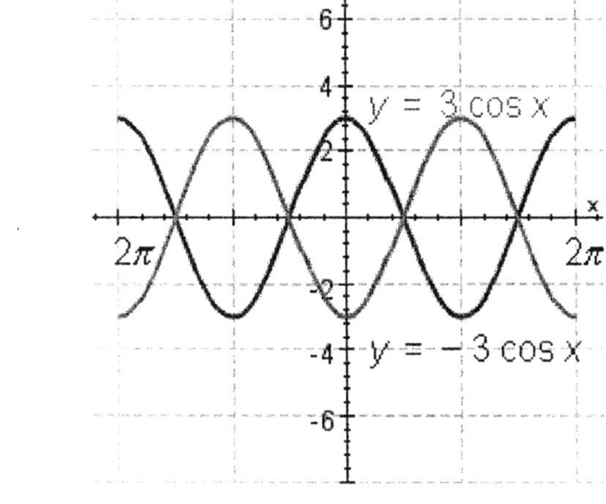

5. Graph one complete cycle of the graph. Label the axes accurately.

$$y = \cos 5x$$

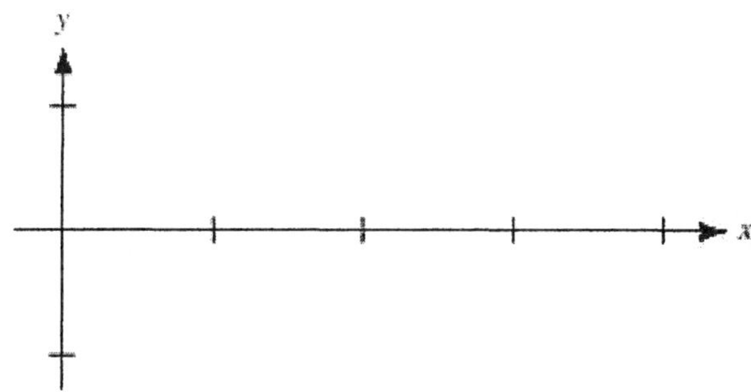

Identify the period.

6. Graph one complete cycle of the graph.

$$y = \sin \frac{\pi}{4} x$$

Select the correct answer.

a.

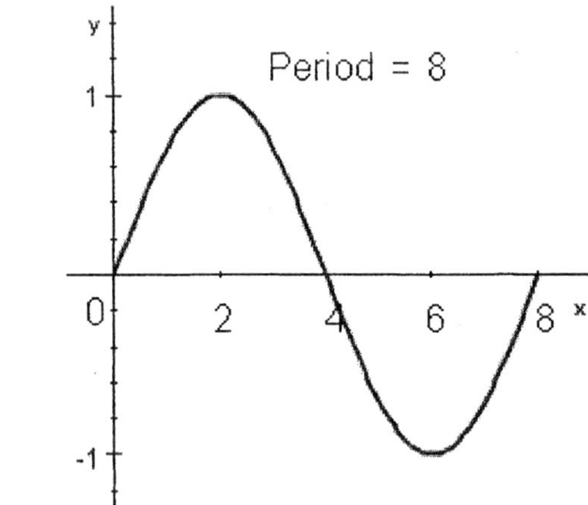

b.

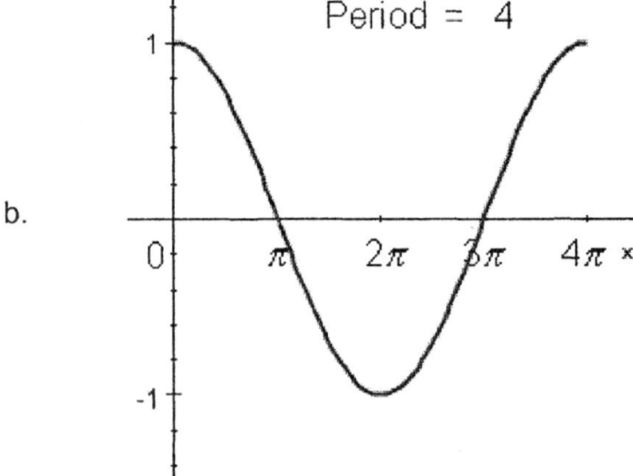

c.

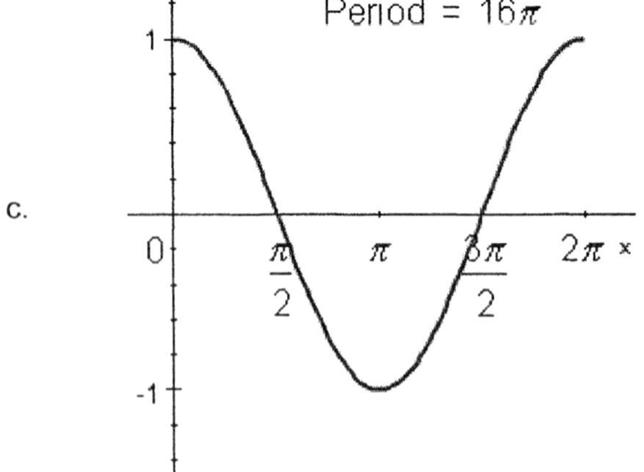

7. Use the graph of $y = \dfrac{3}{2} \cos 3x$ for reference and graph one complete cycle of the equation

$$y = 1 + \dfrac{3}{2} \cos 3x.$$

$$y = \dfrac{3}{2} \cos 3x$$

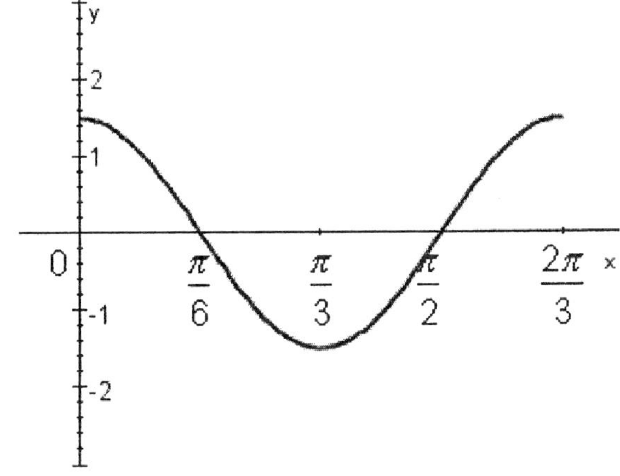

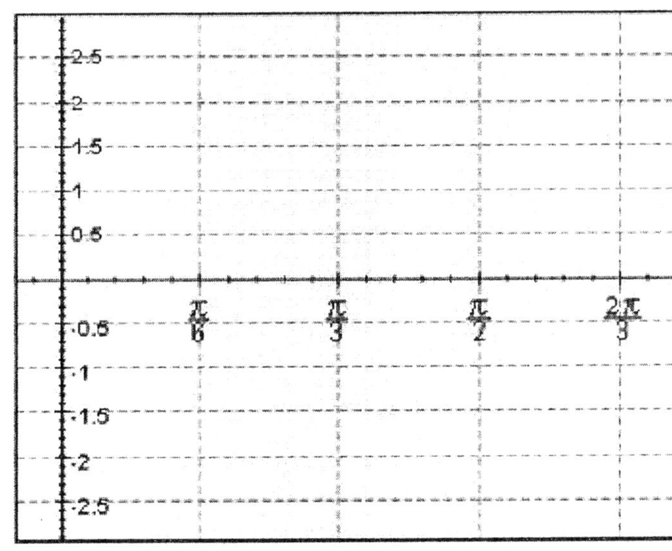

8. Use your graphing calculator to graph each pair of functions together for $-2\pi \le x \le 2\pi$.

$y = \csc x, \ y = 1.5 + \csc x$

$y = \csc x, \ y = -1.5 + \csc x$

$y = \csc x, \ y = -\csc x$

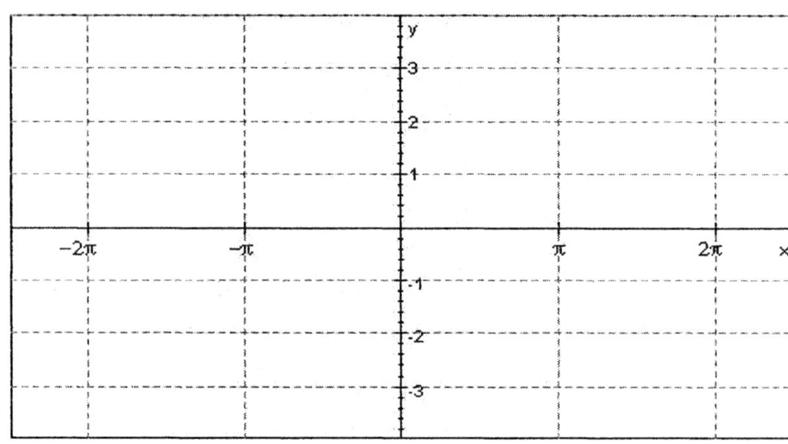

9. Identify the phase shift for the equation.

$$y = \sin\left(x + \frac{\pi}{2}\right)$$

Sketch one complete cycle of the graph. Graph $y = \sin x$ on the same coordinate system.

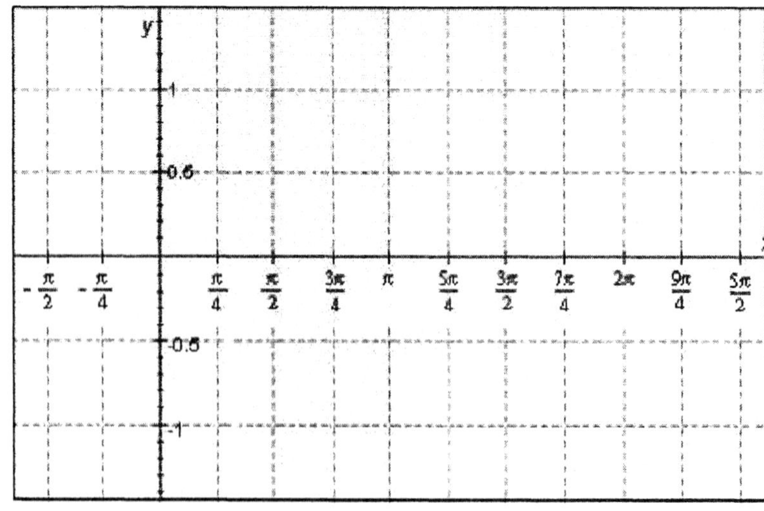

10. For the given equation, first identify the phase shift and then sketch one complete cycle of the graph. Graph $y = \cos x$ on the same coordinate system.

$$y = \cos\left(x + \frac{\pi}{4}\right)$$

Select the correct answer.

a.
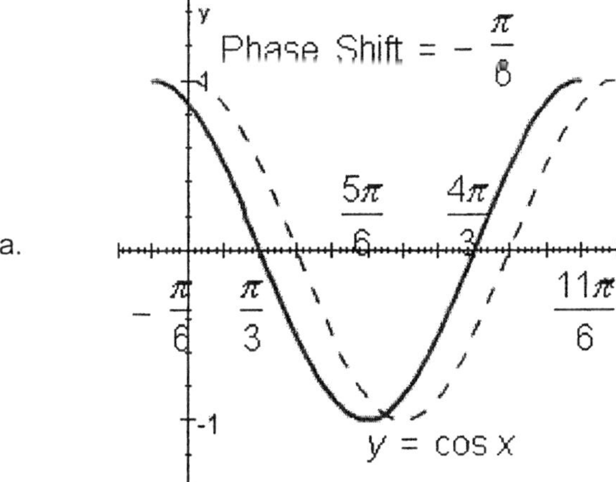

b.
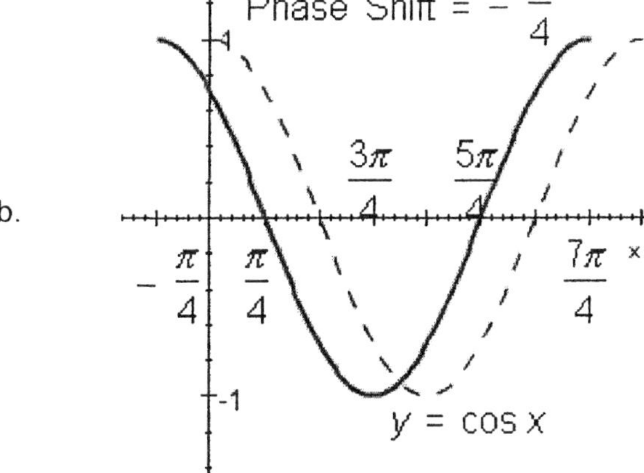

c.
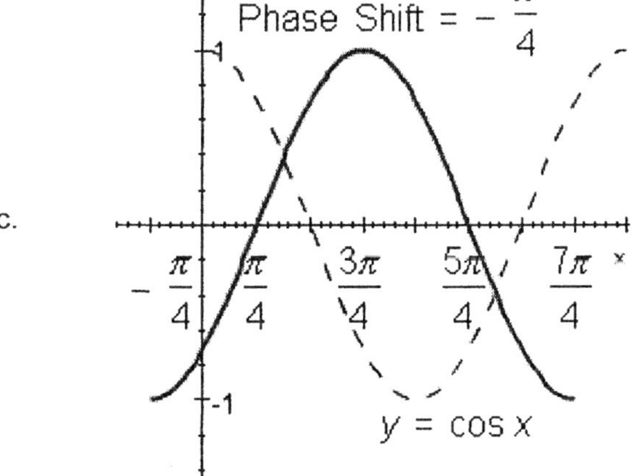

11.

Use the graph of the equation $y = -\sin\left(2x + \dfrac{\pi}{2}\right)$ shown below to graph one

complete cycle of the equation $y = 2 - \sin\left(2x + \dfrac{\pi}{2}\right)$.

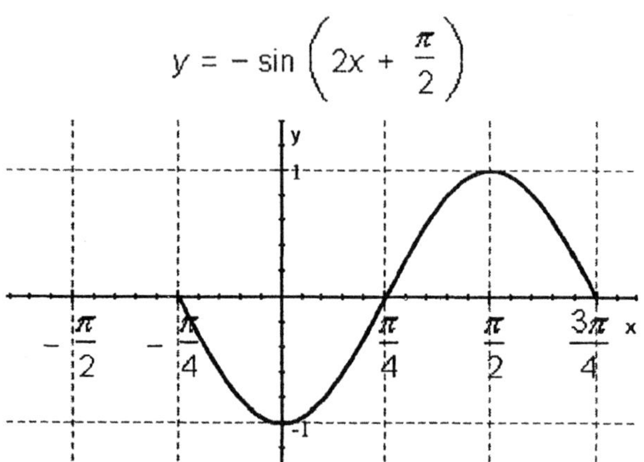

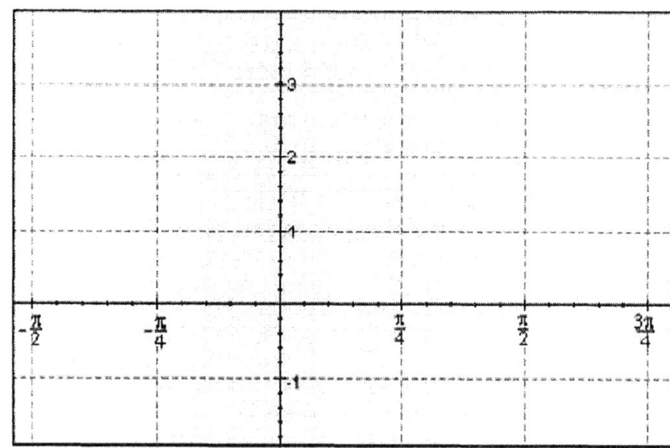

12. Graph the equation over the given interval.

$$y = 3\cos(2x - \pi),\ -\frac{\pi}{4} \le x \le \frac{3\pi}{2}$$

Select the correct answer.

a.

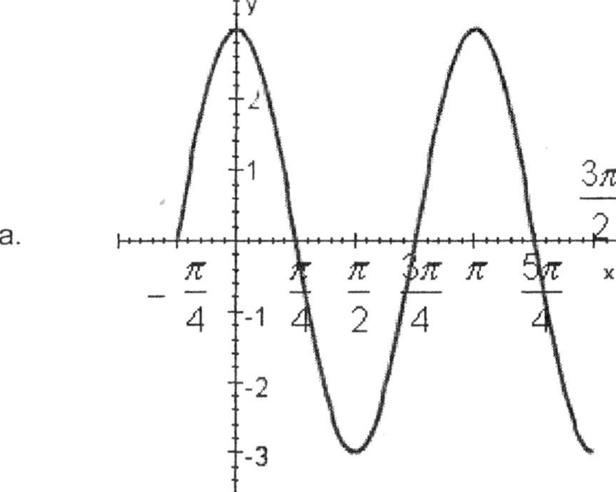

b.

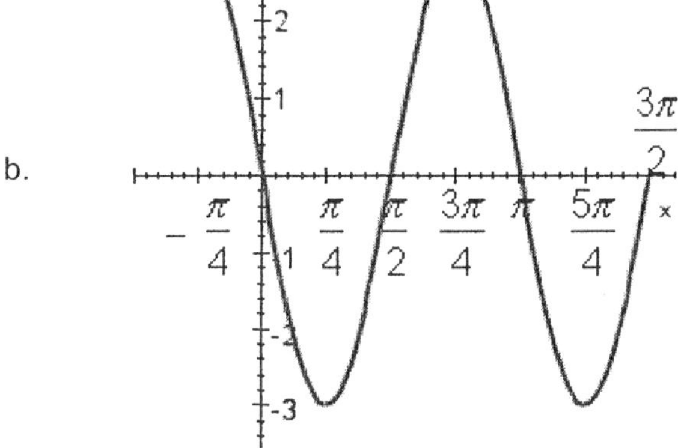

c.

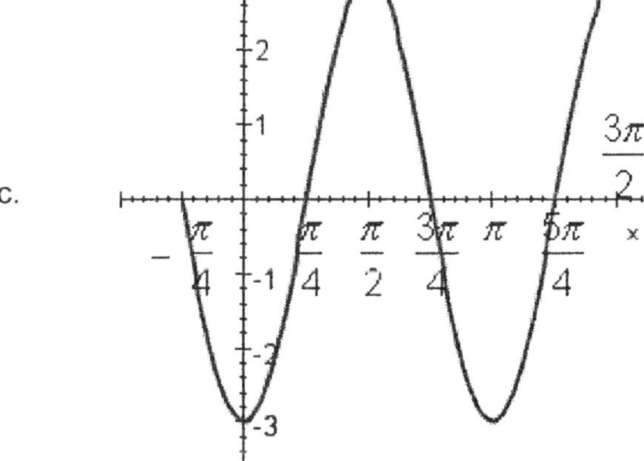

13. Find the equation of the line. Write your answer in slope-intercept form, $y = mx + b$.

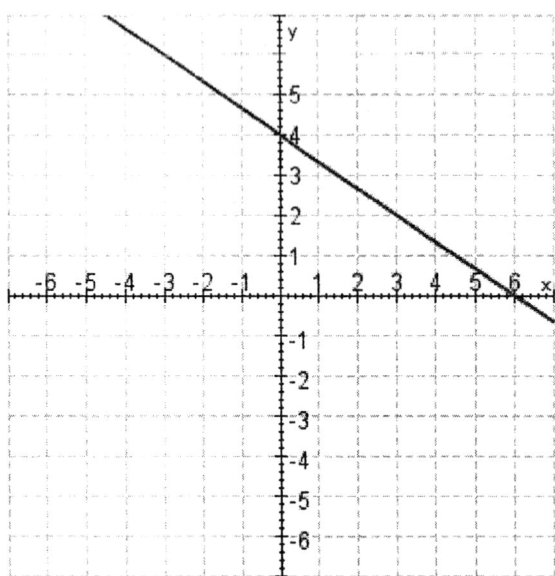

14. The graph below is one complete cycle of the graph of an equation containing a trigonometric function. Find an equation to match the graph.

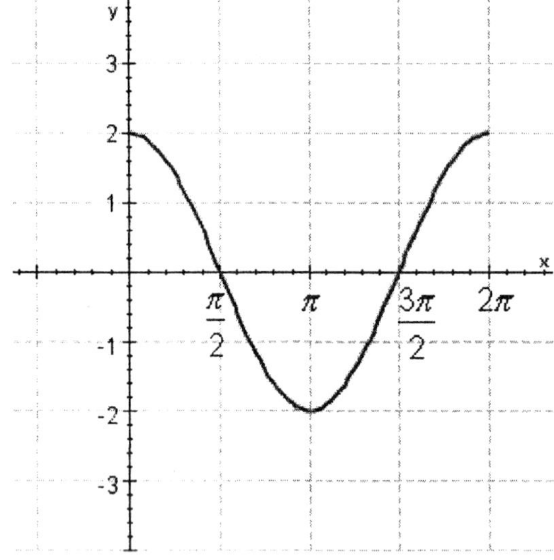

Select the correct answer.

a. $y = 2\cos x$

b. $y = 4\cos x$

c. $y = 6\cos x$

d. $y = 3\cos x$

e. $y = 5\cos x$

15. The graph below is one complete cycle of the graph of an equation containing a trigonometric function. Find an equation to match the graph.

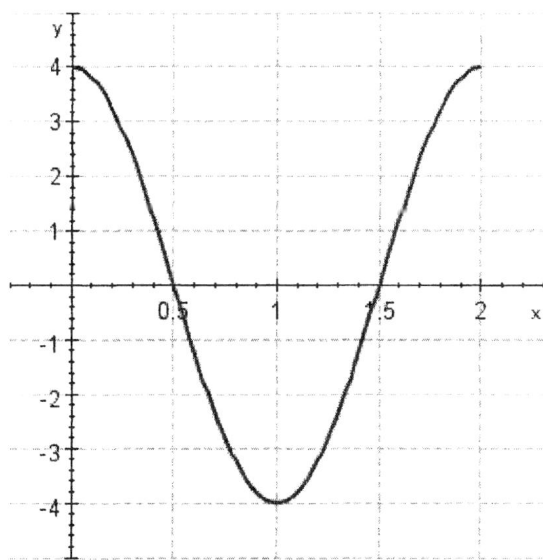

16. The graph below is one complete cycle of the graph of an equation containing a trigonometric function. Find an equation to match the graph.

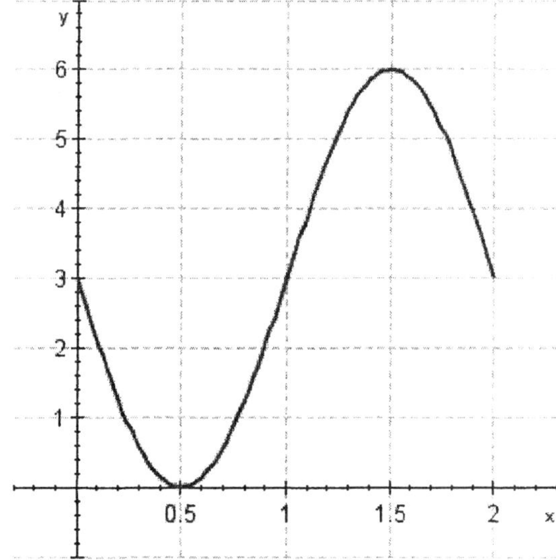

Select the correct answer.

a. $y = 3 - 3\sin \pi x$

b. $y = -1 + \sin \pi x$

c. $y = 1 + 2\sin \pi x$

d. $y = -3 - 4\sin \pi x$

17. Sketch the graph from

$x = 0$ to $x = 4\pi$.

$y = \cos x + \sin \dfrac{x}{2}$

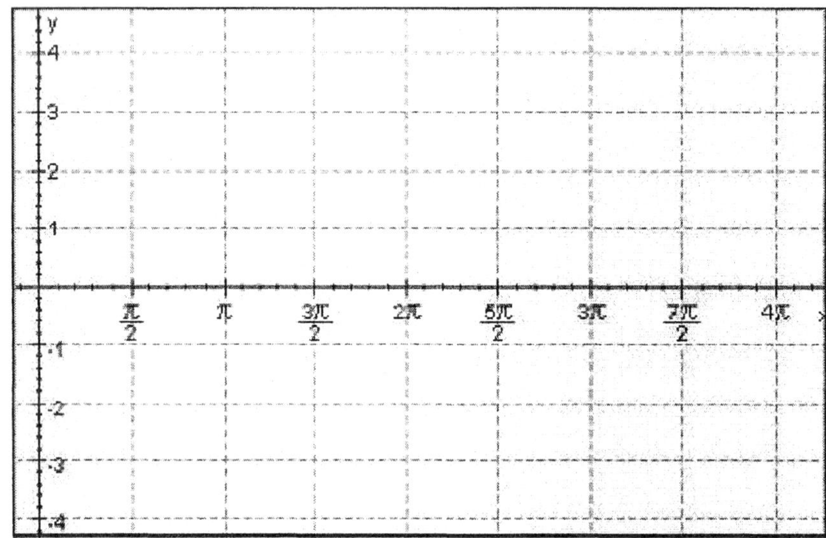

18. Sketch the graph from

$x = 0$ to $x = 4\pi$.

$$y = \sin x - \frac{1}{2}\cos 2x$$

Select the correct answer.

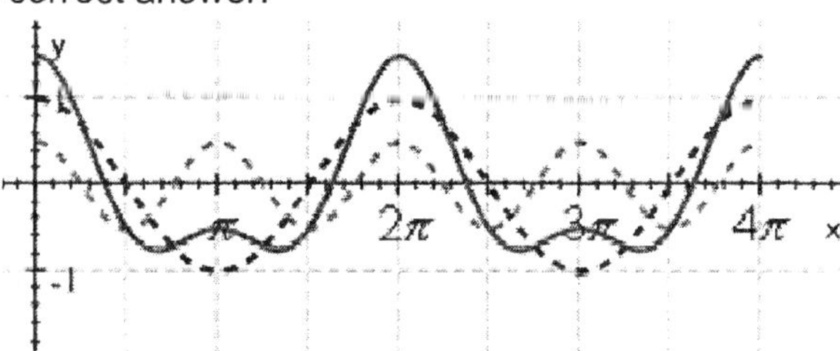

a.

········ $\sin x$ ······ $-\dfrac{1}{2}\cos 2x$

——— $\sin x - \dfrac{1}{2}\cos 2x$

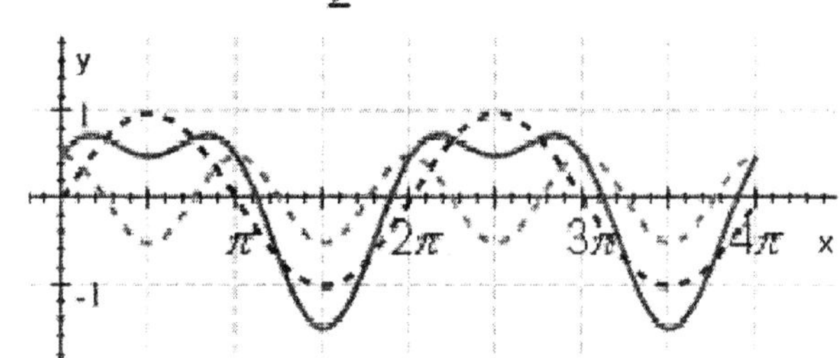

b.

···· $\sin x$ ····· $-\dfrac{1}{2}\cos 2x$

——— $\sin x - \dfrac{1}{2}\cos 2x$

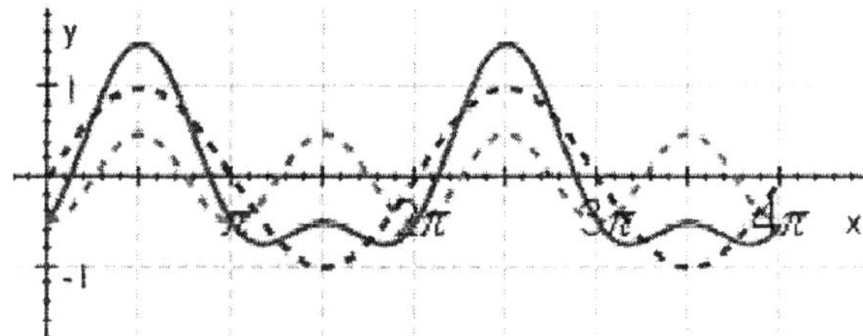

c.

··· $\sin x$ ····· $-\dfrac{1}{2}\cos 2x$

——— $\sin x - \dfrac{1}{2}\cos 2x$

19. Use your graphing calculator to graph the function between $x = 0$ and $x = 4\pi$.

Show the graph of y_1, y_2, and $y = y_1 + y_2$.

$y = 3 + \sin x$

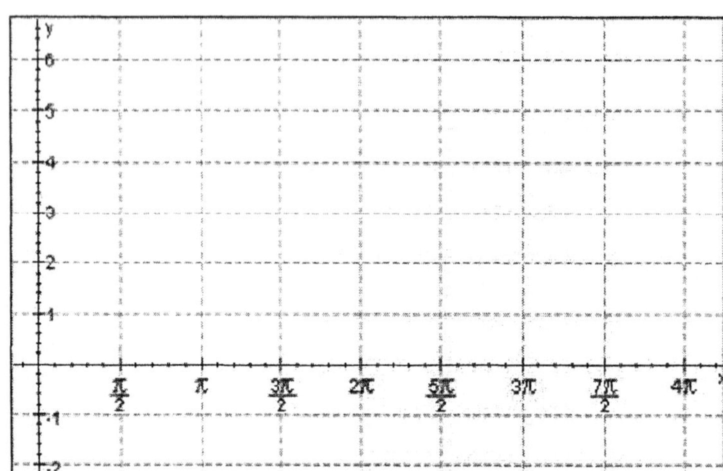

20. Use your graphing calculator to graph the function between $x = 0$ and $x = 4\pi$. Show the graph of y_1, y_2, and $y = y_1 + y_2$.

$y = \sin x + 2\cos x$

Select the correct answer.

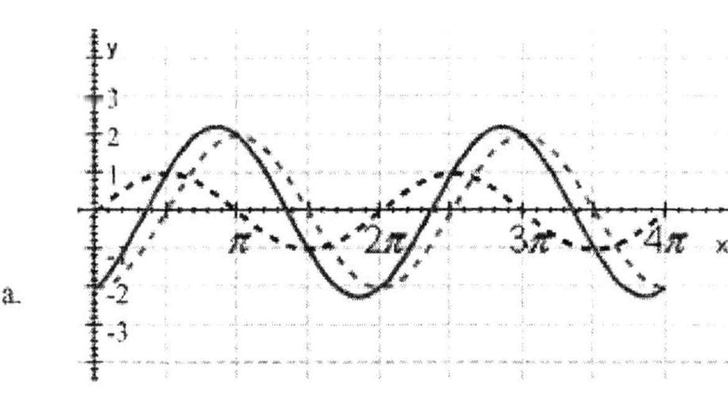

a.

$\cdots\cdot y_1 = \sin x$ $\qquad\cdots\cdot y_2 = 2\cos x$

$\qquad$ $y = \sin x + 2\cos x$

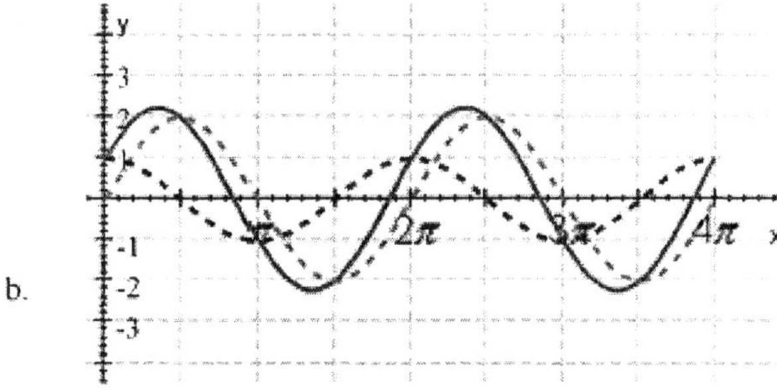

b.

$\cdots\cdot y_1 = \sin x$ $\qquad\cdots\cdot y_2 = 2\cos x$

$\qquad$ $y = \sin x + 2\cos x$

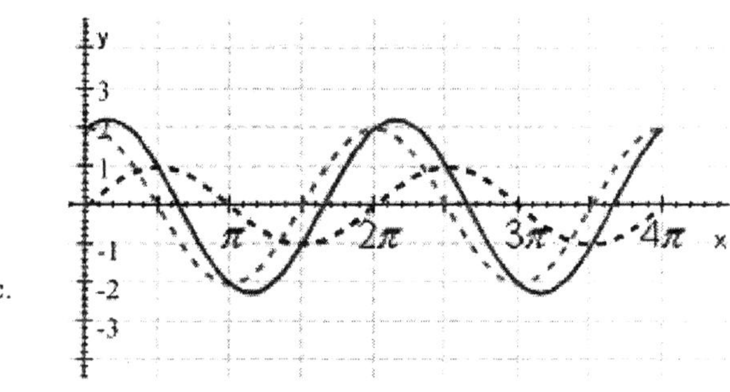

c.

$\cdots\cdot y_1 = \sin x$ $\qquad\cdots\cdot y_2 = 2\cos x$

$\qquad$ $y = \sin x + 2\cos x$

21. Evaluate the expression without using a calculator, and write your answer in radians.

$$\sin^{-1}(1)$$

22. Use a calculator to evaluate the expression to the nearest tenth of a degree.

$$\cos^{-1}(-0.7210)$$

Select the correct answer.

a. 136.1°

b. 130.6°

c. 131.7°

d. 138.4°

e. 137.3°

23. Evaluate without using a calculator.

$$\sin^{-1}(\sin 330^{\circ})$$

Select the correct answer.

a. -45°

b. -180°

c. -90°

d. -30°

e. -150°

24. Evaluate without using a calculator.

$$\tan\left(\cos^{-1}\frac{3}{5}\right)$$

25. Simplify

$$\sin^{-1}(\sin x) \quad \text{if} \quad 0 \le x \le \frac{\pi}{2}.$$

Select the correct answer.

a. $\pi - x$

b. $-x$

c. $x - \pi$

d. $-x - \pi$

e. x

1. c

2. c

3. c

4. b

5.

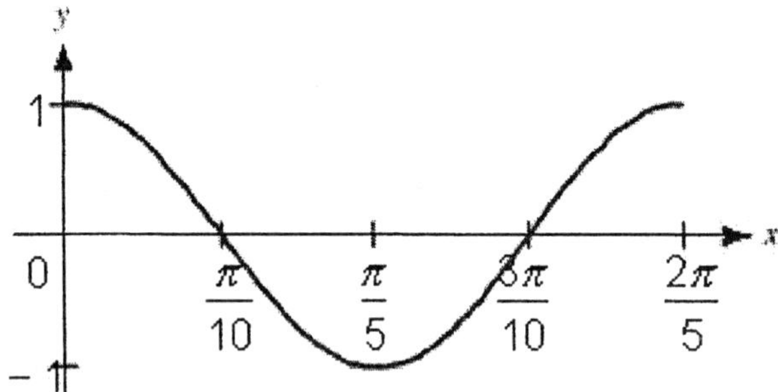

$$\frac{2\pi}{5}$$

6. a

7.

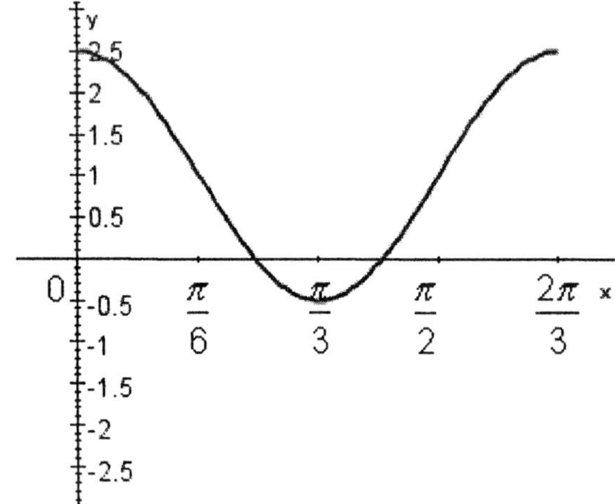

8.

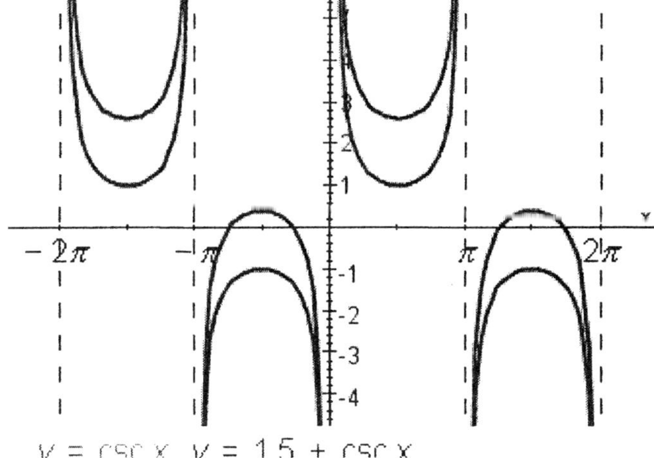

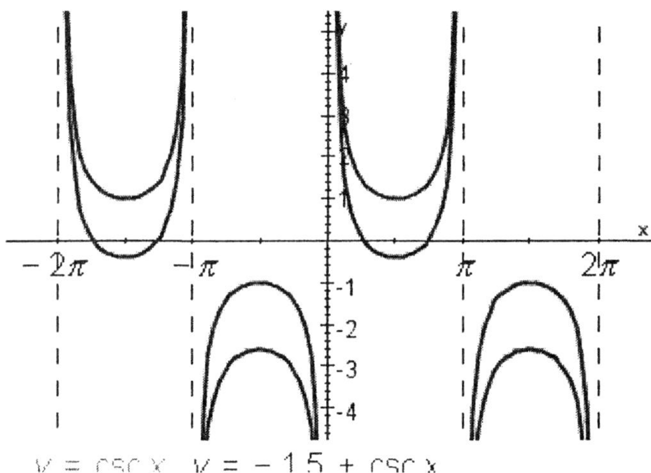

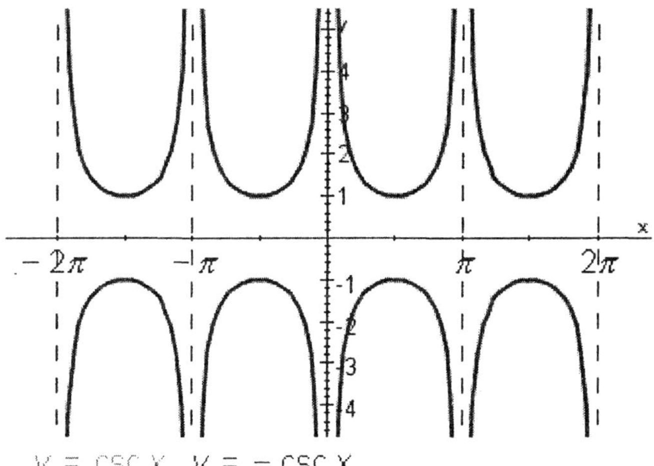

9. $-\dfrac{\pi}{2}$

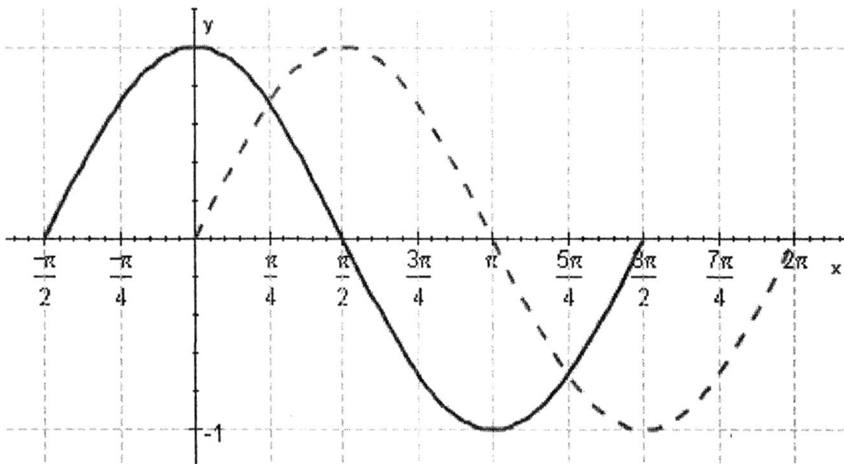

10. b

11.

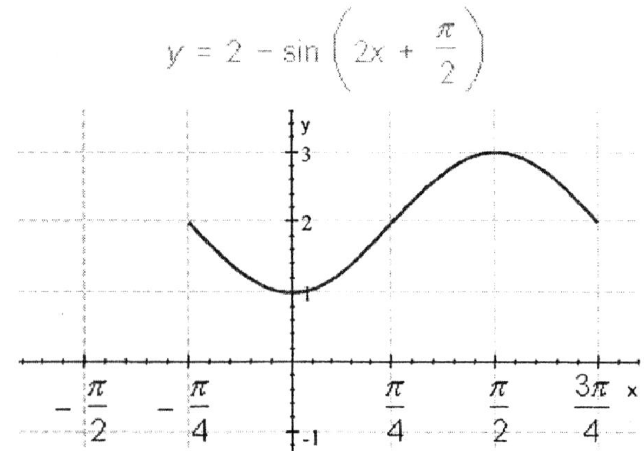

12. c

13. $y = -\dfrac{2}{3} \cdot x + 4$

14. a

15. $y = 4\cos(\pi \cdot x)$

16. a

17.

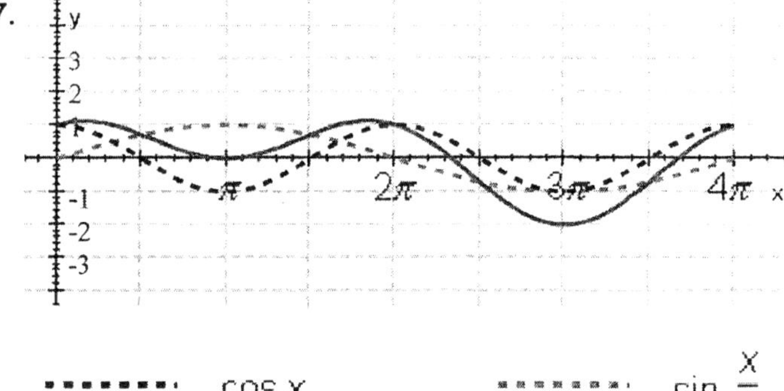

18. c

19.

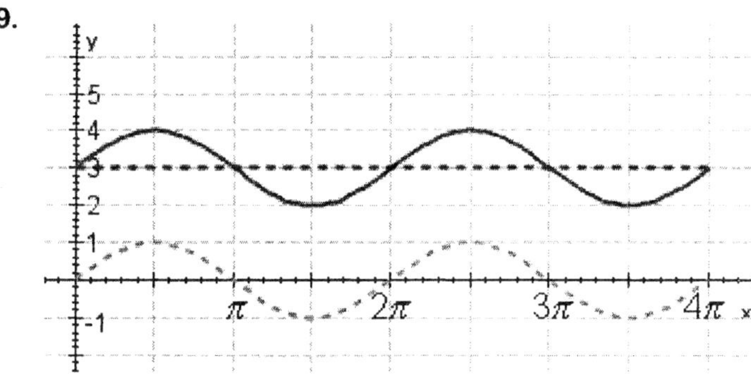

20. c

21. $\dfrac{\pi}{2}$

22. a

23. d

24. $\dfrac{4}{3}$

25. e

1. mctr.04.01.08m_NoAlgs
2. mctr.04.01.38m_NoAlgs
3. mctr.04.01.43m_NoAlgs
4. mctr.04.01.63m_NoAlgs
5. mctr.04.02.01_NoAlgs
6. mctr.04.02.06m_NoAlgs
7. mctr.04.02.36_NoAlgs
8. mctr.04.02.62_NoAlgs
9. mctr.04.03.01_NoAlgs
10. mctr.04.03.09m_NoAlgs
11. mctr.04.03.27_NoAlgs
12. mctr.04.03.33m_NoAlgs
13. mctr.04.04.02_NoAlgs
14. mctr.04.04.07m_NoAlgs
15. mctr.04.04.18_NoAlgs
16. mctr.04.04.21m_NoAlgs
17. mctr.04.05.17_NoAlgs
18. mctr.04.05.22m_NoAlgs
19. mctr.04.05.28_NoAlgs
20. mctr.04.05.32m_NoAlgs
21. mctr.04.06.22_NoAlgs
22. mctr.04.06.38m_NoAlgs
23. mctr.04.06.52m_NoAlgs
24. mctr.04.06.66_NoAlgs
25. mctr.04.06.73m_NoAlgs

1. Use a calculator to evaluate the expression to the nearest tenth of a degree.

$$\cos^{-1}(-0.7210)$$

Select the correct answer.

a. $136.1°$

b. $130.6°$

c. $131.7°$

d. $138.4°$

e. $137.3°$

2. Use your graphing calculator to graph the function between $x = 0$ and $x = 4\pi$. Show the graph of y_1, y_2, and $y = y_1 + y_2$.

$$y = 3 + \sin x$$

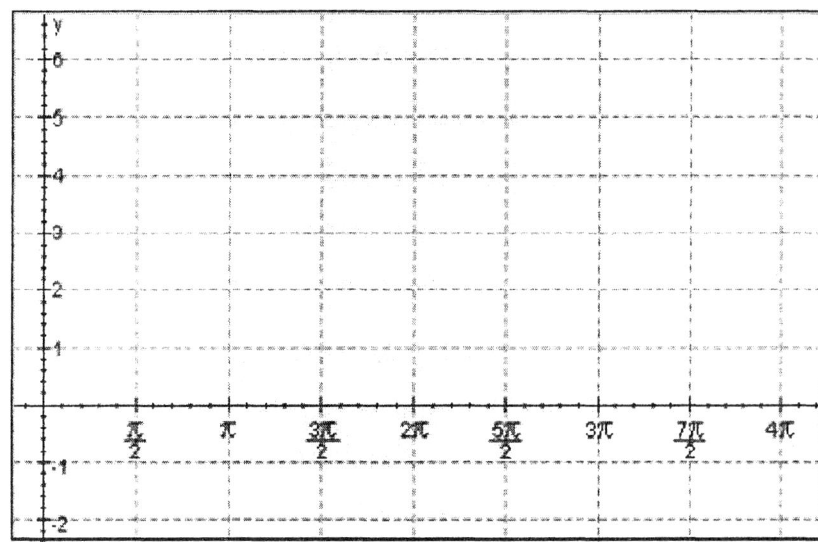

3. Use your graphing calculator to graph the pair of functions for $0 \leq x \leq 4\pi$.

$y = \cos Bx$ for $B = 1, 3$

Select the correct answer.

a.

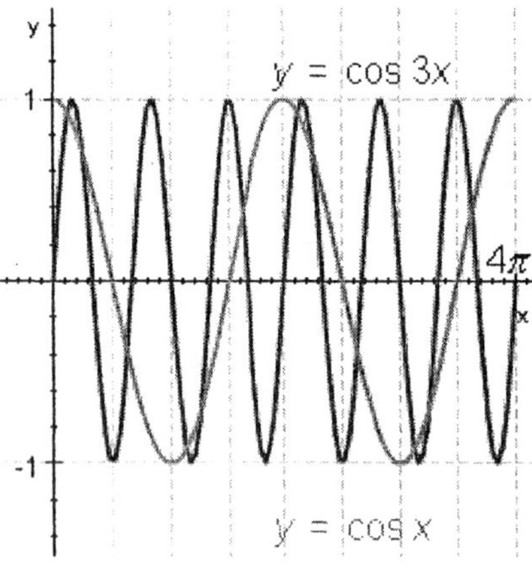

b.

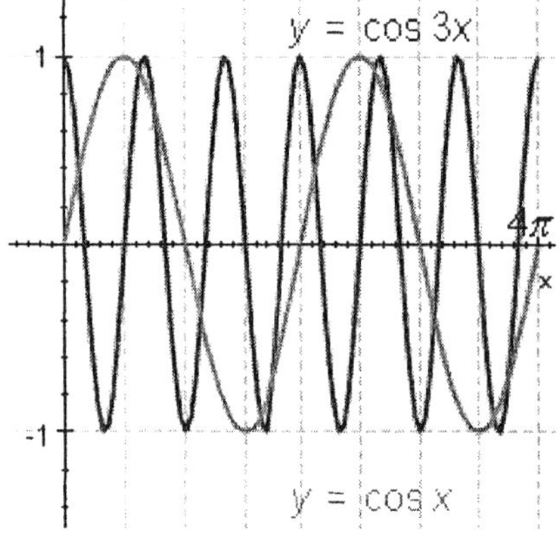

c.

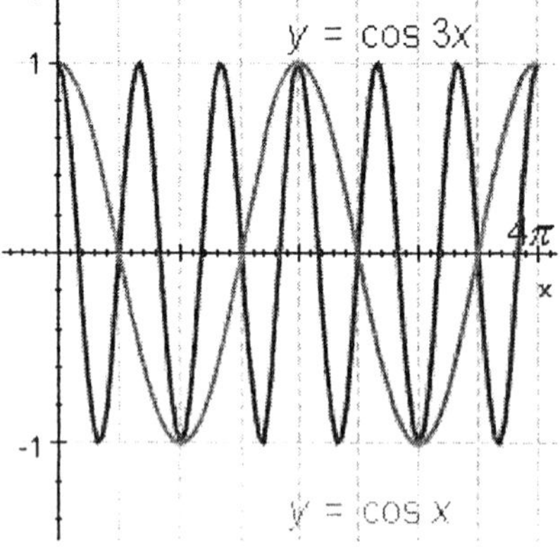

4. Graph the equation over the given interval.

$$y = 3\cos(2x - \pi),\ -\frac{\pi}{4} \le x \le \frac{3\pi}{2}$$

Select the correct answer.

a.

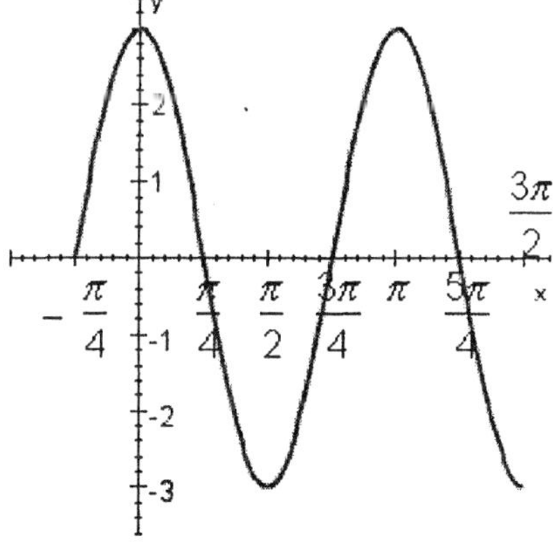

b.

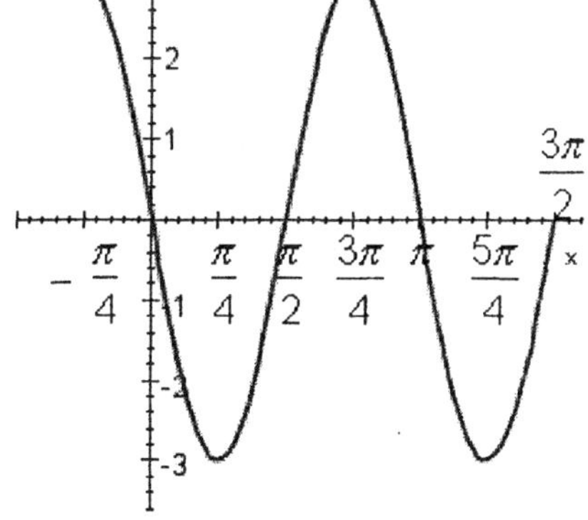

c.

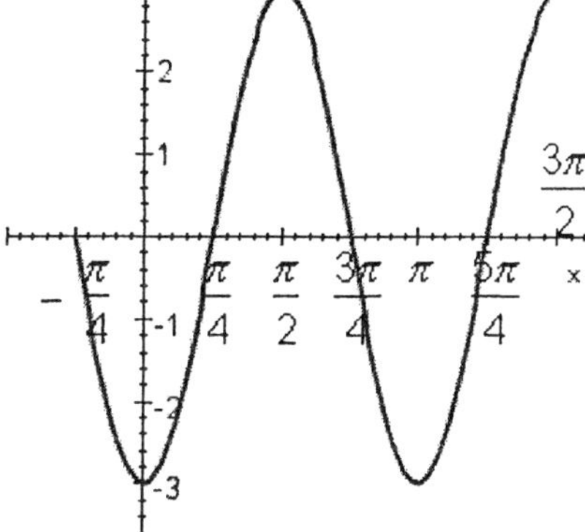

5. Identify the phase shift for the equation.

$$y = \sin\left(x + \frac{\pi}{2}\right)$$

Sketch one complete cycle of the graph. Graph $y = \sin x$ on the same coordinate system.

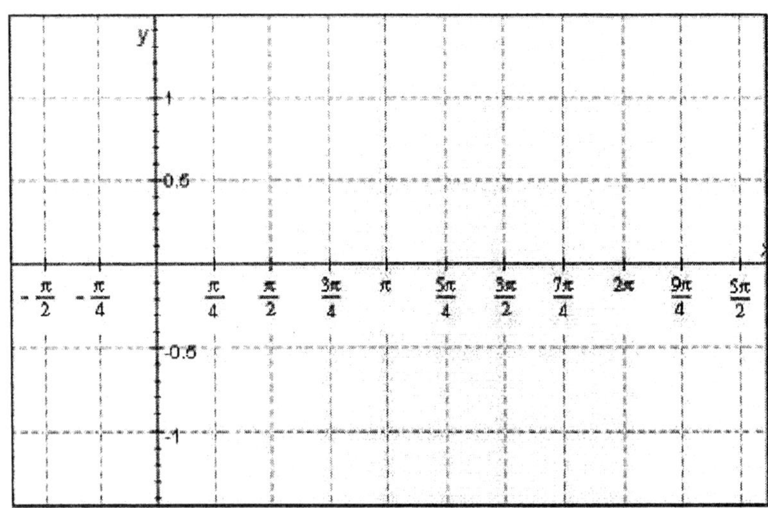

6. Evaluate without using a calculator.

$$\sin^{-1}\left(\sin 330°\right)$$

Select the correct answer.

a. $-45°$

b. $-180°$

c. $-90°$

d. $-30°$

e. $-150°$

7. The graph below is one complete cycle of the graph of an equation containing a trigonometric function. Find an equation to match the graph.

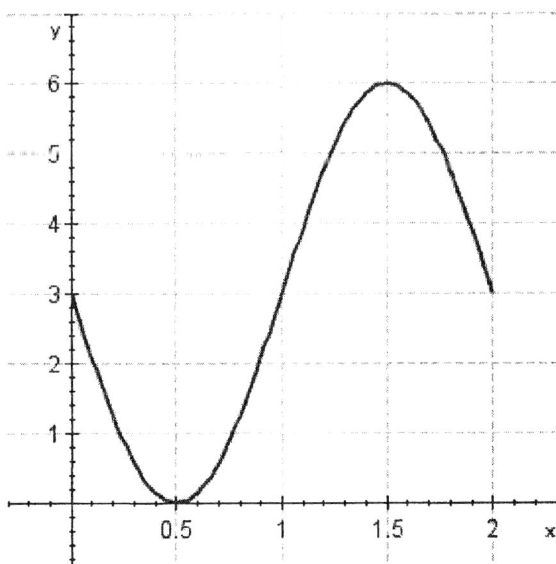

Select the correct answer.

a. $y = 3 - 3\sin \pi x$

b. $y = -1 + \sin \pi x$

c. $y = 1 + 2\sin \pi x$

d. $y = -3 - 4\sin \pi x$

e. $y = 4\sin \pi x$

8. Use your graphing calculator to graph the pair of functions for $-2\pi \le x \le 2\pi$ together on a single coordinate system.

$$y = 3\cos x, \ -3\cos x$$

Select the correct answer.

a.

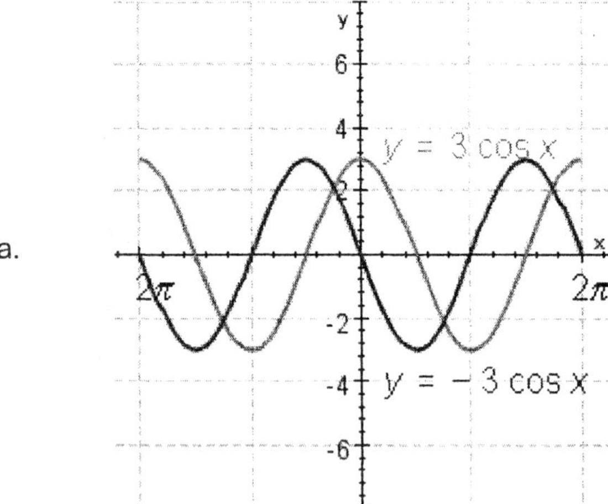

b.

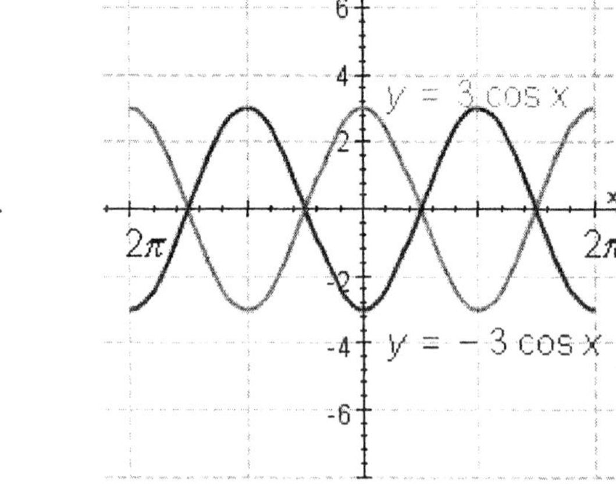

c.

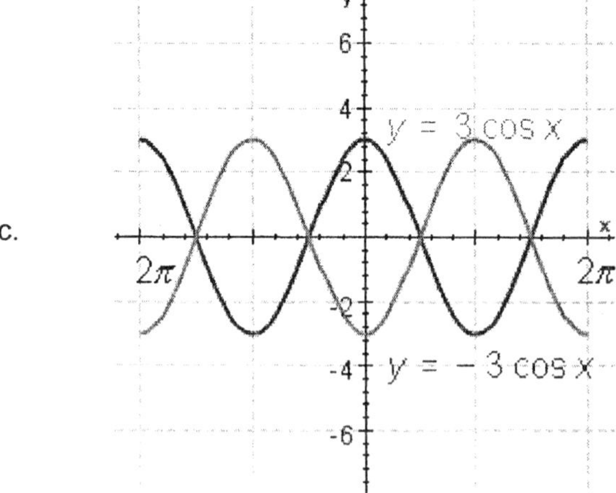

9. For the given equation, first identify the phase shift and then sketch one complete cycle of the graph. Graph $y = \cos x$ on the same coordinate system.

$$y = \cos\left(x + \frac{\pi}{4}\right)$$

Select the correct answer.

a.

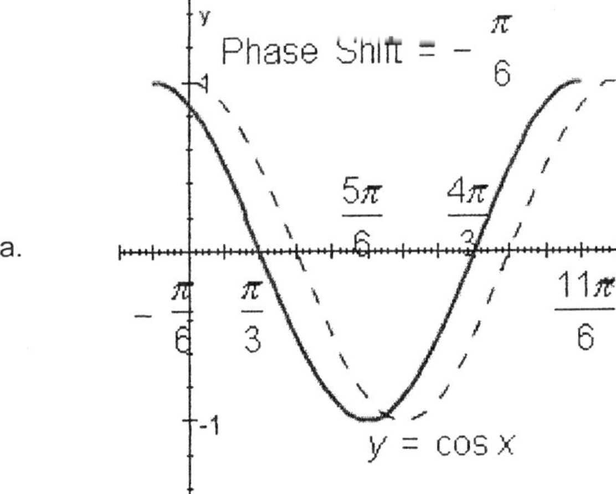

b.

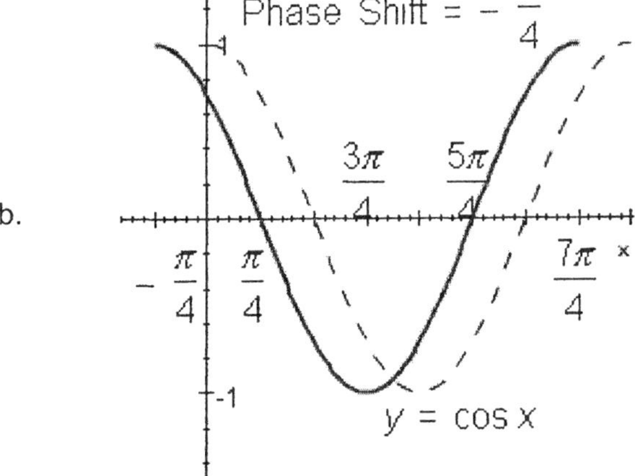

c.

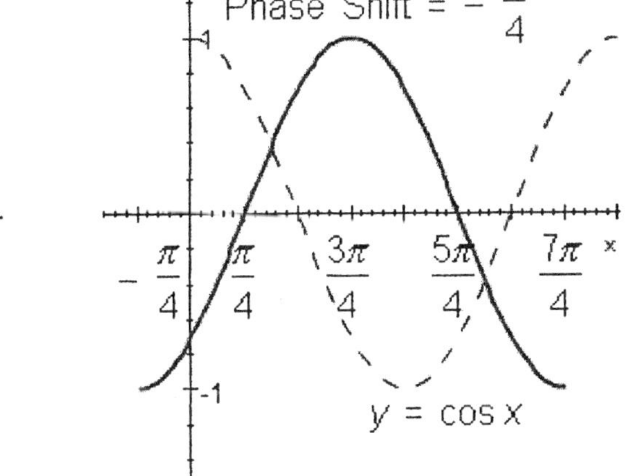

10. Evaluate without using a calculator.

$$\tan\left(\cos^{-1}\frac{3}{5}\right)$$

11. Graph one complete cycle of the graph..

$$y = \sin\frac{\pi}{4}x$$

Select the correct answer.

a.

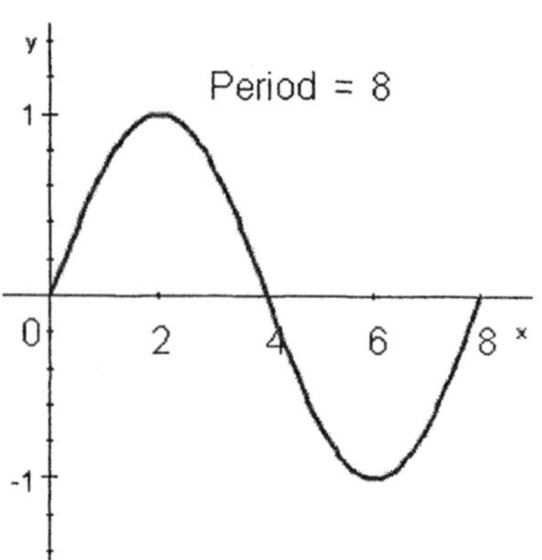

b.

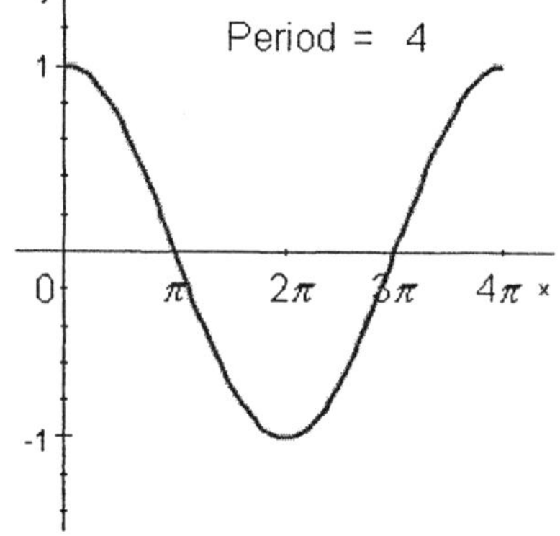

c.

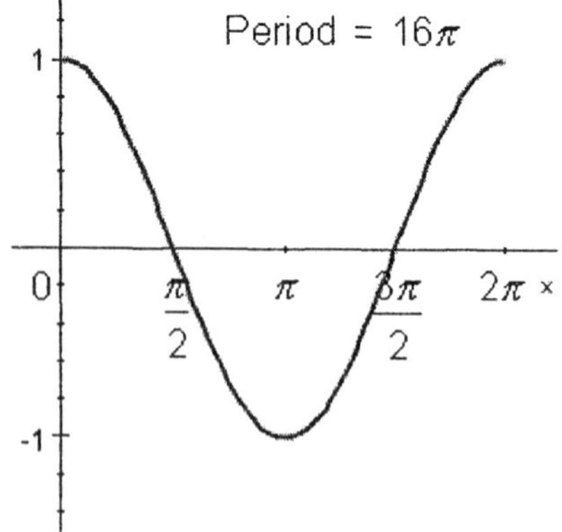

12. Simplify $\sin^{-1}(\sin x)$ if $0 \le x \le \dfrac{\pi}{2}$.

Select the correct answer.

a. $\pi - x$

b. $-x$

c. $x - \pi$

d. $-x - \pi$

e. x

13. Use the graph of the equation $y = -\sin\left(2x + \dfrac{\pi}{2}\right)$ shown below to graph one

complete cycle of the equation $y = 2 - \sin\left(2x + \dfrac{\pi}{2}\right)$.

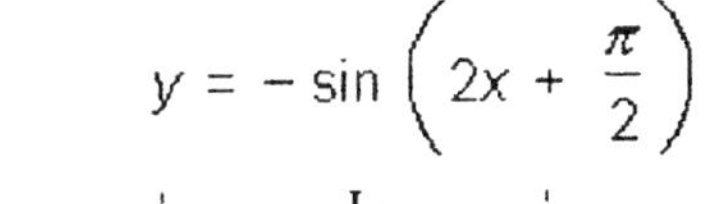

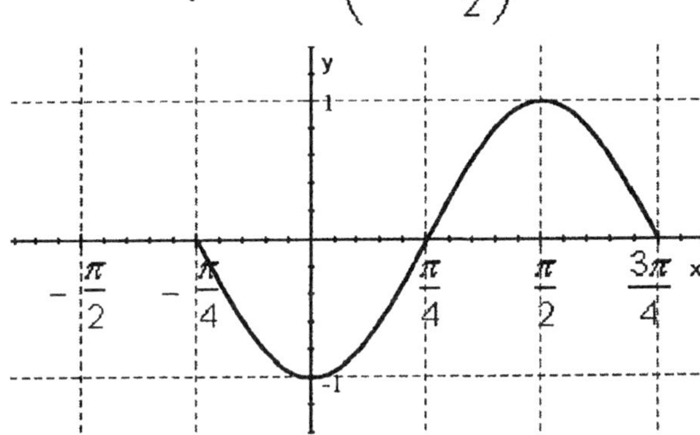

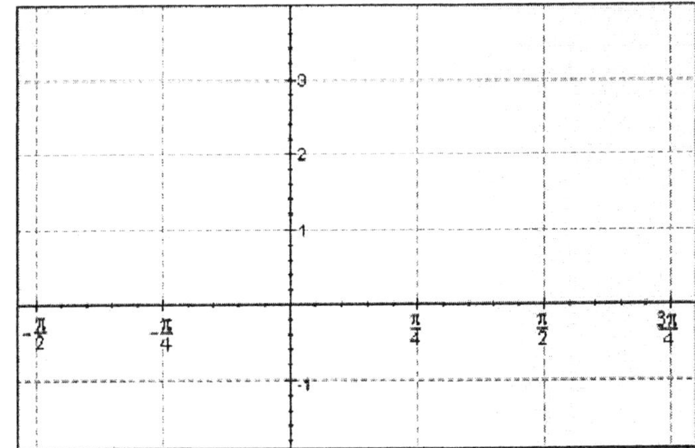

14. The graph below is one complete cycle of the graph of an equation containing a trigonometric function. Find an equation to match the graph.

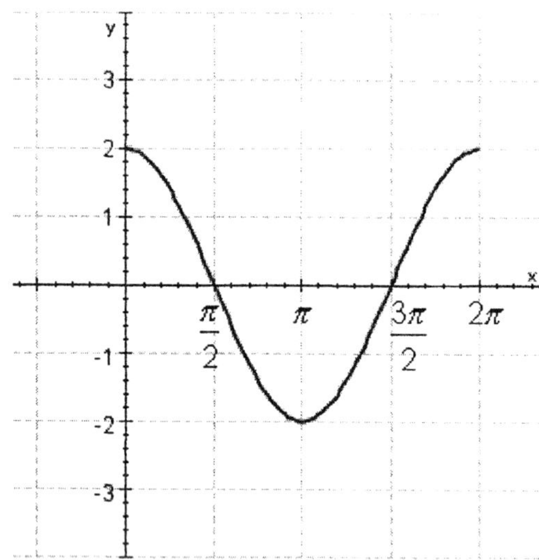

Select the correct answer.

a. $y = 2\cos x$

b. $y = 4\cos x$

c. $y = 6\cos x$

d. $y = 3\cos x$

e. $y = 5\cos x$

15. Sketch the graph from

$x = 0$ to $x = 4\pi$.

$$y = \cos x + \sin \frac{x}{2}$$

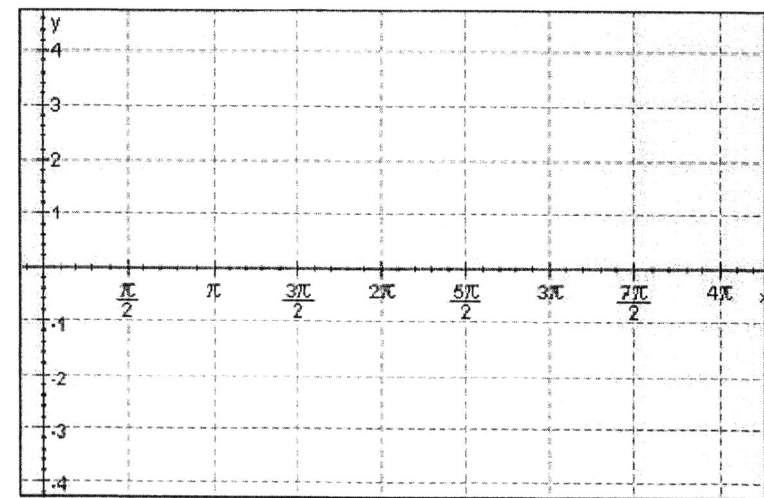

16. Use the graph of $y = \dfrac{3}{2}\cos 3x$ for reference and graph one complete cycle of the equation $y = 1 + \dfrac{3}{2}\cos 3x$.

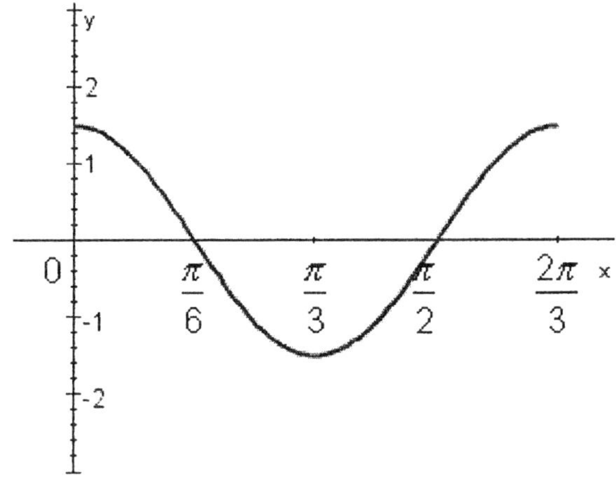

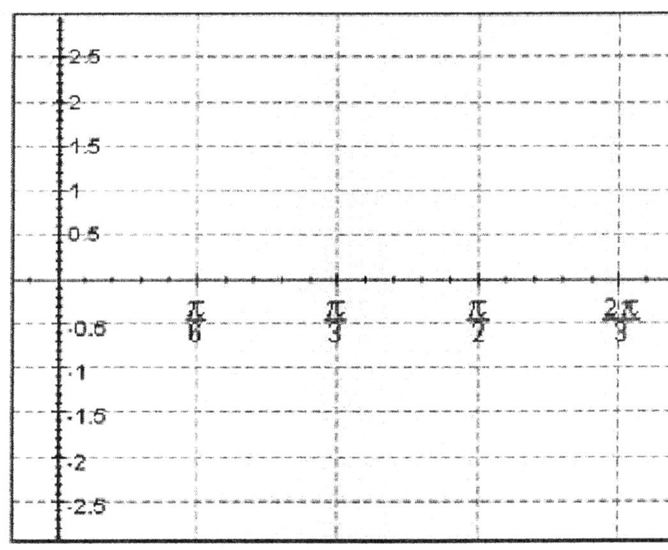

17. Use your graphing calculator to graph each pair of functions together for $-2\pi \leq x \leq 2\pi$.

$y = \csc x, \ y = 1.5 + \csc x$

$y = \csc x, \ y = -1.5 + \csc x$

$y = \csc x, \ y = -\csc x$

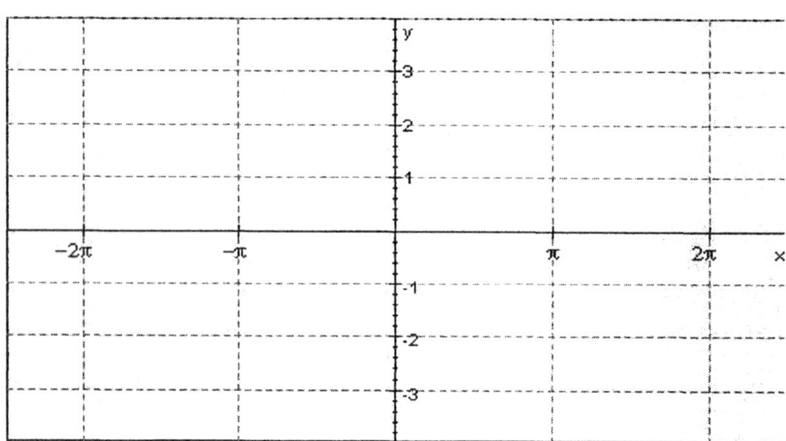

18. Find the equation of the line. Write your answer in slope-intercept form, $y = mx + b$.

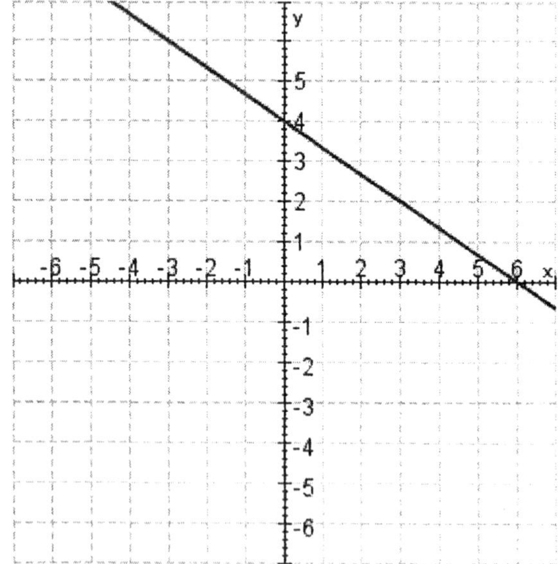

19. Sketch the graph of $y = \cos x$ between $x = -4\pi$ and $x = 4\pi$

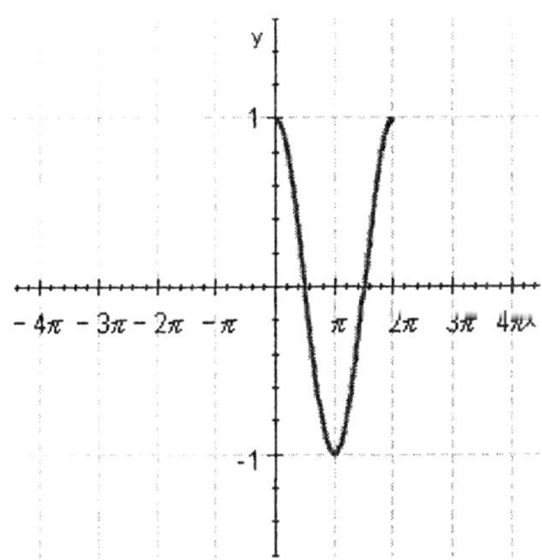

Select the correct answer.

a.

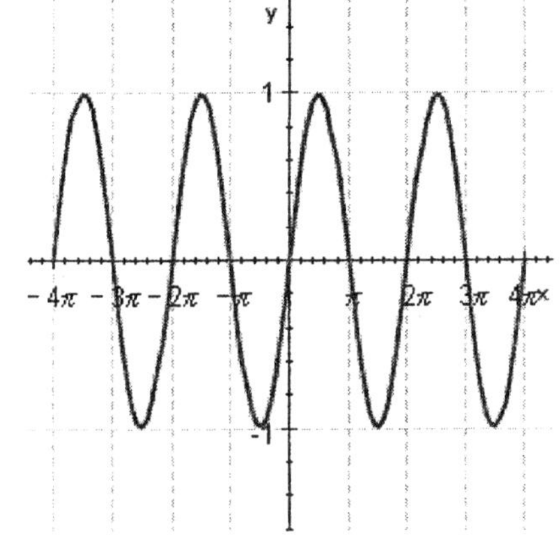

b.

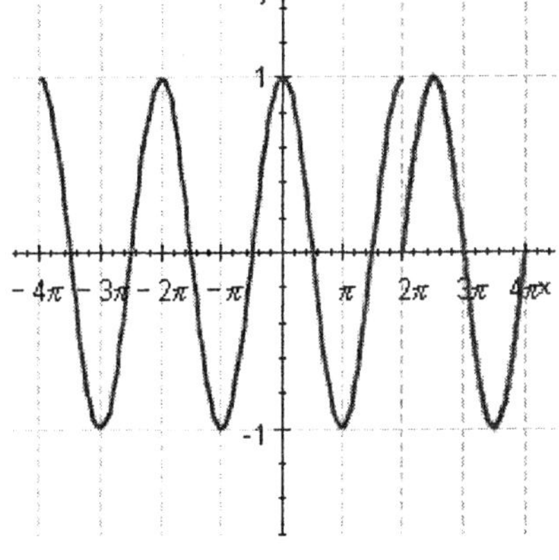

c.

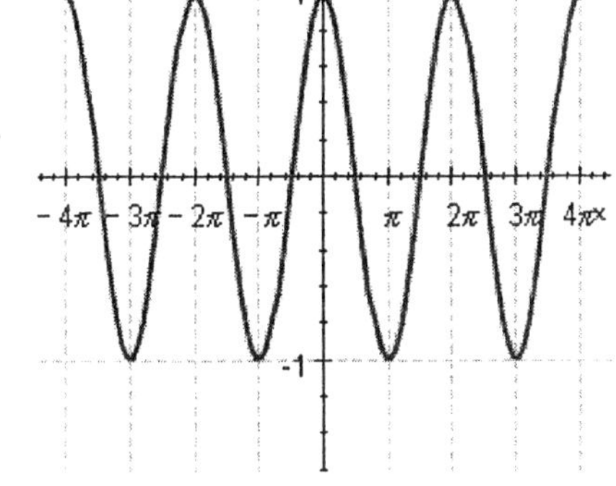

d.

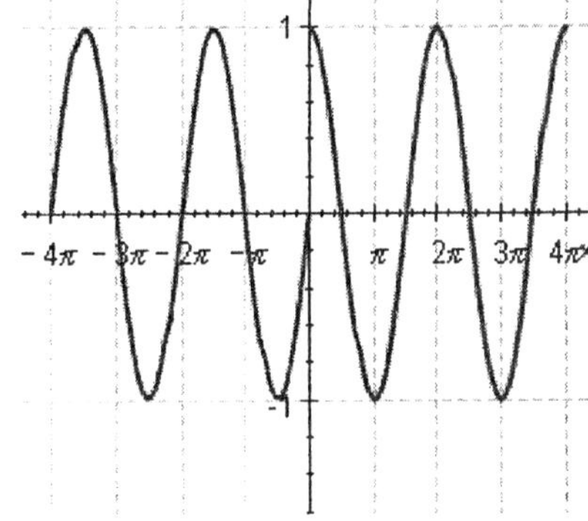

20. Graph one complete cycle of the graph. Label the axes accurately.

$$y = \cos 5x$$

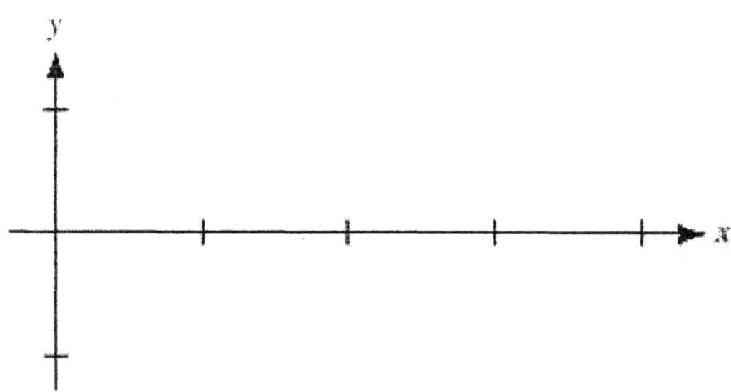

Identify the period.

21. The graph below is one complete cycle of the graph of an equation containing a trigonometric function. Find an equation to match the graph.

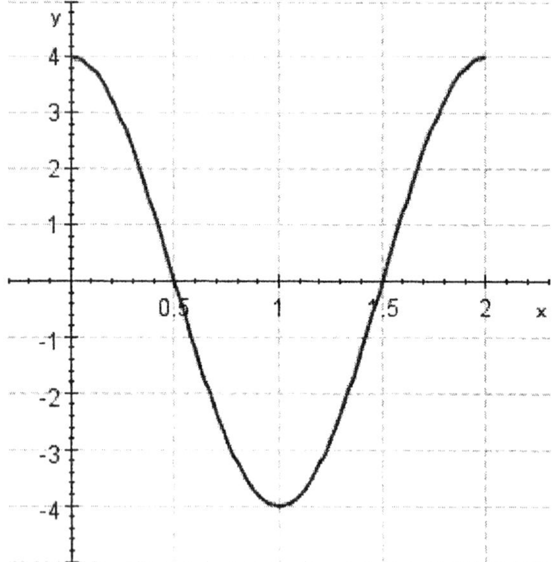

22. Use your graphing calculator to graph the family of functions for $-2\pi \leq x \leq 2\pi$ together on a single coordinate system.

$y = A\cos x$ for $A = 1, 4, 6$

Select the correct answer.

a.

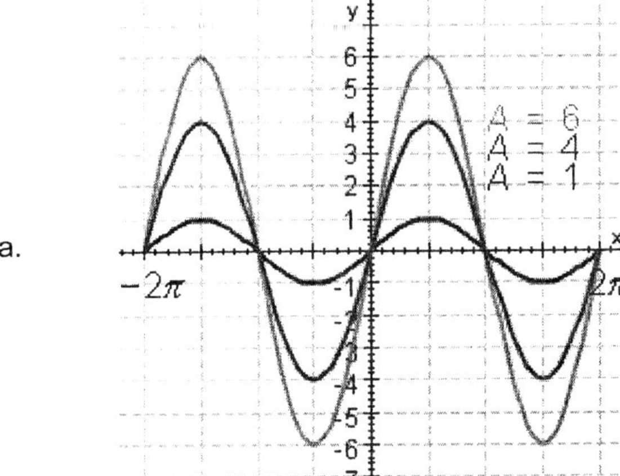

b.

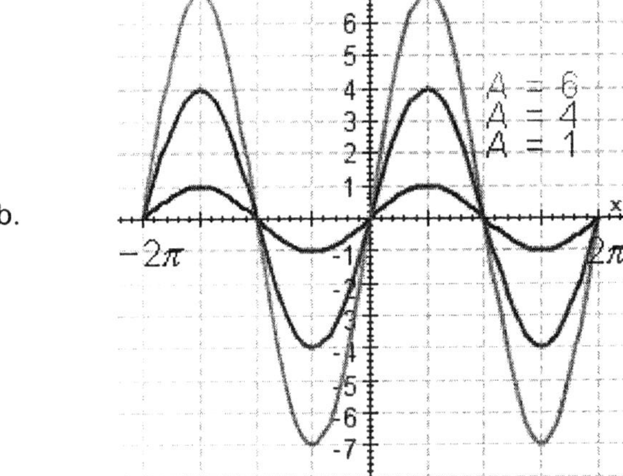

c.

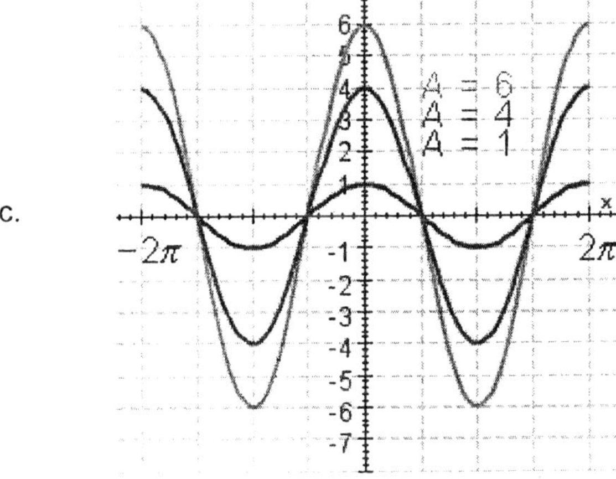

23. Sketch the graph from

$x = 0$ to $x = 4\pi$.

$$y = \sin x - \frac{1}{2}\cos 2x$$

Select the correct answer.

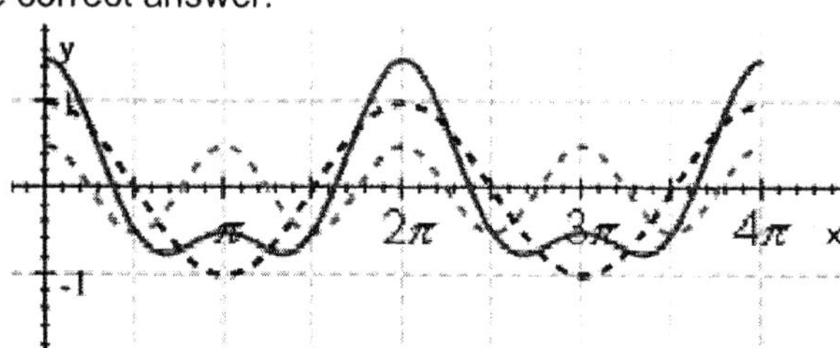

a.

$\cdots\cdots$ $\sin x$ $\cdots\cdots$ $-\frac{1}{2}\cos 2x$

$\overline{}$ $\sin x - \frac{1}{2}\cos 2x$

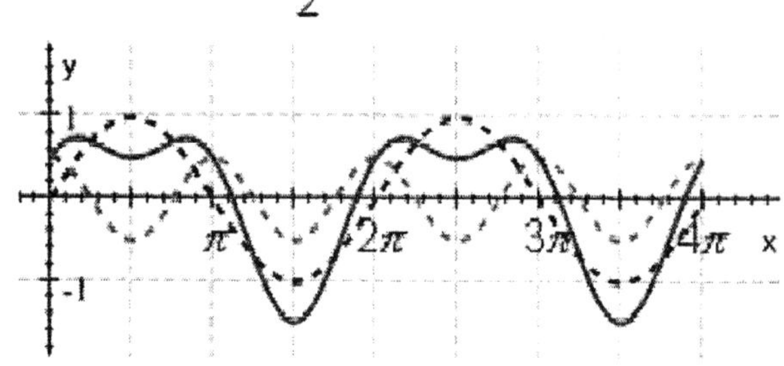

b.

$\cdots\cdots$ $\sin x$ $\cdots\cdots$ $-\frac{1}{2}\cos 2x$

$\overline{}$ $\sin x - \frac{1}{2}\cos 2x$

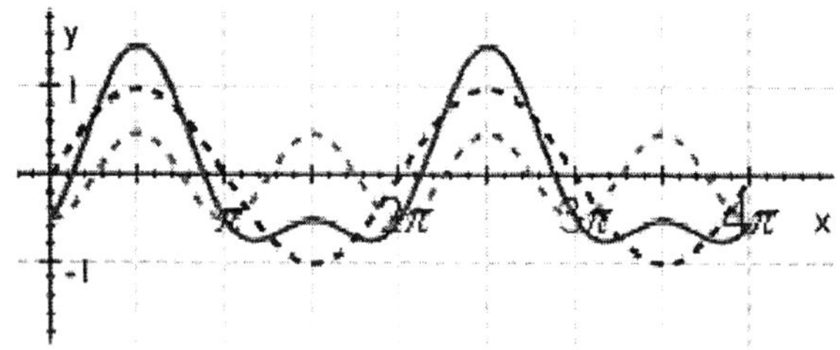

c.

$\cdots\cdots$ $\sin x$ $\cdots\cdots$ $-\frac{1}{2}\cos 2x$

$\overline{}$ $\sin x - \frac{1}{2}\cos 2x$

24. Use your graphing calculator to graph the function between $x = 0$ and $x = 4\pi$. Show the graph of y_1, y_2, and $y = y_1 + y_2$.

$$y = \sin x + 2\cos x$$

Select the correct answer.

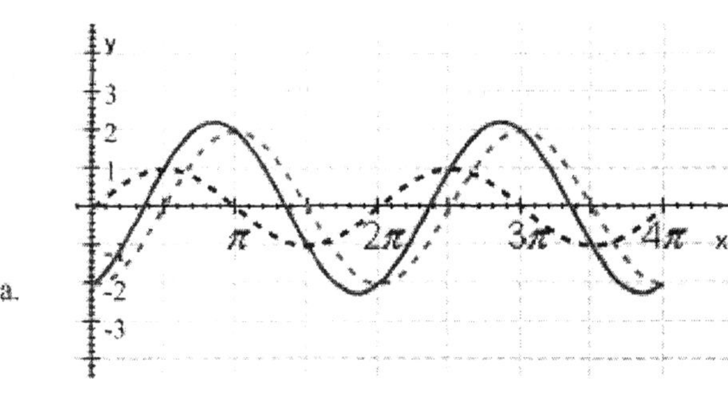

a.

$\cdots y_1 = \sin x$ $\cdots y_2 = 2\cos x$

$—\, y = \sin x + 2\cos x$

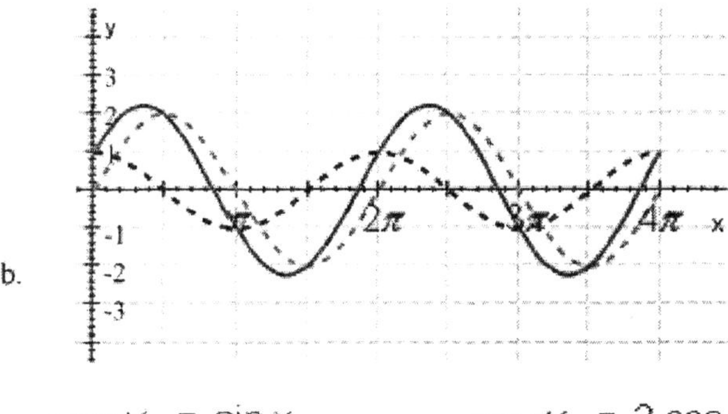

b.

$\cdots y_1 = \sin x$ $\cdots y_2 = 2\cos x$

$—\, y = \sin x + 2\cos x$

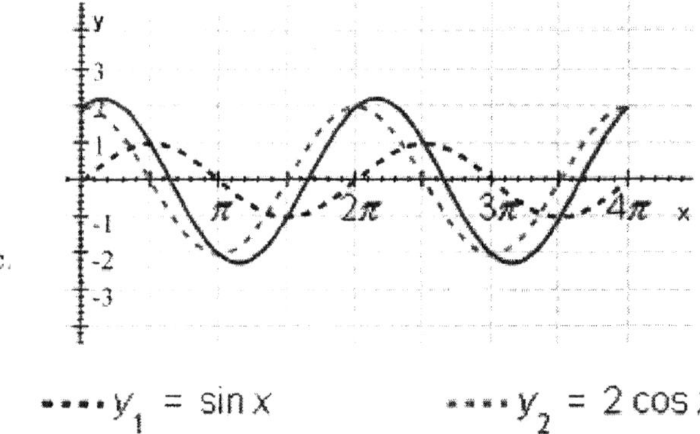

c.

$\cdots y_1 = \sin x$ $\cdots y_2 = 2\cos x$

$—\, y = \sin x + 2\cos x$

25. Evaluate the expression without using a calculator, and write your answer in radians.

$$\sin^{-1}(1)$$

1. a

2. 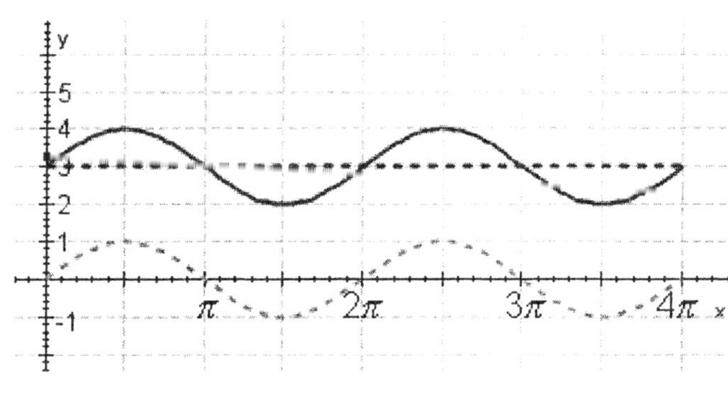

3. c

4. c

5. $-\dfrac{\pi}{2}$

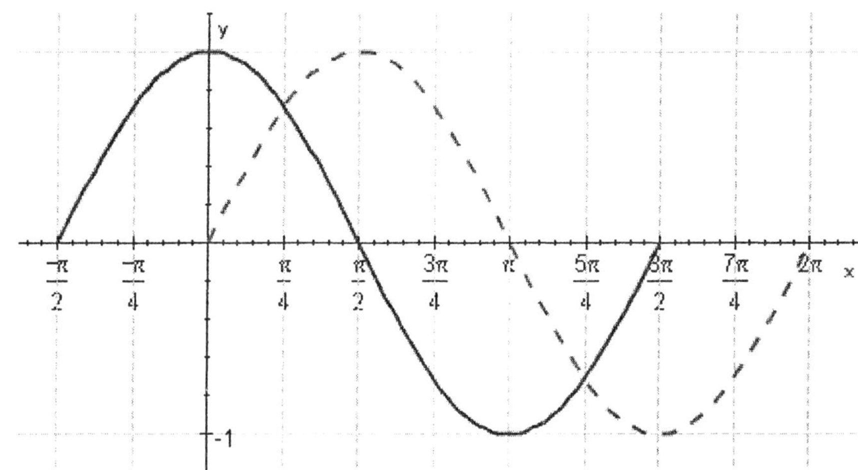

6. d

7. a

8. b

9. b

10. $\dfrac{4}{3}$

11. a

12. e

13.

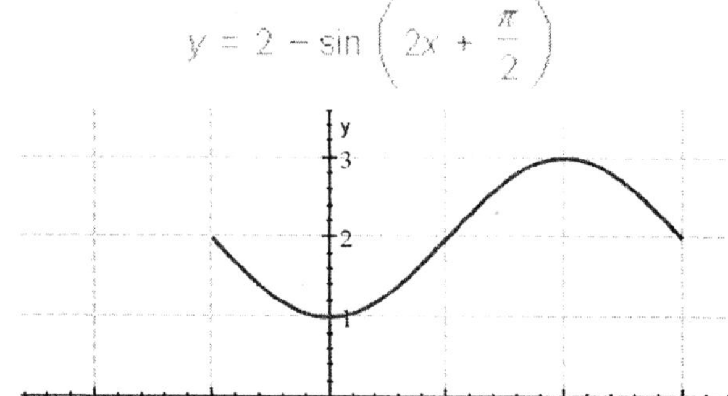

14. a

15.

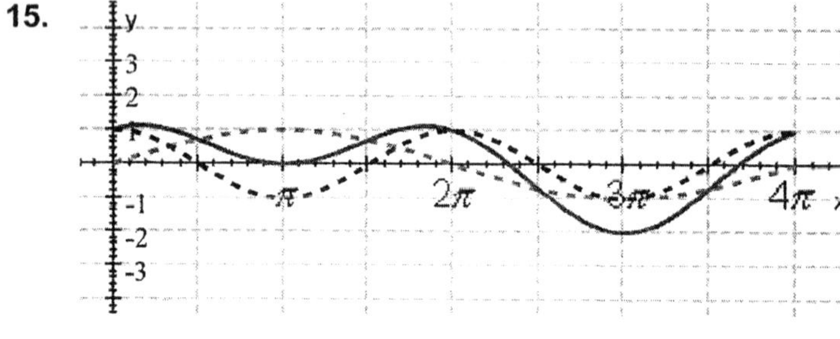

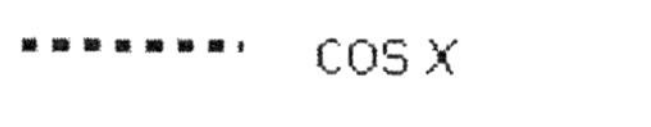

16.

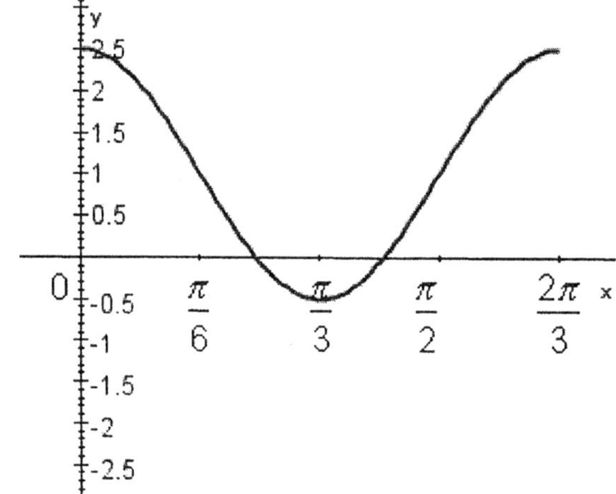

17.

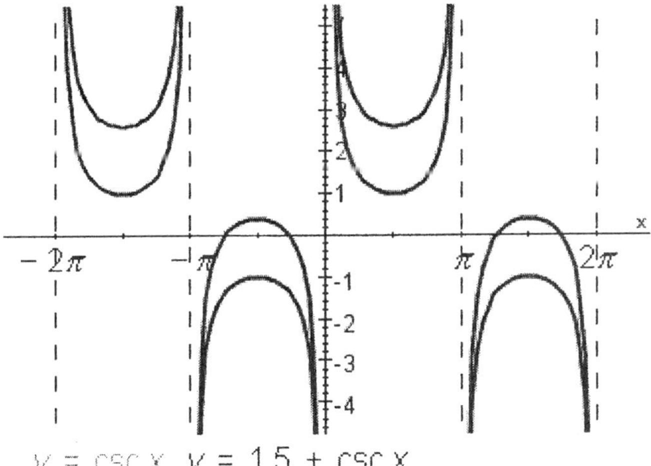

$y = \csc x \quad y = 1.5 + \csc x$

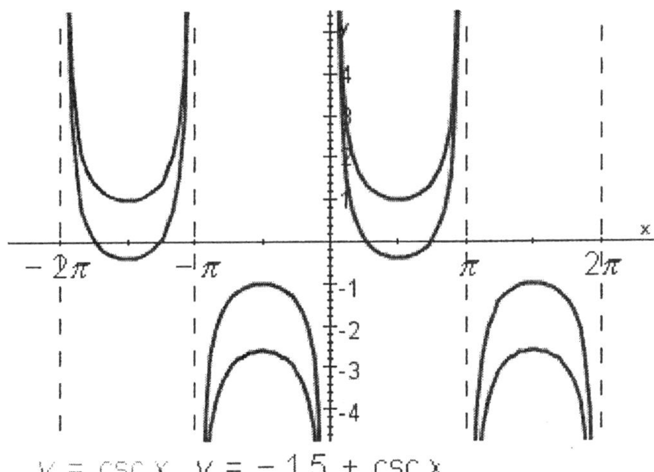

$y = \csc x \quad y = -1.5 + \csc x$

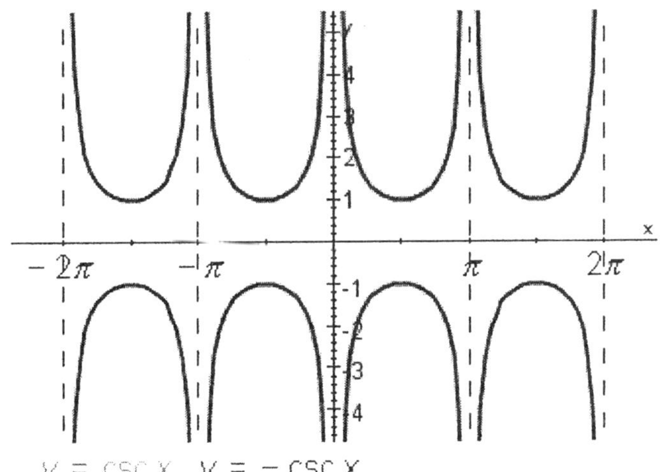

$y = \csc x \quad y = -\csc x$

18. $y = -\dfrac{2}{3} \cdot x + 4$

19. c

20.

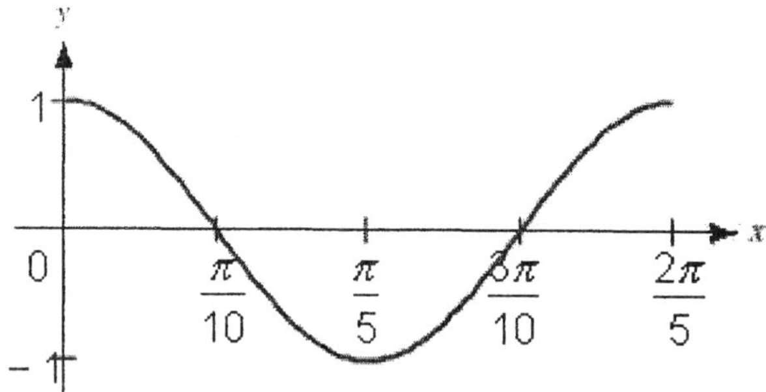

$\dfrac{2\pi}{5}$

21. $y = 4\cos(\pi \cdot x)$

22. c

23. c

24. c

25. $\dfrac{\pi}{2}$

McKeague/Turner - Trigonometry 5e Chapter 4 Form H

1. mctr.04.06.38m_NoAlgs
2. mctr.04.05.28_NoAlgs
3. mctr.04.01.43m_NoAlgs
4. mctr.04.03.33m_NoAlgs
5. mctr.04.03.01_NoAlgs
6. mctr.04.06.52m_NoAlgs
7. mctr.04.04.21m_NoAlgs
8. mctr.04.01.63m_NoAlgs
9. mctr.04.03.09m_NoAlgs
10. mctr.04.06.66_NoAlgs
11. mctr.04.02.06m_NoAlgs
12. mctr.04.06.73m_NoAlgs
13. mctr.04.03.27_NoAlgs
14. mctr.04.04.07m_NoAlgs
15. mctr.04.05.17_NoAlgs
16. mctr.04.02.36_NoAlgs
17. mctr.04.02.62_NoAlgs
18. mctr.04.04.02_NoAlgs
19. mctr.04.01.08m_NoAlgs
20. mctr.04.02.01_NoAlgs
21. mctr.04.04.18_NoAlgs
22. mctr.04.01.38m_NoAlgs
23. mctr.04.05.22m_NoAlgs
24. mctr.04.05.32m_NoAlgs
25. mctr.04.06.22_NoAlgs

1. Prove that the identity is true.

   $\sin x\,(\sec x - \cot x) = \tan x - \cos x$

2. Prove that the identity is true.

   $$\sec^4\theta - \tan^4\theta = \frac{1 + \sin^2\theta}{\cos^2\theta}$$

3. Prove that the identity is true.

   $$\frac{1 + \sec x}{1 - \sec x} = \frac{\cos x + 1}{\cos x - 1}$$

4. Prove that the identity is true.

   $$\sin^4 A - \cos^4 A = 2\sin^2 A - 1$$

5. The identity is from the book *Plane and Spherical Trigonometry with Tables* by Rosenbach, Whitman, and Moskovitz, and published by Ginn and Company in 1937. Verify the identity.

   $$\frac{\tan^2\psi + 2}{1 + \tan^2\psi} = 1 + \cos^2\psi$$

6. Find the exact value for the following.

   $\sin 75^\circ$

7. Show that the following is true.

$$\sin\left(\frac{3\pi}{2} - x\right) = -\cos x$$

8. Graph the following from $x = 0$ to $x = 2\pi$.

$\sin 12x \cos 10x - \cos 12x \sin 10x$

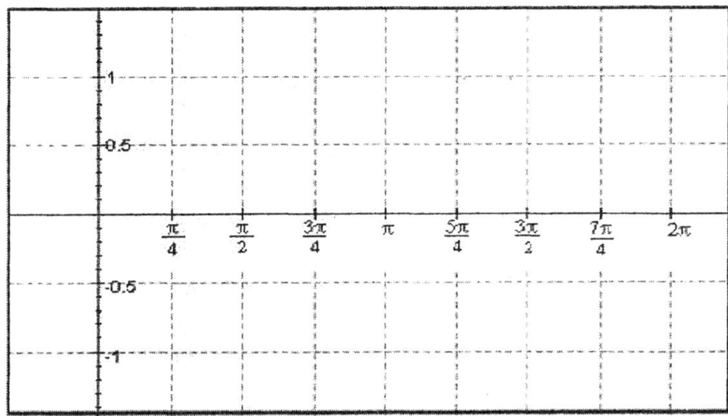

9. If $\tan(A+B) = 3$ and $\tan B = \frac{1}{7}$, find $\tan A$.

$\tan A = $ ________

10. Prove the identity.

$$\cos\left(x + \frac{3\pi}{2}\right) + \cos\left(x - \frac{3\pi}{2}\right) = 0$$

11. Let $\sin A = \dfrac{12}{13}$ with A in QI and find $\sin 2A$.

12. Let $\sin A = \dfrac{1}{\sqrt{5}}$ with A in QII and find $\sec 2A$.

13. If $\tan A = \sqrt{11}$, find $\tan 2A$.

14. Prove the identity.

$$\cos^2 2t = \frac{1 + \cos 4t}{2}$$

15. Prove the identity.

$$\frac{1 - \tan x}{1 + \tan x} = \frac{1 - \sin 2x}{\cos 2x}$$

16. If $\cos A = \dfrac{7}{8}$ with A in QIV, find

$\sin \dfrac{A}{2}$

17. If $\sin B = -\dfrac{1}{5}$ with B in QIII, find

$\sin \dfrac{B}{2}$

18. If $\sin A = \dfrac{3}{5}$ with A in QI, find

$\cos \dfrac{A}{2}$

19. Graph the following function from $x = 0$ to $x = 4\pi$.

$$y = 2\sin^2 \dfrac{x}{2}$$

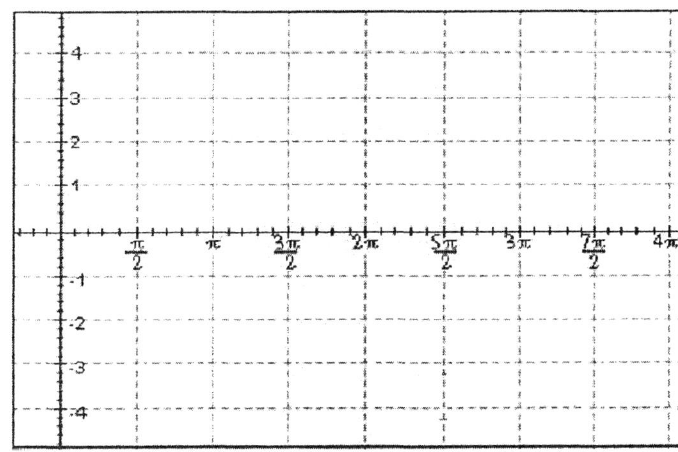

20. Prove the identity.

$$\sin^2 \dfrac{\theta}{2} = \dfrac{\csc \theta - \cot \theta}{2 \csc \theta}$$

21. Evaluate the expression without using a calculator.

$$\sin\left(\arcsin\frac{4}{5} - \arctan 3\right)$$

22. Evaluate the expression without using a calculator. (Assume the variable represents a positive number.)

$$\tan\left(\sin^{-1} 3x\right)$$

23. $10\sin 6x \cos 2x$

Rewrite the expression as a sum or difference using one of the *product to sum formulas*.

Simplify if possible.

24. $\sin 9x + \sin x$

Rewrite the expression as a product using one of the *sum to product formulas*.

Simplify if possible.

25. Verify the identity.

$$-\cot 3x = \frac{\sin 5x - \sin x}{\cos 5x - \cos x}$$

1.
$$\sin x (\sec x - \cot x) = \sin x \left(\frac{1}{\cos x} - \frac{\cos x}{\sin x} \right)$$

$$= \sin x \cdot \frac{1}{\cos x} - \sin x \cdot \frac{\cos x}{\sin x}$$

$$= \frac{\sin x}{\cos x} - \cos x$$

$$= \tan x - \cos x$$

2.
$$\sec^4 \theta - \tan^4 \theta = \frac{1}{\cos^4 \theta} - \frac{\sin^4 \theta}{\cos^4 \theta}$$

$$= \frac{1 - \sin^4 \theta}{\cos^4 \theta}$$

$$= \frac{\left(1 - \sin^2 \theta\right)\left(1 + \sin^2 \theta\right)}{\cos^4 \theta}$$

$$= \frac{\cos^2 \theta \left(1 + \sin^2 \theta\right)}{\cos^4 \theta}$$

$$= \frac{1 + \sin^2 \theta}{\cos^2 \theta}$$

3.
$$\frac{1 + \sec x}{1 - \sec x} = \frac{1 + \frac{1}{\cos x}}{1 - \frac{1}{\cos x}}$$

$$= \frac{\frac{\cos x + 1}{\cos x}}{\frac{\cos x - 1}{\cos x}}$$

$$= \frac{\cos x + 1}{\cos x - 1}$$

4. $\sin^4 A - \cos^4 A = \left(\sin^2 A - \cos^2 A\right)\left(\sin^2 A + \cos^2 A\right)$

$$= \sin^2 A - \cos^2 A = \sin^2 A - \left(1 - \sin^2 A\right)$$

$$= 2\sin^2 A - 1$$

5.

$$\frac{\tan^2 \psi + 2}{1 + \tan^2 \psi} = \frac{\dfrac{\sin^2 \psi}{\cos^2 \psi} + 2}{1 + \dfrac{\sin^2 \psi}{\cos^2 \psi}} = \frac{\sin^2 \psi + 2\cos^2 \psi}{\sin^2 \psi + \cos^2 \psi} = \sin^2 \psi + 2\cos^2 \psi = \left(\sin^2 \psi + \cos^2 \psi\right) + \cos^2 \psi = 1 + \cos^2 \psi$$

6. $\dfrac{\sqrt{6} + \sqrt{2}}{4}$

7. $\sin\left(\dfrac{3\pi}{2} - x\right) = \sin\dfrac{3\pi}{2} \cdot \cos x - \sin x \cdot \cos\dfrac{3\pi}{2} = -1 \cdot \cos x - \sin x \cdot 0 = -\cos x$

8.

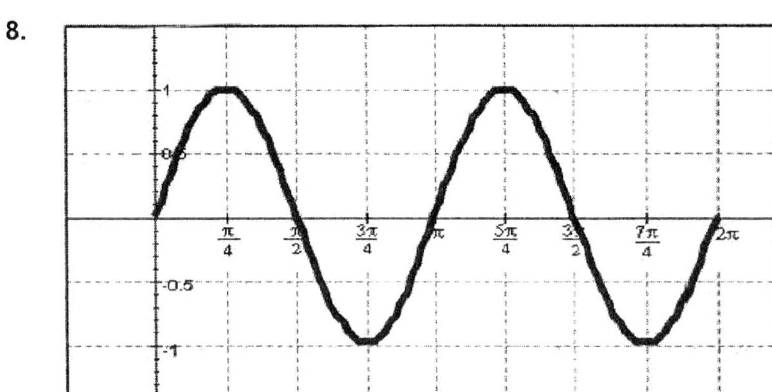

9. 2

10. $\cos\left(x + \dfrac{3\pi}{2}\right) + \cos\left(x - \dfrac{3\pi}{2}\right) = \cos x \cos\dfrac{3\pi}{2} - \sin x \sin\dfrac{3\pi}{2} + \cos x \cos\dfrac{3\pi}{2} + \sin x \sin\dfrac{3\pi}{2} = 0$

11. $\dfrac{120}{169}$

12. $\dfrac{5}{3}$

13. $-\dfrac{\sqrt{11}}{5}$

14. $\dfrac{1 + \cos 4t}{2} = \dfrac{1 - 1 + 2\cos^2 2t}{2} = \cos^2 2t$

15. $\dfrac{1 - \tan x}{1 + \tan x} = \dfrac{\cos x(1 - \tan x)}{\cos x(1 + \tan x)} = \dfrac{\cos x - \sin x}{\cos x + \sin x} =$

$= \dfrac{(\cos x - \sin x)^2}{(\cos x + \sin x)(\cos x - \sin x)} = \dfrac{\cos^2 x + \sin^2 x - 2\sin x \cos x}{\cos^2 x - \sin^2 x} = \dfrac{1 - \sin 2x}{\cos 2x}$

16. $\dfrac{1}{4}$

17. $\sqrt{\dfrac{5 + 2\sqrt{6}}{10}}$

18. $\dfrac{3}{\sqrt{10}}$

19.

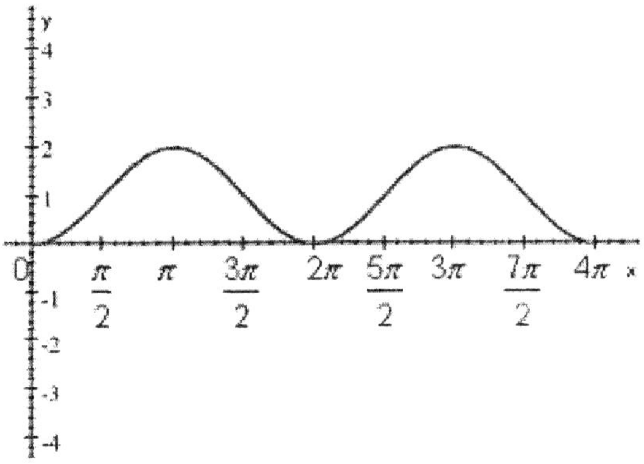

20. $\sin^2\dfrac{\theta}{2} = \dfrac{1 - \left(1 - 2\sin^2\dfrac{\theta}{2}\right)}{2} = \dfrac{1 - \cos\theta}{2} = \dfrac{\dfrac{1}{\sin\theta} - \dfrac{\cos\theta}{\sin\theta}}{2\dfrac{1}{\sin\theta}} = \dfrac{\csc\theta - \cot\theta}{2\csc\theta}$

21. $-\dfrac{1}{\sqrt{10}}$

22. $\dfrac{3x}{\sqrt{1-9x^2}}$

23. $5\left(\sin(8x)+\sin(4x)\right),$

impossible

24. $2\sin(5x)\cdot\cos(4x),$

impossible

25. $\dfrac{\sin 5x - \sin x}{\cos 5x - \cos x} = \dfrac{2\cos 3x\sin 2x}{-2\sin 3x\sin 2x} = -\dfrac{\cos 3x}{\sin 3x} = -\cot 3x$

1. mctr.05.01.09_NoAlgs
2. mctr.05.01.25_NoAlgs
3. mctr.05.01.37_NoAlgs
4. mctr.05.01.51_NoAlgs
5. mctr.05.01.63_NoAlgs
6. mctr.05.02.01_NoAlgs
7. mctr.05.02.19_NoAlgs
8. mctr.05.02.27_NoAlgs
9. mctr.05.02.40_NoAlgs
10. mctr.05.02.52_NoAlgs
11. mctr.05.03.01_NoAlgs
12. mctr.05.03.13_NoAlgs
13. mctr.05.03.28_NoAlgs
14. mctr.05.03.39_NoAlgs
15. mctr.05.03.53_NoAlgs
16. mctr.05.04.01_NoAlgs
17. mctr.05.04.09_NoAlgs
18. mctr.05.04.16_NoAlgs
19. mctr.05.04.30_NoAlgs
20. mctr.05.04.37_NoAlgs
21. mctr.05.05.01_NoAlgs
22. mctr.05.05.07_NoAlgs
23. mctr.05.05.15_NoAlgs
24. mctr.05.05.25_NoAlgs
25. mctr.05.05.31_NoAlgs

1. Evaluate the expression without using a calculator.

$$\sin\left(\arcsin \frac{4}{5} - \arctan 3\right)$$

2. Prove that the identity is true.

$$\sec^4\theta - \tan^4\theta = \frac{1 + \sin^2\theta}{\cos^2\theta}$$

3. If $\sin B = -\frac{1}{5}$ with B in QIII, find

$$\sin \frac{B}{2}$$

4. Prove the identity.

$$\cos\left(x + \frac{3\pi}{2}\right) + \cos\left(x - \frac{3\pi}{2}\right) = 0$$

5. Let $\sin A = \frac{12}{13}$ with A in QI and find sin2A.

6. If $\cos A = \frac{7}{8}$ with A in QIV, find

$$\sin \frac{A}{2}$$

7. Prove that the identity is true.

$$\sin^4 A - \cos^4 A = 2\sin^2 A - 1$$

8. Prove the identity.

$$\frac{1 - \tan x}{1 + \tan x} = \frac{1 - \sin 2x}{\cos 2x}$$

9. Evaluate the expression without using a calculator. (Assume the variable represents a positive number.)

$$\tan\left(\sin^{-1} 3x\right)$$

10. $10 \sin 6x \cos 2x$

   Rewrite the expression as a sum or difference using one of the *product to sum formulas*.

   Simplify if possible.

11. Prove that the identity is true.

$$\frac{1 + \sec x}{1 - \sec x} = \frac{\cos x + 1}{\cos x - 1}$$

12. Prove the identity.

$$\cos^2 2t = \frac{1 + \cos 4t}{2}$$

13. Graph the following function from $x = 0$ to $x = 4\pi$.

$$y = 2\sin^2 \frac{x}{2}$$

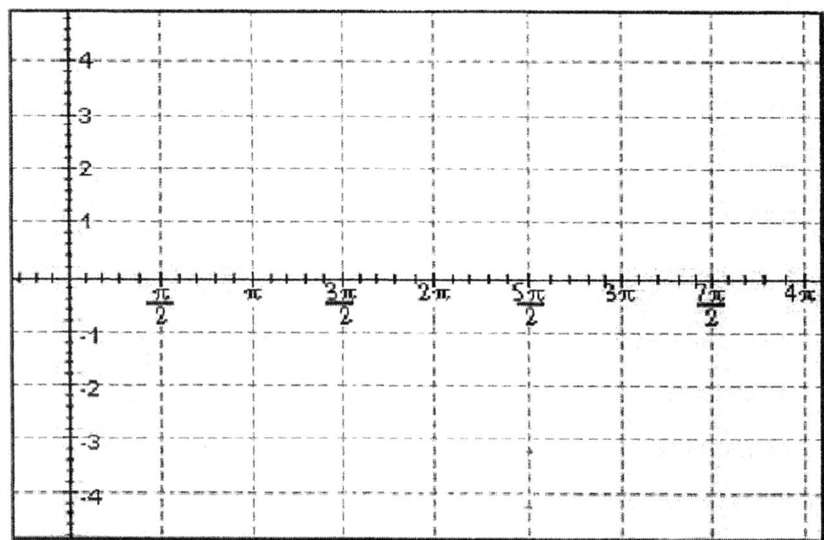

14. Prove the identity.

$$\sin^2 \frac{\theta}{2} = \frac{\csc\theta - \cot\theta}{2\csc\theta}$$

15. Find the exact value for the following.

$$\sin 75°$$

16. Verify the identity.

$$-\cot 3x = \frac{\sin 5x - \sin x}{\cos 5x - \cos x}$$

17. If $\tan A = \sqrt{11}$, find $\tan 2A$.

18. Let $\sin A = \dfrac{1}{\sqrt{5}}$ with A in QII and find sec$2A$.

19. If $\sin A = \dfrac{3}{5}$ with A in QI, find

$\cos \dfrac{A}{2}$

20. Prove that the identity is true.

$\sin x\ (\sec x - \cot x) = \tan x - \cos x$

21. $\sin 9x + \sin x$

Rewrite the expression as a product using one of the *sum to product formulas*.

Simplify if possible.

22. Show that the following is true.

$$\sin \left(\dfrac{3\pi}{2} - x \right) = -\cos x$$

23. The identity is from the book *Plane and Spherical Trigonometry with Tables* by Rosenbach, Whitman, and Moskovitz, and published by Ginn and Company in 1937. Verify the identity.

$$\dfrac{\tan^2 \psi + 2}{1 + \tan^2 \psi} = 1 + \cos^2 \psi$$

24. If $\tan(A+B) = 3$ and $\tan B = \dfrac{1}{7}$, find $\tan A$.

$\tan A = $ _________

25. Graph the following from $x = 0$ to $x = 2\pi$.

$\sin 12x \cos 10x - \cos 12x \sin 10x$

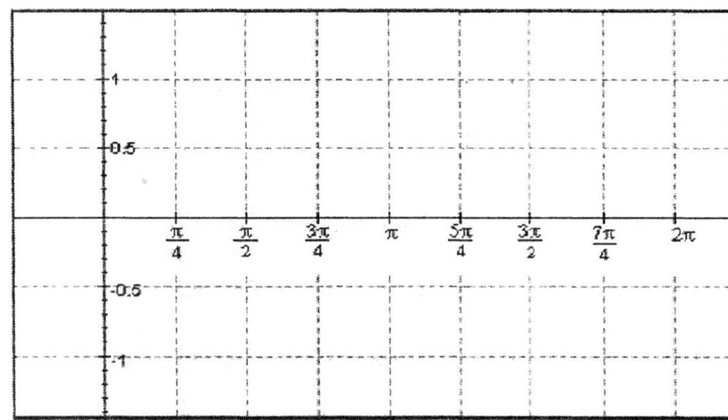

1. $-\dfrac{1}{\sqrt{10}}$

2.
$$\sec^4\theta - \tan^4\theta = \dfrac{1}{\cos^4\theta} - \dfrac{\sin^4\theta}{\cos^4\theta}$$
$$= \dfrac{1 - \sin^4\theta}{\cos^4\theta}$$
$$= \dfrac{\left(1 - \sin^2\theta\right)\left(1 + \sin^2\theta\right)}{\cos^4\theta}$$
$$= \dfrac{\cos^2\theta\left(1 + \sin^2\theta\right)}{\cos^4\theta}$$
$$= \dfrac{1 + \sin^2\theta}{\cos^2\theta}$$

3. $\sqrt{\dfrac{5 + 2\sqrt{6}}{10}}$

4. $\cos\left(x + \dfrac{3\pi}{2}\right) + \cos\left(x - \dfrac{3\pi}{2}\right) = \cos x \cos \dfrac{3\pi}{2} - \sin x \sin \dfrac{3\pi}{2} + \cos x \cos \dfrac{3\pi}{2} + \sin x \sin \dfrac{3\pi}{2} = 0$

5. $\dfrac{120}{169}$

6. $\dfrac{1}{4}$

7.
$$\sin^4 A - \cos^4 A = \left(\sin^2 A - \cos^2 A\right)\left(\sin^2 A + \cos^2 A\right)$$
$$= \sin^2 A - \cos^2 A = \sin^2 A - \left(1 - \sin^2 A\right)$$
$$= 2\sin^2 A - 1$$

8.
$$\frac{1-\tan x}{1+\tan x} = \frac{\cos x(1-\tan x)}{\cos x(1+\tan x)} = \frac{\cos x - \sin x}{\cos x + \sin x} =$$

$$= \frac{(\cos x - \sin x)^2}{(\cos x + \sin x)(\cos x - \sin x)} = \frac{\cos^2 x + \sin^2 x - 2\sin x\cos x}{\cos^2 x - \sin^2 x} = \frac{1 - \sin 2x}{\cos 2x}$$

9.
$$\frac{3x}{\sqrt{1-9x^2}}$$

10. $5\big(\sin(8x) + \sin(4x)\big),$

impossible

11.
$$\frac{1+\sec x}{1-\sec x} = \frac{1 + \dfrac{1}{\cos x}}{1 - \dfrac{1}{\cos x}}$$

$$= \frac{\dfrac{\cos x + 1}{\cos x}}{\dfrac{\cos x - 1}{\cos x}}$$

$$= \frac{\cos x + 1}{\cos x - 1}$$

12.
$$\frac{1+\cos 4t}{2} = \frac{1 - 1 + 2\cos^2 2t}{2} = \cos^2 2t$$

13.

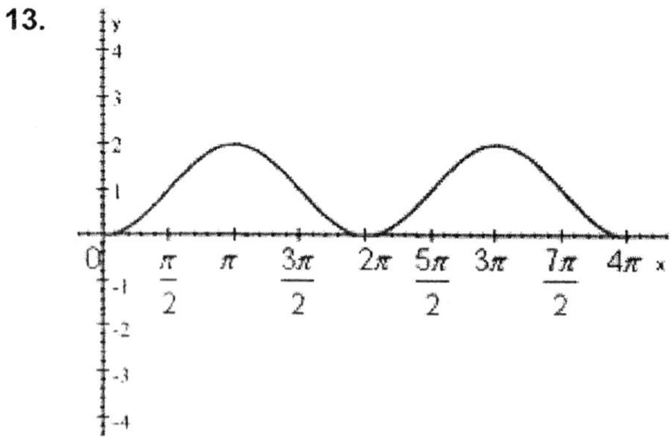

14.

$$\sin^2\frac{\theta}{2} = \frac{1 - \left(1 - 2\sin^2\frac{\theta}{2}\right)}{2} = \frac{1 - \cos\theta}{2} = \frac{\dfrac{1}{\sin\theta} - \dfrac{\cos\theta}{\sin\theta}}{2\,\dfrac{1}{\sin\theta}} = \frac{\csc\theta - \cot\theta}{2\csc\theta}$$

15. $\dfrac{\sqrt{6}+\sqrt{2}}{4}$

16. $\dfrac{\sin 5x - \sin x}{\cos 5x - \cos x} = \dfrac{2\cos 3x\sin 2x}{-2\sin 3x\sin 2x} = -\dfrac{\cos 3x}{\sin 3x} = -\cot 3x$

17. $-\dfrac{\sqrt{11}}{5}$

18. $\dfrac{5}{3}$

19. $\dfrac{3}{\sqrt{10}}$

20.

$$\sin x(\sec x - \cot x) = \sin x\left(\frac{1}{\cos x} - \frac{\cos x}{\sin x}\right)$$

$$= \sin x\cdot\frac{1}{\cos x} - \sin x\cdot\frac{\cos x}{\sin x}$$

$$= \frac{\sin x}{\cos x} - \cos x$$

$$= \tan x - \cos x$$

21. $2\sin(5x)\cdot\cos(4x)$,

impossible

22.

$$\sin\left(\frac{3\pi}{2} - x\right) = \sin\frac{3\pi}{2}\cdot\cos x - \sin x\cdot\cos\frac{3\pi}{2} = -1\cdot\cos x - \sin x\cdot 0 = -\cos x$$

23.

$$\frac{\tan^2\psi + 2}{1 + \tan^2\psi} = \frac{\dfrac{\sin^2\psi}{\cos^2\psi} + 2}{1 + \dfrac{\sin^2\psi}{\cos^2\psi}} = \frac{\sin^2\psi + 2\cos^2\psi}{\sin^2\psi + \cos^2\psi} = \sin^2\psi + 2\cos^2\psi = \left(\sin^2\psi + \cos^2\psi\right) + \cos^2\psi = 1 + \cos^2\psi$$

24. 2

25.

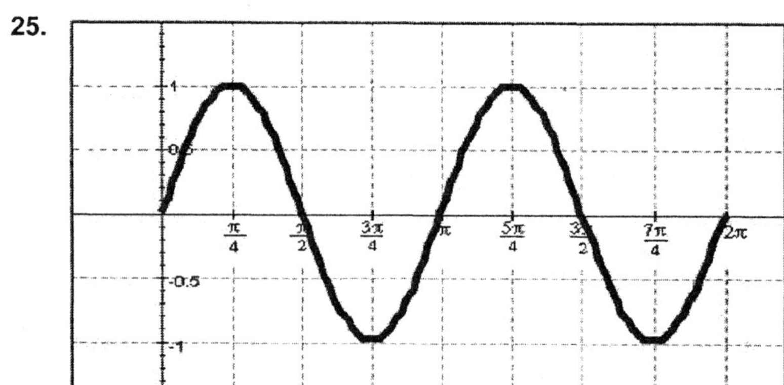

McKeague/Turner - Trigonometry 5e Chapter 5 Form B

1. mctr.05.05.01_NoAlgs
2. mctr.05.01.25_NoAlgs
3. mctr.05.04.09_NoAlgs
4. mctr.05.02.52_NoAlgs
5. mctr.05.03.01_NoAlgs
6. mctr.05.04.01_NoAlgs
7. mctr.05.01.51_NoAlgs
8. mctr.05.03.53_NoAlgs
9. mctr.05.05.07_NoAlgs
10. mctr.05.05.15_NoAlgs
11. mctr.05.01.37_NoAlgs
12. mctr.05.03.39_NoAlgs
13. mctr.05.04.30_NoAlgs
14. mctr.05.04.37_NoAlgs
15. mctr.05.02.01_NoAlgs
16. mctr.05.05.31_NoAlgs
17. mctr.05.03.28_NoAlgs
18. mctr.05.03.13_NoAlgs
19. mctr.05.04.16_NoAlgs
20. mctr.05.01.09_NoAlgs
21. mctr.05.05.25_NoAlgs
22. mctr.05.02.19_NoAlgs
23. mctr.05.01.63_NoAlgs
24. mctr.05.02.40_NoAlgs
25. mctr.05.02.27_NoAlgs

1. Find the identical expression for the following.

$$\frac{\cos^4 t - \sin^4 t}{\cos^2 t}$$

Select the correct answer.

a. $\cot^2 t - 1$

b. $\tan t$

c. $1 - \cot^2 t$

d. $1 - \tan^2 t$

e. $\tan^2 t - 1$

2. Find the identical expression for the following.

$$\frac{\cos x}{1 + \sin x} + \frac{1 + \sin x}{\cos x}$$

Select the correct answer.

a. $2 \csc x$

b. $2 \sec x$

c. $2 \csc^2 x$

d. $2 \tan x$

e. $2 \sec^2 x$

3. Find the identical expression for the following.

$$\frac{1 + \cos x}{1 - \cos x}$$

Select the correct answer.

a. $(\sec x - \cot x)^2$

b. $(\csc x - \cot x)^2$

c. $(\sec x + \cot x)^2$

d. 1

e. $(\csc x + \cot x)^2$

4. Find the identical expression for the following.

$$\frac{\sin^3 A - 1}{\sin A - 1}$$

Select the correct answer.

a. $\tan^2 A + \tan A + 1$

b. -1

c. $\tan^2 A - \tan A + 1$

d. $\sin^2 A + \sin A + 1$

e. $\sin^2 A - \sin A + 1$

5.   In the list below find the right part of the identity.

$$\frac{\cos \beta}{1 - \tan \beta} + \frac{\sin \beta}{1 - \cot \beta} = \ ?$$

Select the correct answer.

   a. $\cot^2 \beta + 1$

   b. $\sin \beta$

   c. $\csc \beta$

   d. $\sec^2 \beta$

   e. $\sin \beta + \cos \beta$

6.   Find the identical expression for the following.

$$\sin \left(x - \frac{\pi}{2} \right)$$

Select the correct answer.

   a. $\tan x$
   b. $-\cos x$
   c. $-\sin x$
   d. $\sin x$
   e. $\cos x$

7.   Write the expression as a single trigonometric function.

$\cos 5x \cos x - \sin 5x \sin x$

Select the correct answer.

   a. $\cos 2x$
   b. $\cos 8x$
   c. $\cos 6x$
   d. $\cos 9x$
   e. $\cos 7x$

8. Graph one complete cycle of $y = \sin x \cos \dfrac{\pi}{6} - \cos x \sin \dfrac{\pi}{6}$.

Select the correct answer.

a. 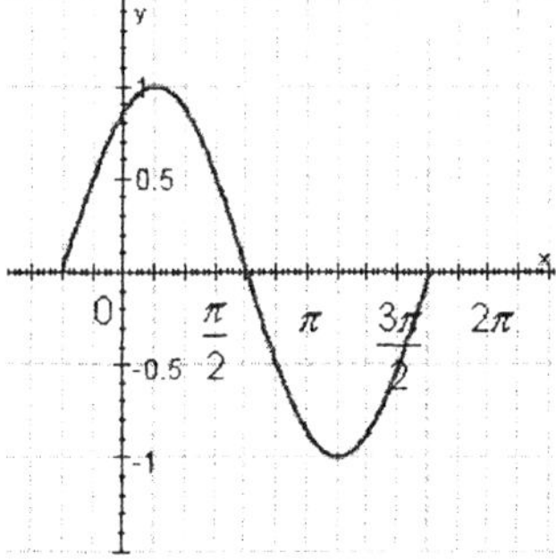

b. 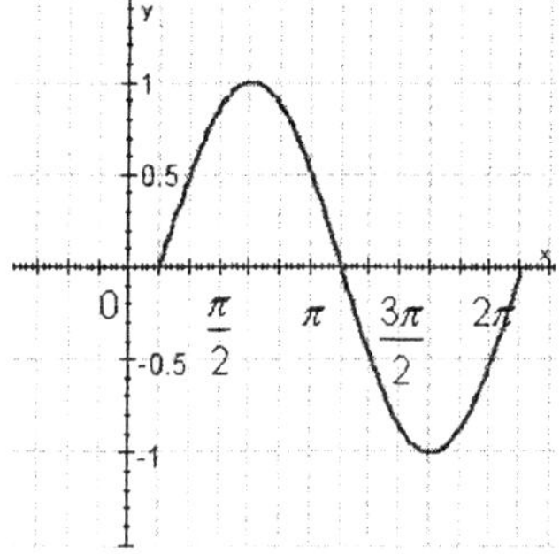

c. 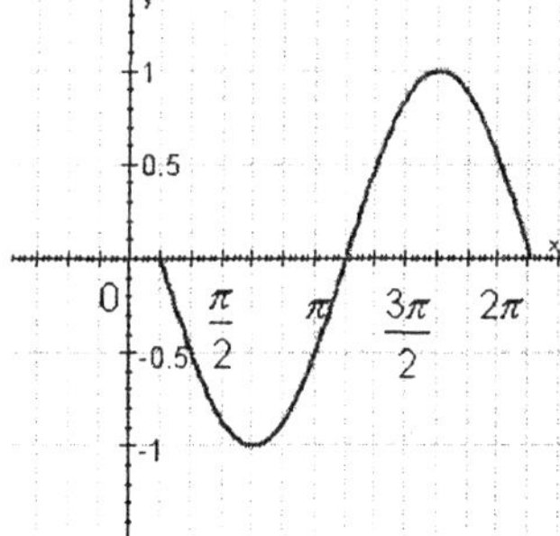

d. 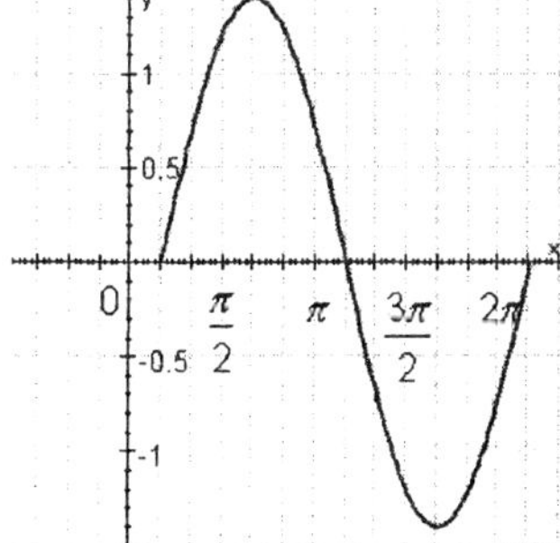

9. Find the identical expression for the following.

$$\sin(90° - x) + \sin(90° + x)$$

Select the correct answer.

a. $2\cos x$

b. 0

c. $-2\cos x$

d. $2\sin x$

e. $-2\sin x$

10. Find the identical expression for the following.

$$\frac{\cos(A - B)}{\sin A \cos B}$$

Select the correct answer.

a. $\cot A + \cot B$

b. $\tan A + \cot B$

c. $\tan A + \tan B$

d. $\cot A - \tan B$

e. $\cot A + \tan B$

11. Let $\cos A = \dfrac{1}{\sqrt{5}}$ with A in QIV and find tan2A.

Select the correct answer.

a. $\dfrac{4}{3}$

b. $-\dfrac{3}{4}$

c. $\dfrac{4}{9}$

d. $\dfrac{8}{9}$

e. $-\dfrac{2}{3}$

12. Graph the function from

$$x = 0 \text{ and } x = 2\pi.$$

$$y = 6\cos^2\frac{x}{2} - 3$$

Select the correct answer.

a.

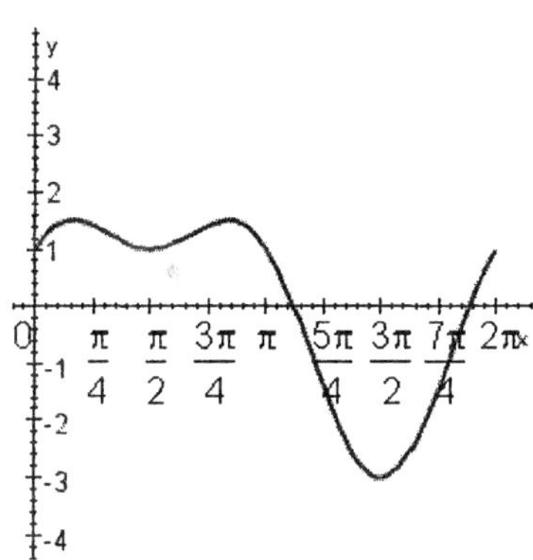

b.

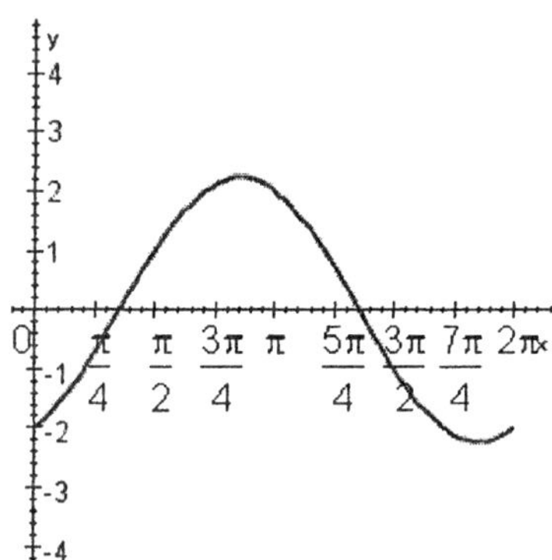

c.

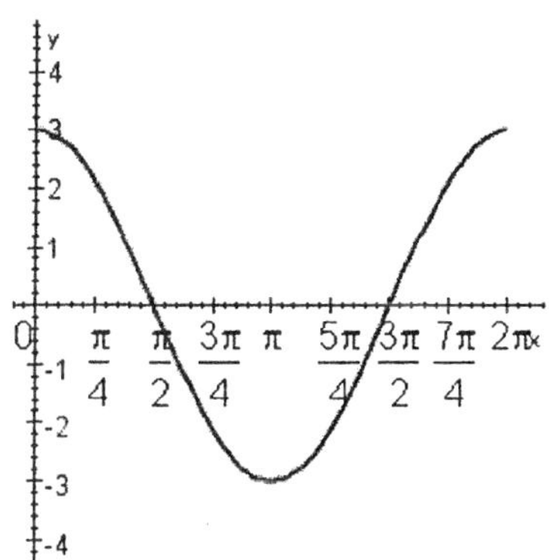

13. Simplify the expression.

$$\cos^2 \frac{\pi}{3} - \sin^2 \frac{\pi}{3}$$

Select the correct answer.

a.　　0

b.　　$\dfrac{1}{3}$

c.　　$-\dfrac{1}{2}$

d.　　$-\dfrac{\sqrt{3}}{4}$

e.　　$-\dfrac{\sqrt{2}}{4}$

14. From the list below find the expression, identical for the given.

$\cos 6\theta$

Select the correct answer.

a.　　$3\cos 2\theta + 4\cos^3 2\theta$

b.　　$4\cos^3 2\theta - 3\cos 2\theta$

c.　　$\cos 2\theta - 4\cos^2 2\theta$

d.　　$3\sin 2\theta - 4\sin^3 2\theta$

e.　　$\sin 2\theta + 4\sin^3 2\theta$

15. From the list below find the expression, identical for the given.

$$\frac{1 - \tan t}{1 + \tan t}$$

Select the correct answer.

a. $\dfrac{1 - \sin 2t}{\cos t}$

b. $\dfrac{1 - \sin 2t}{\cos 2t}$

c. $\dfrac{1 - \cos 2t}{\sin 2t}$

d. $\dfrac{1 + \sin 2t}{\sin t}$

e. $\dfrac{1 - \sin t}{\cos t}$

16. If $\sin A = -\dfrac{3}{5}$ with A in QIV, find

$$\cos \frac{A}{2}$$

Select the correct answer.

a. $\cos \dfrac{A}{2} = -\dfrac{3}{\sqrt{10}}$

b. $\cos \dfrac{A}{2} = -\dfrac{1}{\sqrt{4}}$

c. $\cos \dfrac{A}{2} = -\dfrac{9}{\sqrt{10}}$

d. $\cos \dfrac{A}{2} = \dfrac{3}{\sqrt{10}}$

e. $\cos \dfrac{A}{2} = \dfrac{1}{\sqrt{9}}$

17.

If $\sin B = -\dfrac{1}{3}$ with B in QIII, find

$$\tan \dfrac{B}{2}$$

Select the correct answer.

    a. $\tan \dfrac{B}{2} = 2 - 2\sqrt{5}$

    b. $\tan \dfrac{B}{2} = -3 - 2\sqrt{2}$

    c. $\tan \dfrac{B}{2} = -2 - 2\sqrt{5}$

    d. $\tan \dfrac{B}{2} = -2 - 2\sqrt{2}$

    e. $\tan \dfrac{B}{2} = 2 - 2\sqrt{2}$

18.

If $\sin A = -\dfrac{4}{5}$ with A in QIV, and $\sin B = -\dfrac{3}{5}$ with B in QIII, find

$$\cos(A + B)$$

Select the correct answer.

    a. $\cos(A + B) = -\dfrac{1}{5}$

    b. $\cos(A + B) = \dfrac{\sqrt{2}}{5}$

    c. $\cos(A + B) = -\dfrac{\sqrt{2}}{5}$

    d. $\cos(A + B) = -\dfrac{24}{25}$

    e. $\cos(A + B) = \dfrac{1}{5}$

19. Use half-angle formulas to find the exact value for the following :

$$\tan 165°$$

Select the correct answer.

a. $\tan 165° = \sqrt{\dfrac{1 + \sqrt{2}}{2}}$

b. $\tan 165° = -2 + \sqrt{3}$

c. $\tan 165° = \dfrac{1}{2}$

d. $\tan 165° = -\dfrac{1}{2}$

e. $\tan 165° = -\dfrac{\sqrt{2 + \sqrt{3}}}{2}$

20. Find the expression identical to the one given.

$$\tan \dfrac{B}{2}$$

Select the correct answer.

a. $-\csc^2 B - \cot B$

b. $\csc B + \cot^2 B$

c. $\dfrac{\sec B}{\sec B \csc B - \csc B}$

d. $\csc^2 B - \cot B$

e. $\dfrac{\sec B}{\sec B \csc B + \csc B}$

21. Evaluate the expression without using a calculator.

$$\sin\left(2\sin^{-1}\frac{1}{\sqrt{5}}\right)$$

Select the correct answer.

a. $\quad -\dfrac{4}{5}$

b. $\quad \dfrac{6}{5}$

c. $\quad \dfrac{4}{5}$

d. $\quad -\dfrac{6}{5}$

e. $\quad \dfrac{2}{\sqrt{5}}$

22. Evaluate the expression without using a calculator. (Assume the variable represents a positive number.)

$$\cos\left(2\cos^{-1}9x\right)$$

Select the correct answer.

a. $\quad 162x^2 + 1$

b. $\quad 161x^2 - 1$

c. $\quad 161x^2 + 1$

d. $\quad 160x^2 - 1$

e. $\quad 162x^2 - 1$

23. Rewrite the expression as a sum or difference, then simplify if possible.

$$\cos 180^\circ \cos 360^\circ$$

Select the correct answer.

a. $\dfrac{1}{5}$

b. 0

c. -1

d. $\dfrac{1}{4}$

e. 1

24. Rewrite the expression as a product. Simplify if possible.

$$\sin \frac{7\pi}{12} + \sin \frac{\pi}{12}$$

Select the correct answer.

a. $\dfrac{1}{2\sqrt{6}}$

b. $\dfrac{\sqrt{6}}{2}$

c. $\sqrt{6}$

d. $-\dfrac{\sqrt{6}}{2}$

e. $2\sqrt{6}$

25. From the list below find the expression, identical to the given.

$\tan 11x$

Select the correct answer.

a. $\dfrac{\sin 19x + \sin 3x}{\cos 3x + \cos 19x}$

b. $\dfrac{\sin 20x - \sin 3x}{\cos 3x + \cos 20x}$

c. $\dfrac{\sin 16x + \sin 3x}{\cos 3x + \cos 16x}$

d. $\dfrac{\sin 21x + \sin 3x}{\cos 3x - \cos 21x}$

e. $\dfrac{\sin 15x + \sin 3x}{\cos 3x - \cos 15x}$

McKeague/Turner - Trigonometry 5e Chapter 5 Form C

1. d

2. b

3. e

4. d

5. e

6. b

7. c

8. b

9. a

10. e

11. a

12. c

13. c

14. b

15. b

16. a

17. b

18. d

19. b

20. e

21. c

22. e

23. c

24. b

25. a

McKeague/Turner - Trigonometry 5e Chapter 5 Form C

1. mctr.05.01.18m_NoAlgs
2. mctr.05.01.33m_NoAlgs
3. mctr.05.01.46m_NoAlgs
4. mctr.05.01.57m_NoAlgs
5. mctr.05.01.63m_NoAlgs
6. mctr.05.02.11m_NoAlgs
7. mctr.05.02.23m_NoAlgs
8. mctr.05.02.32m_NoAlgs
9. mctr.05.02.45m_NoAlgs
10. mctr.05.02.55m_NoAlgs
11. mctr.05.03.07m_NoAlgs
12. mctr.05.03.19m_NoAlgs
13. mctr.05.03.33m_NoAlgs
14. mctr.05.03.46m_NoAlgs
15. mctr.05.03.53m_NoAlgs
16. mctr.05.04.05m_NoAlgs
17. mctr.05.04.14m_NoAlgs
18. mctr.05.04.24m_NoAlgs
19. mctr.05.04.35m_NoAlgs
20. mctr.05.04.42m_NoAlgs
21. mctr.05.05.05m_NoAlgs
22. mctr.05.05.11m_NoAlgs
23. mctr.05.05.20m_NoAlgs
24. mctr.05.05.29m_NoAlgs
25. mctr.05.05.35m_NoAlgs

1. From the list below find the expression, identical for the given.

$$\cos 6\theta$$

Select the correct answer.

   a. $3\cos 2\theta + 4\cos^3 2\theta$

   b. $4\cos^3 2\theta - 3\cos 2\theta$

   c. $\cos 2\theta - 4\cos^2 2\theta$

   d. $3\sin 2\theta - 4\sin^3 2\theta$

   e. $\sin 2\theta + 4\sin^3 2\theta$

2. Find the expression identical to the one given.

$$\tan \frac{B}{2}$$

Select the correct answer.

   a. $-\csc^2 B - \cot B$

   b. $\csc B + \cot^2 B$

   c. $\dfrac{\sec B}{\sec B \csc B - \csc B}$

   d. $\csc^2 B - \cot B$

   e. $\dfrac{\sec B}{\sec B \csc B + \csc B}$

3. Find the identical expression for the following.

$$\frac{\sin^3 A - 1}{\sin A - 1}$$

Select the correct answer.

a. $\tan^2 A + \tan A + 1$

b. -1

c. $\tan^2 A - \tan A + 1$

d. $\sin^2 A + \sin A + 1$

e. $\sin^2 A - \sin A + 1$

4. If $\sin A = -\dfrac{3}{5}$ with A in QIV, find

$$\cos \frac{A}{2}$$

Select the correct answer.

a. $\cos \dfrac{A}{2} = -\dfrac{3}{\sqrt{10}}$

b. $\cos \dfrac{A}{2} = -\dfrac{1}{\sqrt{4}}$

c. $\cos \dfrac{A}{2} = -\dfrac{9}{\sqrt{10}}$

d. $\cos \dfrac{A}{2} = \dfrac{3}{\sqrt{10}}$

e. $\cos \dfrac{A}{2} = \dfrac{1}{\sqrt{9}}$

5. Graph one complete cycle of $y = \sin x \cos \dfrac{\pi}{6} - \cos x \sin \dfrac{\pi}{6}$.

Select the correct answer.

a.

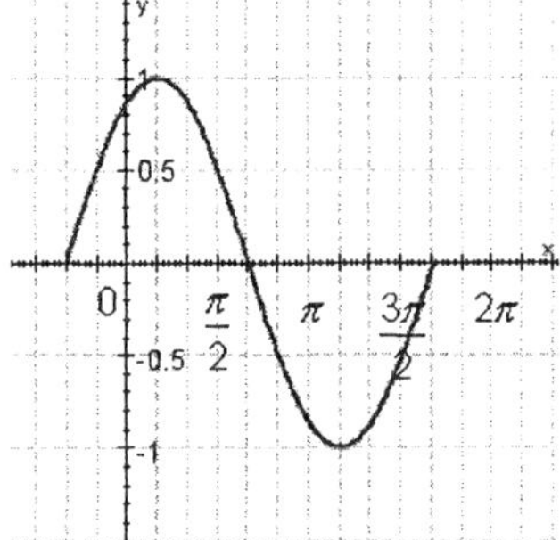

b.

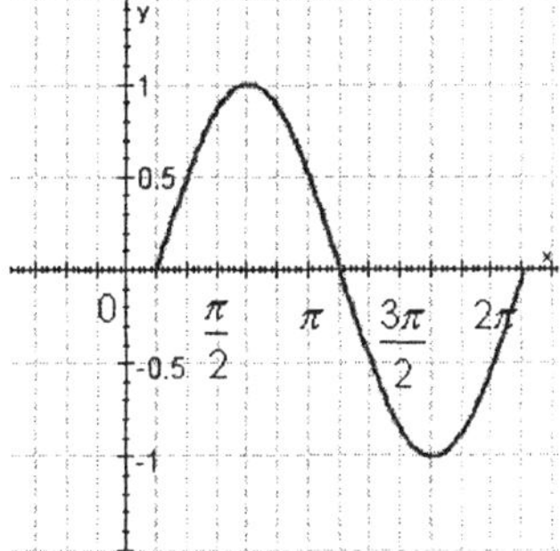

c.

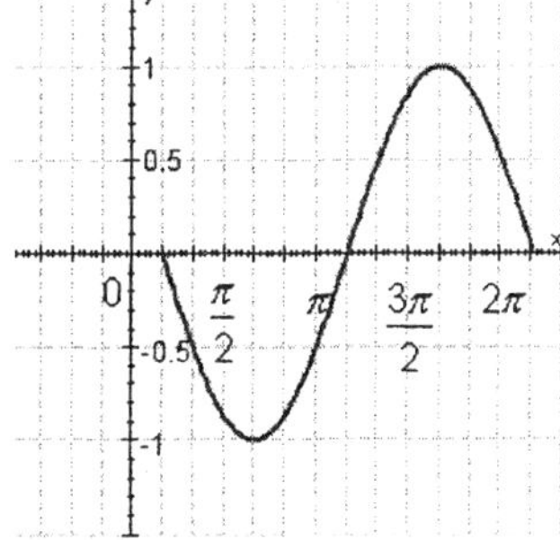

d.

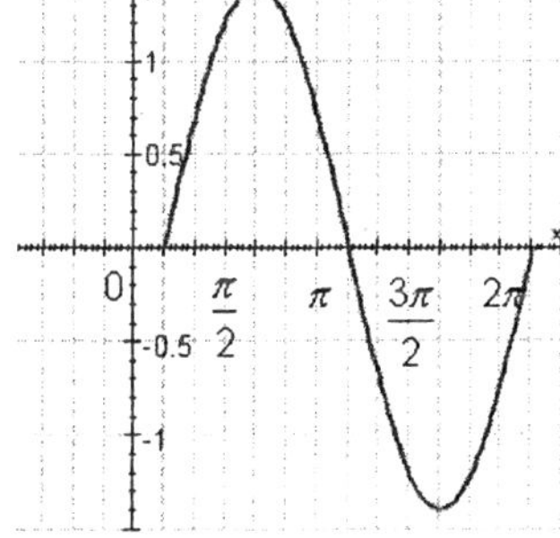

6. Find the identical expression for the following.

$$\frac{\cos x}{1 + \sin x} + \frac{1 + \sin x}{\cos x}$$

Select the correct answer.

a. $2 \csc x$

b. $2 \sec x$

c. $2 \csc^2 x$

d. $2 \tan x$

e. $2 \sec^2 x$

7. Use half-angle formulas to find the exact value for the following :

$\tan 165°$

Select the correct answer.

a. $\tan 165° = \dfrac{\sqrt{1 + \sqrt{2}}}{2}$

b. $\tan 165° = -2 + \sqrt{3}$

c. $\tan 165° = \dfrac{1}{2}$

d. $\tan 165° = -\dfrac{1}{2}$

e. $\tan 165° = -\dfrac{\sqrt{2 + \sqrt{3}}}{2}$

8. Find the identical expression for the following.

$$\frac{\cos^4 t - \sin^4 t}{\cos^2 t}$$

Select the correct answer.

a. $\cot^2 t - 1$

b. $\tan t$

c. $1 - \cot^2 t$

d. $1 - \tan^2 t$

e. $\tan^2 t - 1$

9. Find the identical expression for the following.

$$\frac{\cos(A - B)}{\sin A \cos B}$$

Select the correct answer.

a. $\cot A + \cot B$

b. $\tan A + \cot B$

c. $\tan A + \tan B$

d. $\cot A - \tan B$

e. $\cot A + \tan B$

10. Rewrite the expression as a sum or difference, then simplify if possible.

$$\cos 180° \cos 360°$$

Select the correct answer.

a. $\dfrac{1}{5}$

b. 0

c. -1

d. $\dfrac{1}{4}$

e. 1

11. Find the identical expression for the following.

$$\frac{1 + \cos x}{1 - \cos x}$$

Select the correct answer.

a. $(\sec x - \cot x)^2$

b. $(\csc x - \cot x)^2$

c. $(\sec x + \cot x)^2$

d. 1

e. $(\csc x + \cot x)^2$

12. Find the identical expression for the following.

$$\sin\left(90° - x\right) + \sin\left(90° + x\right)$$

Select the correct answer.

    a. $2\cos x$

    b. 0

    c. $-2\cos x$

    d. $2\sin x$

    e. $-2\sin x$

13. Write the expression as a single trigonometric function.

$$\cos 5x \cos x - \sin 5x \sin x$$

Select the correct answer.

    a. $\cos 2x$
    b. $\cos 8x$
    c. $\cos 6x$
    d. $\cos 9x$
    e. $\cos 7x$

14. From the list below find the expression, identical for the given.

$$\frac{1 - \tan t}{1 + \tan t}$$

Select the correct answer.

    a. $\dfrac{1 - \sin 2t}{\cos t}$

    b. $\dfrac{1 - \sin 2t}{\cos 2t}$

    c. $\dfrac{1 - \cos 2t}{\sin 2t}$

    d. $\dfrac{1 + \sin 2t}{\sin t}$

    e. $\dfrac{1 - \sin t}{\cos t}$

15. In the list below find the right part of the identity.

$$\frac{\cos \beta}{1 - \tan \beta} + \frac{\sin \beta}{1 - \cot \beta} = ?$$

Select the correct answer.

a. $\quad \cot^2 \beta + 1$

b. $\quad \sin \beta$

c. $\quad \csc \beta$

d. $\quad \sec^2 \beta$

e. $\quad \sin \beta + \cos \beta$

16. Evaluate the expression without using a calculator.

$$\sin \left(2 \sin^{-1} \frac{1}{\sqrt{5}} \right)$$

Select the correct answer.

a. $\quad -\dfrac{4}{5}$

b. $\quad \dfrac{6}{5}$

c. $\quad \dfrac{4}{5}$

d. $\quad -\dfrac{6}{5}$

e. $\quad \dfrac{2}{\sqrt{5}}$

17. If $\sin B = -\dfrac{1}{3}$ with B in QIII, find

$\tan \dfrac{B}{2}$

Select the correct answer.

a. $\quad \tan \dfrac{B}{2} = 2 - 2\sqrt{5}$

b. $\quad \tan \dfrac{B}{2} = -3 - 2\sqrt{2}$

c. $\quad \tan \dfrac{B}{2} = -2 - 2\sqrt{5}$

d. $\quad \tan \dfrac{B}{2} = -2 - 2\sqrt{2}$

e. $\quad \tan \dfrac{B}{2} = 2 - 2\sqrt{2}$

18. Let $\cos A = \dfrac{1}{\sqrt{5}}$ with A in QIV and find $\tan 2A$.

Select the correct answer.

a. $\quad \dfrac{4}{3}$

b. $\quad -\dfrac{3}{4}$

c. $\quad \dfrac{4}{9}$

d. $\quad \dfrac{8}{9}$

e. $\quad -\dfrac{2}{3}$

19. Find the identical expression for the following.

$$\sin\left(x - \frac{\pi}{2}\right)$$

Select the correct answer.

a. $\tan x$
b. $-\cos x$
c. $-\sin x$
d. $\sin x$
e. $\cos x$

20. Simplify the expression.

$$\cos^2\frac{\pi}{3} - \sin^2\frac{\pi}{3}$$

Select the correct answer.

a. 0

b. $\dfrac{1}{3}$

c. $-\dfrac{1}{2}$

d. $-\dfrac{\sqrt{3}}{4}$

e. $-\dfrac{\sqrt{2}}{4}$

21. Graph the function from

$$x = 0 \text{ and } x = 2\pi.$$

$$y = 6\cos^2\frac{x}{2} - 3$$

Select the correct answer.

a.

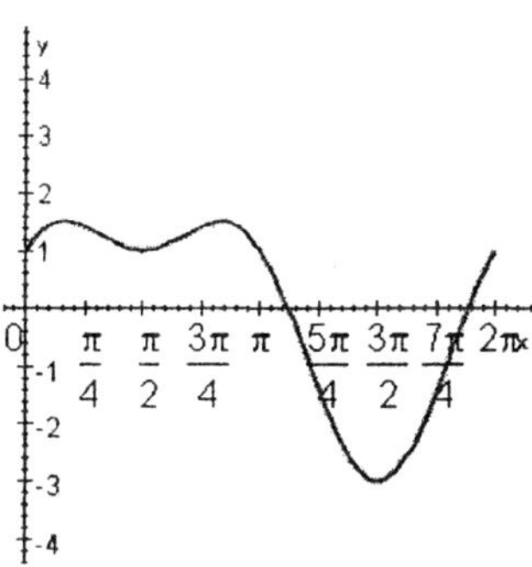

b.

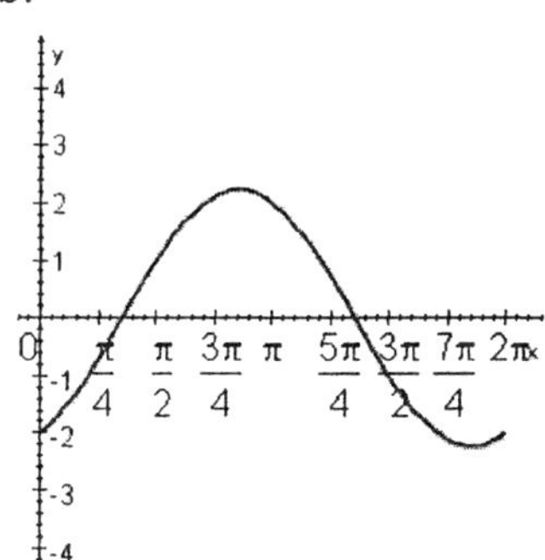

c.

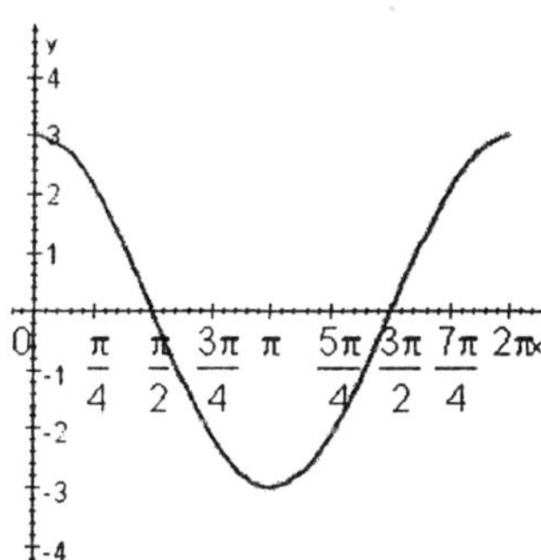

22. Evaluate the expression without using a calculator. (Assume the variable represents a positive number.)

$$\cos\left(2\cos^{-1}9x\right)$$

Select the correct answer.

a. $162x^2 + 1$

b. $161x^2 - 1$

c. $161x^2 + 1$

d. $160x^2 - 1$

e. $162x^2 - 1$

23. If $\sin A = -\dfrac{4}{5}$ with A in QIV, and $\sin B = -\dfrac{3}{5}$ with B in QIII, find

$$\cos(A+B)$$

Select the correct answer.

a. $\cos(A+B) = -\dfrac{1}{5}$

b. $\cos(A+B) = \dfrac{\sqrt{2}}{5}$

c. $\cos(A+B) = -\dfrac{\sqrt{2}}{5}$

d. $\cos(A+B) = -\dfrac{24}{25}$

e. $\cos(A+B) = \dfrac{1}{5}$

24. From the list below find the expression, identical to the given.

$$\tan 11x$$

Select the correct answer.

a. $\dfrac{\sin 19x + \sin 3x}{\cos 3x + \cos 19x}$

b. $\dfrac{\sin 20x - \sin 3x}{\cos 3x + \cos 20x}$

c. $\dfrac{\sin 16x + \sin 3x}{\cos 3x + \cos 16x}$

d. $\dfrac{\sin 21x + \sin 3x}{\cos 3x - \cos 21x}$

e. $\dfrac{\sin 15x + \sin 3x}{\cos 3x - \cos 15x}$

25. Rewrite the expression as a product. Simplify if possible.

$$\sin \frac{7\pi}{12} + \sin \frac{\pi}{12}$$

Select the correct answer.

a. $\dfrac{1}{2\sqrt{6}}$

b. $\dfrac{\sqrt{6}}{2}$

c. $\sqrt{6}$

d. $-\dfrac{\sqrt{6}}{2}$

e. $2\sqrt{6}$

1. b

2. e

3. d

4. a

5. b

6. b

7. b

8. d

9. e

10. c

11. e

12. a

13. c

14. b

15. e

16. c

17. b

18. a

19. b

20. c

21. c

22. e

23. d

24. a

25. b

1. mctr.05.03.46m_NoAlgs
2. mctr.05.04.42m_NoAlgs
3. mctr.05.01.57m_NoAlgs
4. mctr.05.04.05m_NoAlgs
5. mctr.05.02.32m_NoAlgs
6. mctr.05.01.33m_NoAlgs
7. mctr.05.04.35m_NoAlgs
8. mctr.05.01.18m_NoAlgs
9. mctr.05.02.55m_NoAlgs
10. mctr.05.05.20m_NoAlgs
11. mctr.05.01.46m_NoAlgs
12. mctr.05.02.45m_NoAlgs
13. mctr.05.02.23m_NoAlgs
14. mctr.05.03.53m_NoAlgs
15. mctr.05.01.63m_NoAlgs
16. mctr.05.05.05m_NoAlgs
17. mctr.05.04.14m_NoAlgs
18. mctr.05.03.07m_NoAlgs
19. mctr.05.02.11m_NoAlgs
20. mctr.05.03.33m_NoAlgs
21. mctr.05.03.19m_NoAlgs
22. mctr.05.05.11m_NoAlgs
23. mctr.05.04.24m_NoAlgs
24. mctr.05.05.35m_NoAlgs
25. mctr.05.05.29m_NoAlgs

1. Prove that the identity is true.

   $\sin x\,(\sec x - \cot x) = \tan x - \cos x$

2. Find the identical expression for the following.

   $$\frac{\cos^4 t - \sin^4 t}{\cos^2 t}$$

   Select the correct answer.

   a. $\cot^2 t - 1$

   b. $\tan t$

   c. $1 - \cot^2 t$

   d. $1 - \tan^2 t$

   e. $\tan^2 t - 1$

3. Prove that the identity is true.

   $$\frac{1 + \sec x}{1 - \sec x} = \frac{\cos x + 1}{\cos x - 1}$$

4. Find the identical expression for the following.

$$\frac{1 + \cos x}{1 - \cos x}$$

Select the correct answer.

a. $(\sec x - \cot x)^2$

b. $(\csc x - \cot x)^2$

c. $(\sec x + \cot x)^2$

d. 1

e. $(\csc x + \cot x)^2$

5. Find the identical expression for the following.

$$\frac{\sin^3 A - 1}{\sin A - 1}$$

Select the correct answer.

a. $\tan^2 A + \tan A + 1$

b. -1

c. $\tan^2 A - \tan A + 1$

d. $\sin^2 A + \sin A + 1$

e. $\sin^2 A - \sin A + 1$

6. The identity is from the book *Plane and Spherical Trigonometry with Tables* by Rosenbach, Whitman, and Moskovitz, and published by Ginn and Company in 1937. Verify the identity.

$$\frac{\tan^2 \psi + 2}{1 + \tan^2 \psi} = 1 + \cos^2 \psi$$

7. Show that the following is true.

$$\sin\left(\frac{3\pi}{2} - x\right) = -\cos x$$

8. Write the expression as a single trigonometric function.

$\cos 5x \cos x - \sin 5x \sin x$

Select the correct answer.

a. $\cos 2x$
b. $\cos 8x$
c. $\cos 6x$
d. $\cos 9x$
e. $\cos 7x$

9. If $\tan(A+B) = 3$ and $\tan B = \frac{1}{7}$, find $\tan A$.

$\tan A = $ _______

10. Find the identical expression for the following.

$$\sin(90° - x) + \sin(90° + x)$$

Select the correct answer.

a. $2\cos x$

b. 0

c. $-2\cos x$

d. $2\sin x$

e. $-2\sin x$

11. Let $\sin A = \frac{12}{13}$ with A in QI and find $\sin 2A$.

12. Let $\cos A = \dfrac{1}{\sqrt{5}}$ with A in QIV and find tan2A.

Select the correct answer.

a. $\dfrac{4}{3}$

b. $-\dfrac{3}{4}$

c. $\dfrac{4}{9}$

d. $\dfrac{8}{9}$

e. $-\dfrac{2}{3}$

13. If $\tan A = \sqrt{11}$, find tan2A.

14. Simplify the expression.

$$\cos^2 \frac{\pi}{3} - \sin^2 \frac{\pi}{3}$$

Select the correct answer.

a. 0

b. $\dfrac{1}{3}$

c. $-\dfrac{1}{2}$

d. $-\dfrac{\sqrt{3}}{4}$

e. $-\dfrac{\sqrt{2}}{4}$

15. Prove the identity.

$$\cos^2 2t = \frac{1 + \cos 4t}{2}$$

16. From the list below find the expression, identical for the given.

$$\cos 6\theta$$

Select the correct answer.

a. $\quad 3\cos 2\theta + 4\cos^3 2\theta$

b. $\quad 4\cos^3 2\theta - 3\cos 2\theta$

c. $\quad \cos 2\theta - 4\cos^2 2\theta$

d. $\quad 3\sin 2\theta - 4\sin^3 2\theta$

e. $\quad \sin 2\theta + 4\sin^3 2\theta$

17. From the list below find the expression, identical for the given.

$$\frac{1 - \tan t}{1 + \tan t}$$

Select the correct answer.

a. $\quad \dfrac{1 - \sin 2t}{\cos t}$

b. $\quad \dfrac{1 - \sin 2t}{\cos 2t}$

c. $\quad \dfrac{1 - \cos 2t}{\sin 2t}$

d. $\quad \dfrac{1 + \sin 2t}{\sin t}$

e. $\quad \dfrac{1 - \sin t}{\cos t}$

18. If $\sin B = -\dfrac{1}{5}$ with B in QIII, find

$$\sin \frac{B}{2}$$

19.

If $\sin B = -\dfrac{1}{3}$ with B in QIII, find

$$\tan \frac{B}{2}$$

Select the correct answer.

a. $\tan \dfrac{B}{2} = 2 - 2\sqrt{5}$

b. $\tan \dfrac{B}{2} = -3 - 2\sqrt{2}$

c. $\tan \dfrac{B}{2} = -2 - 2\sqrt{5}$

d. $\tan \dfrac{B}{2} = -2 - 2\sqrt{2}$

e. $\tan \dfrac{B}{2} = 2 - 2\sqrt{2}$

20. Graph the following function from $x = 0$ to $x = 4\pi$.

$$y = 2\sin^2 \frac{x}{2}$$

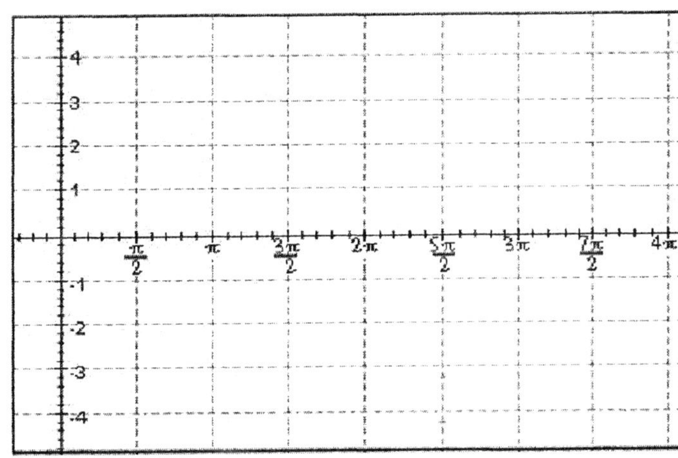

21. Use half-angle formulas to find the exact value for the following :

$$\tan 165°$$

Select the correct answer.

a. $\quad \tan 165° = \dfrac{\sqrt{1 + \sqrt{2}}}{2}$

b. $\quad \tan 165° = -2 + \sqrt{3}$

c. $\quad \tan 165° = \dfrac{1}{2}$

d. $\quad \tan 165° = -\dfrac{1}{2}$

e. $\quad \tan 165° = -\dfrac{\sqrt{2 + \sqrt{3}}}{2}$

22. Evaluate the expression without using a calculator.

$$\sin\left(\arcsin \dfrac{4}{5} - \arctan 3\right)$$

23. Evaluate the expression without using a calculator.

$$\sin\left(2\sin^{-1} \dfrac{1}{\sqrt{5}}\right)$$

Select the correct answer.

a. $\quad -\dfrac{4}{5}$

b. $\quad \dfrac{6}{5}$

c. $\quad \dfrac{4}{5}$

d. $\quad -\dfrac{6}{5}$

e. $\quad \dfrac{2}{\sqrt{5}}$

24. $10 \sin 6x \cos 2x$

Rewrite the expression as a sum or difference using one of the *product to sum formulas*.

Simplify if possible.

25. Rewrite the expression as a sum or difference, then simplify if possible.

$\cos 180° \cos 360°$

Select the correct answer.

a. $\dfrac{1}{5}$

b. 0

c. -1

d. $\dfrac{1}{4}$

e. 1

1.
$$\sin x\,(\sec x - \cot x) = \sin x\left(\frac{1}{\cos x} - \frac{\cos x}{\sin x}\right)$$

$$= \sin x \cdot \frac{1}{\cos x} - \sin x \cdot \frac{\cos x}{\sin x}$$

$$= \frac{\sin x}{\cos x} - \cos x$$

$$= \tan x - \cos x$$

2. d

3.
$$\frac{1 + \sec x}{1 - \sec x} = \frac{1 + \dfrac{1}{\cos x}}{1 - \dfrac{1}{\cos x}}$$

$$= \frac{\dfrac{\cos x + 1}{\cos x}}{\dfrac{\cos x - 1}{\cos x}}$$

$$= \frac{\cos x + 1}{\cos x - 1}$$

4. e

5. d

6.
$$\frac{\tan^2 \psi + 2}{1 + \tan^2 \psi} = \frac{\dfrac{\sin^2 \psi}{\cos^2 \psi} + 2}{1 + \dfrac{\sin^2 \psi}{\cos^2 \psi}} = \frac{\sin^2 \psi + 2\cos^2 \psi}{\sin^2 \psi + \cos^2 \psi} = \sin^2 \psi + 2\cos^2 \psi = \left(\sin^2 \psi + \cos^2 \psi\right) + \cos^2 \psi = 1 + \cos^2 \psi$$

7.
$$\sin\left(\frac{3\pi}{2} - x\right) = \sin\frac{3\pi}{2} \cdot \cos x - \sin x \cdot \cos\frac{3\pi}{2} = -1 \cdot \cos x - \sin x \cdot 0 = -\cos x$$

8. c

9. 2

10. a

11. $\dfrac{120}{169}$

12. a

13. $-\dfrac{\sqrt{11}}{5}$

14. c

15. $\dfrac{1 + \cos 4t}{2} = \dfrac{1 - 1 + 2\cos^2 2t}{2} = \cos^2 2t$

16. b

17. b

18. $\sqrt{\dfrac{5 + 2\sqrt{6}}{10}}$

19. b

20.

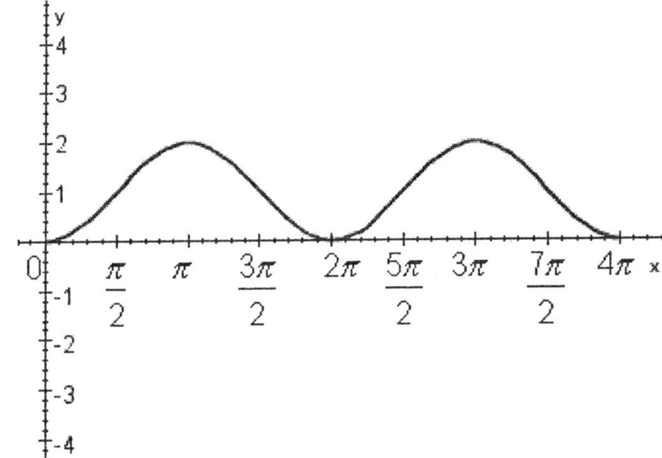

21. b

22. $-\dfrac{1}{\sqrt{10}}$

23. c

24. $5\big(\sin(8x) + \sin(4x)\big),$

impossible

25. c

1. mctr.05.01.09_NoAlgs
2. mctr.05.01.18m_NoAlgs
3. mctr.05.01.37_NoAlgs
4. mctr.05.01.46m_NoAlgs
5. mctr.05.01.57m_NoAlgs
6. mctr.05.01.63_NoAlgs
7. mctr.05.02.19_NoAlgs
8. mctr.05.02.23m_NoAlgs
9. mctr.05.02.40_NoAlgs
10. mctr.05.02.45m_NoAlgs
11. mctr.05.03.01_NoAlgs
12. mctr.05.03.07m_NoAlgs
13. mctr.05.03.28_NoAlgs
14. mctr.05.03.33m_NoAlgs
15. mctr.05.03.39_NoAlgs
16. mctr.05.03.46m_NoAlgs
17. mctr.05.03.53m_NoAlgs
18. mctr.05.04.09_NoAlgs
19. mctr.05.04.14m_NoAlgs
20. mctr.05.04.30_NoAlgs
21. mctr.05.04.35m_NoAlgs
22. mctr.05.05.01_NoAlgs
23. mctr.05.05.05m_NoAlgs
24. mctr.05.05.15_NoAlgs
25. mctr.05.05.20m_NoAlgs

1. Write the expression as a single trigonometric function.

    $\cos 5x \cos x - \sin 5x \sin x$

    Select the correct answer.

    a. $\cos 2x$
    b. $\cos 8x$
    c. $\cos 6x$
    d. $\cos 9x$
    e. $\cos 7x$

2. Show that the following is true.

    $$\sin\left(\frac{3\pi}{2} - x\right) = -\cos x$$

3. $10 \sin 6x \cos 2x$

    Rewrite the expression as a sum or difference using one of the *product to sum formulas*.

    Simplify if possible.

4. Rewrite the expression as a sum or difference, then simplify if possible.

    $\cos 180° \cos 360°$

    Select the correct answer.

    a. $\dfrac{1}{5}$

    b. 0

    c. -1

    d. $\dfrac{1}{4}$

    e. 1

5. From the list below find the expression, identical for the given.

$\cos 6\theta$

Select the correct answer.

    a. $3\cos 2\theta + 4\cos^3 2\theta$

    b. $4\cos^3 2\theta - 3\cos 2\theta$

    c. $\cos 2\theta - 4\cos^2 2\theta$

    d. $3\sin 2\theta - 4\sin^3 2\theta$

    e. $\sin 2\theta + 4\sin^3 2\theta$

6. Let $\sin A = \dfrac{12}{13}$ with A in QI and find sin2A.

7. Let $\cos A = \dfrac{1}{\sqrt{5}}$ with A in QIV and find tan2A.

Select the correct answer.

    a. $\dfrac{4}{3}$

    b. $-\dfrac{3}{4}$

    c. $\dfrac{4}{9}$

    d. $\dfrac{8}{9}$

    e. $-\dfrac{2}{3}$

8. Find the identical expression for the following.

$$\frac{1 + \cos x}{1 - \cos x}$$

Select the correct answer.

a. $(\sec x - \cot x)^2$

b. $(\csc x - \cot x)^2$

c. $(\sec x + \cot x)^2$

d. 1

e. $(\csc x + \cot x)^2$

9. Prove that the identity is true.

$$\frac{1 + \sec x}{1 - \sec x} = \frac{\cos x + 1}{\cos x - 1}$$

10. If $\sin B = -\dfrac{1}{3}$ with B in QIII, find

$$\tan \frac{B}{2}$$

Select the correct answer.

a. $\tan \dfrac{B}{2} = 2 - 2\sqrt{5}$

b. $\tan \dfrac{B}{2} = -3 - 2\sqrt{2}$

c. $\tan \dfrac{B}{2} = -2 - 2\sqrt{5}$

d. $\tan \dfrac{B}{2} = -2 - 2\sqrt{2}$

e. $\tan \dfrac{B}{2} = 2 - 2\sqrt{2}$

11. If $\sin B = -\dfrac{1}{5}$ with B in QIII , find

$\sin \dfrac{B}{2}$

12. Evaluate the expression without using a calculator.

$$\sin \left(\arcsin \frac{4}{5} - \arctan 3 \right)$$

13. Find the identical expression for the following.

$$\frac{\sin^3 A - 1}{\sin A - 1}$$

Select the correct answer.

a. $\tan^2 A + \tan A + 1$

b. -1

c. $\tan^2 A - \tan A + 1$

d. $\sin^2 A + \sin A + 1$

e. $\sin^2 A - \sin A + 1$

14. If $\tan A = \sqrt{11}$, find tan2A.

15. Use half-angle formulas to find the exact value for the following :

$$\tan 165°$$

Select the correct answer.

a. $\quad \tan 165° = \dfrac{\sqrt{1 + \sqrt{2}}}{2}$

b. $\quad \tan 165° = -2 + \sqrt{3}$

c. $\quad \tan 165° = \dfrac{1}{2}$

d. $\quad \tan 165° = -\dfrac{1}{2}$

e. $\quad \tan 165° = -\dfrac{\sqrt{2 + \sqrt{3}}}{2}$

16. Prove that the identity is true.

$$\sin x\,(\sec x - \cot x) = \tan x - \cos x$$

17. Evaluate the expression without using a calculator.

$$\sin\left(2\sin^{-1}\dfrac{1}{\sqrt{5}}\right)$$

Select the correct answer.

a. $\quad -\dfrac{4}{5}$

b. $\quad \dfrac{6}{5}$

c. $\quad \dfrac{4}{5}$

d. $\quad -\dfrac{6}{5}$

e. $\quad \dfrac{2}{\sqrt{5}}$

18. The identity is from the book *Plane and Spherical Trigonometry with Tables* by Rosenbach, Whitman, and Moskovitz, and published by Ginn and Company in 1937. Verify the identity.

$$\frac{\tan^2 \psi + 2}{1 + \tan^2 \psi} = 1 + \cos^2 \psi$$

19. If $\tan(A+B) = 3$ and $\tan B = \dfrac{1}{7}$, find $\tan A$.

$$\tan A = \underline{\hspace{2cm}}$$

20. Simplify the expression.

$$\cos^2 \frac{\pi}{3} - \sin^2 \frac{\pi}{3}$$

Select the correct answer.

a. 0

b. $\dfrac{1}{3}$

c. $-\dfrac{1}{2}$

d. $-\dfrac{\sqrt{3}}{4}$

e. $-\dfrac{\sqrt{2}}{4}$

21. Find the identical expression for the following.

$$\frac{\cos^4 t - \sin^4 t}{\cos^2 t}$$

Select the correct answer.

a. $\cot^2 t - 1$

b. $\tan t$

c. $1 - \cot^2 t$

d. $1 - \tan^2 t$

e. $\tan^2 t - 1$

22. From the list below find the expression, identical for the given.

$$\frac{1 - \tan t}{1 + \tan t}$$

Select the correct answer.

a. $\dfrac{1 - \sin 2t}{\cos t}$

b. $\dfrac{1 - \sin 2t}{\cos 2t}$

c. $\dfrac{1 - \cos 2t}{\sin 2t}$

d. $\dfrac{1 + \sin 2t}{\sin t}$

e. $\dfrac{1 - \sin t}{\cos t}$

23. Prove the identity.

$$\cos^2 2t = \frac{1 + \cos 4t}{2}$$

24. Graph the following function from $x = 0$ to $x = 4\pi$.

$$y = 2\sin^2 \frac{x}{2}$$

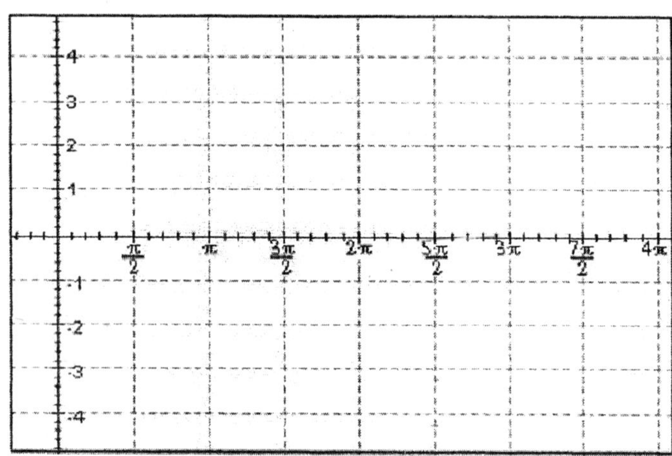

25. Find the identical expression for the following.

$$\sin\left(90° - x\right) + \sin\left(90° + x\right)$$

Select the correct answer.

a. $2\cos x$

b. 0

c. $-2\cos x$

d. $2\sin x$

e. $-2\sin x$

1. c

2. $\sin\left(\dfrac{3\pi}{2} - x\right) = \sin\dfrac{3\pi}{2} \cdot \cos x - \sin x \cdot \cos\dfrac{3\pi}{2} = -1 \cdot \cos x - \sin x \cdot 0 = -\cos x$

3. $5\left(\sin(8x) + \sin(4x)\right),$

   impossible

4. c

5. b

6. $\dfrac{120}{169}$

7. a

8. e

9.
$$\dfrac{1 + \sec x}{1 - \sec x} = \dfrac{1 + \dfrac{1}{\cos x}}{1 - \dfrac{1}{\cos x}}$$

$$= \dfrac{\dfrac{\cos x + 1}{\cos x}}{\dfrac{\cos x - 1}{\cos x}}$$

$$= \dfrac{\cos x + 1}{\cos x - 1}$$

10. b

11. $\sqrt{\dfrac{5 + 2\sqrt{6}}{10}}$

12. $-\dfrac{1}{\sqrt{10}}$

13. d

14. $-\dfrac{\sqrt{11}}{5}$

15. b

16.
$$\sin x\,(\sec x - \cot x) = \sin x\left(\frac{1}{\cos x} - \frac{\cos x}{\sin x}\right)$$

$$= \sin x \cdot \frac{1}{\cos x} - \sin x \cdot \frac{\cos x}{\sin x}$$

$$= \frac{\sin x}{\cos x} - \cos x$$

$$= \tan x - \cos x$$

17. c

18.
$$\frac{\tan^2\psi + 2}{1 + \tan^2\psi} = \frac{\dfrac{\sin^2\psi}{\cos^2\psi} + 2}{1 + \dfrac{\sin^2\psi}{\cos^2\psi}} = \frac{\sin^2\psi + 2\cos^2\psi}{\sin^2\psi + \cos^2\psi} = \sin^2\psi + 2\cos^2\psi = \left(\sin^2\psi + \cos^2\psi\right) + \cos^2\psi = 1 + \cos^2\psi$$

19. 2

20. c

21. d

22. b

23.
$$\frac{1 + \cos 4t}{2} = \frac{1 - 1 + 2\cos^2 2t}{2} = \cos^2 2t$$

24.

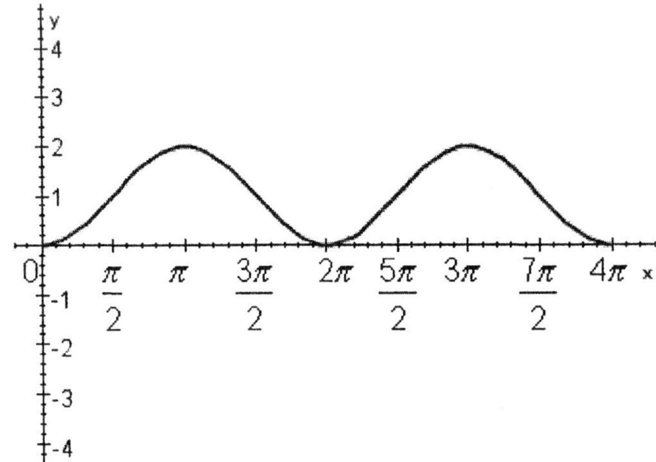

25. a

1. mctr.05.02.23m_NoAlgs
2. mctr.05.02.19_NoAlgs
3. mctr.05.05.15_NoAlgs
4. mctr.05.05.20m_NoAlgs
5. mctr.05.03.46m_NoAlgs
6. mctr.05.03.01_NoAlgs
7. mctr.05.03.07m_NoAlgs
8. mctr.05.01.46m_NoAlgs
9. mctr.05.01.37_NoAlgs
10. mctr.05.04.14m_NoAlgs
11. mctr.05.04.09_NoAlgs
12. mctr.05.05.01_NoAlgs
13. mctr.05.01.57m_NoAlgs
14. mctr.05.03.28_NoAlgs
15. mctr.05.04.35m_NoAlgs
16. mctr.05.01.09_NoAlgs
17. mctr.05.05.05m_NoAlgs
18. mctr.05.01.63_NoAlgs
19. mctr.05.02.40_NoAlgs
20. mctr.05.03.33m_NoAlgs
21. mctr.05.01.18m_NoAlgs
22. mctr.05.03.53m_NoAlgs
23. mctr.05.03.39_NoAlgs
24. mctr.05.04.30_NoAlgs
25. mctr.05.02.45m_NoAlgs

1. Prove that the identity is true.

$$\sec^4\theta - \tan^4\theta = \frac{1 + \sin^2\theta}{\cos^2\theta}$$

2. Find the identical expression for the following.

$$\frac{\cos x}{1 + \sin x} + \frac{1 + \sin x}{\cos x}$$

Select the correct answer.

   a. $2 \csc x$

   b. $2 \sec x$

   c. $2 \csc^2 x$

   d. $2 \tan x$

   e. $2 \sec^2 x$

3. Prove that the identity is true.

$$\sin^4 A - \cos^4 A = 2\sin^2 A - 1$$

4. Use your graphing calculator to determine if the equation appears to be an identity or not by graphing the left expression and right expression together.

$$\sin^4\theta - \cos^4\theta = 2\sin^2\theta - 1$$

Select the correct answer.

   a. is an identity
   b. not an identity

5. Find the exact value for the following.

$$\sin 75°$$

6. Find the identical expression for the following.

$$\sin\left(x - \frac{\pi}{2}\right)$$

Select the correct answer.

a. $\tan x$
b. $-\cos x$
c. $-\sin x$
d. $\sin x$
e. $\cos x$

7. Rewrite the expression as a product. Simplify if possible.

$$\sin \frac{7\pi}{12} + \sin \frac{\pi}{12}$$

Select the correct answer.

a. $\dfrac{1}{2\sqrt{6}}$

b. $\dfrac{\sqrt{6}}{2}$

c. $\sqrt{6}$

d. $-\dfrac{\sqrt{6}}{2}$

e. $2\sqrt{6}$

8. Graph the following from $x = 0$ to $x = 2\pi$.

$\sin 15x \cos 11x - \cos 15x \sin 11x$

Select the correct answer.

a.

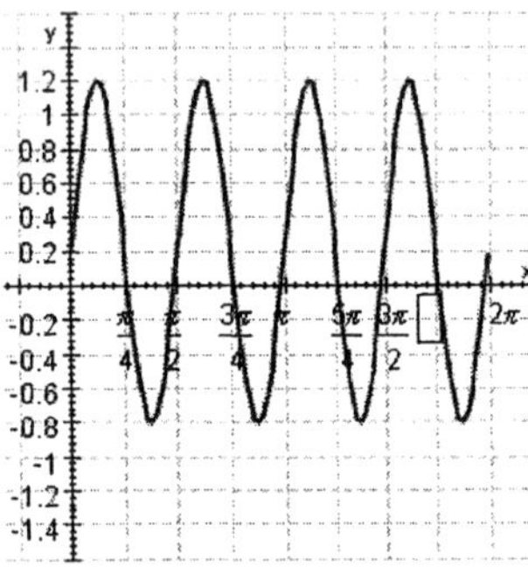

b.

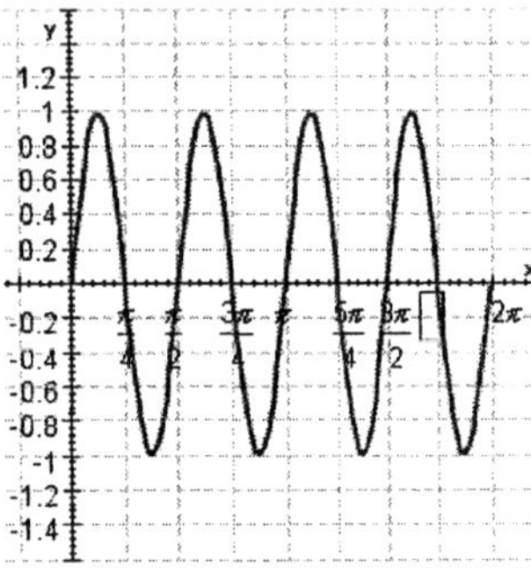

c.

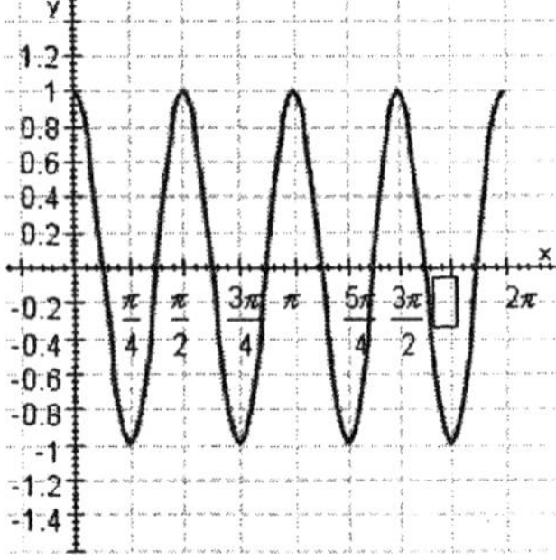

9. Graph one complete cycle of $y = \sin x \cos \dfrac{\pi}{6} - \cos x \sin \dfrac{\pi}{6}$.

Select the correct answer.

a.

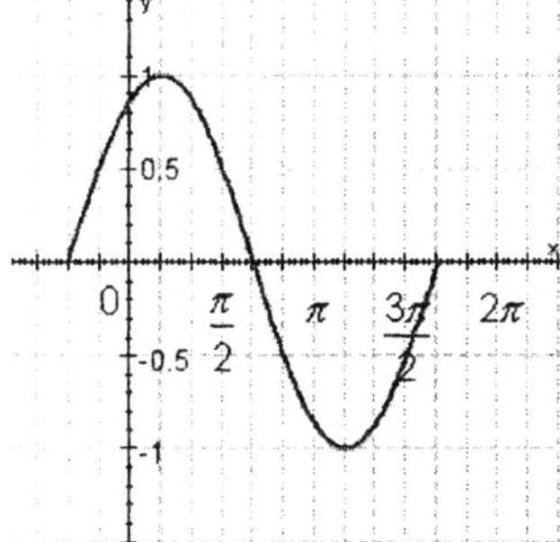

b.

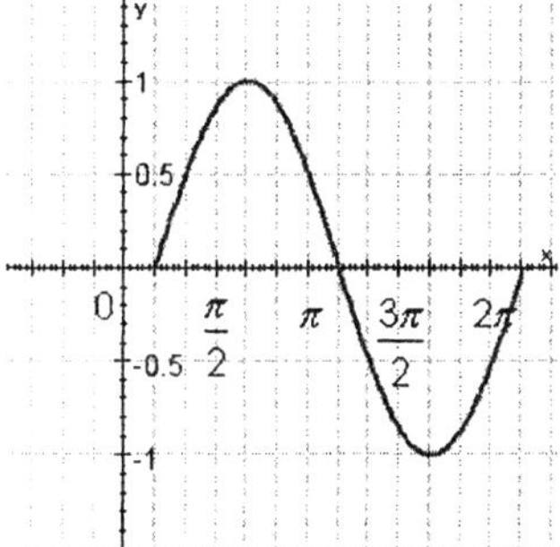

c.

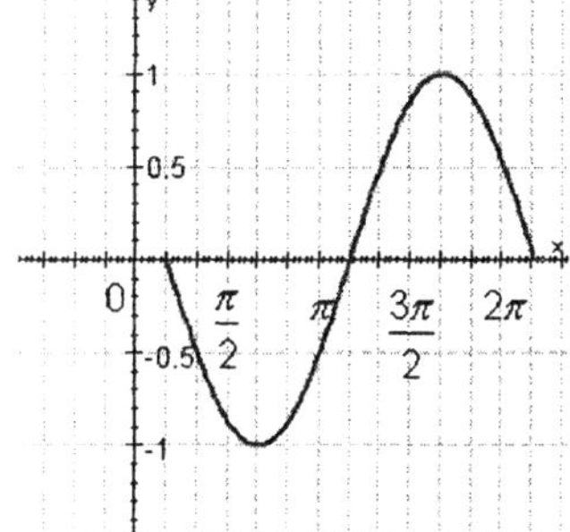

10. Prove the identity.

$$\cos\left(x + \frac{3\pi}{2}\right) + \cos\left(x - \frac{3\pi}{2}\right) = 0$$

11. Find the identical expression for the following.

$$\frac{\cos\left(A - B\right)}{\sin A \cos B}$$

Select the correct answer.

a. $\cot A + \cot B$

b. $\tan A + \cot B$

c. $\tan A + \tan B$

d. $\cot A - \tan B$

e. $\cot A + \tan B$

12. Let $\csc A = \sqrt{10}$ with A in QII and find cos2A.

Select the correct answer.

a. $-\dfrac{4}{5}$

b. $-\dfrac{2}{5}$

c. $-\dfrac{3}{5}$

d. $\dfrac{4}{15}$

e. $\dfrac{4}{5}$

13. Graph the function from

$$x = 0 \text{ and } x = 2\pi.$$

$$y = 6\cos^2 \frac{x}{2} - 3$$

Select the correct answer.

a.

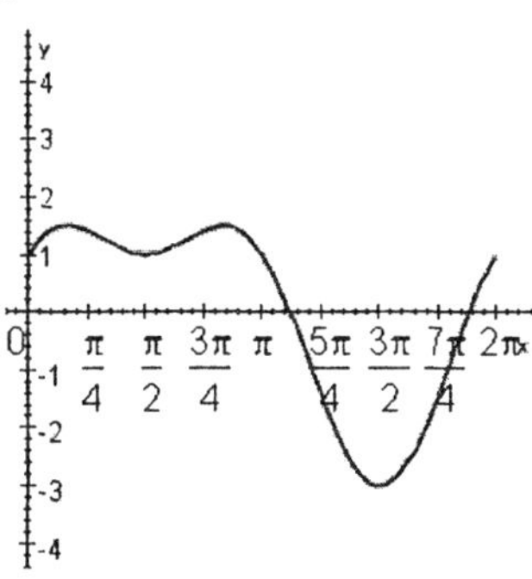

b.

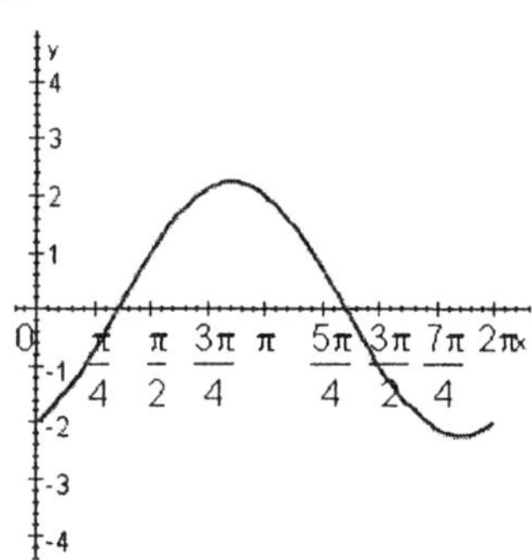

c.

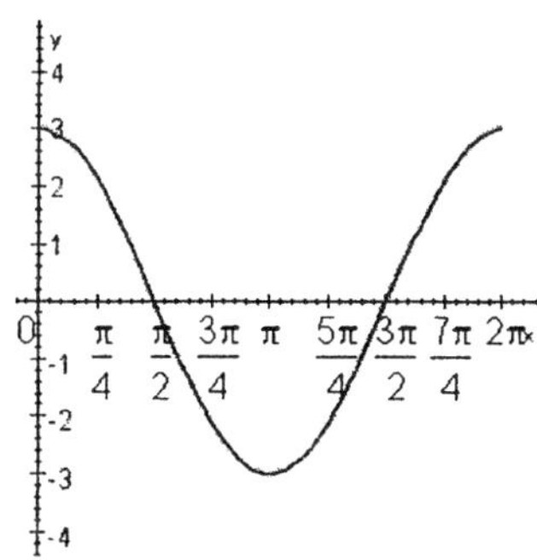

14. From the list below find the expression, identical for the given.

$\cos 6\theta$

Select the correct answer.

a. $\quad 3\cos 2\theta + 4\cos^3 2\theta$

b. $\quad 4\cos^3 2\theta - 3\cos 2\theta$

c. $\quad \cos 2\theta - 4\cos^2 2\theta$

d. $\quad 3\sin 2\theta - 4\sin^3 2\theta$

e. $\quad \sin 2\theta + 4\sin^3 2\theta$

15. Use your graphing calculator to determine if the equation appears to be an identity by graphing the left expression and right expression together.

$$\cot 2t = \frac{\sec^2 t \csc^2 t}{\sec^2 t + \csc^2 t}$$

Select the correct answer.

a.   This is an identity
b.   This is not an identity

16. If $\cos A = \dfrac{7}{8}$ with A in QIV, find

$\sin \dfrac{A}{2}$

17. If $\sin A = -\dfrac{3}{5}$ with A in QIV, find

$$\cos \frac{A}{2}$$

Select the correct answer.

a. $\cos \dfrac{A}{2} = -\dfrac{3}{\sqrt{10}}$

b. $\cos \dfrac{A}{2} = -\dfrac{1}{\sqrt{4}}$

c. $\cos \dfrac{A}{2} = -\dfrac{9}{\sqrt{10}}$

d. $\cos \dfrac{A}{2} = \dfrac{3}{\sqrt{10}}$

e. $\cos \dfrac{A}{2} = \dfrac{1}{\sqrt{9}}$

18. If $\sin A = \dfrac{3}{5}$ with A in QI, find

$$\cos \frac{A}{2}$$

19. If $\sin A = -\dfrac{4}{5}$ with A in QIV, and $\sin B = -\dfrac{3}{5}$ with B in QIII, find $\cos(A + B)$

Select the correct answer.

a. $\cos(A + B) = -\dfrac{1}{5}$

b. $\cos(A + B) = \dfrac{\sqrt{2}}{5}$

c. $\cos(A + B) = -\dfrac{\sqrt{2}}{5}$

d. $\cos(A + B) = -\dfrac{24}{25}$

e. $\cos(A + B) = \dfrac{1}{5}$

20. Prove the identity.

$$\sin^2 \frac{\theta}{2} = \frac{\csc \theta - \cot \theta}{2 \csc \theta}$$

21. Find the expression identical to the one given.

$$\tan \frac{B}{2}$$

Select the correct answer.

a. $\quad -\csc^2 B - \cot B$

b. $\quad \csc B + \cot^2 B$

c. $\quad \dfrac{\sec B}{\sec B \csc B - \csc B}$

d. $\quad \csc^2 B - \cot B$

e. $\quad \dfrac{\sec B}{\sec B \csc B + \csc B}$

22. Evaluate the expression without using a calculator. (Assume the variable represents a positive number.)

$$\tan \left(\sin^{-1} 3x \right)$$

23. Evaluate the expression without using a calculator. (Assume the variable represents a positive number.)

$$\cos\left(2\cos^{-1}9x\right)$$

Select the correct answer.

a. $162x^2 + 1$

b. $161x^2 - 1$

c. $161x^2 + 1$

d. $160x^2 - 1$

e. $162x^2 - 1$

24. Rewrite the expression as a product. Simplify if possible.

$$\sin 5x + \sin 3x$$

Select the correct answer.

a. $2\sin 5x \sin x$

b. $2\cos 6x \sin x$

c. $2\sin 5x \cos x$

d. $2\sin 4x \cos x$

e. $2\sin 6x \cos x$

25. Verify the identity.

$$-\cot 3x = \frac{\sin 5x - \sin x}{\cos 5x - \cos x}$$

1.

$$\sec^4\theta - \tan^4\theta = \frac{1}{\cos^4\theta} - \frac{\sin^4\theta}{\cos^4\theta}$$

$$= \frac{1 - \sin^4\theta}{\cos^4\theta}$$

$$= \frac{\left(1 - \sin^2\theta\right)\left(1 + \sin^2\theta\right)}{\cos^4\theta}$$

$$= \frac{\cos^2\theta\left(1 + \sin^2\theta\right)}{\cos^4\theta}$$

$$= \frac{1 + \sin^2\theta}{\cos^2\theta}$$

2. b

3.

$$\sin^4 A - \cos^4 A = \left(\sin^2 A - \cos^2 A\right)\left(\sin^2 A + \cos^2 A\right)$$

$$= \sin^2 A - \cos^2 A = \sin^2 A - \left(1 - \sin^2 A\right)$$

$$= 2\sin^2 A - 1$$

4. a

5. $\dfrac{\sqrt{6}+\sqrt{2}}{4}$

6. b

7. b

8. b

9.

$$\cos\left(x + \frac{3\pi}{2}\right) + \cos\left(x - \frac{3\pi}{2}\right) = \cos x\cos\frac{3\pi}{2} - \sin x\sin\frac{3\pi}{2} + \cos x\cos\frac{3\pi}{2} + \sin x\,\text{si}$$

10. e

11. e

12. c

13. b

14. b

15. $\dfrac{1}{4}$

16. a

17. $\dfrac{3}{\sqrt{10}}$

18. d

19.
$$\sin^2\frac{\theta}{2} = \frac{1 - \left(1 - 2\sin^2\frac{\theta}{2}\right)}{2} = \frac{1 - \cos\theta}{2} = \frac{\dfrac{1}{\sin\theta} - \dfrac{\cos\theta}{\sin\theta}}{2\,\dfrac{1}{\sin\theta}} = \frac{\csc\theta - \cot\theta}{2\csc\theta}$$

20. e

21. $\dfrac{3x}{\sqrt{1 - 9x^2}}$

22. e

23. d

24. b

25. $\dfrac{\sin 5x - \sin x}{\cos 5x - \cos x} = \dfrac{2\cos 3x \sin 2x}{-2\sin 3x \sin 2x} = -\dfrac{\cos 3x}{\sin 3x} = -\cot 3x$

McKeague/Turner - Trigonometry 5e Chapter 5 Form G

1. mctr.05.01.25_NoAlgs
2. mctr.05.01.33m_NoAlgs
3. mctr.05.01.51_NoAlgs
4. mctr.05.01.72m_NoAlgs
5. mctr.05.02.01_NoAlgs
6. mctr.05.02.11m_NoAlgs
7. mctr.05.05.29m_NoAlgs
8. mctr.05.02.27m_NoAlgs
9. mctr.05.02.32m_NoAlgs
10. mctr.05.02.52_NoAlgs
11. mctr.05.02.55m_NoAlgs
12. mctr.05.03.13m_NoAlgs
13. mctr.05.03.19m_NoAlgs
14. mctr.05.03.46m_NoAlgs
15. mctr.05.03.57m_NoAlgs
16. mctr.05.04.01_NoAlgs
17. mctr.05.04.05m_NoAlgs
18. mctr.05.04.16_NoAlgs
19. mctr.05.04.24m_NoAlgs
20. mctr.05.04.37_NoAlgs
21. mctr.05.04.42m_NoAlgs
22. mctr.05.05.07_NoAlgs
23. mctr.05.05.11m_NoAlgs
24. mctr.05.05.25m_NoAlgs
25. mctr.05.05.31_NoAlgs

1. If $\sin A = -\dfrac{3}{5}$ with A in QIV, find

$\cos \dfrac{A}{2}$

Select the correct answer.

a. $\cos \dfrac{A}{2} = -\dfrac{3}{\sqrt{10}}$

b. $\cos \dfrac{A}{2} = -\dfrac{1}{\sqrt{4}}$

c. $\cos \dfrac{A}{2} = -\dfrac{9}{\sqrt{10}}$

d. $\cos \dfrac{A}{2} = \dfrac{3}{\sqrt{10}}$

e. $\cos \dfrac{A}{2} = \dfrac{1}{\sqrt{9}}$

2. Rewrite the expression as a product. Simplify if possible.

$\sin \dfrac{7\pi}{12} + \sin \dfrac{\pi}{12}$

Select the correct answer.

a. $\dfrac{1}{2\sqrt{6}}$

b. $\dfrac{\sqrt{6}}{2}$

c. $\sqrt{6}$

d. $-\dfrac{\sqrt{6}}{2}$

e. $2\sqrt{6}$

3. Evaluate the expression without using a calculator. (Assume the variable represents a positive number.)

$$\cos\left(2\cos^{-1}9x\right)$$

Select the correct answer.

a. $162x^2 + 1$

b. $161x^2 - 1$

c. $161x^2 + 1$

d. $160x^2 - 1$

e. $162x^2 - 1$

4. If $\sin A = -\dfrac{4}{5}$ with A in QIV, and $\sin B = -\dfrac{3}{5}$ with B in QIII, find

$$\cos(A + B)$$

Select the correct answer.

a. $\cos(A+B) = -\dfrac{1}{5}$

b. $\cos(A+B) = \dfrac{\sqrt{2}}{5}$

c. $\cos(A+B) = -\dfrac{\sqrt{2}}{5}$

d. $\cos(A+B) = -\dfrac{24}{25}$

e. $\cos(A+B) = \dfrac{1}{5}$

5.

Graph one complete cycle of $y = \sin x \cos \dfrac{\pi}{6} - \cos x \sin \dfrac{\pi}{6}$.

Select the correct answer.

a.

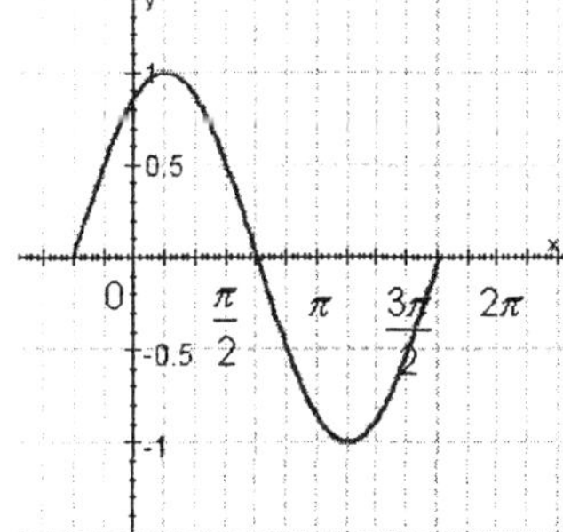

b.

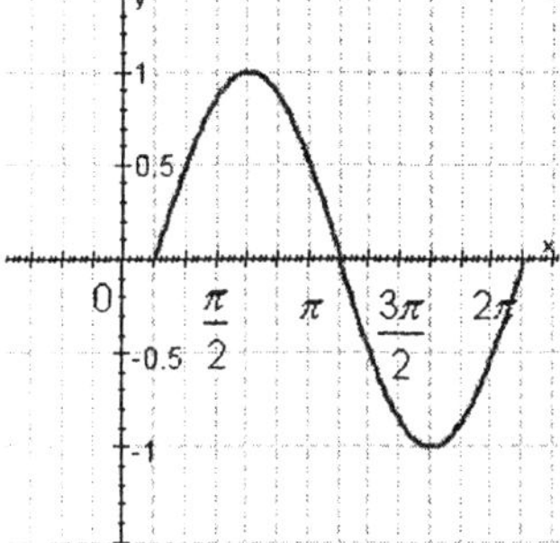

c.

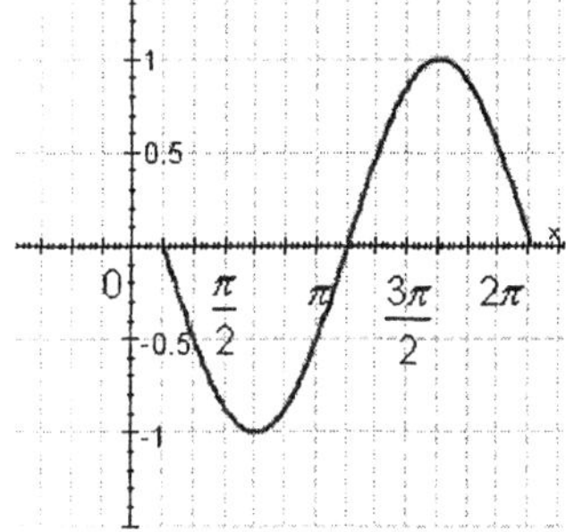

6. Let $\csc A = \sqrt{10}$ with A in QII and find cos2A.

   Select the correct answer.

   a. $-\dfrac{4}{5}$

   b. $-\dfrac{2}{5}$

   c. $-\dfrac{3}{5}$

   d. $\dfrac{4}{15}$

   e. $\dfrac{4}{5}$

7. Graph the following from $x = 0$ to $x = 2\pi$.

    sin15x cos11x - cos15x sin11x

    Select the correct answer.

a.
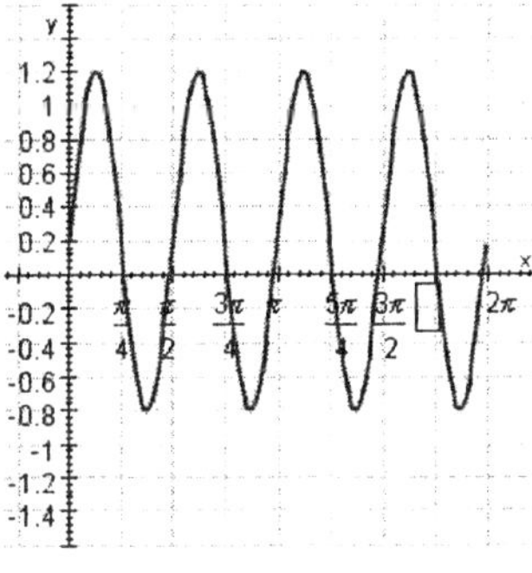

b.
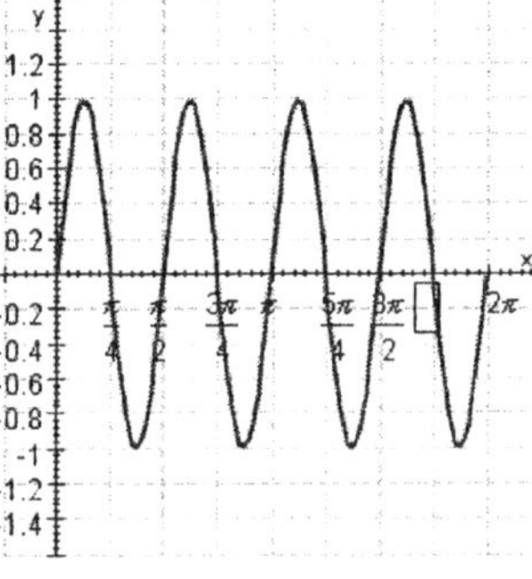

c.
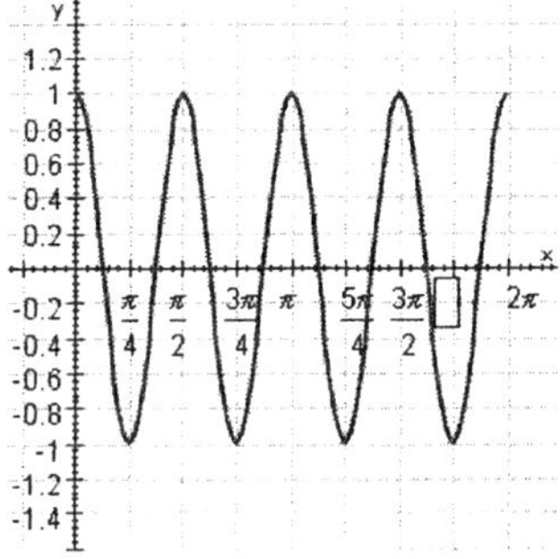

8. From the list below find the expression, identical for the given.

$\cos 6\theta$

Select the correct answer.

a. $3\cos 2\theta + 4\cos^3 2\theta$

b. $4\cos^3 2\theta - 3\cos 2\theta$

c. $\cos 2\theta - 4\cos^2 2\theta$

d. $3\sin 2\theta - 4\sin^3 2\theta$

e. $\sin 2\theta + 4\sin^3 2\theta$

9. Rewrite the expression as a product. Simplify if possible.

$\sin 5x + \sin 3x$

Select the correct answer.

a. $2\sin 5x \sin x$

b. $2\cos 6x \sin x$

c. $2\sin 5x \cos x$

d. $2\sin 4x \cos x$

e. $2\sin 6x \cos x$

10. Graph the function from

$$x = 0 \text{ and } x = 2\pi.$$

$$y = 6\cos^2 \frac{x}{2} - 3$$

Select the correct answer.

a.

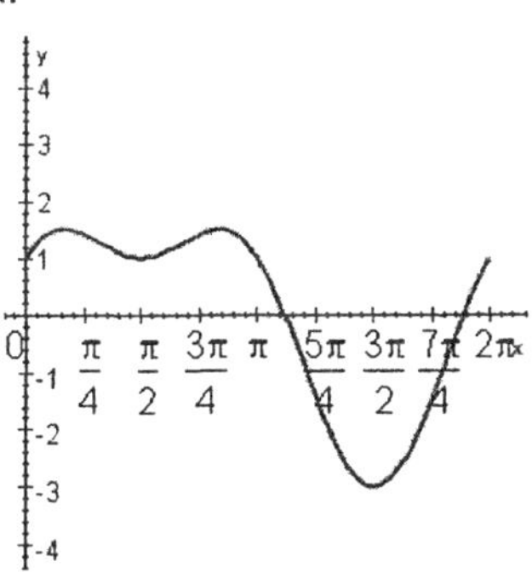

b.

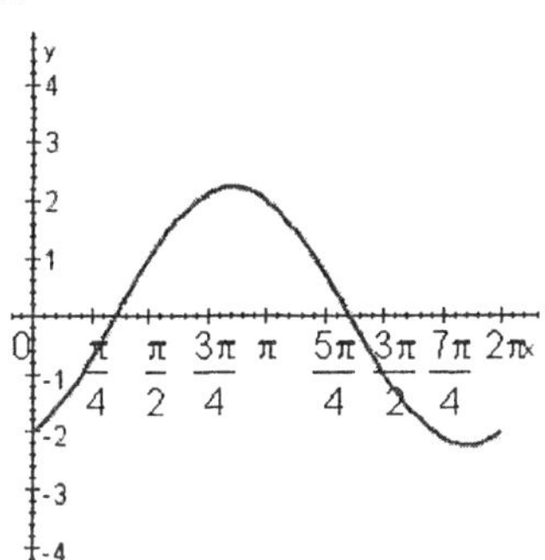

c.

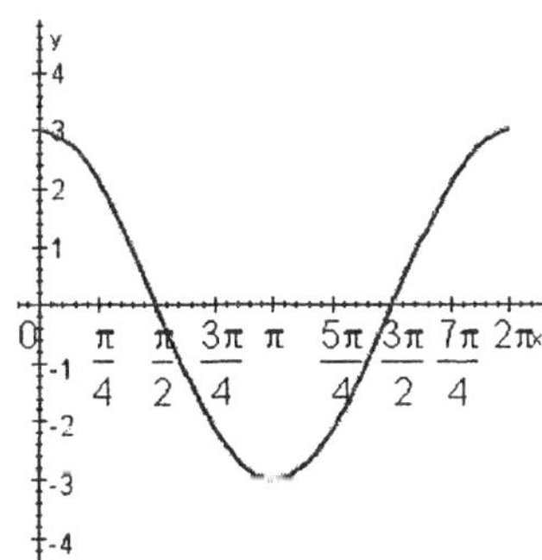

11. Evaluate the expression without using a calculator. (Assume the variable represents a positive number.)

$$\tan\left(\sin^{-1} 3x\right)$$

12. If $\cos A = \dfrac{7}{8}$ with A in QIV, find

$$\sin \dfrac{A}{2}$$

13. Find the exact value for the following.

$$\sin 75°$$

14. If $\sin A = \dfrac{3}{5}$ with A in QI, find

$$\cos \dfrac{A}{2}$$

15. Find the identical expression for the following.

$$\sin\left(x - \dfrac{\pi}{2}\right)$$

Select the correct answer.

a. $\tan x$
b. $- \cos x$
c. $- \sin x$
d. $\sin x$
e. $\cos x$

16. Prove that the identity is true.

$$\sec^4\theta - \tan^4\theta = \frac{1 + \sin^2\theta}{\cos^2\theta}$$

17. Prove the identity.

$$\cos\left(x + \frac{3\pi}{2}\right) + \cos\left(x - \frac{3\pi}{2}\right) = 0$$

18. Find the identical expression for the following.

$$\frac{\cos(A - B)}{\sin A \cos B}$$

Select the correct answer.

a. $\cot A + \cot B$

b. $\tan A + \cot B$

c. $\tan A + \tan B$

d. $\cot A - \tan B$

e. $\cot A + \tan B$

19. Find the identical expression for the following.

$$\frac{\cos x}{1 + \sin x} + \frac{1 + \sin x}{\cos x}$$

Select the correct answer.

    a. $2 \csc x$

    b. $2 \sec x$

    c. $2 \csc^2 x$

    d. $2 \tan x$

    e. $2 \sec^2 x$

20. Prove the identity.

$$\sin^2 \frac{\theta}{2} = \frac{\csc \theta - \cot \theta}{2 \csc \theta}$$

21. Verify the identity.

$$-\cot 3x = \frac{\sin 5x - \sin x}{\cos 5x - \cos x}$$

22. Prove that the identity is true.

$$\sin^4 A - \cos^4 A = 2\sin^2 A - 1$$

23. Use your graphing calculator to determine if the equation appears to be an identity or not by graphing the left expression and right expression together.

$$\sin^4\theta - \cos^4\theta = 2\sin^2\theta - 1$$

Select the correct answer.

a.   is an identity
b.   not an identity

24. Use your graphing calculator to determine if the equation appears to be an identity by graphing the left expression and right expression together.

$$\cot 2t = \frac{\sec^2 t \csc^2 t}{\sec^2 t + \csc^2 t}$$

Select the correct answer.

a.   This is an identity
b.   This is not an identity

25. Find the expression identical to the one given.

$$\tan\frac{B}{2}$$

Select the correct answer.

a. $-\csc^2 B - \cot B$

b. $\csc B + \cot^2 B$

c. $\dfrac{\sec B}{\sec B \csc B - \csc B}$

d. $\csc^2 B - \cot B$

e. $\dfrac{\sec B}{\sec B \csc B + \csc B}$

1. a

2. b

3. e

4. d

5. b

6. e

7. b

8. b

9. d

10. c

11. $\dfrac{3x}{\sqrt{1 - 9x^2}}$

12. $\dfrac{1}{4}$

13. $\dfrac{\sqrt{6} + \sqrt{2}}{4}$

14. $\dfrac{3}{\sqrt{10}}$

15. b

16.
$$\sec^4\theta - \tan^4\theta = \frac{1}{\cos^4\theta} - \frac{\sin^4\theta}{\cos^4\theta}$$

$$= \frac{1 - \sin^4\theta}{\cos^4\theta}$$

$$= \frac{\left(1 - \sin^2\theta\right)\left(1 + \sin^2\theta\right)}{\cos^4\theta}$$

$$= \frac{\cos^2\theta\left(1 + \sin^2\theta\right)}{\cos^4\theta} = \frac{1 + \sin^2\theta}{\cos^2\theta}$$

17. $\cos\left(x + \dfrac{3\pi}{2}\right) + \cos\left(x - \dfrac{3\pi}{2}\right) = \cos x\cos\dfrac{3\pi}{2} - \sin x\sin\dfrac{3\pi}{2} + \cos x\cos\dfrac{3\pi}{2} + \sin x\sin\dfrac{3\pi}{2} = 0$

18. e

19. b

20.
$$\sin^2\dfrac{\theta}{2} = \dfrac{1 - \left(1 - 2\sin^2\dfrac{\theta}{2}\right)}{2} = \dfrac{1 - \cos\theta}{2} = \dfrac{\dfrac{1}{\sin\theta} - \dfrac{\cos\theta}{\sin\theta}}{2\,\dfrac{1}{\sin\theta}} = \dfrac{\csc\theta - \cot\theta}{2\csc\theta}$$

21. $\dfrac{\sin 5x - \sin x}{\cos 5x - \cos x} = \dfrac{2\cos 3x\sin 2x}{-2\sin 3x\sin 2x} = -\dfrac{\cos 3x}{\sin 3x} = -\cot 3x$

22. $\sin^4 A - \cos^4 A = \left(\sin^2 A - \cos^2 A\right)\left(\sin^2 A + \cos^2 A\right)$

$= \sin^2 A - \cos^2 A = \sin^2 A - \left(1 - \sin^2 A\right)$

$= 2\sin^2 A - 1$

23. a

24. b

25. e

1. mctr.05.04.05m_NoAlgs
2. mctr.05.05.29m_NoAlgs
3. mctr.05.05.11m_NoAlgs
4. mctr.05.04.24m_NoAlgs
5. mctr.05.02.32m_NoAlgs
6. mctr.05.03.13m_NoAlgs
7. mctr.05.02.27m_NoAlgs
8. mctr.05.03.46m_NoAlgs
9. mctr.05.05.25m_NoAlgs
10. mctr.05.03.19m_NoAlgs
11. mctr.05.05.07_NoAlgs
12. mctr.05.04.01_NoAlgs
13. mctr.05.02.01_NoAlgs
14. mctr.05.04.16_NoAlgs
15. mctr.05.02.11m_NoAlgs
16. mctr.05.01.25_NoAlgs
17. mctr.05.02.52_NoAlgs
18. mctr.05.02.55m_NoAlgs
19. mctr.05.01.33m_NoAlgs
20. mctr.05.04.37_NoAlgs
21. mctr.05.05.31_NoAlgs
22. mctr.05.01.51_NoAlgs
23. mctr.05.01.72m_NoAlgs
24. mctr.05.03.57m_NoAlgs
25. mctr.05.04.42m_NoAlgs

1. Solve the equation for θ if $0° \leq \theta < 360°$. Do not use a calculator.

$$2\cos\theta - \sqrt{3} = 0$$

2. Find all solutions in the interval $0° \leq \theta < 360°$. Use a calculator on the last step and write all answers to the nearest tenth of a degree.

$$3\sin\theta - 9 = 4\sin\theta - 5$$

3. Solve for x, if $0 \leq x < 2\pi$. Write your answers in exact values only.

$$\sin x + 2\sin x \cos x = 0$$

4. Solve for θ if $0° \leq \theta < 360°$.

$$2\cos^2\theta + 3\cos\theta = 2$$

5. Match each equation in the left column with the corresponding expression representing all solutions in the right column.

$$2\cos\theta = -\sqrt{3}$$

$$\sqrt{2}\sin\theta = 1$$

$150° + 360°k,\ 210° + 360°k$

$45° + 360°k,\ 135° + 360°k$

$30° + 360°k,\ 330° + 360°k$

$225° + 360°k,\ 315° + 360°k$

6. Match each equation in the left column with the corresponding expression representing all solutions in the right column.

$$\sin\left(5A - 40^\circ\right) = -\frac{1}{2}$$

$$\sin\left(5A + 40^\circ\right) = -\frac{1}{2}$$

$20^\circ + 72^\circ k,\ 32^\circ + 72^\circ k$

$50^\circ + 72^\circ k,\ 74^\circ + 72^\circ k$

$34^\circ + 72^\circ k,\ 58^\circ + 72^\circ k$

$4^\circ + 72^\circ k,\ 16^\circ + 72^\circ k$

7. Use your graphing calculator to find the solutions to the equation for θ if $0^\circ \le \theta < 360^\circ$ by graphing the function represented by the left side of the equation and then finding its zeros.

$$\cot\theta + 2\cos\theta\cot\theta = 0$$

8. Solve the equation for θ if $0^\circ \le \theta < 360^\circ$. Give your answer in degrees.

$$\sqrt{3}\csc\theta - 2\cot\theta = 0$$

9. Solve the equation for θ if $0^\circ \le \theta < 360^\circ$. Give your answer in degrees.

$$2\sin\theta + 1 = \csc\theta$$

10. Solve the equation for x if $0 \le x < 2\pi$. Give your answer in radians using exact values only.

$$2\cos x - \tan x - \sec x = 0$$

11. Solve the equation for x if $0 \leq x < 2\pi$. Give your answer in radians using exact values only.

$$2 \sin x + \cot x = \csc x$$

12. Solve the equation for θ if $0° \leq \theta < 360°$. Give your answer in degrees.

$$\sin \theta + \sqrt{3} \cos \theta = -\sqrt{3}$$

13. Write an expression that gives all solutions to the equation.

$$\sin x - \cos x = \sqrt{2}$$

14. Solving the following equation will require you to use the quadratic formula. Solve the equation for θ between $0°$ and $360°$, and round your answers to the nearest tenth of a degree.

$$4 \cos^2 \theta + 2 \sin \theta - 1 = 0$$

15. Find all degree solutions for the following.

$$\cos 4\theta = -1$$

16. Find all solutions if $0 \leq x < 2\pi$. Give your answers in radians using exact values only.

$$\sin 2x \cos x + \cos 2x \sin x = 0$$

17. Find all solutions in radians using exact values only.

$$\sin^2 6x = 1$$

18. Find all degree solutions.

$$2\sin^2 10\theta + 3\sin 10\theta + 1 = 0$$

19. The formula below gives the relationship between the number of sides n, the radius r, and the length of each side l in a regular polygon. Find n, if $l = r\sqrt{2}$.

$$l = 2r \sin \frac{180°}{n}$$

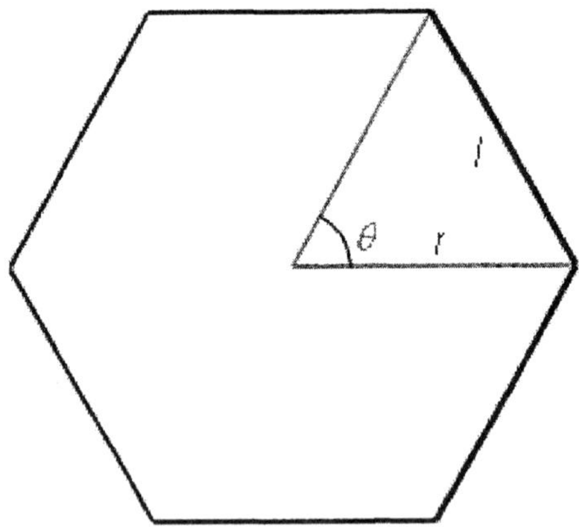

$n = $ ___________

20. Eliminate the parameter t from the following:

$$x = 9\sin t \quad y = 9\cos t$$

Sketch the graph.

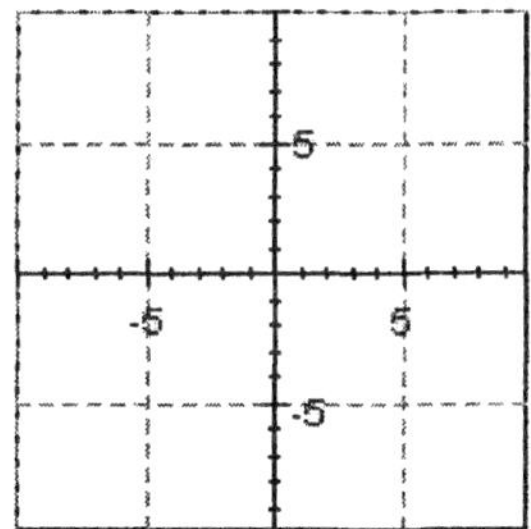

21. Eliminate the parameter t from the following:

$$x = 4\sin t \quad y = 5\cos t$$

Sketch the graph.

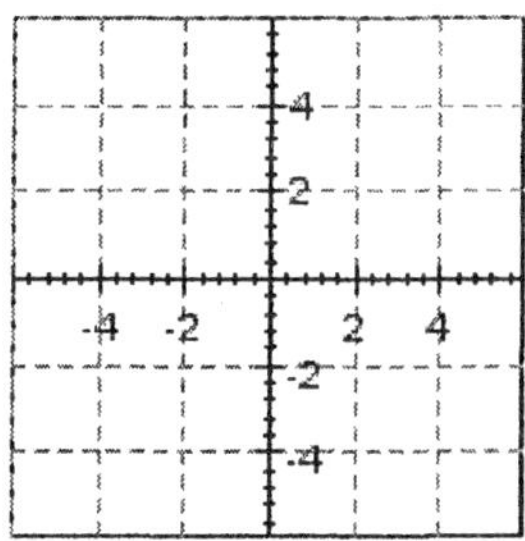

22. Eliminate the parameter t from the following:

$$x = \cos t - 4 \quad y = \sin t + 7$$

Sketch the graph.

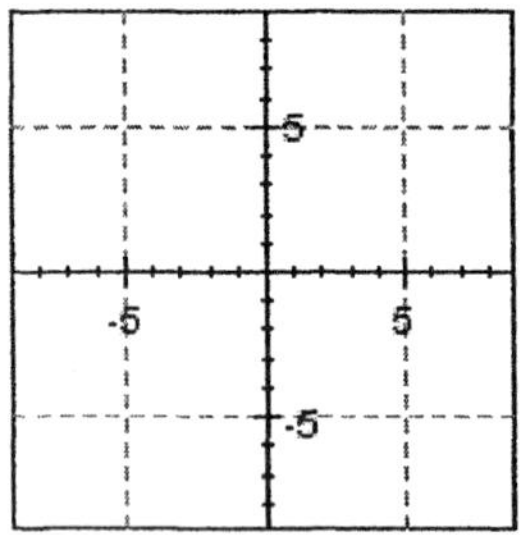

23. Eliminate the parameter t in the following:

$$x = 5\sec t \quad y = 3\tan t$$

24. Eliminate the parameter t in the following:

$$x = 5 + 3\tan t \quad y = 2 + 3\sec t$$

25. Eliminate the parameter t in the following:

$$x = -\cos t \quad y = \cos t$$

1. $30°, 330°$

2. no solution

3. $0, \pi, \dfrac{2\pi}{3}, \dfrac{4\pi}{3}$

4. $60°, 300°$

5. $\sqrt{2}\sin\theta = 1 \rightarrow 45° + 360°k, 135° + 360°k,$

$2\cos\theta = -\sqrt{3} \rightarrow 150° + 360°k, 210° + 360°k$

6. $\sin(5A - 40°) = -\dfrac{1}{2} \rightarrow 50° + 72°k, 74° + 72°k,$

$\sin(5A + 40°) = -\dfrac{1}{2} \rightarrow 34° + 72°k, 58° + 72°k$

7. $90°, 270°, 120°, 240°$

8. $30°, 330°$

9. $30°, 150°, 270°$

10. $\dfrac{\pi}{6}, \dfrac{5\pi}{6}$

11. $\dfrac{2\pi}{3}, \dfrac{4\pi}{3}$

12. $180°, 240°$

13. $\dfrac{3\pi}{4} + 2k\pi$

14. $220.6°, 319.4°$

15. $\theta = 45° + 90°k$

16. $0, \dfrac{\pi}{3}, \dfrac{2\pi}{3}, \pi, \dfrac{4\pi}{3}, \dfrac{5\pi}{3}$

17. $x = \dfrac{\pi}{12} + \dfrac{\pi k}{3}$ or $\dfrac{\pi}{4} + \dfrac{\pi k}{3}$

18. $\theta = 21° + 36°k$ or $\theta = 27° + 36°k$ or $\theta = 33° + 36°k$

19. 4

20. $x^2 + y^2 = 81$

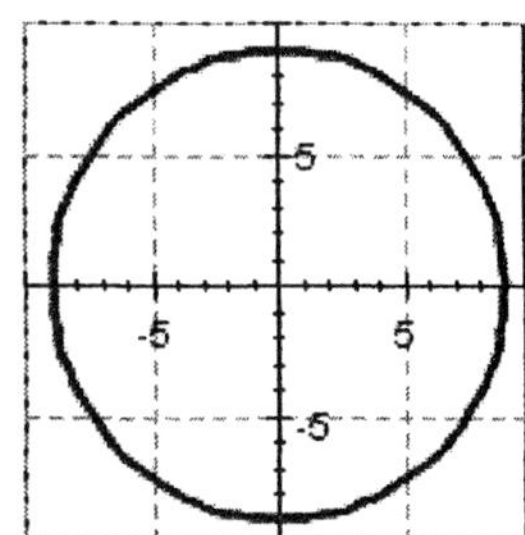

21. $\dfrac{x^2}{16} + \dfrac{y^2}{25} = 1$

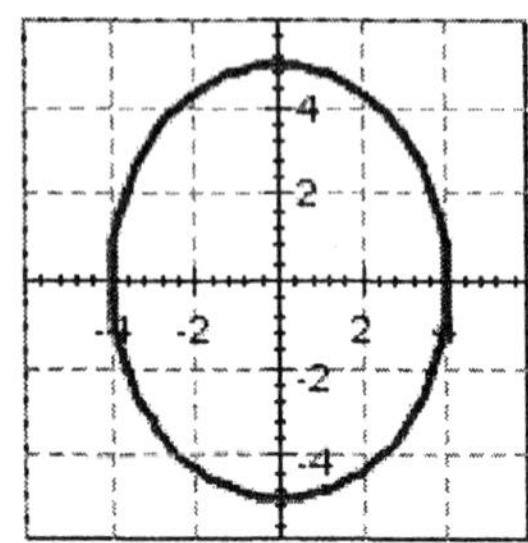

22. $(x + 4)^2 + (y - 7)^2 = 1$

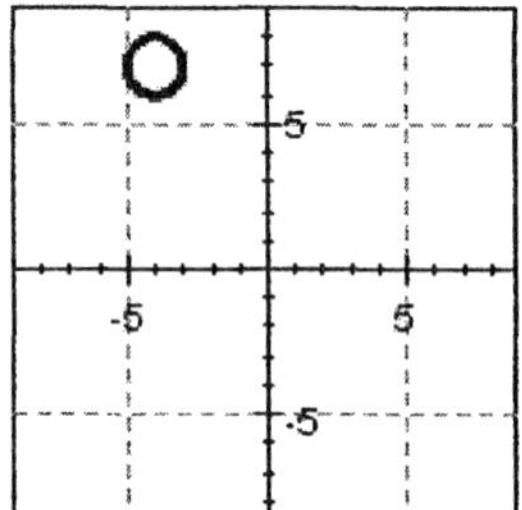

23. $\dfrac{x^2}{25} - \dfrac{y^2}{9} = 1$

24. $\dfrac{(y - 2)^2}{9} - \dfrac{(x - 5)^2}{9} = 1$

25. $y = -x, \ -1 \le x \le 1, \ -1 \le y \le 1$

McKeague/Turner - Trigonometry 5e Chapter 6 Form A

1. mctr.06.01.03_NoAlgs
2. mctr.06.01.15_NoAlgs
3. mctr.06.01.23_NoAlgs
4. mctr.06.01.31_NoAlgs
5. mctr.06.01.39_NoAlgs
6. mctr.06.01.52_NoAlgs
7. mctr.06.01.60_NoAlgs
8. mctr.06.02.07_NoAlgs
9. mctr.06.02.12_NoAlgs
10. mctr.06.02.21_NoAlgs
11. mctr.06.02.22_NoAlgs
12. mctr.06.02.26_NoAlgs
13. mctr.06.02.42_NoAlgs
14. mctr.06.02.48_NoAlgs
15. mctr.06.03.16_NoAlgs
16. mctr.06.03.25_NoAlgs
17. mctr.06.03.31_NoAlgs
18. mctr.06.03.36_NoAlgs
19. mctr.06.03.53_NoAlgs
20. mctr.06.04.01_NoAlgs
21. mctr.06.04.05_NoAlgs
22. mctr.06.04.10_NoAlgs
23. mctr.06.04.15_NoAlgs
24. mctr.06.04.20_NoAlgs
25. mctr.06.04.24_NoAlgs

1. Write an expression that gives all solutions to the equation.

$$\sin x - \cos x = \sqrt{2}$$

2. Solve the equation for θ if $0° \leq \theta < 360°$. Give your answer in degrees.

$$\sin \theta + \sqrt{3}\cos \theta = -\sqrt{3}$$

3. Eliminate the parameter t from the following:

$$x = 4\sin t \quad y = 5\cos t$$

Sketch the graph.

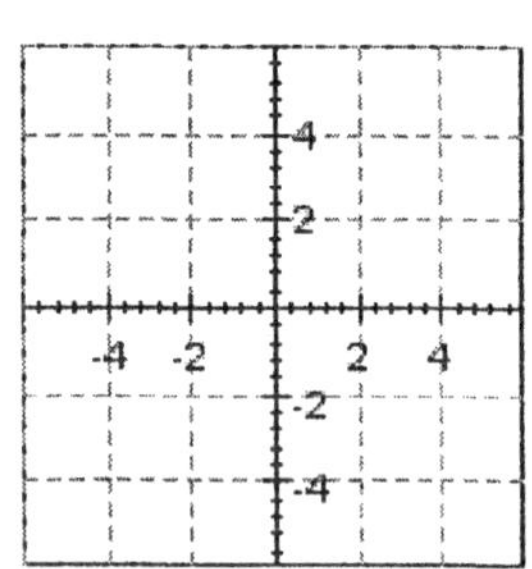

4. Solve the equation for θ if $0° \leq \theta < 360°$. Give your answer in degrees.

$$2\sin \theta + 1 = \csc \theta$$

5. Find all degree solutions.

$$2\sin^2 10\theta + 3\sin 10\theta + 1 = 0$$

6. Solve the equation for θ if $0^\circ \le \theta < 360^\circ$. Do not use a calculator.

$$2\cos\theta - \sqrt{3} = 0$$

7. Solve the equation for x if $0 \le x < 2\pi$. Give your answer in radians using exact values only.

$$2\sin x + \cot x = \csc x$$

8. Use your graphing calculator to find the solutions to the equation for θ if $0^\circ \le \theta < 360^\circ$ by graphing the function represented by the left side of the equation and then finding its zeros.

$$\cot\theta + 2\cos\theta\cot\theta = 0$$

9. Solving the following equation will require you to use the quadratic formula. Solve the equation for θ between 0° and 360°, and round your answers to the nearest tenth of a degree.

$$4\cos^2\theta + 2\sin\theta - 1 = 0$$

10. Find all solutions in the interval $0^\circ \le \theta < 360^\circ$. Use a calculator on the last step and write all answers to the nearest tenth of a degree.

$$3\sin\theta - 9 = 4\sin\theta - 5$$

11. Eliminate the parameter t from the following:

$$x = \cos t - 4 \quad y = \sin t + 7$$

Sketch the graph.

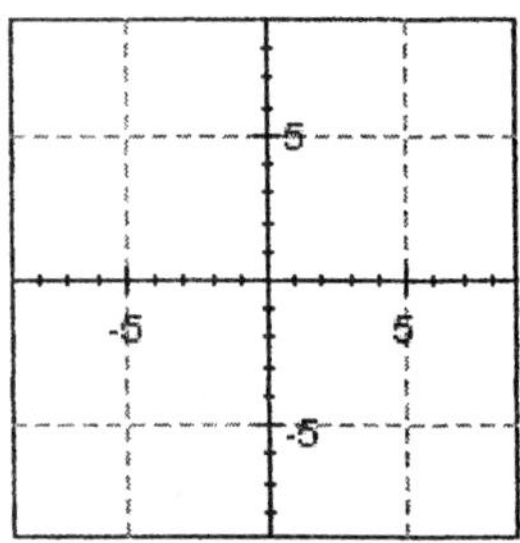

00

12. Eliminate the parameter t in the following:

$$x = 5 + 3\tan t \quad y = 2 + 3\sec t$$

13. Match each equation in the left column with the corresponding expression representing all solutions in the right column.

$$2\cos\theta = -\sqrt{3}$$

$$150° + 360°k, \ 210° + 360°k$$

$$\sqrt{2}\sin\theta = 1$$

$$45° + 360°k, \ 135° + 360°k$$

$$30° + 360°k, \ 330° + 360°k$$

$$225° + 360°k, \ 315° + 360°k$$

14. Solve for x, if $0 \le x < 2\pi$. Write your answers in exact values only.

$$\sin x + 2\sin x \cos x = 0$$

15. Eliminate the parameter t in the following:

$$x = -\cos t \quad y = \cos t$$

16. Find all solutions if $0 \leq x < 2\pi$. Give your answers in radians using exact values only.

$$\sin 2x \cos x + \cos 2x \sin x = 0$$

17. Find all solutions in radians using exact values only.

$$\sin^2 6x = 1$$

18. Match each equation in the left column with the corresponding expression representing all solutions in the right column.

$$\sin\left(5A - 40^\circ\right) = -\frac{1}{2} \qquad\qquad 20^\circ + 72^\circ k,\ 32^\circ + 72^\circ k$$

$$\sin\left(5A + 40^\circ\right) = -\frac{1}{2} \qquad\qquad 50^\circ + 72^\circ k,\ 74^\circ + 72^\circ k$$

$$34^\circ + 72^\circ k,\ 58^\circ + 72^\circ k$$

$$4^\circ + 72^\circ k,\ 16^\circ + 72^\circ k$$

19. Eliminate the parameter t from the following:

$$x = 9\sin t \quad y = 9\cos t$$

Sketch the graph.

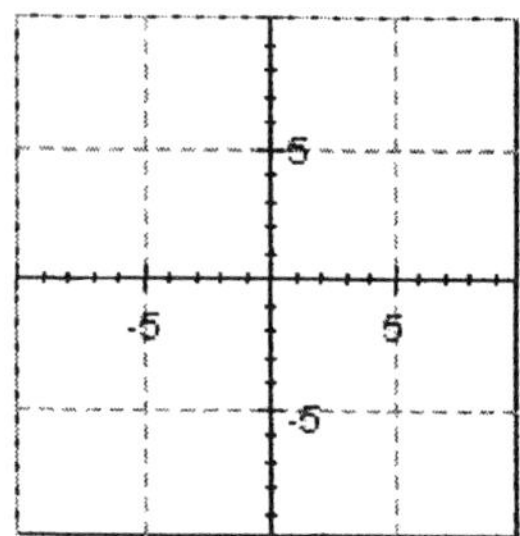

20. The formula below gives the relationship between the number of sides n , the radius r , and the length of each side l in a regular polygon. Find n , if $l = r\sqrt{2}$.

$$l = 2r\sin\frac{180°}{n}$$

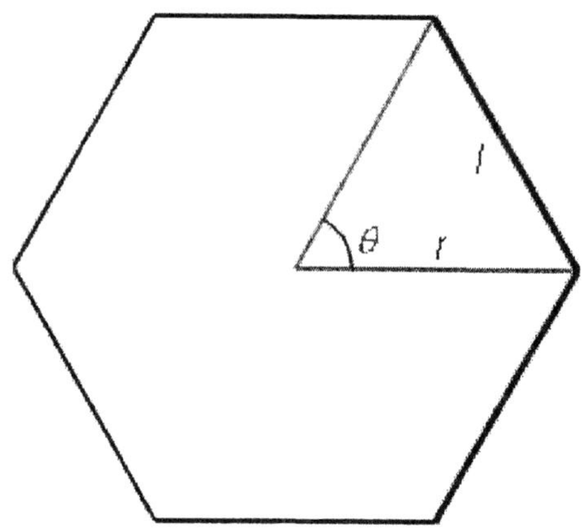

$n = \underline{\qquad}$

21. Eliminate the parameter t in the following:

$$x = 5 \sec t \quad y = 3 \tan t$$

22. Solve the equation for x if $0 \le x < 2\pi$. Give your answer in radians using exact values only.

$$2 \cos x - \tan x - \sec x = 0$$

23. Solve the equation for θ if $0° \le \theta < 360°$. Give your answer in degrees.

$$\sqrt{3} \csc \theta - 2 \cot \theta = 0$$

24. Solve for θ if $0° \le \theta < 360°$.

$$2 \cos^2 \theta + 3 \cos \theta = 2$$
00

25. Find all degree solutions for the following.

$$\cos 4\theta = -1$$

1. $\dfrac{3\pi}{4} + 2k\pi$

2. $180°, 240°$

3. $\dfrac{x^2}{16} + \dfrac{y^2}{25} = 1$

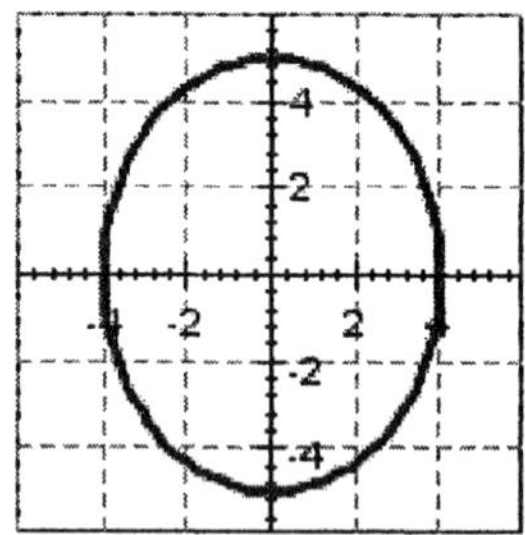

4. $30°, 150°, 270°$

5. $\theta = 21° + 36°k$ or $\theta = 27° + 36°k$ or $\theta = 33° + 36°k$

6. $30°, 330°$

7. $\dfrac{2\pi}{3}, \dfrac{4\pi}{3}$

8. $90°, 270°, 120°, 240°$

9. $220.6°, 319.4°$

10. no solution

11. $(x + 4)^2 + (y - 7)^2 = 1$

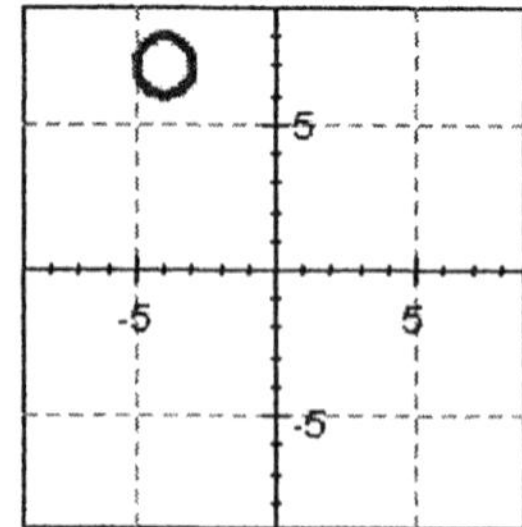

12. $\dfrac{(y-2)^2}{9} - \dfrac{(x-5)^2}{9} = 1$

13. $\sqrt{2}\sin\theta = 1 \;\rightarrow\; 45° + 360°k,\; 135° + 360°k,$

$2\cos\theta = -\sqrt{3} \;\rightarrow\; 150° + 360°k,\; 210° + 360°k$

14. $0,\; \pi,\; \dfrac{2\pi}{3},\; \dfrac{4\pi}{3}$

15. $y = -x,\; -1 \le x \le 1,\; -1 \le y \le 1$

16. $0,\; \dfrac{\pi}{3},\; \dfrac{2\pi}{3},\; \pi,\; \dfrac{4\pi}{3},\; \dfrac{5\pi}{3}$

17. $x = \dfrac{\pi}{12} + \dfrac{\pi k}{3}$ or $\dfrac{\pi}{4} + \dfrac{\pi k}{3}$

18. $\sin(5A - 40°) = -\dfrac{1}{2} \;\rightarrow\; 50° + 72°k,\; 74° + 72°k,$

$\sin(5A + 40°) = -\dfrac{1}{2} \;\rightarrow\; 34° + 72°k,\; 58° + 72°k$

19. $x^2 + y^2 = 81$

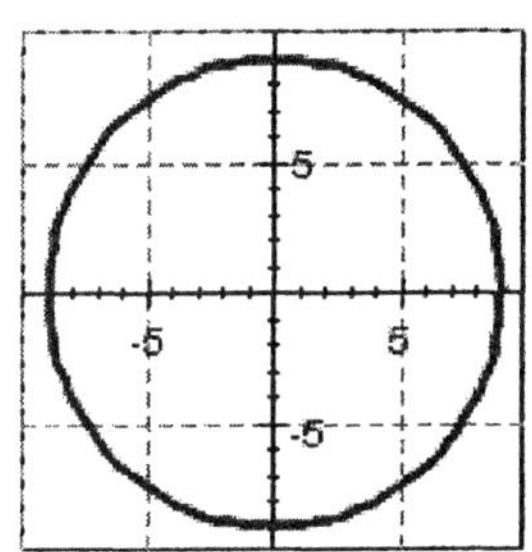

20. 4

21. $\dfrac{x^2}{25} - \dfrac{y^2}{9} = 1$

22. $\dfrac{\pi}{6},\; \dfrac{5\pi}{6}$

23. $30°,\; 330°$

24. $60°,\; 300°$

25. $\theta = 45° + 90°k$

McKeague/Turner - Trigonometry 5e Chapter 6 Form B

1. mctr.06.02.42_NoAlgs
2. mctr.06.02.26_NoAlgs
3. mctr.06.04.05_NoAlgs
4. mctr.06.02.12_NoAlgs
5. mctr.06.03.36_NoAlgs
6. mctr.06.01.03_NoAlgs
7. mctr.06.02.22_NoAlgs
8. mctr.06.01.60_NoAlgs
9. mctr.06.02.48_NoAlgs
10. mctr.06.01.15_NoAlgs
11. mctr.06.04.10_NoAlgs
12. mctr.06.04.20_NoAlgs
13. mctr.06.01.39_NoAlgs
14. mctr.06.01.23_NoAlgs
15. mctr.06.04.24_NoAlgs
16. mctr.06.03.25_NoAlgs
17. mctr.06.03.31_NoAlgs
18. mctr.06.01.52_NoAlgs
19. mctr.06.04.01_NoAlgs
20. mctr.06.03.53_NoAlgs
21. mctr.06.04.15_NoAlgs
22. mctr.06.02.21_NoAlgs
23. mctr.06.02.07_NoAlgs
24. mctr.06.01.31_NoAlgs
25. mctr.06.03.16_NoAlgs

1. Solve the equation for t if $0 \leq t < 2\pi$. Do not use a calculator.

$4 \sin t - 1 = 2 \sin t$

Select the correct answer.

a. $\dfrac{\pi}{4}, \dfrac{3\pi}{4}$

b. $\dfrac{\pi}{6}, \dfrac{5\pi}{6}$

c. $\dfrac{\pi}{3}, \dfrac{2\pi}{3}$

d. $\dfrac{5\pi}{4}, \dfrac{7\pi}{4}$

e. $\dfrac{5\pi}{6}, \dfrac{7\pi}{6}$

2. Solve for x, if $0 \leq x < 2\pi$.

$\cos x - 2 \sin x \cos x = 0$

Select the correct answer.

a. $\dfrac{\pi}{2}, \dfrac{3\pi}{2}, \dfrac{\pi}{6}, \dfrac{5\pi}{6}$

b. $\dfrac{\pi}{2}, \dfrac{3\pi}{2}, \dfrac{\pi}{3}, \dfrac{5\pi}{3}$

c. $0, \pi, \dfrac{\pi}{3}, \dfrac{5\pi}{3}$

d. $\dfrac{\pi}{2}, \dfrac{3\pi}{2}, \dfrac{7\pi}{6}, \dfrac{11\pi}{6}$

e. $0, \pi, \dfrac{\pi}{6}, \dfrac{5\pi}{6}$

3. Solve for θ if $0° \leq \theta < 360°$.

$$2 \sin^2 \theta - 9 \sin \theta = -4$$

Select the correct answer.

a. $60°, 120°$

b. $120°, 240°$

c. $210°, 330°$

d. $30°, 150°$

e. $60°, 300°$

4. Find all degree solutions to the equation.

$$\cos\left(2A - 50°\right) = -\frac{1}{2}$$

Select the correct answer.

a. $85° + 180°k, 145° + 180°k$

b. $35° + 180°k, 95° + 180°k$

c. $35° + 180°k, 145° + 180°k$

d. $85° + 180°k, 95° + 180°k$

e. $55° + 180°k, 85° + 180°k$

5. Find all degree solutions to the equation.

$$\sin\left(4A - 40°\right) = -\frac{1}{2}$$

Select the correct answer.

a. $25° + 90°k, 40° + 90°k$

b. $62.5° + 90°k, 72.5° + 90°k$

c. $42.5° + 90°k, 92.5° + 90°k$

d. $42.5° + 90°k, 72.5° + 90°k$

e. $62.5° + 90°k, 92.5° + 90°k$

6. Use your graphing calculator to find the solutions to the equation for θ if
    $0° \leq \theta < 360°$ by graphing the function represented by the left side of the equation and
    then finding its zeros.

    $\tan \theta + 2 \sin \theta \tan \theta = 0$

    Select the correct answer.

    a. $0°, 180°, 150°, 210°$
    b. $0°, 180°, 210°, 330°$
    c. $90°, 270°, 210°, 330°$
    d. $0°, 180°, 30°, 150°$
    e. $90°, 270°, 30°, 150°$

7. Solve the equation for θ if $0° \leq \theta < 360°$.

    $2 \sin \theta - \sin 2\theta = 0$

    Select the correct answer.

    a. $90°, 180°$
    b. $60°, 300°$
    c. $210°, 330°$
    d. $0°, 180°$
    e. $0°$

8. Solve the equation for x if $0 \leq x < 2\pi$.

$$1 - \cos x - 2\sin^2 x = 0$$

Select the correct answer.

a. $\dfrac{\pi}{6},\ \dfrac{5\pi}{6},\ \dfrac{3\pi}{2}$

b. $0,\ \dfrac{2\pi}{3},\ \dfrac{4\pi}{3}$

c. $\dfrac{\pi}{3},\ \pi,\ \dfrac{5\pi}{3}$

d. $\dfrac{\pi}{2},\ \dfrac{7\pi}{6},\ \dfrac{11\pi}{6}$

e. $\dfrac{\pi}{6},\ \dfrac{\pi}{2},\ \dfrac{5\pi}{6}$

9. Solve the equation for x if $0 \leq x < 2\pi$.

$$2\cos x + \tan x - \sec x = 0$$

Select the correct answer.

a. $\dfrac{7\pi}{6},\ \dfrac{11\pi}{6}$

b. $\dfrac{2\pi}{3},\ \dfrac{4\pi}{3}$

c. $\dfrac{\pi}{6},\ \dfrac{5\pi}{6}$

d. $\dfrac{\pi}{3},\ \dfrac{5\pi}{3}$

e. $\dfrac{\pi}{6},\ \dfrac{5\pi}{6},\ \dfrac{3\pi}{2}$

10. Solve the equation for x if $0 \leq x < 2\pi$.

$$2 \sin x - \cot x = \csc x$$

Select the correct answer.

a. $\dfrac{2\pi}{3}$, $\dfrac{4\pi}{3}$

b. $\dfrac{\pi}{6}$, $\dfrac{5\pi}{6}$

c. $\dfrac{\pi}{3}$, $\dfrac{5\pi}{3}$

d. 0, $\dfrac{2\pi}{3}$, $\dfrac{4\pi}{3}$

e. $\dfrac{7\pi}{6}$, $\dfrac{11\pi}{6}$

11. Solve the equation for θ if $0° \leq \theta < 360°$.

$$\cos \theta - \sqrt{3} \sin \theta = -\sqrt{3}$$

Select the correct answer.

a. $0°$, $60°$

b. $240°$, $300°$

c. $120°$, $180°$

d. $90°$, $150°$

e. $30°$, $330°$

12. Write an expression that gives all solutions to the equation.

$$\cos x - \sin x = -\sqrt{2}$$

Select the correct answer.

a. $\dfrac{5\pi}{4} + 2k\pi$

b. $\dfrac{3\pi}{4} + 2k\pi$

c. $\dfrac{7\pi}{4} + 2k\pi$

d. $\dfrac{5\pi}{4} + k\pi$

e. $\dfrac{3\pi}{4} + k\pi$

13. Solving the following equation will require you to use the quadratic formula. Solve the equation for θ between $0°$ and $360°$, and round your answers to the nearest tenth of a degree.

$$3\cos^2\theta + 2\sin\theta - 1 = 0$$

Select the correct answer.

a. $213.7°, 326.3°$

b. $213.1°, 326.9°$

c. $212.5°, 327.5°$

d. $213.9°, 326.1°$

e. $213.3°, 326.7°$

14. Find all degree solutions for the following.

$$\cos 5\theta = -1$$

Select the correct answer.

a. $\theta = 34° + 72°k$

b. $\theta = 36° + 72°k$

c. $\theta = 41° + 72°k$

d. $\theta = 26° + 72°k$

e. $\theta = 39° + 72°k$

15. Find all solutions if $0 \le x < 2\pi$.

$$\sin 2x \cos x + \cos 2x \sin x = \frac{1}{2}$$

Select the correct answer.

a. $\dfrac{\pi}{18}, \dfrac{5\pi}{18}, \dfrac{13\pi}{18}, \dfrac{17\pi}{18}, \dfrac{25\pi}{18}, \dfrac{29\pi}{18}$

b. $\dfrac{\pi}{6}, \dfrac{5\pi}{6}, \dfrac{3\pi}{2}$

c. $\dfrac{\pi}{9}, \dfrac{2\pi}{9}, \dfrac{7\pi}{9}, \dfrac{8\pi}{9}, \dfrac{13\pi}{9}, \dfrac{14\pi}{9}$

d. $0, \dfrac{\pi}{3}, \dfrac{2\pi}{3}, \pi, \dfrac{4\pi}{3}, \dfrac{5\pi}{3}$

e. $\dfrac{\pi}{12}, \dfrac{\pi}{4}, \dfrac{3\pi}{4}, \dfrac{11\pi}{12}, \dfrac{17\pi}{12}, \dfrac{19\pi}{12}$

16. Find all solutions in radians using exact values only.

$$\sin^2 5x = 1$$

Select the correct answer.

a. $\quad x = \dfrac{\pi}{12} + \dfrac{2\pi k}{5} \text{ or } \dfrac{\pi}{4} + \dfrac{2\pi k}{5}$

b. $\quad x = \dfrac{\pi}{10} + \dfrac{2\pi k}{5} \text{ or } \dfrac{3\pi}{10} + \dfrac{2\pi k}{5}$

c. $\quad x = \dfrac{\pi}{12} + \dfrac{\pi k}{3} \text{ or } \dfrac{\pi}{4} + \dfrac{\pi k}{3}$

d. $\quad x = \dfrac{\pi}{10} + \dfrac{\pi k}{3} \text{ or } \dfrac{3\pi}{10} + \dfrac{\pi k}{3}$

e. $\quad x = \dfrac{\pi}{18} + \dfrac{2\pi k}{9}$

17. Find all degree solutions.

$$2\sin^2 10\theta + \sin 10\theta - 1 = 0$$

Select the correct answer.

a. $\quad \theta = 3° + 60°k \text{ or } \theta = 15° + 60°k \text{ or } \theta = 27° + 36°k$

b. $\quad \theta = 5° + 60°k \text{ or } \theta = 25° + 60°k \text{ or } \theta = 45° + 60°k$

c. $\quad \theta = 60° + 360°k$

d. $\quad \theta = 3° + 36°k \text{ or } \theta = 15° + 36°k \text{ or } \theta = 27° + 36°k$

e. $\quad \theta = 5° + 36°k \text{ or } \theta = 25° + 36°k \text{ or } \theta = 45° + 36°k$

18. Find all solutions if $0° \le \theta < 360°$. Round each of the solutions to the nearest tenth if necessary.

$$\sin^2 2\theta - 7\sin 2\theta - 4 = 0$$

Select the correct answer.

a. $\quad 106.0°, 164.0°, 286.0°, 344.0°$

b. $\quad 106.0°, 164.0°, 282.3°, 349.8°$

c. $\quad 107.2°, 166.1°, 286.0°, 344.0°$

d. $\quad 106.0°, 166.1°, 282.3°, 344.0°$

e. $\quad 107.2°, 164.0°, 286.0°, 349.8°$

19. If central angle θ cuts off a chord of length c in a circle of radius r, then the relationship between θ, c, and r is given by

$$2r \sin \frac{\theta}{2} = c$$

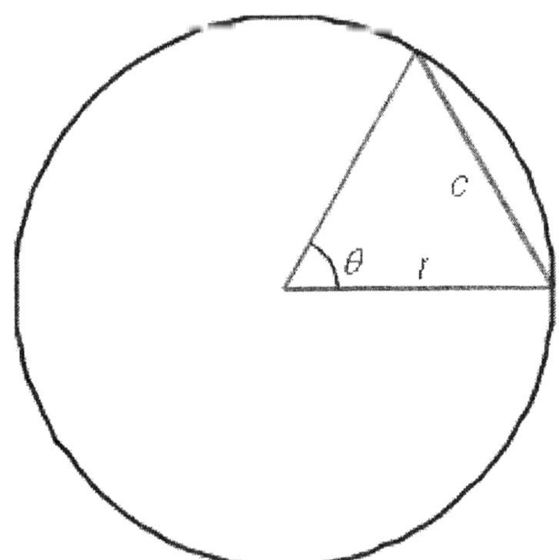

Find θ, if $c = r$.

Select the correct answer.

a. $\theta = 60°, 270°$

b. $\theta = 30°, 270°$

c. $\theta = 30°, 300°$

d. $\theta = 60°, 180°$

e. $\theta = 60°, 300°$

20. Eliminate the parameter t in the following:

$$x = 5\sin t \quad y = 6\sin t$$

Select the correct answer.

a. $6x = -5y, \ -1 \le x \le 1, \ -6 \le y \le 6$

b. $3x = 6y, \ -3 \le x \le 3, \ -1 \le y \le 1$

c. $5x = 6y, \ -5 \le x \le 5, \ -1 \le y \le 1$

d. $6x = 3y, \ -1 \le x \le 1, \ -1 \le y \le 1$

e. $6x = 5y, \ -5 \le x \le 5, \ -6 \le y \le 6$

21. Eliminate the parameter *t* from the following and then sketch the graph:

$$x = 2\cos t \quad y = 2\sin t$$

Select the correct answer.

a. $x^2 - y^2 = 5$

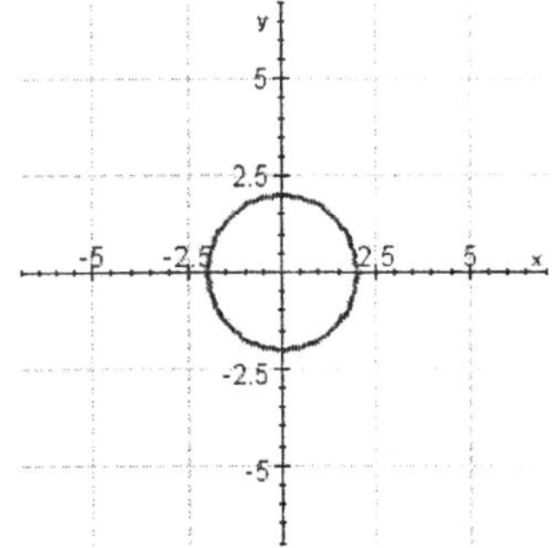

b. $x^2 + y^2 = 5$

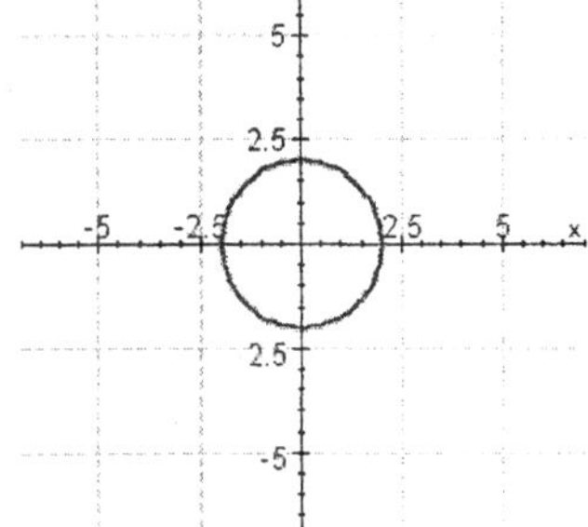

c. $x^2 + y^2 = 4$

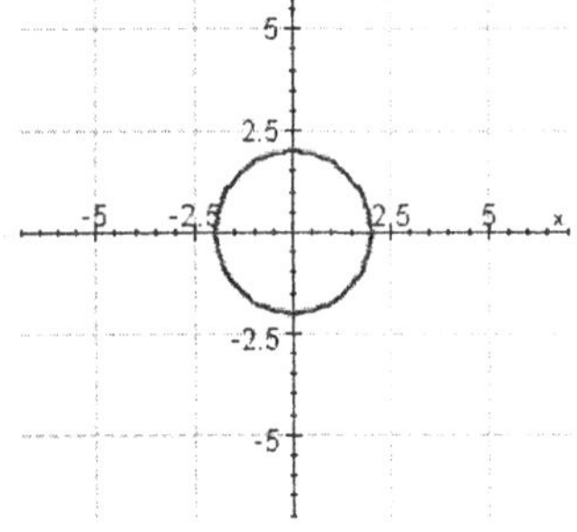

22. Eliminate the parameter t from the following and then sketch the graph:

$$x = 5 + \sin t \quad y = 4 + \cos t$$

Select the correct answer.

a. $(x - 5)^2 + (y - 4)^2 = 1$

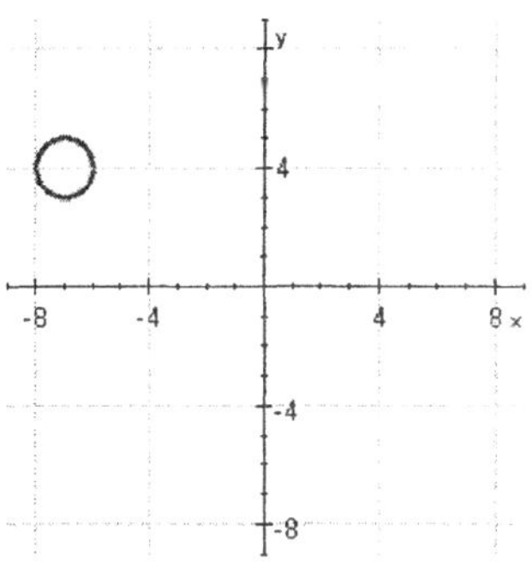

b. $(x + 5)^2 + (y + 4)^2 = 1$

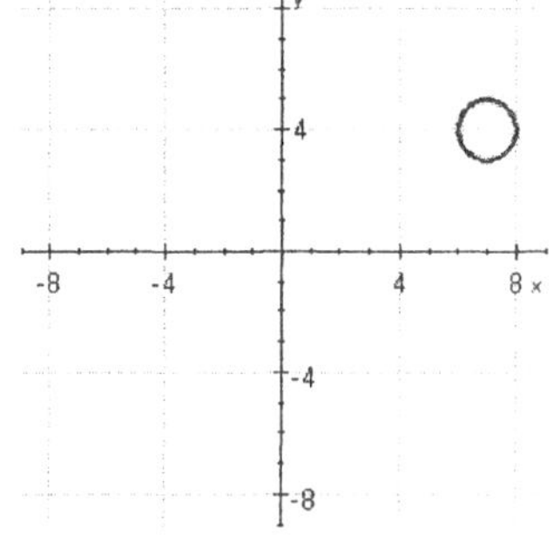

c. $(x - 6)^2 + (y - 4)^2 = 1$

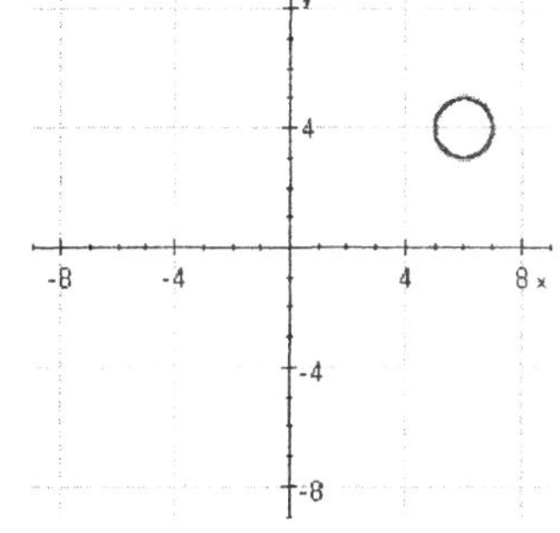

d. $(x - 5)^2 + (y - 4)^2 = 1$

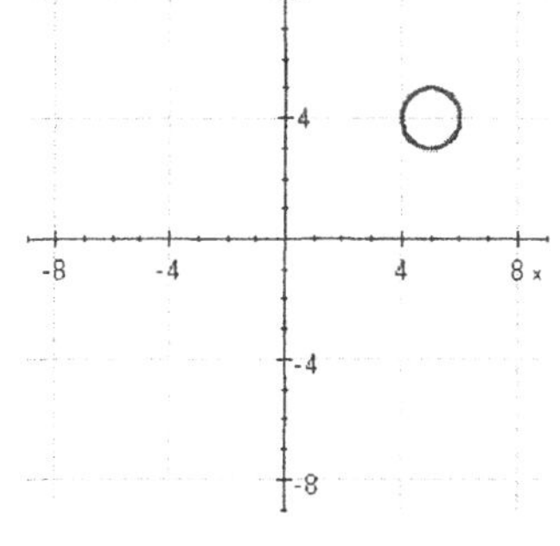

23. Eliminate the parameter t from the following and then sketch the graph:

$$x = 2\cos t - 3 \quad y = 2\sin t + 2$$

Select the correct answer.

a. $\dfrac{(x+4)^2}{4} + \dfrac{(y-2)^2}{4} = 1$

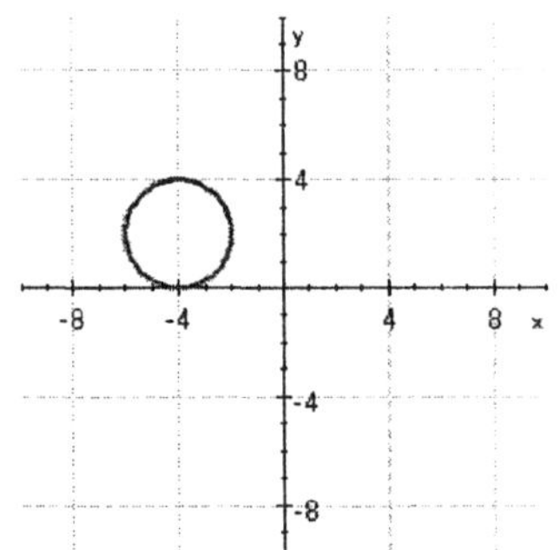

b. $\dfrac{(x+3)^2}{4} - \dfrac{(y-2)^2}{4} = 1$

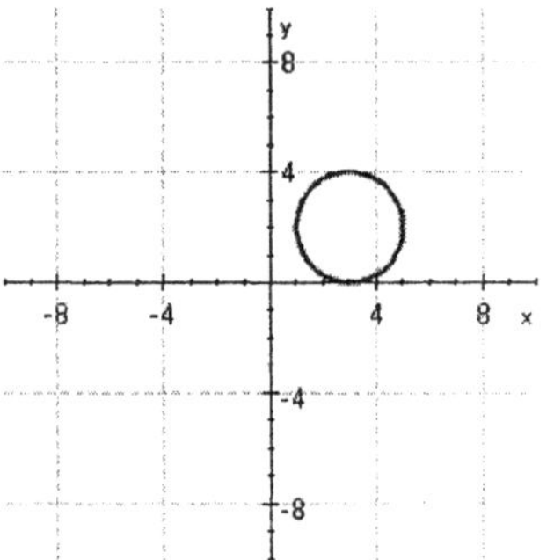

c. $\dfrac{(x+4)^2}{4} + \dfrac{(y-3)^2}{4} = 1$

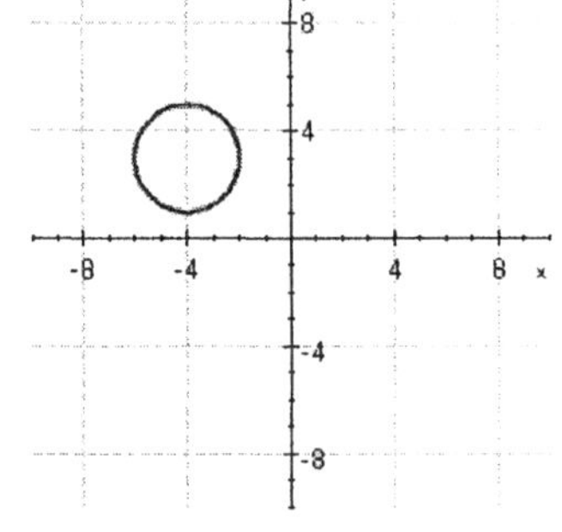

d. $\dfrac{(x+3)^2}{4} + \dfrac{(y-2)^2}{4} = 1$

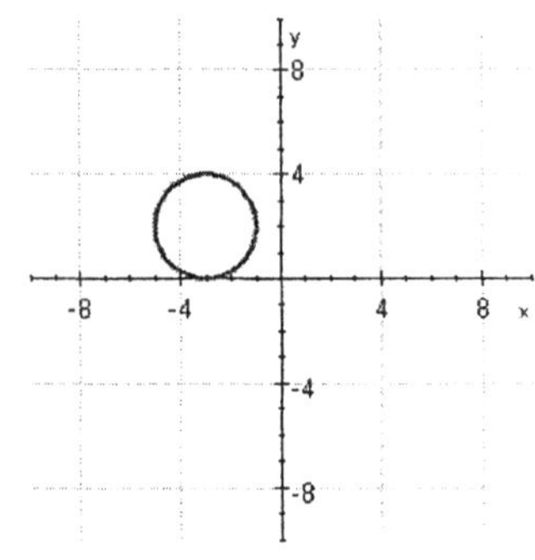

24. Eliminate the parameter t in the following:

$$x = 8 + 3\tan t \quad y = 4 + 3\sec t$$

Select the correct answer.

a. $$\frac{(y-4)^2}{9} - \frac{(x-8)^2}{10} = 1$$

b. $$\frac{(y-4)^2}{9} - \frac{(x-9)^2}{11} = 1$$

c. $$\frac{(y-3)^2}{9} + \frac{(x-8)^2}{9} = 1$$

d. $$\frac{(y-3)^2}{9} + \frac{(x-8)^2}{10} = 1$$

e. $$\frac{(y-4)^2}{9} - \frac{(x-8)^2}{9} = 1$$

25. Eliminate the parameter t in the following:

$$x = -\cos 2t \quad y = \cos t$$

Select the correct answer.

a. $x = 2y^2 - 1, \quad -1 \le x \le 1, \quad -1 \le y \le 1$

b. $x = 2y^2 + 1, \quad -1 \le x \le 1, \quad -1 \le y \le 1$

c. $x = -2y^2 + 1, \quad -1 \le x \le 1, \quad -1 \le y \le 1$

d. $x = 2y + 1, \quad -1 \le x \le 1, \quad -1 \le y \le 1$

e. $x = -2y + 1, \quad -1 \le x \le 1, \quad -1 \le y \le 1$

1. b

2. a

3. d

4. a

5. e

6. b

7. d

8. b

9. a

10. c

11. d

12. b

13. e

14. b

15. a

16. b

17. d

18. a

19. e

20. e

21. c

22. d

23. d

24. e

25. c

1. mctr.06.01.07m_NoAlgs
2. mctr.06.01.23m_NoAlgs
3. mctr.06.01.32m_NoAlgs
4. mctr.06.01.51m_NoAlgs
5. mctr.06.01.52m_NoAlgs
6. mctr.06.01.60m_NoAlgs
7. mctr.06.02.10m_NoAlgs
8. mctr.06.02.17m_NoAlgs
9. mctr.06.02.21m_NoAlgs
10. mctr.06.02.22m_NoAlgs
11. mctr.06.02.26m_NoAlgs
12. mctr.06.02.42m_NoAlgs
13. mctr.06.02.48m_NoAlgs
14. mctr.06.03.16m_NoAlgs
15. mctr.06.03.25m_NoAlgs
16. mctr.06.03.31m_NoAlgs
17. mctr.06.03.35m_NoAlgs
18. mctr.06.03.45m_NoAlgs
19. mctr.06.03.54m_NoAlgs
20. mctr.06.04.25m_NoAlgs
21. mctr.06.04.04m_NoAlgs
22. mctr.06.04.08m_NoAlgs
23. mctr.06.04.13m_NoAlgs
24. mctr.06.04.19m_NoAlgs
25. mctr.06.04.22m_NoAlgs

1. If central angle θ cuts off a chord of length c in a circle of radius r, then the relationship between θ, c, and r is given by

$$2r \sin \frac{\theta}{2} = c$$

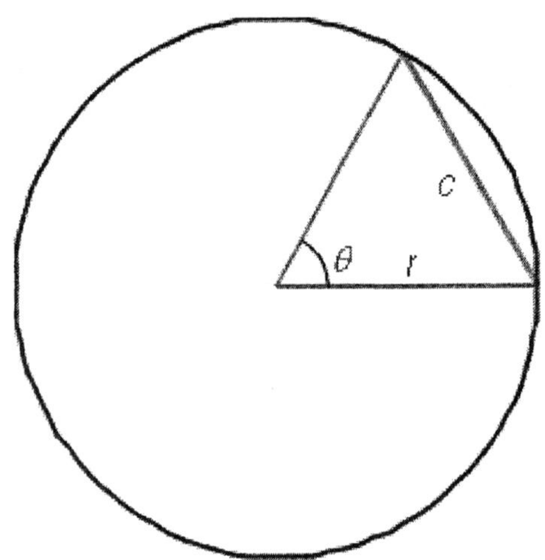

Find θ, if $c = r$.

Select the correct answer.

a. $\theta = 60°, 270°$

b. $\theta = 30°, 270°$

c. $\theta = 30°, 300°$

d. $\theta = 60°, 180°$

e. $\theta = 60°, 300°$

2. Find all degree solutions to the equation.

$$\cos\left(2A - 50°\right) = -\frac{1}{2}$$

Select the correct answer.

a. $85° + 180°k, 145° + 180°k$

b. $35° + 180°k, 95° + 180°k$

c. $35° + 180°k, 145° + 180°k$

d. $85° + 180°k, 95° + 180°k$

e. $55° + 180°k, 85° + 180°k$

3. Eliminate the parameter *t* from the following and then sketch the graph:

$$x = 5 + \sin t \quad y = 4 + \cos t$$

Select the correct answer.

a. $(x - 5)^2 + (y - 4)^2 = 1$

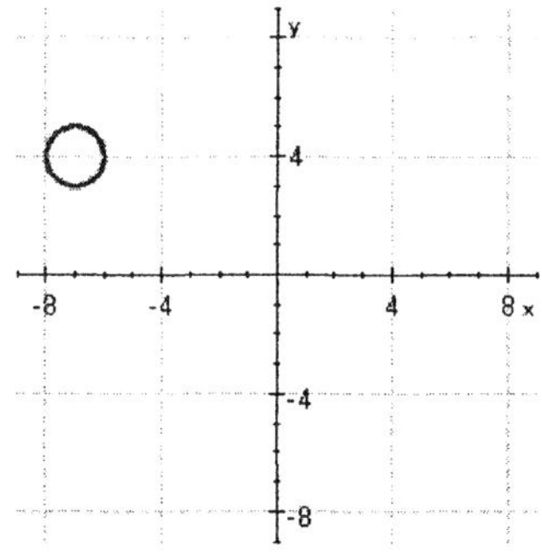

b. $(x + 5)^2 + (y + 4)^2 = 1$

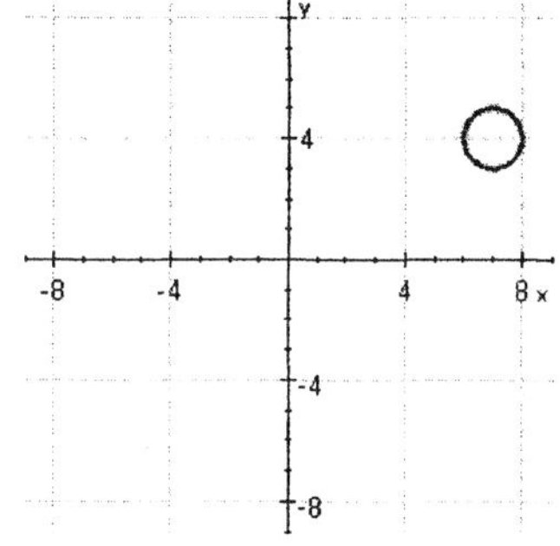

c. $(x - 6)^2 + (y - 4)^2 = 1$

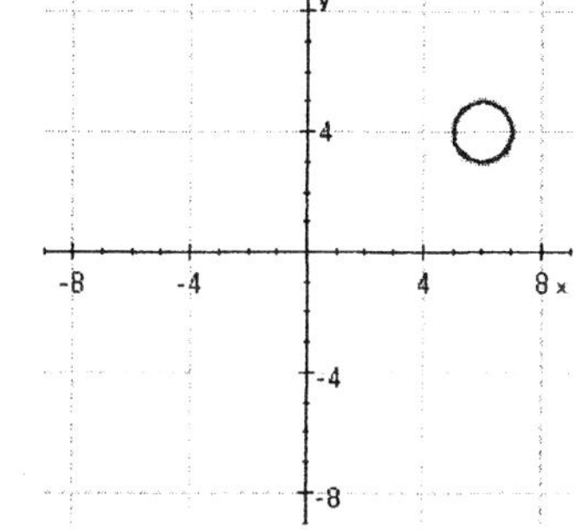

d. $(x - 5)^2 + (y - 4)^2 = 1$

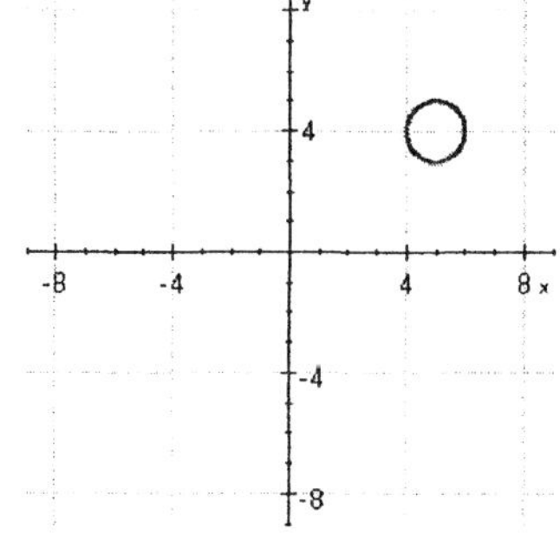

4. Solve the equation for θ if $0° \le \theta < 360°$.

$$2 \sin \theta - \sin 2\theta = 0$$

Select the correct answer.

    a. $90°, 180°$

    b. $60°, 300°$

    c. $210°, 330°$

    d. $0°, 180°$

    e. $0°$

5. Use your graphing calculator to find the solutions to the equation for θ if $0° \le \theta < 360°$ by graphing the function represented by the left side of the equation and then finding its zeros.

$$\tan \theta + 2 \sin \theta \tan \theta = 0$$

Select the correct answer.

    a. $0°, 180°, 150°, 210°$

    b. $0°, 180°, 210°, 330°$

    c. $90°, 270°, 210°, 330°$

    d. $0°, 180°, 30°, 150°$

    e. $90°, 270°, 30°, 150°$

6. Solve for θ if $0° \le \theta < 360°$.

$$2 \sin^2 \theta - 9 \sin \theta = -4$$

Select the correct answer.

    a. $60°, 120°$

    b. $120°, 240°$

    c. $210°, 330°$

    d. $30°, 150°$

    e. $60°, 300°$

7. Write an expression that gives all solutions to the equation.

$$\cos x - \sin x = -\sqrt{2}$$

Select the correct answer.

a. $\dfrac{5\pi}{4} + 2k\pi$

b. $\dfrac{3\pi}{4} + 2k\pi$

c. $\dfrac{7\pi}{4} + 2k\pi$

d. $\dfrac{5\pi}{4} + k\pi$

e. $\dfrac{3\pi}{4} + k\pi$

8. Find all degree solutions.

$$2\sin^2 10\theta + \sin 10\theta - 1 = 0$$

Select the correct answer.

a. $\theta = 3° + 60°k$ or $\theta = 15° + 60°k$ or $\theta = 27° + 36°k$

b. $\theta = 5° + 60°k$ or $\theta = 25° + 60°k$ or $\theta = 45° + 60°k$

c. $\theta = 60° + 360°k$

d. $\theta = 3° + 36°k$ or $\theta = 15° + 36°k$ or $\theta = 27° + 36°k$

e. $\theta = 5° + 36°k$ or $\theta = 25° + 36°k$ or $\theta = 45° + 36°k$

9. Eliminate the parameter *t* from the following and then sketch the graph:

$$x = 2\cos t \quad y = 2\sin t$$

Select the correct answer.

a. $x^2 - y^2 = 5$

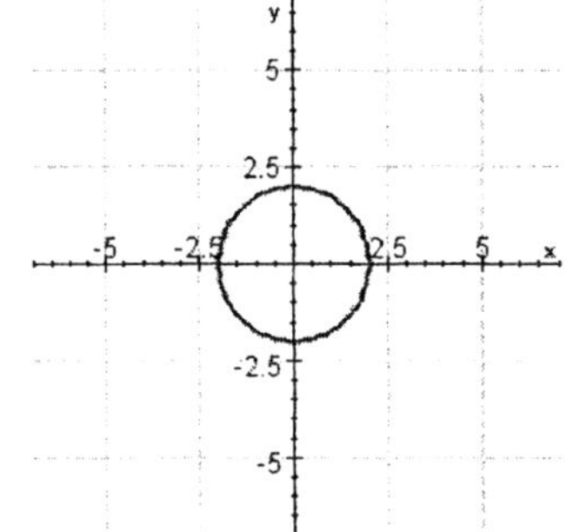

b. $x^2 + y^2 = 5$

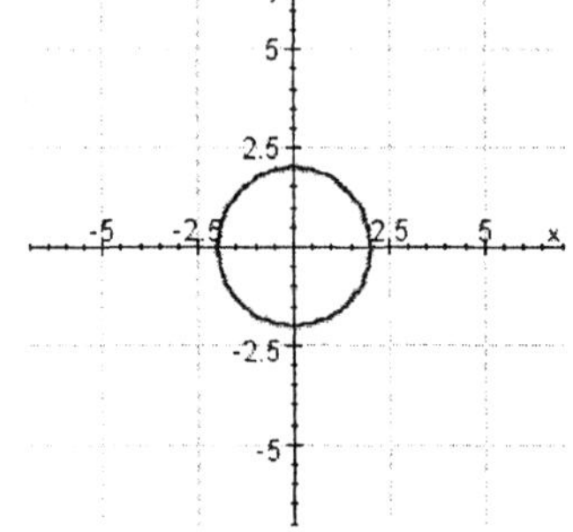

c. $x^2 + y^2 = 4$

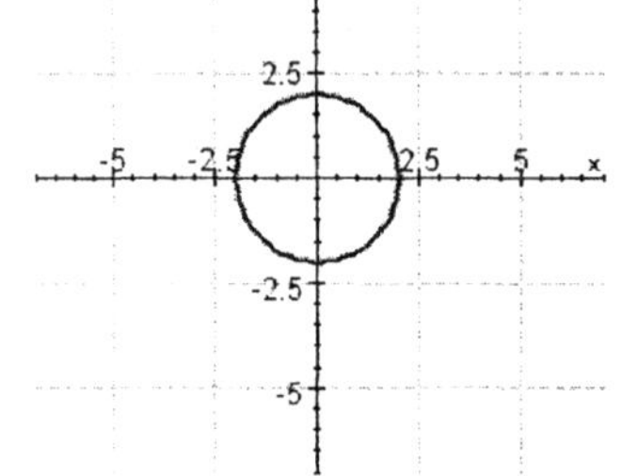

d. $x^2 - y^2 = 9$

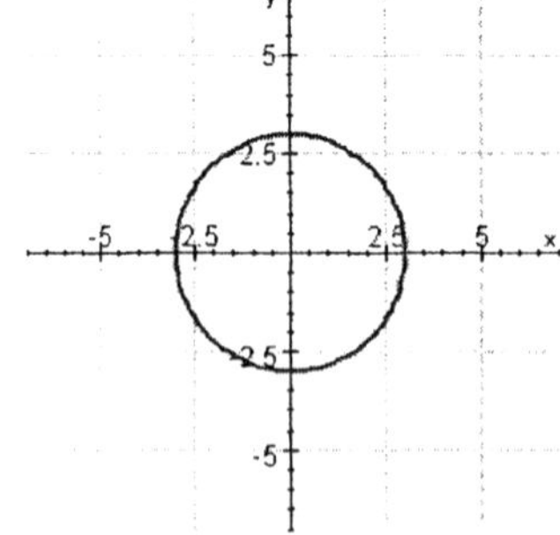

10. Solve the equation for x if $0 \leq x < 2\pi$.

$$2 \sin x - \cot x = \csc x$$

Select the correct answer.

a. $\quad \dfrac{2\pi}{3}, \dfrac{4\pi}{3}$

b. $\quad \dfrac{\pi}{6}, \dfrac{5\pi}{6}$

c. $\quad \dfrac{\pi}{3}, \dfrac{5\pi}{3}$

d. $\quad 0, \dfrac{2\pi}{3}, \dfrac{4\pi}{3}$

e. $\quad \dfrac{7\pi}{6}, \dfrac{11\pi}{6}$

11. Eliminate the parameter t in the following:

$$x = 8 + 3\tan t \quad y = 4 + 3\sec t$$

Select the correct answer.

a. $\quad \dfrac{(y-4)^2}{9} - \dfrac{(x-8)^2}{10} = 1$

b. $\quad \dfrac{(y-4)^2}{9} - \dfrac{(x-9)^2}{11} = 1$

c. $\quad \dfrac{(y-3)^2}{9} + \dfrac{(x-8)^2}{9} = 1$

d. $\quad \dfrac{(y-3)^2}{9} + \dfrac{(x-8)^2}{10} = 1$

e. $\quad \dfrac{(y-4)^2}{9} - \dfrac{(x-8)^2}{9} = 1$

12. Eliminate the parameter t in the following:

$$x = -\cos 2t \quad y = \cos t$$

Select the correct answer.

a. $\quad x = 2y^2 - 1, \quad -1 \leq x \leq 1, \quad -1 \leq y \leq 1$

b. $\quad x = 2y^2 + 1, \quad -1 \leq x \leq 1, \quad -1 \leq y \leq 1$

c. $\quad x = -2y^2 + 1, \quad -1 \leq x \leq 1, \quad -1 \leq y \leq 1$

d. $\quad x = 2y + 1, \quad -1 \leq x \leq 1, \quad -1 \leq y \leq 1$

e. $\quad x = -2y + 1, \quad -1 \leq x \leq 1, \quad -1 \leq y \leq 1$

13. Find all degree solutions for the following.

$$\cos 5\theta = -1$$

Select the correct answer.

a. $\quad \theta = 34° + 72°k$

b. $\quad \theta = 36° + 72°k$

c. $\quad \theta = 41° + 72°k$

d. $\quad \theta = 26° + 72°k$

e. $\quad \theta = 39° + 72°k$

14. Solve the equation for x if $0 \leq x < 2\pi$.

$$1 - \cos x - 2\sin^2 x = 0$$

Select the correct answer.

a. $\dfrac{\pi}{6}, \dfrac{5\pi}{6}, \dfrac{3\pi}{2}$

b. $0, \dfrac{2\pi}{3}, \dfrac{4\pi}{3}$

c. $\dfrac{\pi}{3}, \pi, \dfrac{5\pi}{3}$

d. $\dfrac{\pi}{2}, \dfrac{7\pi}{6}, \dfrac{11\pi}{6}$

e. $\dfrac{\pi}{6}, \dfrac{\pi}{2}, \dfrac{5\pi}{6}$

15. Find all solutions in radians using exact values only.

$$\sin^2 5x = 1$$

Select the correct answer.

a. $x = \dfrac{\pi}{12} + \dfrac{2\pi k}{5}$ or $\dfrac{\pi}{4} + \dfrac{2\pi k}{5}$

b. $x = \dfrac{\pi}{10} + \dfrac{2\pi k}{5}$ or $\dfrac{3\pi}{10} + \dfrac{2\pi k}{5}$

c. $x = \dfrac{\pi}{12} + \dfrac{\pi k}{3}$ or $\dfrac{\pi}{4} + \dfrac{\pi k}{3}$

d. $x = \dfrac{\pi}{10} + \dfrac{\pi k}{3}$ or $\dfrac{3\pi}{10} + \dfrac{\pi k}{3}$

e. $x = \dfrac{\pi}{18} + \dfrac{2\pi k}{9}$

16. Find all solutions if $0 \leq x < 2\pi$.

$$\sin 2x \cos x + \cos 2x \sin x = \frac{1}{2}$$

Select the correct answer.

a. $\dfrac{\pi}{18}, \dfrac{5\pi}{18}, \dfrac{13\pi}{18}, \dfrac{17\pi}{18}, \dfrac{25\pi}{18}, \dfrac{29\pi}{18}$

b. $\dfrac{\pi}{6}, \dfrac{5\pi}{6}, \dfrac{3\pi}{2}$

c. $\dfrac{\pi}{9}, \dfrac{2\pi}{9}, \dfrac{7\pi}{9}, \dfrac{8\pi}{9}, \dfrac{13\pi}{9}, \dfrac{14\pi}{9}$

d. $0, \dfrac{\pi}{3}, \dfrac{2\pi}{3}, \pi, \dfrac{4\pi}{3}, \dfrac{5\pi}{3}$

e. $\dfrac{\pi}{12}, \dfrac{\pi}{4}, \dfrac{3\pi}{4}, \dfrac{11\pi}{12}, \dfrac{17\pi}{12}, \dfrac{19\pi}{12}$

17. Solve the equation for t if $0 \leq t < 2\pi$. Do not use a calculator.

$$4 \sin t - 1 = 2 \sin t$$

Select the correct answer.

a. $\dfrac{\pi}{4}, \dfrac{3\pi}{4}$

b. $\dfrac{\pi}{6}, \dfrac{5\pi}{6}$

c. $\dfrac{\pi}{3}, \dfrac{2\pi}{3}$

d. $\dfrac{5\pi}{4}, \dfrac{7\pi}{4}$

e. $\dfrac{5\pi}{6}, \dfrac{7\pi}{6}$

18. Solve the equation for θ if $0° \leq \theta < 360°$.

$$\cos\theta - \sqrt{3}\sin\theta = -\sqrt{3}$$

Select the correct answer.

a. $0°$, $60°$

b. $240°$, $300°$

c. $120°$, $180°$

d. $90°$, $150°$

e. $30°$, $330°$

19. Eliminate the parameter t in the following:

$$x = 5\sin t \quad y = 6\sin t$$

Select the correct answer.

a. $6x = -5y$, $\ -1 \leq x \leq 1$, $\ -6 \leq y \leq 6$

b. $3x = 6y$, $\ -3 \leq x \leq 3$, $\ -1 \leq y \leq 1$

c. $5x = 6y$, $\ -5 \leq x \leq 5$, $\ -1 \leq y \leq 1$

d. $6x = 3y$, $\ -1 \leq x \leq 1$, $\ -1 \leq y \leq 1$

e. $6x = 5y$, $\ -5 \leq x \leq 5$, $\ -6 \leq y \leq 6$

20. Find all degree solutions to the equation.

$$\sin\left(4A - 40°\right) = -\frac{1}{2}$$

Select the correct answer.

a. $25° + 90°k$, $40° + 90°k$

b. $62.5° + 90°k$, $72.5° + 90°k$

c. $42.5° + 90°k$, $92.5° + 90°k$

d. $42.5° + 90°k$, $72.5° + 90°k$

e. $62.5° + 90°k$, $92.5° + 90°k$

21. Solving the following equation will require you to use the quadratic formula. Solve the equation for θ between $0°$ and $360°$, and round your answers to the nearest tenth of a degree.

$$3\cos^2\theta + 2\sin\theta - 1 = 0$$

Select the correct answer.

a. $213.7°, 326.3°$

b. $213.1°, 326.9°$

c. $212.5°, 327.5°$

d. $213.9°, 326.1°$

e. $213.3°, 326.7°$

22. Solve the equation for x if $0 \le x < 2\pi$.

$$2\cos x + \tan x - \sec x = 0$$

Select the correct answer.

a. $\dfrac{7\pi}{6}, \dfrac{11\pi}{6}$

b. $\dfrac{2\pi}{3}, \dfrac{4\pi}{3}$

c. $\dfrac{\pi}{6}, \dfrac{5\pi}{6}$

d. $\dfrac{\pi}{3}, \dfrac{5\pi}{3}$

e. $\dfrac{\pi}{6}, \dfrac{5\pi}{6}, \dfrac{3\pi}{2}$

23. Find all solutions if $0° \leq \theta < 360°$. Round each of the solutions to the nearest tenth if necessary.

$$\sin^2 2\theta - 7\sin 2\theta - 4 = 0$$

Select the correct answer.

a. $106.0°, 164.0°, 286.0°, 344.0°$

b. $106.0°, 164.0°, 282.3°, 349.8°$

c. $107.2°, 166.1°, 286.0°, 344.0°$

d. $106.0°, 166.1°, 282.3°, 344.0°$

e. $107.2°, 164.0°, 286.0°, 349.8°$

24. Solve for x, if $0 \leq x < 2\pi$.

$$\cos x - 2\sin x \cos x = 0$$

Select the correct answer.

a. $\dfrac{\pi}{2}, \dfrac{3\pi}{2}, \dfrac{\pi}{6}, \dfrac{5\pi}{6}$

b. $\dfrac{\pi}{2}, \dfrac{3\pi}{2}, \dfrac{\pi}{3}, \dfrac{5\pi}{3}$

c. $0, \pi, \dfrac{\pi}{3}, \dfrac{5\pi}{3}$

d. $\dfrac{\pi}{2}, \dfrac{3\pi}{2}, \dfrac{7\pi}{6}, \dfrac{11\pi}{6}$

e. $0, \pi, \dfrac{\pi}{6}, \dfrac{5\pi}{6}$

25. Eliminate the parameter t from the following and then sketch the graph:

$$x = 2\cos t - 3 \quad y = 2\sin t + 2$$

Select the correct answer.

a. $\quad \dfrac{(x+4)^2}{4} + \dfrac{(y-2)^2}{4} = 1$

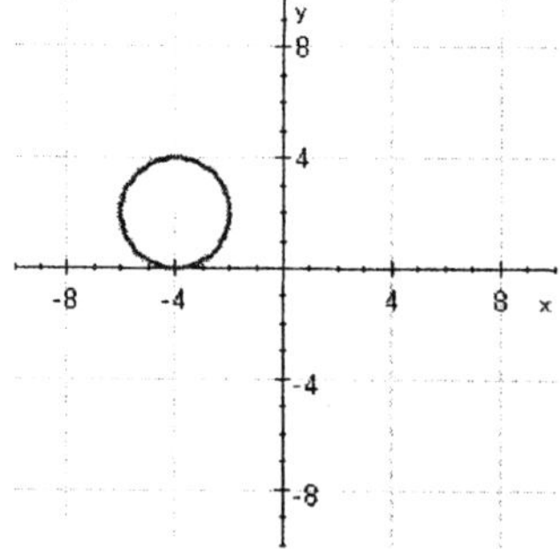

b. $\quad \dfrac{(x+3)^2}{4} - \dfrac{(y-2)^2}{4} = 1$

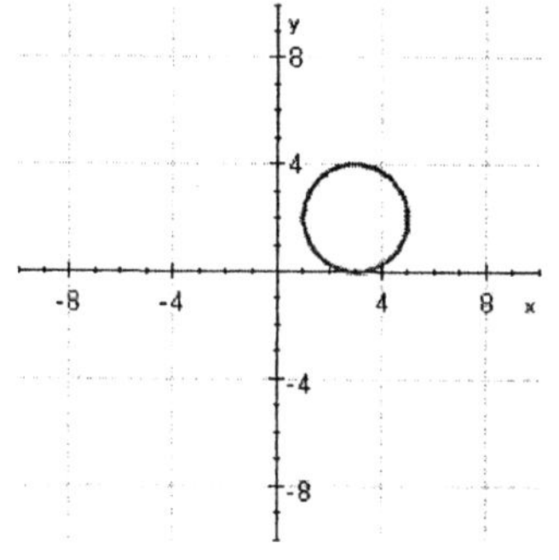

c. $\quad \dfrac{(x+4)^2}{4} + \dfrac{(y-3)^2}{4} = 1$

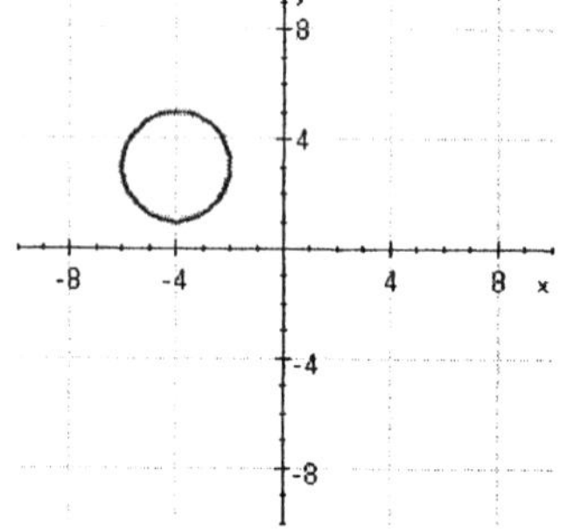

d. $\quad \dfrac{(x+3)^2}{4} + \dfrac{(y-2)^2}{4} = 1$

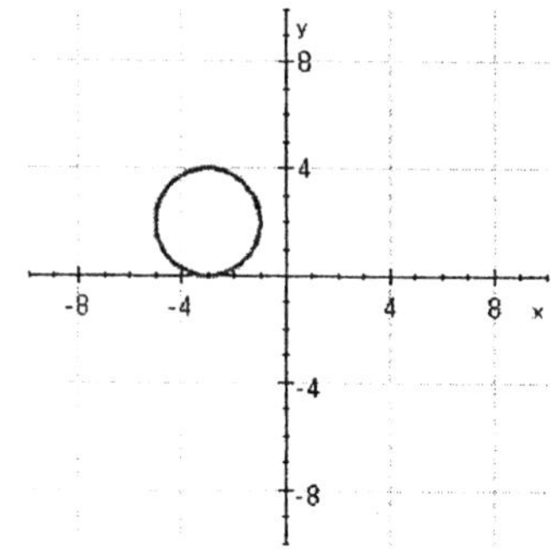

McKeague/Turner - Trigonometry 5e Chapter 6 Form D

1. e

2. a

3. d

4. d

5. b

6. d

7. b

8. d

9. c

10. c

11. e

12. c

13. b

14. b

15. b

16. a

17. b

18. d

19. e

20. e

21. e

22. a

23. a

24. a

25. d

McKeague/Turner - Trigonometry 5e Chapter 6 Form D

1. mctr.06.03.54m_NoAlgs
2. mctr.06.01.51m_NoAlgs
3. mctr.06.04.08m_NoAlgs
4. mctr.06.02.10m_NoAlgs
5. mctr.06.01.60m_NoAlgs
6. mctr.06.01.32m_NoAlgs
7. mctr.06.02.42m_NoAlgs
8. mctr.06.03.35m_NoAlgs
9. mctr.06.04.04m_NoAlgs
10. mctr.06.02.22m_NoAlgs
11. mctr.06.04.19m_NoAlgs
12. mctr.06.04.22m_NoAlgs
13. mctr.06.03.16m_NoAlgs
14. mctr.06.02.17m_NoAlgs
15. mctr.06.03.31m_NoAlgs
16. mctr.06.03.25m_NoAlgs
17. mctr.06.01.07m_NoAlgs
18. mctr.06.02.26m_NoAlgs
19. mctr.06.04.25m_NoAlgs
20. mctr.06.01.52m_NoAlgs
21. mctr.06.02.48m_NoAlgs
22. mctr.06.02.21m_NoAlgs
23. mctr.06.03.45m_NoAlgs
24. mctr.06.01.23m_NoAlgs
25. mctr.06.04.13m_NoAlgs

1. Find all solutions in the interval $0° \leq \theta < 360°$. Use a calculator on the last step and write all answers to the nearest tenth of a degree.

$$3 \sin \theta - 9 = 4 \sin \theta - 5$$

2. Solve for x, if $0 \leq x < 2\pi$.

$$\cos x - 2 \sin x \cos x = 0$$

Select the correct answer.

a. $\dfrac{\pi}{2}, \dfrac{3\pi}{2}, \dfrac{\pi}{6}, \dfrac{5\pi}{6}$

b. $\dfrac{\pi}{2}, \dfrac{3\pi}{2}, \dfrac{\pi}{3}, \dfrac{5\pi}{3}$

c. $0, \pi, \dfrac{\pi}{3}, \dfrac{5\pi}{3}$

d. $\dfrac{\pi}{2}, \dfrac{3\pi}{2}, \dfrac{7\pi}{6}, \dfrac{11\pi}{6}$

e. $0, \pi, \dfrac{\pi}{6}, \dfrac{5\pi}{6}$

3. Solve for θ if $0° \leq \theta < 360°$.

$$2 \sin^2 \theta - 9 \sin \theta = -4$$

Select the correct answer.

a. $60°, 120°$

b. $120°, 240°$

c. $210°, 330°$

d. $30°, 150°$

e. $60°, 300°$

4. Match each equation in the left column with the corresponding expression representing all solutions in the right column.

$$2\cos\theta = -\sqrt{3}$$

$$150° + 360°k,\ 210° + 360°k$$

$$\sqrt{2}\sin\theta = 1$$

$$45° + 360°k,\ 135° + 360°k$$

$$30° + 360°k,\ 330° + 360°k$$

$$225° + 360°k,\ 315° + 360°k$$

5. Find all degree solutions to the equation.

$$\cos\left(2A - 50°\right) = -\frac{1}{2}$$

Select the correct answer.

   a. $85° + 180°k,\ 145° + 180°k$
   b. $35° + 180°k,\ 95° + 180°k$
   c. $35° + 180°k,\ 145° + 180°k$
   d. $85° + 180°k,\ 95° + 180°k$
   e. $55° + 180°k,\ 85° + 180°k$

6. Use your graphing calculator to find the solutions to the equation for θ if $0° \le \theta < 360°$ by graphing the function represented by the left side of the equation and then finding its zeros.

$$\cot\theta + 2\cos\theta\cot\theta = 0$$

7. Solve the equation for θ if $0° \le \theta < 360°$. Give your answer in degrees.

$$\sqrt{3}\csc\theta - 2\cot\theta = 0$$

8. Solve the equation for θ if $0° \leq \theta < 360°$. Give your answer in degrees.

$$2 \sin \theta + 1 = \csc \theta$$

9. Solve the equation for x if $0 \leq x < 2\pi$.

$$1 - \cos x - 2 \sin^2 x = 0$$

Select the correct answer.

a. $\dfrac{\pi}{6}, \dfrac{5\pi}{6}, \dfrac{3\pi}{2}$

b. $0, \dfrac{2\pi}{3}, \dfrac{4\pi}{3}$

c. $\dfrac{\pi}{3}, \pi, \dfrac{5\pi}{3}$

d. $\dfrac{\pi}{2}, \dfrac{7\pi}{6}, \dfrac{11\pi}{6}$

e. $\dfrac{\pi}{6}, \dfrac{\pi}{2}, \dfrac{5\pi}{6}$

10. Solve the equation for x if $0 \leq x < 2\pi$. Give your answer in radians using exact values only.

$$2 \cos x - \tan x - \sec x = 0$$

11. Solve the equation for θ if $0° \leq \theta < 360°$.

$$\cos \theta - \sqrt{3} \sin \theta = -\sqrt{3}$$

Select the correct answer.

    a. $0°, 60°$

    b. $240°, 300°$

    c. $120°, 180°$

    d. $90°, 150°$

    e. $30°, 330°$

12. Write an expression that gives all solutions to the equation.

$$\sin x - \cos x = \sqrt{2}$$

13. Solving the following equation will require you to use the quadratic formula. Solve the equation for θ between $0°$ and $360°$, and round your answers to the nearest tenth of a degree.

$$3\cos^2 \theta + 2\sin \theta - 1 = 0$$

Select the correct answer.

    a. $213.7°, 326.3°$

    b. $213.1°, 326.9°$

    c. $212.5°, 327.5°$

    d. $213.9°, 326.1°$

    e. $213.3°, 326.7°$

14. Find all degree solutions for the following.

$$\cos 4\theta = -1$$

15. Find all solutions if $0 \leq x < 2\pi$.

$$\sin 2x \cos x + \cos 2x \sin x = \frac{1}{2}$$

Select the correct answer.

a. $\dfrac{\pi}{18}, \dfrac{5\pi}{18}, \dfrac{13\pi}{18}, \dfrac{17\pi}{18}, \dfrac{25\pi}{18}, \dfrac{29\pi}{18}$

b. $\dfrac{\pi}{6}, \dfrac{5\pi}{6}, \dfrac{3\pi}{2}$

c. $\dfrac{\pi}{9}, \dfrac{2\pi}{9}, \dfrac{7\pi}{9}, \dfrac{8\pi}{9}, \dfrac{13\pi}{9}, \dfrac{14\pi}{9}$

d. $0, \dfrac{\pi}{3}, \dfrac{2\pi}{3}, \pi, \dfrac{4\pi}{3}, \dfrac{5\pi}{3}$

e. $\dfrac{\pi}{12}, \dfrac{\pi}{4}, \dfrac{3\pi}{4}, \dfrac{11\pi}{12}, \dfrac{17\pi}{12}, \dfrac{19\pi}{12}$

16. Find all solutions in radians using exact values only.

$$\sin^2 6x = 1$$

17. Find all degree solutions.

$$2\sin^2 10\theta + \sin 10\theta - 1 = 0$$

Select the correct answer.

a. $\theta = 3° + 60°k$ or $\theta = 15° + 60°k$ or $\theta = 27° + 36°k$

b. $\theta = 5° + 60°k$ or $\theta = 25° + 60°k$ or $\theta = 45° + 60°k$

c. $\theta = 60° + 360°k$

d. $\theta = 3° + 36°k$ or $\theta = 15° + 36°k$ or $\theta = 27° + 36°k$

e. $\theta = 5° + 36°k$ or $\theta = 25° + 36°k$ or $\theta = 45° + 36°k$

18. The formula below gives the relationship between the number of sides n, the radius r, and the length of each side l in a regular polygon. Find n, if $l = r\sqrt{2}$.

$$l = 2r\sin\frac{180°}{n}$$

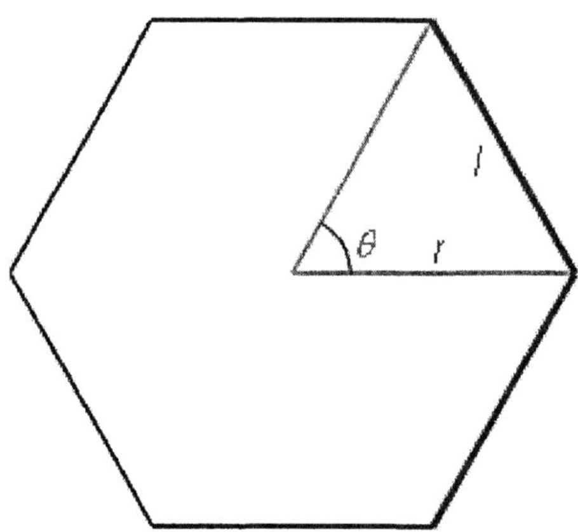

$n =$ _________

19. If central angle θ cuts off a chord of length c in a circle of radius r, then the relationship between θ, c, and r is given by

$$2r \sin \frac{\theta}{2} = c$$

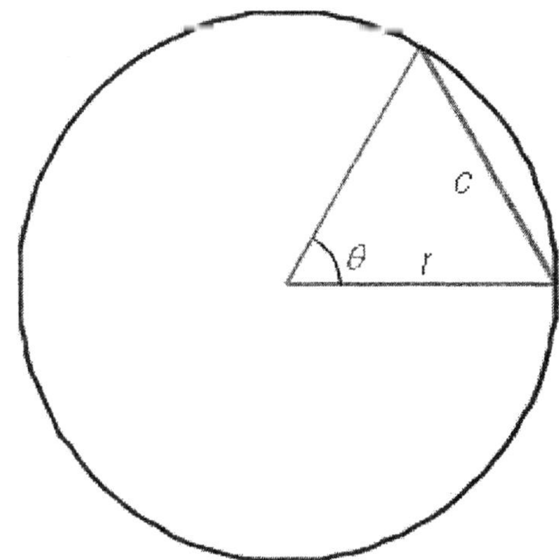

Find θ, if $c = r$.

Select the correct answer.

a. $\theta = 60°, 270°$

b. $\theta = 30°, 270°$

c. $\theta = 30°, 300°$

d. $\theta = 60°, 180°$

e. $\theta = 60°, 300°$

20. Eliminate the parameter t from the following:

$$x = 4\sin t \quad y = 5\cos t$$

Sketch the graph.

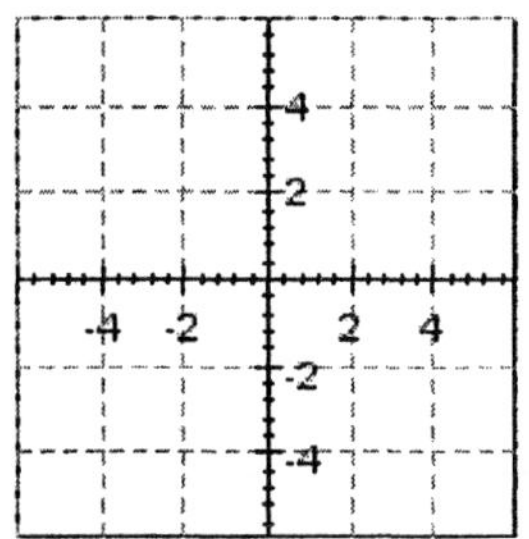

21. Eliminate the parameter t in the following:

$$x = 5\sin t \quad y = 6\sin t$$

Select the correct answer.

a. $\quad 6x = -5y, \quad -1 \le x \le 1, \quad -6 \le y \le 6$

b. $\quad 3x = 6y, \quad -3 \le x \le 3, \quad -1 \le y \le 1$

c. $\quad 5x = 6y, \quad -5 \le x \le 5, \quad -1 \le y \le 1$

d. $\quad 6x = 3y, \quad -1 \le x \le 1, \quad -1 \le y \le 1$

e. $\quad 6x = 5y, \quad -5 \le x \le 5, \quad -6 \le y \le 6$

22. Eliminate the parameter t from the following and then sketch the graph:

$$x = 5 + \sin t \quad y = 4 + \cos t$$

Select the correct answer.

a. $(x - 5)^2 + (y - 4)^2 = 1$

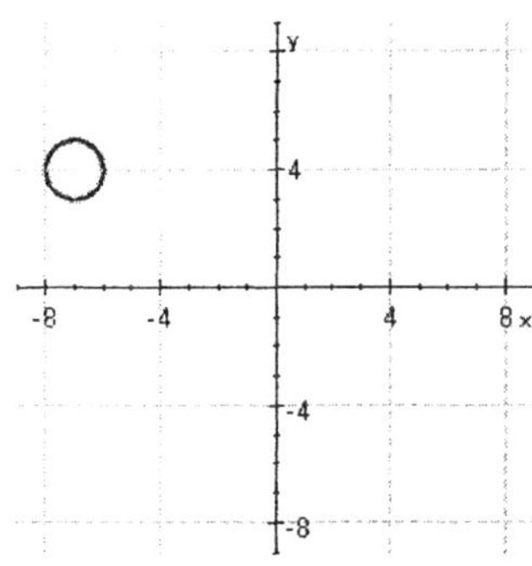

b. $(x + 5)^2 + (y + 4)^2 = 1$

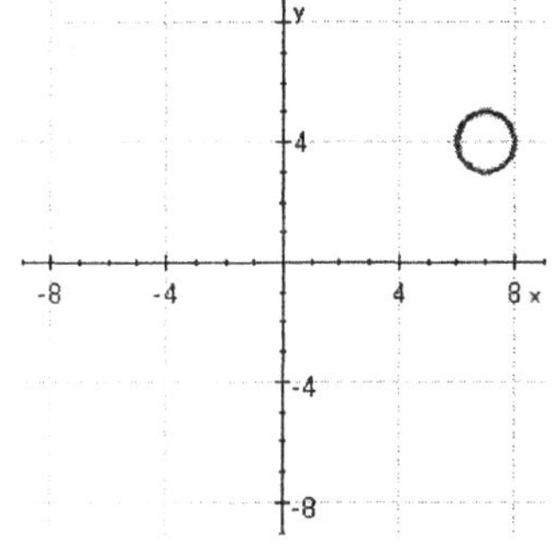

c. $(x - 6)^2 + (y - 4)^2 = 1$

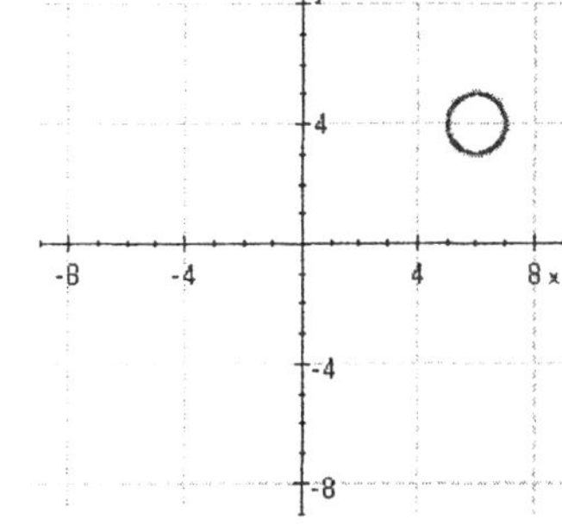

d. $(x - 5)^2 + (y - 4)^2 = 1$

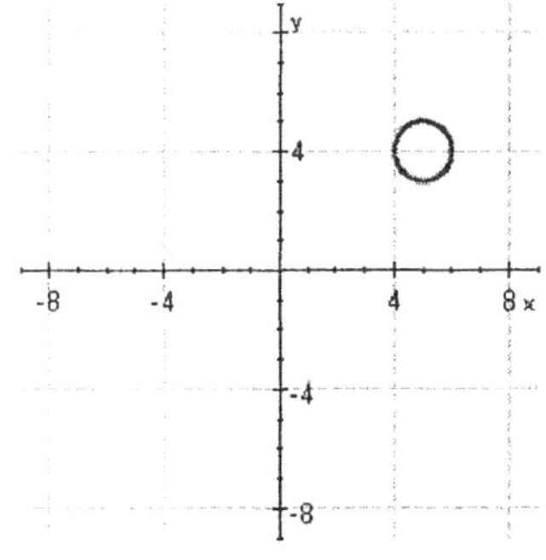

23. Eliminate the parameter t in the following:

$$x = 5\sec t \quad y = 3\tan t$$

24. Eliminate the parameter t in the following:

$$x = 8 + 3\tan t \quad y = 4 + 3\sec t$$

Select the correct answer.

a. $\quad \dfrac{(y-4)^2}{9} - \dfrac{(x-8)^2}{10} = 1$

b. $\quad \dfrac{(y-4)^2}{9} - \dfrac{(x-9)^2}{11} = 1$

c. $\quad \dfrac{(y-3)^2}{9} + \dfrac{(x-8)^2}{9} = 1$

d. $\quad \dfrac{(y-3)^2}{9} + \dfrac{(x-8)^2}{10} = 1$

e. $\quad \dfrac{(y-4)^2}{9} - \dfrac{(x-8)^2}{9} = 1$

25. Eliminate the parameter t in the following:

$$x = -\cos t \quad y = \cos t$$

1. no solution

2. a

3. d

4. $\sqrt{2} \sin \theta = 1 \rightarrow 45° + 360°k,\ 135° + 360°k,$
   $2 \cos \theta = -\sqrt{3} \rightarrow 150° + 360°k,\ 210° + 360°k$

5. a

6. $90°,\ 270°,\ 120°,\ 240°$

7. $30°,\ 330°$

8. $30°,\ 150°,\ 270°$

9. b

10. $\dfrac{\pi}{6},\ \dfrac{5\pi}{6}$

11. d

12. $\dfrac{3\pi}{4} + 2k\pi$

13. e

14. $\theta = 45° + 90°k$

15. a

16. $x = \dfrac{\pi}{12} + \dfrac{\pi k}{3}$ or $\dfrac{\pi}{4} + \dfrac{\pi k}{3}$

17. d

18. 4

19. e

20. $\dfrac{x^2}{16} + \dfrac{y^2}{25} = 1$

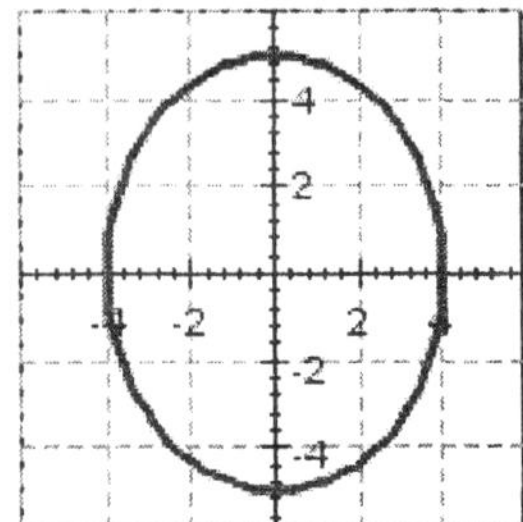

21. e

22. d

23. $\dfrac{x^2}{25} - \dfrac{y^2}{9} = 1$

24. e

25. $y = -x, \ -1 \le x \le 1, \ -1 \le y \le 1$

1. mctr.06.01.15_NoAlgs
2. mctr.06.01.23m_NoAlgs
3. mctr.06.01.32m_NoAlgs
4. mctr.06.01.39_NoAlgs
5. mctr.06.01.51m_NoAlgs
6. mctr.06.01.60_NoAlgs
7. mctr.06.02.07_NoAlgs
8. mctr.06.02.12_NoAlgs
9. mctr.06.02.17m_NoAlgs
10. mctr.06.02.21_NoAlgs
11. mctr.06.02.26m_NoAlgs
12. mctr.06.02.42_NoAlgs
13. mctr.06.02.48m_NoAlgs
14. mctr.06.03.16_NoAlgs
15. mctr.06.03.25m_NoAlgs
16. mctr.06.03.31_NoAlgs
17. mctr.06.03.35m_NoAlgs
18. mctr.06.03.53_NoAlgs
19. mctr.06.03.54m_NoAlgs
20. mctr.06.04.05_NoAlgs
21. mctr.06.04.25m_NoAlgs
22. mctr.06.04.08m_NoAlgs
23. mctr.06.04.15_NoAlgs
24. mctr.06.04.19m_NoAlgs
25. mctr.06.04.24_NoAlgs

1. Solve for x, if $0 \leq x < 2\pi$.

$$\cos x - 2 \sin x \cos x = 0$$

Select the correct answer.

    a. $\dfrac{\pi}{2}, \dfrac{3\pi}{2}, \dfrac{\pi}{6}, \dfrac{5\pi}{6}$

    b. $\dfrac{\pi}{2}, \dfrac{3\pi}{2}, \dfrac{\pi}{3}, \dfrac{5\pi}{3}$

    c. $0, \pi, \dfrac{\pi}{3}, \dfrac{5\pi}{3}$

    d. $\dfrac{\pi}{2}, \dfrac{3\pi}{2}, \dfrac{7\pi}{6}, \dfrac{11\pi}{6}$

    e. $0, \pi, \dfrac{\pi}{6}, \dfrac{5\pi}{6}$

2. Solve for θ if $0° \leq \theta < 360°$.

$$2 \sin^2 \theta - 9 \sin \theta = -4$$

Select the correct answer.

    a. $60°, 120°$

    b. $120°, 240°$

    c. $210°, 330°$

    d. $30°, 150°$

    e. $60°, 300°$

3. Eliminate the parameter t from the following and then sketch the graph:

$$x = 5 + \sin t \quad y = 4 + \cos t$$

Select the correct answer.

a. $(x - 5)^2 + (y - 4)^2 = 1$

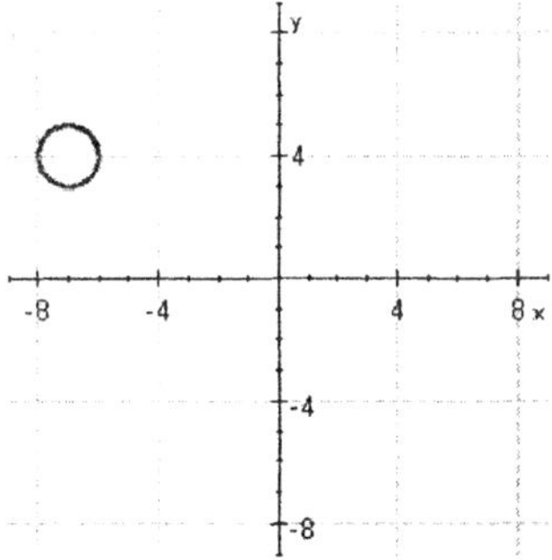

b. $(x + 5)^2 + (y + 4)^2 = 1$

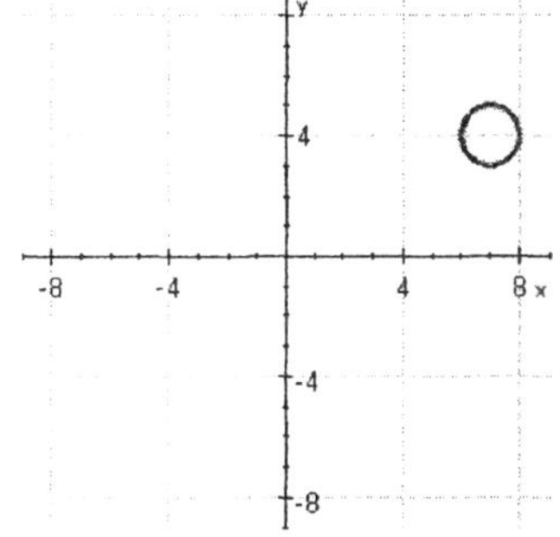

c. $(x - 6)^2 + (y - 4)^2 = 1$

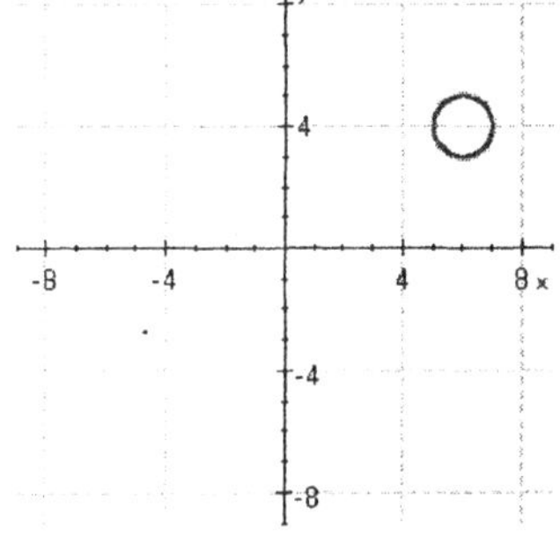

d. $(x - 5)^2 + (y - 4)^2 = 1$

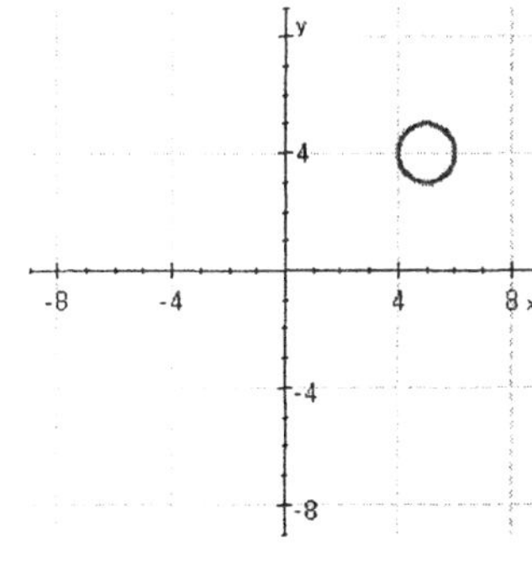

4. Eliminate the parameter t in the following:

$$x = 8 + 3\tan t \quad y = 4 + 3\sec t$$

Select the correct answer.

a. $\dfrac{(y-4)^2}{9} - \dfrac{(x-8)^2}{10} = 1$

b. $\dfrac{(y-4)^2}{9} - \dfrac{(x-9)^2}{11} = 1$

c. $\dfrac{(y-3)^2}{9} + \dfrac{(x-8)^2}{9} = 1$

d. $\dfrac{(y-3)^2}{9} + \dfrac{(x-8)^2}{10} = 1$

e. $\dfrac{(y-4)^2}{9} - \dfrac{(x-8)^2}{9} = 1$

5. Find all degree solutions to the equation.

$$\cos\left(2A - 50°\right) = -\frac{1}{2}$$

Select the correct answer.

a. $85° + 180°k, \ 145° + 180°k$

b. $35° + 180°k, \ 95° + 180°k$

c. $35° + 180°k, \ 145° + 180°k$

d. $85° + 180°k, \ 95° + 180°k$

e. $55° + 180°k, \ 85° + 180°k$

6. Eliminate the parameter t in the following:

$$x = -\cos t \quad y = \cos t$$

7. Eliminate the parameter t in the following:

$$x = 5 \sec t \quad y = 3 \tan t$$

8. Solve the equation for x if $0 \le x < 2\pi$.

$$1 - \cos x - 2 \sin^2 x = 0$$

Select the correct answer.

a. $\quad \dfrac{\pi}{6}, \ \dfrac{5\pi}{6}, \ \dfrac{3\pi}{2}$

b. $\quad 0, \ \dfrac{2\pi}{3}, \ \dfrac{4\pi}{3}$

c. $\quad \dfrac{\pi}{3}, \ \pi, \ \dfrac{5\pi}{3}$

d. $\quad \dfrac{\pi}{2}, \ \dfrac{7\pi}{6}, \ \dfrac{11\pi}{6}$

e. $\quad \dfrac{\pi}{6}, \ \dfrac{\pi}{2}, \ \dfrac{5\pi}{6}$

9. Write an expression that gives all solutions to the equation.

$$\sin x - \cos x = \sqrt{2}$$

10. Find all solutions in radians using exact values only.

$$\sin^2 6x = 1$$

11. Eliminate the parameter t from the following:

$$x = 4 \sin t \quad y = 5 \cos t$$

Sketch the graph.

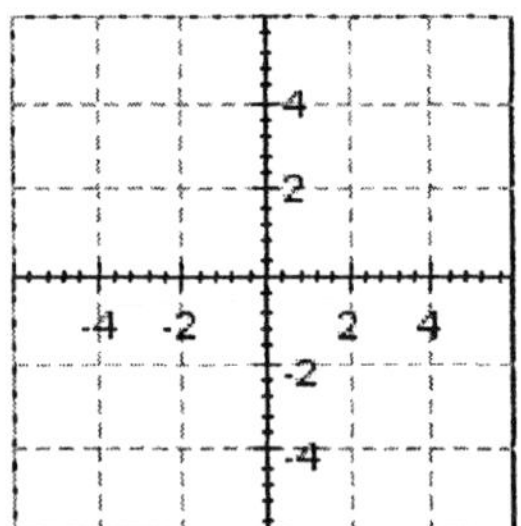

12. If central angle θ cuts off a chord of length c in a circle of radius r, then the relationship between θ, c, and r is given by

$$2r \sin \frac{\theta}{2} = c$$

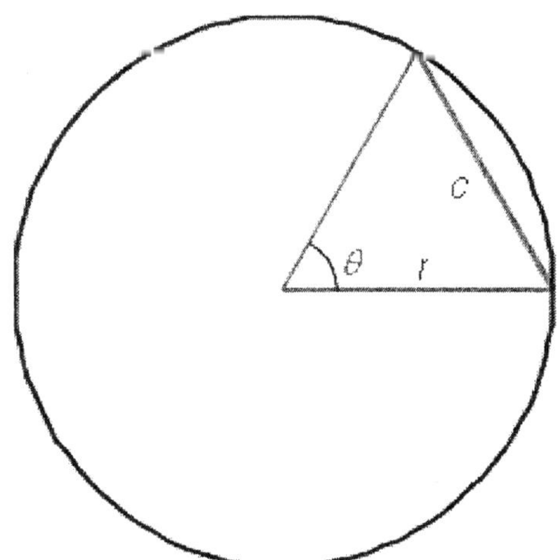

Find θ, if $c = r$.

Select the correct answer.

   a. $\theta = 60°, 270°$

   b. $\theta = 30°, 270°$

   c. $\theta = 30°, 300°$

   d. $\theta = 60°, 180°$

   e. $\theta = 60°, 300°$

13. Match each equation in the left column with the corresponding expression representing all solutions in the right column.

$$2 \cos \theta = -\sqrt{3}$$
$$\sqrt{2} \sin \theta = 1$$

$150° + 360°k,\ 210° + 360°k$

$45° + 360°k,\ 135° + 360°k$

$30° + 360°k,\ 330° + 360°k$

$225° + 360°k,\ 315° + 360°k$

14. Use your graphing calculator to find the solutions to the equation for θ if

$0° \le \theta < 360°$ by graphing the function represented by the left side of the equation and

then finding its zeros.

$$\cot\theta + 2\cos\theta\cot\theta = 0$$

15. The formula below gives the relationship between the number of sides n , the radius r , and

the length of each side l in a regular polygon. Find n , if $l = r\sqrt{2}$.

$$l = 2r\sin\frac{180°}{n}$$

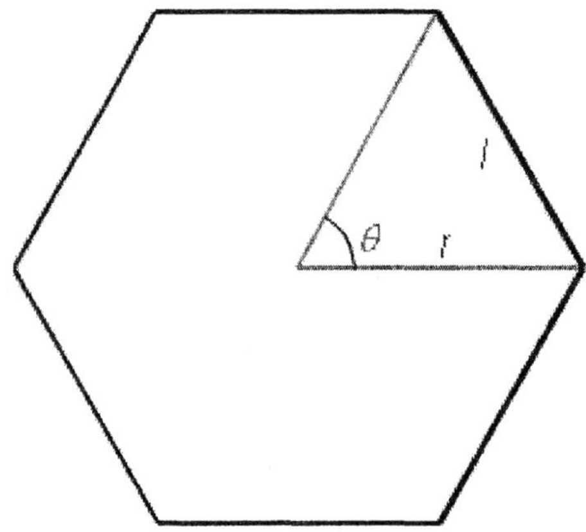

$n = $ ________

16. Solve the equation for θ if $0° \le \theta < 360°$. Give your answer in degrees.

$$\sqrt{3}\,\csc\theta - 2\cot\theta = 0$$

17. Find all solutions in the interval $0° \leq \theta < 360°$. Use a calculator on the last step and write all answers to the nearest tenth of a degree.

$$3 \sin \theta - 9 = 4 \sin \theta - 5$$

18. Solve the equation for θ if $0° \leq \theta < 360°$. Give your answer in degrees.

$$2 \sin \theta + 1 = \csc \theta$$

19. Eliminate the parameter t in the following:

$$x = 5 \sin t \quad y = 6 \sin t$$

Select the correct answer.

a. $6x = -5y, \quad -1 \leq x \leq 1, \quad -6 \leq y \leq 6$

b. $3x = 6y, \quad -3 \leq x \leq 3, \quad -1 \leq y \leq 1$

c. $5x = 6y, \quad -5 \leq x \leq 5, \quad -1 \leq y \leq 1$

d. $6x = 3y, \quad -1 \leq x \leq 1, \quad -1 \leq y \leq 1$

e. $6x = 5y, \quad -5 \leq x \leq 5, \quad -6 \leq y \leq 6$

20. Solve the equation for θ if $0° \leq \theta < 360°$.

$$\cos \theta - \sqrt{3} \sin \theta = -\sqrt{3}$$

Select the correct answer.

a. $0°, 60°$

b. $240°, 300°$

c. $120°, 180°$

d. $90°, 150°$

e. $30°, 330°$

21. Solving the following equation will require you to use the quadratic formula. Solve the equation for θ between $0°$ and $360°$, and round your answers to the nearest tenth of a degree.

$$3\cos^2\theta + 2\sin\theta - 1 = 0$$

Select the correct answer.

    a. $213.7°, 326.3°$

    b. $213.1°, 326.9°$

    c. $212.5°, 327.5°$

    d. $213.9°, 326.1°$

    e. $213.3°, 326.7°$

22. Solve the equation for x if $0 \leq x < 2\pi$. Give your answer in radians using exact values only.

$$2\cos x - \tan x - \sec x = 0$$

23. Find all degree solutions for the following.

$$\cos 4\theta = -1$$

24. Find all solutions if $0 \leq x < 2\pi$.

$$\sin 2x \cos x + \cos 2x \sin x = \frac{1}{2}$$

Select the correct answer.

a. $\dfrac{\pi}{18}, \dfrac{5\pi}{18}, \dfrac{13\pi}{18}, \dfrac{17\pi}{18}, \dfrac{25\pi}{18}, \dfrac{29\pi}{18}$

b. $\dfrac{\pi}{6}, \dfrac{5\pi}{6}, \dfrac{3\pi}{2}$

c. $\dfrac{\pi}{9}, \dfrac{2\pi}{9}, \dfrac{7\pi}{9}, \dfrac{8\pi}{9}, \dfrac{13\pi}{9}, \dfrac{14\pi}{9}$

d. $0, \dfrac{\pi}{3}, \dfrac{2\pi}{3}, \pi, \dfrac{4\pi}{3}, \dfrac{5\pi}{3}$

e. $\dfrac{\pi}{12}, \dfrac{\pi}{4}, \dfrac{3\pi}{4}, \dfrac{11\pi}{12}, \dfrac{17\pi}{12}, \dfrac{19\pi}{12}$

25. Find all degree solutions.

$$2\sin^2 10\theta + \sin 10\theta - 1 = 0$$

Select the correct answer.

a. $\theta = 3° + 60°k$ or $\theta = 15° + 60°k$ or $\theta = 27° + 36°k$

b. $\theta = 5° + 60°k$ or $\theta = 25° + 60°k$ or $\theta = 45° + 60°k$

c. $\theta = 60° + 360°k$

d. $\theta = 3° + 36°k$ or $\theta = 15° + 36°k$ or $\theta = 27° + 36°k$

e. $\theta = 5° + 36°k$ or $\theta = 25° + 36°k$ or $\theta = 45° + 36°k$

1. a

2. d

3. d

4. e

5. a

6. $y = -x, \; -1 \leq x \leq 1, \; -1 \leq y \leq 1$

7. $\dfrac{x^2}{25} - \dfrac{y^2}{9} = 1$

8. b

9. $\dfrac{3\pi}{4} + 2k\pi$

10. $x = \dfrac{\pi}{12} + \dfrac{\pi k}{3}$ or $\dfrac{\pi}{4} + \dfrac{\pi k}{3}$

11. $\dfrac{x^2}{16} + \dfrac{y^2}{25} = 1$

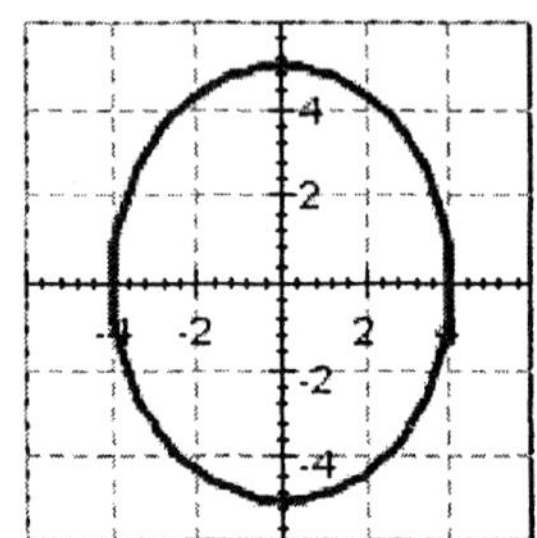

12. e

13. $\sqrt{2}\sin\theta = 1 \rightarrow 45° + 360°k, \; 135° + 360°k,$

$2\cos\theta = -\sqrt{3} \rightarrow 150° + 360°k, \; 210° + 360°k$

14. $90°, 270°, 120°, 240°$

McKeague/Turner - Trigonometry 5e Chapter 6 Form F

15. 4

16. $30°, 330°$

17. no solution

18. $30°, 150°, 270°$

19. e

20. d

21. e

22. $\dfrac{\pi}{6}, \dfrac{5\pi}{6}$

23. $\theta = 45° + 90°k$

24. a

25. d

1. mctr.06.01.23m_NoAlgs
2. mctr.06.01.32m_NoAlgs
3. mctr.06.04.08m_NoAlgs
4. mctr.06.04.19m_NoAlgs
5. mctr.06.01.51m_NoAlgs
6. mctr.06.04.24_NoAlgs
7. mctr.06.04.15_NoAlgs
8. mctr.06.02.17m_NoAlgs
9. mctr.06.02.42_NoAlgs
10. mctr.06.03.31_NoAlgs
11. mctr.06.04.05_NoAlgs
12. mctr.06.03.54m_NoAlgs
13. mctr.06.01.39_NoAlgs
14. mctr.06.01.60_NoAlgs
15. mctr.06.03.53_NoAlgs
16. mctr.06.02.07_NoAlgs
17. mctr.06.01.15_NoAlgs
18. mctr.06.02.12_NoAlgs
19. mctr.06.04.25m_NoAlgs
20. mctr.06.02.26m_NoAlgs
21. mctr.06.02.48m_NoAlgs
22. mctr.06.02.21_NoAlgs
23. mctr.06.03.16_NoAlgs
24. mctr.06.03.25m_NoAlgs
25. mctr.06.03.35m_NoAlgs

1. Solve the equation for θ if $0° \leq \theta < 360°$. Do not use a calculator.

$$2\cos\theta - \sqrt{3} = 0$$
00

2. Solve the equation for t if $0 \leq t < 2\pi$. Do not use a calculator.

$$4\sin t - 1 = 2\sin t$$

Select the correct answer.

   a. $\dfrac{\pi}{4}, \dfrac{3\pi}{4}$

   b. $\dfrac{\pi}{6}, \dfrac{5\pi}{6}$

   c. $\dfrac{\pi}{3}, \dfrac{2\pi}{3}$

   d. $\dfrac{5\pi}{4}, \dfrac{7\pi}{4}$

   e. $\dfrac{5\pi}{6}, \dfrac{7\pi}{6}$

3. Solve for θ if $0° \leq \theta < 360°$.

$$2\cos^2\theta + 3\cos\theta = 2$$

4. Find all degree solutions to the equation.

$$\sin\left(4A - 40°\right) = -\frac{1}{2}$$

Select the correct answer.

   a. $25° + 90°k,\ 40° + 90°k$

   b. $62.5° + 90°k,\ 72.5° + 90°k$

   c. $42.5° + 90°k,\ 92.5° + 90°k$

   d. $42.5° + 90°k,\ 72.5° + 90°k$

   e. $62.5° + 90°k,\ 92.5° + 90°k$

5. Use your graphing calculator to find the solutions to the equation for θ if $0° \le \theta < 360°$ by graphing the function represented by the left side of the equation and then finding its zeros.

$$\tan\theta + 2\sin\theta\tan\theta = 0$$

Select the correct answer.

   a. $0°,\ 180°,\ 150°,\ 210°$

   b. $0°,\ 180°,\ 210°,\ 330°$

   c. $90°,\ 270°,\ 210°,\ 330°$

   d. $0°,\ 180°,\ 30°,\ 150°$

   e. $90°,\ 270°,\ 30°,\ 150°$

6. Use your graphing calculator to find the solutions to the equation to the nearest tenth of a degree in the interval $0° \le \theta < 360°$ by defining the left side and right side of the equation as functions and then finding the intersection points of their graphs.

$$\sin\theta - 5 = -2\sin\theta$$

Select the correct answer.

   a. $27.0°,\ 153.0°$

   b. $\varnothing$

   c. $18.4°,\ 161.6°$

   d. $71.3°,\ 108.7°$

   e. $41.7°,\ 138.3°$

7. Solve the equation for θ if $0° \leq \theta < 360°$. Give your answer in degrees.

$$\sqrt{3}\,\csc\theta - 2\cot\theta = 0$$

8. Solve the equation for θ if $0° \leq \theta < 360°$.

$$2\sin\theta - \sin 2\theta = 0$$

Select the correct answer.

a. $90°,\ 180°$

b. $60°,\ 300°$

c. $210°,\ 330°$

d. $0°,\ 180°$

e. $0°$

9. Solve the equation for x if $0 \leq x < 2\pi$.

$$2\cos x + \tan x - \sec x = 0$$

Select the correct answer.

a. $\dfrac{7\pi}{6},\ \dfrac{11\pi}{6}$

b. $\dfrac{2\pi}{3},\ \dfrac{4\pi}{3}$

c. $\dfrac{\pi}{6},\ \dfrac{5\pi}{6}$

d. $\dfrac{\pi}{3},\ \dfrac{5\pi}{3}$

e. $\dfrac{\pi}{6},\ \dfrac{5\pi}{6},\ \dfrac{3\pi}{2}$

10. Solve the equation for x if $0 \leq x < 2\pi$. Give your answer in radians using exact values only.

$$2 \sin x + \cot x = \csc x$$

11. For the equation, find all degree solutions in the interval $0° \leq \theta < 360°$. If rounding is necessary, round to the nearest tenth of a degree.

$$23 \sec^2 \theta - 22 \tan \theta \sec \theta - 15 = 0$$

Select the correct answer.

a. $41.6°, 53.1°, 126.9°, 138.4°$

b. $41.8°, 52.9°, 127.1°, 138.2°$

c. $41.6°, 52.9°, 127.1°, 138.4°$

d. $41.4°, 53.5°, 126.5°, 138.6°$

e. $41.8°, 53.1°, 126.9°, 138.2°$

12. For the equation, find all degree solutions in the interval $0° \leq \theta < 360°$. If rounding is necessary, round to the nearest tenth of a degree.

$$49 \sin^2 \theta - 16 \cos 2\theta = 0$$

13. Use your graphing calculator to find all radian solutions in the interval $0 \leq x < 2\pi$ for the following equation. Round your answers to four decimal places.

$$\sin^2 x - 3 \sin x - 3 = 0$$

14. Find all solutions if $0 \leq x < 2\pi$.

$$\sin 2x = \frac{1}{\sqrt{2}}$$

Select the correct answer.

a. $\quad 0, \dfrac{\pi}{8}, \dfrac{3\pi}{8}, \dfrac{9\pi}{8}$

b. $\quad \dfrac{\pi}{8}, \dfrac{3\pi}{8}, \dfrac{9\pi}{8}, \dfrac{11\pi}{8}$

c. $\quad \dfrac{3\pi}{8}, \dfrac{9\pi}{8}, \dfrac{11\pi}{8}, \dfrac{15\pi}{8}$

d. $\quad 0, \dfrac{\pi}{8}, \dfrac{3\pi}{8}, \dfrac{11\pi}{8}$

e. $\quad \dfrac{\pi}{8}, \dfrac{3\pi}{8}, \dfrac{9\pi}{8}, \dfrac{15\pi}{8}$

15. Find all degree solutions for the following.

$$\cos 4\theta = -1$$

16. Use your graphing calculator to find all degree solutions in the interval $0° \leq \theta < 360°$ for the following equation.

$$\sin 3x = \frac{\sqrt{3}}{2}$$

Select the correct answer.

a. $\quad 30°, 150°, 270°$

b. $\quad 90°, 210°, 330°$

c. $\quad 20°, 40°, 140°, 160°, 260°, 280°$

d. $\quad 15°, 45°, 135°, 165°, 255°, 285°$

e. $\quad 0°, 60°, 120°, 180°, 240°, 300°$

17. Find all solutions if $0 \leq x < 2\pi$.

$$\sin 2x \cos x + \cos 2x \sin x = \frac{1}{2}$$

Select the correct answer.

a. $\quad \dfrac{\pi}{18}, \dfrac{5\pi}{18}, \dfrac{13\pi}{18}, \dfrac{17\pi}{18}, \dfrac{25\pi}{18}, \dfrac{29\pi}{18}$

b. $\quad \dfrac{\pi}{6}, \dfrac{5\pi}{6}, \dfrac{3\pi}{2}$

c. $\quad \dfrac{\pi}{9}, \dfrac{2\pi}{9}, \dfrac{7\pi}{9}, \dfrac{8\pi}{9}, \dfrac{13\pi}{9}, \dfrac{14\pi}{9}$

d. $\quad 0, \dfrac{\pi}{3}, \dfrac{2\pi}{3}, \pi, \dfrac{4\pi}{3}, \dfrac{5\pi}{3}$

e. $\quad \dfrac{\pi}{12}, \dfrac{\pi}{4}, \dfrac{3\pi}{4}, \dfrac{11\pi}{12}, \dfrac{17\pi}{12}, \dfrac{19\pi}{12}$

18. Find all degree solutions.

$$2\sin^2 10\theta + 3\sin 10\theta + 1 = 0$$

19. Find all solutions if $0° \leq \theta < 360°$. Round each of the solutions to the nearest tenth if necessary.

$$\sin^2 2\theta - 7\sin 2\theta - 4 = 0$$

Select the correct answer.

a. $\quad 106.0°, 164.0°, 286.0°, 344.0°$

b. $\quad 106.0°, 164.0°, 282.3°, 349.8°$

c. $\quad 107.2°, 166.1°, 286.0°, 344.0°$

d. $\quad 106.0°, 166.1°, 282.3°, 344.0°$

e. $\quad 107.2°, 164.0°, 286.0°, 349.8°$

20. Eliminate the parameter t from the following:

$$x = 9\sin t \quad y = 9\cos t$$

Sketch the graph.

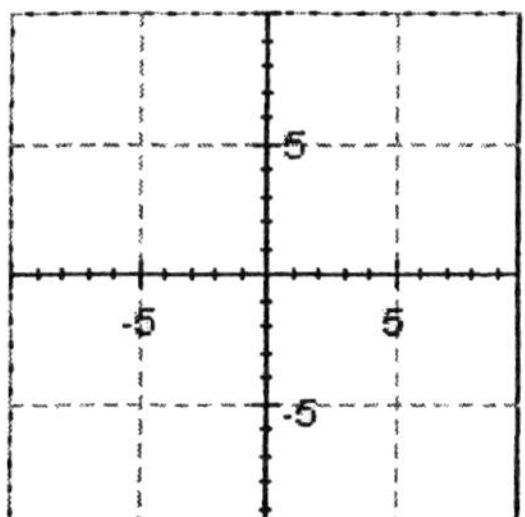

21. Eliminate the parameter t from the following and then sketch the graph:

$$x = 2\cos t \quad y = 2\sin t$$

Select the correct answer.

a. $x^2 - y^2 = 50$

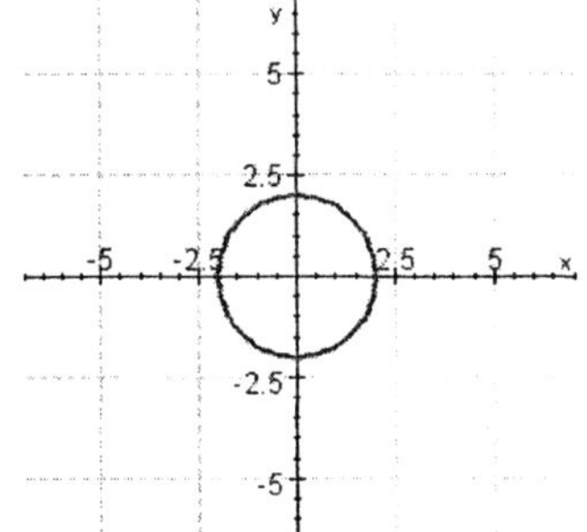

b. $x^2 + y^2 = 5$

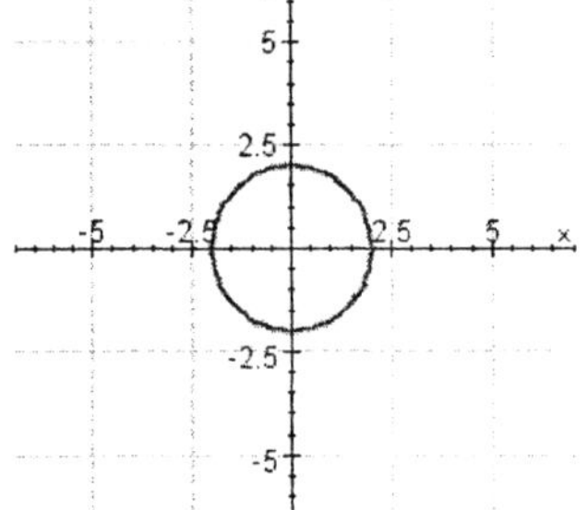

c. $x^2 + y^2 = 4$

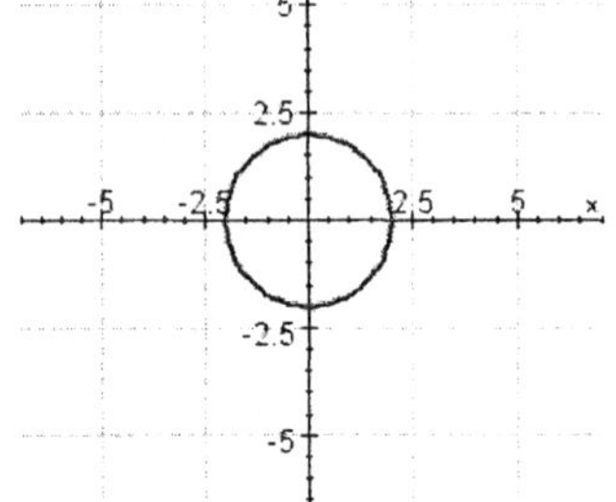

d. $x^2 - y^2 = 9$

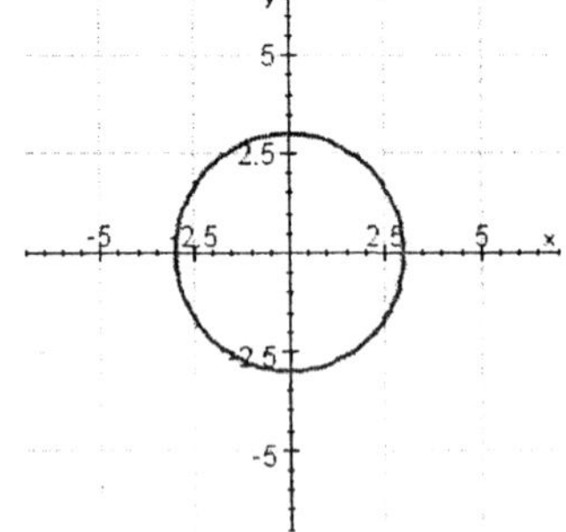

22. Eliminate the parameter t from the following:

$$x = \cos t - 4 \quad y = \sin t + 7$$

Sketch the graph.

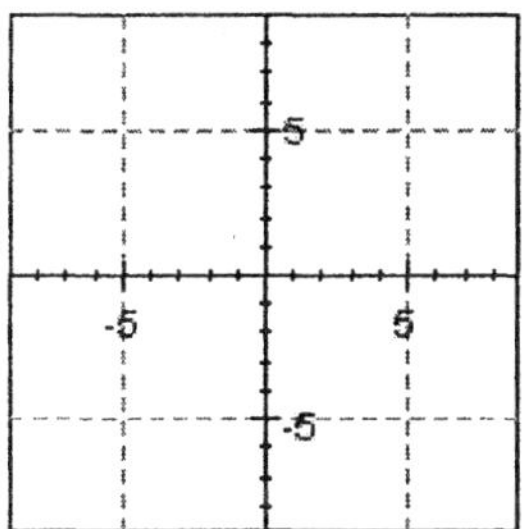

23. Eliminate the parameter t in the following:

$$x = 5 + 3\tan t \quad y = 2 + 3\sec t$$

24. Eliminate the parameter t in the following:

$$x = -\cos 2t \quad y = \cos t$$

Select the correct answer.

a.　　$x = 2y^2 - 1, \quad -1 \le x \le 1, \quad -1 \le y \le 1$

b.　　$x = 2y^2 + 1, \quad -1 \le x \le 1, \quad -1 \le y \le 1$

c.　　$x = -2y^2 + 1, \quad -1 \le x \le 1, \quad -1 \le y \le 1$

d.　　$x = 2y + 1, \quad -1 \le x \le 1, \quad -1 \le y \le 1$

e.　　$x = -2y + 1, \quad -1 \le x \le 1, \quad -1 \le y \le 1$

25. Eliminate the parameter t from the following and then sketch the graph:

$$x = 2\cos t - 3 \quad y = 2\sin t + 2$$

Select the correct answer.

a. $\dfrac{(x+4)^2}{4} + \dfrac{(y-2)^2}{4} = 1$

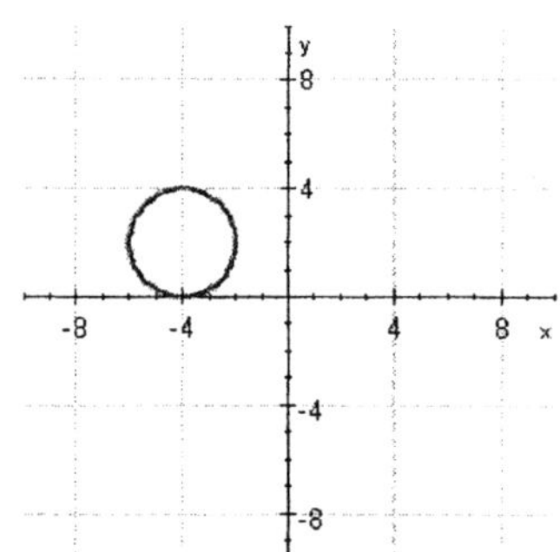

b. $\dfrac{(x+3)^2}{4} - \dfrac{(y-2)^2}{4} = 1$

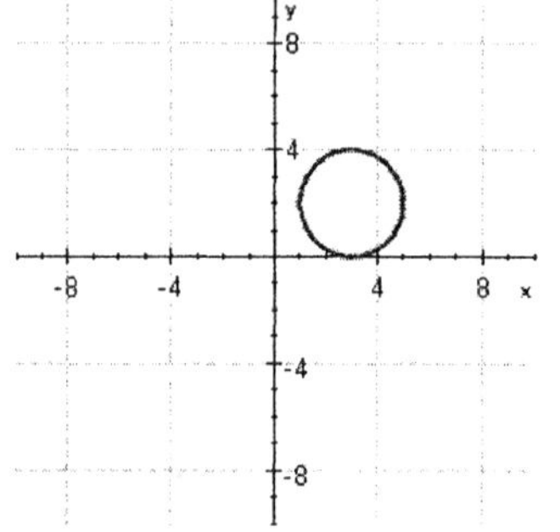

c. $\dfrac{(x+4)^2}{4} + \dfrac{(y-3)^2}{4} = 1$

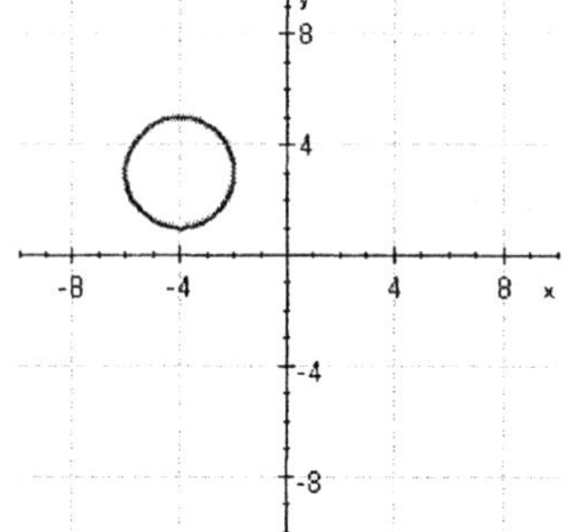

d. $\dfrac{(x+3)^2}{4} + \dfrac{(y-2)^2}{4} = 1$

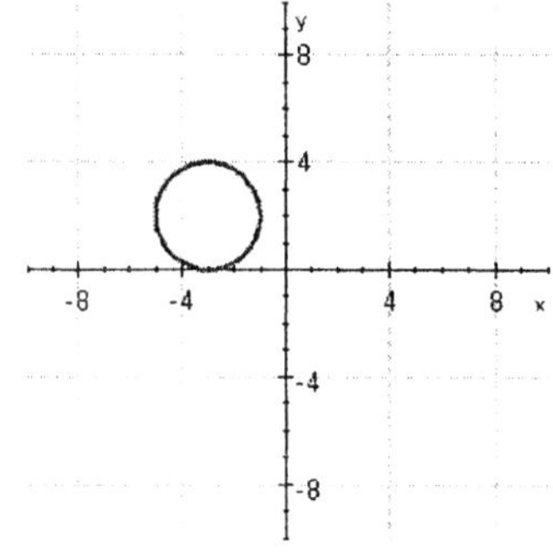

1. $30°, 330°$

2. b

3. $60°, 300°$

4. e

5. b

6. b

7. $30°, 330°$

8. d

9. a

10. $\dfrac{2\pi}{3}, \dfrac{4\pi}{3}$

11. e

12. $26.4°, 153.6°, 206.4°, 333.6°$

13. $4.0545, 5.3703$

14. b

15. $\theta = 45° + 90°k$

16. c

17. a

18. $\theta = 21° + 36°k$ or $\theta = 27° + 36°k$ or $\theta = 33° + 36°k$

19. a

20. $x^2 + y^2 = 81$

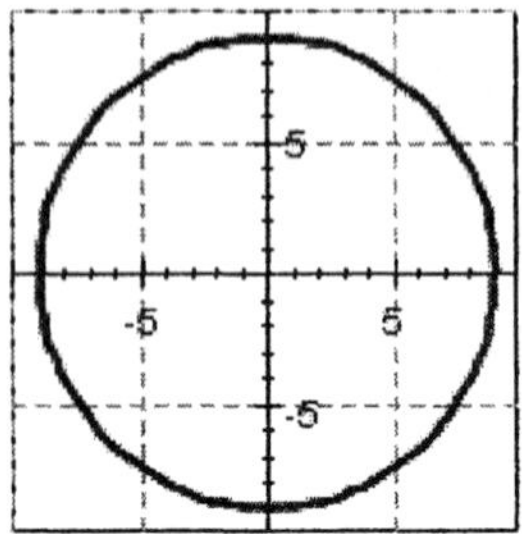

21. c

22. $(x + 4)^2 + (y - 7)^2 = 1$

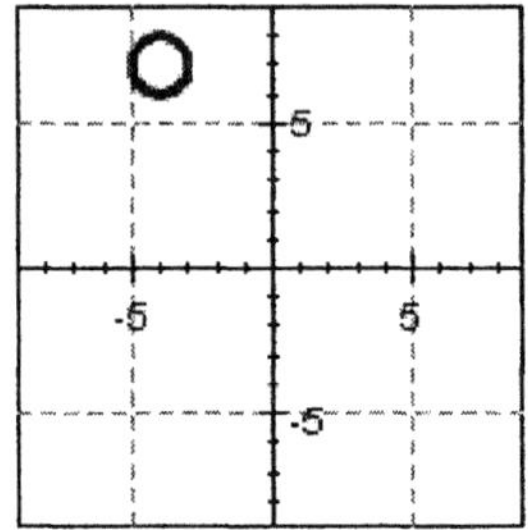

23. $\dfrac{(y - 2)^2}{9} - \dfrac{(x - 5)^2}{9} = 1$

24. c

25. d

1. mctr.06.01.03_NoAlgs
2. mctr.06.01.07m_NoAlgs
3. mctr.06.01.31_NoAlgs
4. mctr.06.01.52m_NoAlgs
5. mctr.06.01.60m_NoAlgs
6. mctr.06.01.00m_NoAlgs
7. mctr.06.02.07_NoAlgs
8. mctr.06.02.10m_NoAlgs
9. mctr.06.02.21m_NoAlgs
10. mctr.06.02.22_NoAlgs
11. mctr.06.02.36m_NoAlgs
12. mctr.06.02.37_NoAlgs
13. mctr.06.02.55_NoAlgs
14. mctr.06.03.07m_NoAlgs
15. mctr.06.03.16_NoAlgs
16. mctr.06.03.22m_NoAlgs
17. mctr.06.03.25m_NoAlgs
18. mctr.06.03.36_NoAlgs
19. mctr.06.03.45m_NoAlgs
20. mctr.06.04.01_NoAlgs
21. mctr.06.04.04m_NoAlgs
22. mctr.06.04.10_NoAlgs
23. mctr.06.04.20_NoAlgs
24. mctr.06.04.22m_NoAlgs
25. mctr.06.04.13m_NoAlgs

1. Solve the equation for x if $0 \leq x < 2\pi$. Give your answer in radians using exact values only.

$$2 \sin x + \cot x = \csc x$$

2. Eliminate the parameter t in the following:

$$x = -\cos 2t \quad y = \cos t$$

Select the correct answer.

   a. $\quad x = 2y^2 - 1, \quad -1 \leq x \leq 1, \quad -1 \leq y \leq 1$

   b. $\quad x = 2y^2 + 1, \quad -1 \leq x \leq 1, \quad -1 \leq y \leq 1$

   c. $\quad x = -2y^2 + 1, \quad -1 \leq x \leq 1, \quad -1 \leq y \leq 1$

   d. $\quad x = 2y + 1, \quad -1 \leq x \leq 1, \quad -1 \leq y \leq 1$

   e. $\quad x = -2y + 1, \quad -1 \leq x \leq 1, \quad -1 \leq y \leq 1$

3. Find all degree solutions.

$$2\sin^2 10\theta + 3\sin 10\theta + 1 = 0$$

4. For the equation, find all degree solutions in the interval $0° \leq \theta < 360°$. If rounding is necessary, round to the nearest tenth of a degree.

$$23\sec^2 \theta - 22\tan \theta \sec \theta - 15 = 0$$

Select the correct answer.

   a. $\quad 41.6°, 53.1°, 126.9°, 138.4°$

   b. $\quad 41.8°, 52.9°, 127.1°, 138.2°$

   c. $\quad 41.6°, 52.9°, 127.1°, 138.4°$

   d. $\quad 41.4°, 53.5°, 126.5°, 138.6°$

   e. $\quad 41.8°, 53.1°, 126.9°, 138.2°$

5. Eliminate the parameter t from the following:

$$x = 9\sin t \quad y = 9\cos t$$

Sketch the graph.

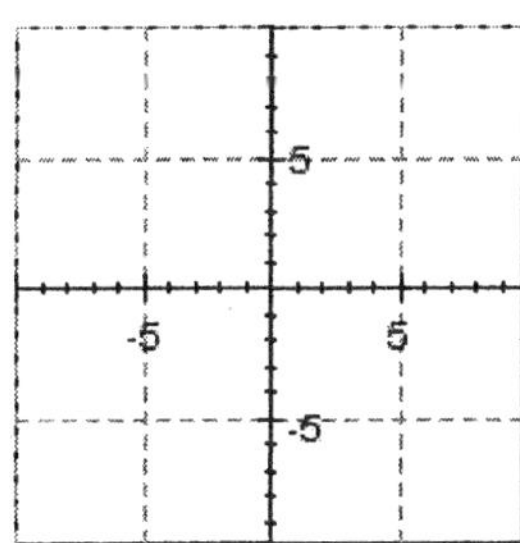

6. Use your graphing calculator to find all degree solutions in the interval $0° \leq \theta < 360°$ for the following equation.

$$\sin 3x = \frac{\sqrt{3}}{2}$$

Select the correct answer.

   a. $30°, 150°, 270°$

   b. $90°, 210°, 330°$

   c. $20°, 40°, 140°, 160°, 260°, 280°$

   d. $15°, 45°, 135°, 165°, 255°, 285°$

   e. $0°, 60°, 120°, 180°, 240°, 300°$

7. Find all solutions if $0 \leq x < 2\pi$. Use exact values only.

$$\sin 2x = \frac{1}{\sqrt{2}}$$

Select the correct answer.

a. $0, \dfrac{\pi}{8}, \dfrac{3\pi}{8}, \dfrac{9\pi}{8}$

b. $\dfrac{\pi}{8}, \dfrac{3\pi}{8}, \dfrac{9\pi}{8}, \dfrac{11\pi}{8}$

c. $\dfrac{3\pi}{8}, \dfrac{9\pi}{8}, \dfrac{11\pi}{8}, \dfrac{15\pi}{8}$

d. $0, \dfrac{\pi}{8}, \dfrac{3\pi}{8}, \dfrac{11\pi}{8}$

e. $\dfrac{\pi}{8}, \dfrac{3\pi}{8}, \dfrac{9\pi}{8}, \dfrac{15\pi}{8}$

8. Use your graphing calculator to find the solutions to the equation to the nearest tenth of a degree in the interval $0° \leq \theta < 360°$ by defining the left side and right side of the equation as functions and then finding the intersection points of their graphs.

$$\sin \theta - 5 = -2 \sin \theta$$

Select the correct answer.

a. $27.0°, 153.0°$

b. $\varnothing$

c. $18.4°, 161.6°$

d. $71.3°, 108.7°$

e. $41.7°, 138.3°$

9. For the equation, find all degree solutions in the interval $0° \leq \theta < 360°$. If rounding is necessary, round to the nearest tenth of a degree.

$$49 \sin^2 \theta - 16 \cos 2\theta = 0$$

10. Solve the equation for t if $0 \leq t < 2\pi$. Do not use a calculator.

$$4 \sin t - 1 = 2 \sin t$$

Select the correct answer.

a. $\dfrac{\pi}{4}, \dfrac{3\pi}{4}$

b. $\dfrac{\pi}{6}, \dfrac{5\pi}{6}$

c. $\dfrac{\pi}{3}, \dfrac{2\pi}{3}$

d. $\dfrac{5\pi}{4}, \dfrac{7\pi}{4}$

e. $\dfrac{5\pi}{6}, \dfrac{7\pi}{6}$

11. Eliminate the parameter t in the following:

$$x = 5 + 3\tan t \quad y = 2 + 3\sec t$$

12. Find all degree solutions to the equation.

$$\sin\left(4A - 40^\circ\right) = -\dfrac{1}{2}$$

Select the correct answer.

a. $25^\circ + 90^\circ k, \ 40^\circ + 90^\circ k$

b. $62.5^\circ + 90^\circ k, \ 72.5^\circ + 90^\circ k$

c. $42.5^\circ + 90^\circ k, \ 92.5^\circ + 90^\circ k$

d. $42.5^\circ + 90^\circ k, \ 72.5^\circ + 90^\circ k$

e. $62.5^\circ + 90^\circ k, \ 92.5^\circ + 90^\circ k$

13. Find all degree solutions for the following.

$$\cos 4\theta = -1$$

14. Eliminate the parameter t from the following and then sketch the graph:

$$x = 2\cos t - 3 \quad y = 2\sin t + 2$$

Select the correct answer.

a. $\dfrac{(x + 4)^2}{4} + \dfrac{(y - 2)^2}{4} = 1$

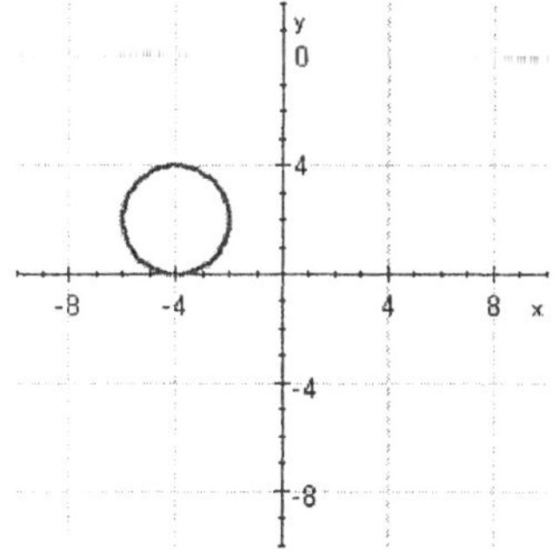

b. $\dfrac{(x + 3)^2}{4} - \dfrac{(y - 2)^2}{4} = 1$

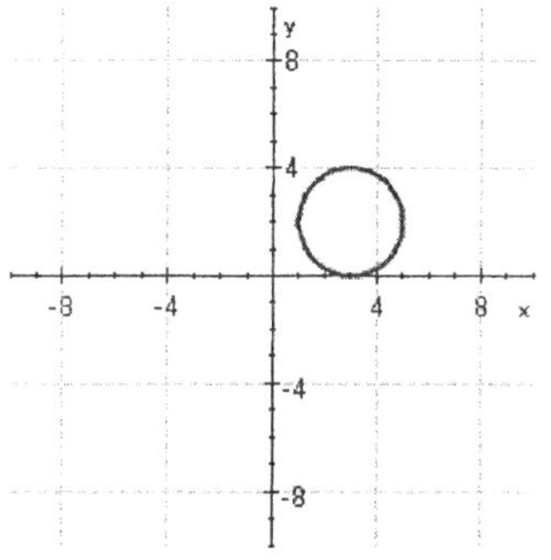

c. $\dfrac{(x + 4)^2}{4} + \dfrac{(y - 3)^2}{4} = 1$

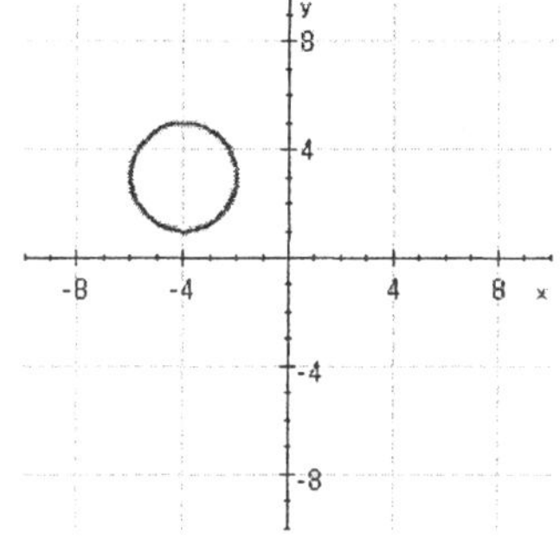

d. $\dfrac{(x + 3)^2}{4} + \dfrac{(y - 2)^2}{4} = 1$

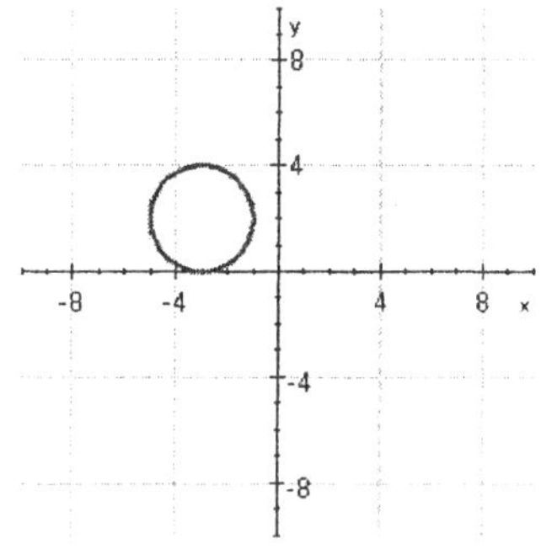

15. Solve the equation for θ if $0° \leq \theta < 360°$. Do not use a calculator.

$$2\cos\theta - \sqrt{3} = 0$$

16. Find all solutions if $0° \leq \theta < 360°$. Round each of the solutions to the nearest tenth if necessary.

$$\sin^2 2\theta - 7\sin 2\theta - 4 = 0$$

Select the correct answer.

a. $106.0°, 164.0°, 286.0°, 344.0°$

b. $106.0°, 164.0°, 282.3°, 349.8°$

c. $107.2°, 166.1°, 286.0°, 344.0°$

d. $106.0°, 166.1°, 282.3°, 344.0°$

e. $107.2°, 164.0°, 286.0°, 349.8°$

17. Eliminate the parameter t from the following:

$$x = \cos t - 4 \quad y = \sin t + 7$$

Sketch the graph.

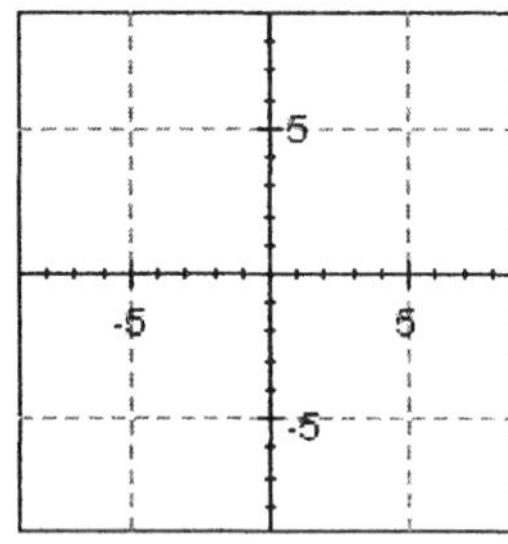

18. Find all solutions if $0 \leq x < 2\pi$.

$$\sin 2x \cos x + \cos 2x \sin x = \frac{1}{2}$$

Select the correct answer.

a. $\dfrac{\pi}{18}, \dfrac{5\pi}{18}, \dfrac{13\pi}{18}, \dfrac{17\pi}{18}, \dfrac{25\pi}{18}, \dfrac{29\pi}{18}$

b. $\dfrac{\pi}{6}, \dfrac{5\pi}{6}, \dfrac{3\pi}{2}$

c. $\dfrac{\pi}{9}, \dfrac{2\pi}{9}, \dfrac{7\pi}{9}, \dfrac{8\pi}{9}, \dfrac{13\pi}{9}, \dfrac{14\pi}{9}$

d. $0, \dfrac{\pi}{3}, \dfrac{2\pi}{3}, \pi, \dfrac{4\pi}{3}, \dfrac{5\pi}{3}$

e. $\dfrac{\pi}{12}, \dfrac{\pi}{4}, \dfrac{3\pi}{4}, \dfrac{11\pi}{12}, \dfrac{17\pi}{12}, \dfrac{19\pi}{12}$

19. Use your graphing calculator to find the solutions to the equation for θ if $0° \leq \theta < 360°$ by graphing the function represented by the left side of the equation and then finding its zeros.

$$\tan \theta + 2 \sin \theta \tan \theta = 0$$

Select the correct answer.

a. $0°, 180°, 150°, 210°$

b. $0°, 180°, 210°, 330°$

c. $90°, 270°, 210°, 330°$

d. $0°, 180°, 30°, 150°$

e. $90°, 270°, 30°, 150°$

20. Solve the equation for x if $0 \leq x < 2\pi$.

$$2\cos x + \tan x - \sec x = 0$$

Select the correct answer.

a. $\quad \dfrac{7\pi}{6} , \dfrac{11\pi}{6}$

b. $\quad \dfrac{2\pi}{3} , \dfrac{4\pi}{3}$

c. $\quad \dfrac{\pi}{6} , \dfrac{5\pi}{6}$

d. $\quad \dfrac{\pi}{3} , \dfrac{5\pi}{3}$

e. $\quad \dfrac{\pi}{6} , \dfrac{5\pi}{6} , \dfrac{3\pi}{2}$

21. Solve for θ if $0° \leq \theta < 360°$.

$$2\cos^2\theta + 3\cos\theta = 2$$

22. Eliminate the parameter *t* from the following and then sketch the graph:

$$x = 2\cos t \quad y = 2\sin t$$

Select the correct answer.

a. $x^2 - y^2 = 5$

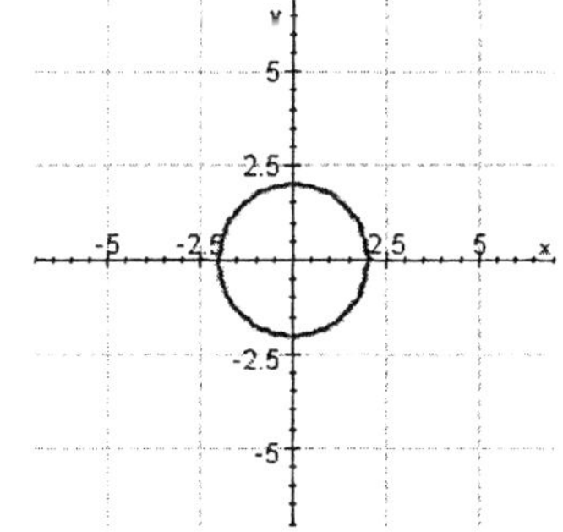

b. $x^2 + y^2 = 5$

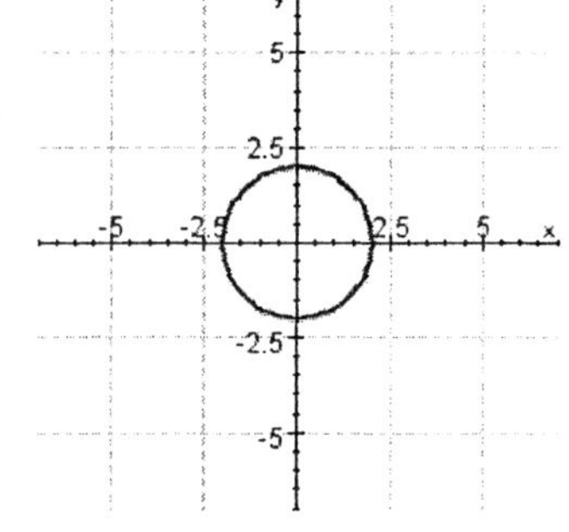

c. $x^2 + y^2 = 4$

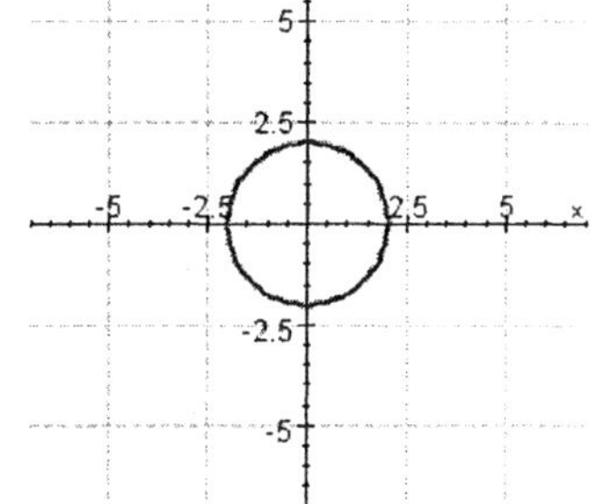

d. $x^2 - y^2 = 9$

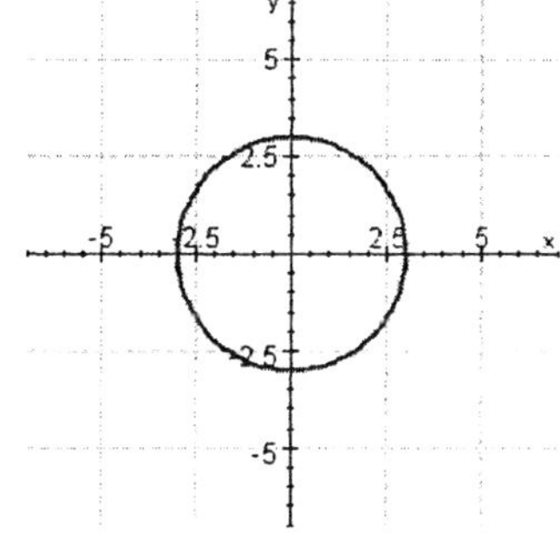

23. Use your graphing calculator to find all radian solutions in the interval $0 \leq x < 2\pi$ for the following equation. Round your answers to four decimal places.

$$\sin^2 x - 3 \sin x - 3 = 0$$

24. Solve the equation for θ if $0° \leq \theta < 360°$. Give your answer in degrees.

$$\sqrt{3} \csc \theta - 2 \cot \theta = 0$$

25. Solve the equation for θ if $0° \leq \theta < 360°$.

$$2 \sin \theta - \sin 2\theta = 0$$

Select the correct answer.

    a. $90°, 180°$

    b. $60°, 300°$

    c. $210°, 330°$

    d. $0°, 180°$

    e. $0°$

1. $\dfrac{2\pi}{3},\ \dfrac{4\pi}{3}$

2. c

3. $\theta = 21° + 36°k$ or $\theta = 27° + 36°k$ or $\theta = 33° + 36°k$

4. e

5. $x^2 + y^2 = 81$

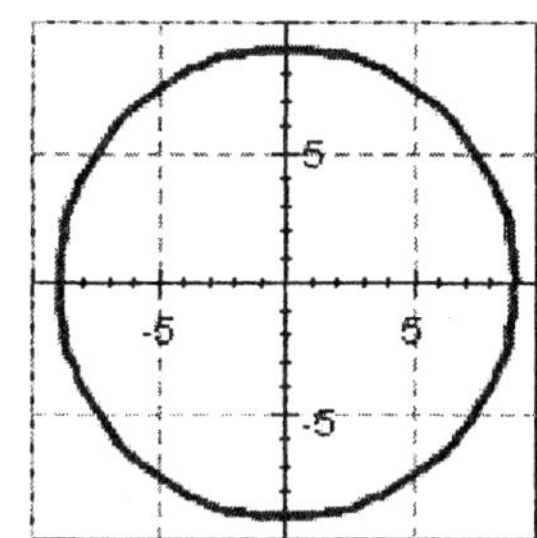

6. c

7. b

8. b

9. $26.4°,\ 153.6°,\ 206.4°,\ 333.6°$

10. b

11. $\dfrac{(y-2)^2}{9} - \dfrac{(x-5)^2}{9} = 1$

12. e

13. $\theta = 45° + 90°k$

14. d

15. $30°,\ 330°$

16. a

17. $(x + 4)^2 + (y - 7)^2 = 1$

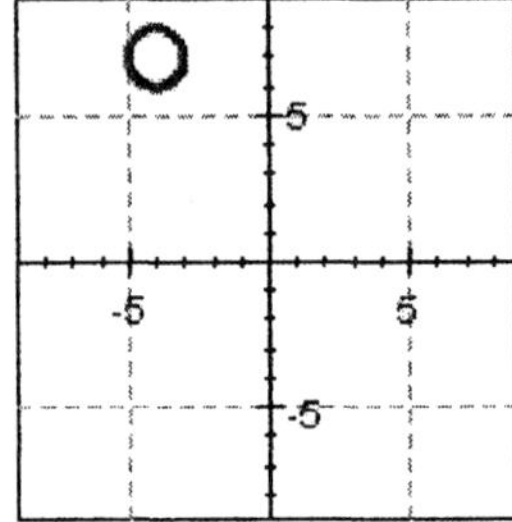

18. a

19. b

20. a

21. $60°, 300°$

22. c

23. $4.0545, 5.3703$

24. $30°, 330°$

25. d

1. mctr.06.02.22_NoAlgs
2. mctr.06.04.22m_NoAlgs
3. mctr.06.03.36_NoAlgs
4. mctr.06.02.36m_NoAlgs
5. mctr.06.04.01_NoAlgs
6. mctr.06.03.22m_NoAlgs
7. mctr.06.03.07m_NoAlgs
8. mctr.06.01.66m_NoAlgs
9. mctr.06.02.37_NoAlgs
10. mctr.06.01.07m_NoAlgs
11. mctr.06.04.20_NoAlgs
12. mctr.06.01.52m_NoAlgs
13. mctr.06.03.16_NoAlgs
14. mctr.06.04.13m_NoAlgs
15. mctr.06.01.03_NoAlgs
16. mctr.06.03.45m_NoAlgs
17. mctr.06.04.10_NoAlgs
18. mctr.06.03.25m_NoAlgs
19. mctr.06.01.60m_NoAlgs
20. mctr.06.02.21m_NoAlgs
21. mctr.06.01.31_NoAlgs
22. mctr.06.04.04m_NoAlgs
23. mctr.06.02.55_NoAlgs
24. mctr.06.02.07_NoAlgs
25. mctr.06.02.10m_NoAlgs

1. The problem that follows refers to triangle *ABC*.

   If $A = 70°$, $B = 40°$ and $b = 8\,\text{cm}$, find a.

   Please round the answer to the nearest centimeter.

   $a =$ _________ cm

2. The problem that follows refers to triangle ABC.

   If $A = 55°$, $B = 48°$ and $c = 16\,\text{cm}$.

   Find C.

   $C =$ _________ °

   Find a. Please round the answer to the nearest centimeter.

   $a =$ _________ cm

3. The information below refers to triangle ABC.

   $B = 12.4°$, $C = 23.8°$, $a = 325\,\text{cm}$

   Find A.

   $A =$ _________ °

   Find b. Please round the answer to the nearest centimeter.

   $b =$ _________ cm

   Find c. Please round the answer to the nearest centimeter.

   $c =$ _________ cm

4. A woman entering an outside glass elevator on the ground floor of a hotel glances up to the top of the building across the street and notices that the angle of elevation is $47°$. She rides the elevator up three floors (60 feet) and finds that the angle of elevation to the top of the building across the street is $30°$. How tall is the building across the street? (Give your answer to the nearest foot.)

________ ft

5. For the triangle described below, solve for B and use the results to explain why the triangle has no solution. Round your answer to the nearest tenth.

$A = 150°$, $b = 60$ ft, $a = 20$ ft

6. Find all solutions to the triangle described below.

$A = 124.3°$, $a = 27.7$ cm, $b = 46.6$ cm

Please round each angle and the length of the missing side to the nearest tenth at the end of computations.

If there is more than one possible triangle, list both.

7. Find all solutions to the triangle described below.

$A = 64°$, $b = 7.7$ yd, $a = 7.1$ yd

Please round each angle to the nearest degree and the length of the missing side to the nearest tenth.

If there is more than one possible triangle, list both.

8. Draw a vector representing the course of a ship that goes 50 miles on a course with direction 225° .

9. For a triangle ABC. If $a = 140$ inches, $b = 63$ inches, and $C = 60^\circ$, find c.

Please round your answer to the nearest integer.

$c =$ _________ inches

10. For a triangle ABC. If $a = 53$ cm, $b = 25$ cm, and $c = 31$ cm, find the largest angle.

Please round your answer to the nearest degree.

_________ $^\circ$

11. Solve the triangle.

$b = 63.4$ km, $c = 75.6$ km, $A = 124^\circ\ 20'$

Please round the length to the nearest tenth and each angle to the nearest minute. Do not round any numbers until the end of each computation.

$a =$ _________ km

$B =$ _________ $^\circ$ _________ '

$C =$ _________ $^\circ$ _________ '

12. Two ships leave a harbor entrance at the same time. The first ship is traveling at a constant 16 miles per hour, while the second is traveling at a constant 21 miles per hour. If the angle between their courses is $121°$, how far apart are they after 2 hours?

Please round your answer to the nearest mile.

__________ mi

13. Referring to triangle ABC, find the triangle's area given

$a = 50$ cm, $b = 70$ cm, $C = 60°$

Please round your answer to the nearest integer.

$S =$ __________ cm^2

14. Referring to triangle ABC, find the triangle's area given

$b = 63.3$ km, $c = 75.4$ km, $A = 125°\ 40'$

Please round your answer to the nearest integer.

$S =$ __________ km^2

15. Referring to triangle ABC, find the triangle's area given

$A = 44°\ 30'$, $C = 115°\ 30'$, $a = 3.46$ ft

Please round your answer to the nearest hundredth.

$S =$ __________ ft^2

16. Referring to triangle ABC, find the triangle's area given

$a = 4.37$ ft, $b = 3.76$ ft, $c = 5.27$ ft

Please round your answer to the nearest hundredth.

$S =$ __________ ft^2

17. Draw the vector **V** that goes from the origin to the given point.

$(-3, 4)$

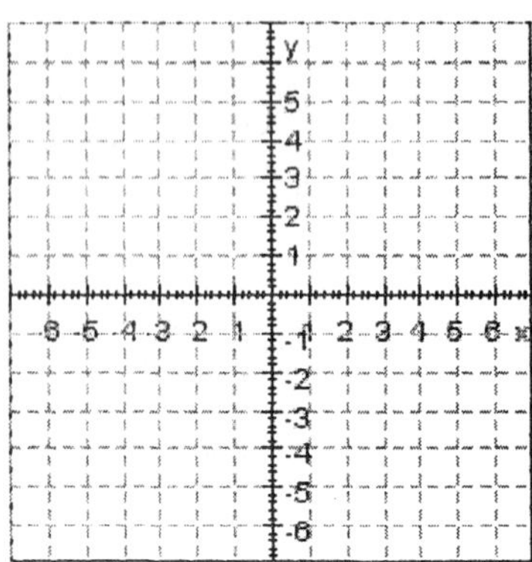

Write the vector **V** in component form $\langle a, b \rangle$.

V = _______

18. Draw the vector **V** that goes from the origin to the given point.

$(4, -6)$

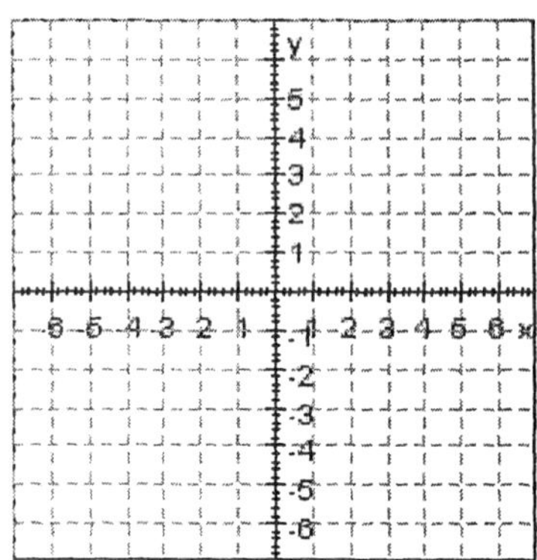

Write the vector **V** in terms of the unit vectors **i** and **j**.

V = _______

19. Find the magnitude of the vector.

$$U = -5i - 12j$$

20. For the pair of vectors, find $U + V$, $U - V$, and $3U + 2V$.

$$U = -2i + 2j, \quad V = 2i + 2j$$

Please enter your answer in terms of the unit vectors i and j.

$U + V =$ ________

$U - V =$ ________

$3U + 2V =$ ________

21. Find the dot product.

$$\langle 4, 5 \rangle \cdot \langle 5, 5 \rangle =$$ ________

22. For the pair of vectors, find $U \cdot V$.

$$U = 6i + 2j \quad V = 2i + 6j$$

$U \cdot V =$ ________

23. Find the angle θ between the given vectors to the nearest tenth of a degree.

$$U = 12i + 7j \quad V = -3i + 5j$$

$\theta =$ ________ $^\circ$

24. Show that the pair of vectors $-\mathbf{i}$ and $-\mathbf{j}$ is perpendicular.

25. Find the work performed when the given force F is applied to an object, whose resulting motion is represented by the displacement vector d. Assume the force is in pounds and the displacement is measured in feet.

$$F = 20\mathbf{i} + 8\mathbf{j} \qquad d = 30\mathbf{i} + 2\mathbf{j}$$

Work = _________ ft-lb

1. 12

2. 77

    13

3. 143.8

    118

    222

4. 130

5. $\sin B = 1.5$; since $\sin B > 1$, there is no triangle.

6. no solution

7. $\left(77^{\circ}, 39^{\circ}, 5\right), \left(103^{\circ}, 13^{\circ}, 1.8\right)$

8.

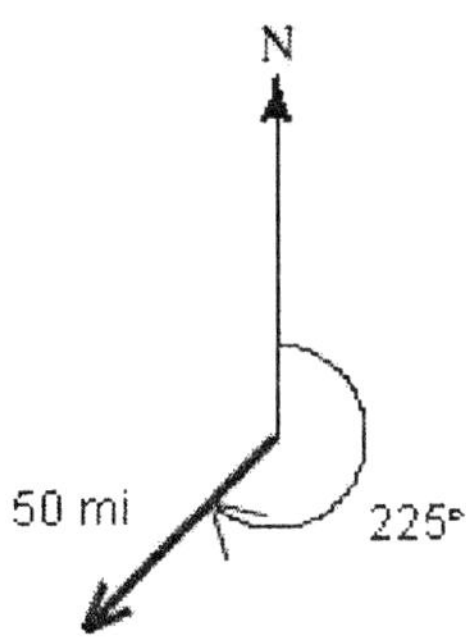

9. 121

10. 142

11. 123.1

    25

    11

    30

    29.2

12. 65

13. 1.516

14. 1.939

15. 2.64

16. 8.10

17.

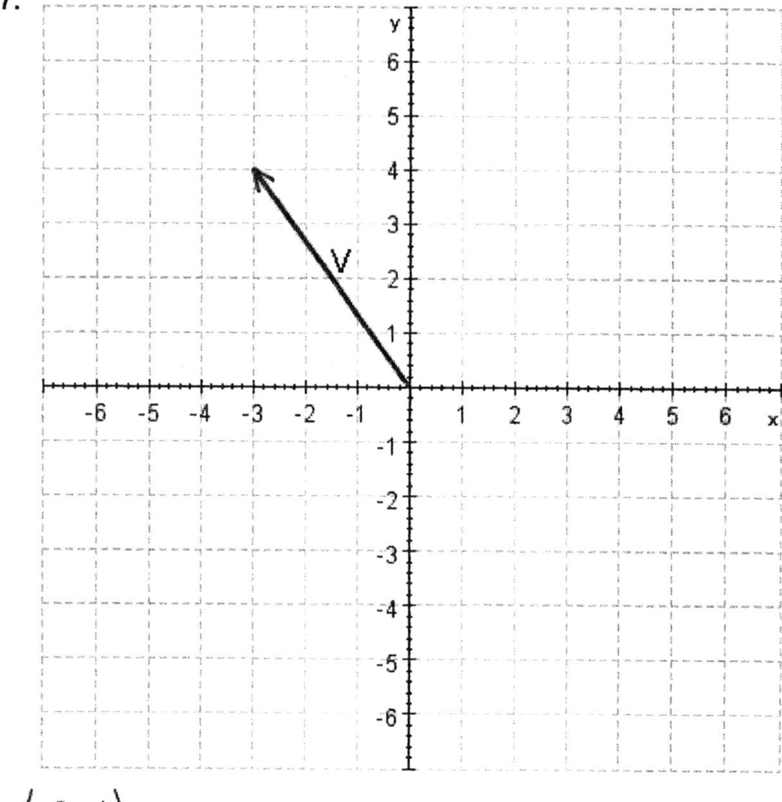

$\langle -3, 4 \rangle$

18.

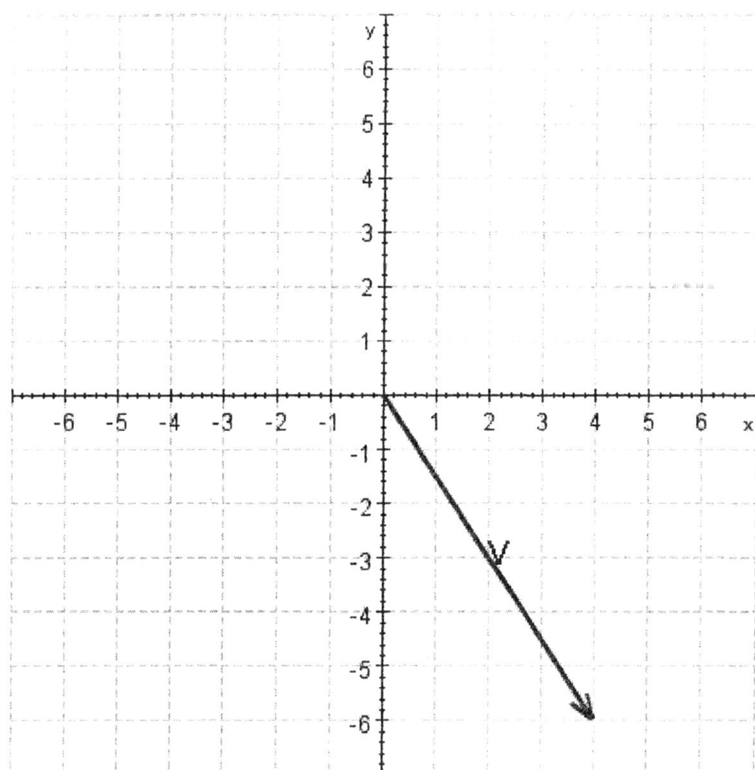

$4\,i - 6\,j$

19. $|U| = 13$

20. $4\,j$

$-4\,i$

$-2i + 10j$

21. 45

22. 24

23. 90.7

24.
$$\cos\theta = \frac{(-1)0 + 0(-1)}{\sqrt{(-1)^2 + 0^2} \cdot \sqrt{0^2 + (-1)^2}} = \frac{0}{\sqrt{1} \cdot \sqrt{1}} = 0$$

$\cos\theta = 0$

$\theta = \cos^{-1}(0) = 90^\circ$

Thus, the pair of vectors is perpendicular.

25. 616

1. mctr.07.01.02_NoAlgs
2. mctr.07.01.09_NoAlgs
3. mctr.07.01.17_NoAlgs
4. mctr.07.01.28_NoAlgs
5. mctr.07.02.02_NoAlgs
6. mctr.07.02.10_NoAlgs
7. mctr.07.02.18_NoAlgs
8. mctr.07.02.26_NoAlgs
9. mctr.07.03.01_NoAlgs
10. mctr.07.03.08_NoAlgs
11. mctr.07.03.14_NoAlgs
12. mctr.07.03.22_NoAlgs
13. mctr.07.04.01_NoAlgs
14. mctr.07.04.06_NoAlgs
15. mctr.07.04.11_NoAlgs
16. mctr.07.04.17_NoAlgs
17. mctr.07.05.03_NoAlgs
18. mctr.07.05.13_NoAlgs
19. mctr.07.05.25_NoAlgs
20. mctr.07.05.36_NoAlgs
21. mctr.07.06.02_NoAlgs
22. mctr.07.06.09_NoAlgs
23. mctr.07.06.16_NoAlgs
24. mctr.07.06.19_NoAlgs
25. mctr.07.06.23_NoAlgs

1. Referring to triangle *ABC,* find the triangle's area given

    $A = 44^\circ\ 30'$, $C = 115^\circ\ 30'$, $a = 3.46$ ft

    Please round your answer to the nearest hundredth.

    $S =$ _________ ft^2

2. The problem that follows refers to triangle ABC.
    If $A = 55^\circ$, $B = 48^\circ$ and $c = 16$ cm.

    Find C.

    $C =$ _________ $^\circ$

    Find a. Please round the answer to the nearest centimeter.

    $a =$ _________ cm

3. The information below refers to triangle ABC.

    $B = 12.4^\circ$, $C = 23.8^\circ$, $a = 325$ cm

    Find A.

    $A =$ _________ $^\circ$

    Find b. Please round the answer to the nearest centimeter.

    $b =$ _________ cm

    Find c. Please round the answer to the nearest centimeter.

    $c =$ _________ cm

4. A woman entering an outside glass elevator on the ground floor of a hotel glances up to the top of the building across the street and notices that the angle of elevation is 47°. She rides the elevator up three floors (60 feet) and finds that the angle of elevation to the top of the building across the street is 30°. How tall is the building across the street? (Give your answer to the nearest foot.)

    __________ ft

5. Find the work performed when the given force F is applied to an object, whose resulting motion is represented by the displacement vector d. Assume the force is in pounds and the displacement is measured in feet.

    $F = 20i + 8j \qquad d = 30i + 2j$

    Work = __________ ft-lb

6. Find the magnitude of the vector.

    $U = -5i - 12j$

7. For the pair of vectors, find $U + V$, $U - V$, and $3U + 2V$.

    $U = -2i + 2j,\ V = 2i + 2j$

    Please enter your answer in terms of the unit vectors i and j.

    $U + V =$ __________

    $U - V =$ __________

    $3U + 2V =$ __________

8. Refer to triangle *ABC*. If $a = 140$ inches, $b = 63$ inches, and $C = 60°$, find c.

   Please round your answer to the nearest integer.

   $c =$ _________ inches

9. For the pair of vectors, find $\mathbf{U} \cdot \mathbf{V}$.

   $\mathbf{U} = 6\mathbf{i} + 2\mathbf{j} \quad \mathbf{V} = 2\mathbf{i} + 6\mathbf{j}$

   $\mathbf{U} \cdot \mathbf{V} =$ _________

10. Show that the pair of vectors $-\mathbf{i}$ and $-\mathbf{j}$ is perpendicular.

11. For the triangle described below, solve for *B* and use the results to explain why the triangle has no solution. Round your answer to the nearest tenth.

    $A = 150°, \ b = 60 \text{ ft}, \ a = 20 \text{ ft}$

12. Find all solutions to the triangle described below.

    $A = 124.3°, \ a = 27.7 \text{ cm}, \ b = 46.6 \text{ cm}$

    Please round each angle and the length of the missing side to the nearest tenth at the end of computations.

    If there is more than one possible triangle, list both.

13. Draw a vector representing the course of a ship that goes 50 miles on a course with direction 225° .

14. Referring to triangle *ABC,* find the triangle's area given

$a = 50$ cm, $b = 70$ cm, $C = 60^\circ$

Please round your answer to the nearest integer.

$S =$ _________ cm^2

15. The problem that follows refers to triangle *ABC*.

If $A = 70^\circ$, $B = 40^\circ$ and $b = 8$ cm, find a.

Please round the answer to the nearest centimeter.

$a =$ _________ cm

16. Find all solutions to the triangle described below.

$A = 64^\circ$, $b = 7.7$ yd, $a = 7.1$ yd

Please round each angle to the nearest degree and the length of the missing side to the nearest tenth.

If there is more than one possible triangle, list both.

17. Solve the triangle.

$b = 63.4$ km, $c = 75.6$ km, $A = 124°$ 20'

Please round the length to the nearest tenth and each angle to the nearest minute. Do not round any numbers until the end of each computation.

$a =$ _________ km

$B =$ _________ ° _________ '

$C =$ _________ ° _________ '

18. Referring to triangle *ABC,* find the triangle's area given

$a = 4.37$ ft, $b = 3.76$ ft, $c = 5.27$ ft

Please round your answer to the nearest hundredth.

$S =$ _________ ft^2

19. Referring to triangle *ABC,* find the triangle's area given

$b = 63.3$ km, $c = 75.4$ km, $A = 125°$ 40'

Please round your answer to the nearest integer.

$S =$ _________ km^2

20. Refer to triangle *ABC*. If $a = 53$ cm, $b = 25$ cm, and $c = 31$ cm, find the largest angle.

Please round your answer to the nearest degree.

_________ °

21. Two ships leave a harbor entrance at the same time. The first ship is traveling at a constant 16 miles per hour, while the second is traveling at a constant 21 miles per hour. If the angle between their courses is $121°$, how far apart are they after 2 hours?

Please round your answer to the nearest mile.

_________ mi

22. Find the dot product.

$$\langle 4, 5 \rangle \cdot \langle 5, 5 \rangle = \underline{\hspace{2cm}}$$

23. Find the angle θ between the given vectors to the nearest tenth of a degree.

$$U = 12i + 7j \qquad V = -3i + 5j$$

$$\theta = \underline{\hspace{2cm}} \,^{\circ}$$

24. Draw the vector **V** that goes from the origin to the given point.

$$(4, -6)$$

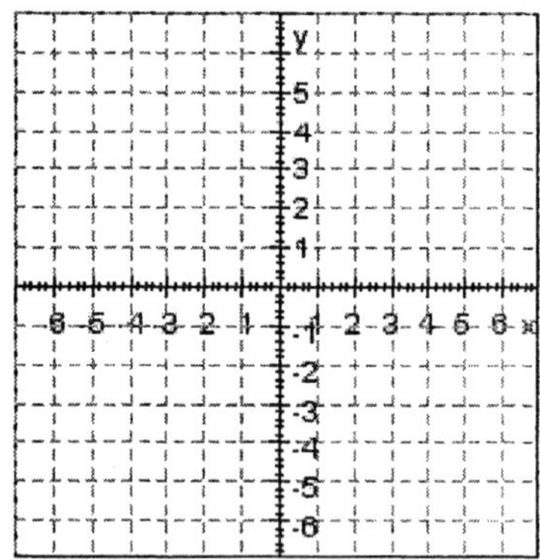

Write the vector **V** in terms of the unit vectors **i** and **j**.

$$V = \underline{\hspace{2cm}}$$

25. Draw the vector **V** that goes from the origin to the given point.

$(-3, 4)$

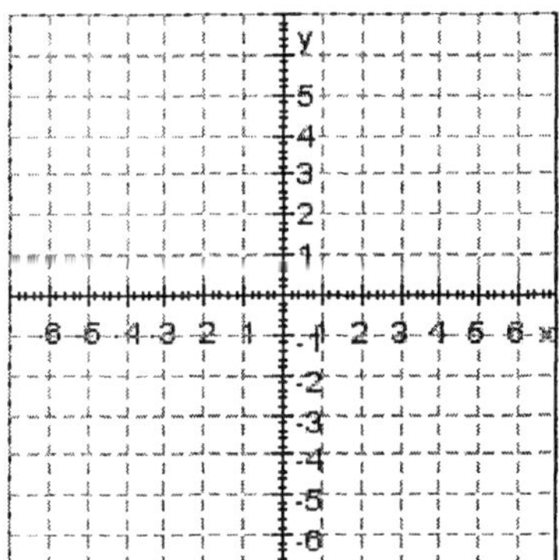

Write the vector **V** in component form (a, b).

V = _______

1. 2.64

2. 77

13

3. 143.8

118

222

4. 130

5. 616

6. $|U| = 13$

7. $4j$

$-4i$

$-2i + 10j$

8. 121

9. 24

10.
$$\cos\theta = \frac{(-1)0 + 0(-1)}{\sqrt{(-1)^2 + 0^2} \cdot \sqrt{0^2 + (-1)^2}} = \frac{0}{\sqrt{1} \cdot \sqrt{1}} = 0$$

$$\cos\theta = 0$$

$$\theta = \cos^{-1}(0) = 90°$$

Thus, the pair of vectors is perpendicular.

11. $\sin B = 1.5$; since $\sin B > 1$, there is no triangle.

12. no solution

13.

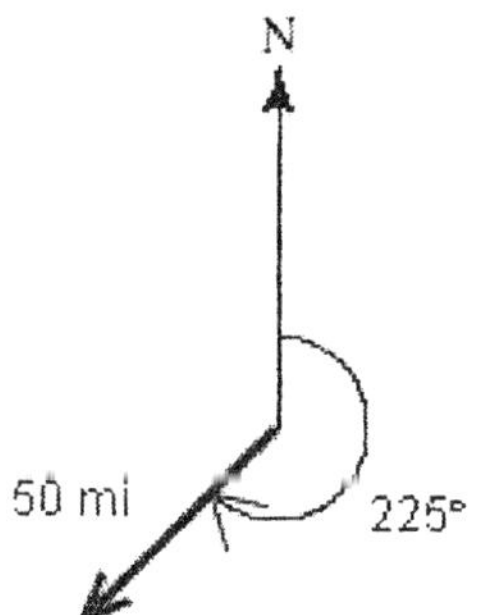

14. 1.516

15. 12

16. $\left(77^\circ, 39^\circ, 5\right), \left(103^\circ, 13^\circ, 1.8\right)$

17. 123.1
   25
   11
   30
   29.2

18. 8.10

19. 1.939

20. 14 2

21. 65

22. 45

23. 90.7

24.

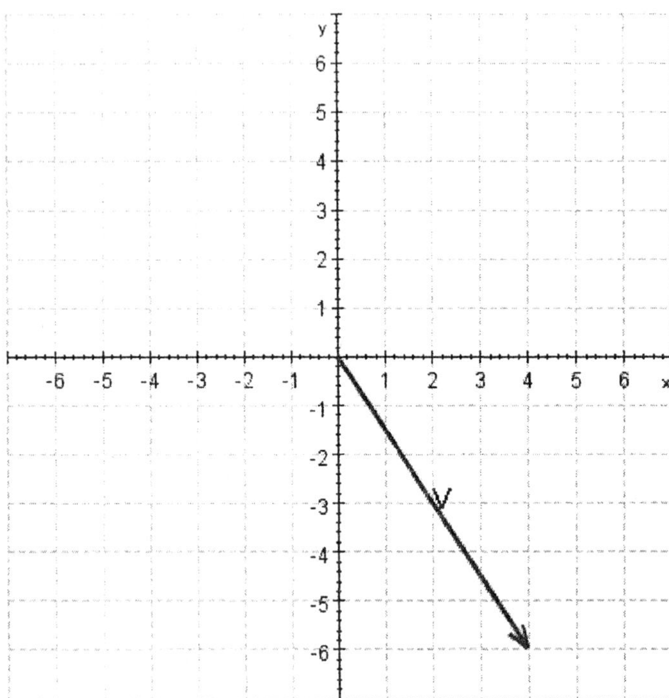

$4i - 6j$

25.

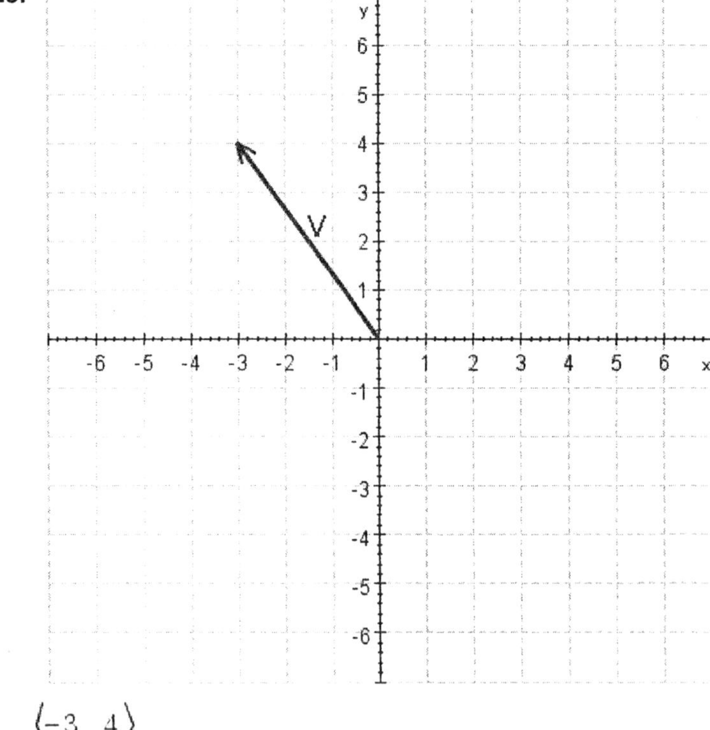

$\langle -3, 4 \rangle$

McKeague/Turner - Trigonometry 5e Chapter 7 Form B

1. mctr.07.04.11_NoAlgs
2. mctr.07.01.09_NoAlgs
3. mctr.07.01.17_NoAlgs
4. mctr.07.01.28_NoAlgs
5. mctr.07.06.23_NoAlgs
6. mctr.07.05.25_NoAlgs
7. mctr.07.05.36_NoAlgs
8. mctr.07.03.01_NoAlgs
9. mctr.07.06.09_NoAlgs
10. mctr.07.06.19_NoAlgs
11. mctr.07.02.02_NoAlgs
12. mctr.07.02.10_NoAlgs
13. mctr.07.02.26_NoAlgs
14. mctr.07.04.01_NoAlgs
15. mctr.07.01.02_NoAlgs
16. mctr.07.02.18_NoAlgs
17. mctr.07.03.14_NoAlgs
18. mctr.07.04.17_NoAlgs
19. mctr.07.04.06_NoAlgs
20. mctr.07.03.08_NoAlgs
21. mctr.07.03.22_NoAlgs
22. mctr.07.06.02_NoAlgs
23. mctr.07.06.16_NoAlgs
24. mctr.07.05.13_NoAlgs
25. mctr.07.05.03_NoAlgs

1. The problem that follows refers to triangle ABC.

   If $B = 100°$, $C = 20°$ and $b = 14$ inches, find c.

   Please round the answer to the nearest inch.

   Select the correct answer.

   a.   5 inches

   b.   6 inches

   c.   7 inches

   d.   8 inches

   e.   1 inches

2. The information below refers to triangle ABC.

   $B = 55°$, $C = 32°$, $a = 7.3$ m.

   Find all the missing parts. Round side lengths to a tenth of a meter.

   Select the correct answer.

   a. $A = 91°$, $b = 6$ m, $c = 3.6$ m

   b. $A = 91°$, $b = 6.2$ m, $c = 3.9$ m

   c. $A = 93°$, $b = 6.2$ m, $c = 3.6$ m

   d. $A = 96°$, $b = 4.5$ m, $c = 9.8$ m

   e. $A = 93°$, $b = 6$ m, $c = 3.9$ m

3. The circle in the figure below has a radius of *r* and center at *C*. The distance from *A* to *B* is *x*, the distance from *A* to *D* is *y*, and the length of arc *BD* is *s*. For the problem, redraw the figure, label it with the values given then solve the problem.

If $A = 40°$, $s = 17$, and $r = 14$, find *y*. Please round the answer to the nearest integer.

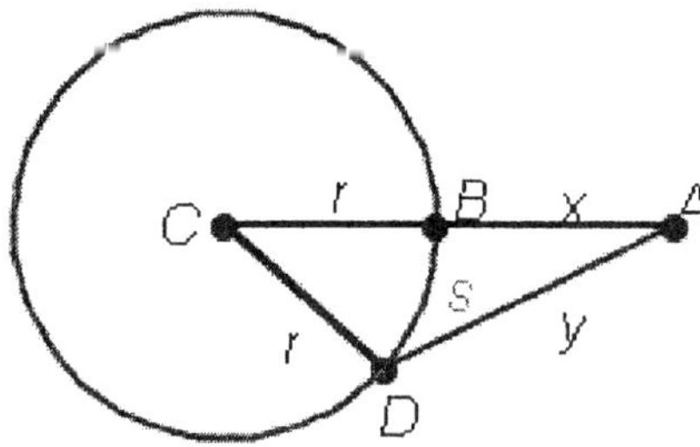

Select the correct answer.

a. $y = 19$

b. $y = 18$

c. $y = 24$

d. $y = 23$

e. $y = 20$

4. A tightrope walker is standing still with one foot on the tightrope as shown in the figure. If the tightrope walker weighs 122 pounds, find the magnitudes of the tension in the rope toward

    each end of the rope. Please round the answers to the nearest pound.

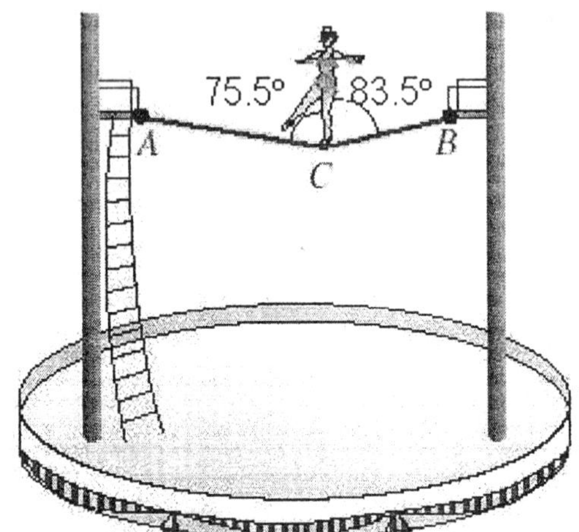

Select the correct answer.

a. $|CB| = 330$ lbs, $|CA| = 338$ lbs

b. $|CB| = 334$ lbs, $|CA| = 336$ lbs

c. $|CB| = 331$ lbs, $|CA| = 340$ lbs

d. $|CB| = 330$ lbs, $|CA| = 340$ lbs

e. $|CB| = 331$ lbs, $|CA| = 338$ lbs

5. For the triangle described below, solve for B. Round your answer to the nearest tenth.

    $A = 60°$, $b = 72$ ft, $a = 64$ ft

Select the correct answer.

a. $B = 103°$ or $B' = 10.9°$

b. $B = 77°$ or $B' = 103°$

c. $B = 77°$

d. $B = 10.9°$

e.   no solutions

6. Find all solutions to the triangle described below.

   $A = 118°$, $a = 0.66$ cm, $b = 0.91$ cm

   Please round each angle to the nearest degree and the length of the missing side to the nearest tenth.

   Select the correct answer.

   a. $B = 19°$, $C = 43°$, $c = 0.51$ cm

   b. $B = 19°$, $C = 43°$, $c = 0.51$ cm or
   $B' = 161°$, $C' = 99°$, $c' = 0.74$ cm

   c. $B = 19°$, $C = 43°$, $c = 0.51$ cm or
   $B' = 22°$, $C' = 40°$, $c' = 0.88$ cm

   d. $B = 22°$, $C = 40°$, $c = 0.88$ cm

   e.   no solution

7. A 50-foot wire running from the top of a tent pole to the ground makes an angle of $55°$ with the ground. If the length of the tent pole is 45 feet, how far is it from the bottom of the tent pole to the point where the wire is fastened to the ground? (The tent pole is not necessarily perpendicular to the ground.)

   Please round your answer to the nearest foot.

   Select the correct answer.

   a.   47 ft
   b.   13 ft
   c.   10 ft or 47 ft
   d.   13 ft or 59 ft
   e.   10 ft

8. A plane headed due north is traveling with an airspeed of 180 miles per hour. The wind currents are moving with constant speed in the direction $150°$. If the ground speed of the plane is 90 miles per hour, what is its true course?

   Select the correct answer.

   a. $43°$

   b. $80°$

   c. $10°$

   d. $15°$

   e. $60°$

9. Refer to triangle ABC. If $b = 4.5$ m, $c = 6.3$ m, and $A = 116°$, find a.

   Please round your answer to the nearest tenth.

   Select the correct answer.

   a. 9.6 m
   b. 9.4 m
   c. 9.2 m
   d. 10.0 m
   e. 9.8 m

10. Solve the triangle.

    $a = 51$ yd, $b = 74$ yd, $c = 64$ yd

    Please round each answer to the nearest integer. Do not round any numbers until the end of each computation.

    Select the correct answer.

    a. $A = 45°, B = 76°, C = 58°$

    b. $A = 43°, B = 79°, C = 59°$

    c. $A = 45°, B = 76°, C = 59°$

    d. $A = 43°, B = 79°, C = 58°$

    e. $A = 39°, B = 73°, C = 68°$

11. Use the law of cosines to find a true statement from the list below, if $A = 120°$.

Select the correct answer.

a. $a^2 = b^2 + c^2 + bc\sqrt{2}$

b. $a^2 = b^2 + c^2 - bc\sqrt{2}$

c. $a^2 = b^2 + c^2 + bc$

d. $a^2 = b^2 + c^2 - bc$

e. $a^2 = b^2 + c^2$

12. A plane is flying with an airspeed of 244 miles per hour with heading 277.2°. The wind currents are running at a constant 45.7 miles per hour in the direction 267.1°. Find the ground speed and true course of the plane.

Please round each answer to the nearest tenth at the end of the computations.

Select the correct answer.

a. 289.1 mph with heading 276.0°

b. 289.1 mph with heading 275.6°

c. 288.7 mph with heading 275.6°

d. 288.3 mph with heading 276.4°

e. 288.7 mph with heading 276.0°

13. Referring to triangle *ABC*, find the triangle's area given

$a = 39.5$ m, $c = 35.5$ m, $B = 154.5°$

Please round your answer to the nearest integer.

Select the correct answer.

a. 308 m^2

b. 302 m^2

c. 298 m^2

d. 300 m^2

e. 294 m^2

14. Referring to triangle *ABC*, find the triangle's area given

$A = 42.2°$, $B = 71.2°$, $a = 220$ inches

Please round your answer to the nearest hundred.

Select the correct answer.

a. 31,500 inches^2

b. 31,000 inches^2

c. 31,300 inches^2

d. 31,200 inches^2

e. 30,900 inches^2

15. Referring to triangle *ABC,* find the triangle's area given

$a = 48$ inches, $b = 68$ inches, $c = 84$ inches

Please round your answer to the nearest ten.

Select the correct answer.

a.   1,690 inches^2

b.   1,710 inches^2

c.   1,630 inches^2

d.   1,650 inches^2

e.   1,590 inches^2

16. The area of a triangle is 50 cm^2. Find the length of the side included between the angles

$A = 30^\circ$ and $B = 40^\circ$.

Please round your answer to the nearest tenth.

Select the correct answer.

a.   16.7 cm
b.   17.3 cm
c.   16.5 cm
d.   16.3 cm
e.   17.1 cm

17. Draw the vector **V** that goes from the origin to the given point. Then write **V** in component form $\langle a, b \rangle$.

$(-6, -4)$

Select the correct answer.

a.

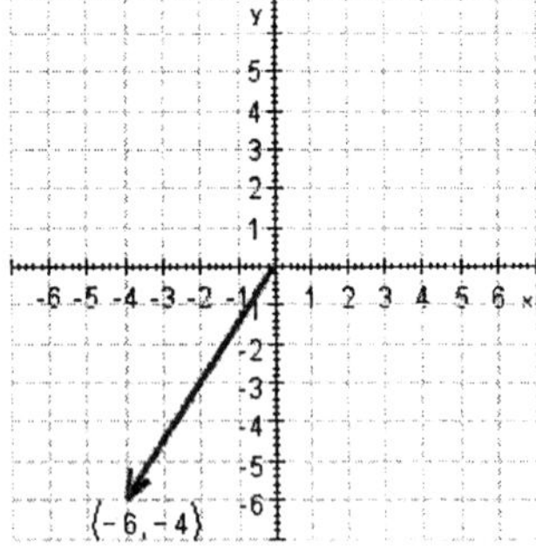

b.

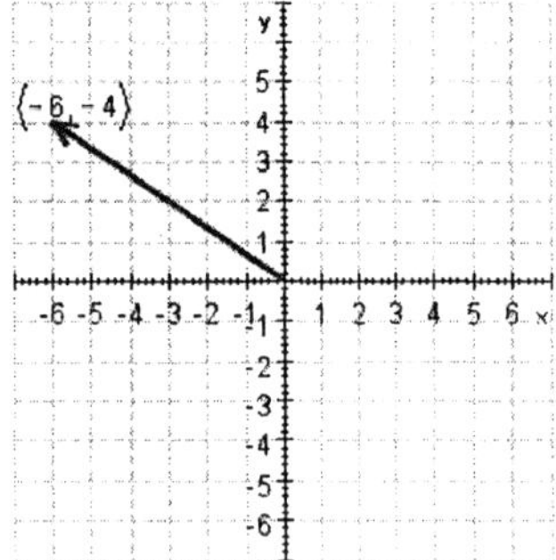

c.

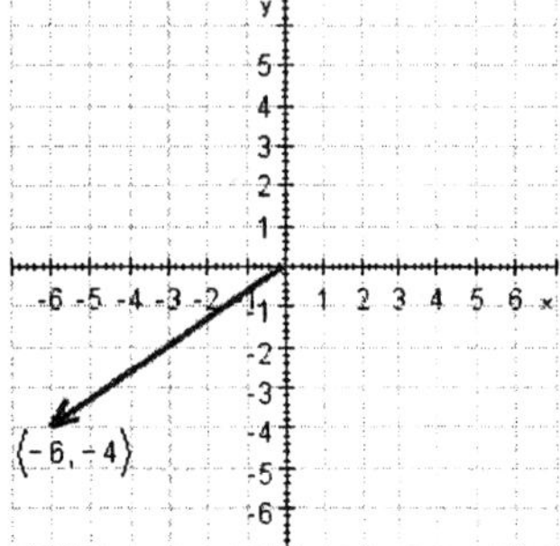

18. Find the magnitude of the vector.

$$\langle 0, 11 \rangle$$

Select the correct answer.

a. $|\mathbf{W}| = 9$

b. $|\mathbf{W}| = 11$

c. $|\mathbf{W}| = 121$

d. $|\mathbf{W}| = 0$

e. $|\mathbf{W}| = 12$

19. For the pair of vectors, find $2\mathbf{U} - 3\mathbf{V}$.

$$\mathbf{U} = \langle 6, 6 \rangle, \ \mathbf{V} = \langle 6, -6 \rangle$$

Select the correct answer.

a. $\langle -6, 30 \rangle$

b. $\langle 30, -6 \rangle$

c. $\langle 29, -7 \rangle$

d. $\langle 6, 30 \rangle$

e. $\langle -7, 29 \rangle$

20. Find the magnitude of the vector and the angle θ, $0° \leq \theta < 360°$, that the vector makes with the positive *x*-axis.

$$\mathbf{W} = \mathbf{i} + \sqrt{3}\,\mathbf{j}$$

Select the correct answer.

a. $\quad |\mathbf{W}| = 2, \ \theta = 120°$

b. $\quad |\mathbf{W}| = 2, \ \theta = 60°$

c. $\quad |\mathbf{W}| = 6, \ \theta = 60°$

d. $\quad |\mathbf{W}| = 3, \ \theta = 300°$

e. $\quad |\mathbf{W}| = 7, \ \theta = 120°$

21. Find the dot product.

$$\left(9, -7\right) \cdot \left(2, -6\right)$$

Select the correct answer.

a. 57
b. 62
c. 60
d. 56
e. 61

22. For the pair of vectors, find $\mathbf{U} \cdot \mathbf{V}$.

$$\mathbf{U} = 2\mathbf{i} + \mathbf{j} \qquad \mathbf{V} = \mathbf{i} + 2\mathbf{j}$$

Select the correct answer.

a. 0
b. 5
c. 2
d. 4
e. 1

23. Find the angle θ between the given vectors to the nearest tenth of a degree.

$$U = -2i + 7j \quad V = 6i + 2j$$

Select the correct answer.

a. $\theta = 83.5°$

b. $\theta = 88.5°$

c. $\theta = 87.5°$

d. $\theta = 89.5°$

e. $\theta = 84.5°$

24. Is the pair of vectors i and $-j$ perpendicular?

Select the correct answer.

a. yes
b. no

25. Find the work performed when the given force F is applied to an object, whose resulting motion is represented by the displacement vector d. Assume the force is in pounds and the displacement is measured in feet.

$$F = 95i \quad d = 8i$$

Select the correct answer.

a. 730 ft-lb
b. 780 ft-lb
c. 760 ft-lb
d. 750 ft-lb
e. 720 ft-lb

1. a
2. e
3. e
4. a
5. b
6. e
7. c
8. e
9. c
10. d
11. c
12. b
13. b
14. c
15. c
16. e
17. c
18. b
19. a
20. b
21. c
22. d
23. c
24. a
25. c

1. mctr.07.01.04m_NoAlgs
2. mctr.07.01.14m_NoAlgs
3. mctr.07.01.23m_NoAlgs
4. mctr.07.01.35m_NoAlgs
5. mctr.07.02.05m_NoAlgs
6. mctr.07.02.15m_NoAlgs
7. mctr.07.02.21m_NoAlgs
8. mctr.07.02.30m_NoAlgs
9. mctr.07.03.05m_NoAlgs
10. mctr.07.03.12m_NoAlgs
11. mctr.07.03.17m_NoAlgs
12. mctr.07.03.26m_NoAlgs
13. mctr.07.04.03m_NoAlgs
14. mctr.07.04.09m_NoAlgs
15. mctr.07.04.13m_NoAlgs
16. mctr.07.04.21m_NoAlgs
17. mctr.07.05.07m_NoAlgs
18. mctr.07.05.20m_NoAlgs
19. mctr.07.05.29m_NoAlgs
20. mctr.07.05.47m_NoAlgs
21. mctr.07.06.04m_NoAlgs
22. mctr.07.06.09m_NoAlgs
23. mctr.07.06.14m_NoAlgs
24. mctr.07.06.19m_NoAlgs
25. mctr.07.06.27m_NoAlgs

1. The problem that follows refers to triangle ABC.

   If $B = 100^\circ$, $C = 20^\circ$ and $b = 14$ inches, find c.

   Please round the answer to the nearest inch.

   Select the correct answer.

   a.   5 inches

   b.   6 inches

   c.   7 inches

   d.   8 inches

   e.   1 inches

2. For the pair of vectors, find $U \cdot V$.

   $$U = 2i + j \quad V = i + 2j$$

   Select the correct answer.

   a.   0
   b.   5
   c.   2
   d.   4
   e.   1

3. Refer to triangle ABC. If $b = 4.5$ m, $c = 6.3$ m, and $A = 116^\circ$, find a.

   Please round your answer to the nearest tenth.

   Select the correct answer.

   a.   9.6 m
   b.   9.4 m
   c.   9.2 m
   d.   10.0 m
   e.   9.8 m

4. Solve the triangle.

    $a = 51$ yd, $b = 74$ yd, $c = 64$ yd

    Please round each answer to the nearest integer. Do not round any numbers until the end of each computation.

    Select the correct answer.

    a. $A = 45°, B = 76°, C = 58°$

    b. $A = 43°, B = 79°, C = 59°$

    c. $A = 45°, B = 76°, C = 59°$

    d. $A = 43°, B = 79°, C = 58°$

    e. $A = 39°, B = 73°, C = 68°$

5. Find the work performed when the given force F is applied to an object, whose resulting motion is represented by the displacement vector d. Assume the force is in pounds and the displacement is measured in feet.

    $F = 95i \qquad d = 8i$

    Select the correct answer.

    a. 730 ft-lb
    b. 780 ft-lb
    c. 760 ft-lb
    d. 750 ft-lb
    e. 720 ft-lb

6. Use the law of cosines to find a true statement from the list below, if $A = 120°$.

    Select the correct answer.

    a. $a^2 = b^2 + c^2 + bc\sqrt{2}$

    b. $a^2 = b^2 + c^2 - bc\sqrt{2}$

    c. $a^2 = b^2 + c^2 + bc$

    d. $a^2 = b^2 + c^2 - bc$

    e. $a^2 = b^2 + c^2$

7. Find the magnitude of the vector and the angle θ, $0° \le \theta < 360°$, that the vector makes with the positive *x*-axis.

$$\mathbf{W} = \mathbf{i} + \sqrt{3}\,\mathbf{j}$$

Select the correct answer.

a. $|\mathbf{W}| = 2,\ \theta = 120°$

b. $|\mathbf{W}| = 2,\ \theta = 60°$

c. $|\mathbf{W}| = 6,\ \theta = 60°$

d. $|\mathbf{W}| = 3,\ \theta = 300°$

e. $|\mathbf{W}| = 7,\ \theta = 120°$

8. A tightrope walker is standing still with one foot on the tightrope as shown in the figure. If the tightrope walker weighs 122 pounds, find the magnitudes of the tension in the rope toward each end of the rope. Please round the answers to the nearest pound.

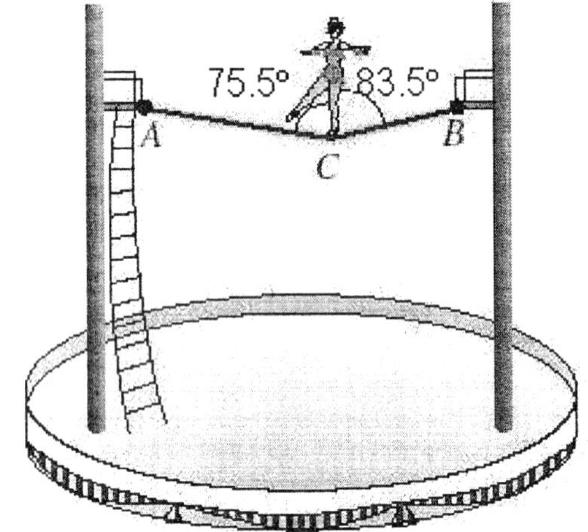

Select the correct answer.

a. $|CB| = 330$ lbs, $|CA| = 338$ lbs

b. $|CB| = 334$ lbs, $|CA| = 336$ lbs

c. $|CB| = 331$ lbs, $|CA| = 340$ lbs

d. $|CB| = 330$ lbs, $|CA| = 340$ lbs

e. $|CB| = 331$ lbs, $|CA| = 338$ lbs

9. For the triangle described below, solve for B. Round your answer to the nearest tenth.

$A = 60°$, $b = 72$ ft, $a = 64$ ft

Select the correct answer.

a. $B = 103°$ or $B' = 10.9°$

b. $B = 77°$ or $B' = 103°$

c. $B = 77°$

d. $B = 10.9°$

e. no solutions

10. Referring to triangle ABC, find the triangle's area given

$A = 42.2°$, $B = 71.2°$, $a = 220$ inches

Please round your answer to the nearest hundred.

Select the correct answer.

a. $31,500$ inches^2

b. $31,000$ inches^2

c. $31,300$ inches^2

d. $31,200$ inches^2

e. $30,900$ inches^2

11. Find the angle θ between the given vectors to the nearest tenth of a degree.

$U = -2i + 7j$ $V = 6i + 2j$

Select the correct answer.

a. $\theta = 83.5°$

b. $\theta = 88.5°$

c. $\theta = 87.5°$

d. $\theta = 89.5°$

e. $\theta = 84.5°$

12. Find the magnitude of the vector.

$$\langle 0, 11 \rangle$$

Select the correct answer.

a. $|\mathbf{W}| = 9$

b. $|\mathbf{W}| = 11$

c. $|\mathbf{W}| = 121$

d. $|\mathbf{W}| = 0$

e. $|\mathbf{W}| = 12$

13. Referring to triangle *ABC,* find the triangle's area given

$a = 39.5$ m, $c = 35.5$ m, $B = 154.5^\circ$

Please round your answer to the nearest integer.

Select the correct answer.

a. 308 m^2

b. 302 m^2

c. 298 m^2

d. 300 m^2

e. 294 m^2

14. Is the pair of vectors $\mathbf{i}$ and $-\mathbf{j}$ perpendicular?

Select the correct answer.

a.   yes

b.   no

15. Find the dot product.

$$\langle 9, -7 \rangle \cdot \langle 2, -6 \rangle$$

Select the correct answer.

a. 57
b. 62
c. C0
d. 56
e. 61

16. The circle in the figure below has a radius of r and center at C. The distance from A to B is x, the distance from A to D is y, and the length of arc BD is s. For the problem, redraw the figure, label it with the values given then solve the problem.
If $A = 40^\circ$, $s = 17$, and $r = 14$, find y. Please round the answer to the nearest integer.

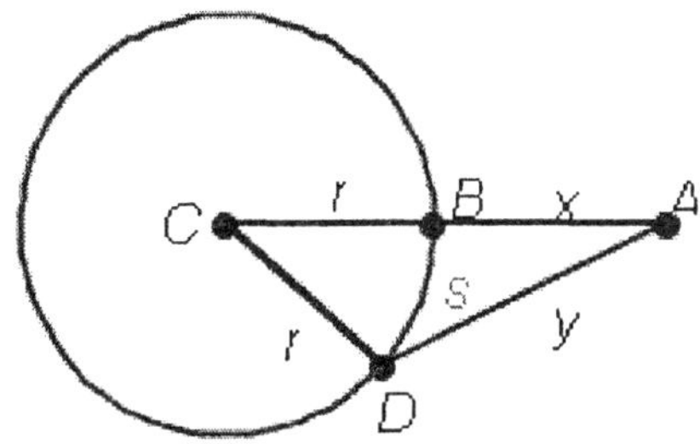

Select the correct answer.

a. $y = 19$

b. $y = 18$

c. $y = 24$

d. $y = 23$

e. $y = 20$

17. A plane headed due north is traveling with an airspeed of 180 miles per hour. The wind currents are moving with constant speed in the direction $150°$. If the ground speed of the plane is 90 miles per hour, what is its true course?

Select the correct answer.

    a. $43°$

    b. $80°$

    c. $10°$

    d. $15°$

    e. $60°$

18. The area of a triangle is 50 cm^2. Find the length of the side included between the angles $A = 30°$ and $B = 40°$.

Please round your answer to the nearest tenth.

Select the correct answer.

    a.   16.7 cm
    b.   17.3 cm
    c.   16.5 cm
    d.   16.3 cm
    e.   17.1 cm

19. For the pair of vectors, find $2\mathbf{U} - 3\mathbf{V}$.

$$\mathbf{U} = \langle 6, 6 \rangle, \ \mathbf{V} = \langle 6, -6 \rangle$$

Select the correct answer.

    a. $\langle -6, 30 \rangle$

    b. $\langle 30, -6 \rangle$

    c. $\langle 29, -7 \rangle$

    d. $\langle 6, 30 \rangle$

    e. $\langle -7, 29 \rangle$

20. A plane is flying with an airspeed of 244 miles per hour with heading 277.2°. The wind currents are running at a constant 45.7 miles per hour in the direction 267.1°. Find the ground speed and true course of the plane.

Please round each answer to the nearest tenth at the end of the computations.

Select the correct answer.

    a.   289.1 mph with heading 276.0°

    b.   289.1 mph with heading 275.6°

    c.   288.7 mph with heading 275.6°

    d.   288.3 mph with heading 276.4°

    e.   288.7 mph with heading 276.0°

21. Find all solutions to the triangle described below.

$$A = 118^\circ, \; a = 0.66 \text{ cm}, \; b = 0.91 \text{ cm}$$

Please round each angle to the nearest degree and the length of the missing side to the nearest tenth.

Select the correct answer.

    a. $B = 19^\circ, \; C = 43^\circ, \; c = 0.51 \text{ cm}$

    b. $B = 19^\circ, \; C = 43^\circ, \; c = 0.51 \text{ cm}$ or
$B' = 161^\circ, \; C' = 99^\circ, \; c' = 0.74 \text{ cm}$

    c. $B = 19^\circ, \; C = 43^\circ, \; c = 0.51 \text{ cm}$ or
$B' = 22^\circ, \; C' = 40^\circ, \; c' = 0.88 \text{ cm}$

    d. $B = 22^\circ, \; C = 40^\circ, \; c = 0.88 \text{ cm}$

    e.   no solution

22. Draw the vector **V** that goes from the origin to the given point. Then write **V** in component form $\langle a, b \rangle$.

$(-6, -4)$

Select the correct answer.

a.

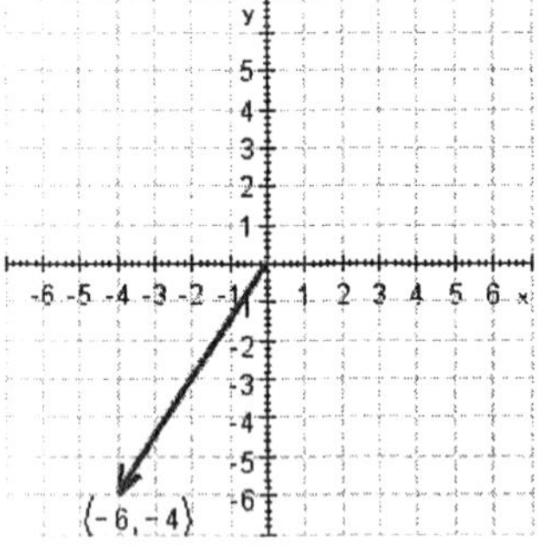

b.

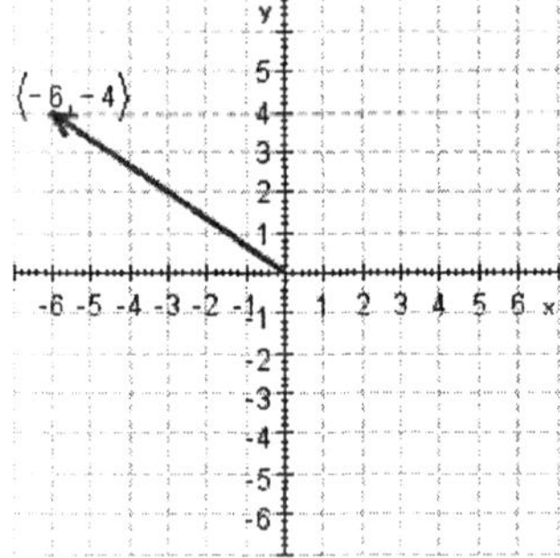

c.

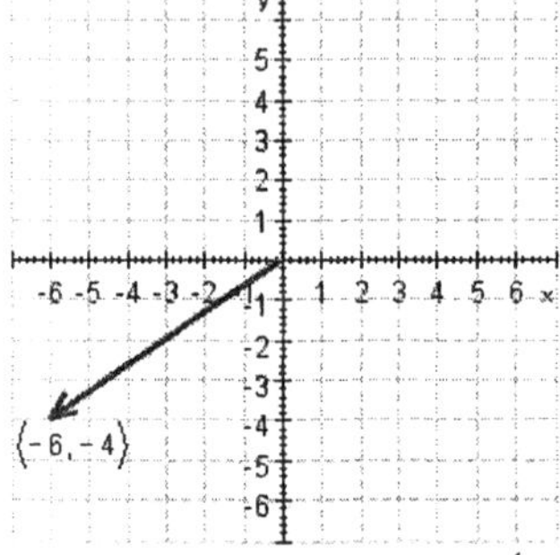

23. The information below refers to triangle ABC.

$B = 55^{\circ}$, $C = 32^{\circ}$, $a = 7.3$ m. Find all the missing parts. Round side lengths to a tenth of a meter.

Select the correct answer.

a. $A = 91^{\circ}$, $b = 6$ m, $c = 3.6$ m

b. $A = 91^{\circ}$, $b = 6.2$ m, $c = 3.9$ m

c. $A = 93^{\circ}$, $b = 6.2$ m, $c = 3.6$ m

d. $A = 96^{\circ}$, $b = 4.5$ m, $c = 9.8$ m

e. $A = 93^{\circ}$, $b = 6$ m, $c = 3.9$ m

24. A 50-foot wire running from the top of a tent pole to the ground makes an angle of 55° with the ground. If the length of the tent pole is 45 feet, how far is it from the bottom of the tent pole to the point where the wire is fastened to the ground? (The tent pole is not necessarily perpendicular to the ground.)

Please round your answer to the nearest foot.

Select the correct answer.

a.   47 ft
b.   13 ft
c.   10 ft or 47 ft
d.   13 ft or 59 ft
e.   10 ft

25. Referring to triangle *ABC*, find the triangle's area given

$a = 48$ inches, $b = 68$ inches, $c = 84$ inches

Please round your answer to the nearest ten.

Select the correct answer.

a.   1,690 inches^2

b.   1,710 inches^2

c.   1,630 inches^2

d.   1,650 inches^2

e.   1,590 inches^2

1. a

2. d

3. c

4. d

5. c

6. c

7. b

8. a

9. b

10. c

11. c

12. b

13. b

14. a

15. c

16. e

17. e

18. e

19. a

20. b

21. e

22. c

23. e

24. c

25. c

McKeague/Turner - Trigonometry 5e Chapter 7 Form D

1. mctr.07.01.04m_NoAlgs
2. mctr.07.06.09m_NoAlgs
3. mctr.07.03.05m_NoAlgs
4. mctr.07.03.12m_NoAlgs
5. mctr.07.06.27m_NoAlgs
6. mctr.07.03.17m_NoAlgs
7. mctr.07.05.47m_NoAlgs
8. mctr.07.01.35m_NoAlgs
9. mctr.07.02.05m_NoAlgs
10. mctr.07.04.09m_NoAlgs
11. mctr.07.06.14m_NoAlgs
12. mctr.07.05.20m_NoAlgs
13. mctr.07.04.03m_NoAlgs
14. mctr.07.06.19m_NoAlgs
15. mctr.07.06.04m_NoAlgs
16. mctr.07.01.23m_NoAlgs
17. mctr.07.02.30m_NoAlgs
18. mctr.07.04.21m_NoAlgs
19. mctr.07.05.29m_NoAlgs
20. mctr.07.03.26m_NoAlgs
21. mctr.07.02.15m_NoAlgs
22. mctr.07.05.07m_NoAlgs
23. mctr.07.01.14m_NoAlgs
24. mctr.07.02.21m_NoAlgs
25. mctr.07.04.13m_NoAlgs

1. The problem that follows refers to triangle ABC.

    If $A = 70°$, $B = 40°$ and $b = 8\,cm$, find a.

    Please round the answer to the nearest centimeter.

    $a =$ _________ cm

2. The problem that follows refers to triangle ABC.

    If $B = 100°$, $C = 20°$ and $b = 14$ inches, find c.

    Please round the answer to the nearest inch.

    Select the correct answer.

    a.   5 inches
    b.   6 inches
    c.   7 inches
    d.   8 inches
    e.   1 inches

3. The information below refers to triangle ABC.

    $B = 12.4°$, $C = 23.8°$, $a = 325\,cm$

    Find A.

    $A =$ _________ °

    Find b. Please round the answer to the nearest centimeter.

    $b =$ _________ cm

    Find c. Please round the answer to the nearest centimeter.

    $c =$ _________ cm

4. The circle in the figure below has a radius of *r* and center at *C*. The distance from *A* to *B* is *x*, the distance from *A* to *D* is *y*, and the length of arc *BD* is *s*. For the problem, redraw the figure, label it with the values given then solve the problem.

    If $A = 40°$, $s = 17$, and $r = 14$, find *y*. Please round the answer to the nearest integer.

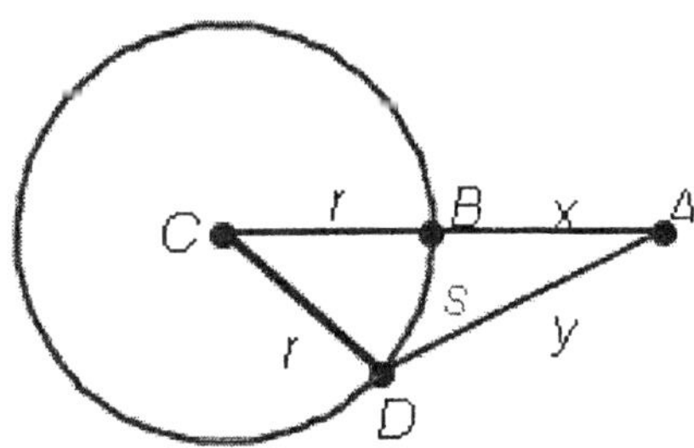

Select the correct answer.

    a. $y = 19$

    b. $y = 18$

    c. $y = 24$

    d. $y = 23$

    e. $y = 20$

5. For the triangle described below, solve for *B* and use the results to explain why the triangle has no solution. Round your answer to the nearest tenth.

    $A = 150°$, $b = 60$ ft, $a = 20$ ft

6. For the triangle described below, solve for *B*. Round your answer to the nearest tenth.

    $A = 60°$, $b = 72$ ft, $a = 64$ ft

    Select the correct answer.

    a. $B = 103°$ or $B' = 10.9°$

    b. $B = 77°$ or $B' = 103°$

    c. $B = 77°$

    d. $B = 10.9°$

    e.   no solutions

7. Find all solutions to the triangle described below.

$A = 64°$, $b = 7.7$ yd, $a = 7.1$ yd

Please round each angle to the nearest degree and the length of the missing side to the nearest tenth.

If there is more than one possible triangle, list both.

8. A 50-foot wire running from the top of a tent pole to the ground makes an angle of $55°$ with the ground. If the length of the tent pole is 45 feet, how far is it from the bottom of the tent pole to the point where the wire is fastened to the ground? (The tent pole is not necessarily perpendicular to the ground.)

Please round your answer to the nearest foot.

Select the correct answer.

a.   47 ft
b.   13 ft
c.   10 ft or 47 ft
d.   13 ft or 59 ft
e.   10 ft

9. Refer to triangle ABC. If $a = 140$ inches, $b = 63$ inches, and $C = 60°$, find c.

Please round your answer to the nearest integer.

$c =$ ________ inches

10. Refer to triangle ABC. If $b = 4.5$ m, $c = 6.3$ m, and $A = 116°$, find a.

Please round your answer to the nearest tenth.

Select the correct answer.

a.   9.6 m
b.   9.4 m
c.   9.2 m
d.   10.0 m
e.   9.8 m

11. Solve the triangle.

$b = 63.4$ km, $c = 75.6$ km, $A = 124°\ 20'$

Please round the length to the nearest tenth and each angle to the nearest minute. Do not round any numbers until the end of each computation.

$a =$ _________ km

$B =$ _________ ° _________ '

$C =$ _________ ° _________ '

12. Use the law of cosines to find a true statement from the list below, if $A = 120°$.

Select the correct answer.

a. $\quad a^2 = b^2 + c^2 + bc\sqrt{2}$

b. $\quad a^2 = b^2 + c^2 - bc\sqrt{2}$

c. $\quad a^2 = b^2 + c^2 + bc$

d. $\quad a^2 = b^2 + c^2 - bc$

e. $\quad a^2 = b^2 + c^2$

13. Referring to triangle *ABC,* find the triangle's area given

$a = 50$ cm, $b = 70$ cm, $C = 60°$

Please round your answer to the nearest integer.

$S =$ _________ cm^2

14. Referring to triangle *ABC,* find the triangle's area given

$a = 39.5$ m, $c = 35.5$ m, $B = 154.5°$

Please round your answer to the nearest integer.

Select the correct answer.

    a.   308 m^2

    b.   302 m^2

    c.   298 m^2

    d.   300 m^2

    e.   294 m^2

15. Referring to triangle *ABC,* find the triangle's area given

$A = 44° \ 30'$, $C = 115° \ 30'$, $a = 3.46$ ft

Please round your answer to the nearest hundredth.

$S = $ _________ ft^2

16. Referring to triangle *ABC,* find the triangle's area given

$a = 48$ inches, $b = 68$ inches, $c = 84$ inches

Please round your answer to the nearest ten.

Select the correct answer.

    a.   1,690 inches^2

    b.   1,710 inches^2

    c.   1,630 inches^2

    d.   1,650 inches^2

    e.   1,590 inches^2

17. Draw the vector **V** that goes from the origin to the given point.

$(-3, 4)$

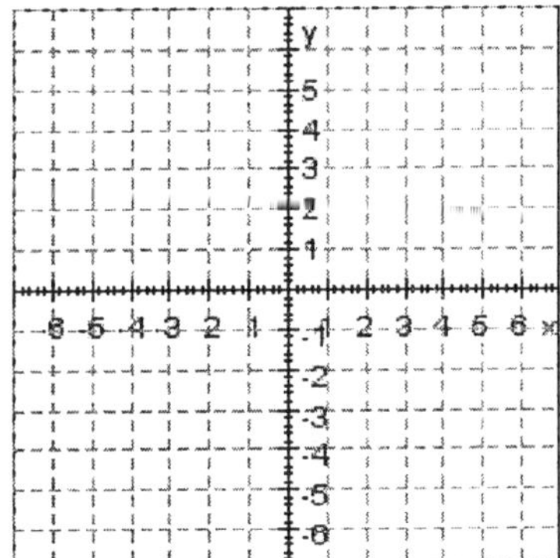

Write the vector **V** in component form (a, b).

$\mathbf{V} = \underline{\hspace{2cm}}$

18. Draw the vector **V** that goes from the origin to the given point. Then write **V** in component form $\langle a, b \rangle$.

$(-6, -4)$

Select the correct answer.

a.
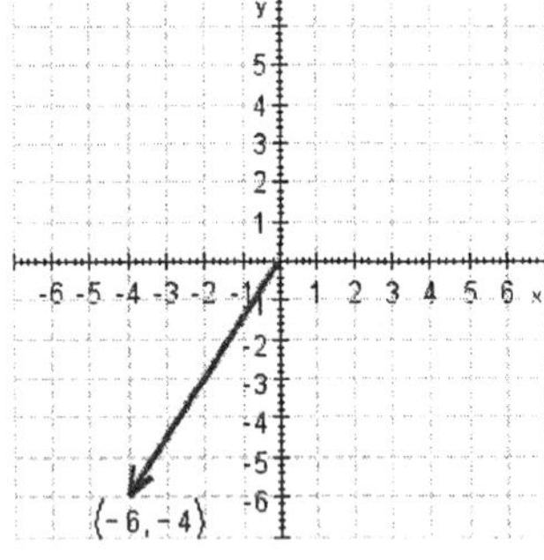

b.
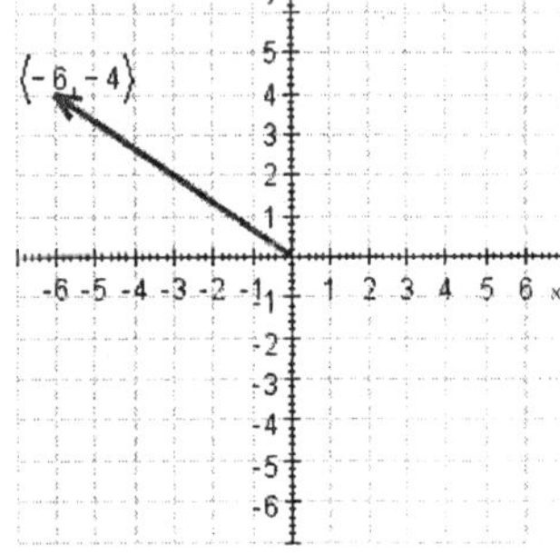

c.
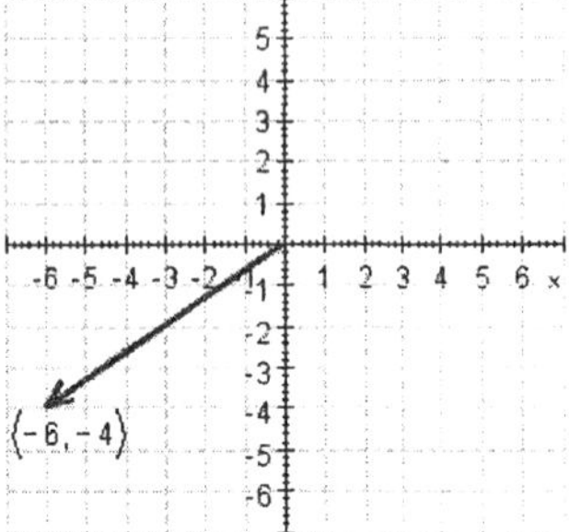

d.
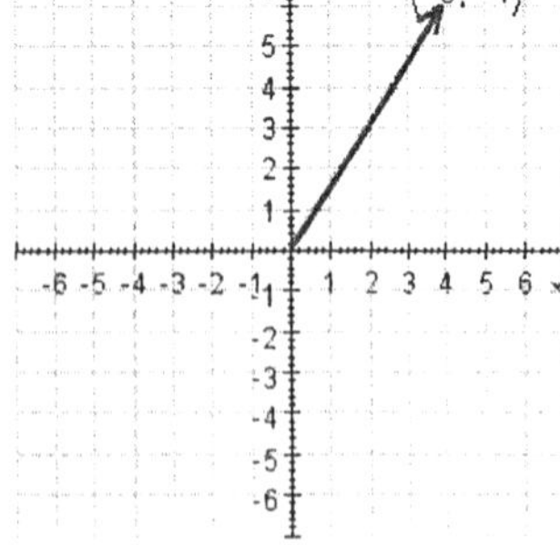

19. Find the magnitude of the vector.

$$U = -5i - 12j$$

20. For the pair of vectors, find $2U - 3V$.

$$U = \langle 6, 6 \rangle, \ V = \langle 6, -6 \rangle$$

Select the correct answer.

a. $\langle -6, 30 \rangle$

b. $\langle 30, -6 \rangle$

c. $\langle 29, -7 \rangle$

d. $\langle 6, 30 \rangle$

e. $\langle -7, 29 \rangle$

21. Find the dot product.

$$\langle 4, 5 \rangle \cdot \langle 5, 5 \rangle = \ \_\_\_\_\_\_$$

22. Find the dot product.

$$\langle 9, -7 \rangle \cdot \langle 2, -6 \rangle$$

Select the correct answer.

a. 57
b. 62
c. 60
d. 56
e. 61

23. Find the angle θ between the given vectors to the nearest tenth of a degree.

$$U = 12i + 7j \qquad V = -3i + 5j$$

$$\theta = \underline{\hspace{2cm}} \,^{\circ}$$

24. Is the pair of vectors i and $-j$ perpendicular?

Select the correct answer.

    a.   yes
    b.   no

25. Find the work performed when the given force F is applied to an object, whose resulting motion is represented by the displacement vector d. Assume the force is in pounds and the displacement is measured in feet.

$$F = 20i + 8j \qquad d = 30i + 2j$$

$$\text{Work} = \underline{\hspace{2cm}} \text{ ft-lb}$$

1. 12

2. a

3. 143.8
   118
   222

4. e

5. $\sin B = 1.5$; since $\sin B > 1$, there is no triangle.

6. b

7. $(77°, 39°, 5)$, $(103°, 13°, 1.8)$

8. c

9. 121

10. c

11. 123.1
    25
    11
    30
    29.2

12. c

13. 1.516

14. b

15. 2.64

16. c

17.

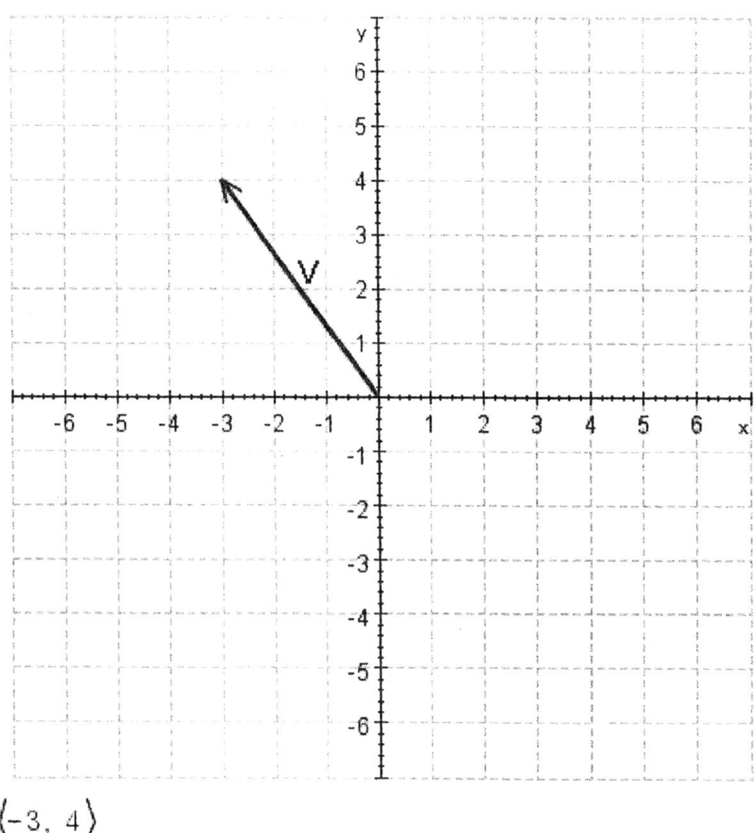

$\langle -3, 4 \rangle$

18. c

19. $|U| = 13$

20. a

21. 45

22. c

23. 90.7

24. a

25. 616

1. mctr.07.01.02_NoAlgs
2. mctr.07.01.04m_NoAlgs
3. mctr.07.01.17_NoAlgs
4. mctr.07.01.23m_NoAlgs
5. mctr.07.02.02_NoAlgs
6. mctr.07.02.05m_NoAlgs
7. mctr.07.02.18_NoAlgs
8. mctr.07.02.21m_NoAlgs
9. mctr.07.03.01_NoAlgs
10. mctr.07.03.05m_NoAlgs
11. mctr.07.03.14_NoAlgs
12. mctr.07.03.17m_NoAlgs
13. mctr.07.04.01_NoAlgs
14. mctr.07.04.03m_NoAlgs
15. mctr.07.04.11_NoAlgs
16. mctr.07.04.13m_NoAlgs
17. mctr.07.05.03_NoAlgs
18. mctr.07.05.07m_NoAlgs
19. mctr.07.05.25_NoAlgs
20. mctr.07.05.29m_NoAlgs
21. mctr.07.06.02_NoAlgs
22. mctr.07.06.04m_NoAlgs
23. mctr.07.06.16_NoAlgs
24. mctr.07.06.19m_NoAlgs
25. mctr.07.06.23_NoAlgs

1. Find the magnitude of the vector.

$$U = -5i - 12j$$

2. For the triangle described below, solve for B and use the results to explain why the triangle has no solution. Round your answer to the nearest tenth.

$$A = 150°, \ b = 60 \text{ ft}, \ a = 20 \text{ ft}$$

3. Refer to triangle ABC. If $a = 140$ inches, $b = 63$ inches, and $C = 60°$, find c.

Please round your answer to the nearest integer.

$c =$ _________ inches

4. The problem that follows refers to triangle ABC.

If $B = 100°$, $C = 20°$ and $b = 14$ inches, find c.

Please round the answer to the nearest inch.

Select the correct answer.

a. 5 inches

b. 6 inches

c. 7 inches

d. 8 inches

e. 1 inches

5. For the triangle described below, solve for B. Round your answer to the nearest tenth.

   $A = 60°$, $b = 72$ ft, $a = 64$ ft

   Select the correct answer.

   a. $B - 103°$ or $B' = 10.9°$
        00
   b. $B = 77°$ or $B' = 103°$

   c. $B = 77°$

   d. $B = 10.9°$

   e.   no solutions

6. Solve the triangle.

   $b = 63.4$ km, $c = 75.6$ km, $A = 124° \ 20'$

   Please round the length to the nearest tenth and each angle to the nearest minute. Do not round any numbers until the end of each computation.

   $a =$ _________ km

   $B =$ _________ ° _________ '

   $C =$ _________ ° _________ '

7. Referring to triangle ABC, find the triangle's area given

   $A = 44° \ 30'$, $C = 115° \ 30'$, $a = 3.46$ ft

   Please round your answer to the nearest hundredth.

   $S =$ _________ ft^2

8. Referring to triangle ABC, find the triangle's area given

   $a = 50$ cm, $b = 70$ cm, $C = 60°$

   Please round your answer to the nearest integer.

   $S =$ _________ cm^2

9. A 50-foot wire running from the top of a tent pole to the ground makes an angle of 55° with the ground. If the length of the tent pole is 45 feet, how far is it from the bottom of the tent pole to the point where the wire is fastened to the ground? (The tent pole is not necessarily perpendicular to the ground.)

    Please round your answer to the nearest foot.

    Select the correct answer.

    a. 47 ft
    b. 13 ft
    c. 10 ft or 47 ft
    d. 13 ft or 59 ft
    e. 10 ft

10. The information below refers to triangle ABC.

    $$B = 12.4^\circ,\ C = 23.8^\circ,\ a = 325\,cm$$

    Find A.

    $A = $ _______ $^\circ$

    Find b. Please round the answer to the nearest centimeter.

    $b = $ _______ cm

    Find c. Please round the answer to the nearest centimeter.

    $c = $ _______ cm
            00

11. Refer to triangle ABC. If $b = 4.5$ m, $c = 6.3$ m, and $A = 116^\circ$, find a.

    Please round your answer to the nearest tenth.

    Select the correct answer.

    a. 9.6 m
    b. 9.4 m
    c. 9.2 m
    d. 10.0 m
    e. 9.8 m

12. Find the angle θ between the given vectors to the nearest tenth of a degree.

$$U = 12i + 7j \quad V = -3i + 5j$$

$$\theta = \underline{\hspace{2cm}}{}^{\circ}$$

13. The problem that follows refers to triangle ABC.

If $A = 70^{\circ}$, $B = 40^{\circ}$ and $b = 8\,cm$, find a.

Please round the answer to the nearest centimeter.

$$a = \underline{\hspace{2cm}} \ cm$$

14. Find the dot product.

$$\langle 4, 5 \rangle \cdot \langle 5, 5 \rangle = \underline{\hspace{2cm}}$$

15. Draw the vector **V** that goes from the origin to the given point.

$(-3, 4)$

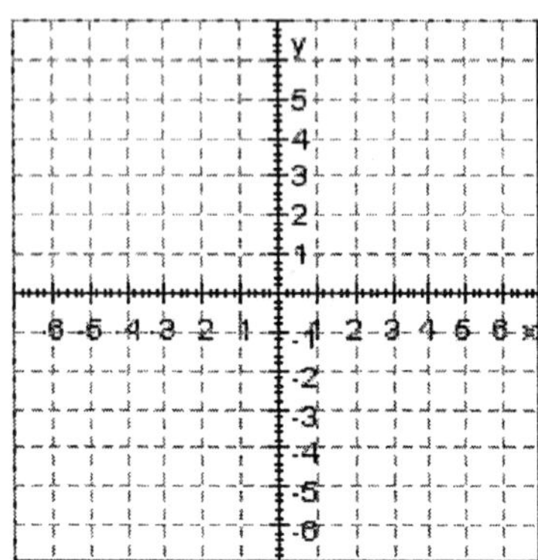

Write the vector **V** in component form $\langle a, b \rangle$.

$$V = \underline{\hspace{2cm}}$$

16. Find the dot product.

$$\langle 9, -7 \rangle \cdot \langle 2, -6 \rangle$$

Select the correct answer.

a.   57
b.   62
c.   60
d.   56
e.   61

17. Referring to triangle *ABC,* find the triangle's area given

a = 48 inches, *b* = 68 inches, *c* = 84 inches

Please round your answer to the nearest ten.

Select the correct answer.

    a.   1,690 inches^2

    b.   1,710 inches^2

    c.   1,630 inches^2

    d.   1,650 inches^2

    e.   1,590 inches^2

18. Find all solutions to the triangle described below.

$$A = 64°,\ b = 7.7\ \text{yd},\ a = 7.1\ \text{yd}$$

Please round each angle to the nearest degree and the length of the missing side to the nearest tenth.

If there is more than one possible triangle, list both.

19. For the pair of vectors, find $2\mathbf{U} - 3\mathbf{V}$.

$$\mathbf{U} = \langle 6, 6 \rangle,\ \mathbf{V} = \langle 6, -6 \rangle$$

Select the correct answer.

    a. $\langle -6, 30 \rangle$

    b. $\langle 30, -6 \rangle$

    c. $\langle 29, -7 \rangle$

    d. $\langle 6, 30 \rangle$

    e. $\langle -7, 29 \rangle$

20. Referring to triangle *ABC,* find the triangle's area given

$a = 39.5$ m, $c = 35.5$ m, $B = 154.5°$

Please round your answer to the nearest integer.

Select the correct answer.

a.   308 m^2

b.   302 m^2

c.   298 m^2

d.   300 m^2

e.   294 m^2

21. Find the work performed when the given force F is applied to an object, whose resulting motion is represented by the displacement vector d . Assume the force is in pounds and the displacement is measured in feet.

$F = 20i + 8j \quad d = 30i + 2j$

Work = ________ ft-lb

22. Is the pair of vectors **i** and $-$**j** perpendicular?

Select the correct answer.

a.   yes
b.   no

23. Use the law of cosines to find a true statement from the list below, if $A = 120°$.

Select the correct answer.

a. $\quad a^2 = b^2 + c^2 + bc\sqrt{2}$

b. $\quad a^2 = b^2 + c^2 - bc\sqrt{2}$

c. $\quad a^2 = b^2 + c^2 + bc$

d. $\quad a^2 = b^2 + c^2 - bc$

e. $\quad a^2 = b^2 + c^2$

24. The circle in the figure below has a radius of r and center at C. The distance from A to B is x, the distance from A to D is y, and the length of arc BD is s. For the problem, redraw the figure, label it with the values given then solve the problem.
If $A = 40°$, $s = 17$, and $r = 14$, find y. Please round the answer to the nearest

integer.

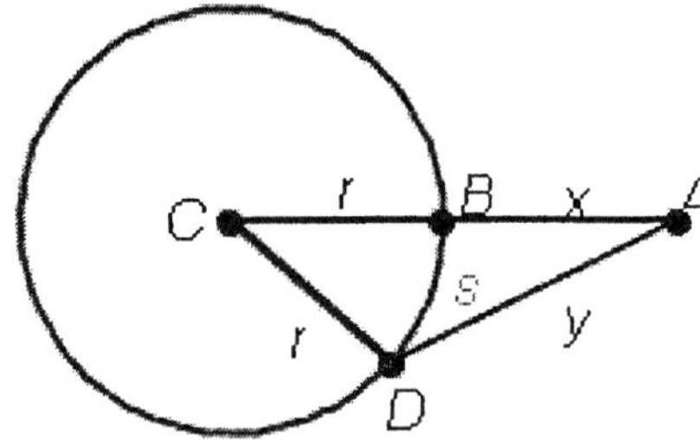

Select the correct answer.

a. $\quad y = 19$

b. $\quad y = 18$

c. $\quad y = 24$

d. $\quad y = 23$

e. $\quad y = 20$

25. Draw the vector **V** that goes from the origin to the given point. Then write **V** in component form $\langle a, b \rangle$.

$$\langle -6, -4 \rangle$$

Select the correct answer.

a.

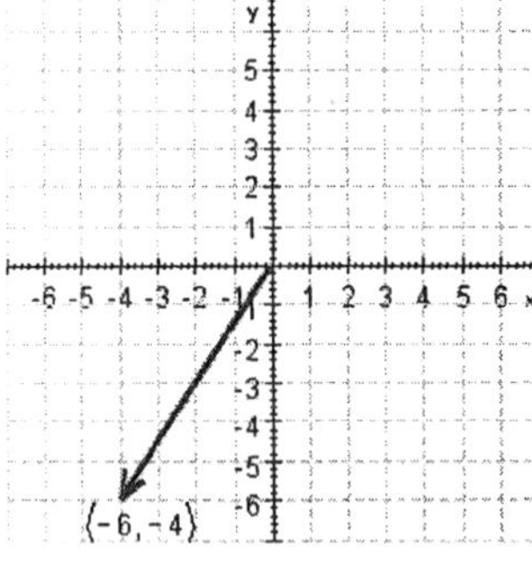

b.

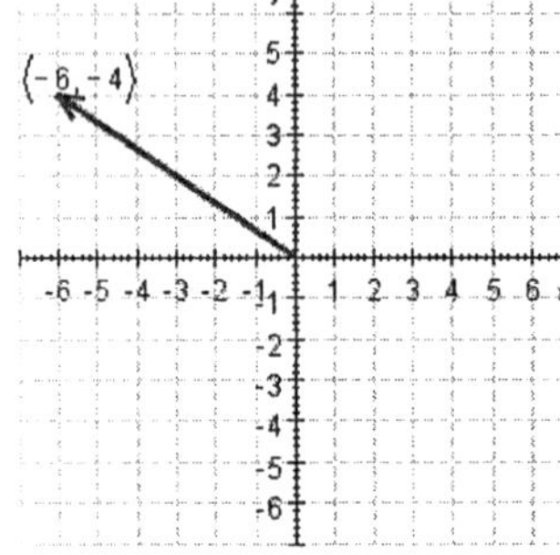

c.

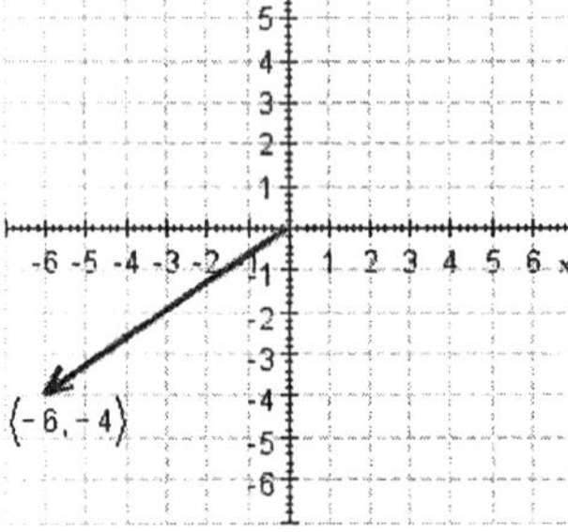

d.

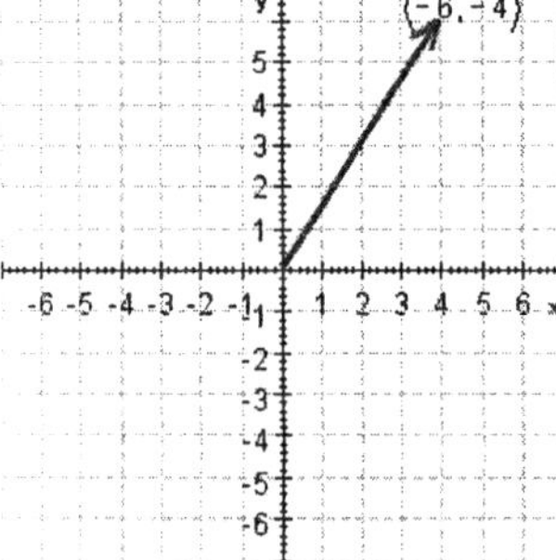

1. $|U| = 13$

2. $\sin B = 1.5$; since $\sin B > 1$, there is no triangle.

3. 121

4. a

5. b

6. 123.1
   25
   11
   30
   29.2

7. 2.64

8. 1,516

9. c

10. 143.8
    118
    222

11. c

12. 90.7

13. 12

14. 45

15.

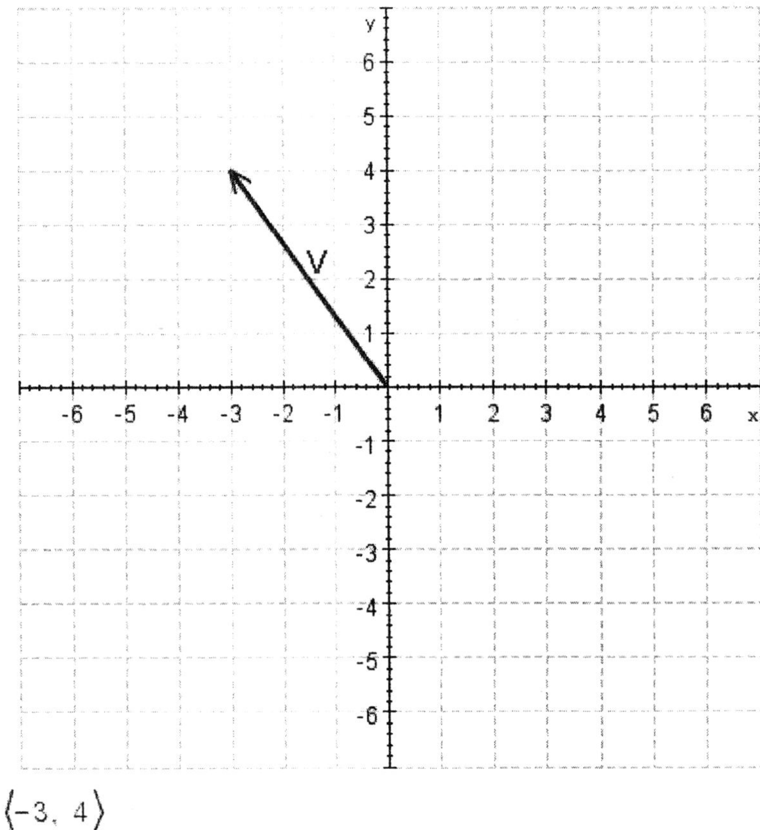

$\langle -3,\ 4 \rangle$

16. c

17. c

18. $\left(77^{\circ},\ 39^{\circ},\ 5\right),\ \left(103^{\circ},\ 13^{\circ},\ 1.8\right)$

19. a

20. b

21. 616

22. a

23. c

24. e

25. c

1. mctr.07.05.25_NoAlgs
2. mctr.07.02.02_NoAlgs
3. mctr.07.03.01_NoAlgs
4. mctr.07.01.04m_NoAlgs
5. mctr.07.02.05m_NoAlgs
6. mctr.07.03.14_NoAlgs
7. mctr.07.04.11_NoAlgs
8. mctr.07.04.01_NoAlgs
9. mctr.07.02.21m_NoAlgs
10. mctr.07.01.17_NoAlgs
11. mctr.07.03.05m_NoAlgs
12. mctr.07.06.16_NoAlgs
13. mctr.07.01.02_NoAlgs
14. mctr.07.06.02_NoAlgs
15. mctr.07.05.03_NoAlgs
16. mctr.07.06.04m_NoAlgs
17. mctr.07.04.13m_NoAlgs
18. mctr.07.02.18_NoAlgs
19. mctr.07.05.29m_NoAlgs
20. mctr.07.04.03m_NoAlgs
21. mctr.07.06.23_NoAlgs
22. mctr.07.06.19m_NoAlgs
23. mctr.07.03.17m_NoAlgs
24. mctr.07.01.23m_NoAlgs
25. mctr.07.05.07m_NoAlgs

1. Refer to triangle *ABC*. If *a* = 53 cm, *b* = 25 cm, and *c* = 31 cm, find the largest angle.

    Please round your answer to the nearest degree.

    ________ º

2. The problem that follows refers to triangle ABC.

    If $A = 55°$, $B = 48°$ and $c = 16\,cm$.

    Find C.

    $C =$ ________ º

    Find a. Please round the answer to the nearest centimeter.

    $a =$ ________ cm

3. Draw the vector **V** that goes from the origin to the given point. Then write **V** in terms of the unit vectors **i** and **j**.

$(3, -4)$

Select the correct answer.

a.

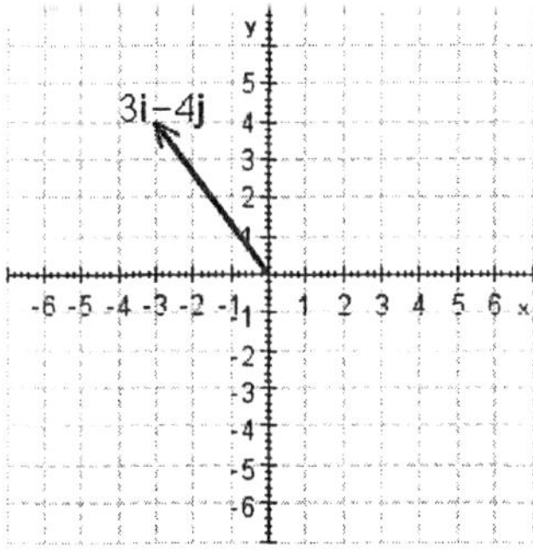

b.

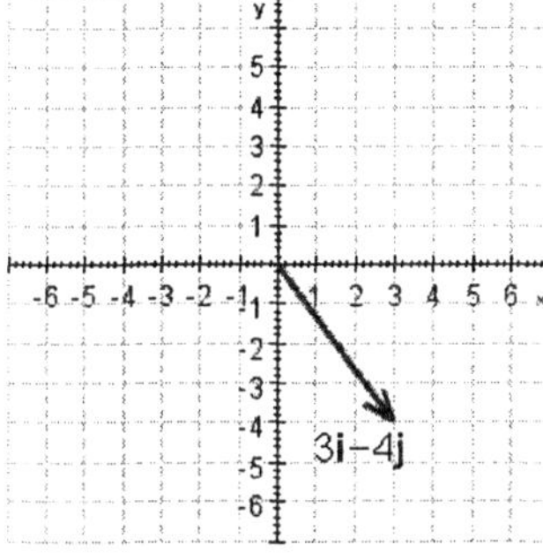

c.

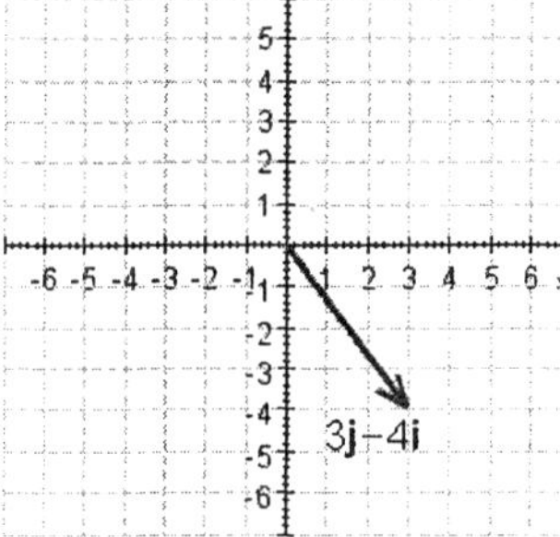

d.

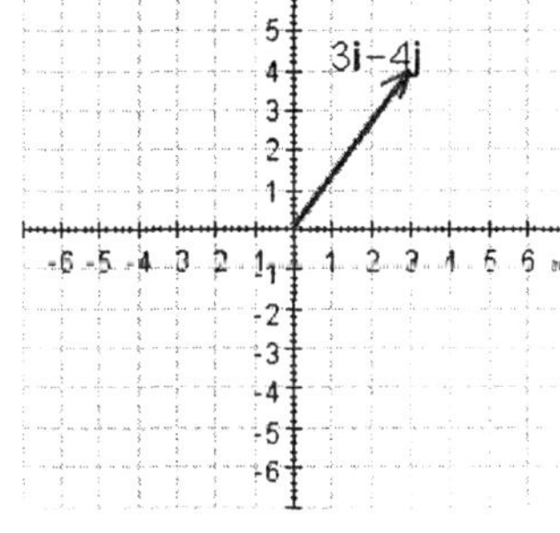

4. Find the magnitude of the vector.

$$\langle 0, 11 \rangle$$

Select the correct answer.

    a. $|\mathbf{W}| = 9$

    b. $|\mathbf{W}| = 11$

    c. $|\mathbf{W}| = 121$

    d. $|\mathbf{W}| = 0$

    e. $|\mathbf{W}| = 12$

5. Find all solutions to the triangle described below.

$$A = 122.5^\circ, \ a = 25.7 \text{ cm}, \ b = 44.6 \text{ cm}$$

Please round each angle and length of the missing side to the nearest tenth at the end of the computations.

Select the correct answer.

a. $B = 14.1^\circ$, $C = 43.4^\circ$, $c = 21$ cm or
$B' = 35.7^\circ$, $C' = 21.8^\circ$, $c' = 25.3$ cm

b. $B = 14.1^\circ$, $C = 43.4^\circ$, $c = 21$ cm or
$B' = 165.9^\circ$, $C' = 108.4^\circ$, $c' = 28.9$ cm

c. $B = 14.1^\circ$, $C = 43.4^\circ$, $c = 21$ cm

d. $B = 35.7^\circ$, $C = 21.8^\circ$, $c = 25.3$ cm

e.   no solution

6. A plane headed due north is traveling with an airspeed of 180 miles per hour. The wind currents are moving with constant speed in the direction $150°$. If the ground speed of the plane is 90 miles per hour, what is its true course?

Select the correct answer.

   a. $43°$

   b. $80°$

   c. $10°$

   d. $15°$

   e. $60°$

7. Solve the triangle.

$a = 51$ yd, $b = 74$ yd, $c = 64$ yd

Please round each answer to the nearest integer. Do not round any numbers until the end of each computation.

Select the correct answer.

   a. $A = 45°, B = 76°, C = 58°$

   b. $A = 43°, B = 79°, C = 59°$

   c. $A = 45°, B = 76°, C = 59°$

   d. $A = 43°, B = 79°, C = 58°$

   e. $A = 39°, B = 73°, C = 68°$

8. Referring to triangle ABC, find the triangle's area given

$A = 42.2°, B = 71.2°, a = 220$ inches

Please round your answer to the nearest hundred.

Select the correct answer.

   a. $31{,}500$ inches^2

   b. $31{,}000$ inches^2

   c. $31{,}300$ inches^2

   d. $31{,}200$ inches^2

   e. $30{,}900$ inches^2

9. Find the work performed when the given force F is applied to an object, whose resulting motion is represented by the displacement vector d. Assume the force is in pounds and the displacement is measured in feet.

$$F = 95i \qquad d = 8i$$

Select the correct answer.

   a.   730 ft-lb
   b.   780 ft-lb
   c.   760 ft-lb
   d.   750 ft-lb
   e.   720 ft-lb

10. For the pair of vectors, find $U \cdot V$.

$$U = 2i + j \qquad V = i + 2j$$

Select the correct answer.

   a.   0
   b.   5
   c.   2
   d.   4
   e.   1

11. A woman entering an outside glass elevator on the ground floor of a hotel glances up to the top of the building across the street and notices that the angle of elevation is $47°$. She rides the elevator up three floors (60 feet) and finds that the angle of elevation to the top of the building across the street is $30°$. How tall is the building across the street? (Give your answer to the nearest foot.)

_________ ft

12. Find all solutions to the triangle described below.

$$A = 118°, \ a = 0.66 \text{ cm}, \ b = 0.91 \text{ cm}$$

Please round each angle to the nearest degree and the length of the missing side to the nearest tenth.

Select the correct answer.

a. $B = 19°, \ C = 43°, \ c = 0.51 \text{ cm}$

b. $B = 19°, \ C = 43°, \ c = 0.51 \text{ cm}$ or
$B' = 161°, \ C' = 99°, \ c' = 0.74 \text{ cm}$

c. $B = 19°, \ C = 43°, \ c = 0.51 \text{ cm}$ or
$B' = 22°, \ C' = 40°, \ c' = 0.88 \text{ cm}$

d. $B = 22°, \ C = 40°, \ c = 0.88 \text{ cm}$

e.   no solution

13. Find the work performed when the given force F is applied to an object, whose resulting motion is represented by the displacement vector **d** . Assume the force is in pounds and the displacement is measured in feet.

$$F = 20i + 8j \qquad d = 30i + 2j$$

Work = __________ ft-lb

14. Find the angle θ between the given vectors to the nearest tenth of a degree.

$$U = -2i + 7j \qquad V = 6i + 2j$$

Select the correct answer.

a. $\theta = 83.5°$

b. $\theta = 88.5°$

c. $\theta = 87.5°$

d. $\theta = 89.5°$

e. $\theta = 84.5°$

15. The information below refers to triangle ABC.

$B = 55°, C = 32°, a = 7.3$ m.

Find all the missing parts. Round side lengths to a tenth of a meter.

Select the correct answer.

a. $A = 91°, b = 6$ m, $c = 3.6$ m

b. $A = 91°, b = 6.2$ m, $c = 3.9$ m

c. $A = 93°, b = 6.2$ m, $c = 3.6$ m

d. $A = 96°, b = 4.5$ m, $c = 9.8$ m

e. $A = 93°, b = 6$ m, $c = 3.9$ m

16. The area of a triangle is 50 cm^2. Find the length of the side included between the angles

$A = 30°$ and $B = 40°$.

Please round your answer to the nearest tenth.

Select the correct answer.

a. 16.7 cm
b. 17.3 cm
c. 16.5 cm
d. 16.3 cm
e. 17.1 cm

17. Referring to triangle ABC, find the triangle's area given

$b = 63.3$ km, $c = 75.4$ km, $A = 125° \ 40'$

Please round your answer to the nearest integer.

$S =$ _________ km^2

18. For the pair of vectors, find $U + V$, $U - V$, and $3U + 2V$.

$$U = -2i + 2j, \ V = 2i + 2j$$

Please enter your answer in terms of the unit vectors i and j.

$U + V =$ __________

$U - V =$ __________

$3U + 2V =$ __________

19. A tightrope walker is standing still with one foot on the tightrope as shown in the figure. If the tightrope walker weighs 122 pounds, find the magnitudes of the tension in the rope toward each end of the rope. Please round the answers to the nearest pound.

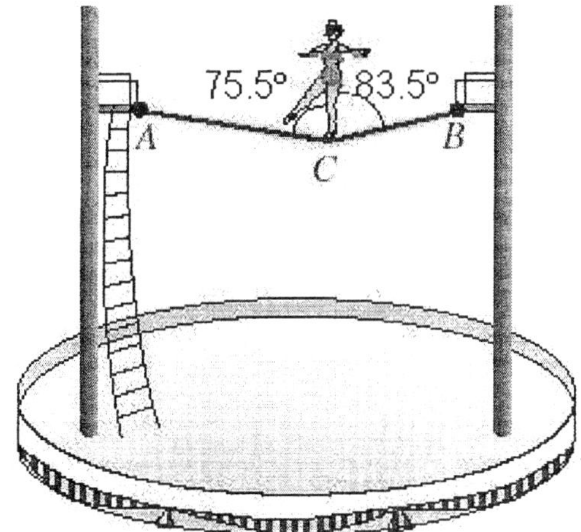

Select the correct answer.

a. $|CB| = 330\,lbs$, $|CA| = 338\,lbs$

b. $|CB| = 334\,lbs$, $|CA| = 336\,lbs$

c. $|CB| = 331\,lbs$, $|CA| = 340\,lbs$

d. $|CB| = 330\,lbs$, $|CA| = 340\,lbs$

e. $|CB| = 331\,lbs$, $|CA| = 338\,lbs$

20. Referring to triangle *ABC,* find the triangle's area given

a = 4.37 ft, b = 3.76 ft, c = 5.27 ft

Please round your answer to the nearest hundredth.

S = _________ ft^2

21. A plane is flying with an airspeed of 244 miles per hour with heading 277.2°. The wind currents are running at a constant 45.7 miles per hour in the direction 267.1°. Find the ground speed and true course of the plane.

Please round each answer to the nearest tenth at the end of the computations.

Select the correct answer.

    a.   289.1 mph with heading 276.0°

    b.   289.1 mph with heading 275.6°

    c.   288.7 mph with heading 275.6°

    d.   288.3 mph with heading 276.4°

    e.   288.7 mph with heading 276.0°

22. Draw a vector representing the course of a ship that goes 35 miles on a course with direction 240⁰ .

Select the correct answer.

a.

b.

c.

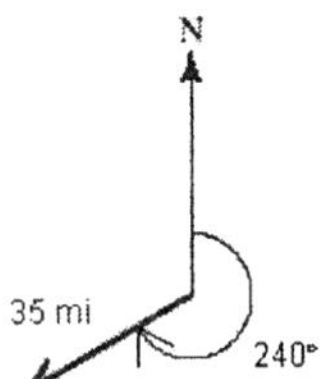

23. Find the dot product.

$$\langle 2, 3 \rangle \cdot \langle 4, 4 \rangle$$

Select the correct answer.

a.   22
b.   16
c.   20
d.   21
e.   23

24. Find the magnitude of the vector and the angle θ, $0° \leq \theta < 360°$, that the vector makes with the positive *x*-axis.

$$\mathbf{W} = \mathbf{i} + \sqrt{3}\,\mathbf{j}$$

Select the correct answer.

a. $|\mathbf{W}| = 2,\ \theta = 120°$

b. $|\mathbf{W}| = 2,\ \theta = 60°$

c. $|\mathbf{W}| = 6,\ \theta = 60°$

d. $|\mathbf{W}| = 3,\ \theta = 300°$

e. $|\mathbf{W}| = 7,\ \theta = 120°$

25. Two ships leave a harbor entrance at the same time. The first ship is traveling at a constant 19 miles per hour, while the second is traveling at a constant 22 miles per hour. If the angle between their courses is $125°$, how far apart are they after 2 hours?

Please round your answer to the nearest mile.

Select the correct answer.

a.   65 mi
b.   79 mi
c.   73 mi
d.   71 mi
e.   69 mi

McKeague/Turner - Trigonometry 5e Chapter 7 Form G

1. 142

2. 77

   13

3. b

4. b

5. e

6. e

7. d

8. c

9. c

10. d

11. 130

12. e

13. 616

14. c

15. e

16. e

17. 1,939

18. 4j

    −4 i

    −2i + 10j

19. a

20. 8.10

21. c

22. c

23. b

24. c

25. b

McKeague/Turner - Trigonometry 5e Chapter 7 Form G

1. mctr.07.03.08_NoAlgs
2. mctr.07.01.09_NoAlgs
3. mctr.07.05.13m_NoAlgs
4. mctr.07.05.20m_NoAlgs
5. mctr.07.02.10m_NoAlgs
6. mctr.07.02.30m_NoAlgs
7. mctr.07.03.12m_NoAlgs
8. mctr.07.04.09m_NoAlgs
9. mctr.07.06.27m_NoAlgs
10. mctr.07.06.09m_NoAlgs
11. mctr.07.01.28_NoAlgs
12. mctr.07.02.15m_NoAlgs
13. mctr.07.06.23_NoAlgs
14. mctr.07.06.14m_NoAlgs
15. mctr.07.01.14m_NoAlgs
16. mctr.07.04.21m_NoAlgs
17. mctr.07.04.06_NoAlgs
18. mctr.07.05.36_NoAlgs
19. mctr.07.01.35m_NoAlgs
20. mctr.07.04.17_NoAlgs
21. mctr.07.03.26m_NoAlgs
22. mctr.07.02.26m_NoAlgs
23. mctr.07.06.02m_NoAlgs
24. mctr.07.05.47m_NoAlgs
25. mctr.07.03.22m_NoAlgs

1. Find the magnitude of the vector.

$$\langle 0, 11 \rangle$$

Select the correct answer.

    a. $|\mathbf{W}| = 9$

    b. $|\mathbf{W}| = 11$

    c. $|\mathbf{W}| = 121$

    d. $|\mathbf{W}| = 0$

    e. $|\mathbf{W}| = 12$

2. A tightrope walker is standing still with one foot on the tightrope as shown in the figure. If the tightrope walker weighs 122 pounds, find the magnitudes of the tension in the rope toward each end of the rope. Please round the answers to the nearest pound.

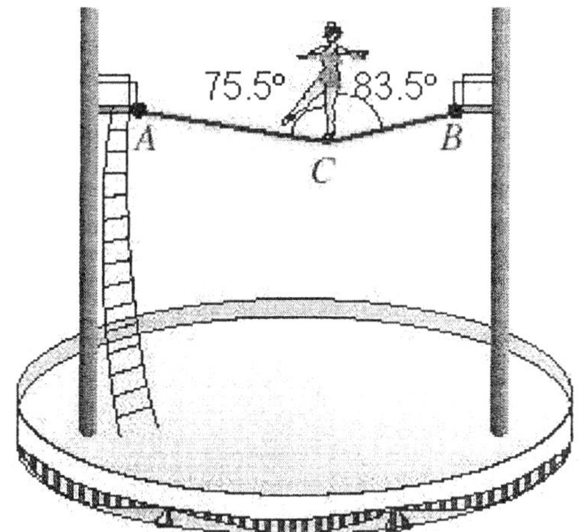

Select the correct answer.

    a. $|CB| = 330\ lbs,\ |CA| = 338\ lbs$

    b. $|CB| = 334\ lbs,\ |CA| = 336\ lbs$

    c. $|CB| = 331\ lbs,\ |CA| = 340\ lbs$

    d. $|CB| = 330\ lbs,\ |CA| = 340\ lbs$

    e. $|CB| = 331\ lbs,\ |CA| = 338\ lbs$

3. The problem that follows refers to triangle ABC.

If $A = 55°$, $B = 48°$ and $c = 16$ cm.

Find C.

$C =$ _________ °

Find a. Please round the answer to the nearest centimeter.

$a =$ _________ cm

4. Find the dot product.

$$\langle 2, 3 \rangle \cdot \langle 4, 4 \rangle$$

Select the correct answer.

a. 22
b. 16
c. 20
d. 21
e. 23

5. Refer to triangle ABC. If $a = 53$ cm, $b = 25$ cm, and $c = 31$ cm, find the largest angle.

Please round your answer to the nearest degree.

_________ °

6. The information below refers to triangle ABC.

$B = 55°$, $C = 32°$, $a = 7.3$ m.

Find all the missing parts. Round side lengths to a tenth of a meter.

Select the correct answer.

a. $A = 91°$, $b = 6$ m, $c = 3.6$ m

b. $A = 91°$, $b = 6.2$ m, $c = 3.9$ m

c. $A = 93°$, $b = 6.2$ m, $c = 3.6$ m

d. $A = 96°$, $b = 4.5$ m, $c = 9.8$ m

e. $A = 93°$, $b = 6$ m, $c = 3.9$ m

7. Referring to triangle *ABC,* find the triangle's area given

   $a = 4.37$ ft, $b = 3.76$ ft, $c = 5.27$ ft

   Please round your answer to the nearest hundredth.

   $S =$ _________ ft^2

8. For the pair of vectors, find $\mathbf{U} \cdot \mathbf{V}$.

   $\mathbf{U} = 2\mathbf{i} + \mathbf{j} \quad \mathbf{V} = \mathbf{i} + 2\mathbf{j}$

   Select the correct answer.

   a.   0
   b.   5
   c.   2
   d.   4
   e.   1

9. Find the work performed when the given force F is applied to an object, whose resulting motion is represented by the displacement vector d. Assume the force is in pounds and the displacement is measured in feet.

   $\mathsf{F} = 95\mathsf{i} \quad \mathsf{d} = 8\mathsf{i}$

   Select the correct answer.

   a.   730 ft-lb
   b.   780 ft-lb
   c.   760 ft-lb
   d.   750 ft-lb
   e.   720 ft-lb

10. The area of a triangle is 50 cm^2. Find the length of the side included between the angles

    $A = 30^\circ$ and $B = 40^\circ$.

    Please round your answer to the nearest tenth.

    Select the correct answer.

    a.   16.7 cm
    b.   17.3 cm
    c.   16.5 cm
    d.   16.3 cm
    e.   17.1 cm

11. Find all solutions to the triangle described below.

$A = 118°$, $a = 0.66$ cm, $b = 0.91$ cm

Please round each angle to the nearest degree and the length of the missing side to the nearest tenth.

Select the correct answer.

a. $B = 19°$, $C = 43°$, $c = 0.51$ cm

b. $B = 19°$, $C = 43°$, $c = 0.51$ cm or
$B' = 161°$, $C' = 99°$, $c' = 0.74$ cm

c. $B = 19°$, $C = 43°$, $c = 0.51$ cm or
$B' = 22°$, $C' = 40°$, $c' = 0.88$ cm

d. $B = 22°$, $C = 40°$, $c = 0.88$ cm

e.   no solution

12. For the pair of vectors, find $\mathbf{U} + \mathbf{V}$, $\mathbf{U} - \mathbf{V}$, and $3\mathbf{U} + 2\mathbf{V}$.

$\mathbf{U} = -2\mathbf{i} + 2\mathbf{j}$, $\mathbf{V} = 2\mathbf{i} + 2\mathbf{j}$

Please enter your answer in terms of the unit vectors $\mathbf{i}$ and $\mathbf{j}$.

$\mathbf{U} + \mathbf{V} = $ _________

$\mathbf{U} - \mathbf{V} = $ _________

$3\mathbf{U} + 2\mathbf{V} = $ _________

13. A plane headed due north is traveling with an airspeed of 180 miles per hour. The wind currents are moving with constant speed in the direction $150°$. If the ground speed of the plane is 90 miles per hour, what is its true course?

Select the correct answer.

a. $43°$

b. $80°$

c. $10°$

d. $15°$

e. $60°$

14. A plane is flying with an airspeed of 244 miles per hour with heading $277.2°$. The wind currents are running at a constant 45.7 miles per hour in the direction $267.1°$. Find the ground speed and true course of the plane.

Please round each answer to the nearest tenth at the end of the computations.

Select the correct answer.

a.   289.1 mph with heading $276.0°$

b.   289.1 mph with heading $275.6°$

c.   288.7 mph with heading $275.6°$

d.   288.3 mph with heading $276.4°$

e.   288.7 mph with heading $276.0°$

15. Find all solutions to the triangle described below.

$$A = 122.5°, \ a = 25.7 \text{ cm}, \ b = 44.6 \text{ cm}$$

Please round each angle and length of the missing side to the nearest tenth at the end of the computations.

Select the correct answer.

a. $B = 14.1°$, $C = 43.4°$, $c = 21$ cm or
$B' = 35.7°$, $C' = 21.8°$, $c' = 25.3$ cm

b. $B = 14.1°$, $C = 43.4°$, $c = 21$ cm or
$B' = 165.9°$, $C' = 108.4°$, $c' = 28.9$ cm

c. $B = 14.1°$, $C = 43.4°$, $c = 21$ cm

d. $B = 35.7°$, $C = 21.8°$, $c = 25.3$ cm

e. no solution

16. Referring to triangle *ABC,* find the triangle's area given

$$A = 42.2°, \ B = 71.2°, \ a = 220 \text{ inches}$$

Please round your answer to the nearest hundred.

Select the correct answer.

a. 31,500 inches^2

b. 31,000 inches^2

c. 31,300 inches^2

d. 31,200 inches^2

e. 30,900 inches^2

17. Draw a vector representing the course of a ship that goes 35 miles on a course with direction 240° .

Select the correct answer.

a.

b.

c.

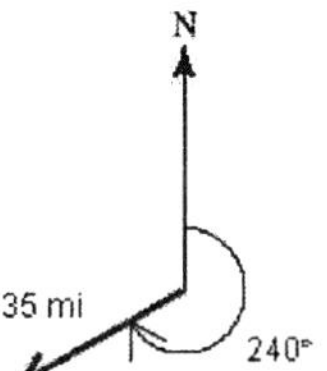

d.

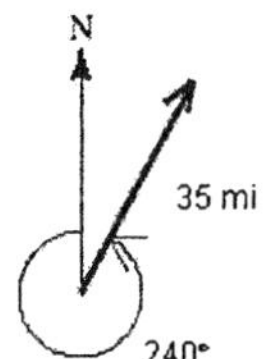

18. Two ships leave a harbor entrance at the same time. The first ship is traveling at a constant 19 miles per hour, while the second is traveling at a constant 22 miles per hour. If the angle between their courses is $125°$, how far apart are they after 2 hours?

Please round your answer to the nearest mile.

Select the correct answer.

a.   65 mi
b.   79 mi
c.   73 mi
d.   71 mi
e.   69 mi

19. Find the magnitude of the vector and the angle θ, $0° \leq \theta < 360°$, that the vector makes with the positive *x*-axis.

$$\mathbf{W} = \mathbf{i} + \sqrt{3}\,\mathbf{j}$$

Select the correct answer.

a. $|\mathbf{W}| = 2, \theta = 120°$

b. $|\mathbf{W}| = 2, \theta = 60°$

c. $|\mathbf{W}| = 6, \theta = 60°$

d. $|\mathbf{W}| = 3, \theta = 300°$

e. $|\mathbf{W}| = 7, \theta = 120°$

20. Find the angle θ between the given vectors to the nearest tenth of a degree.

$$\mathbf{U} = -2\mathbf{i} + 7\mathbf{j} \qquad \mathbf{V} = 6\mathbf{i} + 2\mathbf{j}$$

Select the correct answer.

a. $\theta = 83.5°$

b. $\theta = 88.5°$

c. $\theta = 87.5°$

d. $\theta = 89.5°$

e. $\theta = 84.5°$

21. Draw the vector $\mathbf{V}$ that goes from the origin to the given point. Then write $\mathbf{V}$ in terms of the unit vectors $\mathbf{i}$ and $\mathbf{j}$.

$(3, -4)$

Select the correct answer.

a. 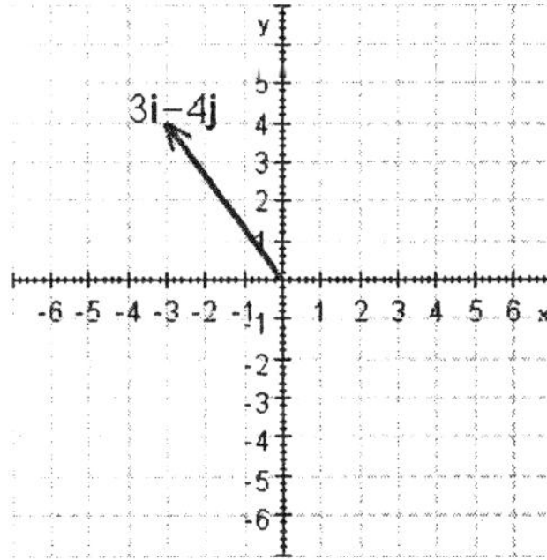

b. 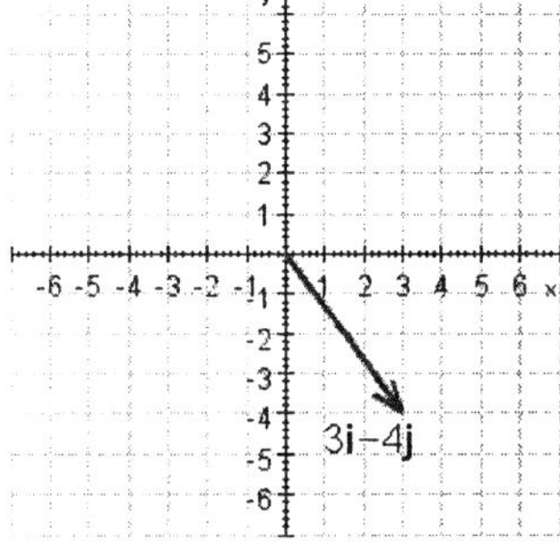

c. 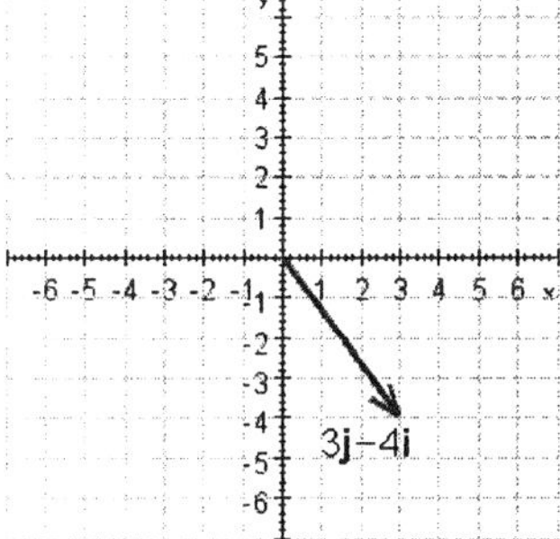

22. Referring to triangle *ABC,* find the triangle's area given

$b = 63.3$ km, $c = 75.4$ km, $A = 125° \ 40'$

Please round your answer to the nearest integer.

$S =$ _________ km^2

23. Find the work performed when the given force F is applied to an object, whose resulting motion is represented by the displacement vector d . Assume the force is in pounds and the displacement is measured in feet.

$$F = 20i + 8j \qquad d = 30i + 2j$$

Work = _________ ft-lb

24. A woman entering an outside glass elevator on the ground floor of a hotel glances up to the top of the building across the street and notices that the angle of elevation is $47°$. She rides the elevator up three floors (60 feet) and finds that the angle of elevation to the top of the building across the street is $30°$. How tall is the building across the street? (Give your answer to the nearest foot.)

_________ ft

25. Solve the triangle.

$a = 51$ yd, $b = 74$ yd, $c = 64$ yd

Please round each answer to the nearest integer. Do not round any numbers until the end of each computation.

Select the correct answer.

a. $A = 45°, B = 76°, C = 58°$

b. $A = 43°, B = 79°, C = 59°$

c. $A = 45°, B = 76°, C = 59°$

d. $A = 43°, B = 79°, C = 58°$

e. $A = 39°, B = 73°, C = 68°$

McKeague/Turner - Trigonometry 5e Chapter 7 Form H

1. b

2. a

3. 77
   13

4. c

5. 142

6. e

7. 8.10

8. d

9. c

10. e

11. e

12. 4j
    −4i
    −2i + 10j

13. e

14. b

15. e

16. c

17. c

18. c

19. b

20. c

21. b

22. 1,939

23. 616

24. 130

25. d

1. mctr.07.05.20m_NoAlgs
2. mctr.07.01.35m_NoAlgs
3. mctr.07.01.09_NoAlgs
4. mctr.07.06.02m_NoAlgs
5. mctr.07.03.08_NoAlgs
6. mctr.07.01.14m_NoAlgs
7. mctr.07.04.17_NoAlgs
8. mctr.07.06.09m_NoAlgs
9. mctr.07.06.27m_NoAlgs
10. mctr.07.04.21m_NoAlgs
11. mctr.07.02.15m_NoAlgs
12. mctr.07.05.36_NoAlgs
13. mctr.07.02.30m_NoAlgs
14. mctr.07.03.26m_NoAlgs
15. mctr.07.02.10m_NoAlgs
16. mctr.07.04.09m_NoAlgs
17. mctr.07.02.26m_NoAlgs
18. mctr.07.03.22m_NoAlgs
19. mctr.07.05.47m_NoAlgs
20. mctr.07.06.14m_NoAlgs
21. mctr.07.05.13m_NoAlgs
22. mctr.07.04.06_NoAlgs
23. mctr.07.06.23_NoAlgs
24. mctr.07.01.28_NoAlgs
25. mctr.07.03.12m_NoAlgs

1. Write the expression $\sqrt{-8}$ in terms of i.

2. Combine the complex numbers.

   $(9 + 8i) - (3 + i)$

   Enter your answer in standard form for complex numbers.

   ____

3. Find the product.

   $(2 + 3i)(3 - i)$

   Enter your answer in standard form for complex numbers.

   ____

4. Let $z_1 = 5 + 6i$ and $z_2 = 5 - 6i$ and find

   $z_1 z_2$.

   Enter your answer in standard form for complex numbers.

   $z_1 z_2 =$ ________

5. Graph the complex number.

$- 3i$

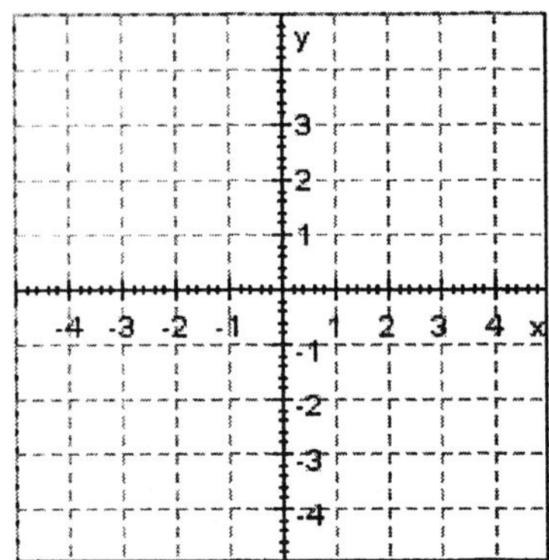

Give the absolute value of the number.

____

6. Graph the complex number along with its opposite and conjugate.

2

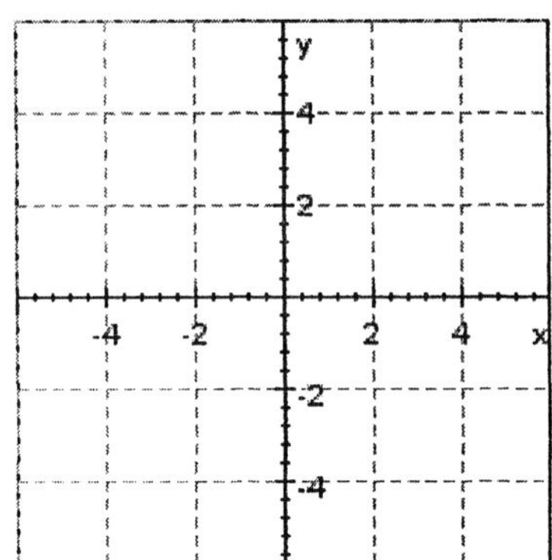

7. Write the complex number in trigonometric form.

$1 - i$

Please enter the argument in degrees and use the degree symbol.

8. Write the complex number in trigonometric form.
   Round the angle to the nearest hundredth of a degree.

   $$-24 - 7i$$

   Please enter the argument in degrees with the degree symbol.

9. Multiply. Leave the answer in trigonometric form.

   $$6(\cos 20° + i \sin 20°) \cdot 6(\cos 35° + i \sin 35°)$$

10. Find the product $z_1 z_2$ in standard form.

    $$z_1 = \sqrt{3} + i, \; z_2 = -1 + i\sqrt{3}$$

    Write z_1 and z_2 in trigonometric form.

    Please enter your answer in the form: $z_1 = $ _____ , $z_2 = $ _____

    Find the product $z_1 z_2$ in trigonometric form.

    Convert the answer that is in trigonometric form to standard form to show that the two products are equal.

11. Use DeMoivre's Theorem to find the following. Write your answer in standard form.

    $$\left(\sqrt{3} + i\right)^6$$

12.

Find the quotient $\dfrac{z_1}{z_2}$ in standard form.

$$z_1 = 3\sqrt{3} - 3i, \ z_2 = 6i$$

Write z_1 and z_2 in trigonometric form.

Please enter your answer in the form: $z_1 = $ _____ , $z_2 = $ _____

Find the quotient $\dfrac{z_1}{z_2}$ in trigonometric form.

Convert the answer that is in trigonometric form to standard form to show that the two products are equal.

13. Change the equation to rectangular coordinates and then graph.

$$r(1 + \cos\theta) = 1$$

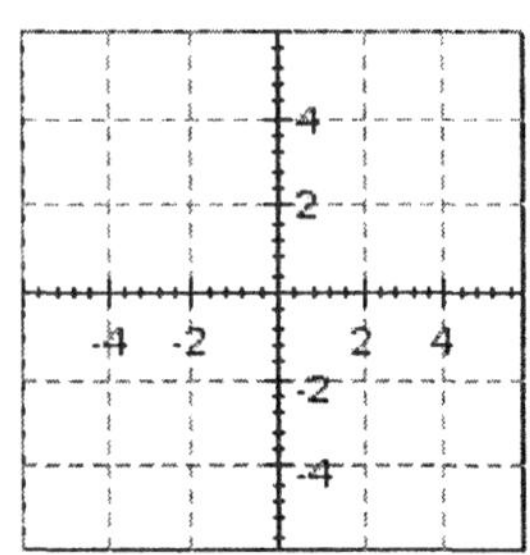

14. Find two square roots for the complex number. Leave your answers in trigonometric form.

$$16\left(\cos 120^\circ + i\sin 120^\circ\right)$$

Graph the two roots.

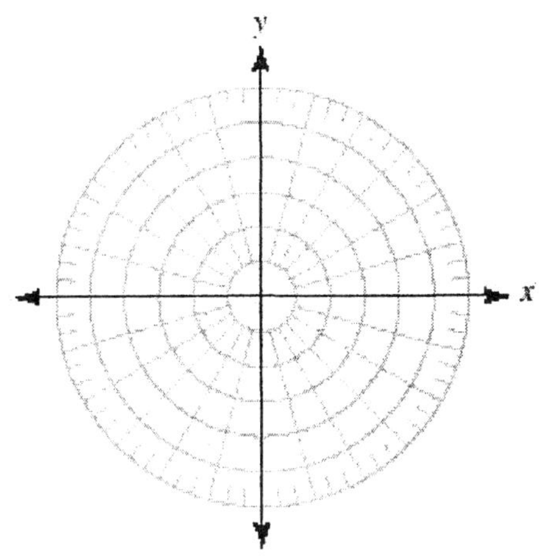

15. Find two square roots for the complex number. Write your answer in standard form.

$$8 - 8i\sqrt{3}$$

16. Find three cube roots for the complex number. Leave your answers in trigonometric form.

$$8\left(\cos 291^\circ + i\sin 291^\circ\right)$$

17. Solve the equation.

$$x^4 + 1{,}296 = 0$$

Please enter the answer in standard form.

18. Graph the ordered pair on a polar coordinate system.

$$(4 , 150°)$$

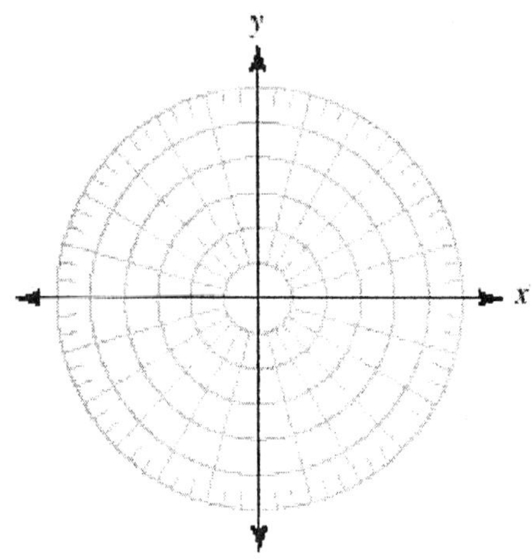

19. For the ordered pair, give three other ordered pairs with degree values between $-360°$ and $360°$ that name the same point.

$$(6 , 125°)$$

20. Convert to rectangular coordinates. Use exact values.

$$\left(5\sqrt{2} , -45°\right)$$

21. Convert to polar coordinates with $r \geq 0$ and θ between $0°$ and $360°$.

$$\left(4, \ 4\sqrt{3}\right)$$

$$\left(\underline{\qquad} , \ \underline{\qquad}° \right)$$

22. Graph the equation using your graphing calculator.

$$r = 4 \sin 3\theta$$

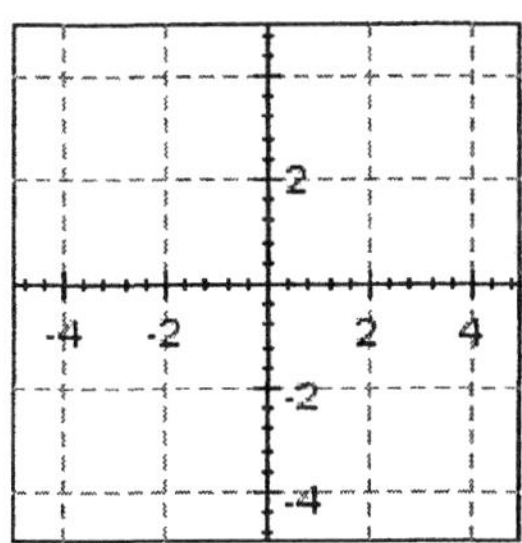

23. Graph the equation.

$$\theta = 225°$$

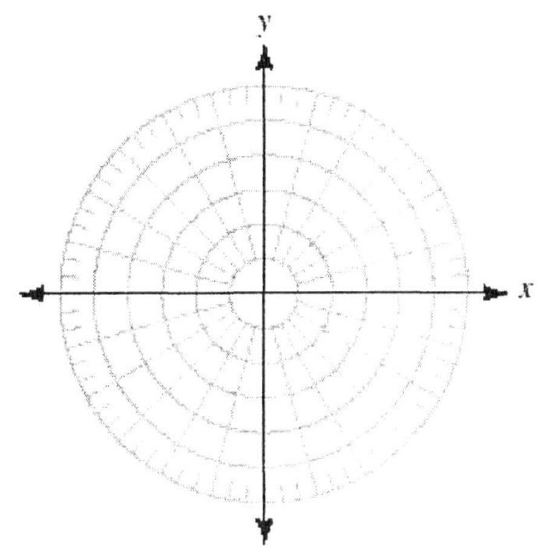

24. Graph the equation.

$$r = 4 + 5\cos\theta$$

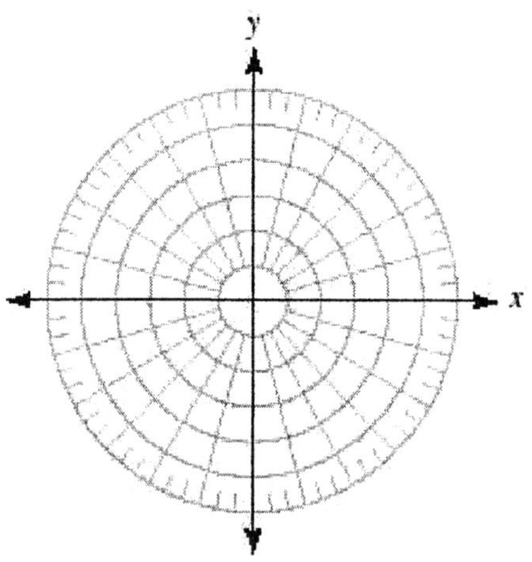

25. Graph the equation.

$$r = 2\cos 2\theta$$

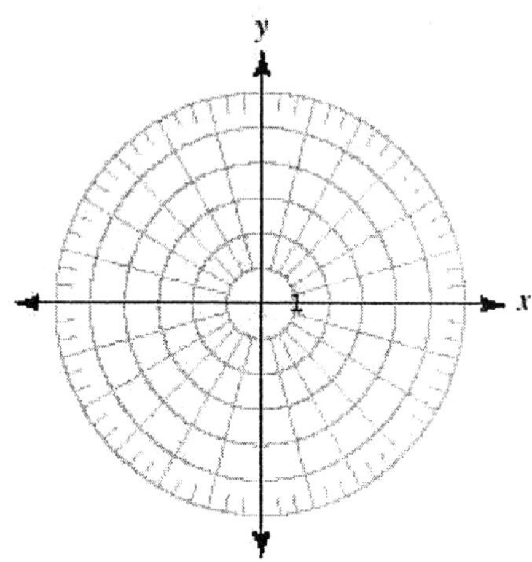

1. $2i \cdot \sqrt{2}$

2. $6 + 7i$

3. $9 + 7i$

4. $6\,i$

5.

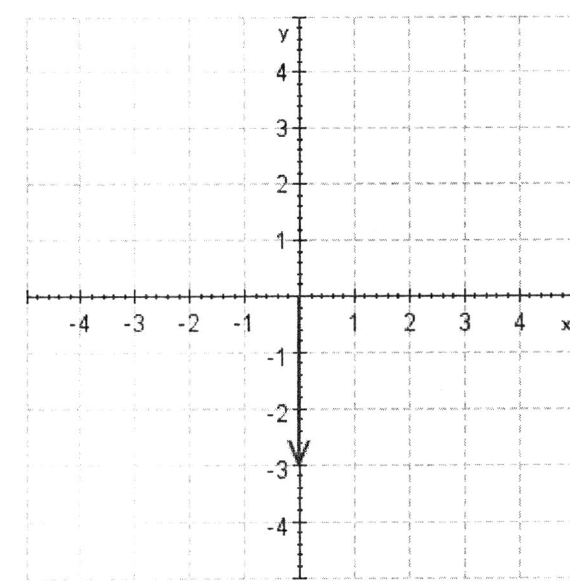

3

6.

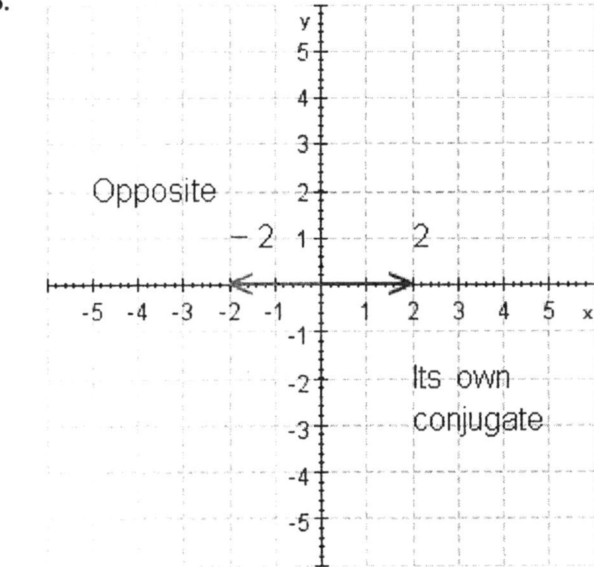

7. $\sqrt{2} \cdot (\cos(315°) + i \cdot \sin(315°))$

8. $25(\cos(196.26°) + i \cdot \sin(196.26°))$

9. $36(\cos(55°) + i \cdot \sin(55°))$

10. $-2\sqrt{3} + 2i$

$z_1 = 2(\cos(30°) + i \cdot \sin(30°)), \; z_2 = 2(\cos(120°) + i \cdot \sin(120°))$

$4(\cos(150°) + i \cdot \sin(150°))$

$z_1 z_2 = 4(\cos 150° + i\sin 150°) = 4\left(-\dfrac{\sqrt{3}}{2} + \dfrac{1}{2}i\right) = -2\sqrt{3} + 2i$

11. -64

12. $-\dfrac{1}{2} - \dfrac{\sqrt{3}}{2} \cdot i$

$z_1 = 6(\cos(330°) + i \cdot \sin(330°)), \; z_2 = 6(\cos(90°) + i \cdot \sin(90°))$

$\cos(240°) + i \cdot \sin(240°)$

$\dfrac{z_1}{z_2} = \cos 240° + i\sin 240° = -\dfrac{1}{2} - \dfrac{\sqrt{3}}{2}i$

13.

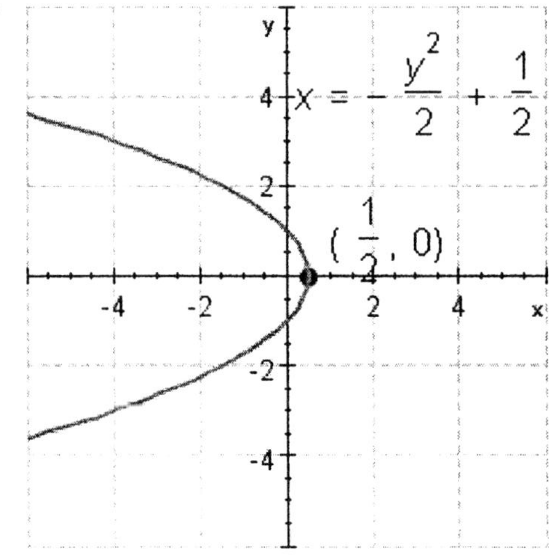

14. $4\left(\cos(60°)+i\cdot\sin(60°)\right)$, $4\left(\cos(240°)+i\cdot\sin(240°)\right)$

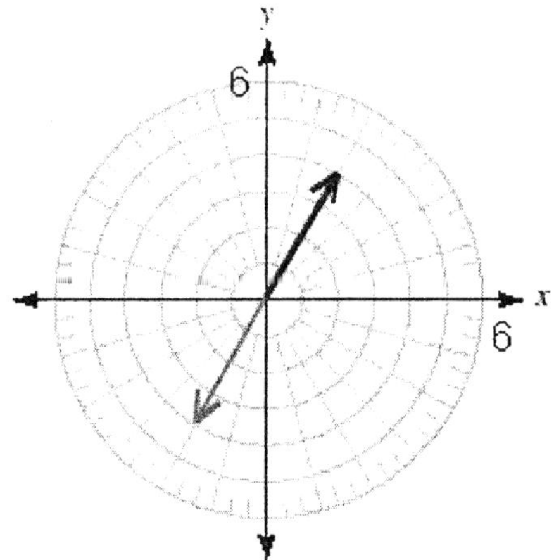

15. $-2\sqrt{3}+2i$, $2\sqrt{3}-2i$

16. $2\left(\cos(97°)+i\cdot\sin(97°)\right)$, $2\left(\cos(217°)+i\cdot\sin(217°)\right)$, $2\left(\cos(337°)+i\cdot\sin(337°)\right)$

17. $3\sqrt{2}+3\sqrt{2}\cdot i$, $-3\sqrt{2}+3\sqrt{2}\cdot i$, $-3\sqrt{2}-3\sqrt{2}\cdot i$, $3\sqrt{2}-3\sqrt{2}\cdot i$

18.

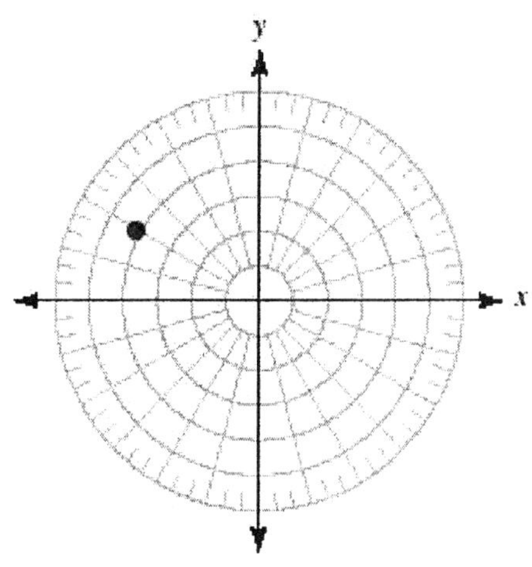

19. $(6,\ -235°)$, $(-6,\ 305°)$, $(-6,\ -55°)$

20. $(5,\ -5)$

21. 8

60

22.

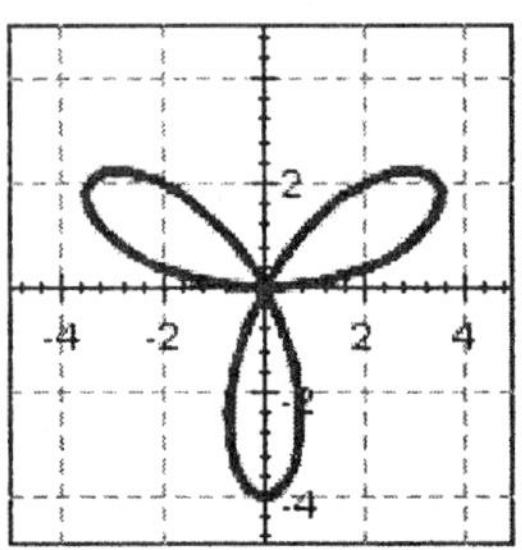

23.

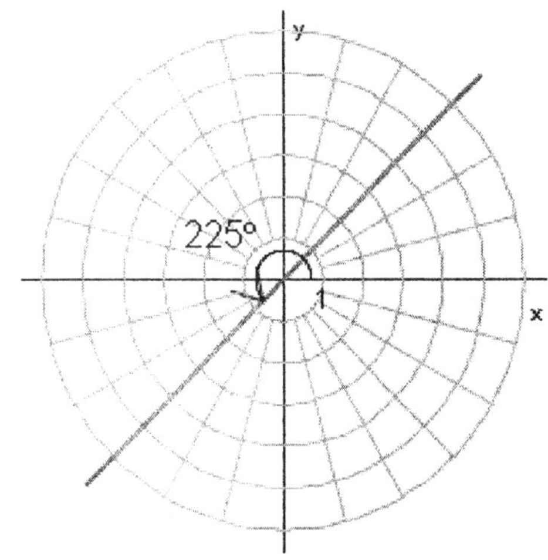

24.

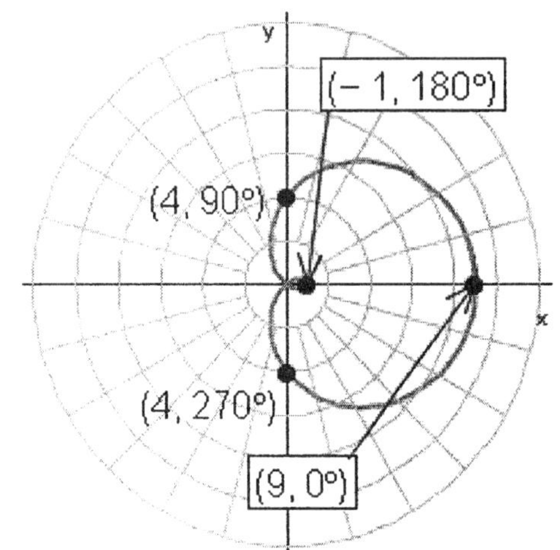

McKeague/Turner - Trigonometry 5e Chapter 8 Form A

25.

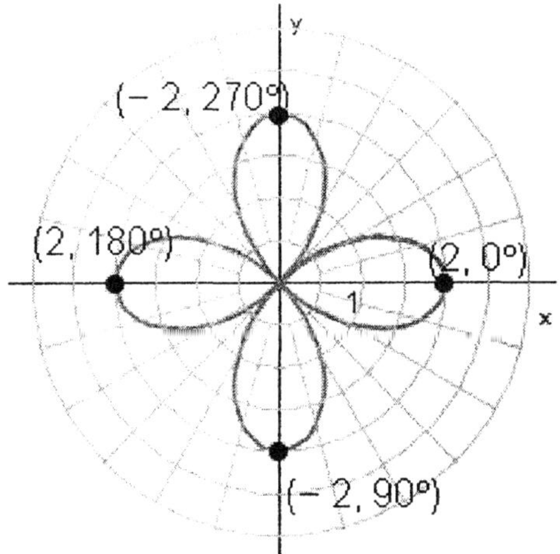

1. mctr.08.01.07_NoAlgs
2. mctr.08.01.25_NoAlgs
3. mctr.08.01.46_NoAlgs
4. mctr.08.01.65_NoAlgs
5. mctr.08.02.05_NoAlgs
6. mctr.08.02.16_NoAlgs
7. mctr.08.02.35_NoAlgs
8. mctr.08.02.51_NoAlgs
9. mctr.08.03.02_NoAlgs
10. mctr.08.03.10_NoAlgs
11. mctr.08.03.28_NoAlgs
12. mctr.08.03.41_NoAlgs
13. mctr.08.06.23_NoAlgs
14. mctr.08.04.03_NoAlgs
15. mctr.08.04.08_NoAlgs
16. mctr.08.04.16_NoAlgs
17. mctr.08.04.26_NoAlgs
18. mctr.08.05.03_NoAlgs
19. mctr.08.05.16_NoAlgs
20. mctr.08.05.24_NoAlgs
21. mctr.08.05.34_NoAlgs
22. mctr.08.06.28_NoAlgs
23. mctr.08.06.12_NoAlgs
24. mctr.08.06.17_NoAlgs
25. mctr.08.06.47_NoAlgs

1. Let $z_1 = 5 + 6i$ and $z_2 = 5 - 6i$ and find

   $z_1 z_2$.

   Enter your answer in standard form for complex numbers.

   $z_1 z_2 = $ ________

2.
   Find the quotient $\dfrac{z_1}{z_2}$ in standard form.

   $z_1 = 3\sqrt{3} - 3i, \ z_2 = 6i$

   Write z_1 and z_2 in trigonometric form.

   Please enter your answer in the form: $z_1 = $ ____ , $z_2 = $ ____

   Find the quotient $\dfrac{z_1}{z_2}$ in trigonometric form.

   Convert the answer that is in trigonometric form to standard form to show that the two products are equal.

3. Graph the complex number.

$- 3i$

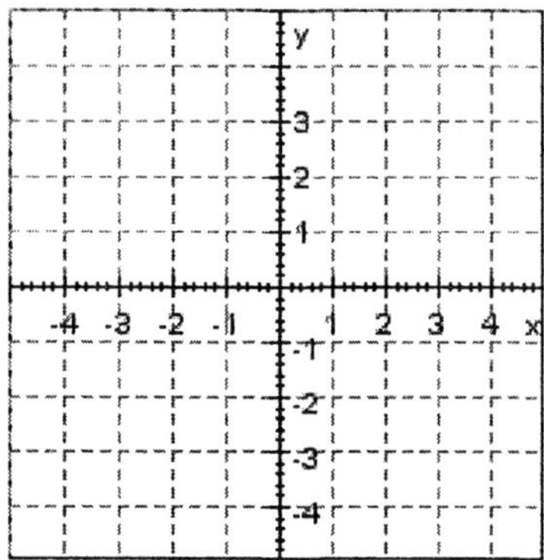

Give the absolute value of the number.

____

4. Solve the equation.

$$x^4 + 1,296 = 0$$

Please enter the answer in standard form.

5. Convert to rectangular coordinates. Use exact values.

$$\left(5\sqrt{2},\ -45°\right)$$

____

6. For the ordered pair, give three other ordered pairs with degree values between $-360°$ and $360°$ that name the same point.

$$(6,\ 125°)$$

7. Convert to polar coordinates with $r \geq 0$ and θ between $0°$ and $360°$.

$$\left(4,\ 4\sqrt{3}\right)$$

$$(\ \underline{\hspace{2cm}},\ \underline{\hspace{2cm}}°)$$

8. Find two square roots for the complex number. Leave your answers in trigonometric form.

$$16\left(\cos 120° + i\sin 120°\right)$$

Graph the two roots.

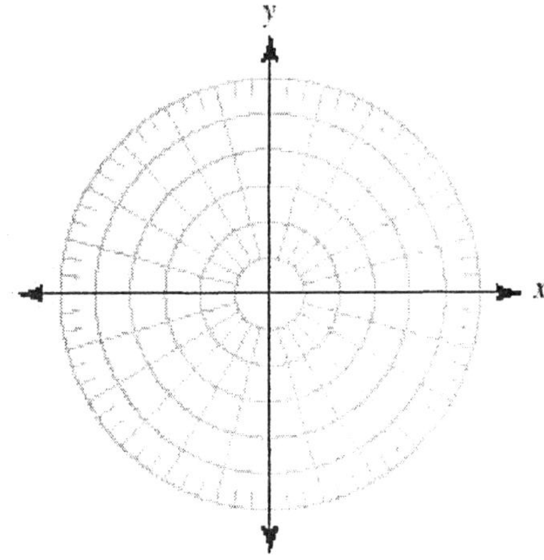

9. Find two square roots for the complex number. Write your answer in standard form.

$$8 - 8i\sqrt{3}$$

10. Graph the equation.

$$\theta = 225°$$

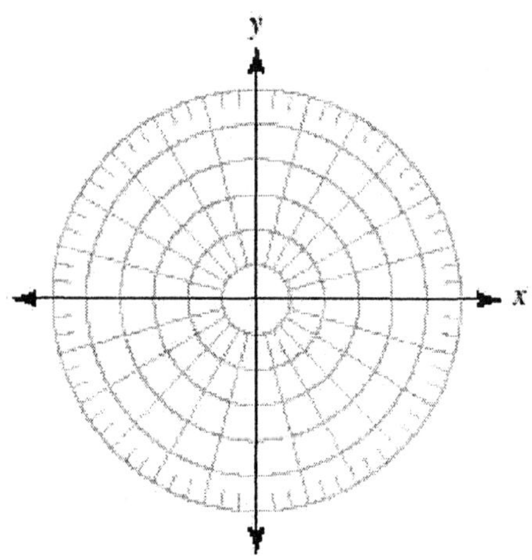

11. Graph the equation.

$$r = 2\cos 2\theta$$

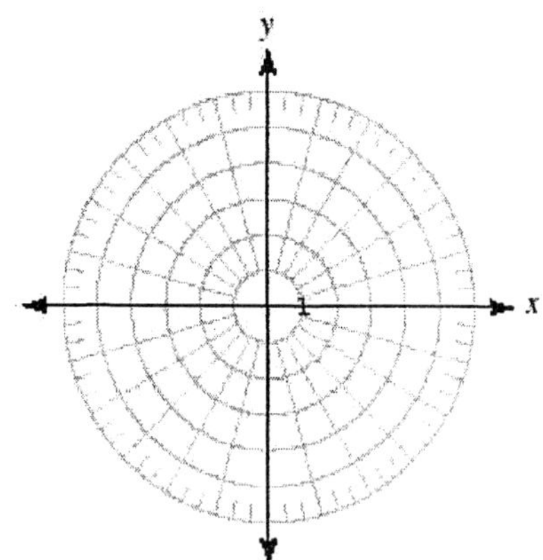

12. Find the product $z_1 z_2$ in standard form.

$$z_1 = \sqrt{3} + i, \; z_2 = -1 + i\sqrt{3}$$

Write z_1 and z_2 in trigonometric form.

Please enter your answer in the form: $z_1 = \underline{\quad}, \; z_2 = \underline{\quad}$

Find the product $z_1 z_2$ in trigonometric form.

Convert the answer that is in trigonometric form to standard form to show that the two products are equal.

13. Graph the ordered pair on a polar coordinate system.

$$(4, 150^\circ)$$

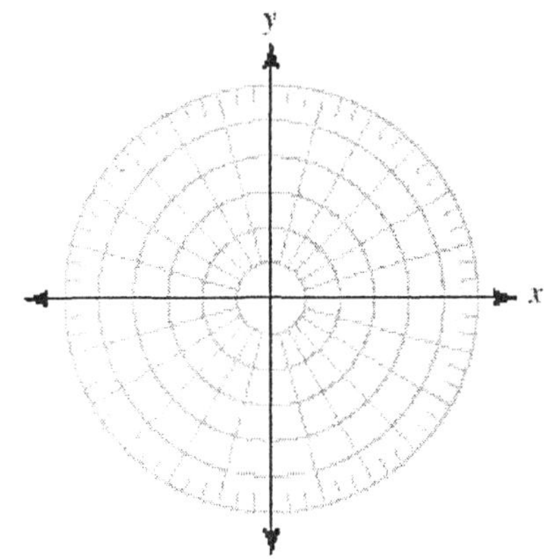

14. Write the complex number in trigonometric form.
Round the angle to the nearest hundredth of a degree.

$-24 - 7i$

Please enter the argument in degrees with the degree symbol.

15. Graph the complex number along with its opposite and conjugate.

2

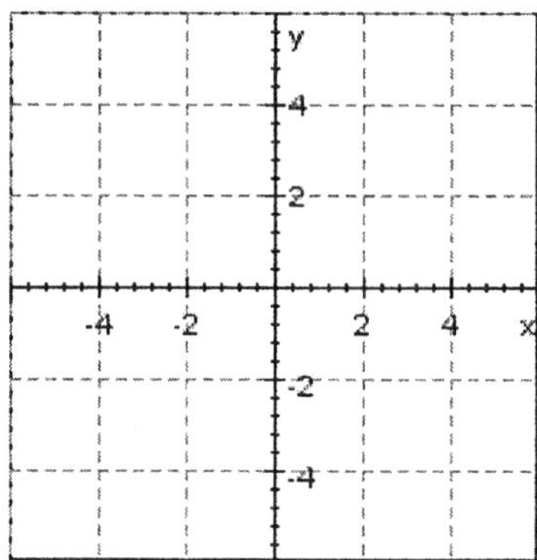

16. Find the product.

$(2 + 3i)(3 - i)$

Enter your answer in standard form for complex numbers.

____

17. Graph the equation using your graphing calculator in polar mode.

$r = 4 \sin 3\theta$

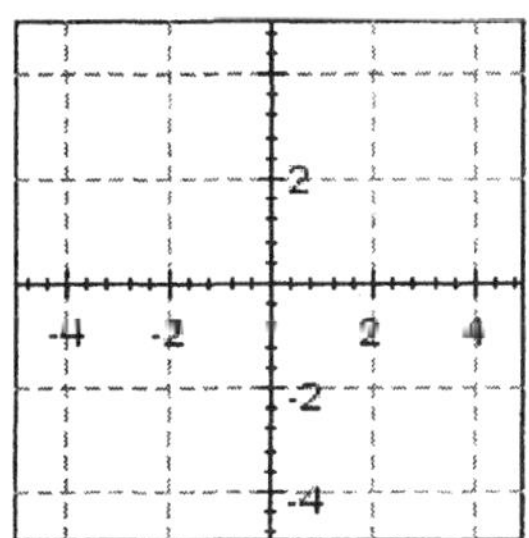

18. Combine the complex numbers.

$(9 + 8i) - (3 + i)$

Enter your answer in standard form for complex numbers.

____

19. Write the expression $\sqrt{-8}$ in terms of i.

20. Graph the equation.

$$r = 4 + 5\cos\theta$$

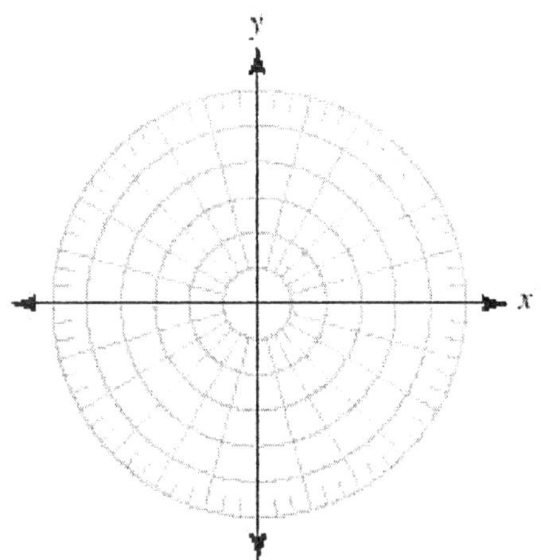

21. Write the complex number in trigonometric form.

$$1 - i$$

Please enter the argument in degrees and use the degree symbol.

22. Change the equation to rectangular coordinates and then graph.

$$r(1 + \cos\theta) = 1$$

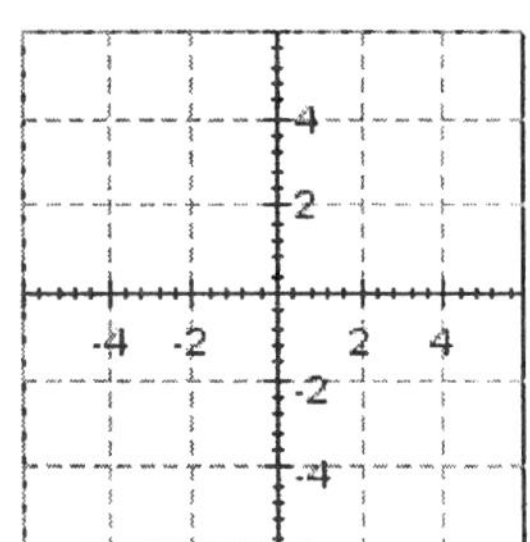

23. Find three cube roots for the complex number. Leave your answers in trigonometric form.

$$8(\cos 291° + i\sin 291°)$$

24. Multiply. Leave the answer in trigonometric form.

$$6(\cos 20° + i\sin 20°) \cdot 6(\cos 35° + i\sin 35°)$$

25. Use DeMoivre's Theorem to find the following. Write your answer in standard form.

$$\left(\sqrt{3} + i\right)^6$$

00

1. 61

2. $-\dfrac{1}{2} - \dfrac{\sqrt{3}}{2} \cdot i$

$z_1 = 6\left(\cos(330^\circ) + i \cdot \sin(330^\circ)\right),\ z_2 = 6\left(\cos(90^\circ) + i \cdot \sin(90^\circ)\right)$

$\cos(240^\circ) + i \cdot \sin(240^\circ)$

$\dfrac{z_1}{z_2} = \cos 240^\circ + i\sin 240^\circ = -\dfrac{1}{2} - \dfrac{\sqrt{3}}{2} i$

3.

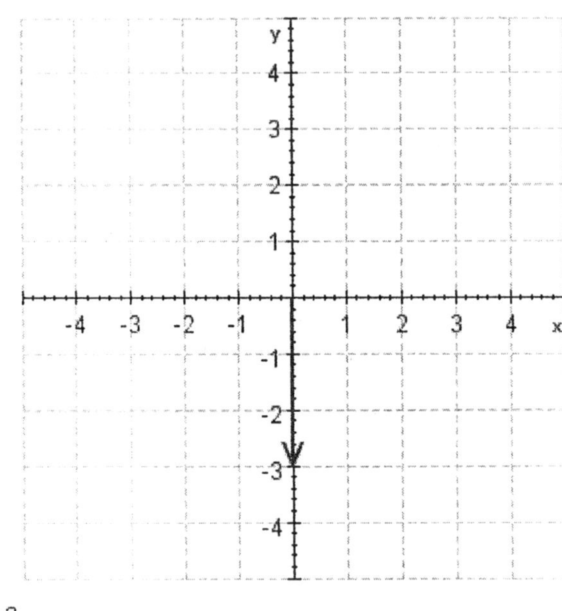

3

4. $3\sqrt{2} + 3\sqrt{2}\cdot i,\ -3\sqrt{2} + 3\sqrt{2}\cdot i,\ -3\sqrt{2} - 3\sqrt{2}\cdot i,\ 3\sqrt{2} - 3\sqrt{2}\cdot i$

5. $\left(5,\ -5\right)$

6. $\left(6,\ -235^\circ\right),\ \left(-6,\ 305^\circ\right),\ \left(-6,\ -55^\circ\right)$

7. 8

60

8. $4(\cos(60°)+i \cdot \sin(60°))$, $4(\cos(240°)+i \cdot \sin(240°))$

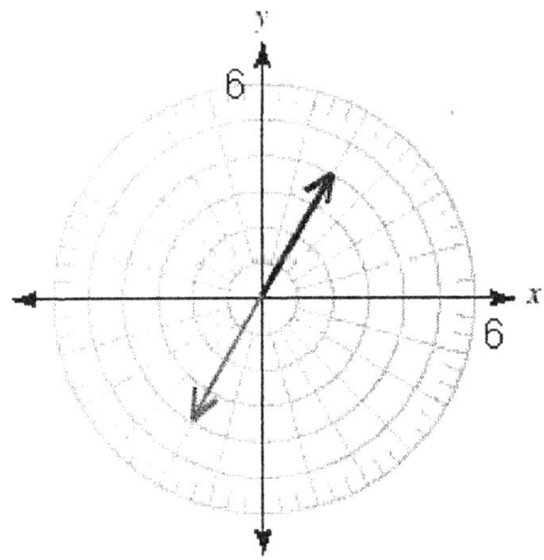

9. $-2\sqrt{3}+2i$, $2\sqrt{3}-2i$

10.

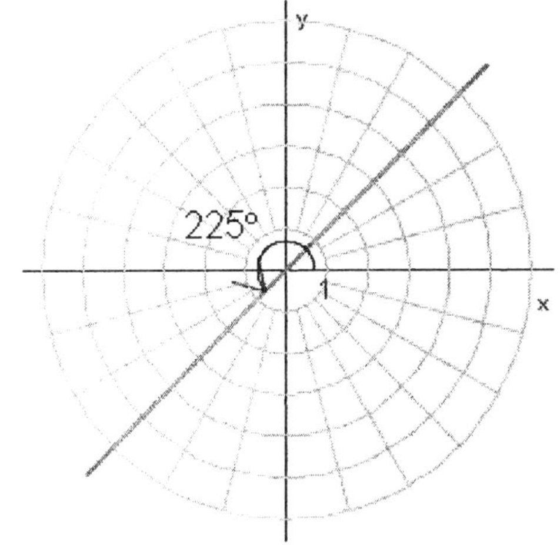

$4(\cos(60°)+i \cdot \sin(60°))$, $4(\cos(240°)+i \cdot \sin(240°))$

11.

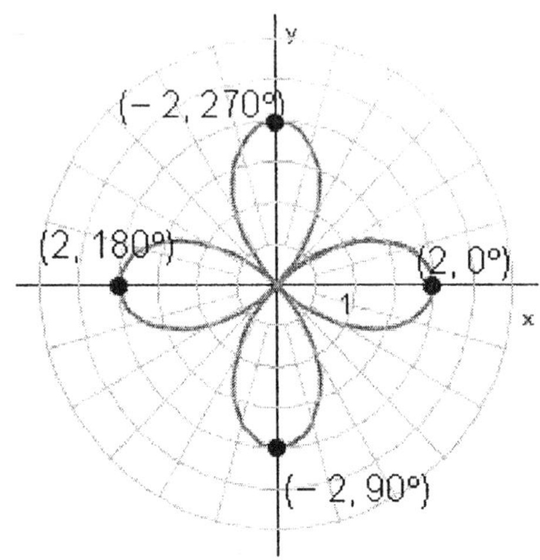

12. $-2\sqrt{3}+2i$

$z_1 = 2\big(\cos(30°)+i \cdot \sin(30°)\big),\ z_2 = 2\big(\cos(120°)+i \cdot \sin(120°)\big)$

$4\big(\cos(150°)+i \cdot \sin(150°)\big)$

$$z_1 z_2 = 4\big(\cos 150° + i\sin 150°\big) = 4\left(-\frac{\sqrt{3}}{2} + \frac{1}{2}i\right) = -2\sqrt{3} + 2i$$

13.

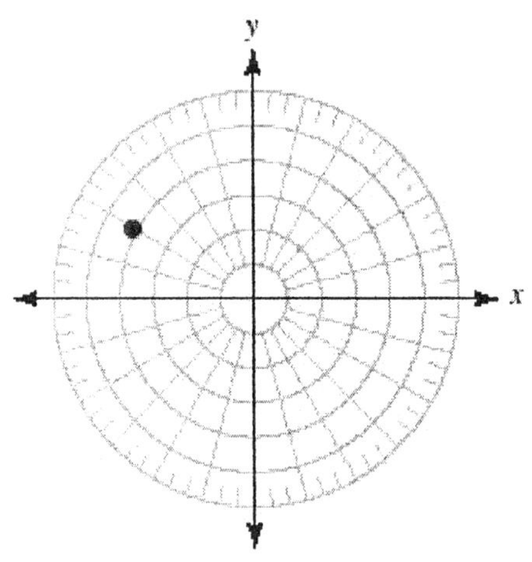

14. $25\big(\cos(196.26°)+i \cdot \sin(196.26°)\big)$

15.

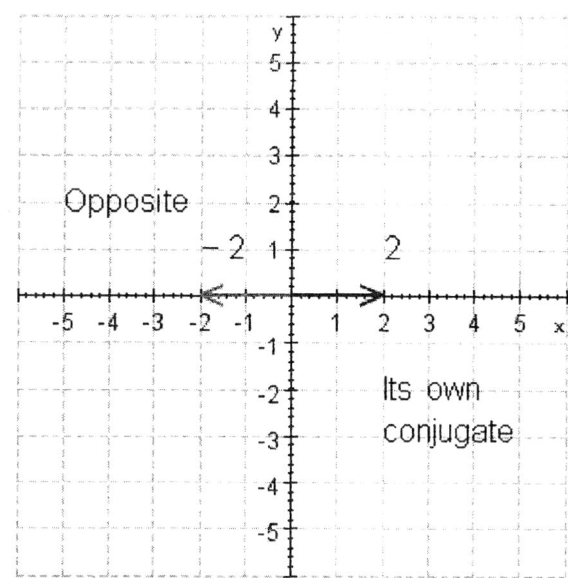

16. $9 + 7i$

17.

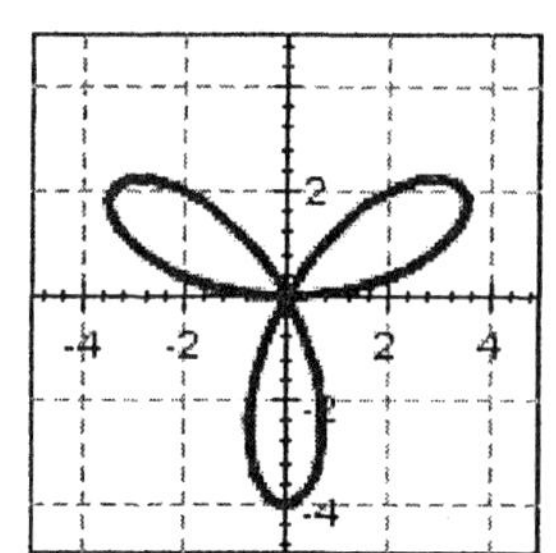

18. $6 + 7i$

19. $2i \cdot \sqrt{2}$

20.

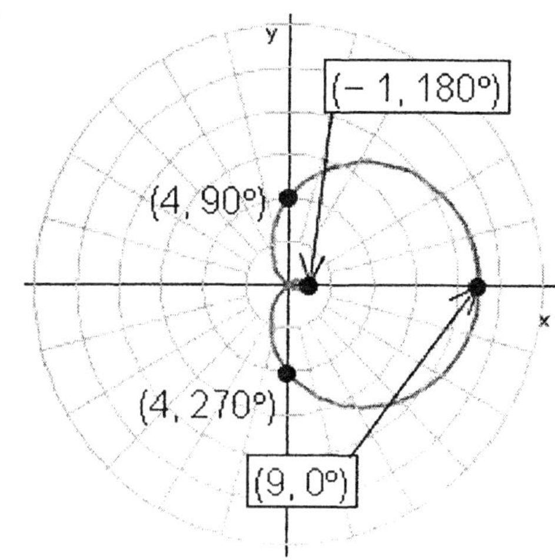

21. $\sqrt{2} \cdot (\cos(315°) + i \cdot \sin(315°))$

22.

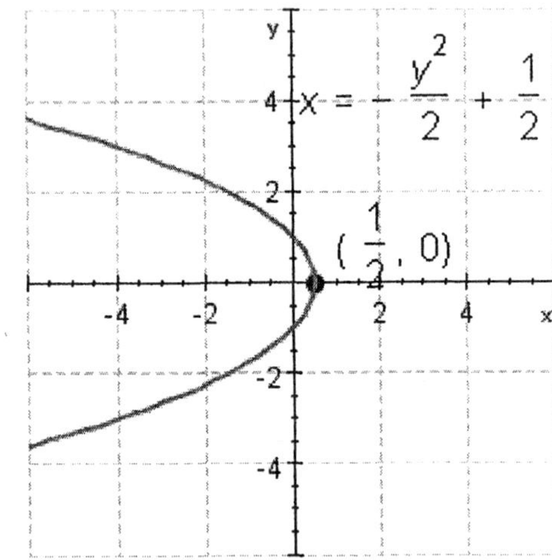

23. $2(\cos(97°) + i \cdot \sin(97°)), \ 2(\cos(217°) + i \cdot \sin(217°)), \ 2(\cos(337°) + i \cdot \sin(337°))$

24. $36(\cos(55°) + i \cdot \sin(55°))$

25. -64

McKeague/Turner - Trigonometry 5e Chapter 8 Form B

1. mctr.08.01.65_NoAlgs
2. mctr.08.03.41_NoAlgs
3. mctr.08.02.05_NoAlgs
4. mctr.08.04.26_NoAlgs
5. mctr.08.05.24_NoAlgs
6. mctr.08.05.10_NoAlgs
7. mctr.08.05.34_NoAlgs
8. mctr.08.04.03_NoAlgs
9. mctr.08.04.08_NoAlgs
10. mctr.08.06.12_NoAlgs
11. mctr.08.06.23_NoAlgs
12. mctr.08.03.10_NoAlgs
13. mctr.08.05.03_NoAlgs
14. mctr.08.02.51_NoAlgs
15. mctr.08.02.16_NoAlgs
16. mctr.08.01.46_NoAlgs
17. mctr.08.06.28_NoAlgs
18. mctr.08.01.25_NoAlgs
19. mctr.08.01.07_NoAlgs
20. mctr.08.06.17_NoAlgs
21. mctr.08.02.35_NoAlgs
22. mctr.08.06.47_NoAlgs
23. mctr.08.04.16_NoAlgs
24. mctr.08.03.02_NoAlgs
25. mctr.08.03.28_NoAlgs

1. Write the expression $\sqrt{-16} \cdot \sqrt{-9}$ in terms of i and then simplify.

   Select the correct answer.

   a. 12

   b. −15

   c. 8

   d. 15

   e. −12

2. Simplify the power of i.

   i^{17}

   Select the correct answer.

   a. − i
   b. i
   c. − 1
   d. 1
   e. 0

3. Find the product.

   $(3 + 7i)(3 - 7i)$

   Select the correct answer.

   a. 58
   b. 62
   c. 56 - 42i
   d. 56
   e. 58 - 42i

4. Read the method of solving the system of equations.

$$x + y = 10$$
$$xy = 40$$

Solving the first equation for y, we have $y = 10 - x$. Substituting this value of y into the second equation, we have

$$x(10 - x) = 40$$

The equation is quadratic. We write it in standard form and apply the quadratic formula.

$$0 = x^2 - 10x + 40$$

$$x = \frac{10 \pm \sqrt{100 - 4(1)(40)}}{2}$$

$$= \frac{10 \pm \sqrt{100 - 160}}{2}$$

$$= \frac{10 \pm \sqrt{-60}}{2}$$

$$= \frac{10 \pm 2\sqrt{-15}}{2}$$

$$= 5 \pm \sqrt{-15}$$

$$= 5 \pm i\sqrt{15}$$

$$y = 5 \mp i\sqrt{15}$$

Use the method shown above to solve the system of equations.

$$x + y = 4$$
$$xy = 68$$

Select the correct answer.

a. $x = 2 + 8i,\ y = 2 - 8i;\ x = 2 - 8i,\ y = 2 + 8i$

b. $x = 6 + 6i,\ y = 6 - 6i;\ x = 6 - 6i,\ y = 6 + 6i$

c. $x = 2 + 8i,\ y = 2 + 8i;\ x = 2 - 8i,\ y = 2 - 8i$

d. $x = 6 + 8i,\ y = 6 - 8i;\ x = 6 - 8i,\ y = 6 + 8i$

e. $x = 2 + 6i,\ y = 2 - 6i;\ x = 8 - 6i,\ y = 2 + 6i$

5. Graph the complex number and give the absolute value of it.

$6 - 8i$

Select the correct answer.

a.

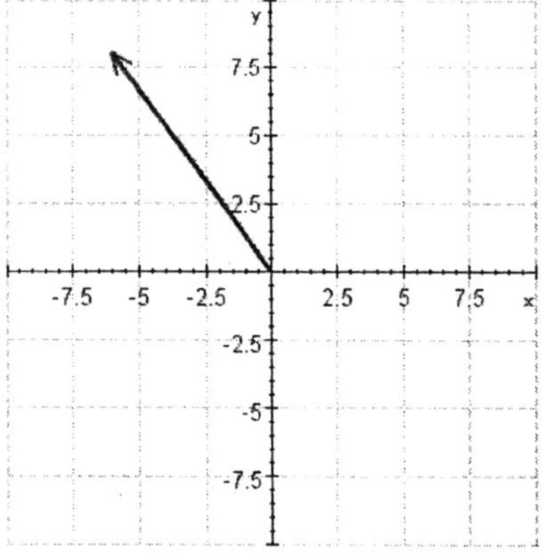

The absolute value is 10

b.

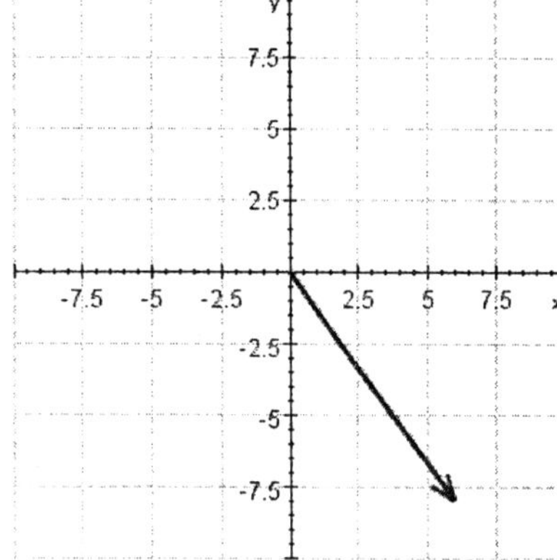

The absolute value is 10

c.

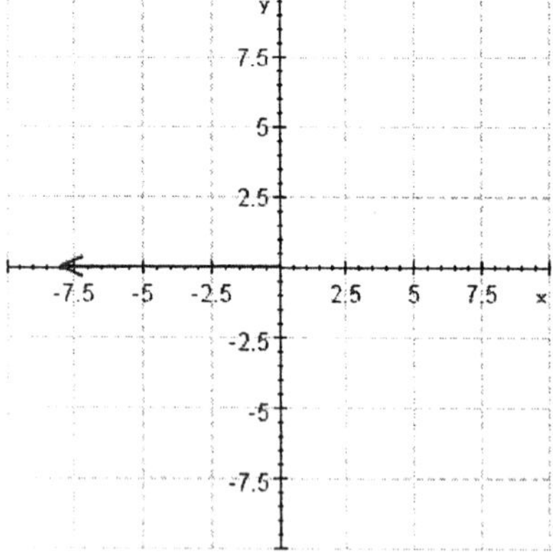

The absolute value is 6

6. Use a calculator to help write the complex number in standard form. Round the numbers in your answer to the nearest hundredth.

$$100\left(\cos 144° + i\sin 144°\right)$$

Select the correct answer.

a. - 80.90 + 58.78i
b. - 80.90 + 50.74i
c. - 80.94 + 58.78i
d. - 80.98 + 58.86i
e. - 80.94 + 58.74i

7. Write the complex number in trigonometric form.

$$3 + 3i\sqrt{3}$$

Select the correct answer.

a. $6\operatorname{cis}150°$

b. $6\operatorname{cis}300°$

c. $6\operatorname{cis}60°$

d. $6\operatorname{cis}120°$

e. $3\operatorname{cis}60°$

8. Write the complex number in trigonometric form.
    Round the angle to the nearest hundredth of a degree.

$24 + 7i$

Select the correct answer.

a. $25\operatorname{cis}16.38°$

b. $25\operatorname{cis}16.35°$

c. $25\operatorname{cis}16.23°$

d. $25\operatorname{cis}16.20°$

e. $25\operatorname{cis}16.26°$

9. Multiply.

$$4 \operatorname{cis} 150° \cdot 2 \operatorname{cis} 30°$$

Select the correct answer.

  a. $8 \operatorname{cis} 180°$

  b. $4 \operatorname{cis} 180°$

  c. $8 \operatorname{cis} 190°$

  d. $4 \operatorname{cis} 190°$

  e. $8 \operatorname{cis} 185°$

10. Use DeMoivre's Theorem to find the following.

$$\left(\sqrt{2} \operatorname{cis} 60°\right)^{10}$$

Select the correct answer.

  a. $-16\sqrt{3} + 16i$

  b. $-4 + 4i$

  c. $-16 - 16i\sqrt{3}$

  d. $-4 + 4i\sqrt{3}$

  e. $4 + 4i\sqrt{3}$

11. Divide.

$$\frac{18(\cos 50° + i \sin 50°)}{12(\cos 35° + i \sin 35°)}$$

Select the correct answer.

  a. $2.5(\cos 20° + i\sin 20°)$

  b. $0.5(\cos 85° + i\sin 85°)$

  c. $1.5(\cos 5° + i\sin 5°)$

  d. $1.5(\cos 15° + i\sin 15°)$

  e. $1.5(\cos 85° + i\sin 85°)$

12. DeMoivre's Theorem can be used to find reciprocals of complex numbers. Recall from algebra that the reciprocal of x is $\dfrac{1}{x}$, which can be expressed as x^{-1}.

Use this fact, along with DeMoivre's Theorem, to find the reciprocal of the number below.

$$-\sqrt{3} + i$$

Select the correct answer.

a. $\quad -\dfrac{\sqrt{3}}{4} - \dfrac{1}{4}i$

b. $\quad \dfrac{1}{2} + \dfrac{1}{2}i$

c. $\quad \dfrac{1}{4} + \dfrac{\sqrt{3}}{4}i$

d. $\quad \dfrac{\sqrt{3}}{4} + \dfrac{1}{4}i$

e. $\quad -\dfrac{1}{2} - \dfrac{1}{2}i$

13. Find two square roots for the complex number.

$$64 \operatorname{cis} 0°$$

Select the correct answer.

a. $\quad 8 \operatorname{cis} 0°,\ 8 \operatorname{cis} 270°$

b. $\quad 8 \operatorname{cis} 0°,\ 8 \operatorname{cis} 180°$

c. $\quad 64 \operatorname{cis} 90°,\ 64 \operatorname{cis} 180°$

d. $\quad 8 \operatorname{cis} 90°,\ 8 \operatorname{cis} 180°$

e. $\quad 8 \operatorname{cis} 90°,\ 8 \operatorname{cis} 270°$

14. Find two square roots for the complex number.

16

Select the correct answer.

a. $4, -4i$

b. $4i, -4$

c. $4i, -4i$

d. $4, -4$

e. $16i, -16i$

15. Find three cube roots for the complex number.

$$-32\sqrt{3} + 32i$$

Select the correct answer.

a. $4 \operatorname{cis} 50°, 4 \operatorname{cis} 170°, 4 \operatorname{cis} 290°$

b. $4 \operatorname{cis} 55°, 4 \operatorname{cis} 175°, 4 \operatorname{cis} 295°$

c. $4 \operatorname{cis} 40°, 4 \operatorname{cis} 160°, 4 \operatorname{cis} 280°$

d. $2 \operatorname{cis} 40°, 2 \operatorname{cis} 160°, 2 \operatorname{cis} 280°$

e. $2 \operatorname{cis} 50°, 2 \operatorname{cis} 170°, 2 \operatorname{cis} 290°$

16. Solve the equation.

$$x^4 - 2\sqrt{3}\,x^2 + 4 = 0$$

Select the correct answer.

a. $\sqrt{2} \operatorname{cis} 10°, \sqrt{2} \operatorname{cis} 170°, \sqrt{2} \operatorname{cis} 200°, \sqrt{2} \operatorname{cis} 350°$

b. $\sqrt{2} \operatorname{cis} 15°, \sqrt{2} \operatorname{cis} 165°, \sqrt{2} \operatorname{cis} 195°, \sqrt{2} \operatorname{cis} 345°$

c. $\sqrt{2} \operatorname{cis} 10°, \sqrt{2} \operatorname{cis} 200°$

d. $\sqrt{2} \operatorname{cis} 170°, \sqrt{2} \operatorname{cis} 350°$

e. $\sqrt{2} \operatorname{cis} 15°, \sqrt{2} \operatorname{cis} 195°$

17. Graph the ordered pair on a polar coordinate system.

$$(-3, \ 45°)$$

Select the correct answer.

a.

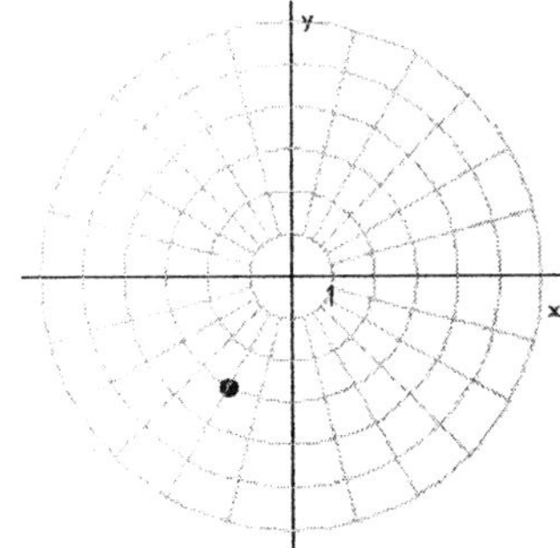

b.

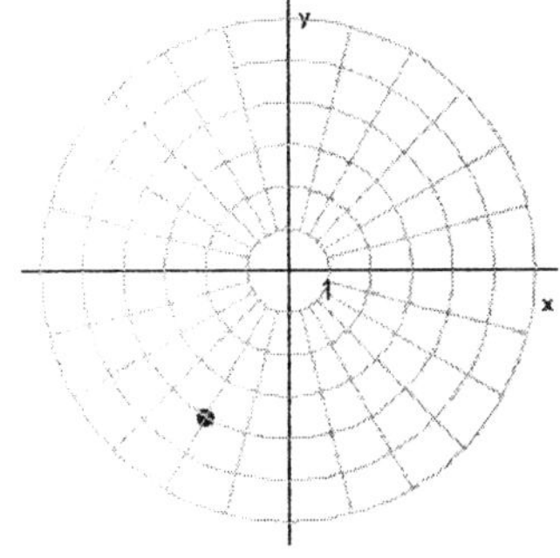

c.

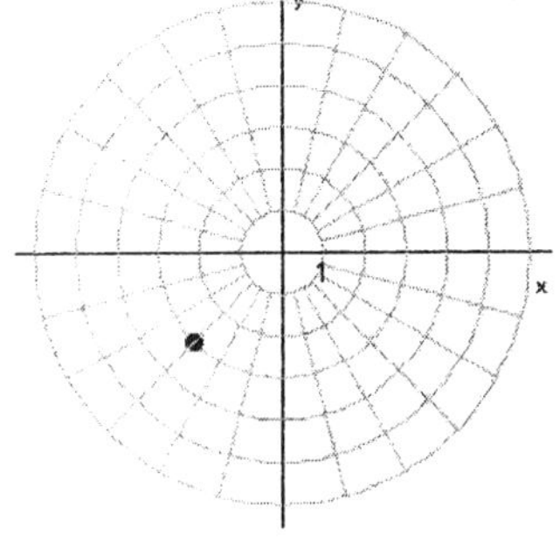

18. Convert to rectangular coordinates. Use exact values.

$$\left(6\sqrt{2},\ -45^\circ\right)$$

Select the correct answer.

a. $(-4, -4)$
b. $(6, -6)$
c. $(-6, 6)$
d. $(-4, 6)$
e. $(-7, -7)$

19. Convert to polar coordinates with $r \geq 0$ and θ between 0° and 360°.

$$\left(-1,\ -\sqrt{3}\right)$$

Select the correct answer.

a. $\left(1,\ 235^\circ\right)$

b. $\left(4,\ 250^\circ\right)$

c. $\left(2,\ 260^\circ\right)$

d. $\left(1,\ 225^\circ\right)$

e. $\left(2,\ 240^\circ\right)$

20. Write the equation in polar coordinates.

$$x^2 + y^2 = 5$$

Select the correct answer.

a. $r^2 = 1$

b. $r^2 = 2$

c. $r^2 = 7$

d. $r^2 = 5$

e. $r^2 = 4$

21. Use your graphing calculator to generate solutions for the equation using values of θ that are multiples of 15°. Sketch the graph of the equation using the values from your table.

$r = 2 + 2\sin\theta$

Select the correct answer.

a.

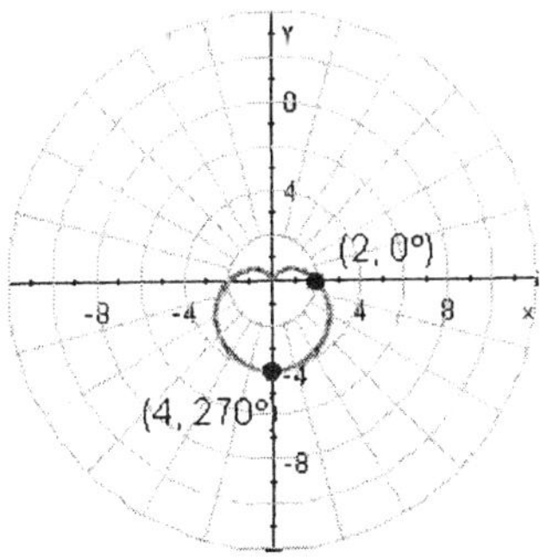

b.

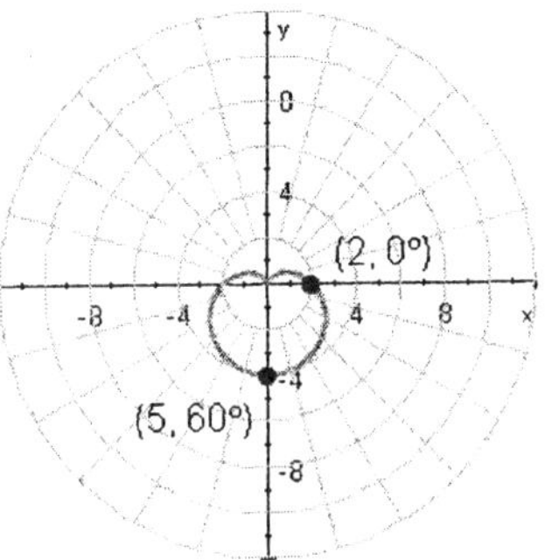

c.

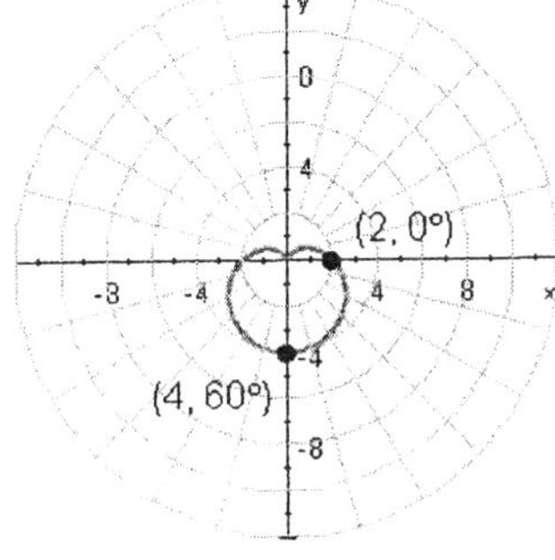

d.

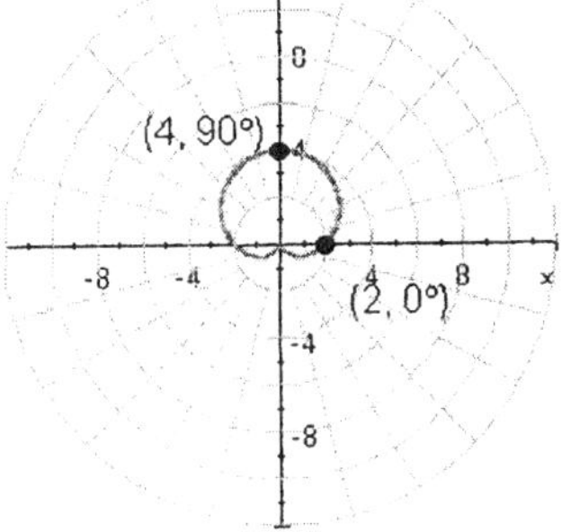

22. Graph the equation.

$$\theta = 225°$$

Select the correct answer.

a.

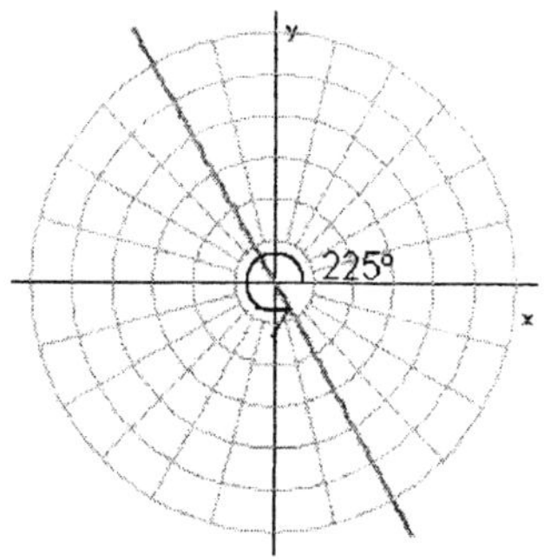

b.

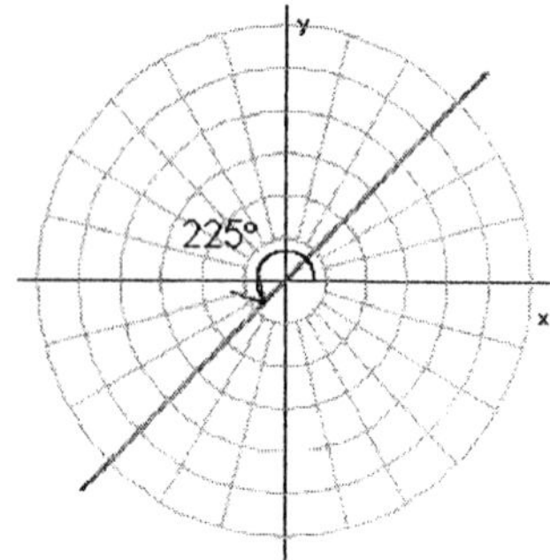

c.

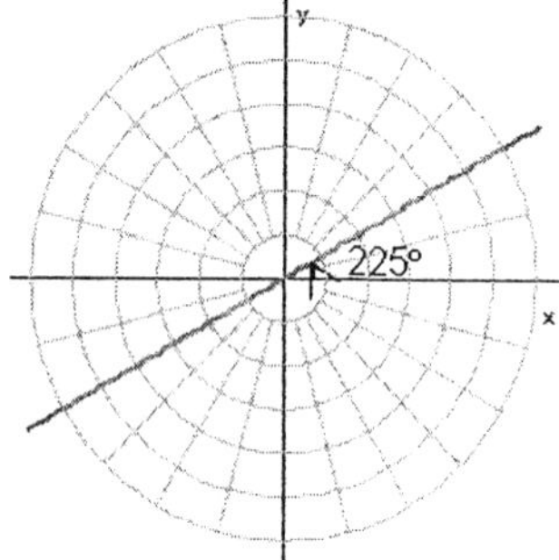

d.

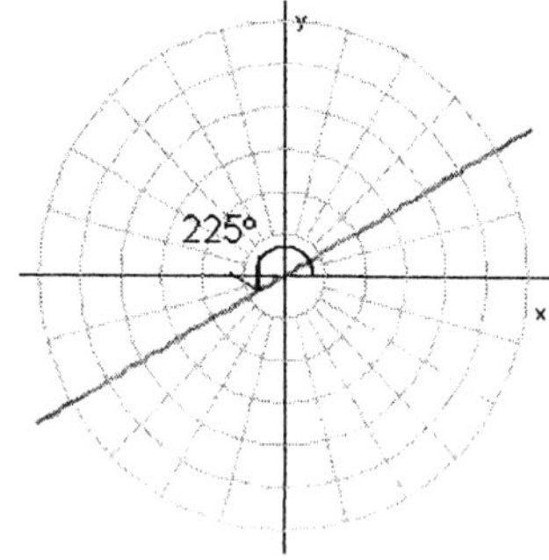

23. Graph the equation.

$$r = 4 + 5\cos\theta$$

Select the correct answer.

a.

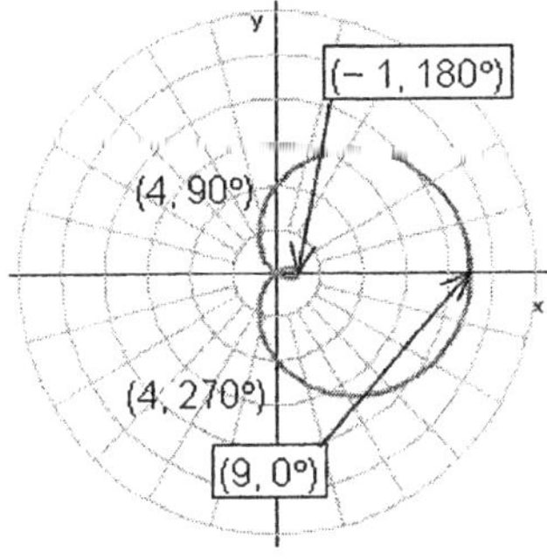

b.

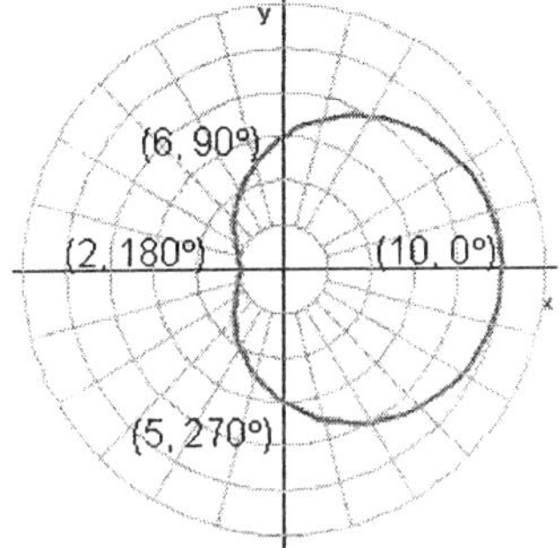

c.

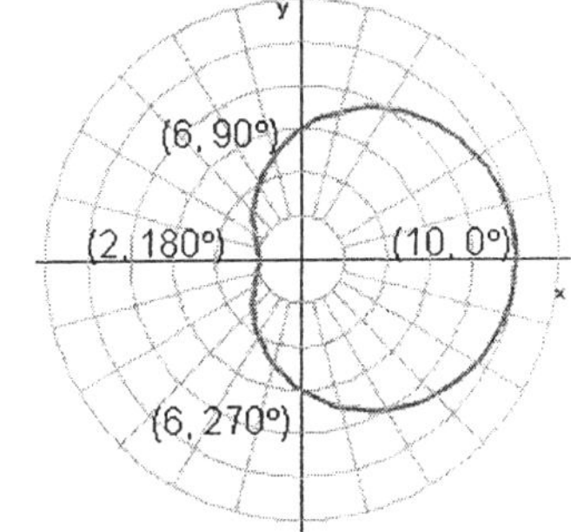

d.

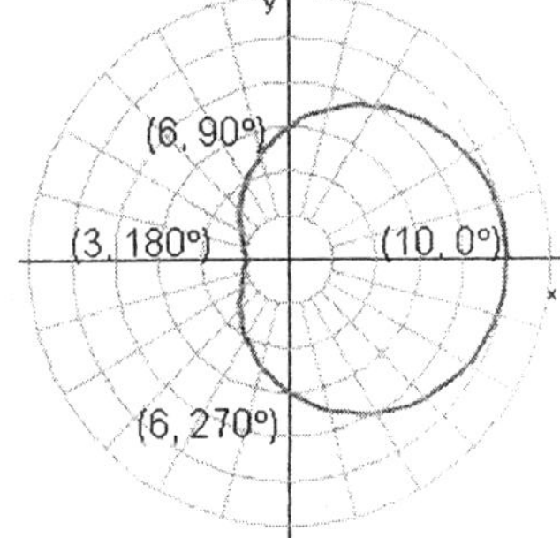

24. Graph the equation.

$$r = 2\sin 2\theta$$

Select the correct answer.

a.

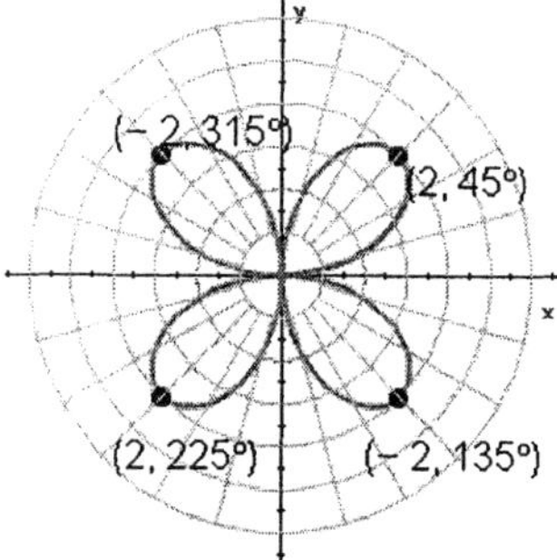

b.

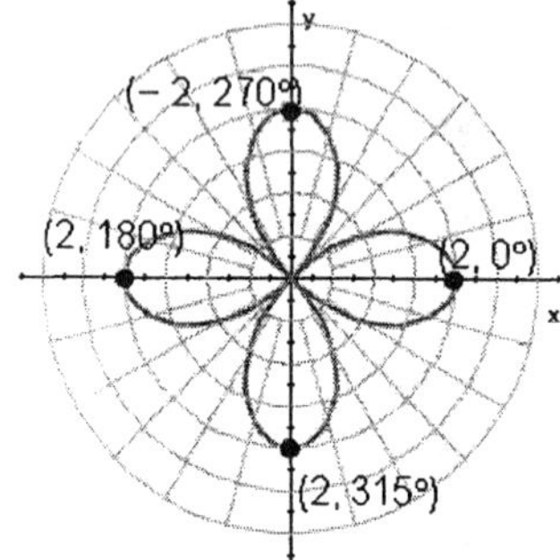

c.

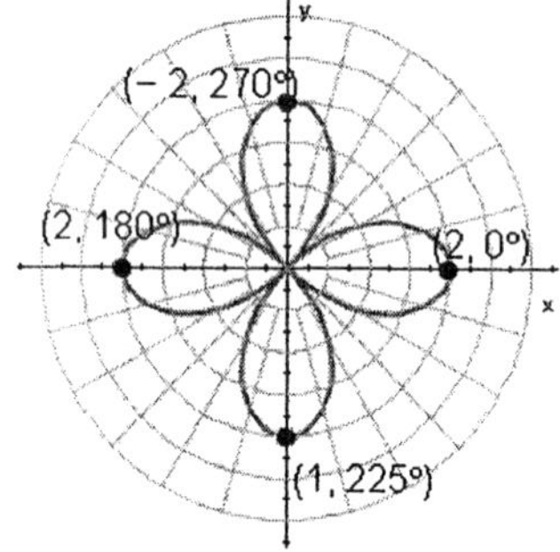

d.

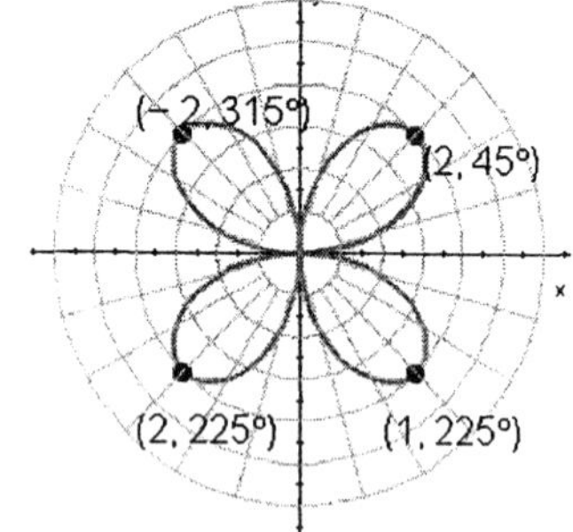

25. Change the equation to rectangular coordinates and then graph.

$$r(1 + \cos\theta) = 3$$

Select the correct answer.

a.

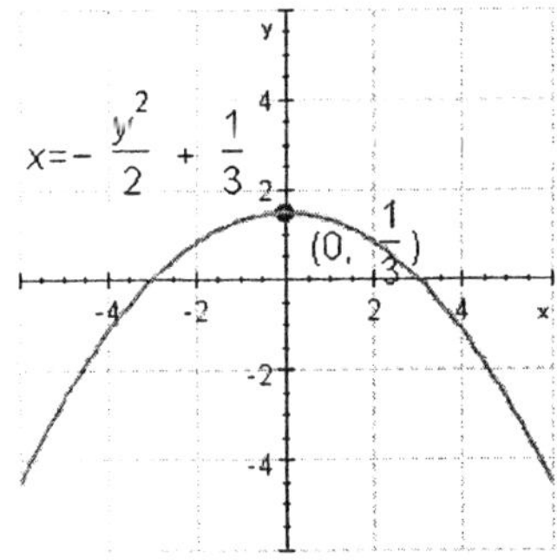

b.

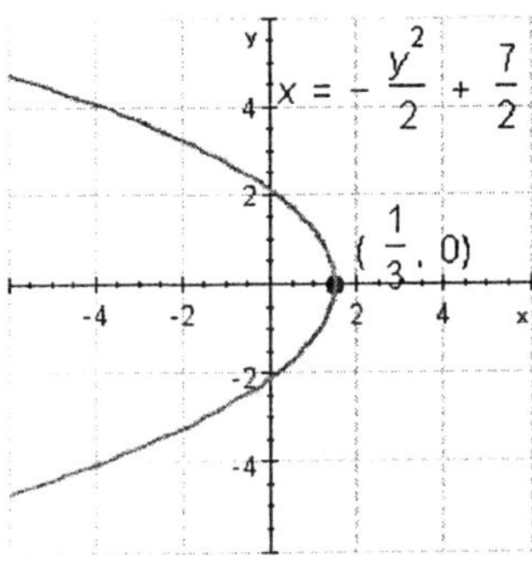

c.

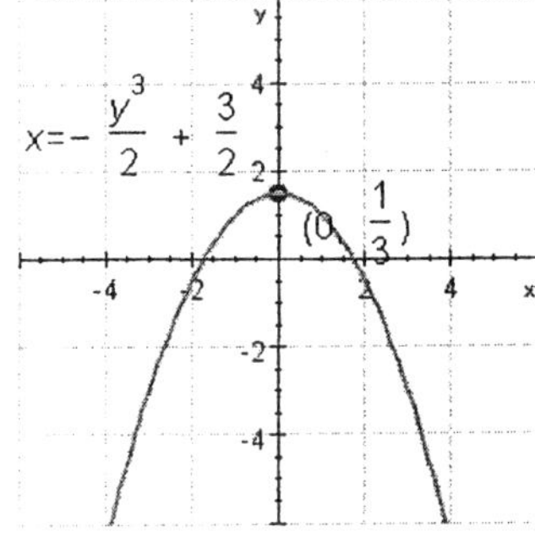

d.

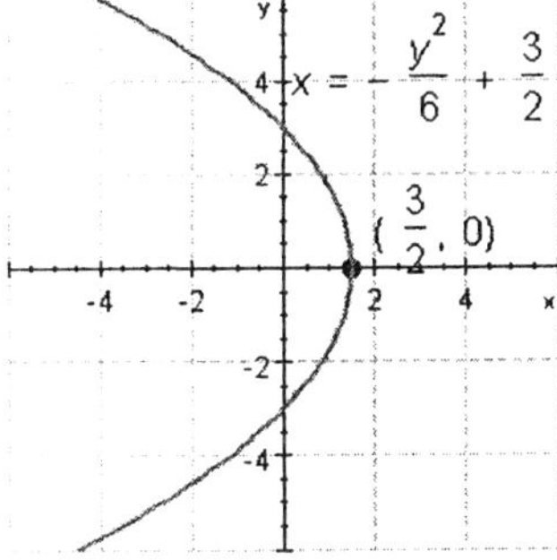

1. e
2. b
3. a
4. a
5. b
6. a
7. c
8. e
9. a
10. c
11. d
12. a
13. b
14. d
15. a
16. b
17. c
18. b
19. e
20. d
21. d
22. b
23. a
24. a
25. d

McKeague/Turner - Trigonometry 5e Chapter 8 Form C

McKeague/Turner - Trigonometry 5e Chapter 8 Form C

1. mctr.08.01.12m_NoAlgs
2. mctr.08.01.38m_NoAlgs
3. mctr.08.01.50m_NoAlgs
4. mctr.08.01.79m_NoAlgs
5. mctr.08.02.10m_NoAlgs
6. mctr.08.02.29m_NoAlgs
7. mctr.08.02.45m_NoAlgs
8. mctr.08.02.51m_NoAlgs
9. mctr.08.03.05m_NoAlgs
10. mctr.08.03.24m_NoAlgs
11. mctr.08.03.35m_NoAlgs
12. mctr.08.03.57m_NoAlgs
13. mctr.08.04.05m_NoAlgs
14. mctr.08.04.11m_NoAlgs
15. mctr.08.04.18m_NoAlgs
16. mctr.08.04.33m_NoAlgs
17. mctr.08.05.08m_NoAlgs
18. mctr.08.05.24m_NoAlgs
19. mctr.08.05.38m_NoAlgs
20. mctr.08.05.59m_NoAlgs
21. mctr.08.06.07m_NoAlgs
22. mctr.08.06.12m_NoAlgs
23. mctr.08.06.17m_NoAlgs
24. mctr.08.06.23m_NoAlgs
25. mctr.08.06.47m_NoAlgs

1. Find the product.

    $(3 + 7i)(3 - 7i)$

    Select the correct answer.

    a.   58
    b.   62
    c. $56 - 42i$
    d.   56
    e. $58 - 42i$

2. Write the complex number in trigonometric form.
    Round the angle to the nearest hundredth of a degree.

    $24 + 7i$

    Select the correct answer.

    a. $25 \operatorname{cis} 16.38°$

    b. $25 \operatorname{cis} 16.35°$

    c. $25 \operatorname{cis} 16.23°$

    d. $25 \operatorname{cis} 16.20°$

    e. $25 \operatorname{cis} 16.26°$

3. DeMoivre's Theorem can be used to find reciprocals of complex numbers. Recall from
    algebra that the reciprocal of x is $\dfrac{1}{x}$, which can be expressed as x^{-1}.

    Use this fact, along with DeMoivre's Theorem, to find the reciprocal of the number below.

    $-\sqrt{3} + i$

    Select the correct answer.

    a. $-\dfrac{\sqrt{3}}{4} - \dfrac{1}{4}i$

    b. $\dfrac{1}{2} + \dfrac{1}{2}i$

    c. $\dfrac{1}{4} + \dfrac{\sqrt{3}}{4}i$

    d. $\dfrac{\sqrt{3}}{4} + \dfrac{1}{4}i$

    e. $-\dfrac{1}{2} - \dfrac{1}{2}i$

4. Change the equation to rectangular coordinates and then graph.

$$r(1 + \cos\theta) = 3$$

Select the correct answer.

a. 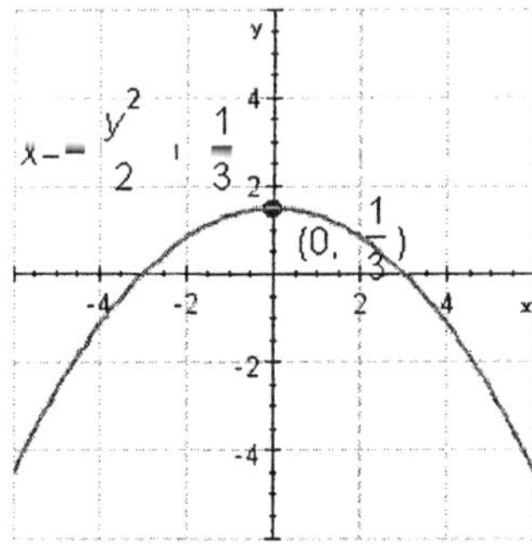

b. 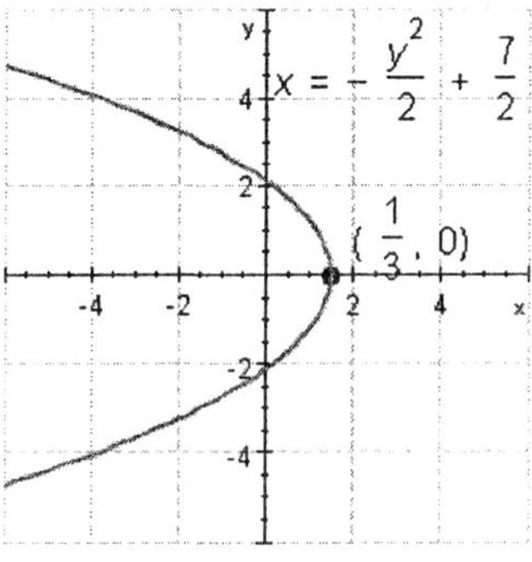

c. 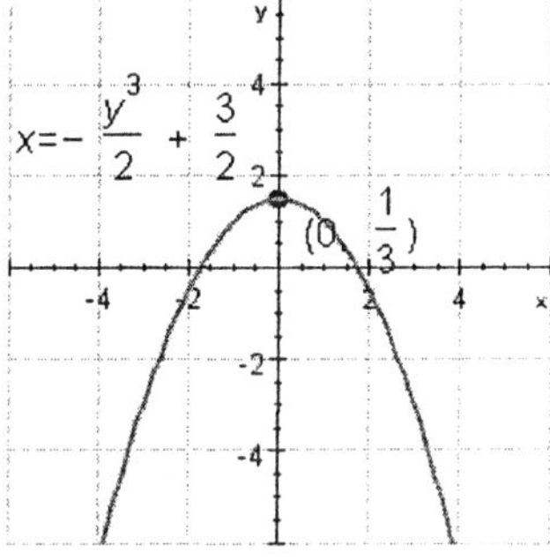

d. 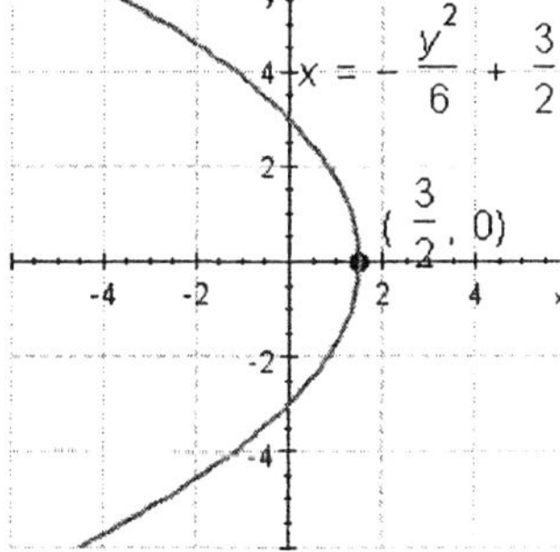

5. Convert to rectangular coordinates. Use exact values.

$$\left(6\sqrt{2},\ -45^\circ\right)$$

Select the correct answer.

a. (- 4, -4)
b. (6, -6)
c. (- 6,6)
d. (- 4,6)
e. (- 7, -7)

6. Convert to polar coordinates with $r \geq 0$ and θ between 0° and 360°.

$$\left(-1,\ -\sqrt{3}\right)$$

Select the correct answer.

a. $\left(1,\ 235^\circ\right)$

b. $\left(4,\ 250^\circ\right)$

c. $\left(2,\ 260^\circ\right)$

d. $\left(1,\ 225^\circ\right)$

e. $\left(2,\ 240^\circ\right)$

7. Graph the equation.

$$r = 4 + 5\cos\theta$$

Select the correct answer.

a.

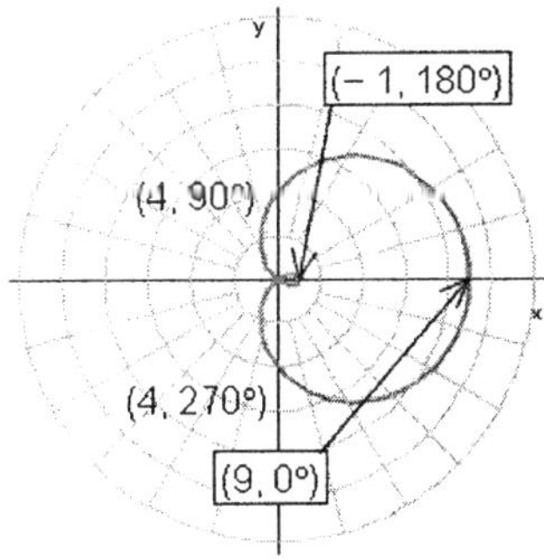

b.

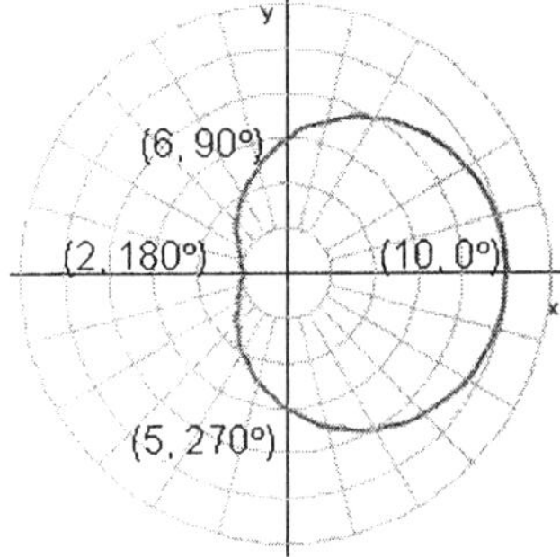

c.

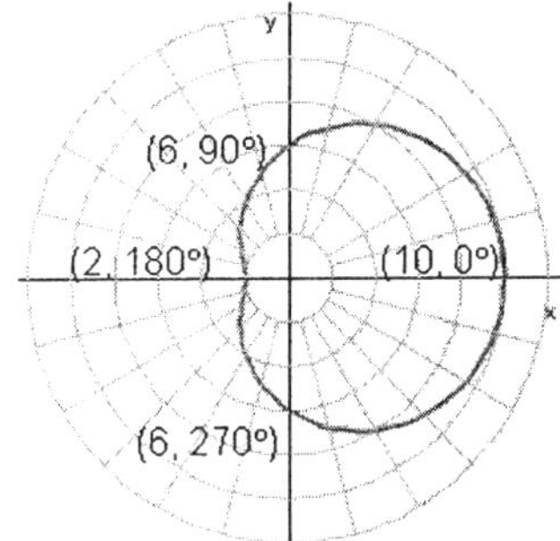

d.

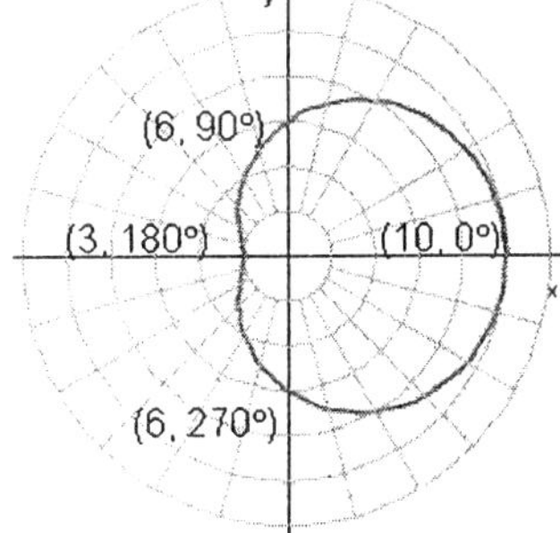

8. Write the expression $\sqrt{-16} \cdot \sqrt{-9}$ in terms of i and then simplify.

Select the correct answer.

a. 12

b. -15

c. 8

d. 15

e. -12

9. Solve the equation.

$$x^4 - 2\sqrt{3}\,x^2 + 4 = 0$$

Select the correct answer.

a. $\sqrt{2}\operatorname{cis}10°,\ \sqrt{2}\operatorname{cis}170°,\ \sqrt{2}\operatorname{cis}200°,\ \sqrt{2}\operatorname{cis}350°$

b. $\sqrt{2}\operatorname{cis}15°,\ \sqrt{2}\operatorname{cis}165°,\ \sqrt{2}\operatorname{cis}195°,\ \sqrt{2}\operatorname{cis}345°$

c. $\sqrt{2}\operatorname{cis}10°,\ \sqrt{2}\operatorname{cis}200°$

d. $\sqrt{2}\operatorname{cis}170°,\ \sqrt{2}\operatorname{cis}350°$

e. $\sqrt{2}\operatorname{cis}15°,\ \sqrt{2}\operatorname{cis}195°$

10. Find two square roots for the complex number.

16

Select the correct answer.

a. $4,\ -4i$

b. $4i,\ -4$

c. $4i,\ -4i$

d. $4,\ -4$

e. $16i,\ -16i$

11. Simplify the power of i.

$$i^{17}$$

Select the correct answer.

a. $-i$
b. i
c. -1
d. 1
e. 0

12. Use a calculator to help write the complex number in standard form. Round the numbers in your answer to the nearest hundredth.

$$100\left(\cos 144^\circ + i\sin 144^\circ\right)$$

Select the correct answer.

a. $-80.90 + 58.78i$
b. $-80.90 + 58.74i$
c. $-80.94 + 58.78i$
d. $-80.98 + 58.86i$
e. $-80.94 + 58.74i$

13. Use your graphing calculator to generate solutions for the equation using values of θ that are multiples of $15°$. Sketch the graph of the equation using the values from your table.

$$r = 2 + 2\sin\theta$$

Select the correct answer.

a.

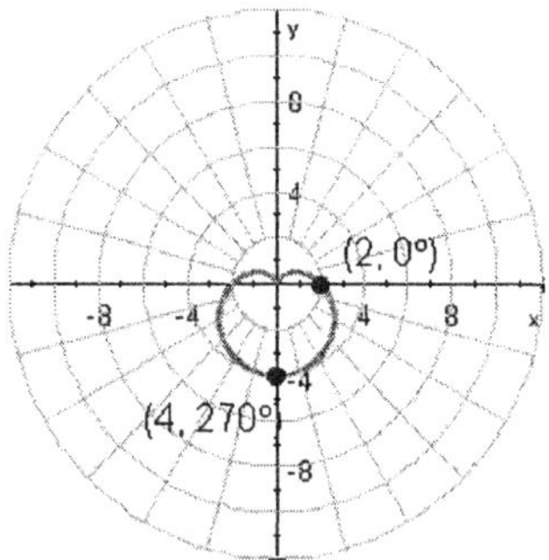

b.

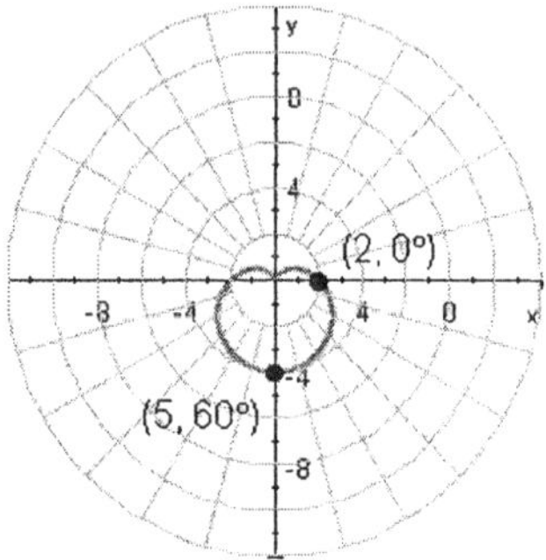

c.

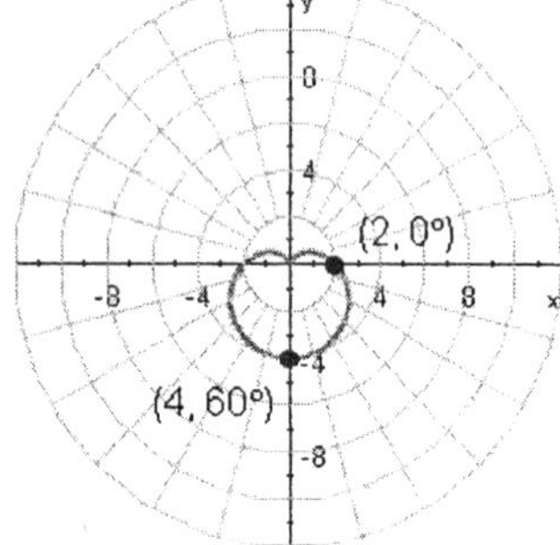

d.

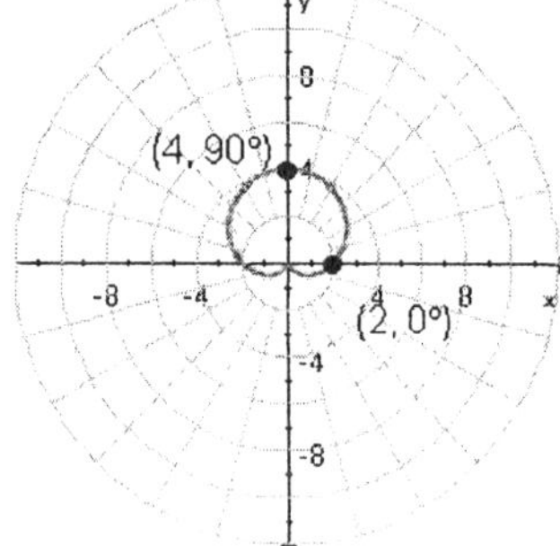

14. Find two square roots for the complex number.

$$64 \text{ cis } 0°$$

Select the correct answer.

a. $8 \text{ cis } 0°$, $8 \text{ cis } 270°$

b. $8 \text{ cis } 0°$, $8 \text{ cis } 180°$

c. $64 \text{ cis } 90°$, $64 \text{ cis } 180°$

d. $8 \text{ cis } 90°$, $8 \text{ cis } 180°$

e. $8 \text{ cis } 90°$, $8 \text{ cis } 270°$

15. Read the method of solving the system of equations.

$$x + y = 10$$
$$xy = 40$$

Solving the first equation for y, we have $y = 10 - x$. Substituting this value of y into the second equation, we have

$$x(10 - x) = 40$$

The equation is quadratic. We write it in standard form and apply the quadratic formula.

$$0 = x^2 - 10x + 40$$

$$x = \frac{10 \pm \sqrt{100 - 4(1)(40)}}{2}$$

$$= \frac{10 \pm \sqrt{100 - 160}}{2}$$

$$= \frac{10 \pm \sqrt{-60}}{2}$$

$$= \frac{10 \pm 2\sqrt{-15}}{2}$$

$$= 5 \pm \sqrt{-15}$$

$$= 5 \pm i\sqrt{15}$$

$$y = 5 \mp i\sqrt{15}$$

Use the method shown above to solve the system of equations.

$$x + y = 4$$
$$xy = 68$$

Select the correct answer.

    a. $x = 2 + 8i,\ y = 2 - 8i;\ x = 2 - 8i,\ y = 2 + 8i$

    b. $x = 6 + 6i,\ y = 6 - 6i;\ x = 6 - 6i,\ y = 6 + 6i$

    c. $x = 2 + 8i,\ y = 2 + 8i;\ x = 2 - 8i,\ y = 2 - 8i$

    d. $x = 6 + 8i,\ y = 6 - 8i;\ x = 6 - 8i,\ y = 6 + 8i$

    e. $x = 2 + 6i,\ y = 2 - 6i;\ x = 2 - 6i,\ y = 2 + 6i$

16. Graph the equation.

$$\theta = 225^\circ$$

Select the correct answer.

a.

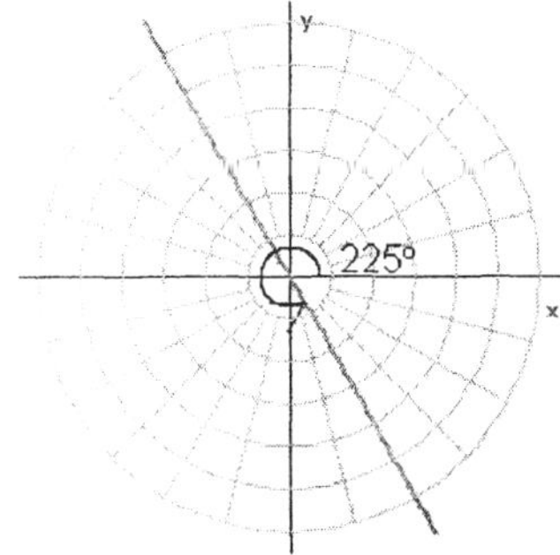

b.

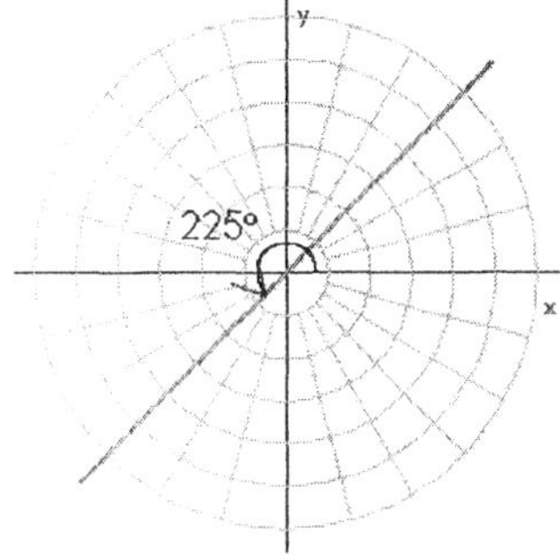

c.

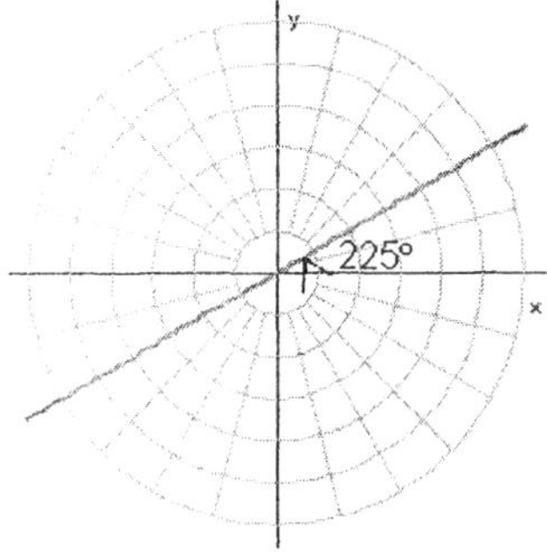

d.

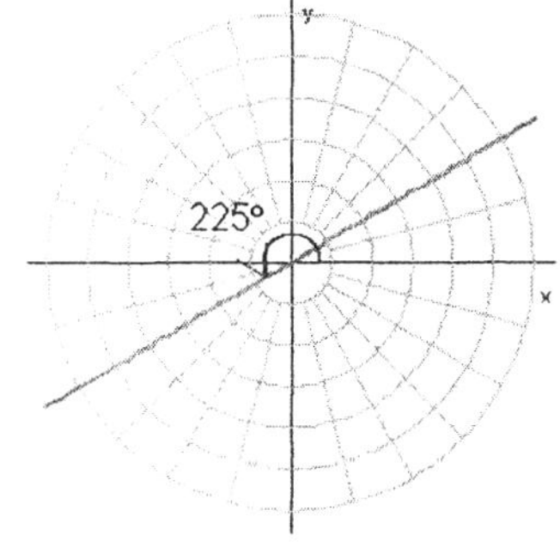

17. Graph the equation.

$$r = 2\sin 2\theta$$

Select the correct answer.

a.

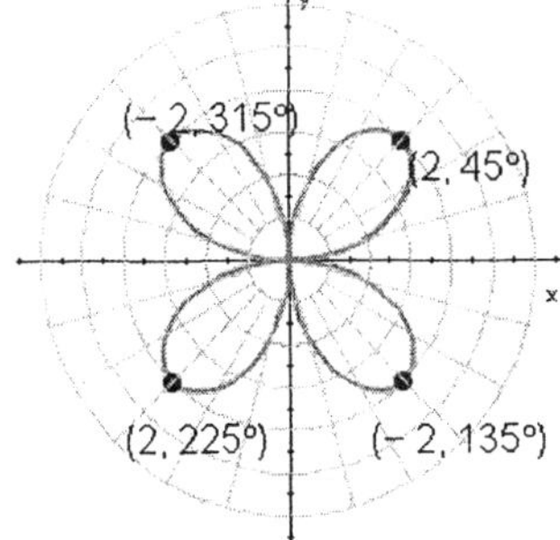

b.

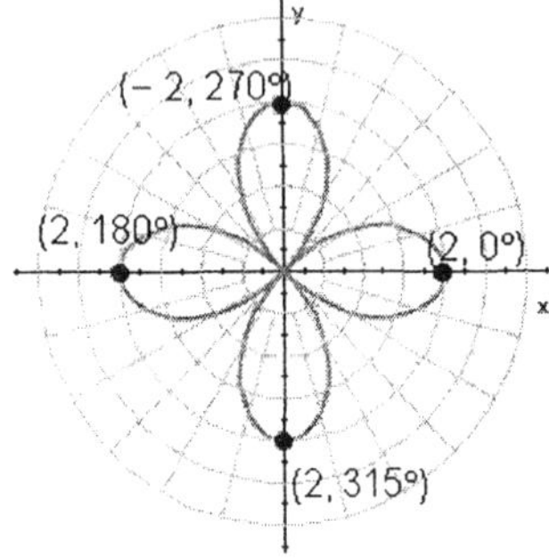

c.

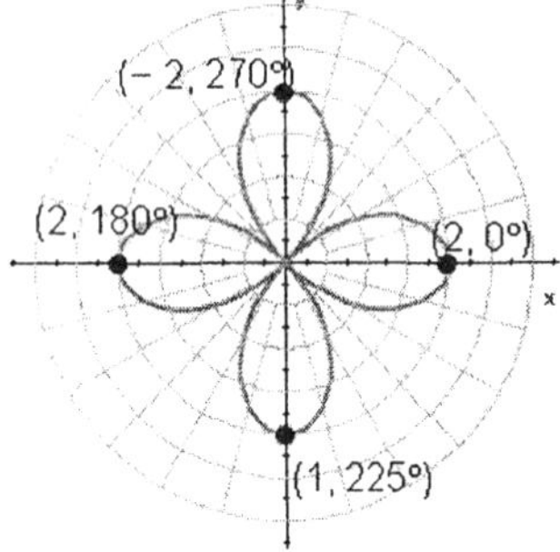

d.

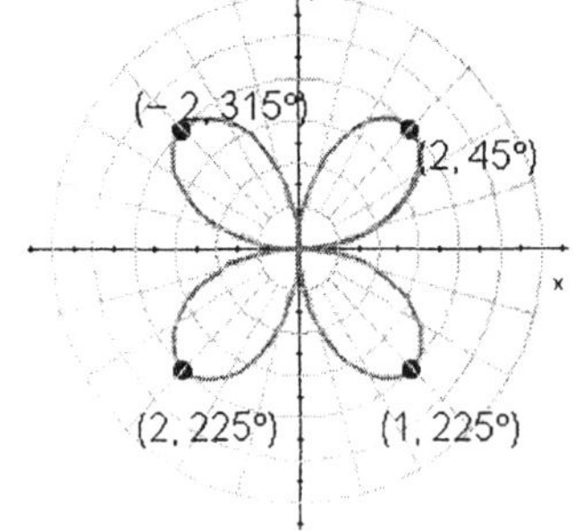

18. Write the equation in polar coordinates.

$$x^2 + y^2 = 5$$

Select the correct answer.

a. $r^2 = 1$

b. $r^2 = 2$

c. $r^2 = 7$

d. $r^2 = 5$

e. $r^2 = 4$

19. Write the complex number in trigonometric form.

$$3 + 3i\sqrt{3}$$

Select the correct answer.

a. $6\operatorname{cis}150°$

b. $6\operatorname{cis}300°$

c. $6\operatorname{cis}60°$

d. $6\operatorname{cis}120°$

e. $3\operatorname{cis}60°$

20. Divide.

$$\frac{18(\cos 50° + i\sin 50°)}{12(\cos 35° + i\sin 35°)}$$

Select the correct answer.

a. $2.5(\cos 20° + i\sin 20°)$

b. $0.5(\cos 85° + i\sin 85°)$

c. $1.5(\cos 5° + i\sin 5°)$

d. $1.5(\cos 15° + i\sin 15°)$

e. $1.5(\cos 85° + i\sin 85°)$

21. Graph the ordered pair on a polar coordinate system.

$$(-3, 45°)$$

Select the correct answer.

a.

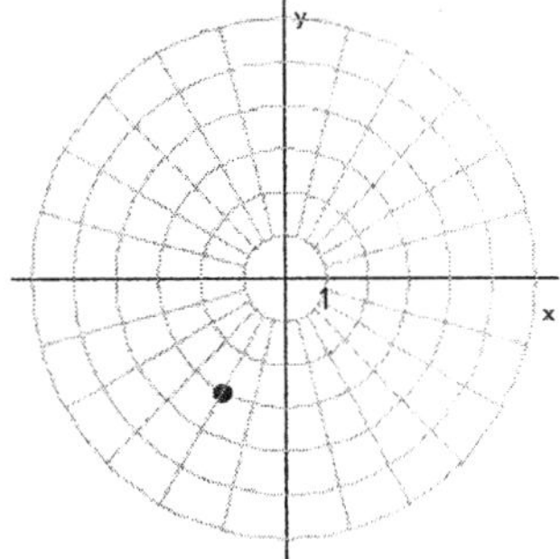

b.

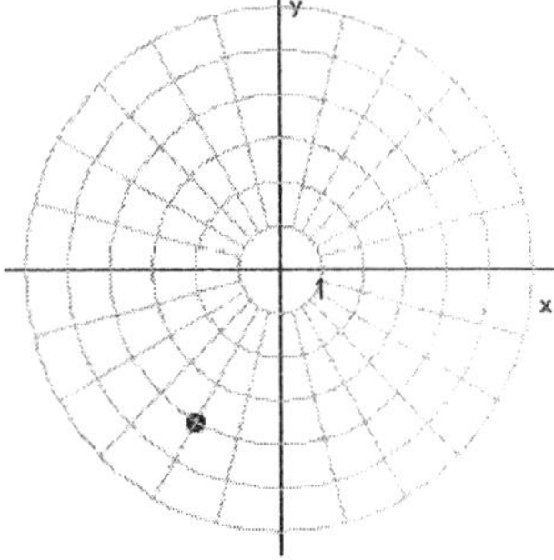

c.

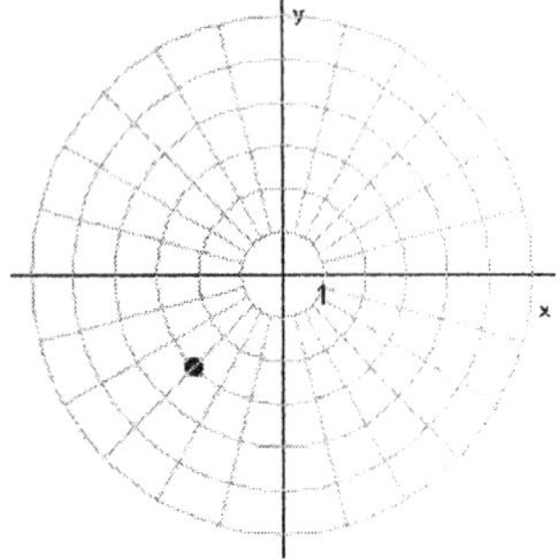

22. Graph the complex number and give the absolute value of it.

$6 - 8i$

Select the correct answer.

a.

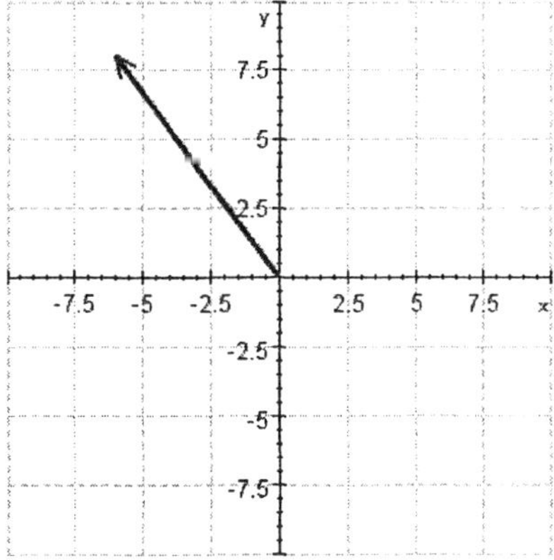

The absolute value is 10

b.

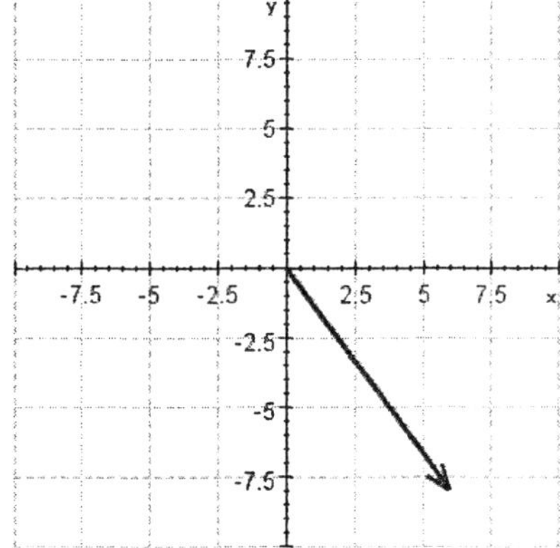

The absolute value is 10

c.

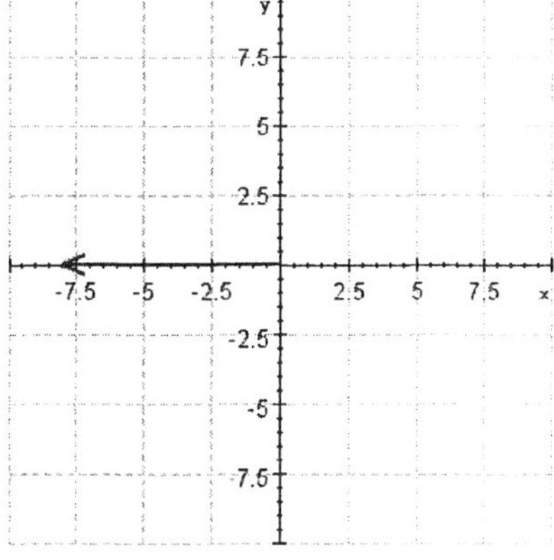

The absolute value is 6

23. Use DeMoivre's Theorem to find the following.

$$\left(\sqrt{2}\,\text{cis}\,60°\right)^{10}$$

Select the correct answer.

a. $-16\sqrt{3} + 16i$

b. $-4 + 4i$

c. $-16 - 16i\sqrt{3}$

d. $-4 + 4i\sqrt{3}$

e. $4 + 4i\sqrt{3}$

24. Multiply.

$$4\,\text{cis}\,150° \cdot 2\,\text{cis}\,30°$$

Select the correct answer.

a. $8\,\text{cis}\,180°$

b. $4\,\text{cis}\,180°$

c. $8\,\text{cis}\,190°$

d. $4\,\text{cis}\,190°$

e. $8\,\text{cis}\,185°$

25. Find three cube roots for the complex number.

$$-32\sqrt{3} + 32i$$

Select the correct answer.

a. $4\,\text{cis}\,50°,\ 4\,\text{cis}\,170°,\ 4\,\text{cis}\,290°$

b. $4\,\text{cis}\,55°,\ 4\,\text{cis}\,175°,\ 4\,\text{cis}\,295°$

c. $4\,\text{cis}\,40°,\ 4\,\text{cis}\,160°,\ 4\,\text{cis}\,280°$

d. $2\,\text{cis}\,40°,\ 2\,\text{cis}\,160°,\ 2\,\text{cis}\,280°$

e. $2\,\text{cis}\,50°,\ 2\,\text{cis}\,170°,\ 2\,\text{cis}\,290°$

1. a

2. e

3. a

4. d

5. h

6. e

7. a

8. e

9. b

10. d

11. b

12. a

13. d

14. b

15. a

16. b

17. a

18. d

19. c

20. d

21. c

22. b

23. c

24. a

25. a

1. mctr.08.01.50m_NoAlgs
2. mctr.08.02.51m_NoAlgs
3. mctr.08.03.57m_NoAlgs
4. mctr.08.06.47m_NoAlgs
5. mctr.08.05.24m_NoAlgs
6. mctr.08.05.38m_NoAlgs
7. mctr.08.06.17m_NoAlgs
8. mctr.08.01.12m_NoAlgs
9. mctr.08.04.33m_NoAlgs
10. mctr.08.04.11m_NoAlgs
11. mctr.08.01.38m_NoAlgs
12. mctr.08.02.29m_NoAlgs
13. mctr.08.06.07m_NoAlgs
14. mctr.08.04.05m_NoAlgs
15. mctr.08.01.79m_NoAlgs
16. mctr.08.06.12m_NoAlgs
17. mctr.08.06.23m_NoAlgs
18. mctr.08.05.59m_NoAlgs
19. mctr.08.02.45m_NoAlgs
20. mctr.08.03.35m_NoAlgs
21. mctr.08.05.08m_NoAlgs
22. mctr.08.02.10m_NoAlgs
23. mctr.08.03.24m_NoAlgs
24. mctr.08.03.05m_NoAlgs
25. mctr.08.04.18m_NoAlgs

1. Write the expression $\sqrt{-8}$ in terms of i.

2. Write the expression $\sqrt{-16} \cdot \sqrt{-9}$ in terms of i and then simplify.

   Select the correct answer.

   a. 12

   b. -15

   c. 8

   d. 15

   e. -12

3. Find the product.

   $(2 + 3i)(3 - i)$

   Enter your answer in standard form for complex numbers.

   ____

4. Find the product.

   $(3 + 7i)(3 - 7i)$

   Select the correct answer.

   a. 58
   b. 62
   c. 56 - 42i
   d. 56
   e. 58 - 42i

5. Graph the complex number.

$-3i$

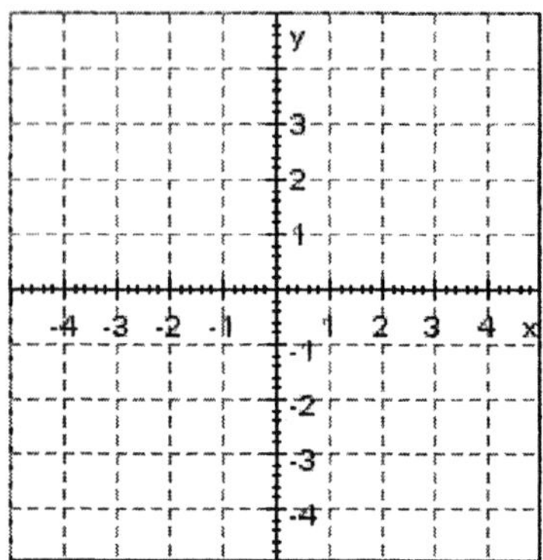

Give the absolute value of the number.

____

6. Graph the complex number and give the absolute value of it.

   $6 - 8i$

   Select the correct answer.

a. 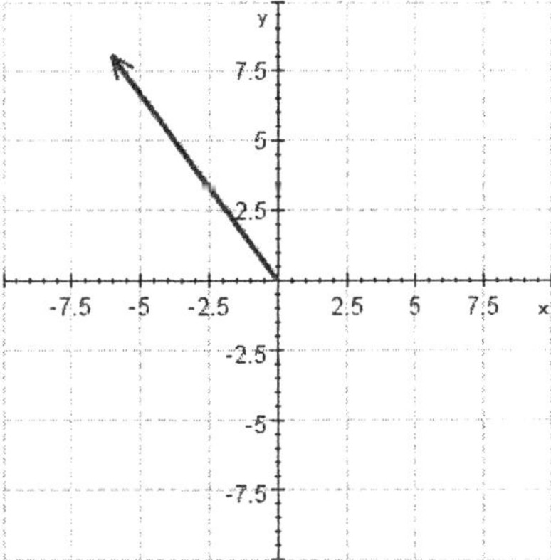 The absolute value is 10

b. 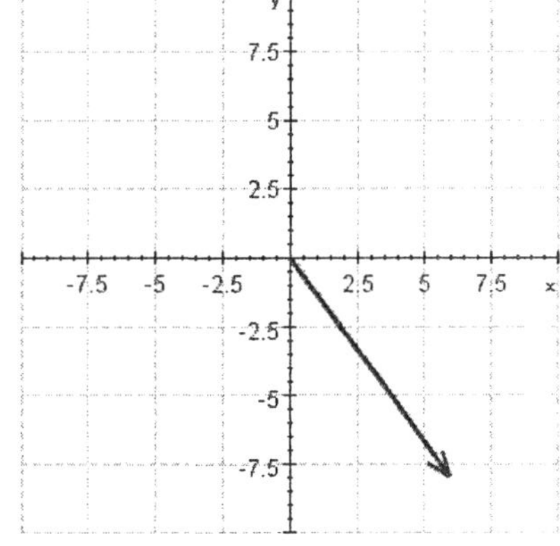 The absolute value is 10

c. 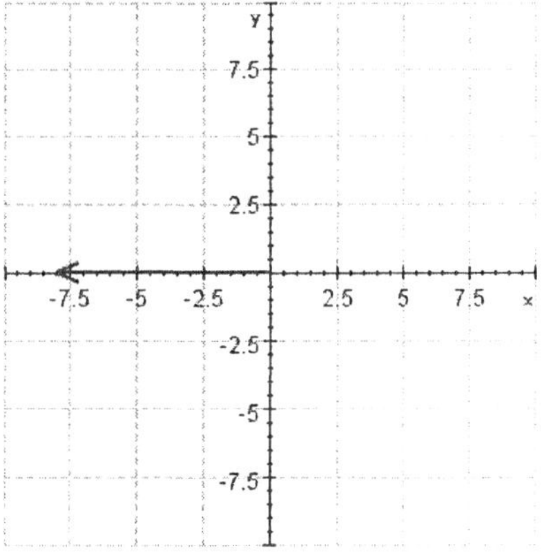 The absolute value is 6

7. Write the complex number in trigonometric form.

   $1 - i$

   Please enter the argument in degrees and use the degree symbol.

8. Write the complex number in trigonometric form.

   $3 + 3i\sqrt{3}$

   Select the correct answer.

   a. $6\operatorname{cis}150°$

   b. $6\operatorname{cis}300°$

   c. $6\operatorname{cis}60°$

   d. $6\operatorname{cis}120°$

   e. $3\operatorname{cis}60°$

9. Multiply. Leave the answer in trigonometric form.

   $6(\cos 20° + i\sin 20°) \cdot 6(\cos 35° + i\sin 35°)$

10. Multiply.

    $4\operatorname{cis}150° \cdot 2\operatorname{cis}30°$

    Select the correct answer.

    a. $8\operatorname{cis}180°$

    b. $4\operatorname{cis}180°$

    c. $8\operatorname{cis}190°$

    d. $4\operatorname{cis}190°$

    e. $8\operatorname{cis}185°$

11. Use DeMoivre's Theorem to find the following. Write your answer in standard form.

$$\left(\sqrt{3} + i\right)^{6}$$

12. Divide.

$$\frac{18\left(\cos 50^{\circ} + i \sin 50^{\circ}\right)}{12\left(\cos 35^{\circ} + i \sin 35^{\circ}\right)}$$

Select the correct answer.

a. $2.5\left(\cos 20^{\circ} + i\sin 20^{\circ}\right)$

b. $0.5\left(\cos 85^{\circ} + i\sin 85^{\circ}\right)$

c. $1.5\left(\cos 5^{\circ} + i\sin 5^{\circ}\right)$

d. $1.5\left(\cos 15^{\circ} + i\sin 15^{\circ}\right)$

e. $1.5\left(\cos 85^{\circ} + i\sin 85^{\circ}\right)$

13. Find two square roots for the complex number. Leave your answers in trigonometric form.

$$16\left(\cos 120^{\circ} + i\sin 120^{\circ}\right)$$

Graph the roots.

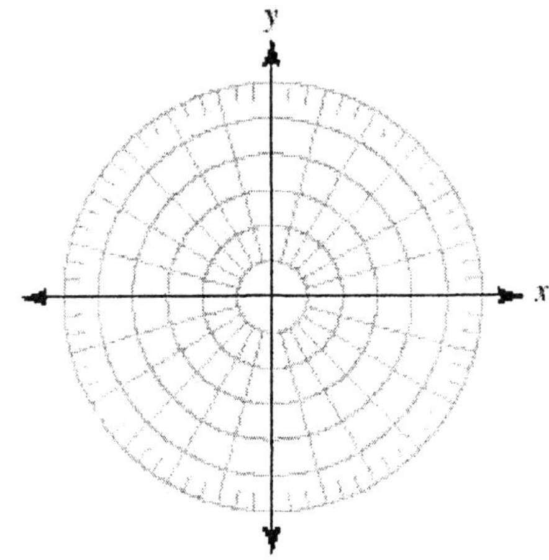

14. Find two square roots for the complex number.

$64 \operatorname{cis} 0°$

Select the correct answer.

a. $8 \operatorname{cis} 0°, 8 \operatorname{cis} 270°$

b. $8 \operatorname{cis} 0°, 8 \operatorname{cis} 180°$

c. $64 \operatorname{cis} 90°, 64 \operatorname{cis} 180°$

d. $8 \operatorname{cis} 90°, 8 \operatorname{cis} 180°$

e. $8 \operatorname{cis} 90°, 8 \operatorname{cis} 270°$

15. Find three cube roots for the complex number. Leave your answers in trigonometric form.

$8\left(\cos 291° + i \sin 291°\right)$

16. Find three cube roots for the complex number.

$-32\sqrt{3} + 32i$

Select the correct answer.

a. $4 \operatorname{cis} 50°, 4 \operatorname{cis} 170°, 4 \operatorname{cis} 290°$

b. $4 \operatorname{cis} 55°, 4 \operatorname{cis} 175°, 4 \operatorname{cis} 295°$

c. $4 \operatorname{cis} 40°, 4 \operatorname{cis} 160°, 4 \operatorname{cis} 280°$

d. $2 \operatorname{cis} 40°, 2 \operatorname{cis} 160°, 2 \operatorname{cis} 280°$

e. $2 \operatorname{cis} 50°, 2 \operatorname{cis} 170°, 2 \operatorname{cis} 290°$

17. Graph the ordered pair on a polar coordinate system.

$$(4, 150°)$$

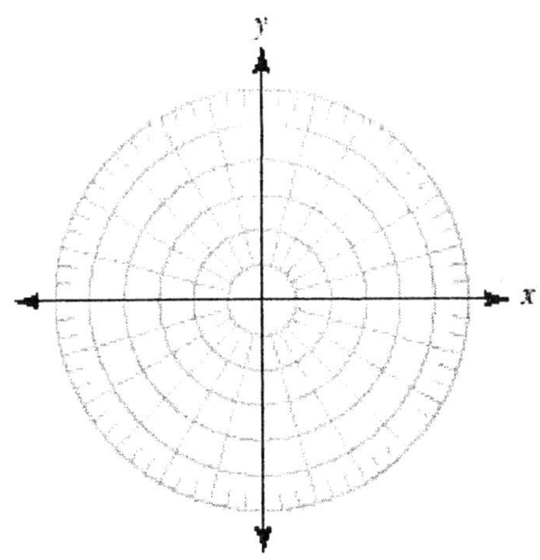

18. Graph the ordered pair on a polar coordinate system.

$$(-3, 45°)$$

Select the correct answer.

a.

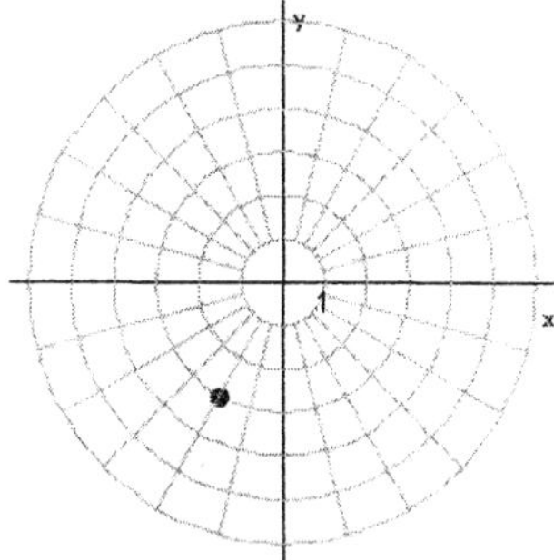

b.

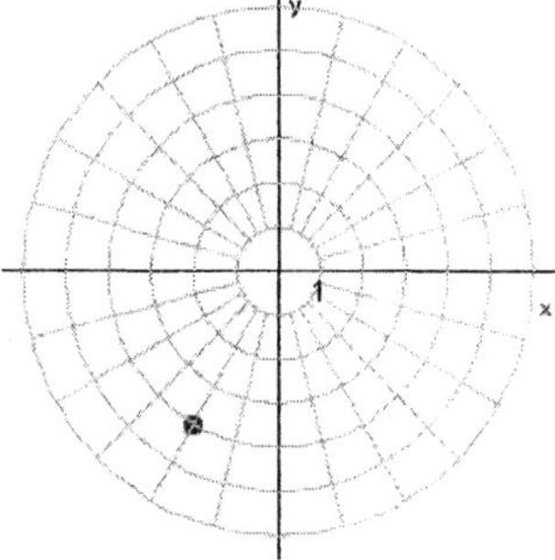

c.

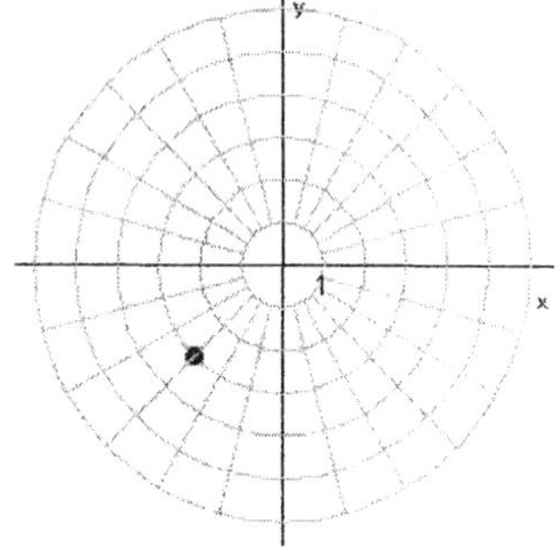

19. For the ordered pair, give three other ordered pairs with degree values between $-360°$ and $360°$ that name the same point.

$$(6, 125°)$$

20. Convert to polar coordinates with $r \geq 0$ and θ between $0°$ and $360°$.

$$\left(-2\sqrt{3},\ 2\right)$$

Select the correct answer.

a. $\left(4,\ 130°\right)$

b. $\left(7,\ 165°\right)$

c. $\left(2,\ 160°\right)$

d. $\left(3,\ 145°\right)$

e. $\left(4,\ 150°\right)$

21. Convert to polar coordinates with $r \geq 0$ and θ between $0°$ and $360°$.

$$\left(-1,\ -\sqrt{3}\right)$$

Select the correct answer.

a. $\left(1,\ 235°\right)$

b. $\left(4,\ 250°\right)$

c. $\left(2,\ 260°\right)$

d. $\left(1,\ 225°\right)$

e. $\left(2,\ 240°\right)$

22. Graph the equation using your graphing calculator.

$$r = 4 \sin 3\theta$$

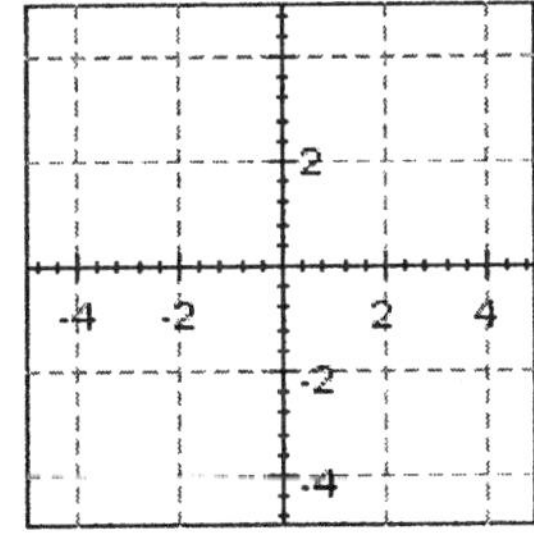

23. Use your graphing calculator to generate solutions for the equation using values of θ that are multiples of $15°$. Sketch the graph of the equation using the values from your table.

$$r = 2 + 2\sin\theta$$

Select the correct answer.

a.

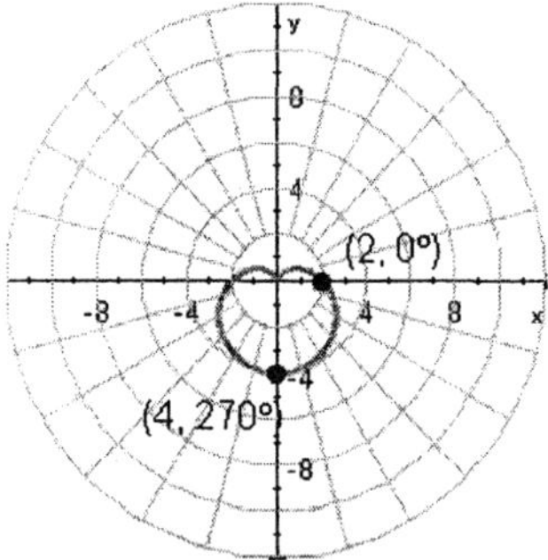

b.

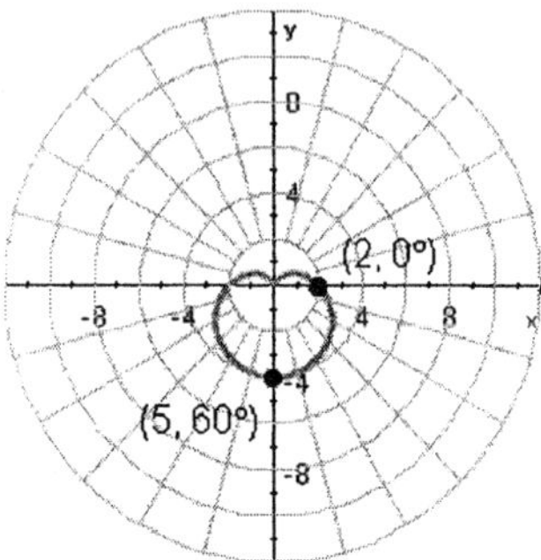

c.

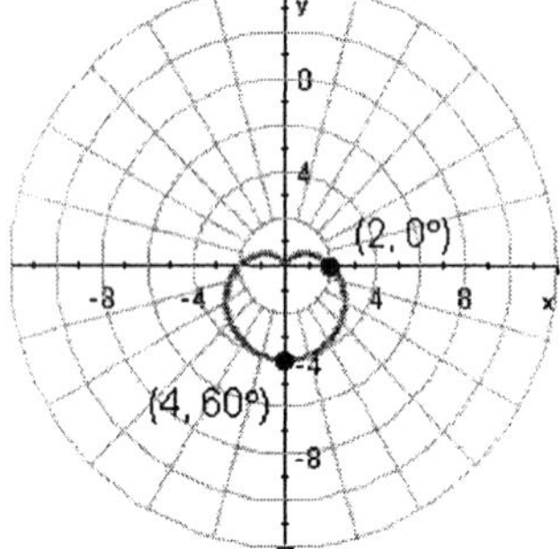

d.

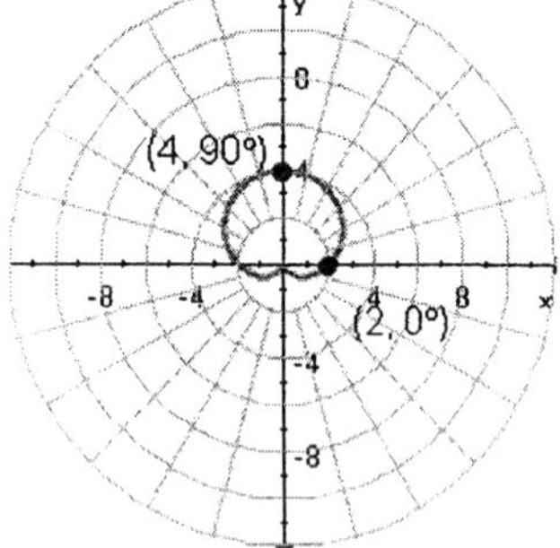

24. Graph the equation.

$$r = 2\cos 2\theta$$

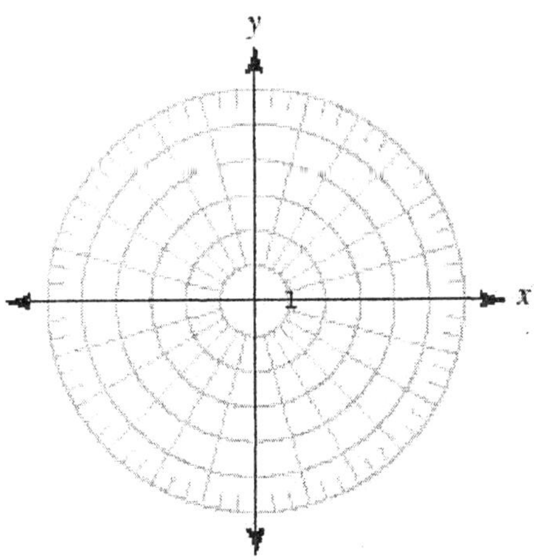

25. Change the equation to rectangular coordinates and then graph.

$$r(1 + \cos\theta) = 3$$

Select the correct answer.

a. 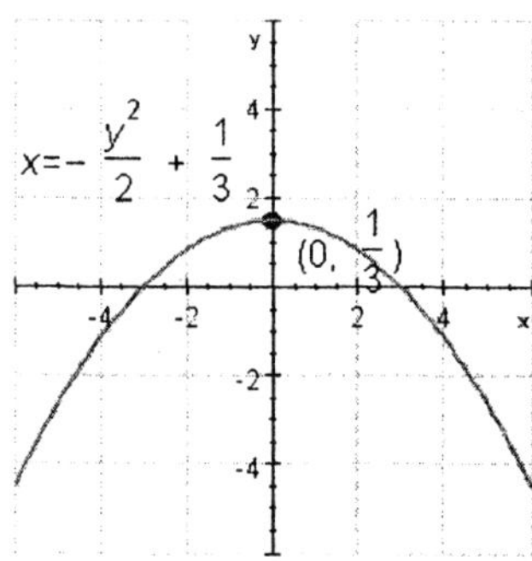

b. 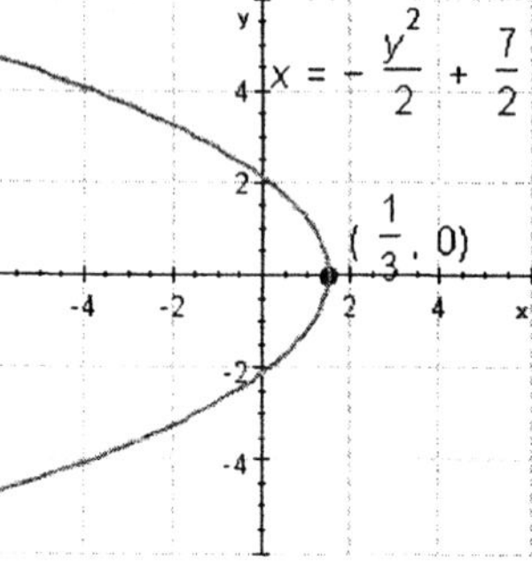

c. 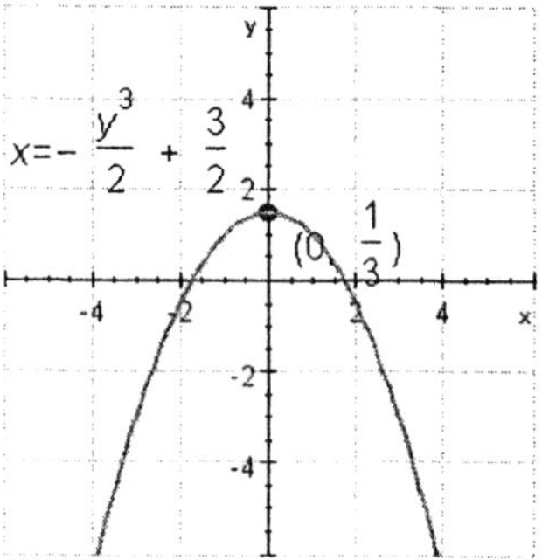

d. 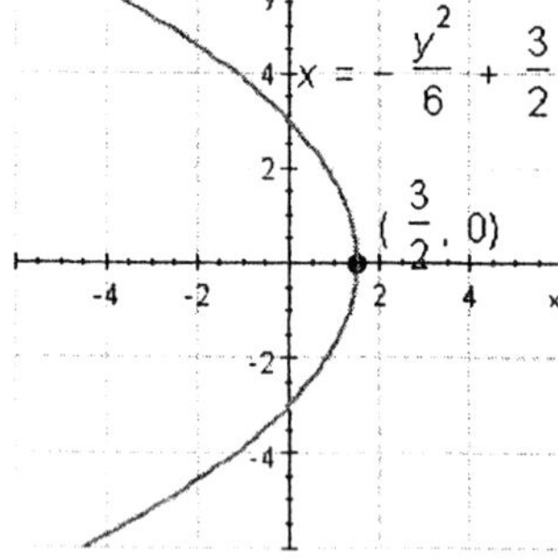

1. $2i \cdot \sqrt{2}$

2. e

3. $9 + 7i$

4. a

5.

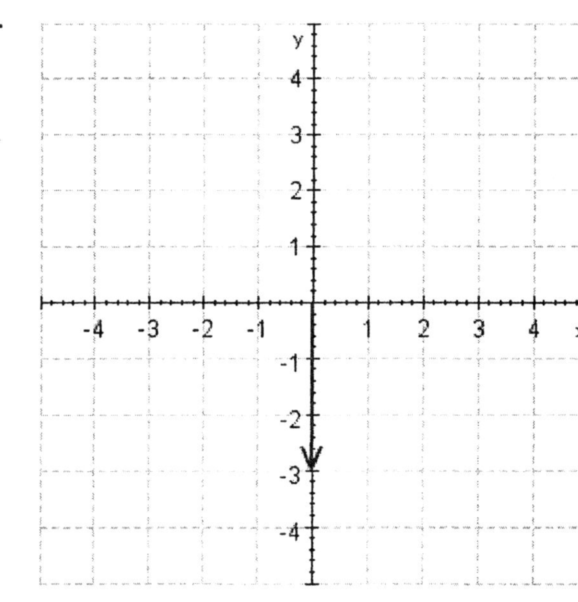

3

6. b

7. $\sqrt{2} \cdot \left(\cos\left(315^\circ\right) + i \cdot \sin\left(315^\circ\right)\right)$

8. c

9. $36\left(\cos\left(55^\circ\right) + i \cdot \sin\left(55^\circ\right)\right)$

10. a

11. -64

12. d

13. $4\left(\cos\left(60^\circ\right)+i\cdot\sin\left(60^\circ\right)\right),\ 4\left(\cos\left(240^\circ\right)+i\cdot\sin\left(240^\circ\right)\right)$

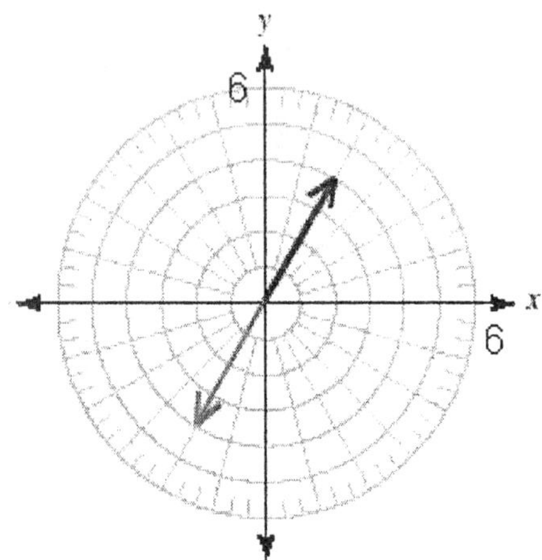

14. b

15. $2\left(\cos\left(97^\circ\right)+i\cdot\sin\left(97^\circ\right)\right),\ 2\left(\cos\left(217^\circ\right)+i\cdot\sin\left(217^\circ\right)\right),\ 2\left(\cos\left(337^\circ\right)+i\cdot\sin\left(337^\circ\right)\right)$

16. a

17.

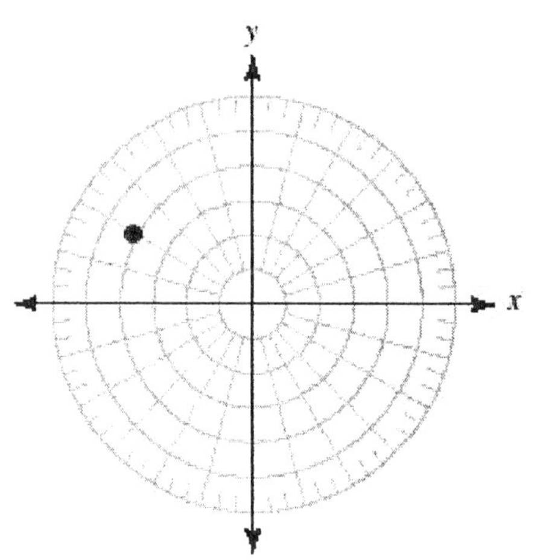

18. e

19. $\left(6,\ -235^\circ\right),\ \left(-6,\ 305^\circ\right),\ \left(-6,\ -55^\circ\right)$

20. e

21. e

22.

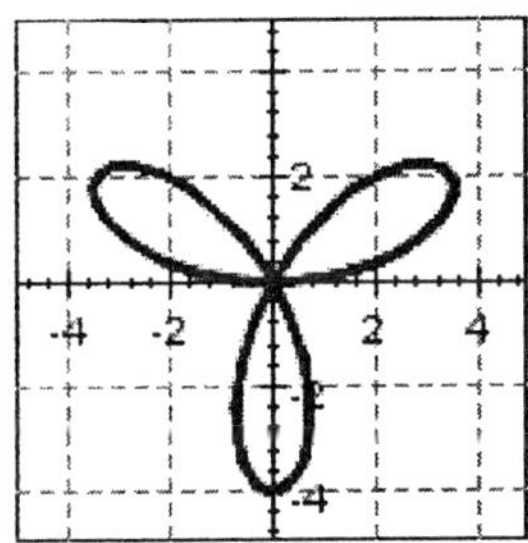

23. d

24.

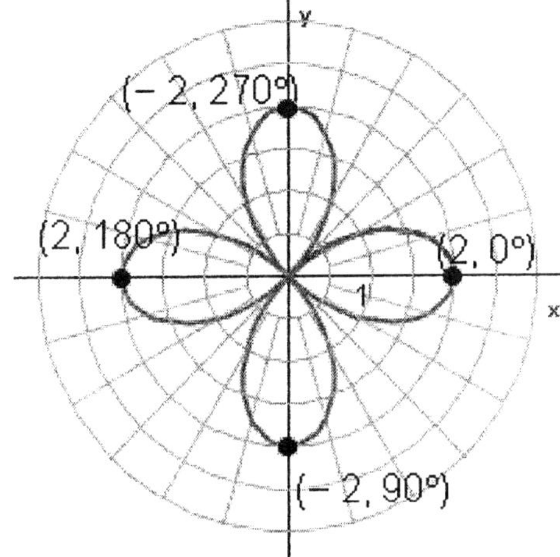

25. d

1. mctr.08.01.07_NoAlgs
2. mctr.08.01.12m_NoAlgs
3. mctr.08.01.46_NoAlgs
4. mctr.08.01.50m_NoAlgs
5. mctr.08.02.05_NoAlgs
6. mctr.08.02.10m_NoAlgs
7. mctr.08.02.35_NoAlgs
8. mctr.08.02.45m_NoAlgs
9. mctr.08.03.02_NoAlgs
10. mctr.08.03.05m_NoAlgs
11. mctr.08.03.28_NoAlgs
12. mctr.08.03.35m_NoAlgs
13. mctr.08.04.03_NoAlgs
14. mctr.08.04.05m_NoAlgs
15. mctr.08.04.16_NoAlgs
16. mctr.08.04.18m_NoAlgs
17. mctr.08.05.03_NoAlgs
18. mctr.08.05.08m_NoAlgs
19. mctr.08.05.16_NoAlgs
20. mctr.08.05.34m_NoAlgs
21. mctr.08.05.38m_NoAlgs
22. mctr.08.06.28_NoAlgs
23. mctr.08.06.07m_NoAlgs
24. mctr.08.06.23_NoAlgs
25. mctr.08.06.47m_NoAlgs

1.   Convert to polar coordinates with $r \geq 0$ and θ between $0°$ and $360°$.

$$\left(-1, -\sqrt{3}\right)$$

Select the correct answer.

   a. $\left(1, 235°\right)$

   b. $\left(4, 250°\right)$

   c. $\left(2, 260°\right)$

   d. $\left(1, 225°\right)$

   e. $\left(2, 240°\right)$

2.   Divide.

$$\frac{18\left(\cos 50° + i \sin 50°\right)}{12\left(\cos 35° + i \sin 35°\right)}$$

Select the correct answer.

   a. $2.5\left(\cos 20° + i\sin 20°\right)$

   b. $0.5\left(\cos 85° + i\sin 85°\right)$

   c. $1.5\left(\cos 5° + i\sin 5°\right)$

   d. $1.5\left(\cos 15° + i\sin 15°\right)$

   e. $1.5\left(\cos 85° + i\sin 85°\right)$

3.   For the ordered pair, give three other ordered pairs with degree values between $-360°$ and $360°$ that name the same point.

$$\left(6, 125°\right)$$

4. Multiply.

$$4\,\mathrm{cis}\,150° \cdot 2\,\mathrm{cis}\,30°$$

Select the correct answer.

   a. $8\,\mathrm{cis}\,180°$

   b. $4\,\mathrm{cis}\,180°$

   c. $8\,\mathrm{cis}\,190°$

   d. $4\,\mathrm{cis}\,190°$

   e. $8\,\mathrm{cis}\,185°$

5. Use DeMoivre's Theorem to find the following. Write your answer in standard form.

$$\left(\sqrt{3} + i\right)^{6}$$

6. Find the product.

$$(3 + 7i)(3 - 7i)$$

Select the correct answer.

   a.   58
   b.   62
   c. $56 - 42i$
   d.   56
   e. $58 - 42i$

7. Use your graphing calculator to generate solutions for the equation using values of θ that are multiples of $15°$. Sketch the graph of the equation using the values from your table.

$r = 2 + 2\sin\theta$

Select the correct answer.

a.

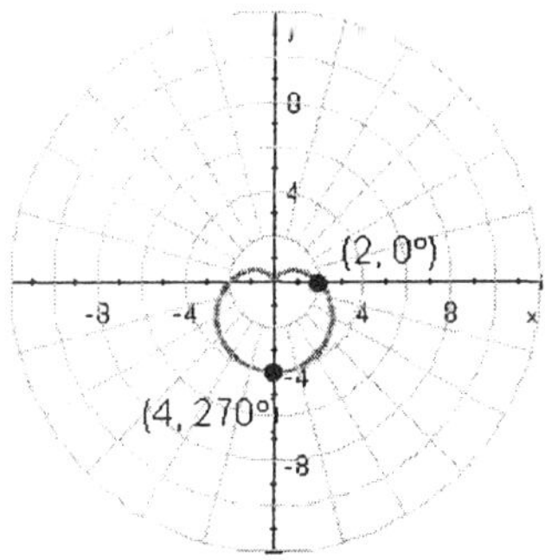

b.

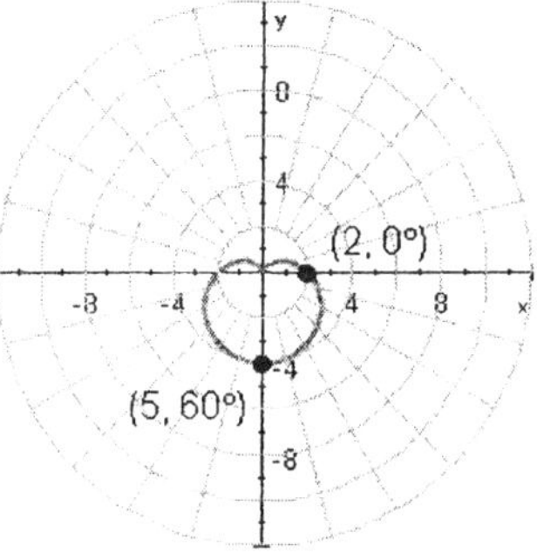

c.

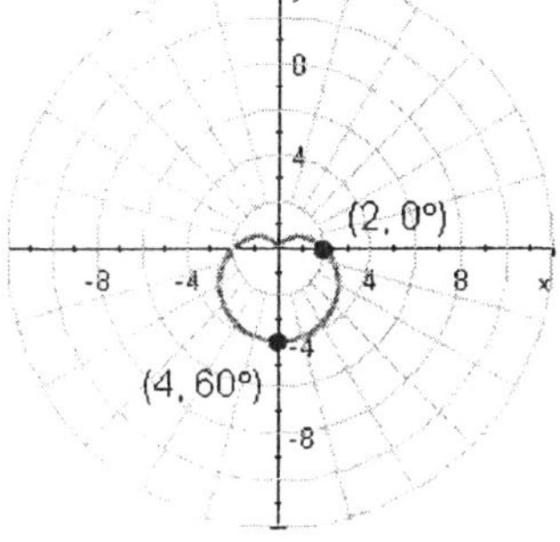

d.

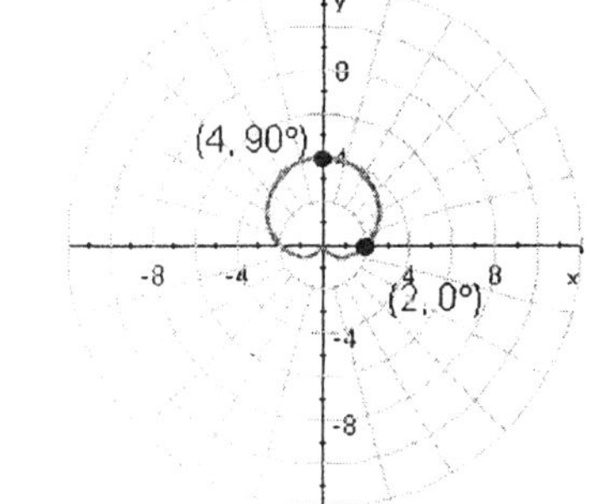

8. Graph the complex number.

$- 3i$

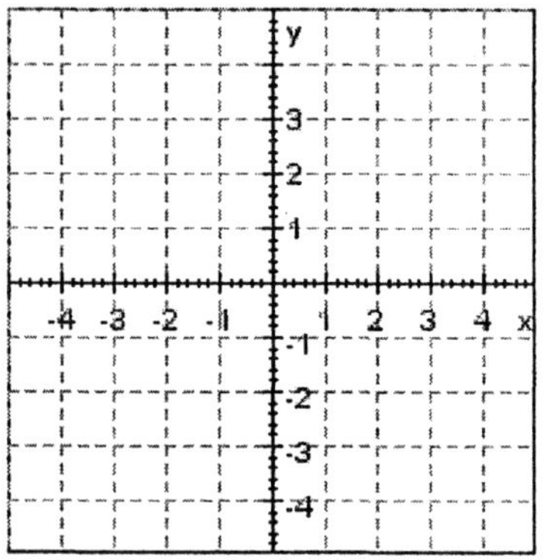

Give the absolute value of the number.

____

9. Graph the equation using your graphing calculator.

$r = 4 \sin 3\theta$

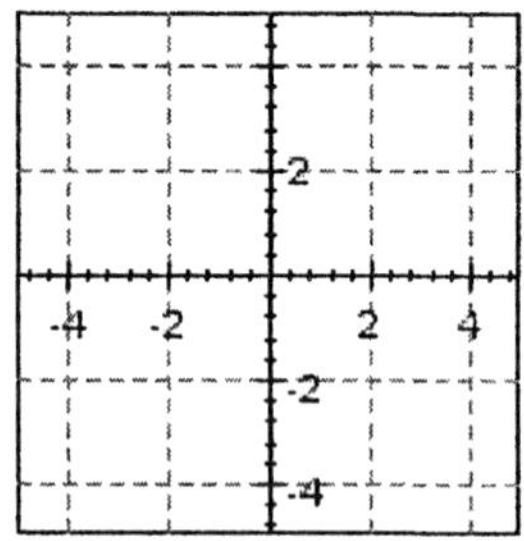

10. Change the equation to rectangular coordinates and then graph.

$$r(1 + \cos\theta) = 3$$

Select the correct answer.

a. 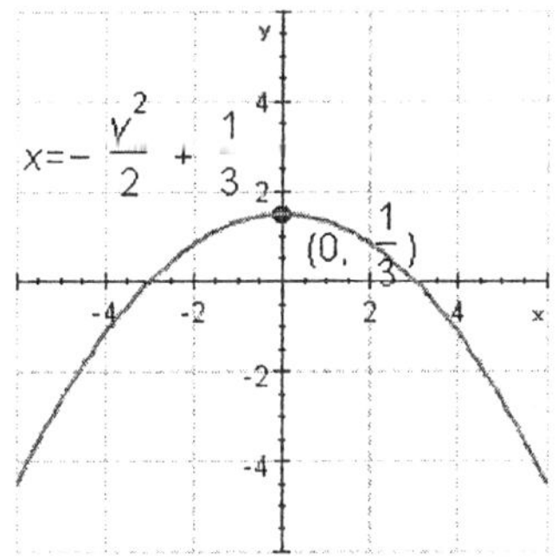

b. 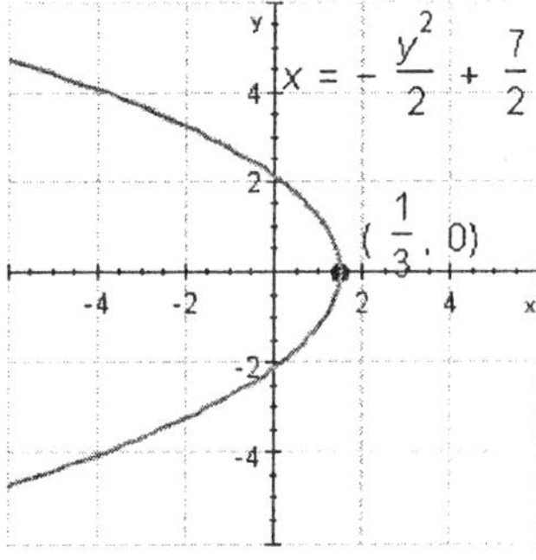

c. 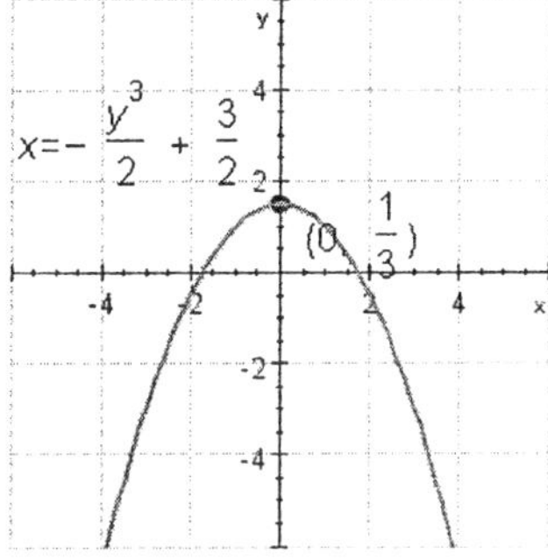

d. 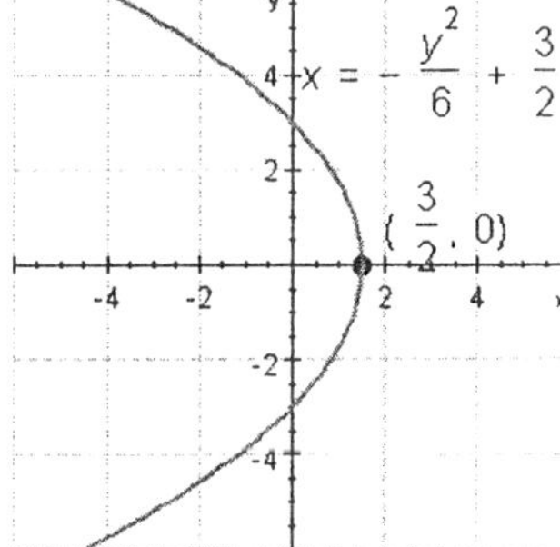

11. Find two square roots for the complex number.

$64 \text{ cis } 0°$

Select the correct answer.

a. $8 \text{ cis } 0°, 8 \text{ cis } 270°$

b. $8 \text{ cis } 0°, 8 \text{ cis } 180°$

c. $64 \text{ cis } 90°, 64 \text{ cis } 180°$

d. $8 \text{ cis } 90°, 8 \text{ cis } 180°$

e. $8 \text{ cis } 90°, 8 \text{ cis } 270°$

12. Find three cube roots for the complex number.

$-32\sqrt{3} + 32i$

Select the correct answer.

a. $4 \text{ cis } 50°, 4 \text{ cis } 170°, 4 \text{ cis } 290°$

b. $4 \text{ cis } 55°, 4 \text{ cis } 175°, 4 \text{ cis } 295°$

c. $4 \text{ cis } 40°, 4 \text{ cis } 160°, 4 \text{ cis } 280°$

d. $2 \text{ cis } 40°, 2 \text{ cis } 160°, 2 \text{ cis } 280°$

e. $2 \text{ cis } 50°, 2 \text{ cis } 170°, 2 \text{ cis } 290°$

13. Write the expression $\sqrt{-16} \cdot \sqrt{-9}$ in terms of i and then simplify.

Select the correct answer.

a. 12

b. -15

c. 8

d. 15

e. -12

14. Graph the ordered pair on a polar coordinate system.

$$(4, 150°)$$

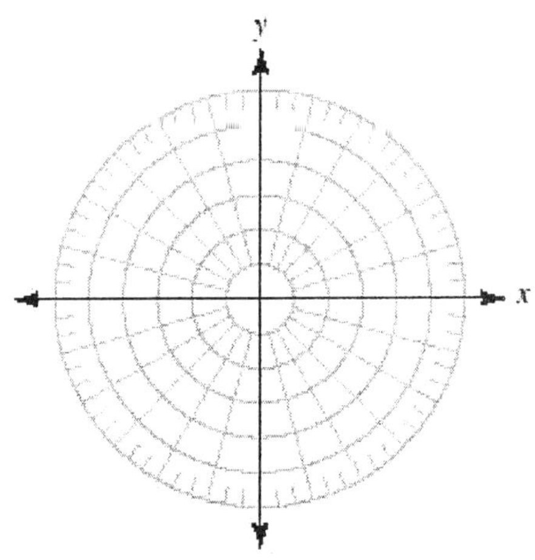

15. Write the complex number in trigonometric form.

$$3 + 3i\sqrt{3}$$

Select the correct answer.

    a. $6\,\text{cis}\,150°$

    b. $6\,\text{cis}\,300°$

    c. $6\,\text{cis}\,60°$

    d. $6\,\text{cis}\,120°$

    e. $3\,\text{cis}\,60°$

16. Find two square roots for the complex number. Leave your answers in trigonometric form.

$$16(\cos 120^\circ + i\sin 120^\circ)$$

Graph the two roots.

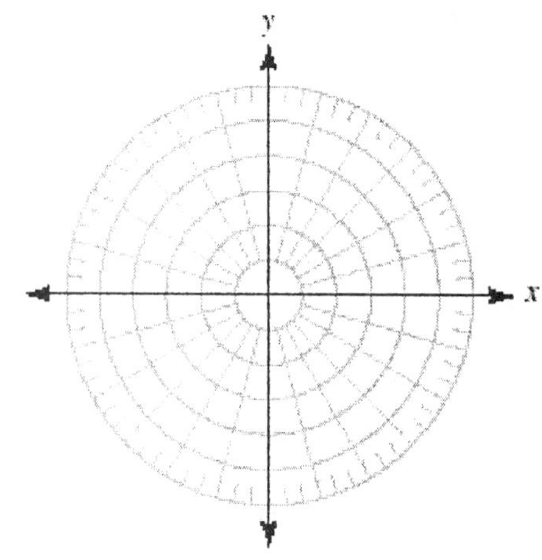

17. Find three cube roots for the complex number. Leave your answers in trigonometric form.

$$8(\cos 291^\circ + i\sin 291^\circ)$$

18. Write the complex number in trigonometric form.

$$1 - i$$

Please enter the argument in degrees and use the degree symbol.

19. Graph the equation.

$$r = 2\cos 2\theta$$

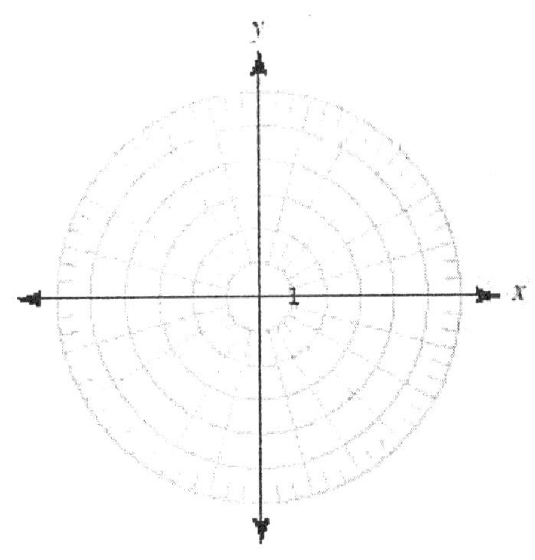

20. Convert to polar coordinates with $r \geq 0$ and θ between $0°$ and $360°$.

$$\left(-2\sqrt{3},\ 2\right)$$

Select the correct answer.

a. $\left(4,\ 130°\right)$
b. $\left(7,\ 165°\right)$
c. $\left(2,\ 160°\right)$
d. $\left(3,\ 145°\right)$
e. $\left(4,\ 150°\right)$

21. Find the product.

$(2 + 3i)(3 - i)$

Enter your answer in standard form for complex numbers.

____

22. Graph the complex number and give the absolute value of it.

6 - 8*i*

Select the correct answer.

a. 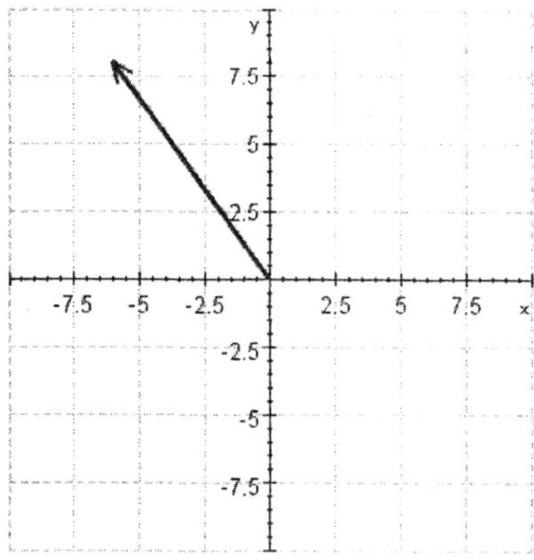 The absolute value is 10

b. 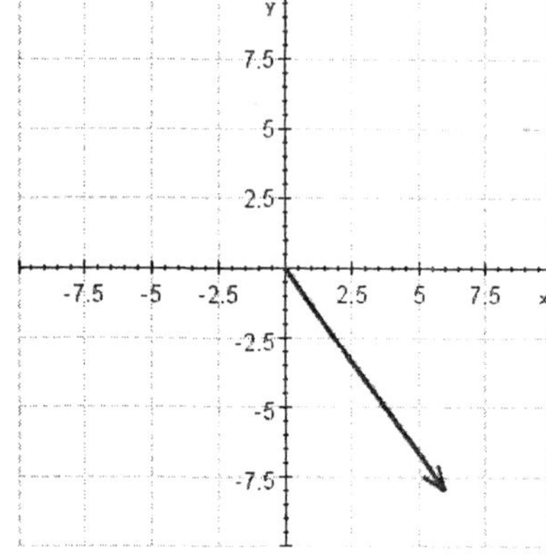 The absolute value is 10

c. 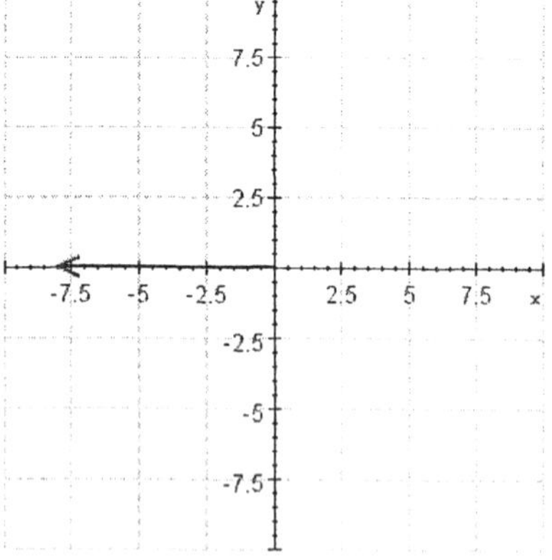 The absolute value is 6

23. Multiply. Leave the answer in trigonometric form.

$$6(\cos 20^\circ + i \sin 20^\circ) \cdot 6(\cos 35^\circ + i \sin 35^\circ)$$

24. Write the expression $\sqrt{-8}$ in terms of i.

25. Graph the ordered pair on a polar coordinate system.

$$(-3, 45^\circ)$$

Select the correct answer.

a.

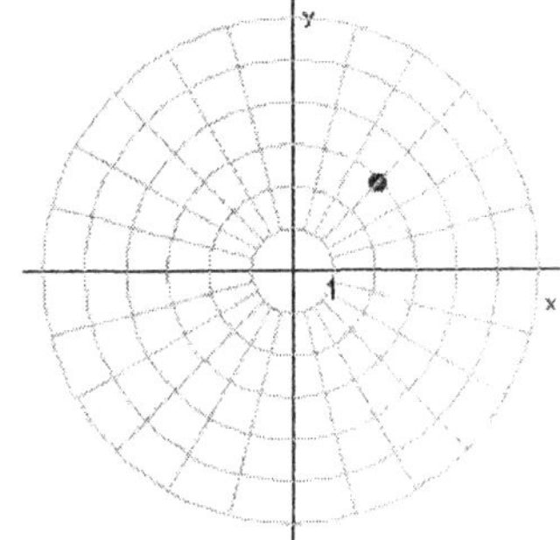

b.

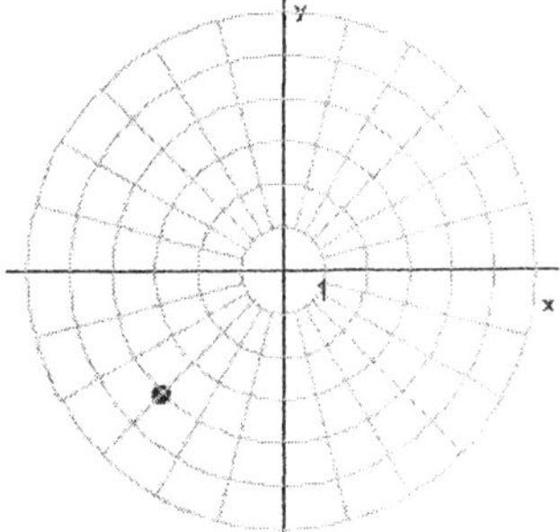

c.

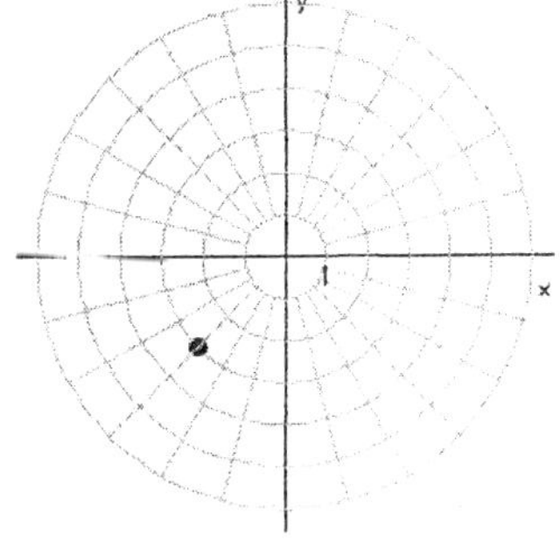

1. e

2. d

3. $\left(6,\ -235°\right),\ \left(-6,\ 305°\right),\ \left(-6,\ -55°\right)$

4. a

5. -64

6. a

7. d

8.

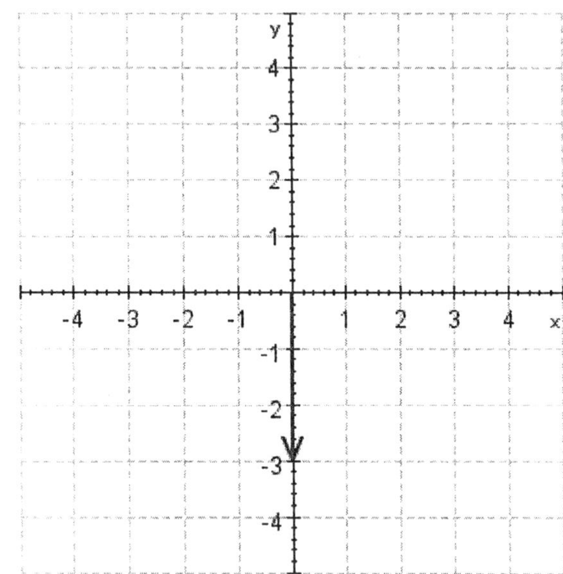

3

9.

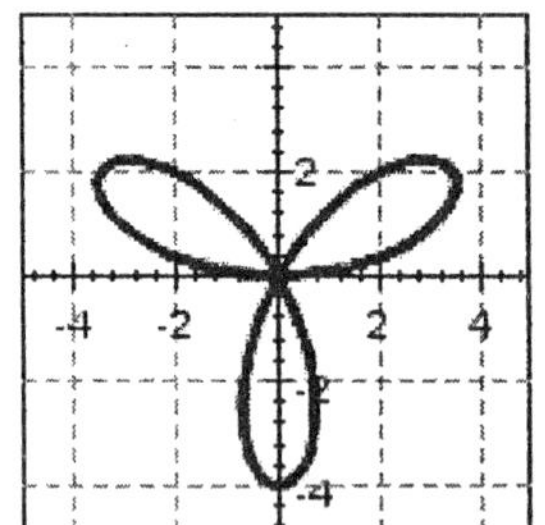

10. d

11. b

12. a

13. e

14.

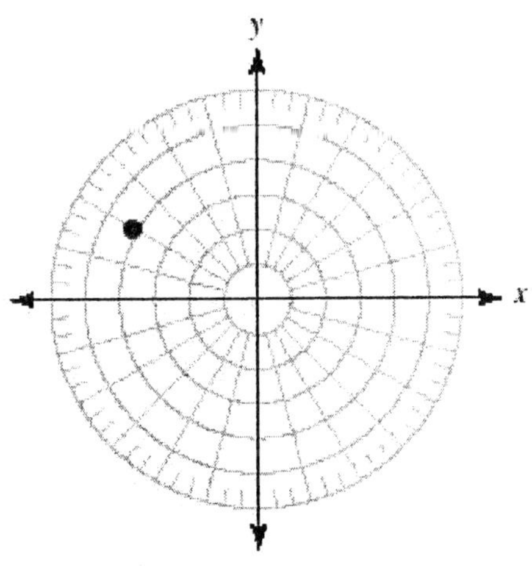

15. c

16. $4\left(\cos\left(60°\right)+i\cdot\sin\left(60°\right)\right),\ 4\left(\cos\left(240°\right)+i\cdot\sin\left(240°\right)\right)$

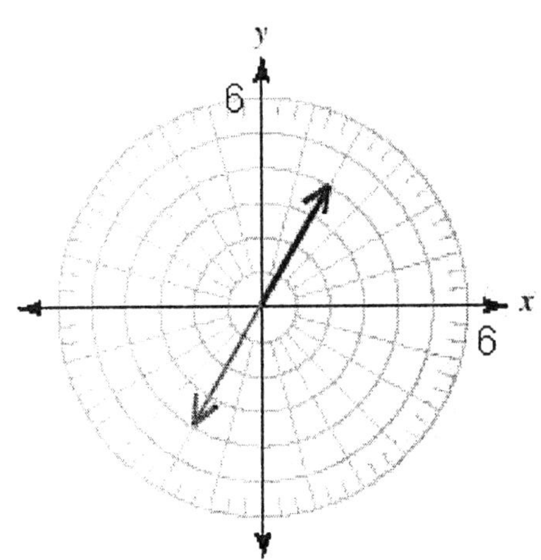

17. $2\left(\cos\left(97°\right)+i\cdot\sin\left(97°\right)\right),\ 2\left(\cos\left(217°\right)+i\cdot\sin\left(217°\right)\right),\ 2\left(\cos\left(337°\right)+i\cdot\sin\left(337°\right)\right)$

18. $\sqrt{2}\cdot\left(\cos\left(315°\right)+i\cdot\sin\left(315°\right)\right)$

19.

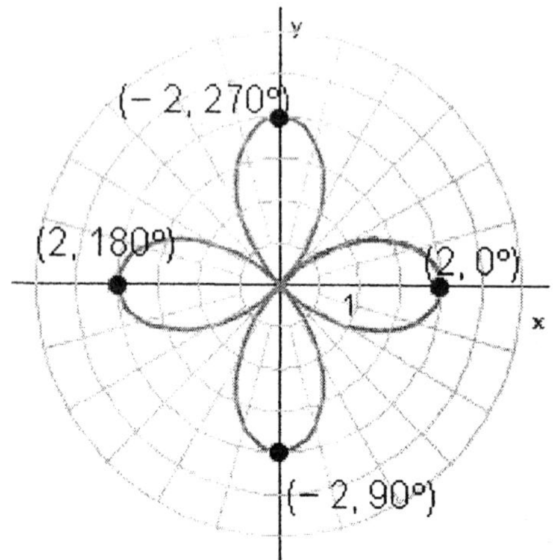

20. e

21. $9 + 7i$

22. b

23. $36\left(\cos\left(55°\right)+i \cdot \sin\left(55°\right)\right)$

24. $2i \cdot \sqrt{2}$

25. e

1. mctr.08.05.38m_NoAlgs
2. mctr.08.03.35m_NoAlgs
3. mctr.08.05.16_NoAlgs
4. mctr.08.03.05m_NoAlgs
5. mctr.08.03.28_NoAlgs
6. mctr.08.01.50m_NoAlgs
7. mctr.08.06.07m_NoAlgs
8. mctr.08.02.05_NoAlgs
9. mctr.08.06.28_NoAlgs
10. mctr.08.06.47m_NoAlgs
11. mctr.08.04.05m_NoAlgs
12. mctr.08.04.18m_NoAlgs
13. mctr.08.01.12m_NoAlgs
14. mctr.08.05.03_NoAlgs
15. mctr.08.02.45m_NoAlgs
16. mctr.08.04.03_NoAlgs
17. mctr.08.04.16_NoAlgs
18. mctr.08.02.35_NoAlgs
19. mctr.08.06.23_NoAlgs
20. mctr.08.05.34m_NoAlgs
21. mctr.08.01.46_NoAlgs
22. mctr.08.02.10m_NoAlgs
23. mctr.08.03.02_NoAlgs
24. mctr.08.01.07_NoAlgs
25. mctr.08.05.08m_NoAlgs

1. Combine the complex numbers.

    $(9 + 8i) - (3 + i)$

    Enter your answer in standard form for complex numbers.

    ____

2. Simplify the power of i.

    i^{17}

    Select the correct answer.

    a. $-i$
    b. i
    c. -1
    d. 1
    e. 0

3. Let $z_1 = 5 + 6i$ and $z_2 = 5 - 6i$ and find

    $z_1 z_2$.

    Enter your answer in standard form for complex numbers.

    $z_1 z_2 =$ ________

4. Read the method of solving the system of equations.

$$x + y = 10$$
$$xy = 40$$

Solving the first equation for y, we have $y = 10 - x$. Substituting this value of y into the second equation, we have

$$x(10 - x) = 40$$

The equation is quadratic. We write it in standard form and apply the quadratic formula

$$0 = x^2 - 10x + 40$$

$$x = \frac{10 \pm \sqrt{100 - 4(1)(40)}}{2}$$

$$= \frac{10 \pm \sqrt{100 - 160}}{2}$$

$$= \frac{10 \pm \sqrt{-60}}{2}$$

$$= \frac{10 \pm 2\sqrt{-15}}{2}$$

$$= 5 \pm \sqrt{-15}$$

$$= 5 \pm i\sqrt{15}$$

$$y = 5 \mp i\sqrt{15}$$

Use the method shown above to solve the system of equations.

$$x + y = 4$$
$$xy = 68$$

Select the correct answer.

    a. $x = 2 + 8i,\ y = 2 - 8i;\ x = 2 - 8i,\ y = 2 + 8i$

    b. $x = 6 + 6i,\ y = 6 - 6i;\ x = 6 - 6i,\ y = 6 + 6i$

    c. $x = 2 + 8i,\ y = 2 + 8i;\ x = 2 - 8i,\ y = 2 - 8i$

    d. $x = 6 + 8i,\ y = 6 - 8i;\ x = 6 - 8i,\ y = 6 + 8i$

    e. $x = 2 + 6i,\ y = 2 - 6i;\ x = 2 - 6i,\ y = 2 + 6i$

5. Graph the complex number along with its opposite and conjugate.

$- 3$

Select the correct answer.

a.

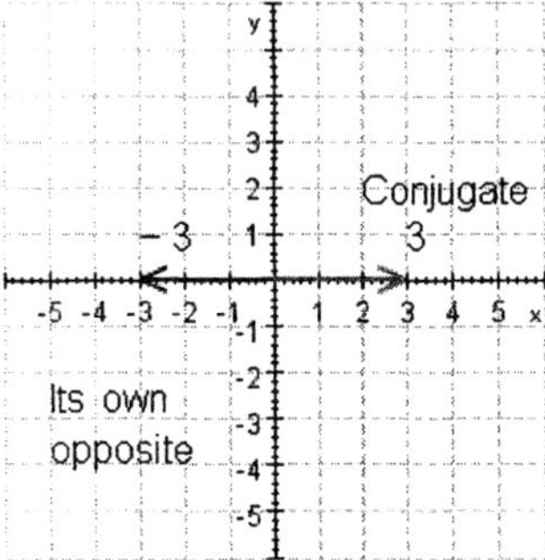

b.

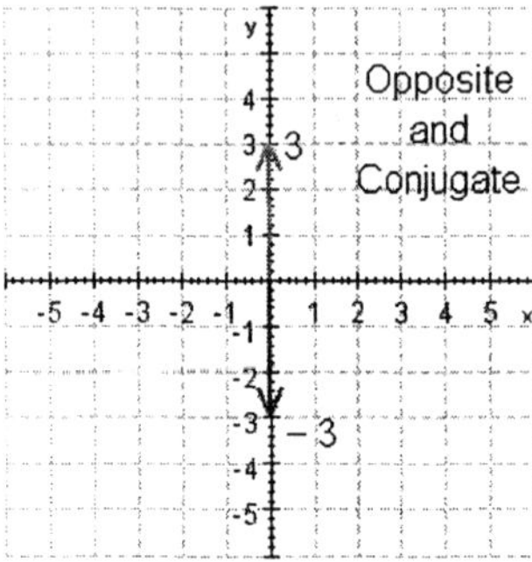

c.

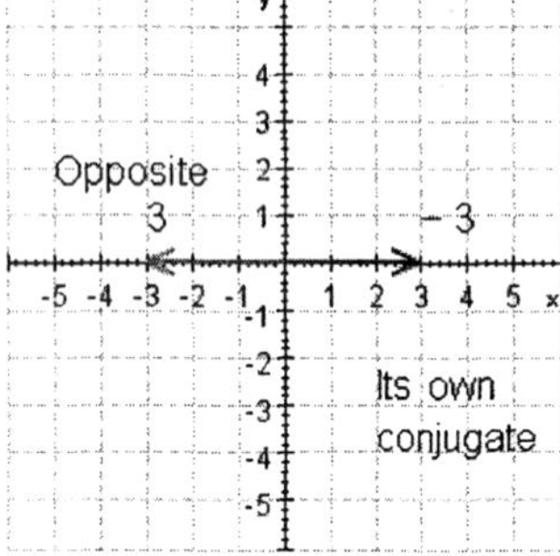

d.

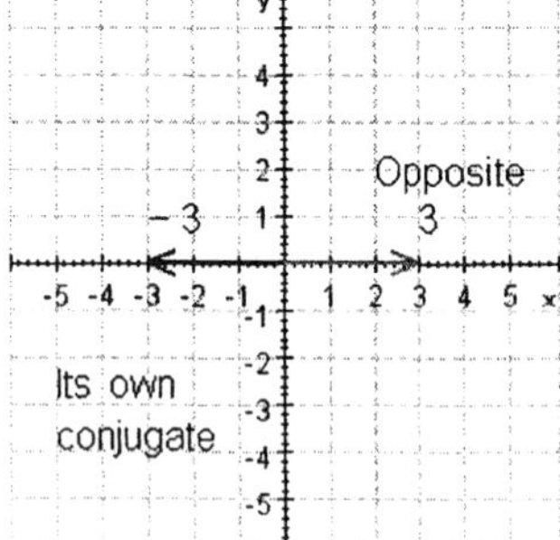

6. Use a calculator to help write the complex number in standard form. Round the numbers in your answer to the nearest hundredth.

$$100\left(\cos 144° + i \sin 144°\right)$$

Select the correct answer.

a.   - 80.90 + 58.78i
b.   - 80.90 + 58.74i
c.   - 80.94 + 58.78i
d.   - 80.98 + 58.86i
e.   - 80.94 + 58.74i

7. Write the complex number in trigonometric form.
   Round the angle to the nearest hundredth of a degree.

24 + 7i

Select the correct answer.

a.   25 cis 1 6.38°

b.   25 cis 1 6.35°

c.   25 cis 1 6.23°

d.   25 cis 1 6.20°

e.   25 cis 1 6.26°

8. Use your graphing calculator to convert the complex number to trigonometric form. Round the angle to the nearest hundredth of a degree.

- 21 - 20i

Select the correct answer.

a.   29 cis 223.50°

b.   29 cis 223.60°

c.   29 cis 223.40°

d.   29 cis 223.90°

e.   29 cis 223.20°

9. Find the product $z_1 z_2$ in standard form.

$$z_1 = \sqrt{3} + i, \ z_2 = -1 + i\sqrt{3}$$

Write z_1 and z_2 in trigonometric form.

Please enter your answer in the form: $z_1 = $ ____ , $z_2 = $ ____

Find the product $z_1 z_2$ in trigonometric form.

Convert the answer that is in trigonometric form to standard form to show that the two products are equal.

10. Use DeMoivre's Theorem to find the following.

$$\left(\sqrt{2} \text{ cis } 60°\right)^{10}$$

Select the correct answer.

a. $-16\sqrt{3} + 16i$

b. $-4 + 4i$

c. $-16 - 16i\sqrt{3}$

d. $-4 + 4i\sqrt{3}$

e. $4 + 4i\sqrt{3}$

11.

Find the quotient $\dfrac{z_1}{z_2}$ in standard form.

$$z_1 = 3\sqrt{3} - 3i, \quad z_2 = 6i$$

Write z_1 and z_2 in trigonometric form.

Please enter your answer in the form: $z_1 = $ _____ , $z_2 = $ _____

Find the quotient $\dfrac{z_1}{z_2}$ in trigonometric form.

Convert the answer that is in trigonometric form to standard form to show that the two products are equal.

12. DeMoivre's Theorem can be used to find reciprocals of complex numbers. Recall from algebra that the reciprocal of x is $\dfrac{1}{x}$, which can be expressed as x^{-1}.

Use this fact, along with DeMoivre's Theorem, to find the reciprocal of the number below.

$$-\sqrt{3} + i$$

Select the correct answer.

a. $\quad -\dfrac{\sqrt{3}}{4} - \dfrac{1}{4}i$

b. $\quad \dfrac{1}{2} + \dfrac{1}{2}i$

c. $\quad \dfrac{1}{4} + \dfrac{\sqrt{3}}{4}i$

d. $\quad \dfrac{\sqrt{3}}{4} + \dfrac{1}{4}i$

e. $\quad -\dfrac{1}{2} - \dfrac{1}{2}i$

13. Find two square roots for the complex number. Write your answer in standard form.

$$8 - 8i\sqrt{3}$$

14. Find two square roots for the complex number.

$$16$$

Select the correct answer.

a. $4, -4i$

b. $4i, -4$

c. $4i, -4i$

d. $4, -4$

e. $16i, -16i$

15. Solve the equation.

$$x^4 + 256 = 0$$

Select the correct answer.

a. $\pm 2, \pm 2i$

b. $-2\sqrt{3} \pm 2i, 2\sqrt{3} \pm 2i$

c. $\pm 3, \pm 3i$

d. $-2\sqrt{2} \pm 2i\sqrt{2}, 2\sqrt{2} \pm 2i\sqrt{2}$

e. $-3\sqrt{2} \pm 3i\sqrt{2}, 3\sqrt{2} \pm 3i\sqrt{2}$

16. Solve the equation.

$$x^4 - 2\sqrt{3}\,x^2 + 4 = 0$$

Select the correct answer.

a.　$\sqrt{2}\,\text{cis}\,10°,\ \sqrt{2}\,\text{cis}\,170°,\ \sqrt{2}\,\text{cis}\,200°,\ \sqrt{2}\,\text{cis}\,350°$

b.　$\sqrt{2}\,\text{cis}\,15°,\ \sqrt{2}\,\text{cis}\,165°,\ \sqrt{2}\,\text{cis}\,195°,\ \sqrt{2}\,\text{cis}\,345°$

c.　$\sqrt{2}\,\text{cis}\,10°,\ \sqrt{2}\,\text{cis}\,200°$

d.　$\sqrt{2}\,\text{cis}\,170°,\ \sqrt{2}\,\text{cis}\,350°$

e.　$\sqrt{2}\,\text{cis}\,15°,\ \sqrt{2}\,\text{cis}\,195°$

17. For the ordered pair, give three other ordered pairs with degree values between $-360°$ and $360°$ that name the same point.

$$(6,\ 125°)$$

18. Convert to rectangular coordinates. Use exact values.

$$\left(6\sqrt{2},\ -45°\right)$$

Select the correct answer.

a.　$(-4, -4)$
b.　$(6, -6)$
c.　$(-6, 6)$
d.　$(-4, 6)$
e.　$(-7, -7)$

19. Use your graphing calculator to convert to polar coordinates. Round all values to four significant digits.

(6, - 3)

Select the correct answer.

a. $(10.708, -22.57°)$

b. $(6.708, -26.57°)$

c. $(8.708, -28.57°)$

d. $(7.708, -25.57°)$

e. $(3.708, -23.57°)$

20. Write the equation in polar coordinates.

$$x^2 + y^2 = 5$$

Select the correct answer.

a. $r^2 = 1$

b. $r^2 = 2$

c. $r^2 = 7$

d. $r^2 = 5$

e. $r^2 = 4$

21. Graph the equation.

$\theta = 225°$

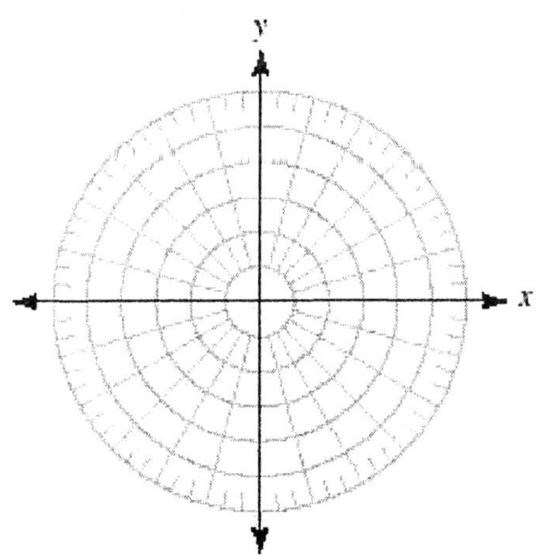

22. Graph the equation using your graphing calculator.

$r = 4 \sin 3\theta$

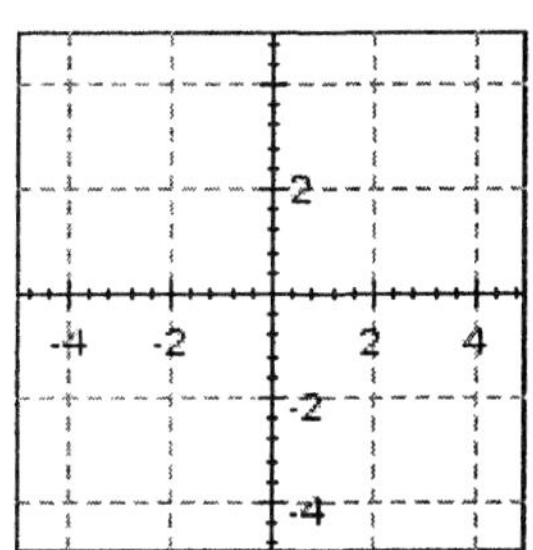

23. Graph the equation.

$$r = 4 + 5\cos\theta$$

Select the correct answer.

a.

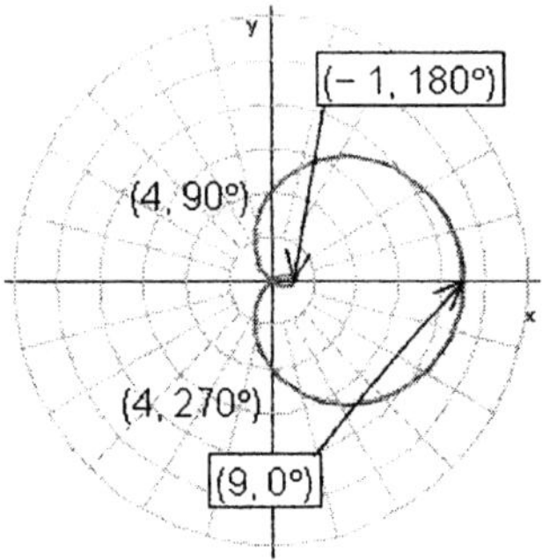

b.

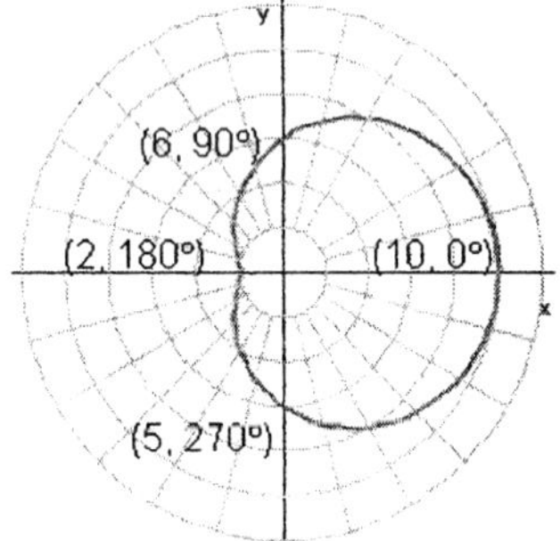

c.

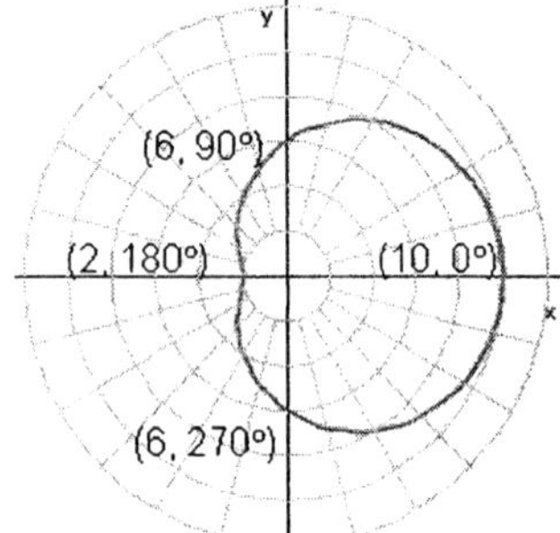

d.

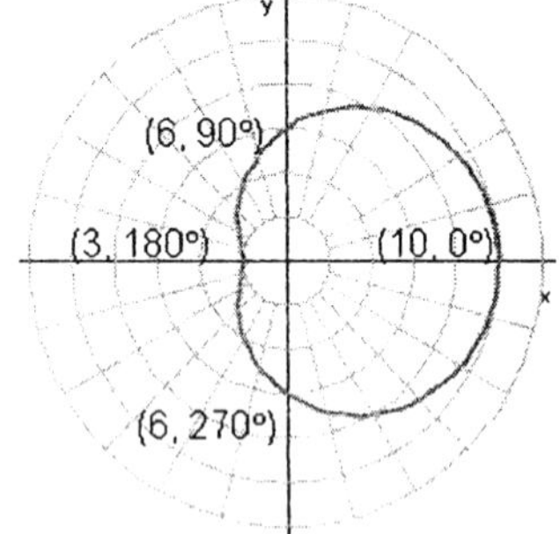

24. Graph the equation using your graphing calculator.

$$r = 4 - 6 \sin \theta$$

Select the correct answer.

a.

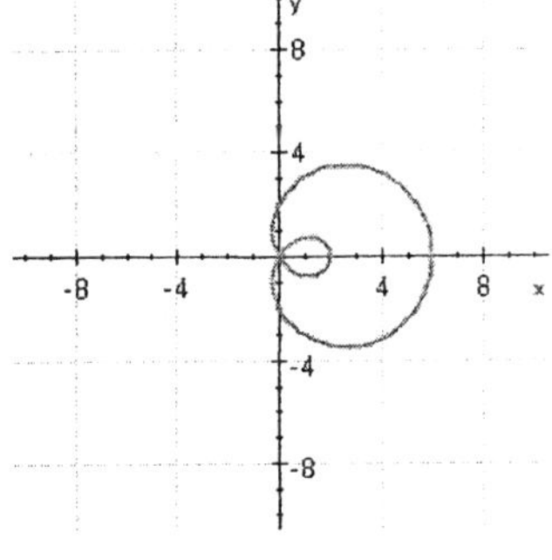

b.

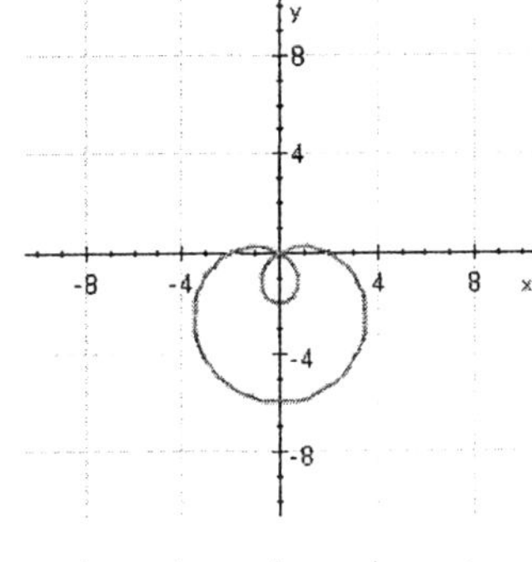

c.

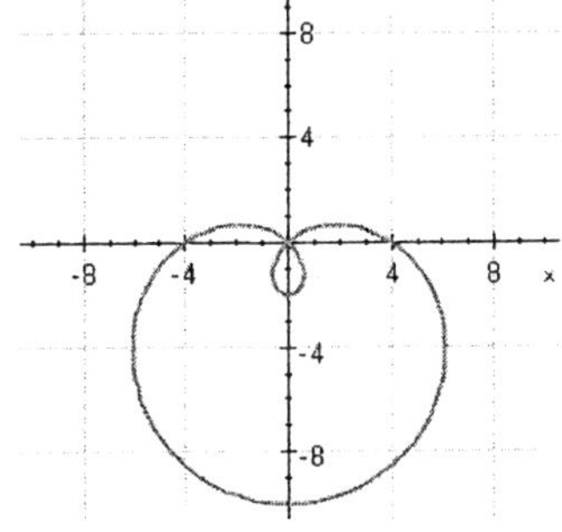

25. Change the equation to rectangular coordinates and then graph.

$$r(1 + \cos\theta) = 3$$

Select the correct answer.

a. 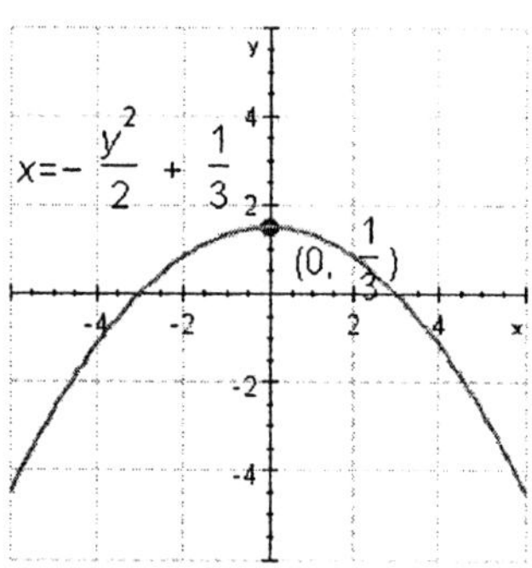

b. 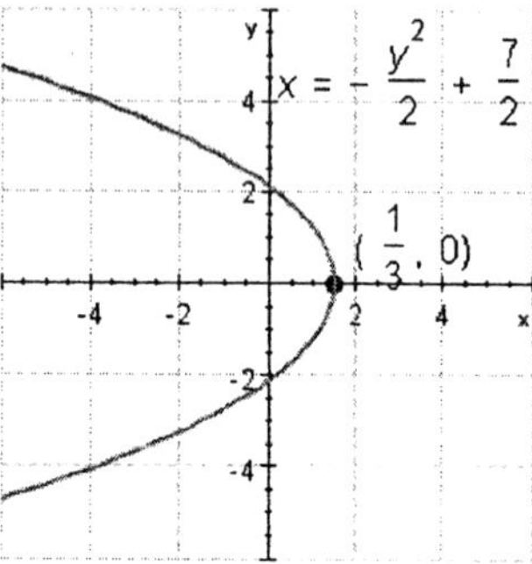

c. 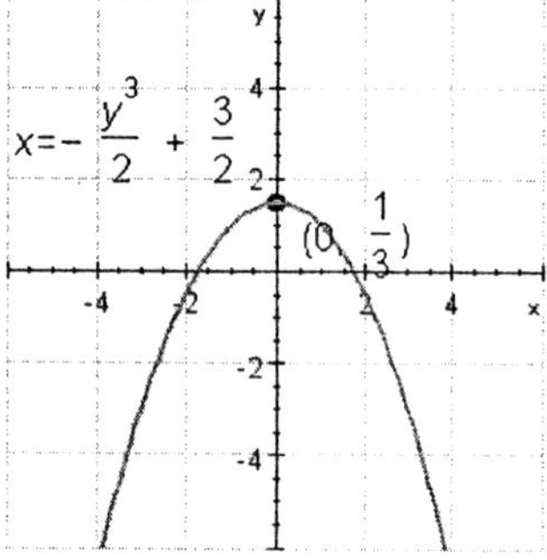

d. 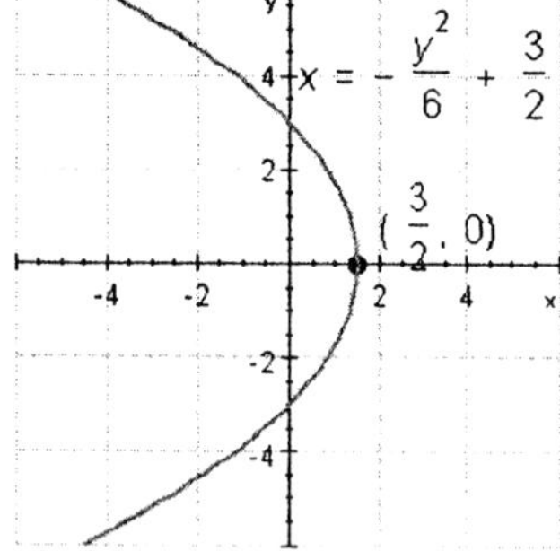

1. $6 + 7i$

2. b

3. 61

4. d

5. d

6. a

7. e

8. b

9. $-2\sqrt{3}+2i$

$z_1 = 2\left(\cos(30°)+i \cdot \sin(30°)\right), z_2 = 2\left(\cos(120°)+i \cdot \sin(120°)\right)$

$4\left(\cos(150°)+i \cdot \sin(150°)\right)$

$z_1 z_2 = 4\left(\cos 150° + i\sin 150°\right) = 4\left(-\dfrac{\sqrt{3}}{2} + \dfrac{1}{2}i\right) = -2\sqrt{3} + 2i$

10. c

11. $-\dfrac{1}{2} - \dfrac{\sqrt{3}}{2} \cdot i$

$z_1 = 6\left(\cos(330°)+i \cdot \sin(330°)\right), z_2 = 6\left(\cos(90°)+i \cdot \sin(90°)\right)$

$\cos(240°)+i \cdot \sin(240°)$

$\dfrac{z_1}{z_2} = \cos 240° + i\sin 240° = -\dfrac{1}{2} - \dfrac{\sqrt{3}}{2}i$

12. a

13. $-2\sqrt{3}+2i, \ 2\sqrt{3}-2i$

14. d

15. d

16. b

17. $(6, -235°), (-6, 305°), (-6, -55°)$

18. b

19. b

20. d

21.

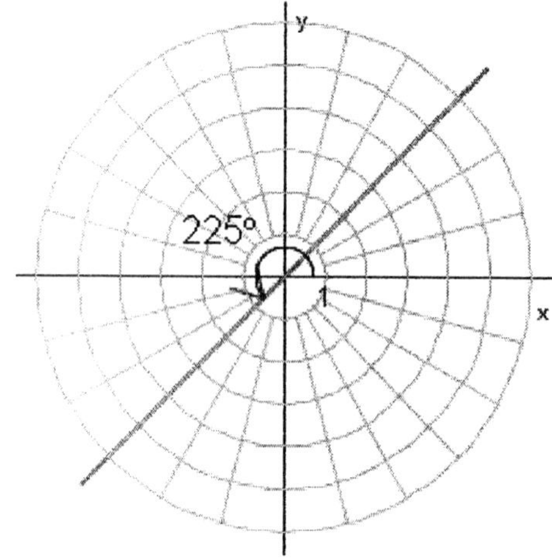

22.

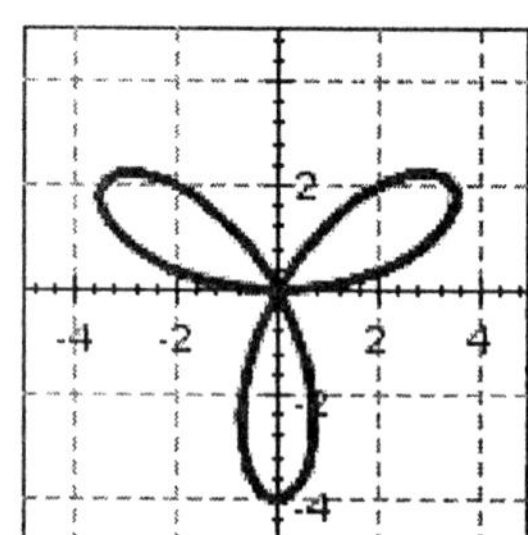

23. a

24. c

25. d

1. mctr.08.01.25_NoAlgs
2. mctr.08.01.38m_NoAlgs
3. mctr.08.01.65_NoAlgs
4. mctr.08.01.79m_NoAlgs
5. mctr.08.02.16m_NoAlgs
6. mctr.08.02.29m_NoAlgs
7. mctr.08.02.51m_NoAlgs
8. mctr.08.02.58m_NoAlgs
9. mctr.08.03.10_NoAlgs
10. mctr.08.03.24m_NoAlgs
11. mctr.08.03.41_NoAlgs
12. mctr.08.03.57m_NoAlgs
13. mctr.08.04.08_NoAlgs
14. mctr.08.04.11m_NoAlgs
15. mctr.08.04.26m_NoAlgs
16. mctr.08.04.33m_NoAlgs
17. mctr.08.05.16_NoAlgs
18. mctr.08.05.24m_NoAlgs
19. mctr.08.05.48m_NoAlgs
20. mctr.08.05.59m_NoAlgs
21. mctr.08.06.12_NoAlgs
22. mctr.08.06.28_NoAlgs
23. mctr.08.06.17m_NoAlgs
24. mctr.08.06.34m_NoAlgs
25. mctr.08.06.47m_NoAlgs

1. Read the method of solving the system of equations.

$$x + y = 10$$
$$xy = 40$$

Solving the first equation for y, we have $y = 10 - x$. Substituting this value of y into the second equation, we have

$x(10 - x) = 40$

The equation is quadratic. We write it in standard form and apply the quadratic formula.

$$0 = x^2 - 10x + 40$$

$$x = \frac{10 \pm \sqrt{100 - 4(1)(40)}}{2}$$

$$= \frac{10 \pm \sqrt{100 - 160}}{2}$$

$$= \frac{10 \pm \sqrt{-60}}{2}$$

$$= \frac{10 \pm 2\sqrt{-15}}{2}$$

$$= 5 \pm \sqrt{-15}$$

$$= 5 \pm i\sqrt{15}$$

$$y = 5 \mp i\sqrt{15}$$

Use the method shown above to solve the system of equations.

$$x + y = 4$$
$$xy = 68$$

Select the correct answer.

a. $x = 2 + 8i,\ y = 2 - 8i;\ x = 2 - 8i,\ y = 2 + 8i$

b. $x = 6 + 6i,\ y = 6 - 6i;\ x = 6 - 6i,\ y = 6 + 6i$

c. $x = 2 + 8i,\ y = 2 + 8i;\ x = 2 - 8i,\ y = 2 - 8i$

d. $x = 6 + 8i,\ y = 6 - 8i;\ x = 6 - 8i,\ y = 6 + 8i$

e. $x = 2 + 6i,\ y = 2 - 6i;\ x = 2 - 6i,\ y = 2 + 6i$

2. Write the equation in polar coordinates.

$$x^2 + y^2 = 5$$

Select the correct answer.

a. $r^2 = 1$

b. $r^2 = 2$

c. $r^2 = 7$

d. $r^2 = 5$

e. $r^2 = 4$

3. Find the quotient $\dfrac{z_1}{z_2}$ in standard form.

$$z_1 = 3\sqrt{3} - 3i, \quad z_2 = 6i$$

Write z_1 and z_2 in trigonometric form.

Please enter your answer in the form: $z_1 = $ _____, $z_2 = $ _____

Find the quotient $\dfrac{z_1}{z_2}$ in trigonometric form.

Convert the answer that is in trigonometric form to standard form to show that the two products are equal.

4. DeMoivre's Theorem can be used to find reciprocals of complex numbers. Recall from algebra that the reciprocal of x is $\dfrac{1}{x}$, which can be expressed as x^{-1}.

    Use this fact, along with DeMoivre's Theorem, to find the reciprocal of the number below.

    $$-\sqrt{3} + i$$

    Select the correct answer.

    a. $-\dfrac{\sqrt{3}}{4} - \dfrac{1}{4}i$

    b. $\dfrac{1}{2} + \dfrac{1}{2}i$

    c. $\dfrac{1}{4} + \dfrac{\sqrt{3}}{4}i$

    d. $\dfrac{\sqrt{3}}{4} + \dfrac{1}{4}i$

    e. $-\dfrac{1}{2} - \dfrac{1}{2}i$

5. Use a calculator to help write the complex number in standard form. Round the numbers in your answer to the nearest hundredth.

    $$100\left(\cos 144^\circ + i\sin 144^\circ\right)$$

    Select the correct answer.

    a.   - 80.90 + 58.78i
    b.   - 80.90 + 58.74i
    c.   - 80.94 + 58.78i
    d.   - 80.98 + 58.86i
    e.   - 80.94 + 58.74i

6. Graph the equation using your graphing calculator.

   $r = 4 \sin 3\theta$

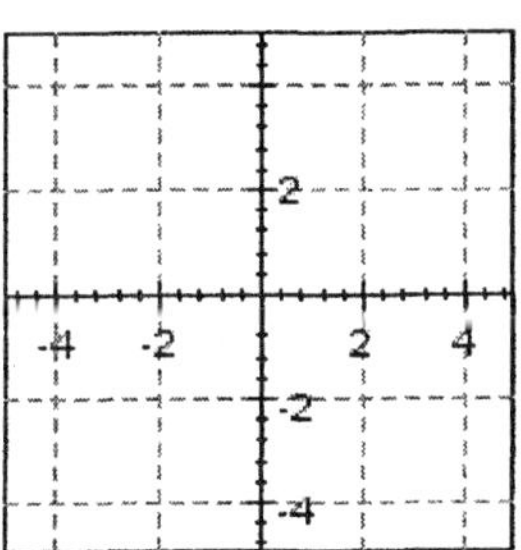

7. Use your graphing calculator to convert to polar coordinates. Round all values to four significant digits.

   $(6, -3)$

   Select the correct answer.

   a. $(10.708, -22.57°)$
   b. $(6.708, -26.57°)$
   c. $(8.708, -28.57°)$
   d. $(7.708, -25.57°)$
   e. $(3.708, -23.57°)$

8. Change the equation to rectangular coordinates and then graph.

$$r(1 + \cos\theta) = 3$$

Select the correct answer.

a.

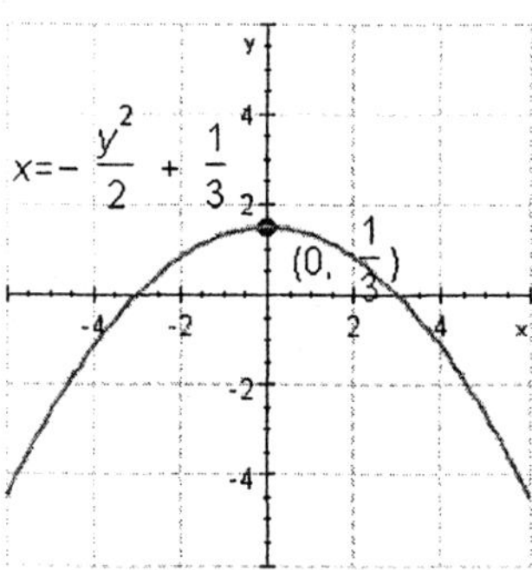

b.

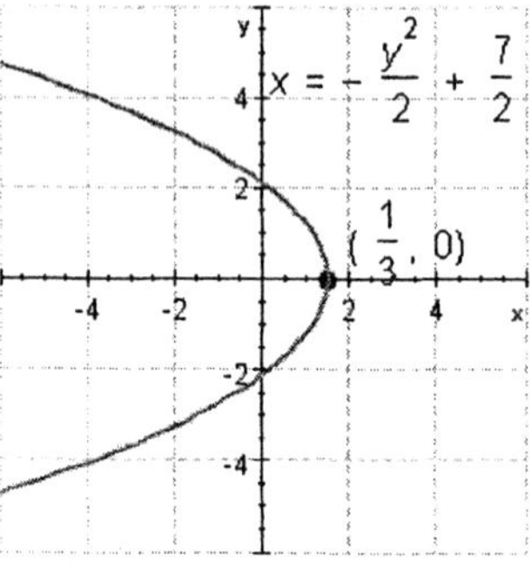

c.

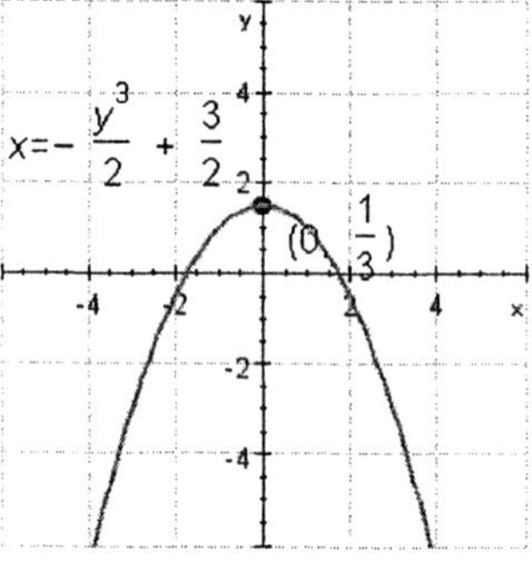

d.

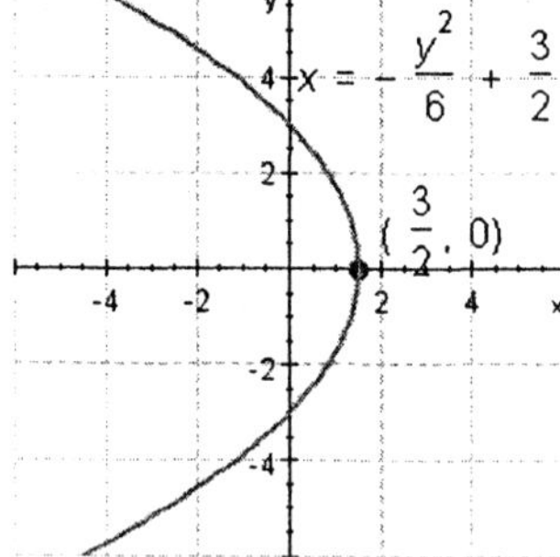

9. Graph the equation using your graphing calculator.

$$r = 4 - 6 \sin \theta$$

Select the correct answer.

a.

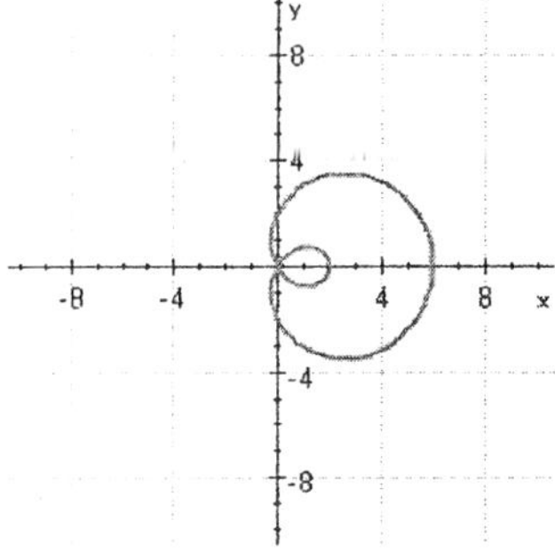

b.

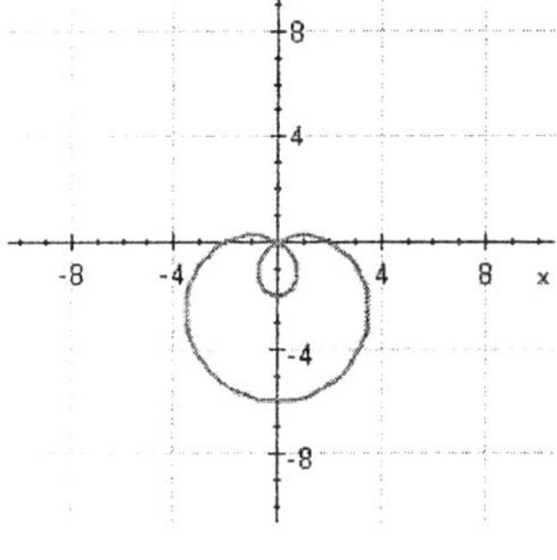

c.

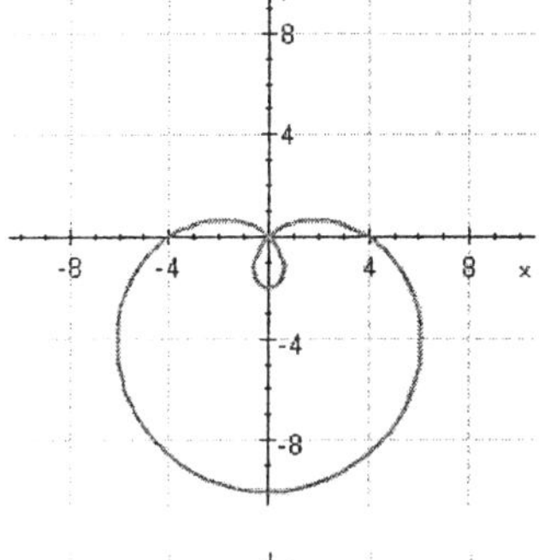

d.

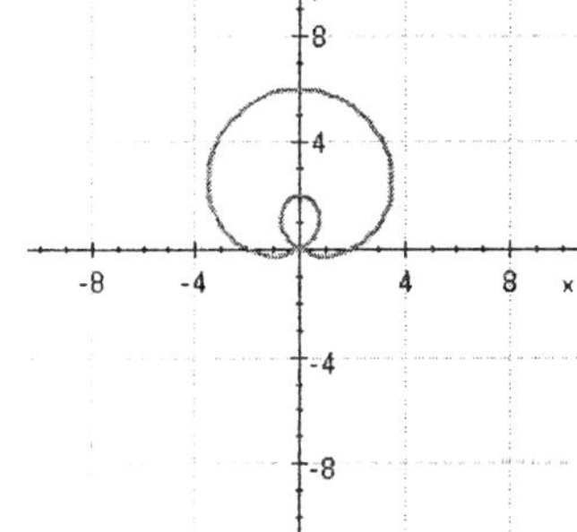

10. Use DeMoivre's Theorem to find the following.

$$\left(\sqrt{2}\ \text{cis}\ 60°\right)^{10}$$

Select the correct answer.

a. $-16\sqrt{3} + 16i$

b. $-4 + 4i$

c. $-16 - 16i\sqrt{3}$

d. $-4 + 4i\sqrt{3}$

e. $4 + 4i\sqrt{3}$

11. Simplify the power of i.

$$i^{17}$$

Select the correct answer.

a. $-i$
b. i
c. -1
d. 1
e. 0

12. Graph the equation.

$$r = 4 + 5\cos\theta$$

Select the correct answer.

a.
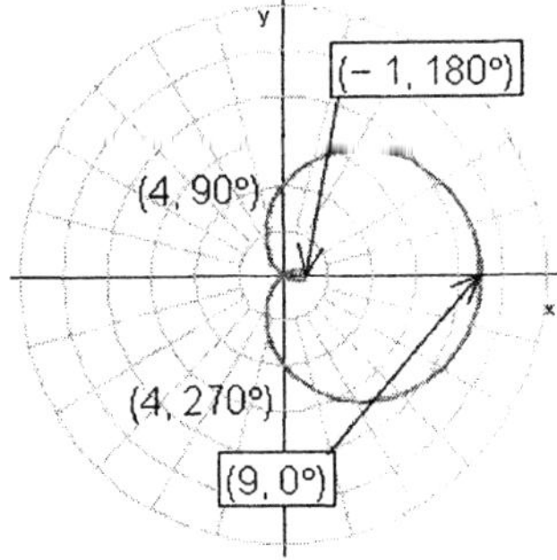

b.
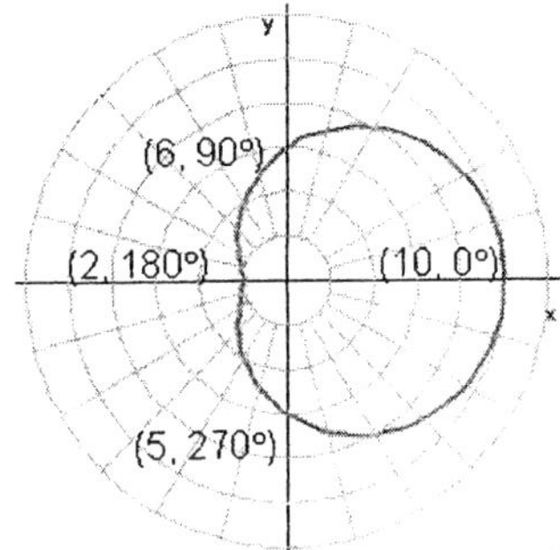

c.
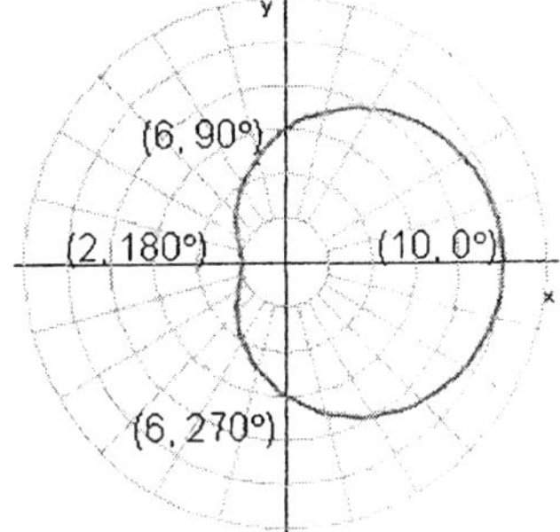

d.
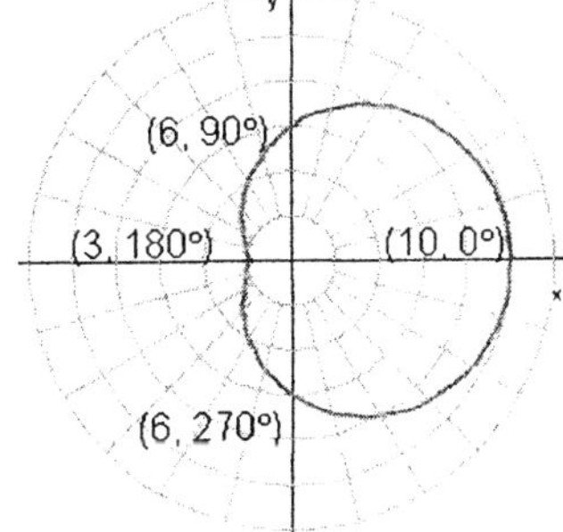

13. Combine the complex numbers.

$(9 + 8i) - (3 + i)$

Enter your answer in standard form for complex numbers.

————

14. Graph the equation.

$\theta = 225°$

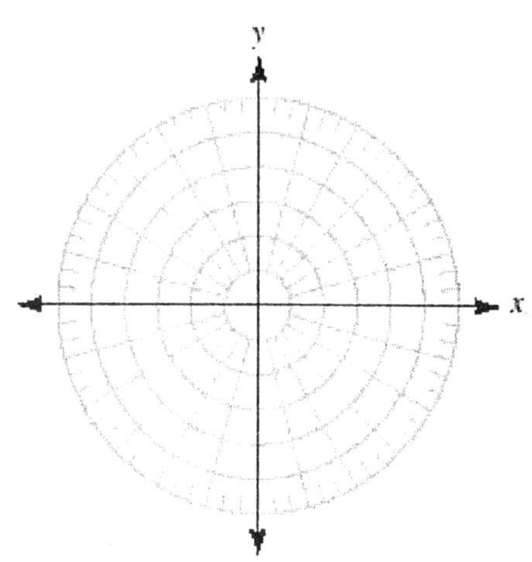

15. Write the complex number in trigonometric form.
Round the angle to the nearest hundredth of a degree.

$24 + 7i$

Select the correct answer.

    a. $25 \operatorname{cis} 16.38°$

    b. $25 \operatorname{cis} 16.35°$

    c. $25 \operatorname{cis} 16.23°$

    d. $25 \operatorname{cis} 16.20°$

    e. $25 \operatorname{cis} 16.26°$

16. Graph the complex number along with its opposite and conjugate.

-3

Select the correct answer.

a.

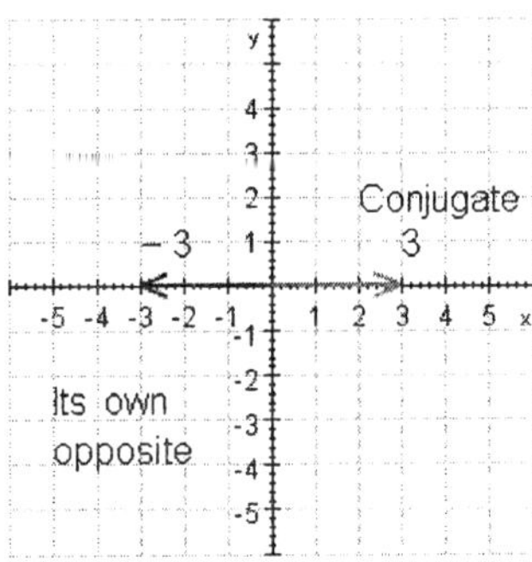

b.

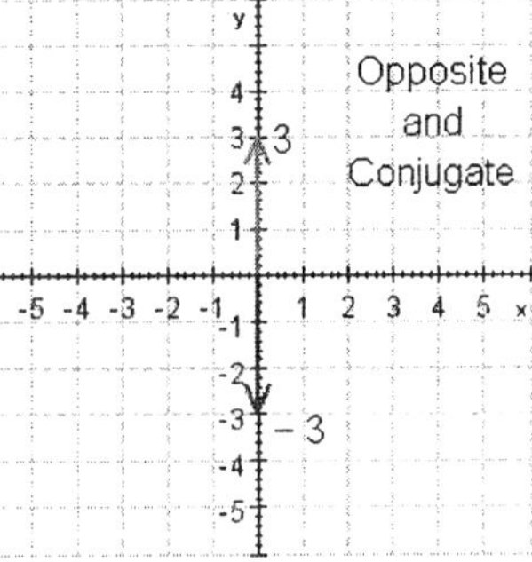

c.

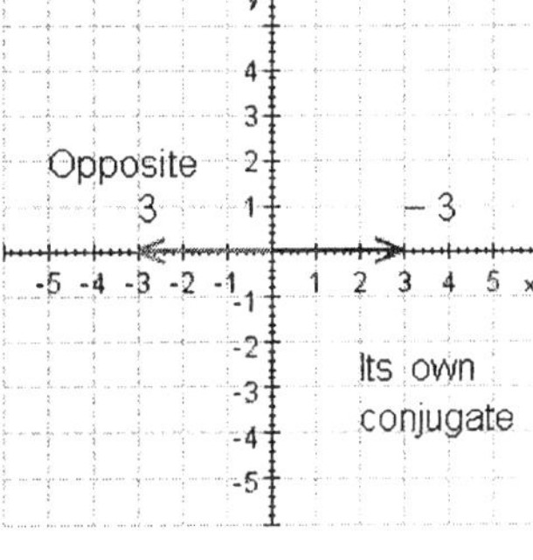

d.

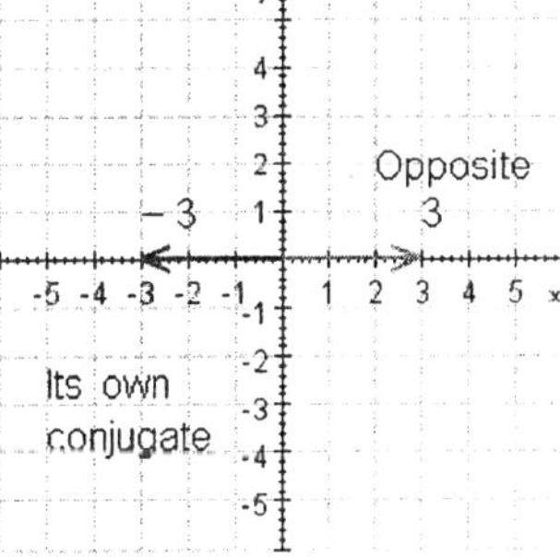

17. Solve the equation.

$$x^4 - 2\sqrt{3}\,x^2 + 4 = 0$$

Select the correct answer.

    a. $\sqrt{2}\,\text{cis}\,10°,\ \sqrt{2}\,\text{cis}\,170°,\ \sqrt{2}\,\text{cis}\,200°,\ \sqrt{2}\,\text{cis}\,350°$

    b. $\sqrt{2}\,\text{cis}\,15°,\ \sqrt{2}\,\text{cis}\,165°,\ \sqrt{2}\,\text{cis}\,195°,\ \sqrt{2}\,\text{cis}\,345°$

    c. $\sqrt{2}\,\text{cis}\,10°,\ \sqrt{2}\,\text{cis}\,200°$

    d. $\sqrt{2}\,\text{cis}\,170°,\ \sqrt{2}\,\text{cis}\,350°$

    e. $\sqrt{2}\,\text{cis}\,15°,\ \sqrt{2}\,\text{cis}\,195°$

18. Use your graphing calculator to convert the complex number to trigonometric form. Round the angle to the nearest hundredth of a degree.

- 21 - 20i

Select the correct answer.

    a. $29\,\text{cis}\,223.50°$

    b. $29\,\text{cis}\,223.60°$

    c. $29\,\text{cis}\,223.40°$

    d. $29\,\text{cis}\,223.90°$

    e. $29\,\text{cis}\,223.20°$

19. For the ordered pair, give three other ordered pairs with degree values between $-360°$ and $360°$ that name the same point.

$$(6,\ 125°)$$

20. Let $z_1 = 5 + 6i$ and $z_2 = 5 - 6i$ and find

$z_1 z_2$.

Enter your answer in standard form for complex numbers.

$z_1 z_2 = $ _______

21. Convert to rectangular coordinates. Use exact values:

$$\left(6\sqrt{2},\ -45^\circ\right)$$

Select the correct answer.

a. $(-4, -4)$
b. $(6, -6)$
c. $(-6, 6)$
d. $(-4, 6)$
e. $(-7, -7)$

22. Solve the equation.

$$x^4 + 256 = 0$$

Select the correct answer.

a. $\pm 2,\ \pm 2i$

b. $-2\sqrt{3} \pm 2i,\ 2\sqrt{3} \pm 2i$

c. $\pm 3,\ \pm 3i$

d. $-2\sqrt{2} \pm 2i\sqrt{2},\ 2\sqrt{2} \pm 2i\sqrt{2}$

e. $-3\sqrt{2} \pm 3i\sqrt{2},\ 3\sqrt{2} \pm 3i\sqrt{2}$

23. Find the product $z_1 z_2$ in standard form.

$$z_1 = \sqrt{3} + i, \; z_2 = -1 + i\sqrt{3}$$

Write z_1 and z_2 in trigonometric form.

Please enter your answer in the form: $z_1 = $ ____ , $z_2 = $ ____

Find the product $z_1 z_2$ in trigonometric form.

Convert the answer that is in trigonometric form to standard form to show that the two products are equal.

24. Find two square roots for the complex number. Write your answer in standard form.

$$8 - 8i\sqrt{3}$$

25. Find two square roots for the complex number.

16

Select the correct answer.

a. $4, -4i$

b. $4i, -4$

c. $4i, -4i$

d. $4, -4$

e. $16i, -16i$

1. a

2. d

3. $-\dfrac{1}{2} - \dfrac{\sqrt{3}}{2} \cdot i$

$z_1 = 6\left(\cos(330°) + i \cdot \sin(330°)\right),\ z_2 = 6\left(\cos(90°) + i \cdot \sin(90°)\right)$

$\cos(240°) + i \cdot \sin(240°)$

$$\dfrac{z_1}{z_2} = \cos 240° + i\sin 240° = -\dfrac{1}{2} - \dfrac{\sqrt{3}}{2}i$$

4. a

5. a

6. 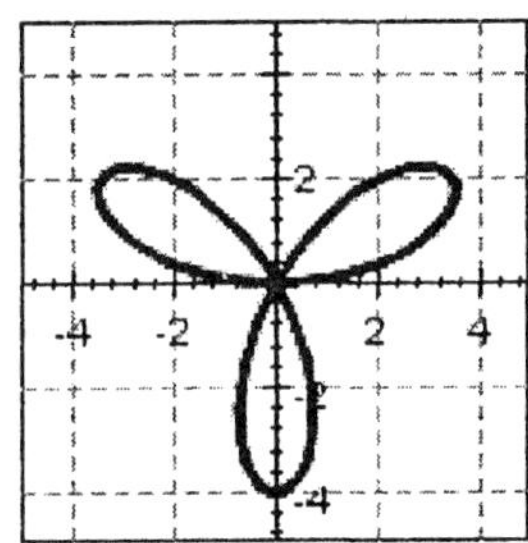

7. b

8. d

9. c

10. c

11. b

12. a

13. $6 + 7i$

14.

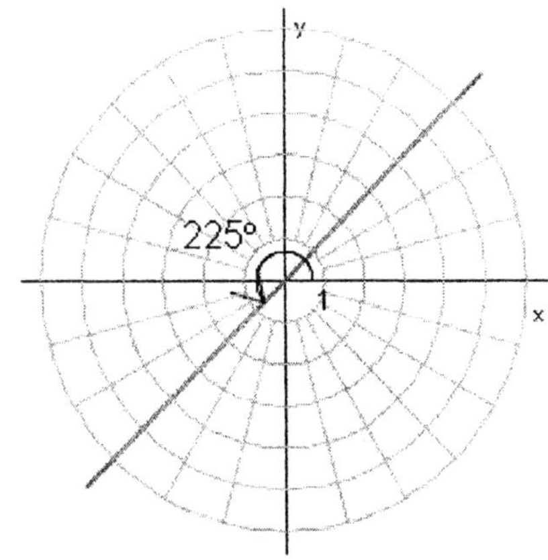

15. e

16. d

17. b

18. b

19. $(6, -235°), (-6, 305°), (-6, -55°)$

20. 61

21. b

22. d

23. $-2\sqrt{3}+2i$

$z_1 = 2(\cos(30°)+i \cdot \sin(30°)), \ z_2 = 2(\cos(120°)+i \cdot \sin(120°))$

$4(\cos(150°)+i \cdot \sin(150°))$

$z_1 z_2 = 4(\cos 150° + i\sin 150°) = 4\left(-\dfrac{\sqrt{3}}{2} + \dfrac{1}{2}i\right) = -2\sqrt{3} + 2i$

24. $-2\sqrt{3}+2i, \ 2\sqrt{3}-2i$

25. d

McKeague/Turner - Trigonometry 5e Chapter 8 Form H

1. mctr.08.01.79m_NoAlgs
2. mctr.08.05.59m_NoAlgs
3. mctr.08.03.41_NoAlgs
4. mctr.08.03.57m_NoAlgs
5. mctr.08.02.29m_NoAlgs
6. mctr.08.06.28_NoAlgs
7. mctr.08.05.48m_NoAlgs
8. mctr.08.06.47m_NoAlgs
9. mctr.08.06.34m_NoAlgs
10. mctr.08.03.24m_NoAlgs
11. mctr.08.01.38m_NoAlgs
12. mctr.08.06.17m_NoAlgs
13. mctr.08.01.25_NoAlgs
14. mctr.08.06.12_NoAlgs
15. mctr.08.02.51m_NoAlgs
16. mctr.08.02.16m_NoAlgs
17. mctr.08.04.33m_NoAlgs
18. mctr.08.02.58m_NoAlgs
19. mctr.08.05.16_NoAlgs
20. mctr.08.01.65_NoAlgs
21. mctr.08.05.24m_NoAlgs
22. mctr.08.04.26m_NoAlgs
23. mctr.08.03.10_NoAlgs
24. mctr.08.04.08_NoAlgs
25. mctr.08.04.11m_NoAlgs

1. Suppose ABC is a right triangle with $C = 90°$.

   If $a = 9$ and $c = 15$, find b.

2. Draw $-45°$ in standard position. Name a point on the terminal side of the angle. Find the distance from the origin to the point.

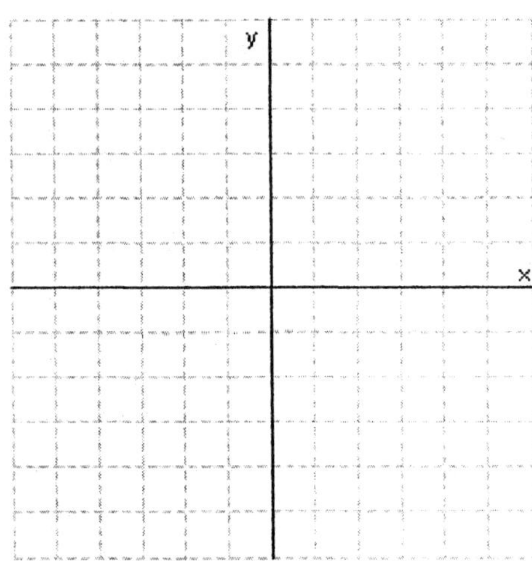

Name another angle that is coterminal with the angle you have drawn.

3. Draw the following angle in standard position.

   $315°$

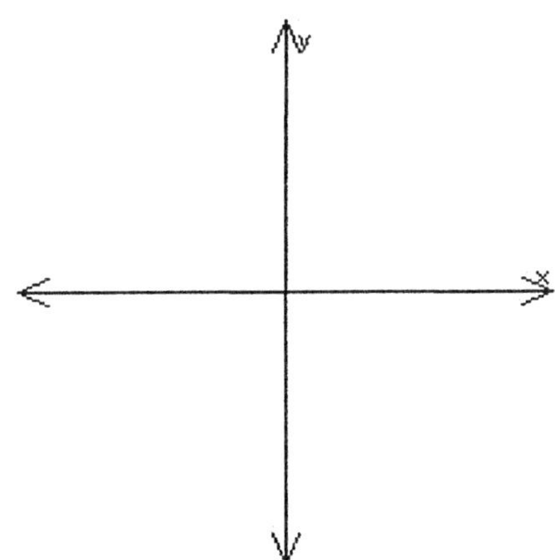

   Find a point on the terminal side.

   Find the sine of the angle.

   Find the cosine of the angle.

   Find the tangent of the angle.

4. Use a ratio identity to find $\cot \theta$ if

   $$\sin \theta = \frac{5}{\sqrt{61}} \quad \text{and} \quad \cos \theta = \frac{6}{\sqrt{61}}.$$

5. Write the following in terms of $\sin \theta$ and $\cos \theta$ and then simplify if possible.

   $\sec \theta - \tan \theta \sin \theta$

6. Show that the following statement is true by transforming the left side into the right side.

$$\sin\alpha\cot\alpha = \cos\alpha$$

7. Simplify the expression by first substituting values from the table of exact values and then simplifying the resulting expression.

$$\sin^2 30° + \cos^2 30°$$

Table of Exact Values

x	$\sin x$	$\cos x$
$0°$	0	1
$30°$	$\dfrac{1}{2}$	$\dfrac{\sqrt{3}}{2}$
$45°$	$\dfrac{1}{\sqrt{2}}$	$\dfrac{1}{\sqrt{2}}$
$60°$	$\dfrac{\sqrt{3}}{2}$	$\dfrac{1}{2}$
$90°$	1	0

8. Use a calculator to find a value of θ between $0°$ and $90°$ that satisfies the statement below. Write your answer in degrees and minutes rounded to the nearest minute.

$$\csc\theta = 7.5493$$

$$\theta =$$

9. Refer to right triangle ABC with $C = 90°$. Solve for all the missing parts using the given information.

$$b = 377.6 \text{ inches}, \quad c = 588.7 \text{ inches}$$

Please round the answers for the angles to two decimal places, and for the side, to one decimal place.

$A = $ _________ °

$B = $ _________ °

$A = $ _________ inches

10. A person standing 160 centimeters from a mirror notices that the angle of depression from his eyes to the bottom of the mirror is $13°$, while the angle of elevation to the top of the mirror is $12°$. Find the vertical dimension of the mirror (see the figure). Please round your answer to the nearest centimeter.

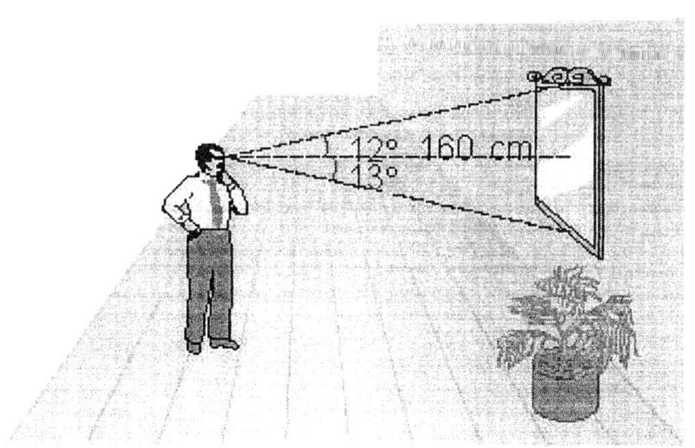

__________ cm

11. The problem refers to a vector **V** with magnitude | **V** | that forms an angle with the positive x-axis. Give the magnitude of the horizontal and vertical vector components of **V**, namely V_x and V_y, respectively. Round your answers to the nearest integer.

$$|\mathbf{V}| = 60,\ \theta = 150°$$

$$\left|\mathbf{V}_x\right| = \underline{\hspace{2cm}}$$

$$\left|\mathbf{V}_y\right| = \underline{\hspace{2cm}}$$

12. The horizontal and vertical components of the velocity of an arrow shot into the air are 15.0 feet per second and 27.0 feet per second, respectively. Find the velocity of the arrow. Round your answers to the nearest tenth.

__________ ft/sec

13. Use a calculator to find $\sec 100.9^\circ$.

Please round the answer to the nearest ten-thousandth.

____

14. Convert to degree measure.

$$\theta = \frac{8\pi}{9}$$

$\theta =$ __________ $^\circ$

Draw the angle in standard position.

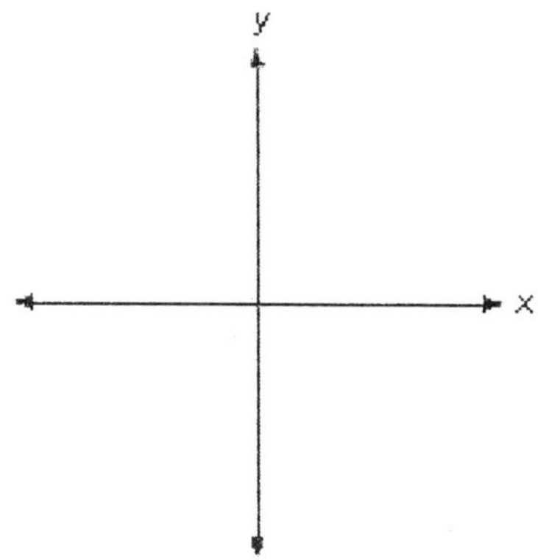

Label the reference angle in both degrees and radians.

15. If we start at the point (1, 0) and travel once around the unit circle, we travel a distance of 2π units and arrive back where we started at the point (1, 0). If we continue around the unit circle a second time, we will repeat all the values of x and y that occurred during our first trip around. Use this discussion to evaluate the expression.

$$\sin\left(2\pi + \frac{\pi}{2}\right)$$

16. The pendulum on a grandfather clock swings from side to side once every second. If the length of the pendulum is 5 feet and the angle through which it swings is $16°$, how far does the tip of the pendulum travel in 1 second?

If the answer needs rounding, round it to three significant digits.

__________ feet

17. Find the area of the sector formed by central angle $\theta = 2.2$ in a circle of radius $r = 5$ inches.

If the answer needs rounding, round it to three significant digits.

$A =$ __________ in^2

18. Point P moves with angular velocity w on a circle of radius r. Find the distance s traveled by the point in time t.

$w = 5 \text{ rad/sec}, r = 3 \text{ ft}, t = 5 \text{ min}$

If the answer needs rounding, round it to three significant digits.

__________ ft

19.

Graph one complete cycle of $y = \dfrac{3}{10} \cos x$. Label the axes accurately.

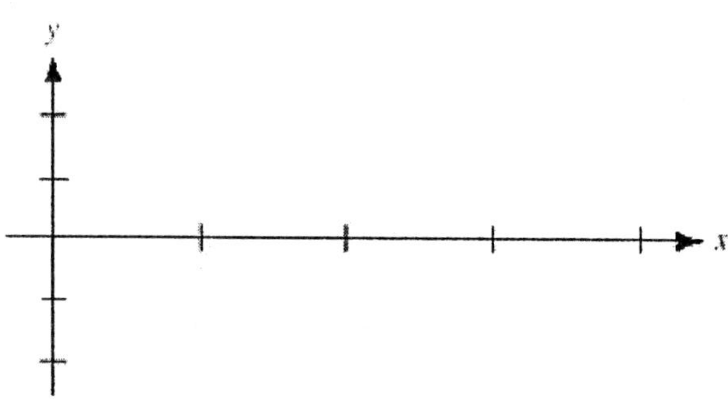

Identify the amplitude for the graph (if defined)

20. Graph one complete cycle for the function. Label the axes so that the amplitude (if defined) and period are easy to read.

$$y = 2 \csc \dfrac{1}{2} x$$

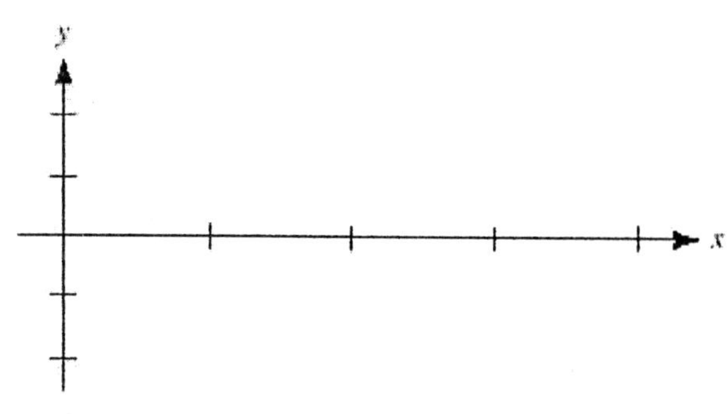

21. Identify the amplitude for the equation.

$$y = 2\sin\left(\pi x + \frac{\pi}{3}\right)$$

____

Identify the period for the equation.

Identify the phase shift for the equation.

Label the axes accordingly and sketch one complete cycle of the curve.

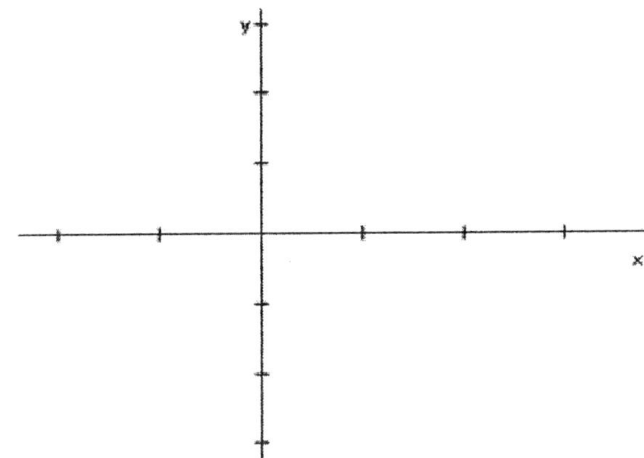

22.

Use the graph of the equation $y = -\sin\left(2x + \dfrac{\pi}{2}\right)$ shown below to graph one complete cycle of the equation $y = 2 - \sin\left(2x + \dfrac{\pi}{2}\right)$.

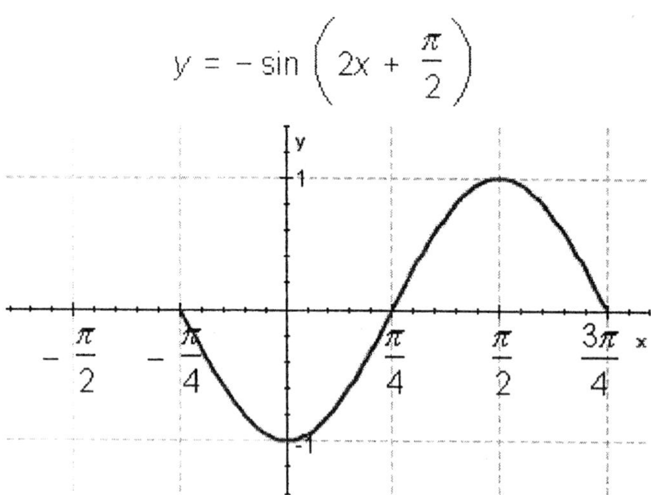

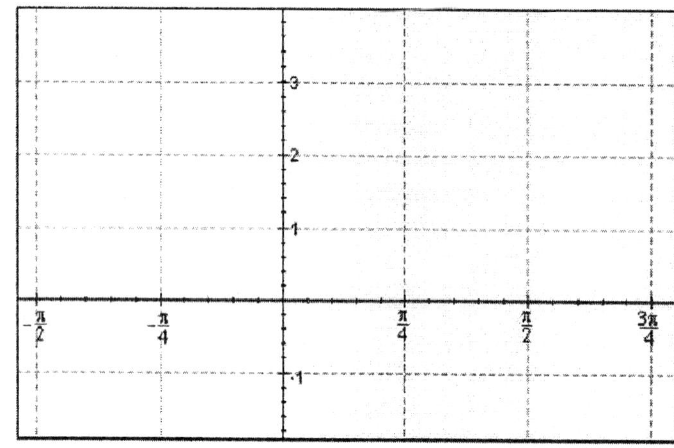

23. The graph below is one complete cycle of the graph of an equation containing a trigonometric function. Find an equation to match the graph.

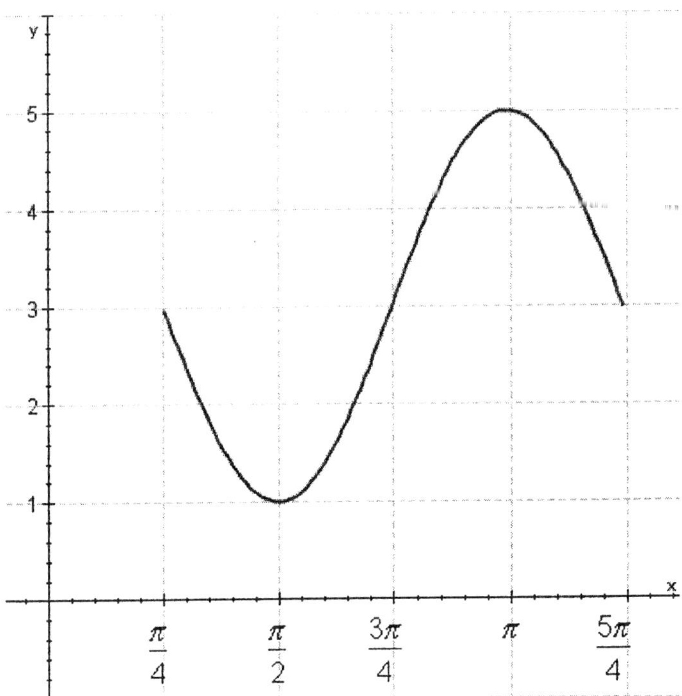

24. Sketch the graph from $x = 0$ to $x = 4\pi$.

$y = 3\sin x + \sin 2x$

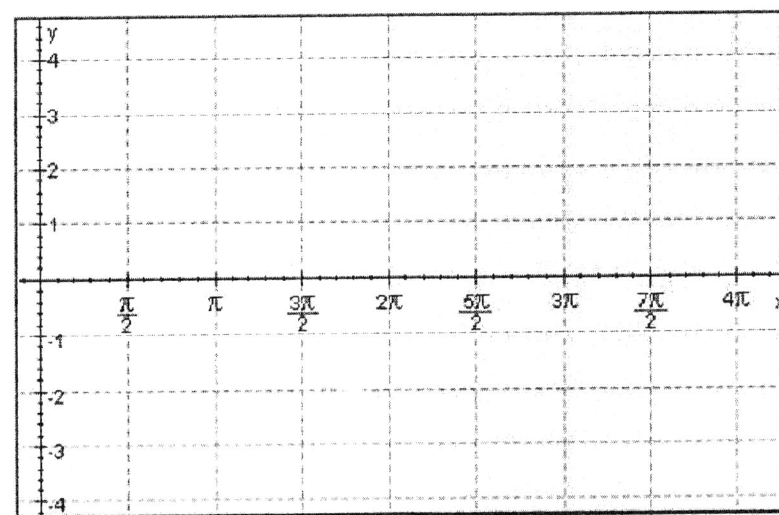

25. Evaluate without using a calculator.

$$\sin^{-1}\left(\sin 210^{\circ}\right)$$

26. Prove that the identity is true.

$$\frac{1 + \sec x}{1 - \sec x} = \frac{\cos x + 1}{\cos x - 1}$$

27. Match the expression with a single trigonometric function.

$\cos 9x \cos 4x - \sin 9x \sin 4x$ $\cos 13x$

$\cos 5x \cos x - \sin 5x \sin x$ $\cos 6x$

28. Let $\cos A = -\dfrac{2}{\sqrt{13}}$ with A in QII and find $\tan 2A$.

29. Graph the function from $x = 0$ and $x = 2\pi$.

$$y = 4\cos^2 2x - 2$$

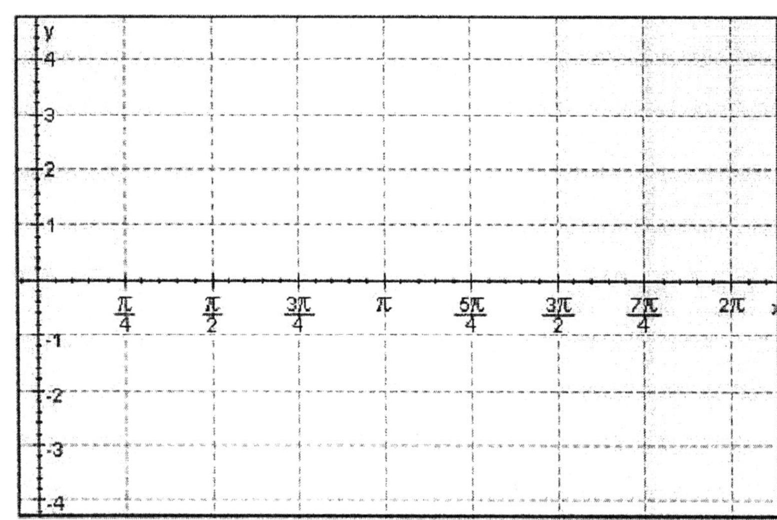

30. If $\sin B = -\dfrac{1}{6}$ with B in QIII, find

$$\tan \dfrac{B}{2}$$

31. Evaluate the expression without using a calculator. (Assume the variable represents a positive number.)

$$\cos\left(2\cos^{-1}2x\right)$$

32. Solve for θ if $0° \leq \theta < 360°$.

$$2\cos^2\theta + 3\cos\theta = 2$$

33. Solve the equation for x if $0 \leq x < 2\pi$. Give your answer in radians using exact values only.

$$2\cos^2 x - \sin x - 1 = 0$$

34. Write an expression that gives all solutions to the equation.

$$\sin x - \cos x = \sqrt{2}$$

35. Find all degree solutions.

$$2\sin^2 10\theta + 3\sin 10\theta + 1 = 0$$

36. Eliminate the parameter t from the following:

$x = \cos t - 4 \quad y = \sin t + 7$

Enter your answer as an equation in standard form if possible.

Sketch the graph.

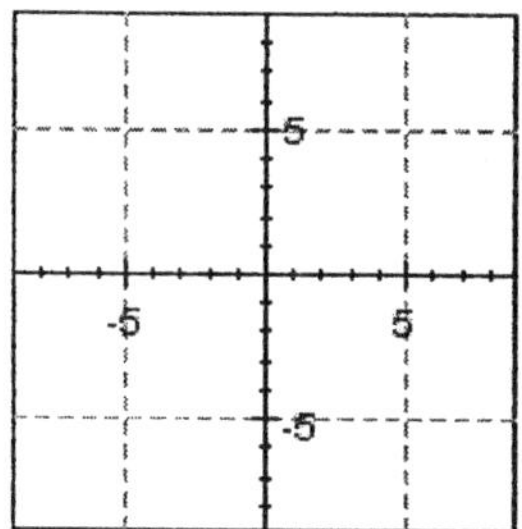

37. The information below refers to triangle ABC.

$B = 58°, C = 31°, a = 7.5$ m.

Find A.

$A = \underline{\hspace{2cm}}$ °

Find b. Please round the answer to a tenth of a meter.

$b = \underline{\hspace{2cm}}$ m

Find c. Please round the answer to a tenth of a meter.

$c = \underline{\hspace{2cm}}$ m

38. Find all solutions to the triangle described below.

$A = 64°, b = 7.7$ yd, $a = 7.1$ yd

Please round each angle to the nearest degree and the length of the missing side to the nearest tenth.

If there is more than one possible triangle, list both.

39. Solve the triangle.

$A = 48$ yd, $b = 75$ yd, $c = 60$ yd

Please round your answers to the nearest integer. Do not round any numbers until the end of each computation.

$A = $ _________ $^\circ$

$B = $ _________ $^\circ$

$C = $ _________ $^\circ$

40. Referring to triangle *ABC,* find the triangle's area given

$A = 44^\circ\ 30'$, $C = 115^\circ\ 30'$, $a = 3.46$ ft

Please round your answer to the nearest hundredth.

$S = $ _________ ft^2

41. For the pair of vectors, find $U + V$, $U - V$, and $2U - 3V$.

$U = \langle 3, 3 \rangle$, $V = \langle 3, -3 \rangle$

$U + V = \langle$ _________ , _________ $\rangle$

$U - V = \langle$ _________ , _________ $\rangle$

$2U - 3V = \langle$ _________ , _________ $\rangle$

42. For the pair of vectors, find $U \cdot V$.

$U = 6i + 2j \quad V = 2i + 6j$

$U \cdot V = $ _________

43. Find the angle θ between the given vectors to the nearest tenth of a degree.

$$U = 12i + 7j \quad V = -3i + 5j$$

$\theta = $ _________ °

44. Let $z_1 = 5 + 6i$ and $z_2 = 5 - 6i$ and find $z_1 z_2$.

Enter your answer in standard form for complex numbers.

$z_1 z_2 = $ _________

45. Write the complex number in trigonometric form.

$1 - i$

Please enter the argument in degrees and use the degree symbol.

46. Use DeMoivre's Theorem to find the following. Write your answer in standard form.

$$\left(\sqrt{2} \text{ cis } 81° \right)^5$$

47. Find two square roots for the complex number. Write your answer in standard form.

$$8 - 8i\sqrt{3}$$

48. Convert to rectangular coordinates. Use exact values.

$$\left(5\sqrt{2},\ -45^\circ\right)$$

____

49. Write the equation in polar coordinates.

$$x^2 + y^2 = 2$$

50. Graph the equation.

$$r = 4 + 5\cos\theta$$

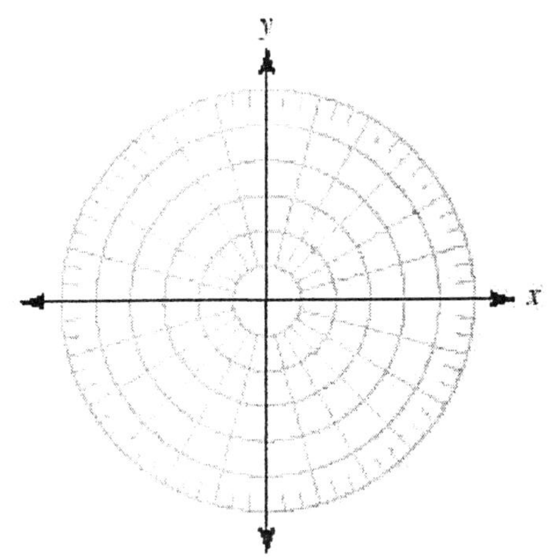

McKeague/Turner - Trigonometry 5e Final Exam Form A

1. 12

2.

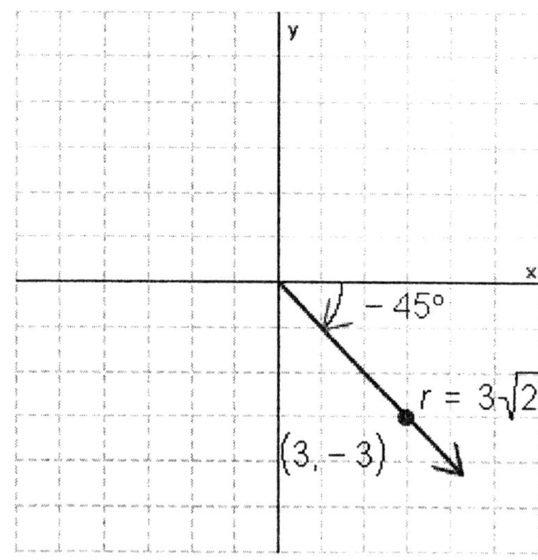

Coterminal angle is $315°$.

3.

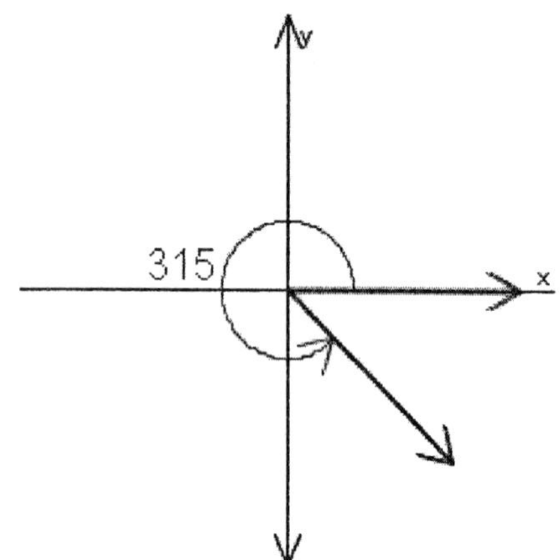

$(1, -1)$ for example

$-\dfrac{\sqrt{2}}{2}$

$\dfrac{\sqrt{2}}{2}$

-1

4. $\cot(\theta) = \dfrac{6}{5}$

5. $\cos(\theta)$

6. We begin by writing everything on the left side in terms of $\sin\alpha$ and $\cos\alpha$.

$$\sin\alpha \cot\alpha = \sin\alpha\,\frac{\cos\alpha}{\sin\alpha} = \frac{\sin\alpha\,\cos\alpha}{\sin\alpha} = \cos\alpha$$

Since we have succeeded in transforming the left side into the right side, we have shown that the statement $\sin\alpha \cot\alpha = \cos\alpha$ is an identity.

7. 1

8. 7
   37

9. 50.10
   39.90
   451.6

10. 71

11. 52
    30

12. 30.9

13. -5.2883

14. 160

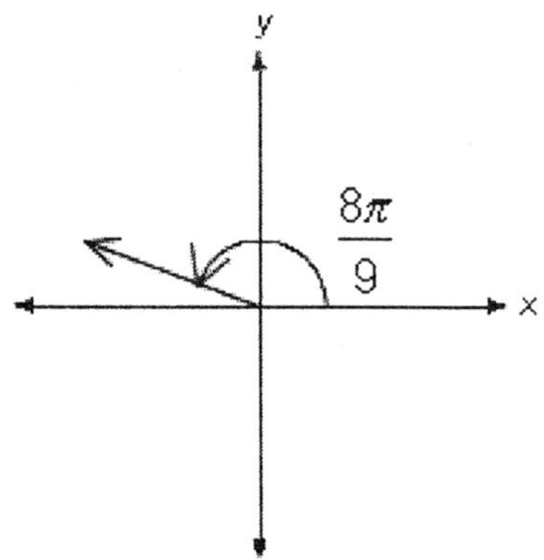

$\dfrac{\pi}{9}$, $20°$

15. 1

16. 1.40

17. 27.5

18. 4.500

19.

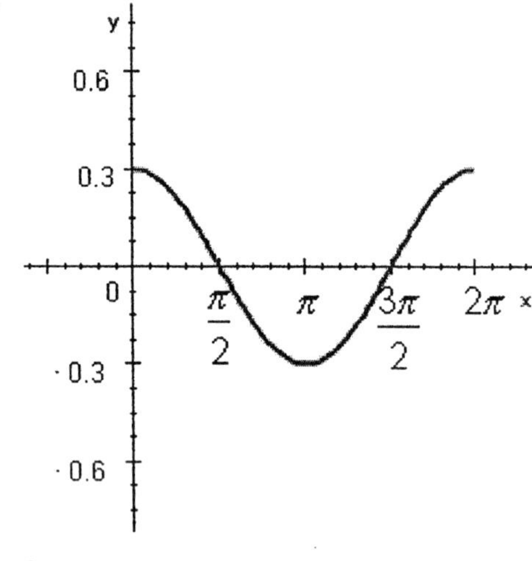

$\dfrac{3}{10}$

20.

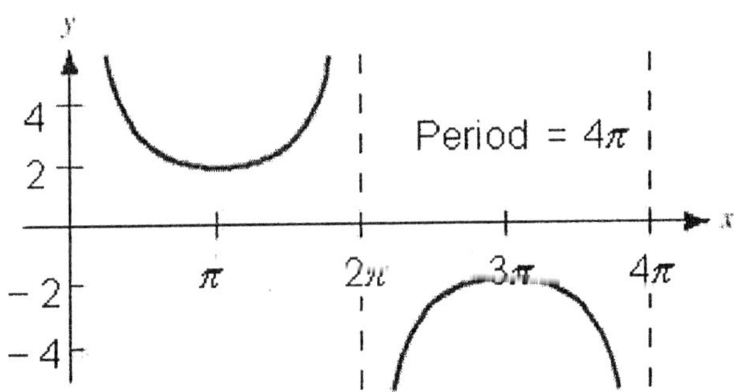

21. 2

2

$-\dfrac{1}{3}$

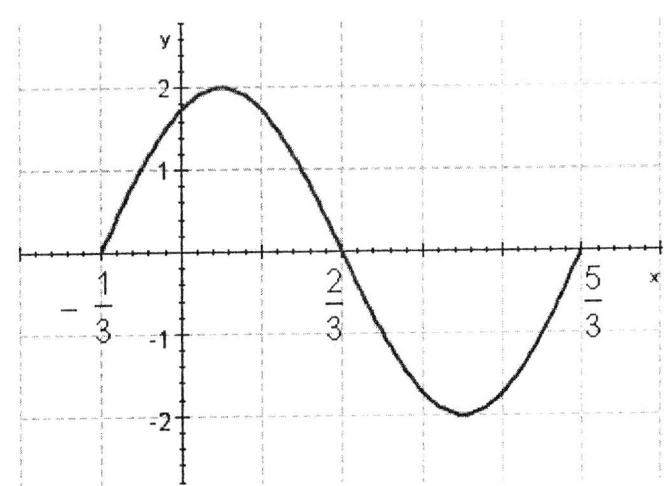

22.

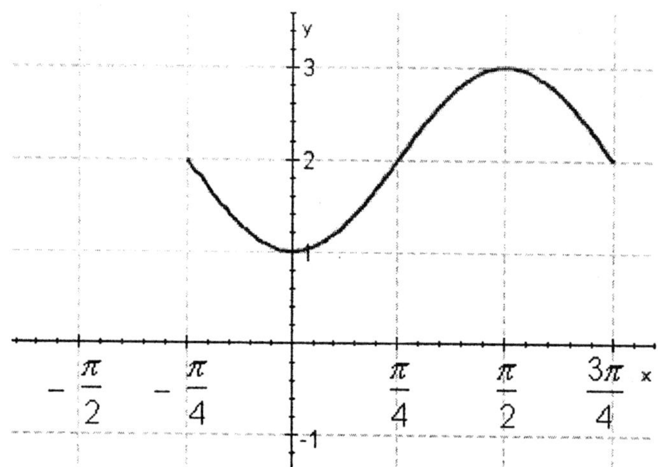

23.

$$y = 3 - 2\sin\left(2x - \frac{\pi}{2}\right)$$

24.

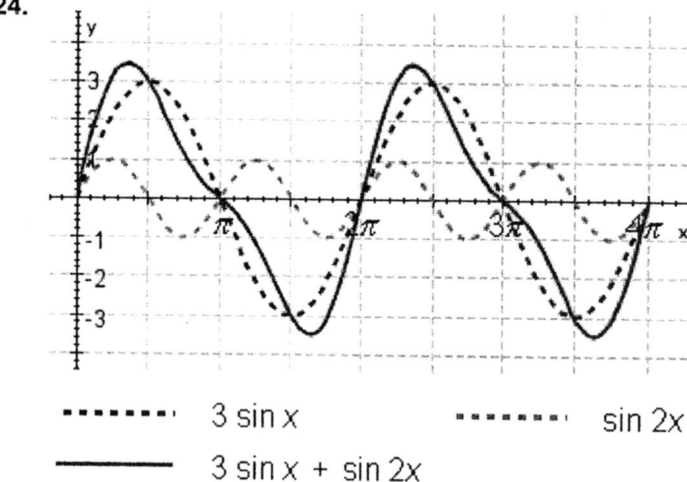

25. -30°

26.

$$\frac{1 + \sec x}{1 - \sec x} = \frac{1 + \dfrac{1}{\cos x}}{1 - \dfrac{1}{\cos x}}$$

$$= \frac{\dfrac{\cos x + 1}{\cos x}}{\dfrac{\cos x - 1}{\cos x}} = \frac{\cos x + 1}{\cos x - 1}$$

27. $\cos 5x \cos x - \sin 5x \sin x \rightarrow \cos 6x$,

$\cos 9x \cos 4x - \sin 9x \sin 4x \rightarrow \cos 13x$

28. $\dfrac{12}{5}$

29.

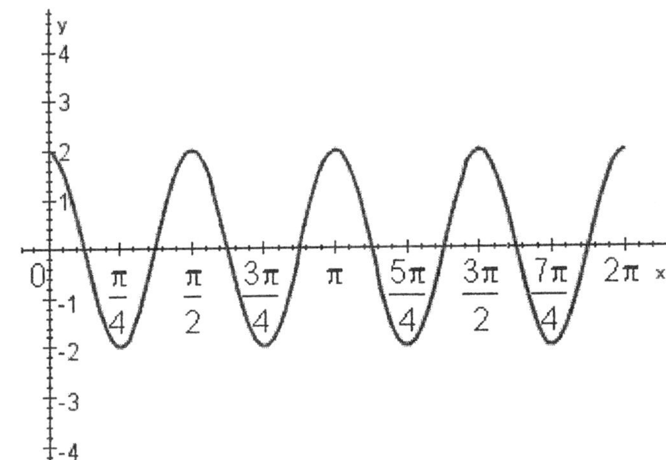

30. $-6 - \sqrt{35}$

31. $8x^2 - 1$

32. $60°, 300°$

33. $\dfrac{\pi}{6}, \dfrac{5\pi}{6}, \dfrac{3\pi}{2}$

34. $\dfrac{3\pi}{4} + 2k\pi$

35. $\theta = 21° + 36°k$ or $\theta = 27° + 36°k$ or $\theta = 33° + 36°k$

36. $(x+4)^2+(y-7)^2=1$

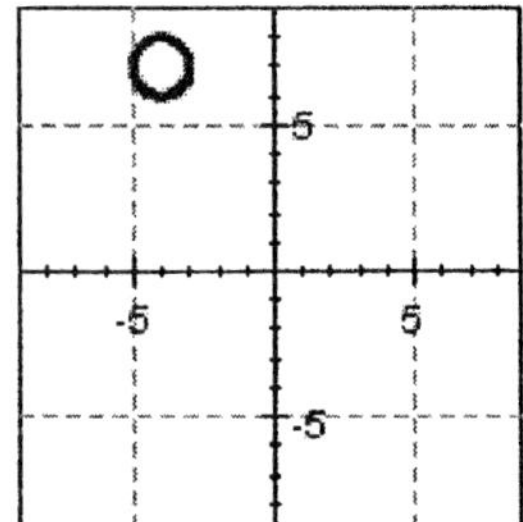

37. 91

   6.4

   3.9

38. $(77°, 39°, 5), (103°, 13°, 1.8)$

39. 40

   87

   53

40. 2.64

41. 6

   0

   0

   6

   -3

   15

42. 24

43. 90.7

44. 61

45. $\sqrt{2}\cdot(\cos(315°)+i\cdot\sin(315°))$

46. $4 + 4i$

47. $-2\sqrt{3} + 2i,\ 2\sqrt{3} - 2i$

48. $\left(5,\ -5\right)$

49. $r^2 - 2$

50.

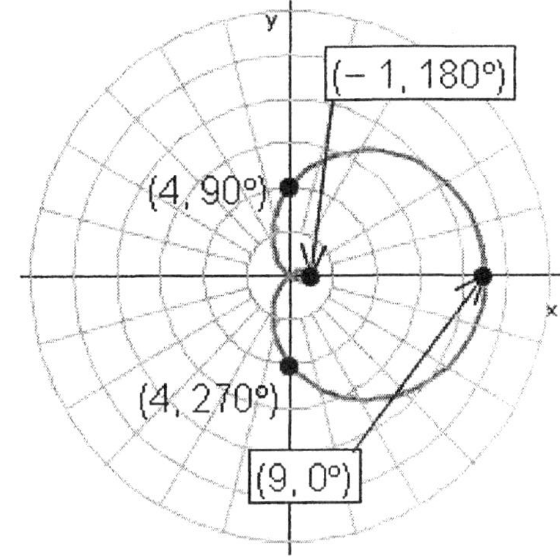

McKeague/Turner - Trigonometry 5e Final Exam Form A

---

1. mctr.01.01.27_NoAlgs
2. mctr.01.02.77_NoAlgs
3. mctr.01.03.30_NoAlgs
4. mctr.01.04.18_NoAlgs
5. mctr.01.05.25_NoAlgs
6. mctr.01.05.59_NoAlgs
7. mctr.02.01.32_NoAlgs
8. mctr.02.02.74_NoAlgs
9. mctr.02.03.32_NoAlgs
10. mctr.02.04.11_NoAlgs
11. mctr.02.05.17_NoAlgs
12. mctr.02.05.28_NoAlgs
13. mctr.03.01.31_NoAlgs
14. mctr.03.02.33_NoAlgs
15. mctr.03.03.43_NoAlgs
16. mctr.03.04.15_NoAlgs
17. mctr.03.04.41_NoAlgs
18. mctr.03.05.28_NoAlgs
19. mctr.04.01.53_NoAlgs
20. mctr.04.02.28_NoAlgs
21. mctr.04.03.13_NoAlgs
22. mctr.04.03.27_NoAlgs
23. mctr.04.04.30_NoAlgs
24. mctr.04.05.14_NoAlgs
25. mctr.04.06.52_NoAlgs
26. mctr.05.01.37_NoAlgs
27. mctr.05.02.23_NoAlgs
28. mctr.05.03.07_NoAlgs
29. mctr.05.03.19_NoAlgs
30. mctr.05.04.14_NoAlgs
31. mctr.05.05.11_NoAlgs
32. mctr.06.01.31_NoAlgs
33. mctr.06.02.17_NoAlgs
34. mctr.06.02.42_NoAlgs
35. mctr.06.03.36_NoAlgs

---

36. mctr.06.04.10_NoAlgs
37. mctr.07.01.14_NoAlgs
38. mctr.07.02.18_NoAlgs
39. mctr.07.03.12_NoAlgs
40. mctr.07.04.11_NoAlgs
41. mctr.07.05.29_NoAlgs
42. mctr.07.06.09_NoAlgs
43. mctr.07.06.16_NoAlgs
44. mctr.08.01.65_NoAlgs
45. mctr.08.02.35_NoAlgs
46. mctr.08.03.24_NoAlgs
47. mctr.08.04.08_NoAlgs
48. mctr.08.05.24_NoAlgs
49. mctr.08.05.59_NoAlgs
50. mctr.08.06.17_NoAlgs

1.
Suppose ABC is a right triangle with $C = 90^\circ$.

If $a = 4$ and $c = 5$, find b.

Select the correct answer.

   a. $b = 14$

   b. $b = 12$

   c. $b = 3$

   d. $b = 6$

   e. $b = 4$

2.   Name another angle that is coterminal with -60°.

Select the correct answer(s).

   a. 300°

   b. 60°

   c. 420°

   d. -420°

   e. -300°

3.   Find the right part of the identity.

$\cos\alpha \tan\alpha =$ ______

Select the correct answer.

   a. $\cos\alpha$

   b. $\sec\alpha$

   c. $\tan\alpha$

   d. $\csc\alpha$

   e. $\sin\alpha$

4. Draw the following angle in standard position; find the sine, the cosine, and the tangent of the angle below.

$315°$

Select the correct answer.

a.

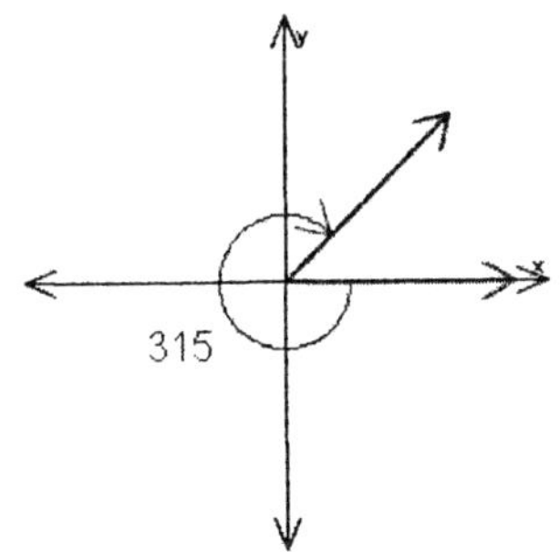

$$\sin 315° = -\frac{\sqrt{2}}{2}, \ \cos 315° = -1, \ \tan 315° = \frac{\sqrt{2}}{2}$$

b.

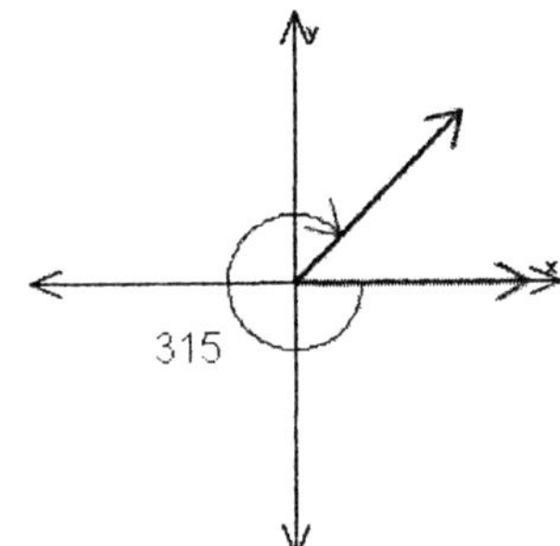

$$\sin 315° = \frac{\sqrt{2}}{2}, \ \cos 315° = -\frac{\sqrt{2}}{2}, \ \tan 315° = -1$$

c.

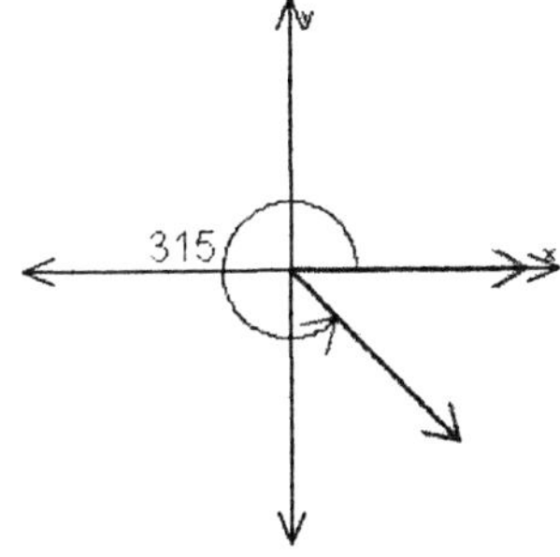

$$\sin 315° = -\frac{\sqrt{2}}{2}, \ \cos 315° = -1, \ \tan 315° = \frac{\sqrt{2}}{2}$$

d.

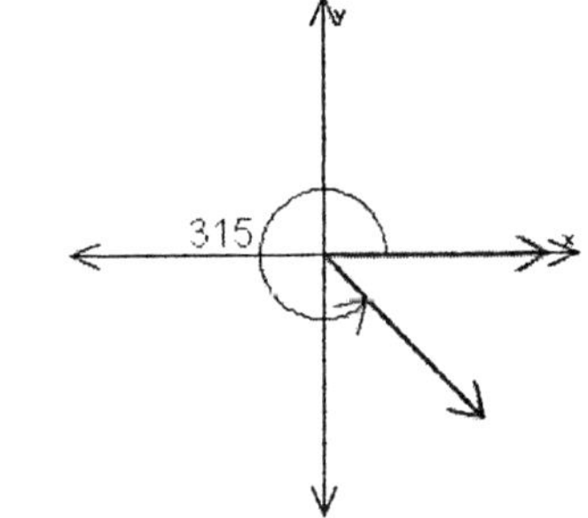

$$\sin 315° = -\frac{\sqrt{2}}{2}, \ \cos 315° = \frac{\sqrt{2}}{2}, \ \tan 315° = -1$$

5. Use a ratio identity to find $\cot\theta$ if

$$\sin\theta = \frac{9}{\sqrt{181}} \quad \text{and} \quad \cos\theta = \frac{10}{\sqrt{181}}.$$

Select the correct answer.

a. $\cot\theta = \dfrac{9}{10\sqrt{181}}$

00

b. $\cot\theta = \dfrac{10}{9\sqrt{181}}$

c. $\cot\theta = \dfrac{10}{9}$

d. $\cot\theta = \dfrac{9}{10}$

e. $\cot\theta = \dfrac{10\sqrt{181}}{9}$

6. Write the following in terms of $\sin\theta$ and $\cos\theta$ and then simplify if possible.

$$\csc\theta - \sin\theta\cot^2\theta$$

Select the correct answer.

a. $\sin\theta$

b. $\dfrac{\sin\theta}{\cos\theta}$

c. $2\sin\theta$

d. $\dfrac{\cos\theta}{\sin\theta}$

e. $\cos\theta$

7. Simplify expression by first substituting values from the table of exact values and then simplifying the resulting expression.

$$\left(\sin 60^\circ + \cos 60^\circ\right)^2$$

Table of Exact Values

x	$\sin x$	$\cos x$
0°	0	1
30°	$\dfrac{1}{2}$	$\dfrac{\sqrt{3}}{2}$
45°	$\dfrac{1}{\sqrt{2}}$	$\dfrac{1}{\sqrt{2}}$
60°	$\dfrac{\sqrt{3}}{2}$	$\dfrac{1}{2}$
90°	1	0

Select the correct answer.

a. $\dfrac{2 + \sqrt{3}}{2}$

b. $\dfrac{2 + \sqrt{2}}{2}$

c. $\dfrac{1 - \sqrt{3}}{2}$

d. $\dfrac{2 - \sqrt{2}}{2}$

e. $\dfrac{2 - \sqrt{3}}{2}$

8. Use a calculator to find a value of θ between $0°$ and $90°$ that satisfies the statement below.

   $$\csc \theta = 9.4856$$
   00

   Select the correct answer.

   a. $\theta = 7°03'$

   b. $\theta = 6°05'$

   c. $\theta = 3°03'$

   d. $\theta = 7°05'$

   e. $\theta = 6°03'$

9. Refer to right triangle ABC with $C = 90°$. Solve for all the missing parts using the given information.

   $b = 377.6$ inches, $c = 588.5$ inches

   Please round the answers for the angles to two decimal places, and for the side, to one decimal place.

   Select the correct answer.

   a. $A = 50.09°$, $B = 39.91°$, $a = 451.2$ in

   b. $A = 50.13°$, $B = 39.87°$, $a = 451.8$ in

   c. $A = 50.09°$, $B = 39.91°$, $a = 451.4$ in

   d. $A = 50.11°$, $B = 39.89°$, $a = 451.2$ in

   e. $A = 50.11°$, $B = 39.89°$, $a = 451.4$ in

10. A person standing 130 centimeters from a mirror notices that the angle of depression from his eyes to the bottom of the mirror is 14°, while the angle of elevation to the top of the mirror is 12°. Find the vertical dimension of the mirror (see the figure). Please round your answer to the nearest centimeter.

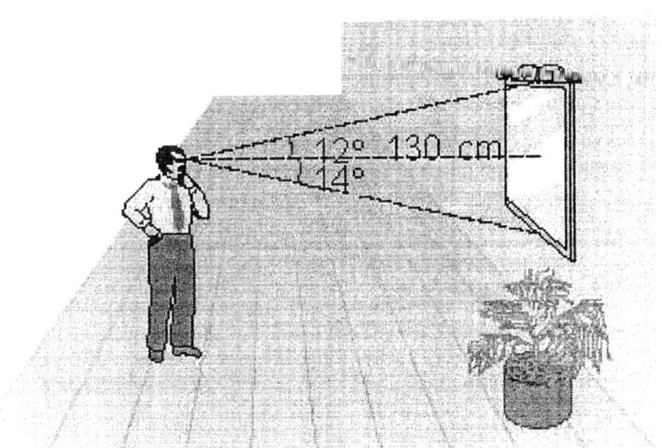

Select the correct answer.

a. 60 cm
b. 62 cm
c. 50 cm
d. 57 cm
e. 61 cm

11. The problem refers to a vector **V** with magnitude | **V** | that forms an angle with the positive x-axis. Give the magnitude of the horizontal and vertical vector components of **V**, namely V_x and V_y, respectively. Round your answers to the nearest integer.

$$|\mathbf{V}| = 66,\ \theta = 120^\circ$$

Select the correct answer.

a. $\left|\mathbf{V}_x\right| = -36,\ \left|\mathbf{V}_y\right| = -56$

b. $\left|\mathbf{V}_x\right| = -29,\ \left|\mathbf{V}_y\right| = 57$

c. $\left|\mathbf{V}_x\right| = 33,\ \left|\mathbf{V}_y\right| = 59$

d. $\left|\mathbf{V}_x\right| = 33,\ \left|\mathbf{V}_y\right| = 57$

e. $\left|\mathbf{V}_x\right| = 29,\ \left|\mathbf{V}_y\right| = 59$

12. The horizontal and vertical components of the velocity of an arrow shot into the air are 18.0 feet per second and 21.0 feet per second, respectively. Find the velocity of the arrow. Please round your answers to the nearest tenth.

    Select the correct answer.

    a. 32.3 ft/sec
    b. 36.1 ft/sec
    c. 25 ft/sec
    d. 27.7 ft/sec
    e. 41 ft/sec

13. Use a calculator to find $\sec 101.4^\circ$.

    Please round the answer to the nearest ten-thousandth.

    Select the correct answer.

    a. -5.0593
    b. -5.1093
    c. -5.0594
    d. -5.0893
    e. -5.0373

14. Convert to degree measure.

$$\theta = \frac{8\pi}{9}$$

    Select the correct answer.

    a. 135°
    b. 140°
    c. 160°
    d. 150°
    e. 170°
       00

15. If we start at the point (1, 0) and travel once around the unit circle, we travel a distance of 2π units and arrive back where we started at the point (1, 0). If we continue around the unit circle a second time, we will repeat all the values of x and y that occurred during our first trip around. Use this discussion to evaluate the expression.

$$\sin\left(2\pi - \frac{\pi}{2}\right)$$

Select the correct answer.

a.   1

b. $\dfrac{1}{2}$

c.   0

d. $\dfrac{\sqrt{2}}{2}$

e. -1

16. The pendulum on a grandfather clock swings from side to side once every second. If the length of the pendulum is 4 feet and the angle through which it swings is 21°, how far does the tip of the pendulum travel in 1 second?

Select the correct answer.

a.   1.67 feet
b.   1.87 feet
c.   1.47 feet
d.   1.77 feet
e.   1.57 feet

17. Find the area of the sector formed by central angle $\theta = 2.4$ in a circle of radius $r = 6$ inches.

Select the correct answer.

a.   43.4 inches^2

b.   42.9 inches^2

c.   43.2 inches^2

d.   43.6 inches^2

e.   43.1 inches^2

18. Point P moves with angular velocity ω on a circle of radius r. Find the distance s traveled by the point in time t.

$$\omega = 10 \text{ rad/sec}, \ r = 5 \text{ ft}, \ t = 3 \text{ min}$$

Select the correct answer.

a.   9,500 ft
b.   5,790 ft
c.   5,400 ft
d.   9,000 ft
e.   6,300 ft

19.

Graph one complete cycle of $y = \dfrac{2}{5}\cos x$. Label the axes accurately.

Select the correct answer.

a.

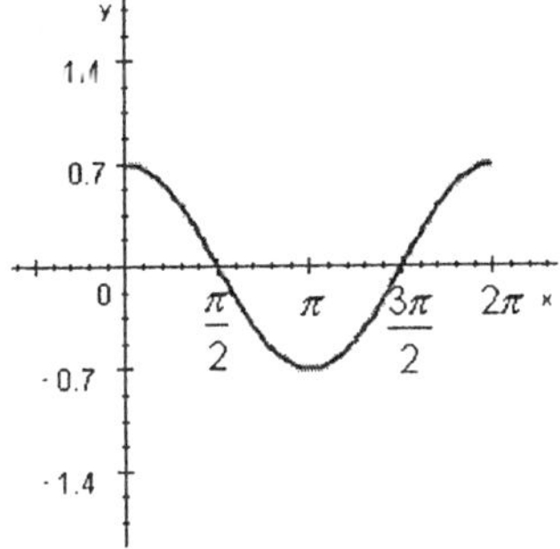

b.

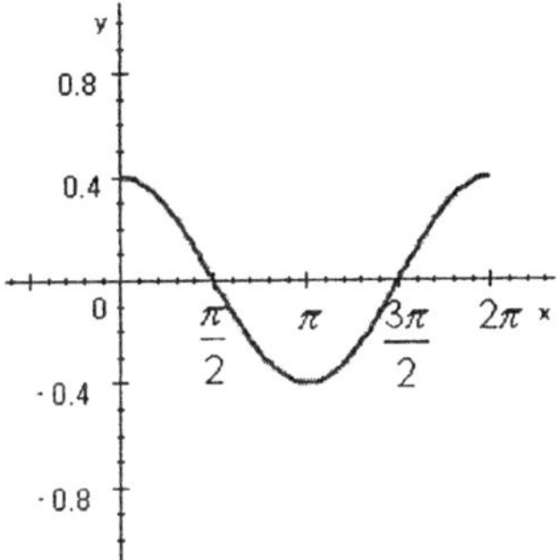

c.

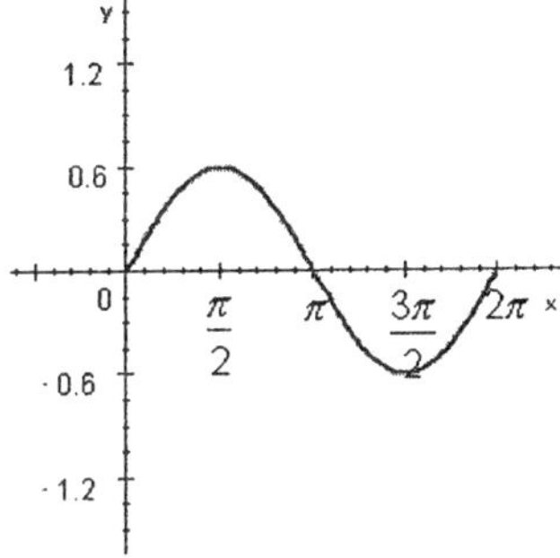

d.

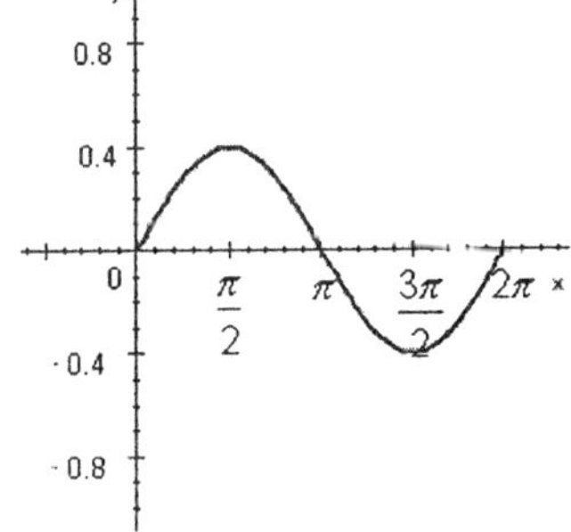

20. Graph one complete cycle for the function. Label the axes so that the amplitude (if defined) and period are easy to read.

$$y = 4\csc \frac{1}{6} x$$

Select the correct answer.

a.

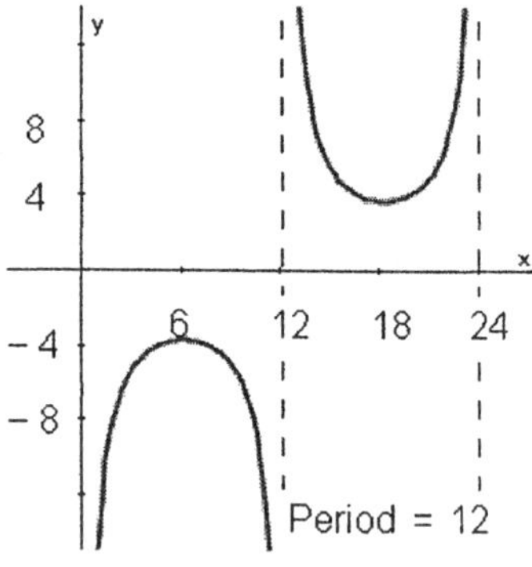

b.

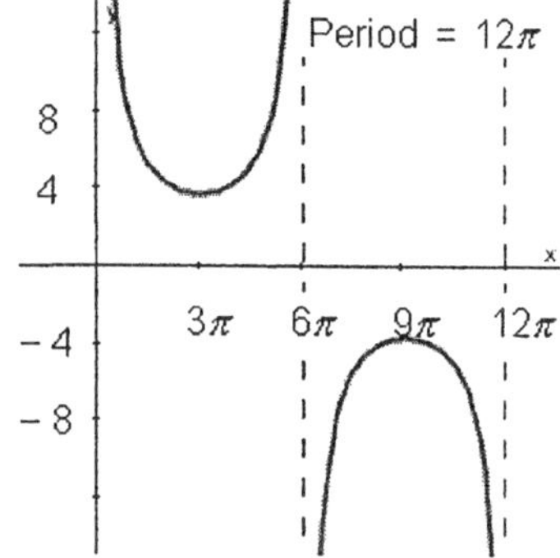

c.

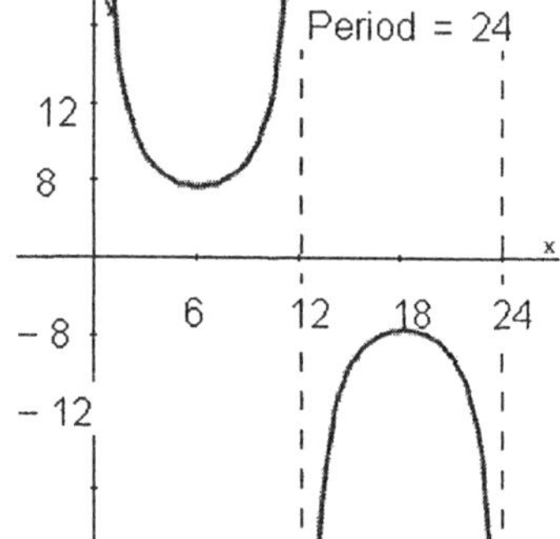

21. Identify the period for the equation. Then label the axes accordingly and sketch one complete cycle of the curve.

$$y = 4\sin\left(3\pi x + \frac{\pi}{3}\right)$$

Select the correct answer.

a.

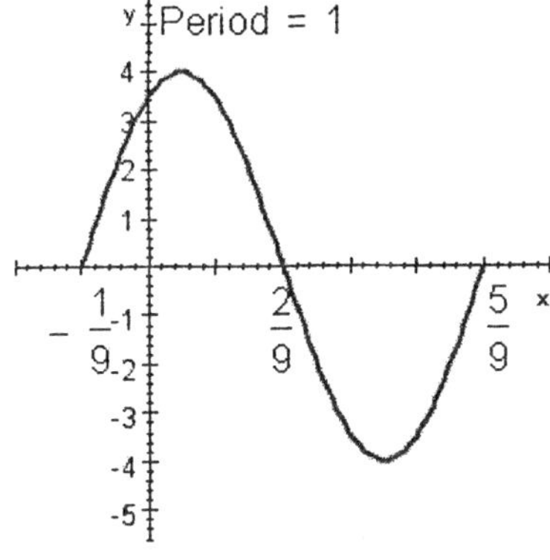

b.

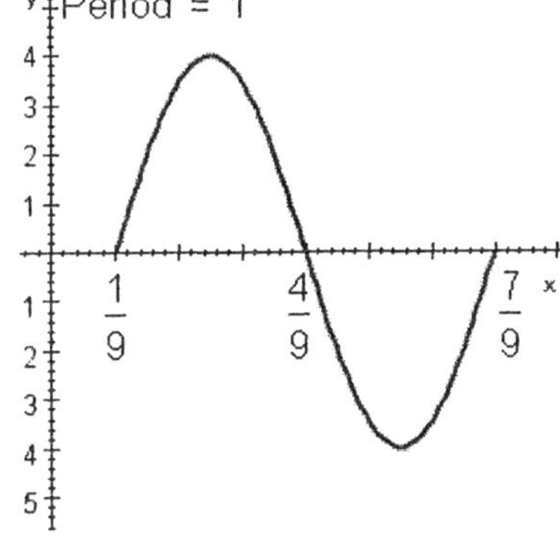

c.

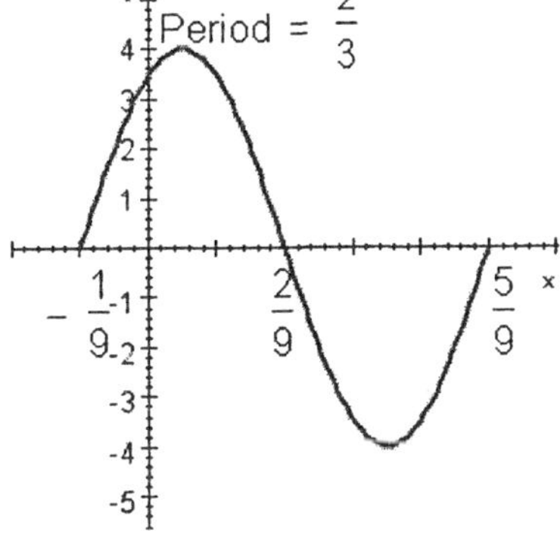

22.

Use the graph of the equation $y = -\cos\left(2x + \dfrac{\pi}{2}\right)$ shown below to graph one complete cycle of the equation $y = 3 - \cos\left(2x + \dfrac{\pi}{2}\right)$.

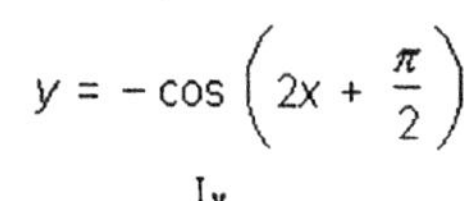

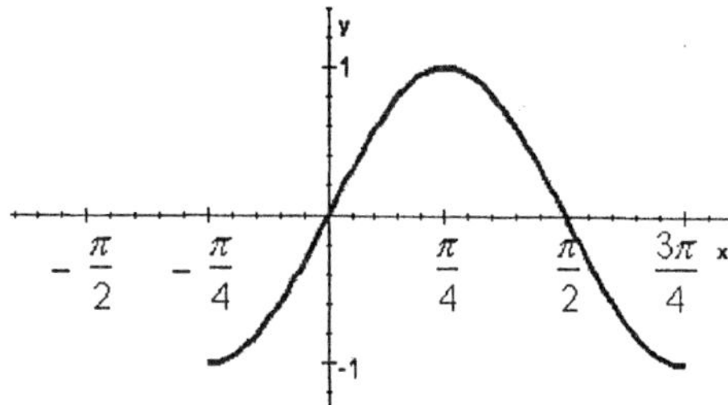

Select the correct answer.

a.

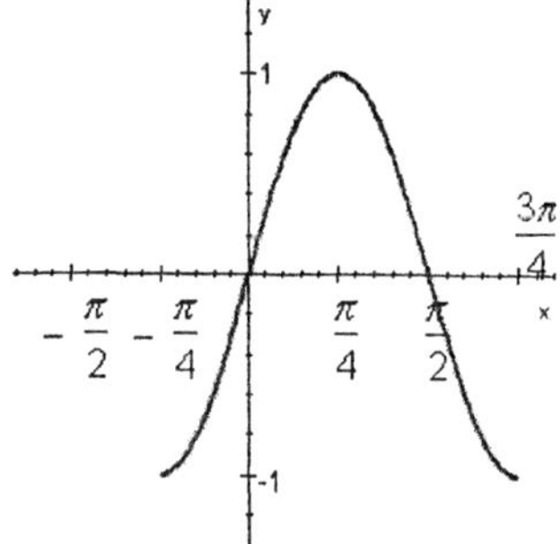

b.

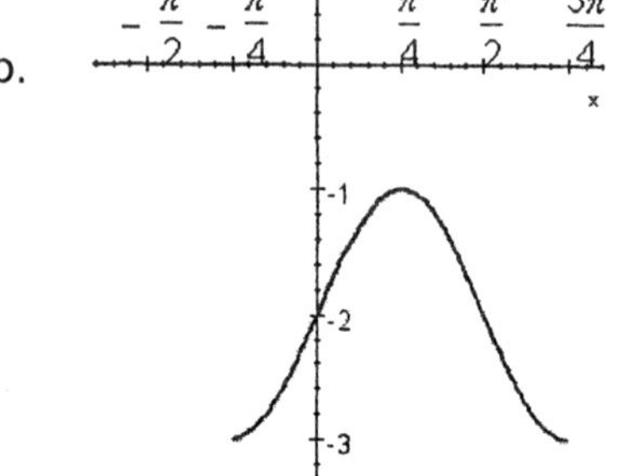

c.

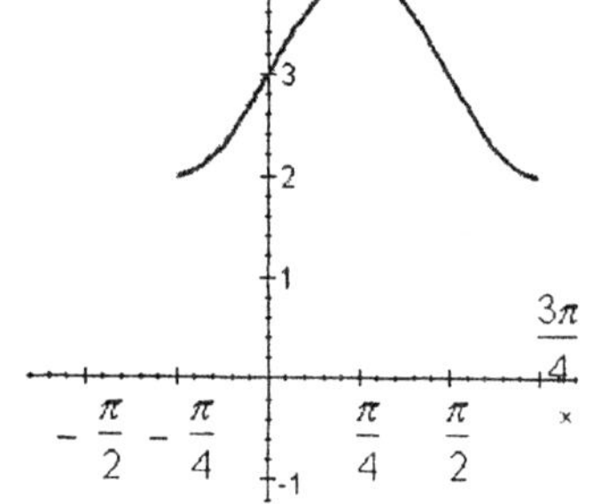

25. Evaluate without using a calculator.

$$\sin^{-1}\left(\sin 330^\circ\right)$$

Select the correct answer.

a. -45°

b. -180°

c. -90°

d. -30°

e. -150°

26. Find the identical expression for the following.

$$\frac{\sec x + 1}{\sec x - 1}$$

Select the correct answer.

a. $\dfrac{\cos x + 1}{\cos x - 1}$

b. $\dfrac{1 + \sin x}{1 - \sin x}$

c. 0

d. $\dfrac{1 + \cos x}{1 - \cos x}$

e. $\dfrac{\sin x + 1}{\sin x - 1}$

27. Write the expression as a single trigonometric function.

$\cos 5x \cos x - \sin 5x \sin x$

Select the correct answer.

a. $\cos 2x$
b. $\cos 8x$
c. $\cos 6x$
d. $\cos 9x$
e. $\cos 7x$

28.

Let $\cos A = \dfrac{1}{\sqrt{5}}$ with A in QIV and find $\tan 2A$.

Select the correct answer.

a. $\dfrac{4}{3}$

b. $-\dfrac{3}{4}$

c. $\dfrac{4}{9}$

d. $\dfrac{8}{9}$

e. $-\dfrac{2}{3}$

29. Graph the function from $x = 0$ and $x = 2\pi$.

$$y = 6\cos^2\frac{x}{2} - 3$$

Select the correct answer.

a.

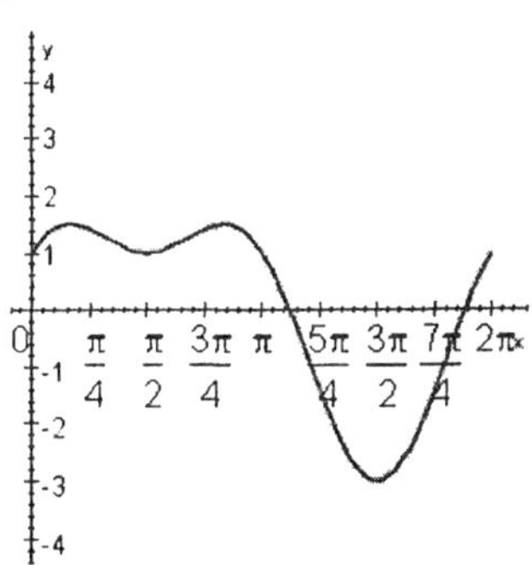

b.

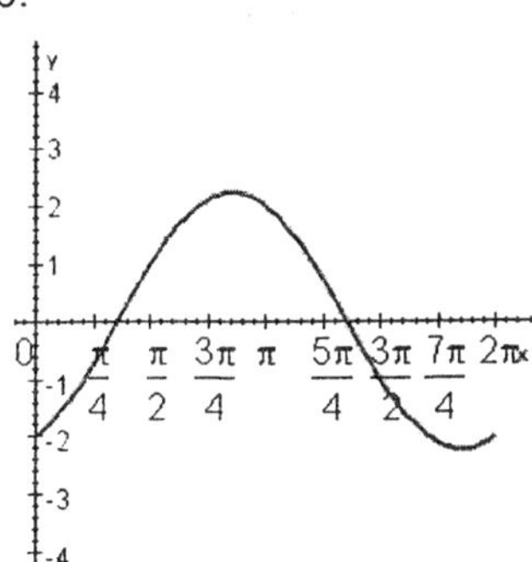

c.

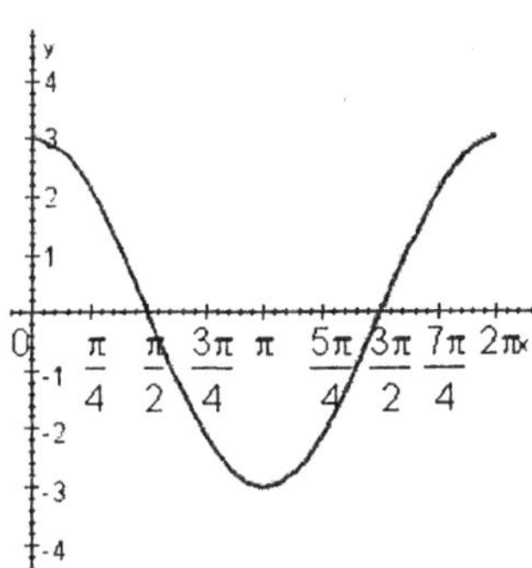

30.

If $\sin B = -\dfrac{1}{3}$ with B in QIII, find

$\tan \dfrac{B}{2}$

Select the correct answer.

a. $\tan \dfrac{B}{2} = 2 - 2\sqrt{5}$

b. $\tan \dfrac{B}{2} = -3 - 2\sqrt{2}$

c. $\tan \dfrac{B}{2} = -2 - 2\sqrt{5}$

d. $\tan \dfrac{B}{2} = -2 - 2\sqrt{2}$

e. $\tan \dfrac{B}{2} = 2 - 2\sqrt{2}$

31. Evaluate the expression without using a calculator. (Assume the variable represents a positive number.)

$\cos\left(2\cos^{-1} 9x\right)$

Select the correct answer.

a. $162x^2 + 1$

b. $161x^2 - 1$

c. $161x^2 + 1$

d. $160x^2 - 1$

e. $162x^2 - 1$

32. Solve for θ if $0° \leq \theta < 360°$.

$$2\cos^2\theta + 11\cos\theta = -5$$

Select the correct answer.

a. $30°, 150°$

b. $120°, 240°$

c. $60°, 120°$

d. $60°, 300°$

e. $30°, 330°$

33. Solve the equation for x if $0 \leq x < 2\pi$. Give your answer in radians using exact values only.

$$1 - \cos x - 2\sin^2 x = 0$$

Select the correct answer.

a. $\dfrac{\pi}{6}, \dfrac{5\pi}{6}, \dfrac{3\pi}{2}$

b. $0, \dfrac{2\pi}{3}, \dfrac{4\pi}{3}$

c. $\dfrac{\pi}{3}, \pi, \dfrac{5\pi}{3}$

d. $\dfrac{\pi}{2}, \dfrac{7\pi}{6}, \dfrac{11\pi}{6}$

e. $\dfrac{\pi}{6}, \dfrac{\pi}{2}, \dfrac{5\pi}{6}$

34. Write an expression that gives all solutions to the equation.

$$\cos x - \sin x = -\sqrt{2}$$

Select the correct answer.

a. $\dfrac{5\pi}{4} + 2k\pi$

b. $\dfrac{3\pi}{4} + 2k\pi$

c. $\dfrac{7\pi}{4} + 2k\pi$

d. $\dfrac{5\pi}{4} + k\pi$

e. $\dfrac{3\pi}{4} + k\pi$

35. Find all degree solutions.

$$2\sin^2 6\theta + 3\sin 6\theta + 1 = 0$$

Select the correct answer.

a. $\theta = 105° + 180°k$ or $\theta = 135° + 180°k$ or $\theta = 165° + 180°k$

b. $\theta = 100° + 120°k$

c. $\theta = 35° + 180°k$ or $\theta = 45° + 180°k$ or $\theta = 55° + 180°k$

d. $\theta = 35° + 60°k$ or $\theta = 45° + 60°k$ or $\theta = 55° + 60°k$

e. $\theta = 105° + 60°k$ or $\theta = 135° + 60°k$ or $\theta = 165° + 60°k$

36. Eliminate the parameter t from the following and then sketch the graph:

$$x = \cos t - 8 \quad y = \sin t + 4$$

Select the correct answer.

a. $(x + 8)^2 + (y + 4)^2 = 1$

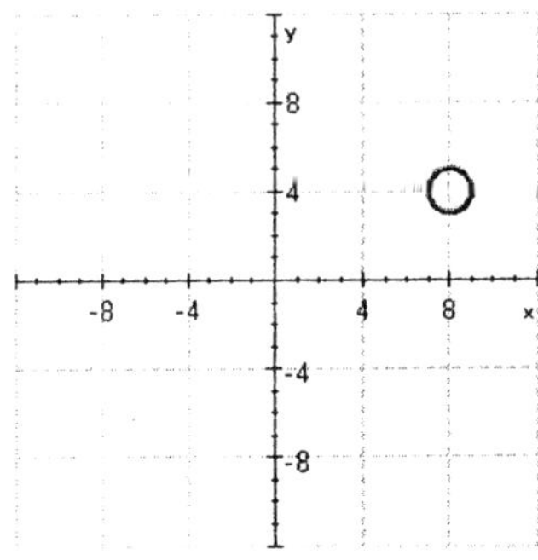

b. $(x + 8)^2 + (y - 4)^2 = 1$

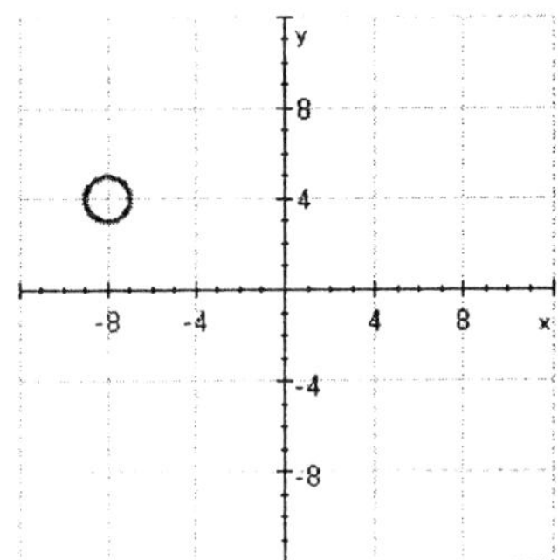

c. $(x + 7)^2 + (y - 4)^2 = 1$

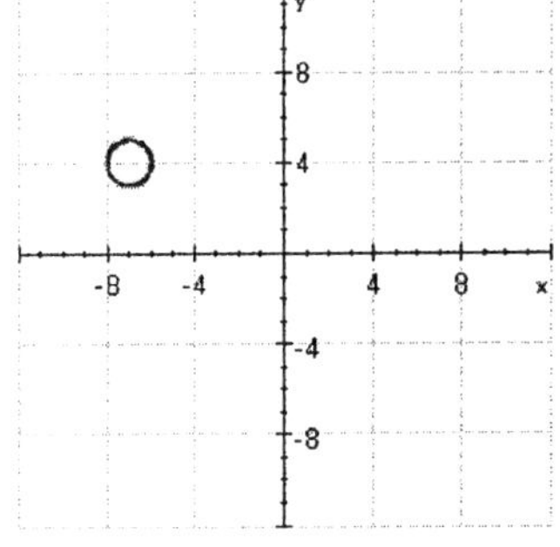

d. $(x + 8)^2 + (y - 3)^2 = 1$

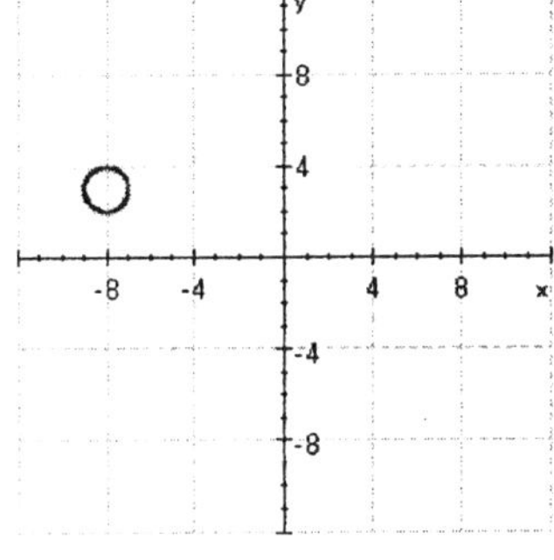

37. The information below refers to triangle ABC.

$B = 55°, C = 32°, a = 7.3$ m. Find all the missing parts. Round side lengths to a tenth of a meter.

Select the correct answer.

a. $A = 91°, b = 6$ m, $c = 3.6$ m

b. $A = 91°, b = 6.2$ m, $c = 3.9$ m

c. $A = 93°, b = 6.2$ m, $c = 3.6$ m

d. $A = 96°, b = 4.5$ m, $c = 9.8$ m

e. $A = 93°, b = 6$ m, $c = 3.9$ m

38. Find all solutions to the triangle described below.

$A = 64°, b = 7.6$ yd, $a = 6.9$ yd

Please round each angle to the nearest degree and the length of the missing side to the nearest tenth.

Select the correct answer.

a. $B = 82°, C = 34°, c = 4.3$ yd

b. $B = 39°, C = 77°, c = 2$ yd

c. $B = 82°, C = 34°, c = 4.3$ yd or $B' = 98°, C' = 77°, c' = 2$ yd

d. $B = 82°, C = 34°, c = 4.3$ yd or $B' = 98°, C' = 18°, c' = 2.4$ yd

e. no solution

39. Solve the triangle.

$A = 51$ yd, $b = 74$ yd, $c = 64$ yd

Please round each answer to the nearest integer. Do not round any numbers until the end of each computation.

Select the correct answer.

a. $A = 45°, B = 76°, C = 58°$

b. $A = 43°, B = 79°, C = 59°$

c. $A = 45°, B = 76°, C = 59°$

d. $A = 43°, B = 79°, C = 58°$

e. $A = 39°, B = 73°, C = 68°$

40. Referring to triangle ABC, find the triangle's area given

$A = 46° \, 30'$, $C = 110° \, 30'$, $a = 3.48$ ft

Please round your answer to the nearest hundredth.

Select the correct answer.

a. 3.02 ft^2

b. 3.06 ft^2

c. 2.98 ft^2

d. 2.90 ft^2

e. 2.94 ft^2

41. For the pair of vectors, find $2\mathbf{U} - 3\mathbf{V}$.

$\mathbf{U} = \langle 6,6 \rangle, \mathbf{V} = \langle 6,-6 \rangle$

Select the correct answer.

a. $\langle -6,30 \rangle$

b. $\langle 30,-6 \rangle$

c. $\langle 29,-7 \rangle$

d. $\langle 6,30 \rangle$

e. $\langle -7,29 \rangle$

42. For the pair of vectors, find $U \cdot V$.

$$U = 2i + j \quad V = i + 2j$$

Select the correct answer.

a.   0
b.   5
c.   2
d.   4
e.   1

43. Find the angle θ between the given vectors to the nearest tenth of a degree.

$$U = 14i + 7j \quad V = -4i + 6j$$

Select the correct answer.

a. $\theta = 100.1°$

b. $\theta = 99.1°$

c. $\theta = 93.1°$

d. $\theta = 98.1°$

e. $\theta = 97.1°$

44. Let $z_1 = 4 + 7i$ and $z_2 = 4 - 7i$ and find $z_1 z_2$.

Select the correct answer.

a.   65 - 56i
b.   61
c.   67 - 56i
d.   65
e.   67

45. Write the complex number in trigonometric form.

$-1 + i$

Select the correct answer.

a. $\sqrt{2} \text{ cis } 315°$

b. $\sqrt{2} \text{ cis } 45°$

c. $\sqrt{3} \text{ cis } 80°$

d. $\sqrt{2} \text{ cis } 225°$

e. $\sqrt{2} \text{ cis } 135°$

46. Use DeMoivre's Theorem to find the following.

$$\left(\sqrt{2} \text{ cis } 60° \right)^{10}$$

Select the correct answer.

a. $-16\sqrt{3} + 16i$

b. $-4 + 4i$

c. $-16 - 16i\sqrt{3}$

d. $-4 + 4i\sqrt{3}$

e. $4 + 4i\sqrt{3}$

47. Find two square roots for the complex number.

$32 - 32i\sqrt{3}$

Select the correct answer.

a. $-4\sqrt{3} + 4i,\ 4\sqrt{3} - 4i$

b. $4\sqrt{3} + 4i,\ -4\sqrt{3} - 4i$

c. $-4\sqrt{2} + 4i\sqrt{2},\ 4\sqrt{2} - 4i\sqrt{2}$

d. $4 + 4i\sqrt{3},\ -4 - 4i\sqrt{3}$

e. $-4 + 4i\sqrt{3},\ 4 - 4i\sqrt{3}$

48. Convert to rectangular coordinates. Use exact values.

$$\left(6\sqrt{2},\ -45^\circ\right)$$

Select the correct answer.

    a. $(-4, -4)$
    b. $(6, -6)$
    c. $(-6, 6)$
    d. $(-4, 6)$
    e. $(-7, -7)$

49. Write the equation in polar coordinates.

$$x^2 + y^2 = 5$$

Select the correct answer.

    a. $r^2 = 1$

    b. $r^2 = 2$

    c. $r^2 = 7$

    d. $r^2 = 5$

    e. $r^2 = 4$

50. Graph the equation.

$$r = 4 + 5\cos\theta$$

Select the correct answer.

a.

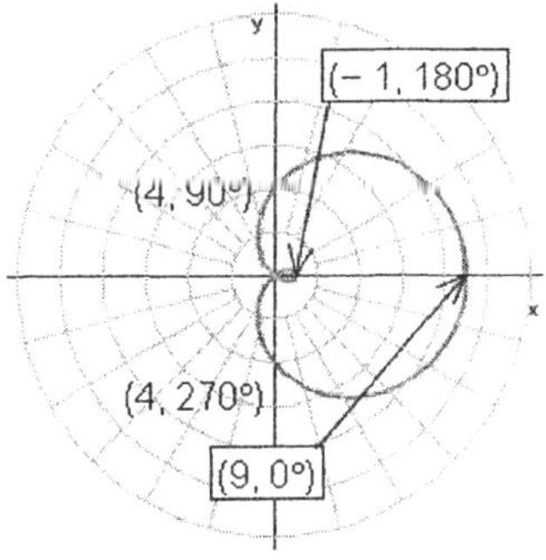

b.

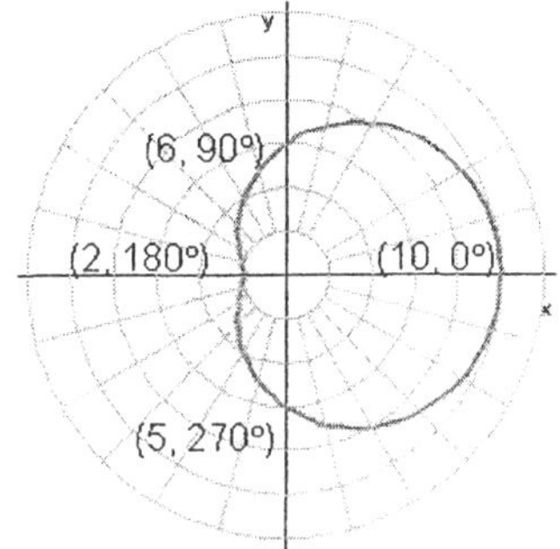

c.

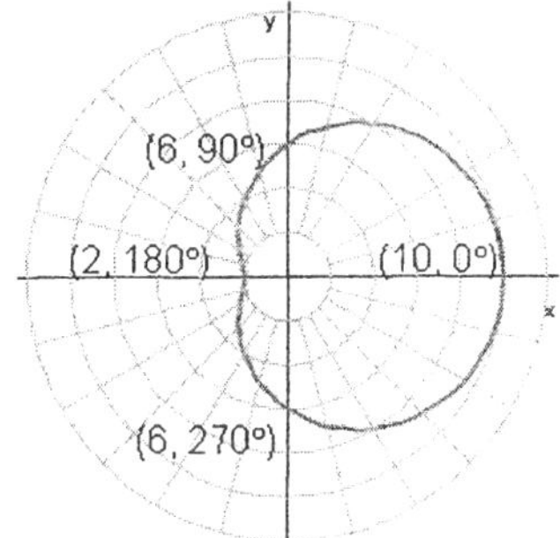

d.

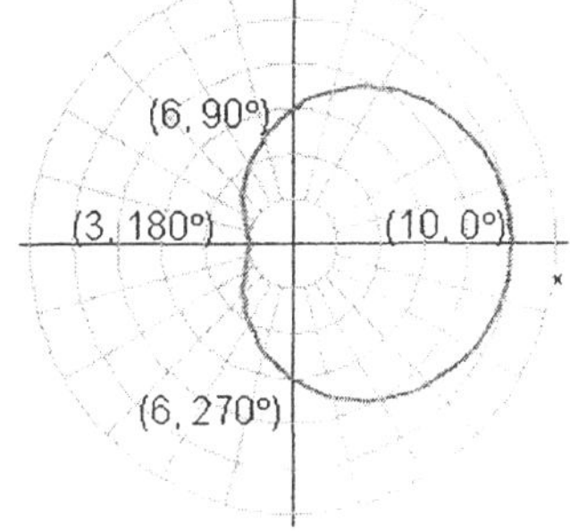

1. c

2. a,d

3. e

4. d

5. c

6. a

7. a

8. e

9. c

10. a

11. d

12. d

13. a

14. c

15. e

16. c

17. c

18. d

19. b

20. b

21. c

22. c

23. c

24. c

25. d

26. d

27. c

McKeague/Turner - Trigonometry 5e Final Exam Form B

28. a

29. c

30. b

31. e

32. b

33. b

34. b

35. d

36. b

37. e

38. d

39. d

40. b

41. a

42. d

43. e

44. d

45. e

46. c

47. a

48. b

49. d

50. a

McKeague/Turner - Trigonometry 5e Final Exam Form B

1. mctr.01.01.27m_NoAlgs
2. mctr.01.02.77m_NoAlgs
3. mctr.01.05.59m_NoAlgs
4. mctr.01.03.30m_NoAlgs
5. mctr.01.04.18m_NoAlgs
6. mctr.01.05.25m_NoAlgs
7. mctr.02.01.32m_NoAlgs
8. mctr.02.02.74m_NoAlgs
9. mctr.02.03.32m_NoAlgs
10. mctr.02.04.11m_NoAlgs
11. mctr.02.05.17m_NoAlgs
12. mctr.02.05.28m_NoAlgs
13. mctr.03.01.31m_NoAlgs
14. mctr.03.02.33m_NoAlgs
15. mctr.03.03.43m_NoAlgs
16. mctr.03.04.15m_NoAlgs
17. mctr.03.04.41m_NoAlgs
18. mctr.03.05.28m_NoAlgs
19. mctr.04.01.53m_NoAlgs
20. mctr.04.02.28m_NoAlgs
21. mctr.04.03.13m_NoAlgs
22. mctr.04.03.27m_NoAlgs
23. mctr.04.04.30m_NoAlgs
24. mctr.04.05.14m_NoAlgs
25. mctr.04.06.52m_NoAlgs
26. mctr.05.01.37m_NoAlgs
27. mctr.05.02.23m_NoAlgs
28. mctr.05.03.07m_NoAlgs
29. mctr.05.03.19m_NoAlgs
30. mctr.05.04.14m_NoAlgs
31. mctr.05.05.11m_NoAlgs
32. mctr.06.01.31m_NoAlgs
33. mctr.06.02.17m_NoAlgs
34. mctr.06.02.42m_NoAlgs
35. mctr.06.03.36m_NoAlgs

McKeague/Turner - Trigonometry 5e Final Exam Form B

36. mctr.06.04.10m_NoAlgs
37. mctr.07.01.14m_NoAlgs
38. mctr.07.02.18m_NoAlgs
39. mctr.07.03.12m_NoAlgs
40. mctr.07.04.11m_NoAlgs
41. mctr.07.05.29m_NoAlgs
42. mctr.07.06.09m_NoAlgs
43. mctr.07.06.16m_NoAlgs
44. mctr.08.01.65m_NoAlgs
45. mctr.08.02.35m_NoAlgs
46. mctr.08.03.24m_NoAlgs
47. mctr.08.04.08m_NoAlgs
48. mctr.08.05.24m_NoAlgs
49. mctr.08.05.59m_NoAlgs
50. mctr.08.06.17m_NoAlgs

1. Evaluate the expression without using a calculator. (Assume the variable represents a positive number.)

$$\cos\left(2\cos^{-1}2x\right)$$

2. Point *P* moves with angular velocity w on a circle of radius *r*. Find the distance *s* traveled by the point in time *t*.

$$w = 10 \text{ rad/sec}, \ r = 5 \text{ ft}, \ t = 3 \text{ min}$$

Select the correct answer.

a. 5,790 ft
b. 9,500 ft
c. 6,300 ft
d. 9,000 ft
e. 5,400 ft

3. If $\sin B = -\dfrac{1}{3}$ with B in QIII, find $\tan\dfrac{B}{2}$

Select the correct answer.

a. $\tan\dfrac{B}{2} = -2 - 2\sqrt{5}$

b. $\tan\dfrac{B}{2} = -3 - 2\sqrt{2}$

c. $\tan\dfrac{B}{2} = -2 - 2\sqrt{2}$

d. $\tan\dfrac{B}{2} = 2 - 2\sqrt{5}$

e. $\tan\dfrac{B}{2} = 2 - 2\sqrt{2}$

4. For the pair of vectors, find $U \cdot V$.

$$U = 2i + j \qquad V = i + 2j$$

Select the correct answer.

a. 2
b. 5
c. 1
d. 4
e. 0

5. Solve the equation for x if $0 \leq x < 2\pi$. Give your answer in radians using exact values only.

$$2\cos^2 x - \sin x - 1 = 0$$

Enter $x_1, x_2 \cdots$, where, x_1 etc. are the solutions of the equation.

6. Graph one complete cycle of $y = \dfrac{3}{10}\cos x$. Label the axes accurately.

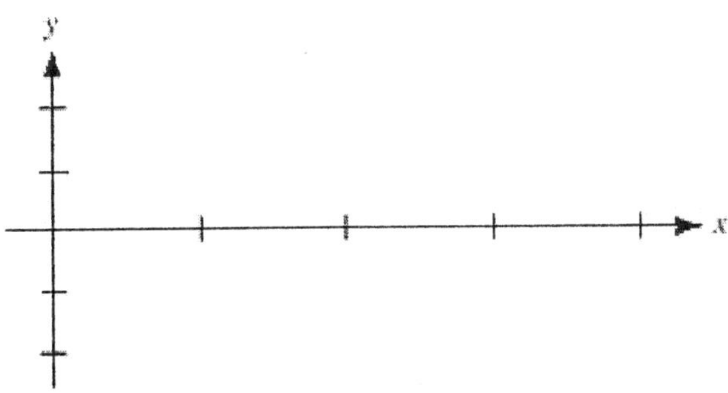

Identify the amplitude for the graph (if defined). If the amplitude cannot not be defined enter *undefined*.

7. The graph below is one complete cycle of the graph of an equation containing a trigonometric function. Find an equation to match the graph. If you are using a graphing calculator, graph your equation to verify that it is correct.

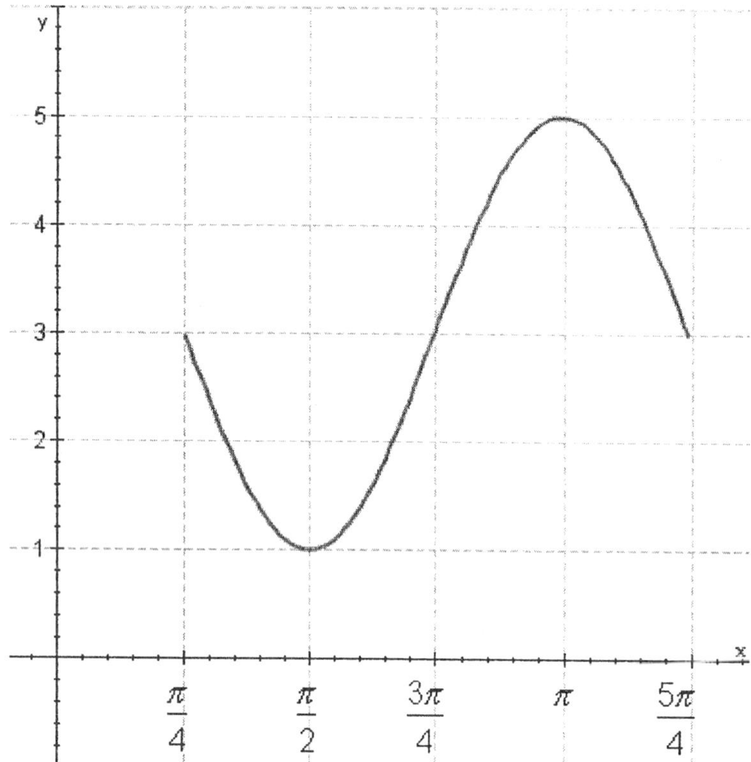

Please enter your answer in the form $y = f(x)$.

8. Find the angle θ between the given vectors to the nearest tenth of a degree.

$$U = 12\mathbf{i} + 7\mathbf{j} \qquad V = -3\mathbf{i} + 5\mathbf{j}$$

$\theta = $ _________ °

9. Let $\cos A = \dfrac{1}{\sqrt{5}}$ with *A* in QIV and find tan 2*A*.

Select the correct answer.

a. $-\dfrac{2}{3}$

b. $\dfrac{4}{9}$

c. $\dfrac{8}{9}$

d. $-\dfrac{3}{4}$

e. $\dfrac{4}{3}$

10. Graph one complete cycle for the function. Label the axes so that the amplitude (if defined) and period are easy to read.

$$y = 4 \csc \frac{1}{6} x$$

Select the correct answer.

a.

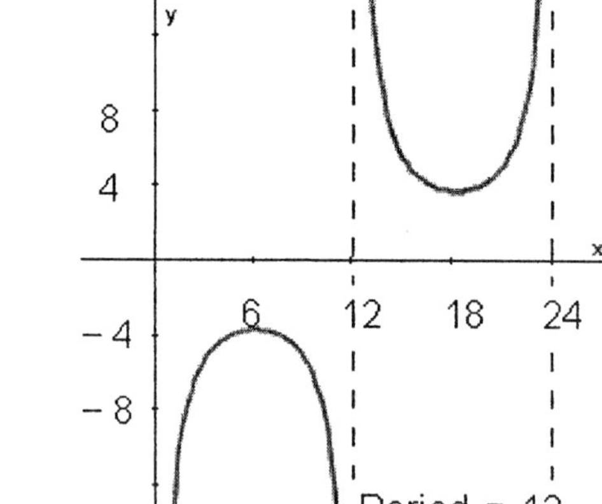

b.

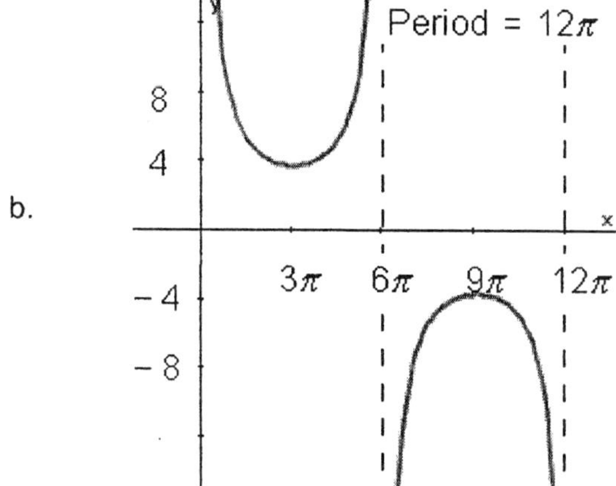

c.

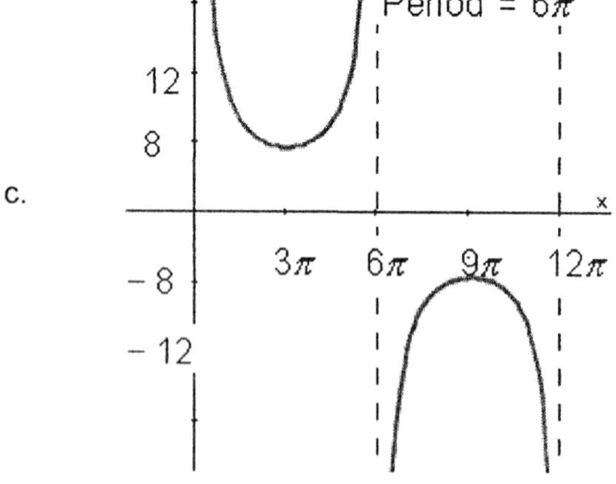

11. Convert to rectangular coordinates. Use exact values.

$$\left(6\sqrt{2}, \; -45°\right)$$

Select the correct answer.

    a. $(-4, -4)$
    b. $(-6, 6)$
    c. $(-7, -7)$
    d. $(6, -6)$
    e. $(-4, 8)$

12. The horizontal and vertical components of the velocity of an arrow shot into the air are 18.0 feet per second and 21.0 feet per second, respectively. Find the velocity of the arrow. Please round your answers to the nearest tenth.

    Select the correct answer.

    a.   27.7 ft/sec
    b.   41 ft/sec
    c.   32.3 ft/sec
    d.   36.1 ft/sec
    e.   25 ft/sec

13. A person standing 130 centimeters from a mirror notices that the angle of depression from his eyes to the bottom of the mirror is $14°$, while the angle of elevation to the top of the mirror is $12°$. Find the vertical dimension of the mirror (see the figure). Please round your answer to the nearest centimeter.

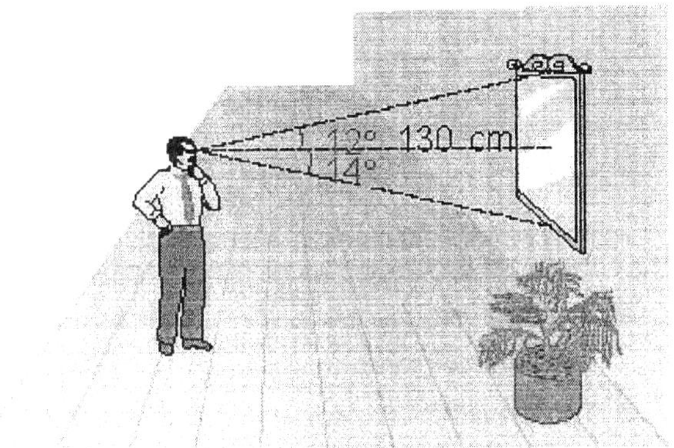

    Select the correct answer.

    a.   50 cm
    b.   60 cm
    c.   61 cm
    d.   62 cm
    e.   57 cm

14. Write an expression that gives all solutions to the equation.

$$\cos x - \sin x = -\sqrt{2}$$

Select the correct answer.

a. $\dfrac{5\pi}{4} + k\pi$

b. $\dfrac{5\pi}{4} + 2k\pi$

c. $\dfrac{7\pi}{4} + 2k\pi$

d. $\dfrac{3\pi}{4} + 2k\pi$

e. $\dfrac{3\pi}{4} + k\pi$

15. Draw the following angle in standard position.

$315°$

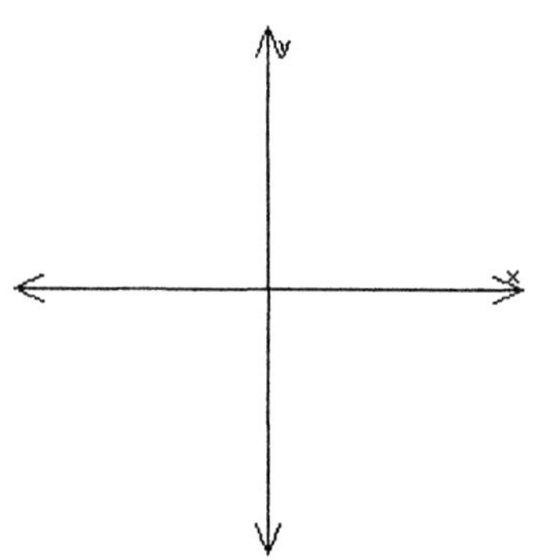

Find a point on the terminal side.

Find the sine of the angle.

Find the cosine of the angle.

Find the tangent of the angle.

16. Match the expression with a single trigonometric function.

$\cos 9x \cos 4x - \sin 9x \sin 4x$  $\cos 6x$

$\cos 5x \cos x - \sin 5x \sin x$  $\cos 13x$

17. Find all degree solutions.

$$2\sin^2 10\theta + 3\sin 10\theta + 1 = 0$$

18. Use DeMoivre's Theorem to find the following. Write your answer in standard form.

$$\left(\sqrt{2}\ \text{cis}\ 60^\circ\right)^{10}$$

Select the correct answer.

a. $-16 - 16i\sqrt{3}$

b. $4 + 4i\sqrt{3}$

c. $-4 + 4i$

d. $-16\sqrt{3} + 16i$

e. $-4 + 4i\sqrt{3}$

19. Identify the amplitude for the equation.

$$y = 2\sin\left(\pi x + \frac{\pi}{3}\right)$$

____

Identify the period for the equation.

Identify the phase shift for the equation.

Label the axes accordingly and sketch one complete cycle of the curve.

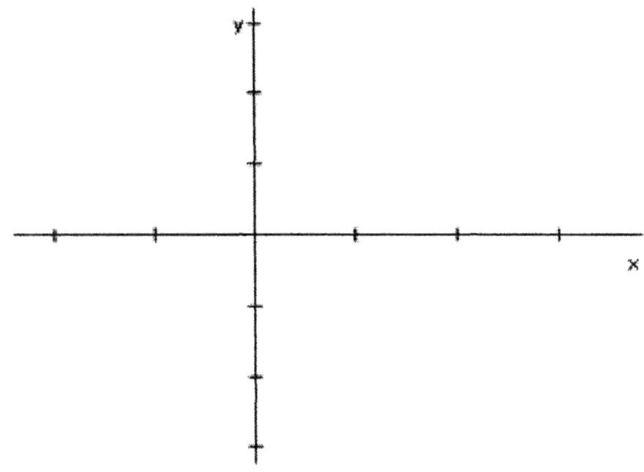

20. Solve for θ if $0° \le \theta < 360°$.

$$2\cos^2\theta + 11\cos\theta = -5$$

Select the correct answer.

a.   30°, 150°

b.   120°, 240°

c.   30°, 330°

d.   60°, 120°

e.   60°, 300°

21. Use a calculator to find a value of θ between $0°$ and $90°$ that satisfies the statement below.

Write your answer in degrees and minutes rounded to the nearest minute.

$\csc \theta = 9.4856$

Select the correct answer.

a. $\theta = 3°03'$

b. $\theta = 7°05'$

c. $\theta = 6°05'$

d. $\theta = 7°03'$

e. $\theta = 6°03'$

22. Graph the equation.

$$r = 4 + 5\cos\theta$$

Select the correct answer.

a.

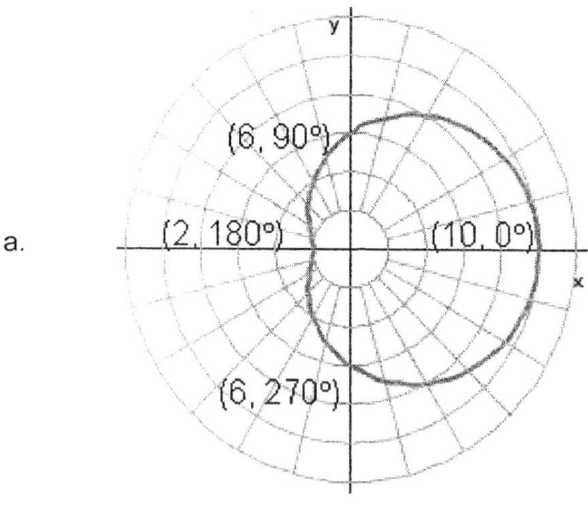

b.

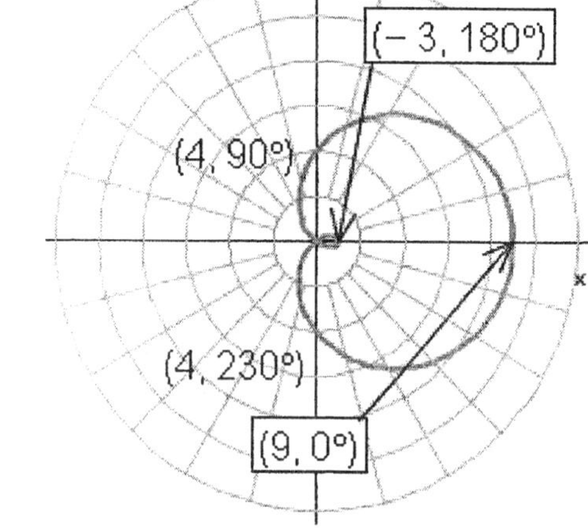

c.

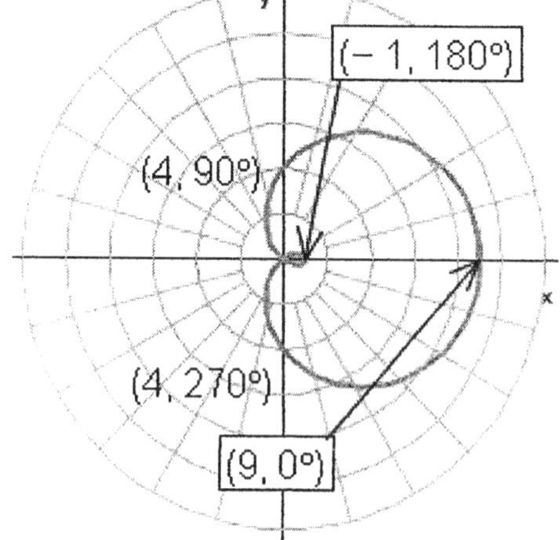

23. Use a calculator to find $\sec 100.9^{\circ}$.

Please round the answer to the nearest ten-thousandth.

____

24. Suppose ABC is a right triangle with $C = 90^{\circ}$.

If $a = 9$ and $c = 15$, find b.

25. Use a ratio identity to find $\cot \theta$ if

$$\sin \theta = \frac{9}{\sqrt{181}} \quad \text{and} \quad \cos \theta = \frac{10}{\sqrt{181}}.$$

Select the correct answer.

a. $\quad \cot \theta = \dfrac{10}{9}$

b. $\quad \cot \theta = \dfrac{9}{10\sqrt{181}}$

c. $\quad \cot \theta = \dfrac{10}{9\sqrt{181}}$

d. $\quad \cot \theta = \dfrac{9}{10}$

e. $\quad \cot \theta = \dfrac{10\sqrt{181}}{9}$

26. The problem refers to a vector **V** with magnitude $|\,\mathbf{V}\,|$ that forms an angle with the positive x-axis. Give the magnitude of the horizontal and vertical vector components of **V**, namely $\mathbf{V}_x$ and $\mathbf{V}_y$, respectively. Round your answers to the nearest integer.

$|\mathbf{V}| = 60, \ \theta = 150^{\circ}$

$\left|\mathbf{V}_x\right| = $ __________

$\left|\mathbf{V}_y\right| = $ __________

27. Let $z_1 = 4 + 7i$ and $z_2 = 4 - 7i$ and find $z_1 z_2$.

Select the correct answer.

   a.   61
   b.   65
   c.   67
   d.   65 - 56i
   e.   67 - 56i

28. Convert to degree measure.

$$\theta = \frac{8\pi}{9}$$

Select the correct answer.

   a.   170°

   b.   160°

   c.   135°

   d.   140°

   e.   150°

29. The information below refers to triangle ABC.

$B = 58°, C = 31°, a = 7.5$ m.

Find A.

$A =$ _________ °

Find b. Please round the answer to a tenth of a meter.

$b =$ _________ m

Find c. Please round the answer to a tenth of a meter.

$c =$ _________ m

30. Find the area of the sector formed by central angle $\theta = 2.2$ in a circle of radius $r = 5$ inches.

If the answer needs rounding, round it to three significant digits.

$A=$ _________ in^2

31. Refer to right triangle ABC with $C= 90^\circ$. Solve for all the missing parts using the given information.

$b = 377.6$ inches, $c = 588.7$ inches

Please round the answers for the angles to two decimal places, and for the side, to one decimal place.

$A=$ _________ $^\circ$

$B=$ _________ $^\circ$

$a=$ _________ inches

32. Evaluate without using a calculator.

$$\sin^{-1}\left(\sin 210^\circ\right)$$

33. Write the equation in polar coordinates.

$$x^2 + y^2 = 2$$

34. Eliminate the parameter t from the following and then sketch the graph:

$$x = \cos t - 8 \quad y = \sin t + 4$$

Select the correct answer.

a. $(x + 8)^2 + (y + 4)^2 = 1$

b. $(x + 8)^2 + (y - 3)^2 = 1$

c. $(x + 8)^2 + (y - 4)^2 = 1$

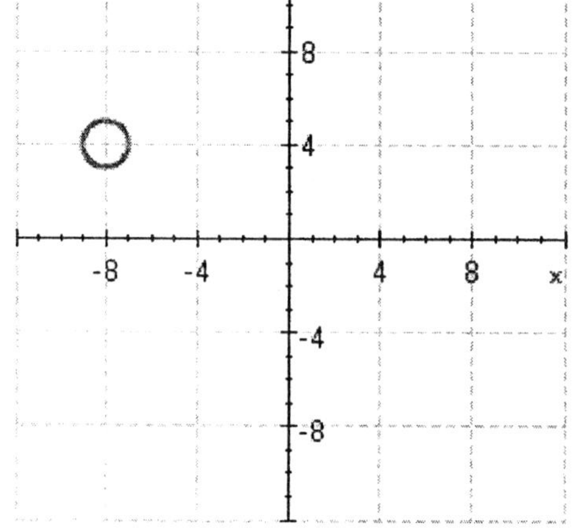

35. Name another angle that is coterminal with $-60°$.

Select the correct answer(s).

a. $-300°$

b. $300°$

c. $-420°$

d. $60°$

e. $420°$

36. Write the following in terms of $\sin\theta$ and $\cos\theta$ and then simplify if possible.

$\sec\theta - \tan\theta \sin\theta$

37. Graph the function from $x = 0$ and $x = 2\pi$.

$$y = 4\cos^2 2x - 2$$

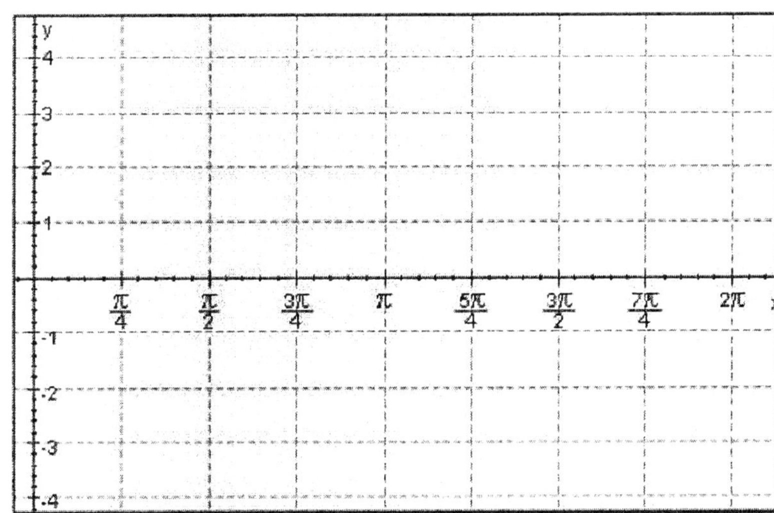

38. Find all solutions to the triangle described below.

$A = 64°$, $b = 7.6$ yd, $a = 6.9$ yd

Please round each angle to the nearest degree and the length of the missing side to the nearest tenth.

Select the correct answer.

a. $B = 82°$, $C = 34°$, $c = 4.3$ yd

b. $B = 82°$, $C = 34°$, $c = 4.3$ yd or $B' = 98°$, $C' = 18°$, $c' = 2.4$ yd

c. $B = 82°$, $C = 34°$, $c = 4.3$ yd or $B' = 98°$, $C' = 77°$, $c' = 2$ yd

d. $B = 39°$, $C = 77°$, $c = 2$ yd

e. no solution

39. For the pair of vectors, find $U + V$, $U - V$, and $2U - 3V$.

$U = \langle 3, 3 \rangle$, $V = \langle 3, -3 \rangle$

$U + V = \langle$ _______ , _______ $\rangle$

$U - V = \langle$ _______ , _______ $\rangle$

$2U - 3V = \langle$ _______ , _______ $\rangle$

40. The pendulum on a grandfather clock swings from side to side once every second. If the length of the pendulum is 4 feet and the angle through which it swings is $21°$, how far does the tip of the pendulum travel in 1 second?

Select the correct answer.

a. 1.77 feet
b. 1.47 feet
c. 1.67 feet
d. 1.87 feet
e. 1.57 feet

41. Find the right part of the identity.

$$\cos\alpha\,\tan\alpha = \underline{\hspace{3cm}}$$

Select the correct answer.

a. $\csc\alpha$

b. $\tan\alpha$

c. $\sin\alpha$

d. $\sec\alpha$

e. $\cos\alpha$

42. Simplify expression by first substituting values from the table of exact values and then simplifying the resulting expression.

$$\sin^2 30° + \cos^2 30°$$

Table of Exact Values

x	$\sin x$	$\cos x$
0°	0	1
30°	$\dfrac{1}{2}$	$\dfrac{\sqrt{3}}{2}$
45°	$\dfrac{1}{\sqrt{2}}$	$\dfrac{1}{\sqrt{2}}$
60°	$\dfrac{\sqrt{3}}{2}$	$\dfrac{1}{2}$
90°	1	0

43. Solve the triangle.

$a= 48$ yd, $b= 75$ yd, $c= 60$ yd

Please round your answers to the nearest integer. Do not round any numbers until the end of each computation.

$A=$ _________ °

$B=$ _________ °

$C=$ _________ °

44. If we start at the point (1, 0) and travel once around the unit circle, we travel a distance of 2π units and arrive back where we started at the point (1, 0). If we continue around the unit circle a second time, we will repeat all the values of x and y that occurred during our first trip around. Use this discussion to evaluate the expression.

$$\sin\left(2\pi + \frac{\pi}{2}\right)$$

45. Write the complex number in trigonometric form. Begin by sketching the graph to help find the argument θ.

$1 - i$

Please enter the argument in degrees and use the degree symbol.

Enter your answer in the form $r(\cos\theta + i\sin\theta)$.

46. Find two square roots for the complex number. Write your answer in standard form.

$8 - 8i\sqrt{3}$

Please enter your answer as two numbers, separated with a comma.

47.

Use the graph of the equation $y = -\cos\left(2x + \dfrac{\pi}{2}\right)$ shown below to graph one complete cycle of the equation $y = 3 - \cos\left(2x + \dfrac{\pi}{2}\right)$.

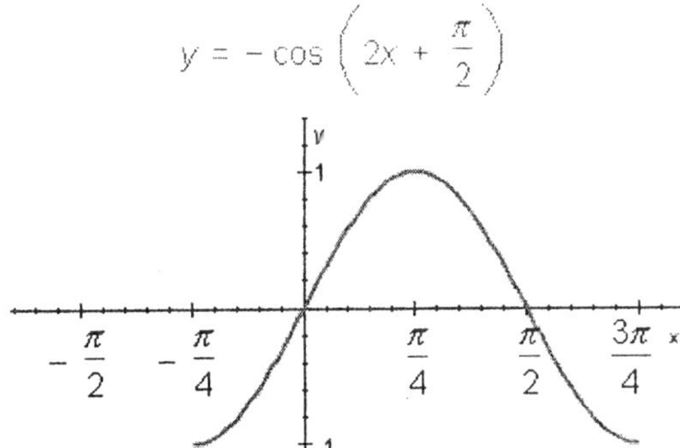

Select the correct answer.

a.

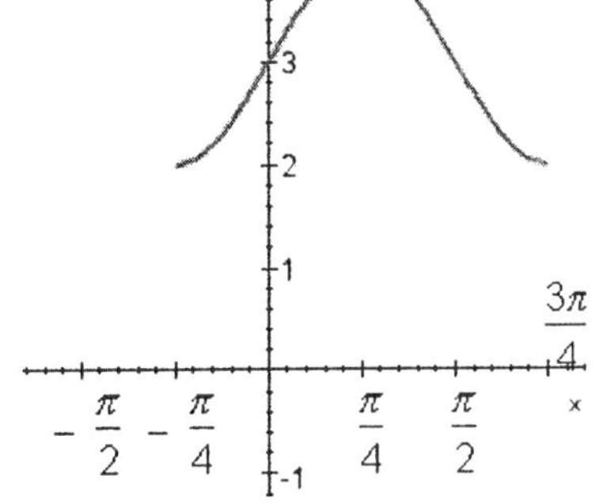

b.

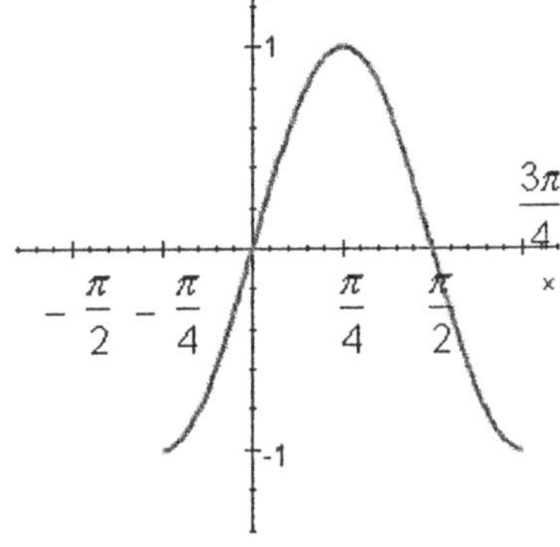

c.

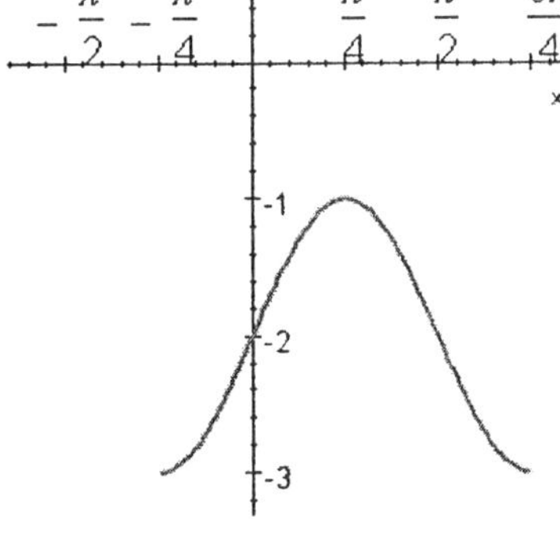

d.

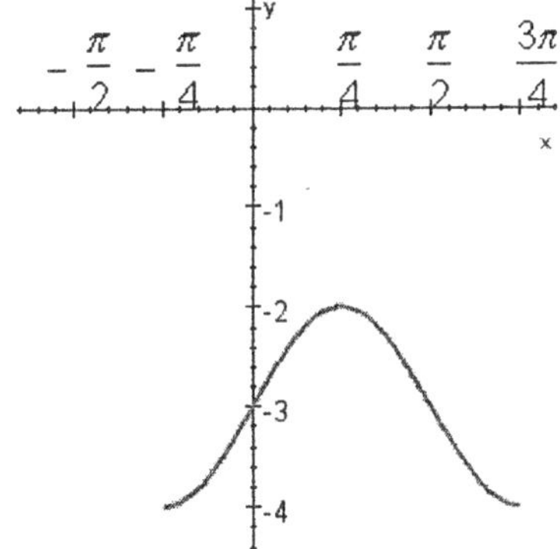

48. Sketch the graph from $x = 0$ to $x = 4\pi$.

$$y = 2\sin x + \sin 2x$$

Select the correct answer.

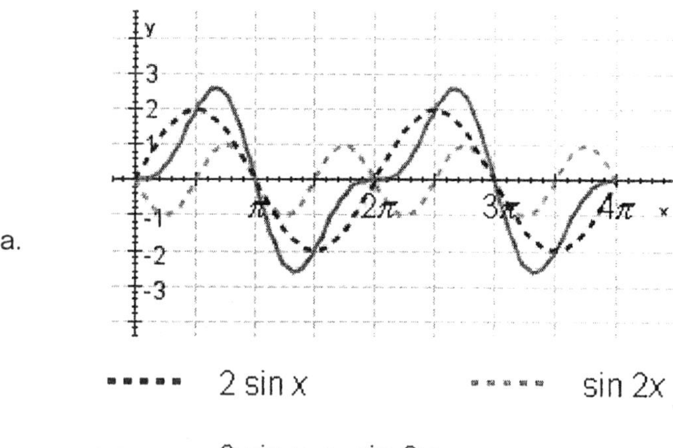

a.

$\cdots\cdots$ $2 \sin x$ $\cdots\cdots$ $\sin 2x$

$\underline{\hspace{2cm}}$ $2 \sin x + \sin 2x$

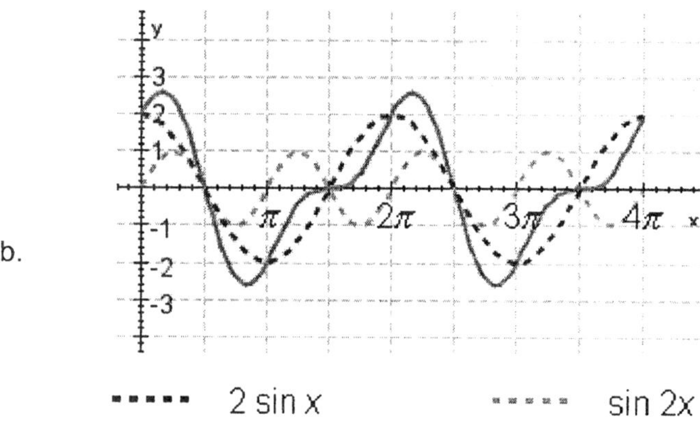

b.

$\cdots\cdots$ $2 \sin x$ $\cdots\cdots$ $\sin 2x$

$\underline{\hspace{2cm}}$ $2 \sin x + \sin 2x$

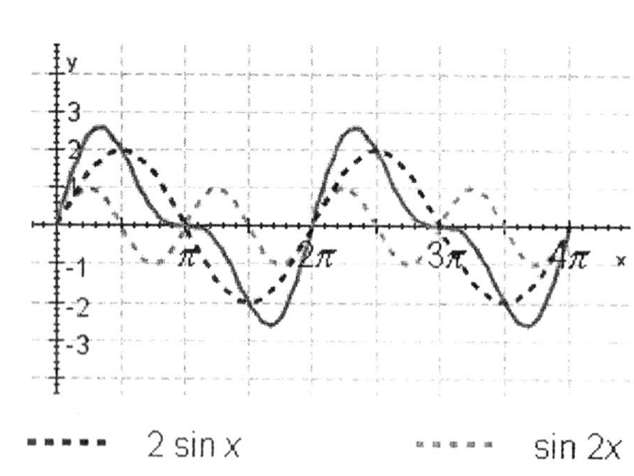

c.

$\cdots\cdots$ $2 \sin x$ $\cdots\cdots$ $\sin 2x$

$\underline{\hspace{2cm}}$ $2 \sin x + \sin 2x$

49. Referring to triangle *ABC*, find the triangle's area given

$A= 46^{\circ}\ 30'$, $C= 110^{\circ}\ 30'$, $a= 3.48$ ft

Please round your answer to the nearest hundredth.

Select the correct answer.

a.   3.06 ft^2

b.   2.94 ft^2

c.   3.02 ft^2

d.   2.90 ft^2

e.   2.98 ft^2

50. Find the identical expression for the following.

$$\frac{\sec x + 1}{\sec x - 1}$$

Select the correct answer.

a.   0

b. $\dfrac{\cos x + 1}{\cos x - 1}$

c. $\dfrac{1 + \sin x}{1 - \sin x}$

d. $\dfrac{\sin x + 1}{\sin x - 1}$

e. $\dfrac{1 + \cos x}{1 - \cos x}$

1. $8x^2 - 1$

2. d

3. b

4. d

5. $\dfrac{\pi}{6},\ \dfrac{5\pi}{6},\ \dfrac{3\pi}{2}$

6. 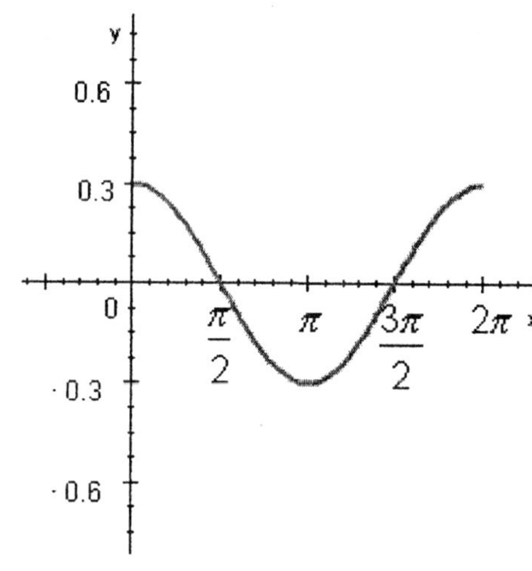

$\dfrac{3}{10}$

7. $y = 3 - 2\sin\left(2x - \dfrac{\pi}{2}\right)$

8. 90.7

9. e

10. b

11. d

12. a

13. b

14. d

15.

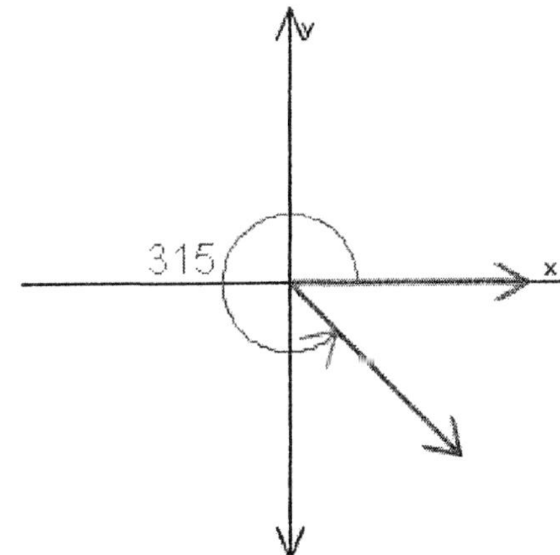

$(1, -1)$ for example

$-\dfrac{\sqrt{2}}{2}$

$\dfrac{\sqrt{2}}{2}$

-1

16. $\cos 5x \cos x - \sin 5x \sin x \rightarrow \cos 6x,$

$\cos 9x\cos 4x - \sin 9x\sin 4x \rightarrow \cos 13x$

17. $\theta = 21° + 36°k$ or $\theta = 27° + 36°k$ or $\theta = 33° + 36°k$

18. a

19. 2

2

$-\dfrac{1}{3}$

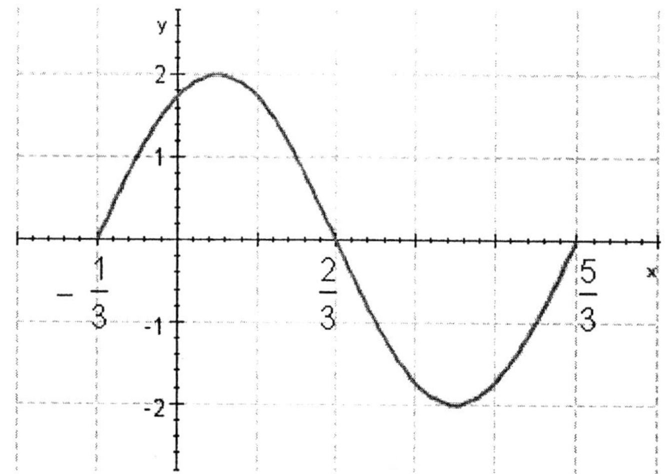

20. b

21. e

22. c

23. -5.2883

24. 12

25. a

26. 52

30

27. b

28. b

29. 91

6.4

3.9

30. 27.5

31. 50.10

39.90

451.6

32. -30°

33. $r^2 = 2$

34. c

35. b, c

36. $\cos(\theta)$

37.

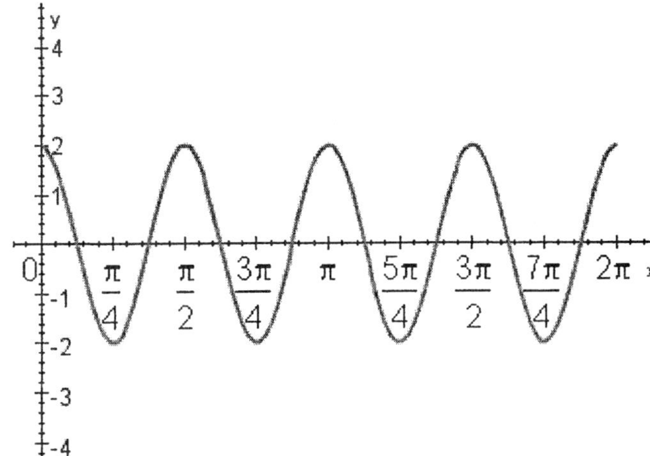

38. b

39. 6

0

0

6

-3

15

40. b

41. c

42. 1

43. 40

87

53

44. 1

45. $\sqrt{2}\cdot\left(\cos\left(315^\circ\right)+i\cdot\sin\left(315^\circ\right)\right)$

46. $-2\sqrt{3}+2i,\ 2\sqrt{3}-2i$

47. a

48. c

49. a

50. e

McKeague/Turner - Trigonometry 5e Final Exam Form C

1. mctr.05.05.11_NoAlgs
2. mctr.03.05.28m_NoAlgs
3. mctr.05.04.14m_NoAlgs
4. mctr.07.06.09m_NoAlgs
5. mctr.06.02.17_NoAlgs
6. mctr.04.01.53_NoAlgs
7. mctr.04.04.30_NoAlgs
8. mctr.07.06.16_NoAlgs
9. mctr.05.03.07m_NoAlgs
10. mctr.04.02.28m_NoAlgs
11. mctr.08.05.24m_NoAlgs
12. mctr.02.05.28m_NoAlgs
13. mctr.02.04.11m_NoAlgs
14. mctr.06.02.42m_NoAlgs
15. mctr.01.03.30_NoAlgs
16. mctr.05.02.23_NoAlgs
17. mctr.06.03.36_NoAlgs
18. mctr.08.03.24m_NoAlgs
19. mctr.04.03.13_NoAlgs
20. mctr.06.01.31m_NoAlgs
21. mctr.02.02.74m_NoAlgs
22. mctr.08.06.17m_NoAlgs
23. mctr.03.01.31_NoAlgs
24. mctr.01.01.27_NoAlgs
25. mctr.01.04.18m_NoAlgs
26. mctr.02.05.17_NoAlgs
27. mctr.08.01.65m_NoAlgs
28. mctr.03.02.33m_NoAlgs
29. mctr.07.01.14_NoAlgs
30. mctr.03.04.41_NoAlgs
31. mctr.02.03.32_NoAlgs
32. mctr.04.06.52_NoAlgs
33. mctr.08.05.59_NoAlgs
34. mctr.06.04.10m_NoAlgs
35. mctr.01.02.77m_NoAlgs
36. mctr.01.05.25_NoAlgs

McKeague/Turner - Trigonometry 5e Final Exam Form C

37. mctr.05.03.19_NoAlgs
38. mctr.07.02.18m_NoAlgs
39. mctr.07.05.29_NoAlgs
40. mctr.03.04.15m_NoAlgs
41. mctr.01.05.59m_NoAlgs
42. mctr.02.01.32_NoAlgs
43. mctr.07.03.12_NoAlgs
44. mctr.03.03.43_NoAlgs
45. mctr.08.02.35_NoAlgs
46. mctr.08.04.08_NoAlgs
47. mctr.04.03.27m_NoAlgs
48. mctr.04.05.14m_NoAlgs
49. mctr.07.04.11m_NoAlgs
50. mctr.05.01.37m_NoAlgs